T0135267

Lecture Notes in Computational Science and Engineering 104

Editors:

More information about this series at
http://www.springer.com/series/3527

Thomas Dickopf • Martin J. Gander •
Laurence Halpern • Rolf Krause • Luca F. Pavarino
Editors

Domain Decomposition Methods in Science and Engineering XXII

Editors
Thomas Dickopf
Fakultät für Mathematik
Technische Universität München
Garching, Germany

Martin J. Gander
Section de Mathématiques
Université de Genève
Genève, Switzerland

Laurence Halpern
Laboratoire Analyse
Geométrie & Applications
Université Paris XIII
Villetaneuse, France

Rolf Krause
Institute of Computational Science
University of Lugano
Lugano, Switzerland

Luca F. Pavarino
Dipartimento di Matematica
Università degli Studi di Milano
Milano, Italy

ISSN 1439-7358 ISSN 2197-7100 (electronic)
Lecture Notes in Computational Science and Engineering
ISBN 978-3-319-79260-6 ISBN 978-3-319-18827-0 (eBook)
DOI 10.1007/978-3-319-18827-0

Mathematics Subject Classification (2010): 65F10, 65Y05, 65N30, 65N55

Springer Cham Heidelberg New York Dordrecht London

Softcover reprint of the hardcover 1st edition 2016

Cover Photo: Università della Svizzera italiana, Lugano

Printed on acid-free paper

Springer International Publishing AG Switzerland is part of Springer Science+Business Media (www.springer.com)

Preface of DD22 Book of Proceedings

This volume contains the proceedings of the 22nd International Conference on Domain Decomposition Methods, which was hosted by the Institute for Computational Science (ICS) at the Università della Svizzera italiana (USI), Lugano, Switzerland, September 16–20, 2013.

Background of the Conference Series

The International Conference on Domain Decomposition Methods has been held in fourteen countries throughout Asia, Europe, and North America, beginning in Paris in 1987. Held annually for the first fourteen meetings, it has been spaced out since DD15 at roughly 18-month intervals. A complete list of the past meetings appears below. The twenty-second International Conference on Domain Decomposition Methods was the first one held in Switzerland and it took place in the Italian speaking part of Switzerland in Lugano.

The main technical content of the DD conference series has always been mathematical, but the principal motivation was and is to make efficient use of distributed memory computers for complex applications arising in science and engineering. As we approach the dawn of exascale computing, where we will command 10^{18} floating point operations per second, clearly efficient and mathematically well-founded methods for the solution of large-scale systems become more and more important—as does their sound realization in the framework of modern HPC architectures. In fact, the massive parallelism, which makes exascale computing possible, requires the development of new solutions methods, which are capable of efficiently exploiting this large number of cores as well as the connected hierarchies for memory access. Ongoing developments such as parallelization in time, asynchronous iterative methods, or nonlinear domain decomposition methods show that this massive parallelism does not only demand for new solution and discretization methods, but also allows to foster the development of new approaches.

The progress obtained in domain decomposition techniques during the last decades has led to a broadening of the conference program in terms of methods and applications. Multi-physics, nonlinear problems, and space-time decomposition methods are more prominent these days than they have been previously. Domain decomposition has always been an active and vivid field, and this conference series is representing well the highly active and fast advancing scientific community behind it. This is also due to the fact that there is basically no alternative to domain decomposition methods as a general approach for massively parallel simulations at a large scale. Thus, with growing scale and growing hardware capabilities, also the methods can—and have to–improve.

However, even if domain decomposition methods are motivated historically by the need for efficient simulation tools for large scale applications, there are also many interesting aspects of domain decomposition, which are not necessarily motivated by the need for massive parallelism. Examples are the choice of transmission conditions between sub-domains, new coupling strategies, or the principal handling of interface conditions in problem classes such as fluid structure interaction or contact problems in elasticity.

While research in domain decomposition methods is presented at numerous venues, the International Conference on Domain Decomposition Methods is the only regularly occurring international forum dedicated to interdisciplinary technical interactions between theoreticians and practitioners working in the development, analysis, software implementation, and application of domain decomposition methods.

The list of previous Domain Decomposition Conferences is the following:

1. Paris, France, January 7–9, 1987
2. Los Angeles, USA, January 14–16, 1988
3. Houston, USA, March 20–22, 1989
4. Moscow, USSR, May 21–25, 1990
5. Norfolk, USA, May 6–8, 1991
6. Como, Italy, June 15–19, 1992
7. University Park, Pennsylvania, USA, October 27–30, 1993
8. Beijing, China, May 16–19, 1995
9. Ullensvang, Norway, June 3–8, 1996
10. Boulder, USA, August 10–14, 1997
11. Greenwich, UK, July 20–24, 1998
12. Chiba, Japan, October 25–29, 1999
13. Lyon, France, October 9–12, 2000
14. Cocoyoc, Mexico, January 6–11, 2002
15. Berlin, Germany, July 21–25, 2003
16. New York, USA, January 12–15, 2005
17. St. Wolfgang-Strobl, Austria, July 3–7, 2006
18. Jerusalem, Israel, January 12–17, 2008
19. Zhangjiajie, China, August 17–22, 2009
20. San Diego, California, USA, February 7–11, 2011

21. Rennes, France, June 25–29, 2012
22. Lugano, Switzerland, September 16–20, 2013

International Scientific Committee on Domain Decomposition Methods

- Petter Bjørstad, University of Bergen, Norway
- Susanne Brenner, Louisiana State University, USA
- Xiao-Chuan Cai, CU Boulder, USA
- Martin Gander, University of Geneva, Switzerland
- Roland Glowinski, University of Houston, USA
- Laurence Halpern, University Paris 13, France
- Ronald Hoppe, Universities of Augsburg, Germany, and Houston, USA
- David Keyes, KAUST, Saudi Arabia
- Hyea Hyun Kim, Kyung Hee University, Korea
- Axel Klawonn, Universität zu Köln, Germany
- Ralf Kornhuber, Freie Universität Berlin, Germany
- Ulrich Langer, University of Linz, Austria
- Alfio Quarteroni, EPFL, Switzerland
- Olof Widlund, Courant Institute, USA
- Jinchao Xu, Penn State, USA
- Jun Zou, Chinese University of Hong Kong

About the Twenty-Second Conference

The twenty-second International Conference on Domain Decomposition Methods had 172 participants from over 24 countries. It was the first one to be held in Switzerland. It was hosted by the Institute of Computational Science (ICS) at USI. The ICS was founded in 2009 towards realizing the vision of USI to become a new scientific and educational node for computational science in Switzerland. ICS has since then grown into a place with strong competences in mathematical modeling, numerical simulation, and high-performance computing. Research areas range from numerical simulation in science, medicine, and engineering, through computational time series analysis and computational shape analysis, to computational cardiology and the (multi-scale) simulation of physical and biological systems and processes.

As in previous meetings, DD22 featured a well-balanced mixture of established and new topics, such as the manifold theory of Schwarz methods, isogeometric analysis, discontinuous Galerkin methods, exploitation of modern HPC architectures, and industrial applications. From the conference program, it is evident that the growing capabilities in terms of theory and available hardware allow for increasingly complex nonlinear and multi-physics simulations, confirming the huge potential and flexibility of the domain decomposition idea.

The conference, which was organized over an entire week, featured presentations of three different types: The conference contained

- 14 invited presentations, fostering also younger scientists and their scientific development, selected by the International Scientific Committee,
- a poster session, which also gave rise to intense discussions with the mostly younger presenting scientists,
- 13 minisymposia, arranged around a special topic,
- 14 sessions of contributed talks.

The present proceedings volume contains a selection of 66 papers, split into 5 plenary papers, 35 minisymposia papers, and 26 contributed papers.

Sponsoring Organizations

- Swiss National Science Foundation
- Nvidia
- Fondazione Cardiocentro Ticino
- Swiss Mathematical Society
- Swiss National Supercomputing Centre
- CRUS via the PhD school FOMICS *Foundations of Mathematics and Informatics for Computer Simulations in Science and Engineering* located at ICS/USI.

Local Organizing Committee Members

- Rolf Krause (ICS/USI Lugano; Chair)
- Thomas Dickopf (ICS/USI Lugano)
- Martin Gander (U Genève)
- Ralf Hiptmair (ETH Zürich)
- Luca F. Pavarino (U Milano)
- Alfio Quarteroni (EPF Lausanne)
- William Sawyer (CSCS Lugano)
- Olaf Schenk (ICS/USI Lugano)

The organizing committee would like to thank USI staff for their invaluable support and the sponsors for the financial support.

Research Activity in Domain Decomposition According to DD22 and Its Proceedings

The conference and the proceedings contain three parts: the plenary presentations, the minisymposia presentations, and the contributed talks.

Plenary Presentations

The plenary presentations of the conference have been dealing with established topics in Domain Decomposition as well as with new approaches, including Domain Decomposition for multiphysics problems and nonlinear problems.

- Nonlinear FETI-DP methods. Oliver Rheinbach (TU Freiberg, Germany)
- Domain decomposition methods for high-order discontinuous Galerkin discretizations. Paola F. Antonietti (MOX Milano, Italy)
- Numerical treatment of tensors and new discretisation paradigms. Wolfgang Hackbusch (MPI Leipzig, Germany)
- Domain decomposition methods in isogeometric analysis. Lourenço Beirão da Veiga (University of Milano, Italy)
- Auxiliary space multigrid based on domain decomposition. Johannes Kraus (RICAM Linz, Austria)
- Domain decomposition in nonlinear function spaces. Oliver Sander (RWTH Aachen, Germany)
- Numerical solution of PDE eigenvalue problems in acoustic field computation. Volker Mehrmann (TU Berlin, Germany)
- Applications of the Voronoi implicit interface method to domain decomposition. James A. Sethian (UC Berkeley, USA)
- Robin-Neumann explicit schemes in fluid-structure interaction problems. Marina Vidrascu (INRIA Le Chesnay, France)
- An assembled inexact Schur-complement preconditioner. Joachim Schöberl (TU Wien, Austria)
- Local simplification of Darcy's equations with pressure dependent permeability. Christine Bernardi (LJLL Paris, France)
- BDDC deluxe domain decomposition algorithms. Olof B. Widlund (NYU, USA)
- Coupling Stokes and Darcy equations: modeling and numerical methods. Marco Discacciati (UPC Barcelona, Spain)
- Robust discretization and iterative methods for multi-physics systems. Jinchao Xu (Penn State University, USA)

Minisymposia

There were 13 minisymposia organized within DD22:

1. Advances in FETI-DP and BDDC methods (Axel Klawonn, Olof B. Widlund)
   This minisymposium focuses on recent developments of the closely related families of BDDC and FETI-DP domain decomposition algorithms. These algorithms have proven very effective in a variety of applications. Talks are offered on applications to nonlinear problems, discontinuous Galerkin methods, mixed finite elements for the Stokes equations with continuous pressures, and on adaptive coarse spaces based on the solution of suitable eigenvalue problems. Recently, there has also been considerable activity in the development of a new variant of BDDC, which is due to Clark Dohrmann. Among the applications of these new ideas are algorithms for H(div) in 3D and for new special discontinuous approximations of H(curl) problems in 2D.
2. Achieving scalability in domain decomposition methods: advances in coarse spaces and alternatives (Felix Kwok, Kevin Santugini)
   With the increasing availability of massively parallel machines, scalability becomes a crucial factor in the design of domain decomposition algorithms. To be scalable, an iterative algorithm must have a convergence rate that does not depend on the number of subdomains. This precludes methods in which subdomains send information only to their direct neighbors, since they cannot converge in fewer iterations than the diameter of the connectivity graph of the decomposition. A traditional way of introducing long-range communication is to add a coarse space component; there are also other methods inspired by multilevel decompositions and interpolation. Speakers present their work on either innovative coarse spaces or new alternatives to coarse spaces.
3. Non-overlapping discretization methods and how to achieve the DDM-paradigm (Ismael Herrera, Luis Miguel de La Cruz)
   The DDM-paradigm is to obtain the global solution by solving local problems exclusively. The introduction of non-overlapping DDMs represented an important step towards achieving this paradigm. However, in non-overlapping DDMs, the interface-nodes are shared by two or more subdomains of the coarse-mesh. In this minisymposium, we present the non-overlapping discretization methods, which use systems of nodes with the property that each node belongs to one and only one subdomain of the coarse mesh. Then, it is explained how using non-overlapping discretization methods the DDM-paradigm can be achieved.
4. Solution techniques for discontinuous Galerkin methods (Blanca Ayuso de Dios, Susanne Brenner)
   Based on the discontinuous finite element spaces, DG methods are extremely versatile and have many attractive features: local conservation properties; flexibility in designing hp-adaptivity strategies, and built-in parallelism. DG methods can deal robustly with PDEs of almost any kind. However, their use in many real applications is still limited by the larger number of degrees-of-freedom required compared with other classical discretization methods. The

aim of this minisymposium is to bring together experts in the field to discuss and identify the most relevant aspects of the state of the art for DG methods, including design, theoretical analysis, and issues related to the implementation and applications of the various solution techniques.

5. Solvers for isogeometric analysis and applications (Lourenço Beirão da Veiga, Luca F. Pavarino, Simone Scacchi)

   Isogeometric analysis (IGA) is a novel and fast developing technology for the numerical solution of PDEs, that integrates CAD geometric parametrization and finite element analysis. Since its introduction in 2005 by T.J.R. Hughes and co-workers, IGA is having a strong impact on the engineering and scientific computing community, producing a large amount of journal publications and developing advanced computer codes. In recent years, researchers in this quickly growing field have started to focus on the design and analysis of efficient solvers for IGA discrete systems, and in particular of multilevel domain decomposition methods yielding parallel and scalable preconditioners. The high (global) regularity and polynomial degree of the NURBS spaces employed in IGA discretizations introduce both new difficulties and opportunities for the construction and analysis of novel solution techniques. The aim of the minisymposium is to bring together researchers in both fields of IGA and domain decomposition methods, focusing on the latest developments and on the new research pathways and applications.

6. Efficient solvers for heterogeneous nonlinear problems (Juan Galvis, Lisandro Dalcin, Nathan Collier, Victor Calo)

   Multiple scales and nonlinearities are present in many applications, such as porous media and material sciences. Heterogeneities and disparity in media properties make it difficult to design robust preconditioning techniques and coarse multiscale approximations. Certainly, the presence of nonlinearities or many possible (properly parametrized) scenarios make this task even more challenging. In particular, the design and analysis of iterative solvers with good convergence properties with respect to physical parameters and nonlinearities is important for applications. A main interest of this minisymposium is to develop techniques and algorithms to approach efficiently heterogeneous and nonlinear problems such as Richard's equation for heterogeneous porous media and other nonlinear models. In this session, we bring together experts working on domain decomposition methods for multiscale and nonlinear problems.

7. Domain decomposition techniques in practical flow applications (Menno Genseberger, Mart Borsboom)

   Last decade's domain decomposition techniques have been incorporated in large computer codes for real-life applications. By bringing together some of them, this minisymposium aims to (1) illustrate the importance of domain decomposition (for modeling flexibility, parallel performance, etc.) in the application field and (2) highlight the applied domain decomposition techniques, to discuss these approaches and, reconsider or further improve them. Application area is restricted to hydrodynamics to have a good basis for further discussion. The presentations consider domain decomposition techniques in large computer codes being used worldwide for shallow water flow in coastal areas, lakes, rivers, ocean flow, and climate modeling.

8. Domain decomposition methods in implementations (Christian Engwer, Guido Kanschat)

   Domain decomposition and subspace correction methods are tools with potential for high impact on practical applications. They yield efficient solvers for high performance simulations of multi-physics applications or multi-scale problems, way beyond the realm of currently available theoretical analysis. They are in particular suitable for generic implementations in tool-boxes and programming libraries, since they replicate structures existing on the whole computational domain on subdomains, and their mathematical structure coincides with parallel implementation. Thus, it is possible to implement these methods in a way that their optimal performance can be evaluated for the provable problems, but application of the very same code structures to more advanced problems is straightforward. We bring together experts on the development of software frameworks for high performance computing and on challenging applications to discuss possible approaches for generic implementations as well as demands posed by advanced applications and performance results. By incorporating improved domain decomposition algorithms into high-level frameworks, they can be made readily accessible to a wide audience without particular knowledge of their technical details. The talks focus on different challenges in the context of domain decomposition methods, e.g., multi-physics simulations, construction of preconditioners or generic parallel simulations, and discuss how such topics can be incorporated into a general purpose framework and made available to the application level.

9. Parallel multigrid methods (Karsten Kahl, Matthias Bolten)

   Modern simulation codes must solve extremely large systems of equations—up to billions of equations. Hence, there is an acute need for scalable parallel linear solvers, i.e., algorithms for which the time to solution (or number of iterations) remains constant as both problem size and number of processors increase. Multigrid (MG), known to be an optimal serial algorithm, is often scalable when implemented on a parallel computer. However, newly emerging many-core architectures present several new challenges that must be addressed if these methods are to be competitive on such computing platforms. Here, we discuss new techniques for parallelizing MG solvers for various problems.

10. Efficient solvers for frequency domain wave problems (Victorita Dolean, Martin J. Gander, Ivan Graham)

   In this minisymposium, we explore iterative methods for frequency domain wave problems such as the Helmholtz and Maxwell equations. Driven by important technological applications, considerable recent progress in this topic aims towards obtaining a wavenumber robust efficient scalable solver, accompanied by a rigorous analysis. The minisymposium discusses several areas of recent progress including sweeping and source transfer preconditioners, techniques based on the principle of limited absorption, and new advances in optimized Schwarz methods.

11. Domain decomposition methods for environmental modeling (Florian Lemarié, Antoine Rousseau)
    Many applications in geophysical fluid dynamics and natural hazards prediction require the development of domain decomposition methods (DDMs) either to optimally use the increasing computational power or to accurately simulate multi-physics phenomena. Due to the complexity of such numerical codes, additional constraints arise in the design of the numerical methods as, for example, in space-time discretizations, subgrid scale parameterizations, physical/numerical interfaces, etc. In this context, a compromise between efficient numerical methods and their according constraints imposed by the target applications must be found. The aim of this minisymposium is to bring together theoretical and applied scientists working on realistic environmental simulations. Work presented explores a range of applications from hydrological, oceanic, and atmospheric modeling to earthquake dynamics.
12. Efficient solvers (Sébastien Loisel)
    Solving large problems is a core interest in domain decomposition. In order to be useful, an algorithm should be efficient—whether from high parallelization, ease of implementation, or low floating point operation counts. One may improve the efficiency of algorithms by carefully choosing artificial interface boundary conditions (Dirichlet, Neumann, or Robin); this choice then impacts the design and implementation of algorithms. A further issue is the physical nature of the problem (e.g., elliptic or parabolic, with possible heterogeneities). In this minisymposium, we discuss algorithms related to the optimized Schwarz and FETI methods and consider especially their performance advantages.
13. Space-time parallel methods (Daniel Ruprecht, Robert Speck)
    The number of cores in modern supercomputers increases rapidly, requiring new inherently parallel algorithms in order to actually harness their computational capacities. This fact leads to increasing need for methods that provide levels of concurrency in addition to already ubiquitous spatial parallelization. For time-dependent problems, algorithms that replace classical serial time-stepping with typically iterative approaches more amenable to parallelization have been demonstrated to be promising. The minisymposium features four talks on recent methodological and application-related developments for three different methods: Parareal, revisionist deferred corrections (RIDC), and the parallel full approximation scheme in space and time (PFASST).

Contributed Presentations

The contributed talks have been distributed over 14 different sessions:

1. Helmholtz equation
2. Implementation strategies
3. Flow and porous media
4. Adaptivity in HPC simulations

5. Additive Schwarz methods
6. Optimized Schwarz methods
7. Parallelization in time
8. Maxwell's equation
9. Inverse problems
10. Preconditioners and solvers
11. Non-matching meshes
12. Multiphysics problems
13. Parallelization in time
14. FETI and BDD methods

Acknowledgements

In closing, we would like to thank all the participants gathered here in Lugano for their contributions to the scientific success of this conference. Moreover, it is our pleasure to express our sincere thanks to everybody who has supported this conference on the administrative side. This includes the chairs of the conference sessions, the students helping on the practical and technical issues, and last but not least, the dean's office of the faculty of informatics, whose members have provided invaluable support.

R. Krause
University of Lugano, Lugano, Switzerland

T. Dickopf
Technische Universität München, Garching, Germany

M.J. Gander
Université de Genève, Genève, Switzerland

L. Halpern
Université Paris XIII, Villetaneuse, France

L.F. Pavarino
Università degli Studi di Milano, Milano, Italy

30 September 2015

Contents

Part I
Plenary Talks (PT)

Multigrid Algorithms for High Order Discontinuous Galerkin Methods

Paola F. Antonietti, Marco Sarti, and Marco Verani

1 Introduction

In the framework of multigrid solvers for Discontinuous Galerkin (DG) schemes, the first contributions are due to [10, 16]. In [16] a V-cycle preconditioner for a Symmetric Interior Penalty (SIP) discretization of an elliptic problem is analyzed. They prove that the condition number of the preconditioned system is uniformly bounded with respect to the mesh size and the number of levels. In [10] V-cycle, F-cycle and W-cycle multigrid schemes for SIP discretizations are presented and analyzed, employing the additive theory developed in [8, 9]. A uniform bound for the error propagation operator is shown provided the number of smoothing steps is large enough. All the previously cited works focus on low order, i.e., linear, DG approximations. With regard to high order DG discretizations, h- and p-multigrid schemes are successfully employed for the numerical solution of many different kinds of problems, see e.g. [6, 14, 20–22, 24], even if only few theoretical results are available that show the convergence properties of the underlying algorithms. In the context of fast solution techniques for high order DG methods, we mention [1, 11, 12], see also [3] were a substructuring preconditioner is analyzed for an hp domain decomposition method with interior penalty mortaring. Recently, in [2] a convergence analysis of W-cycle h- and p-multigrid algorithms for a wide

P.F. Antonietti (✉) • M. Sarti • M. Verani
MOX, Dipartimento di Matematica, Politecnico di Milano, Piazza Leonardo da Vinci 32, 20133 Milano, Italy
e-mail: paola.antonietti@polimi.it; marco.sarti@polimi.it; marco.verani@polimi.it

T. Dickopf et al. (eds.), *Domain Decomposition Methods in Science and Engineering XXII*, Lecture Notes in Computational Science and Engineering 104, DOI 10.1007/978-3-319-18827-0_1

class of high order DG schemes is provided. More precisely, it is shown that, if a Richardson smoother is employed, the W-cycle algorithms converge uniformly with respect the granularity of the underlying mesh and the number of levels; but the contraction factor of the scheme deteriorates when increasing the polynomial order. As a further development of the results contained in [2], the aim of this paper is to investigate the performance of *h*- and *p*-multigrid algorithms for high order DG methods, considering a wide class of smoothers and addressing both two- and three-dimensional test cases. The paper is organized as follows. In Sect. 2 we briefly introduce the model problem and its DG discretization. The *h*- and *p*-multigrid methods are described in Sect. 3. The numerical experiments are presented in Sect. 4, where the W-cycle schemes are tested on two- and three-dimensional problems.

2 Model Problem and DG Methods

Given an open, bounded polygonal/polyhedral domain Ω and a given function $f \in L^2(\Omega)$, we consider the weak formulation of the Poisson problem with homogeneous boundary conditions: find $u \in H_0^1(\Omega)$ such that

$$(\nabla u, \nabla v)_\Omega = (f, v)_\Omega \qquad \forall v \in H_0^1(\Omega), \tag{1}$$

where $(\cdot,\cdot)_\Omega$ denotes the standard L^2 product. We consider a sequence of discontinuous finite dimensional spaces V_k, $k = 1, \dots, K$, defined as

$$V_k = \{v \in L^2(\Omega) : v|_E \in \mathbb{M}^{p_k}(T) \quad \forall\, T \in \mathcal{T}_k\} \quad k = 1, \dots, K,$$

where $\mathbb{M}^{p_k}$ is a suitable space of polynomials of degree $p_k \geq 1$ and $\mathcal{T}_k$ is a partition of Ω with granularity h_k. The sequence of spaces V_k is generated with two different approaches, depending on whether we are interested in *h*- or *p*-multigrid algorithms. In the *h*-multigrid algorithm, we fix the polynomial approximation degree $p_k = p$ for all $k = 1, \dots, K$, and the spaces V_k are associated to a sequence of nested partitions $\{\mathcal{T}_k\}_{k=1,\dots,K}$ obtained from successive uniform refinements of an initial (coarse) shape regular and quasi-uniform partition $\mathcal{T}_1$, cf. Fig. 1 (left). In *p*-multigrid

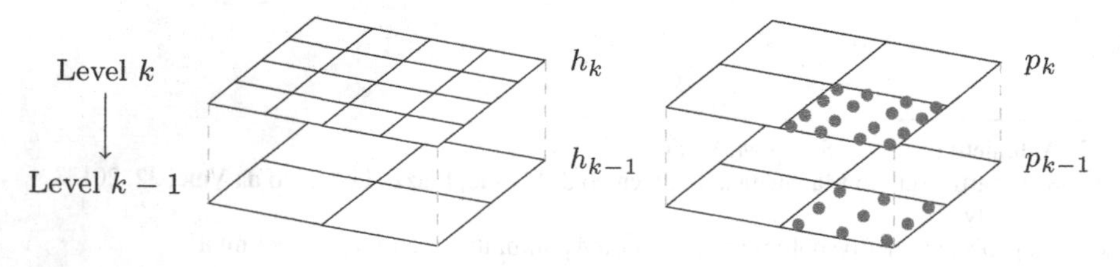

Fig. 1 Sample of the space V_k and V_{k-1} in the *h*- (*left*) and *p*- (*right*) multigrid schemes

schemes, the grid is kept fixed on all the levels and from the level k to the level $k-1$ the polynomial degree is lowered down, i.e., $p_{k-1} \leq p_k$ for any $k = 2, \ldots, K$, cf. Fig. 1 (right). Notice that, with such a construction the spaces V_k are nested, i.e., $V_1 \subseteq V_2 \subseteq \cdots \subseteq V_K$. For the sake of simplicity, we will also suppose that the polynomial degrees p_k satisfy the following local bounded variation among levels: there exists a constant $C > 0$ such that $p_k \leq C p_{k-1}$, for any $k = 2, \ldots, K$.

For any level k, we denote by $\mathcal{F}_k^I$ and $\mathcal{F}_k^B$ the sets of interior and boundary faces of $\mathcal{T}_k$, respectively, set $\mathcal{F}_k = \mathcal{F}_k^I \cup \mathcal{F}_k^B$, and define the lifting operators

$$(\mathcal{R}_k(\boldsymbol{\tau}), \boldsymbol{\eta})_\Omega = -\sum_{F \in \mathcal{F}_k} (\boldsymbol{\tau}, \{\boldsymbol{\eta}\})_F \quad \forall \boldsymbol{\eta} \in [V_k]^d, \quad k = 1, \ldots, K,$$

$$(\mathcal{L}_k(v), \boldsymbol{\eta})_\Omega = -\sum_{F \in \mathcal{F}_k^I} (v, [\![\boldsymbol{\eta}]\!])_F \quad \forall \boldsymbol{\eta} \in [V_k]^d, \quad k = 1, \ldots, K,$$

where the *jump* and *average* trace operators are defined as in [5].

We next define the DG bilinear forms $\mathcal{A}_k(\cdot, \cdot) : V_k \times V_k \to \mathbb{R}$, $k = 1, \ldots, K$, as

$$\begin{aligned}\mathcal{A}_k(w, v) = (\nabla w + \mathcal{R}_k([\![w]\!]) + \mathcal{L}_k(\boldsymbol{\beta} \cdot [\![w]\!]), \nabla v + \mathcal{R}_k([\![v]\!]) + \mathcal{L}_k(\boldsymbol{\beta} \cdot [\![v]\!]))_\Omega \\ - \theta(\mathcal{R}_k([\![w]\!]), \mathcal{R}_k([\![v]\!]))_\Omega + \sum_{F \in \mathcal{F}_k} (\sigma_k [\![w]\!], [\![v]\!])_F \qquad (2)\end{aligned}$$

where, for a constant $\alpha_k > 0$, the stabilization function σ_k is defined as

$$\sigma_k|_F = \frac{\alpha_k p_k^2}{\min\{\mathrm{diam}(T^+), \mathrm{diam}(T^-)\}} \quad F \in \mathcal{F}_k^I, \quad \sigma_k|_F = \frac{\alpha_k p_k^2}{\mathrm{diam}(T)} \quad F \in \mathcal{F}_k^B,$$

$T^{\pm}$ being the two neighboring elements sharing the face $F \in \mathcal{F}_k^I$. For $\theta = 1$ and $\boldsymbol{\beta} = \mathbf{0}$, the bilinear form (3) correspond to the SIP method [4], whereas for $\theta = 1$ and $\boldsymbol{\beta}$ a uniformly bounded (and possibly null) vector in $\mathbb{R}^d$ we obtain the LDG bilinear form [13].

We are interested in solving the following problem on the finest level K:

$$\text{find } u_K \in V_K \text{ such that } \quad \mathcal{A}_K(u_K, v_K) = (f, v_K)_\Omega \quad \forall v_K \in V_K, \qquad (3)$$

with a W-cycle multigrid method. Fixing a basis for V_K, Eq. (3) is equivalent to the following linear system of equations

$$A_K u_K = F_K, \qquad (4)$$

where A_K and F_K are the matrix representations of the bilinear form $\mathcal{A}_K(\cdot, \cdot)$ and of the right hand side in (3), respectively, and where, with a slight abuse of notation, we used to the same symbol to denote both a function in the finite element space V_K and the vector of its expansion coefficients in a given basis.

It can be shown that the bilinear form $\mathcal{A}_K(\cdot,\cdot)$ defined in (3) is continuous and coercive with respect to the following (mesh-dependent) DG norm

$$\|v\|_{\mathrm{DG},K}^2 = \sum_{T\in\mathcal{T}_K} \|\nabla v\|_{L^2(T)}^2 + \sum_{F\in\mathcal{F}_K} \|\sigma_K^{1/2}[\![v]\!]\|_{L^2(F)}^2, \tag{5}$$

and that the following error estimates are satisfied, cf. [18, 23, 25] for example.

Theorem 1 *Let u be the exact solution of problem* (1) *such that $u \in H^{s+1}(\mathcal{T}_K)$, $s \geq 1$, and let $u_K \in V_K$ be the DG solution of problem* (3)*. Then,*

$$\|u - u_K\|_{\mathrm{DG},K} \lesssim \frac{h_K^{\min(p_K,s)}}{p_K^{s-\mu/2}} \|u\|_{H^{s+1}(\mathcal{T}_K)}, \tag{6}$$

with $\mu = 0$ whenever a continuous interpolant can be built, cf. [25], or the projector of [15] can be employed and $\mu = 1$ otherwise.

3 W-Cycle *h*- and *p*-Multigrid Algorithms

As usual in the multigrid framework, we will employ a recursive algorithm to describe the multigrid scheme. To this aim, we define on each level k the problem

$$A_k z_k = b_k,$$

where A_k is the matrix representation of the bilinear form (3), and z_k, b_k are vectors of dimension $\dim(V_k)$. The first ingredient to build a multigrid algorithm are the intergrid transfer operators, which we denoted by R_{k-1}^k (prolongation from V_{k-1} to V_k) and by R_k^{k-1} (restriction from V_k to V_{k-1}). Given we are considering nested spaces, we can simply take R_{k-1}^k as the classical embedding operator and R_k^{k-1} as its adjoint with respect to the L^2 scalar product. The second ingredient is a suitable smoother, which we denote by B_k. Denoting by $u_k^{(0)} \in V_k$ the initial guess, and by m_1 and m_2 the number of pre- and post-smoothing steps, respectively, the W-cycle multigrid algorithm $u_k = \mathsf{MG}_{\mathcal{W}}(k, b_k, u_k^{(0)}, m_1, m_2)$ is defined recursively as shown in Algorithm 1. We then employ Algorithm 1 to solve the linear system (4), i.e.,

$$u_K = \mathsf{MG}_{\mathcal{W}}(K, b_K, u_K^{(0)}, m_1, m_2).$$

Notice that if the spaces V_k are associated to a sequence of grids $\mathcal{T}_k$ with variable mesh size and the polynomial degree is kept fixed on all the levels we obtain the

Algorithm 1 $u_k = \mathsf{MG}_{\mathcal{W}}(k, b_k, u_k^{(0)}, m_1, m_2)$

```
if k=1 then                                                    ▷ Solution on the coarsest level
    Solve A_k u_k = b_k
else
    for ℓ = 1,...,m_1 do                                       ▷ Pre-smoothing
        Set B̃_k = B_k, if ℓ is odd and B̃_k = B_k^T if ℓ is even.
        u_k^(ℓ) = u_k^(ℓ-1) + B̃_k^{-1}(b_k − A_k u_k^(ℓ-1));
    end for
    Set r_{k-1} = R_k^{k-1}(b_k − A_k u_k^(m_1));             ▷ Restriction of the residual
    Set u_{k-1}^(0) = 0;
    Call ē_{k-1} = MG_W(k−1, r_{k-1}, u_{k-1}^(0), m_1, m_2);  ▷ Recursion
    Call e_{k-1} = MG_W(k−1, r_{k-1}, ē_{k-1}, m_1, m_2);
    Set u_k^(m_1+1) = u_k^(m_1) + R_{k-1}^k e_{k-1};
    for ℓ = m_1+2,...,m_1+m_2+1 do                             ▷ Post-smoothing
        Set B̃_k = B_k, if ℓ is odd and B̃_k = B_k^T if ℓ is even.
        u_k^(ℓ) = u_k^(ℓ-1) + B̃_k^{-1}(b_k − A_k u_k^(ℓ-1));
    end for
    Set u_k = u_k^(m_1+m_2+1);
end if
```

W-cycle h-multigrid scheme, whereas if the mesh is kept fixed and the polynomial degree is lower down from one level to a coarser one we then have a W-cycle p-multigrid algorithm.

We next introduce the following operator $P_k^{k-1} : V_k \to V_{k-1}$

$$\mathcal{A}_{k-1}(P_k^{k-1} v, w) = \mathcal{A}_k(v, R_{k-1}^k w) \quad \forall v \in V_k, w \in V_{k-1},$$

and the following discrete norm

$$|||v|||_{1,k}^2 = (A_k v, v)_k = \mathcal{A}_k(v, v) \quad \forall v \in V_k.$$

The error propagation operator associated to the W-cycle multigrid scheme is given by

$$E_{k,m_1,m_2} v = \begin{cases} 0 & k = 1, \\ G_k^{m_2}(\mathrm{I}_k - R_{k-1}^k(\mathrm{I}_k - E_{k-1,m_1,m_2}^2)P_k^{k-1})G_k^{m_1} v & k > 1, \end{cases}$$

where I_k is the identity operator, and $G_k = \mathrm{I}_k - B_k^{-1} A_k$, cf. [7, 17]. The following result, which is proved in [2], state that, whenever a Richardson smoother is employed, the W-cycle algorithms converge uniformly with respect to the granularity of the underlying mesh and the number of levels, provided the number of smoothing steps is chosen sufficiently large, but the contraction factor of the scheme deteriorates when increasing the approximation order.

Theorem 2 *For any k, let B_k be the Richardson smoother, i.e., $B_k = \Lambda_k I_k$, where Λ_k is an upper bound for the maximum eigenvalue of A_k. Then, there exist a constant $C_W > 0$ and an integer m_W that are independent of the mesh size, but dependent on the polynomial degree, such that*

$$\|E_{k,m_1,m_2} v\|_{1,k} \leq C_W \frac{p_k^{2+\mu}}{(1+m_1)^{1/2}(1+m_2)^{1/2}} \|v\|_{1,k} \quad \forall v \in V_k, \quad k = 2, \ldots, K,$$

provided $m_1 + m_2 \geq m_W = m_W(p_k)$.

4 Numerical Results

In this section we test the performance of the W-cycle h- and p-multigrid schemes in both two- and three-dimensional test cases and with different choices of smoothers. We compute the convergence factor as

$$\rho = \exp\left(\frac{1}{N} \ln \frac{\|\mathbf{r}_N\|_2}{\|\mathbf{r}_0\|_2}\right),$$

with N denoting the iteration counts needed to achieve convergence up to a relative tolerance of 10^{-8} and $\mathbf{r}_N$ and $\mathbf{r}_0$ denoting the final and initial computed residuals, respectively. Throughout the section we have employed an equal number of pre- and post-smoothing steps, i.e., $m_1 = m_2 = m$, and we have set the penalty parameter α_k appearing in the definition of the DG bilinear form as $\alpha_k = 10$, for any level $k = 1, \ldots, K$.

We first consider a two-dimensional example with $\Omega = (0,1)^2$ and focus on the performance of the h-multigrid algorithm. To this aim, we fix a coarse (triangular/Cartesian) grid $\mathcal{T}_1$ with granularity $h_1 = 0.25$ and consider a sequence of nested grids $\mathcal{T}_k$, $k = 2, \ldots, K$, obtained from successive uniform refinements of $\mathcal{T}_1$. In Table 1 we report the computed convergence factors as a function of the number of smoothing steps m and the number of levels K, fixing the polynomial degree $p_k = p = 1, 2$ for all the levels $k = 1, \ldots, K$. The results reported in Table 1 have been obtained with the SIP method on structured triangular grids and with the LDG scheme on Cartesian grids, and employing a Richardson smoother. The symbol "-" means that the maximum number of 1000 iterations has been reached without achieving the desired tolerance. We have repeated the same set of experiments employing $p = 3, 4$, and the same behavior as been observed; for brevity these results have been omitted. As expected from Theorem 2, the convergence factor is independent of the number of levels K, decreases when m increases, and the performance of the algorithm deteriorates as p grows up.

We next fix the number of pre- and post-smoothing steps $m = 6$, and investigate how the performance of the h-multigrid algorithm depends on the polynomial degree, always employing a Richardson smoother. Table 2 shows the computed

Table 1 2D test case, SIP and LDG methods, h-multigrid scheme

	SIP, triangular grids				LDG, cartesian grids			
	$K=2$	$K=3$	$K=4$	$K=5$	$K=2$	$K=3$	$K=4$	$K=5$
	$p=1$							
$m=2$	0.77	0.78	0.78	0.78	–	–	–	–
$m=4$	0.60	0.62	0.62	0.62	0.86	0.88	0.87	0.87
$m=10$	0.38	0.40	0.40	0.39	0.74	0.76	0.76	0.75
	$p=2$							
$m=2$	0.93	0.94	0.93	0.78	0.96	0.96	0.96	0.96
$m=4$	0.87	0.88	0.88	0.62	0.93	0.93	0.93	0.92
$m=10$	0.76	0.77	0.77	0.39	0.88	0.88	0.88	0.87

Convergence factor as a function of the number of levels K, the polynomial approximation degree p, and the number of smoothing steps m. Richardson smoother

Table 2 2D test case, SIP and LDG methods, h-multigrid scheme

	SIP, triangular grids			LDG, cartesian grids		
	$K=2$	$K=3$	$K=4$	$K=2$	$K=3$	$K=4$
$p=1$	0.50	0.51	0.50	0.81	0.82	0.82
$p=2$	0.83	0.84	0.84	0.91	0.91	0.91
$p=3$	0.91	0.92	0.91	0.94	0.94	0.93
$p=4$	0.95	0.94	0.93	0.96	0.95	0.95
$p=5$	0.96	0.95	0.94	0.97	0.95	0.96
$p=6$	0.95	0.96	0.96	0.98	0.96	0.97

Convergence factor as a function of the number of levels K and the polynomial approximation order p. Richardson smoother ($m = 6$)

convergence factors as a function of the polynomial degree $p = 1, 2, \ldots, 6$ and the number of levels $K = 2, 3, 4$, for both the SIP and LDG methods. We observe that, as predicted by Theorem 2, the performance of the h-multigrid algorithm are independent of the number of levels but deteriorates as p increases.

We next test the performance of the h-multigrid scheme employing different smoothers as the Gauss-Seidel smoother of [16], an (elementwise) block Gauss-Seidel smoother and the polynomial smoother proposed in [19]. In Table 3 we report the computed convergence factors as a function of the number of pre- and post-smoothing steps $m = 2, 4, 10$, the number of levels $K = 2, 3, 4$ and the polynomial approximation degree $p = 1, 2, 3, 4$. These results have been obtained with the SIP method and employing triangular grids. In all the cases the performance of the h-multigrid algorithm seems to be fairly independent of the number of levels. Moreover, as expected, the convergence factor decreases as the number of smoothing steps increases, but still deteriorates as p grows up (even if the dependence of the convergence factor on p seems to be weaker than for the Richardson smoother). Moreover, all the smoothers outperform the Richardson smoother and the polynomial smoother seems to provide the best convergence factors. The extension of the convergence analysis presented in [2] to

Table 3 2D test case, SIP method (triangular grids), h-multigrid scheme

	$p=1$			$p=2$			$p=3$			$p=4$		
$K \rightarrow$	2	3	4	2	3	4	2	3	4	2	3	4
	Gauss-Seidel smoother											
$m=2$	0.55	0.56	0.56	0.80	0.80	0.80	0.88	0.87	0.86	0.92	0.92	0.93
$m=4$	0.40	0.41	0.41	0.68	0.68	0.68	0.79	0.78	0.77	0.86	0.86	0.86
$m=10$	0.20	0.21	0.21	0.44	0.44	0.44	0.61	0.59	0.58	0.71	0.71	0.70
	Block Gauss-Seidel smoother											
$m=2$	0.55	0.56	0.56	0.71	0.72	0.72	0.82	0.82	0.82	0.84	0.84	0.84
$m=4$	0.40	0.42	0.41	0.54	0.56	0.55	0.70	0.70	0.70	0.73	0.73	0.73
$m=10$	0.20	0.21	0.21	0.27	0.31	0.29	0.47	0.47	0.46	0.51	0.50	0.50
	Polynomial smoother											
$m=2$	0.30	0.31	0.31	0.68	0.69	0.68	0.80	0.80	0.78	0.89	0.88	0.87
$m=4$	0.17	0.17	0.17	0.50	0.50	0.49	0.66	0.65	0.63	0.80	0.79	0.78
$m=10$	0.07	0.07	0.06	0.21	0.21	0.21	0.40	0.38	0.37	0.60	0.59	0.59

Convergence factors as a function of the number of levels K, the polynomial approximation degree p, and the number of smoothing steps m. Gauss-Seidel, block Gauss-Seidel and polynomial smoothers

Table 4 2D test case, SIP and LDG methods, p-multigrid scheme

	SIP, triangular grid			LDG, cartesian grid		
	$K=2$	$K=3$	$K=4$	$K=2$	$K=3$	$K=4$
$m=2$	0.91	0.91	0.94	0.95	0.95	0.97
$m=4$	0.85	0.85	0.90	0.88	0.89	0.92
$m=10$	0.78	0.77	0.80	0.86	0.86	0.89

Convergence factor as a function of the number of levels K and the number of smoothing steps m. Richardson smoother, $p_K=5$

h-multigrid algorithms based on these (more effective) smoothers is currently under investigation.

We next turn our attention to the performance of the p-multigrid scheme. To this aim, we fix the finest computational level K, the mesh $\mathcal{T}_K$ and the polynomial approximation degree $p_K \geq K$ employed therein. Then, for each level k, we set $p_{k-1} = p_k - 1$, $k = K, K-1, \ldots, 2$. In Table 4 we report the computed convergence factors obtained with $p_K = 5$ and varying the number of smoothing steps m and the number of levels K. The results reported in Table 4 have been obtained with the LDG and SIP methods and employing a Richardson smoother. Next, we fix the number of smoothing steps $m = 6$ and vary the polynomial approximation degree p_K employed on the finest level. The results obtained with the SIP method and employing the Richardson smoother are reported in Table 5. From the results reported in Table 4 and in Table 5, we can conclude that the p-multigrid scheme seems to be asymptotically uniform with respect to the number of levels (notice that in this case the ratio p_k/p_{k-1} is not constant from one level to the other), and that, as expected, the performance of the algorithm improves as m increases. We finally

Table 5 2D test case, SIP and LDG methods, p-multigrid scheme

	SIP, triangular grid			LDG, Cartesian grid		
	$K=2$	$K=3$	$K=4$	$K=2$	$K=3$	$K=4$
$p_K=2$	0.62	–	–	0.83	–	–
$p_K=3$	0.77	0.77	–	0.89	0.90	–
$p_K=4$	0.79	0.80	0.86	0.86	0.89	0.90
$p_K=5$	0.83	0.82	0.87	0.89	0.89	0.92
$p_K=6$	0.86	0.86	0.86	0.91	0.91	0.90

Convergence factor as a function of the number of levels K and the polynomial degree p_K. Richardson smoother ($m = 6$)

Table 6 2D test case, SIP method (triangular grid), p-multigrid scheme

	$p_K=2$	$p_K=3$		$p_K=4$			$p_K=5$			$p_K=6$		
$K\rightarrow$	2	2	3	2	3	4	2	3	4	2	3	4
$m=2$	0.76	0.79	0.79	0.84	0.84	0.85	0.85	0.85	0.85	0.88	0.87	0.86
$m=4$	0.60	0.66	0.66	0.73	0.73	0.73	0.75	0.75	0.75	0.79	0.78	0.77
$m=6$	0.48	0.57	0.56	0.63	0.63	0.63	0.67	0.67	0.67	0.71	0.71	0.70
$m=10$	0.34	0.44	0.44	0.49	0.49	0.49	0.56	0.56	0.56	0.59	0.58	0.58

Convergence factor as a function of the number of levels K, the polynomial degree p_K, and the number of smoothing steps m. Gauss-Seidel smoother

address the performance of the p-multigrid method when employing a different smoother. For this set of experiments we have considered the SIP formulation and tested the Gauss-Seidel smoother. The results reported in Table 6 show the computed convergence factors as a function of the number of levels K, the number of smoothing steps m and the polynomial degree p_K employed on the finest level. The computed convergence factor seems to be fairly insensitive to the number of levels employed in the algorithm and it improves as the number of pre- and post-smoothing steps increases (notice that, no restriction on the minimum number of smoothing steps seems to be needed in this case). Nevertheless, the convergence factor still depends on the polynomial degree even if such a dependence seems to be weaker than that observed employing the Richardson smoother (cf. Table 5). Finally, comparing these results with the ones reported in Table 5 it is clear that, as for the h- multigrid algorithm, the Gauss-Seidel smoother outperforms the Richardson smoother.

We next present some three-dimensional numerical experiments. We have employed an h-multigrid scheme to solve the linear system of equations arising from the SIP discretization of model problem (1) posed on $\Omega = (0, 1)^3$. We employ a sequence of tetrahedral meshed obtained by successive uniform refinements of an initial coarse grid with granularity $h_1 = 0.25$. As before, we fix $p_k = p$ for all the levels $k = 1, 2, \ldots, K$ and consider the Richardson, the Gauss-Seidel and the symmetric Gauss-Seidel smoothers. The computed convergence factors varying the number of levels K, the number of pre-and post-smoothing steps m as well as the polynomial degree p are reported in Table 7. We observe that the performance of

Table 7 3D test case, SIP method (tetrahedral grids), h-multigrid scheme

	$p=1$			$p=2$			$p=3$	
	$K=2$	$K=3$	$K=4$	$K=2$	$K=3$	$K=4$	$K=2$	$K=3$
	Richardson smoother							
$m=2$	0.57	0.55	0.53	0.82	0.81	0.80	0.90	0.90
$m=4$	0.71	0.71	0.69	0.91	0.90	0.90	0.95	0.95
$m=10$	0.46	0.44	0.41	0.79	0.78	0.77	0.88	0.88
	Gauss-Seidel smoother							
$m=2$	0.57	0.55	0.53	0.82	0.81	0.79	0.89	0.89
$m=4$	0.35	0.33	0.30	0.68	0.67	0.65	0.81	0.80
$m=10$	0.13	0.15	0.12	0.43	0.41	0.40	0.61	0.60
	Symmetric Gauss-Seidel smoother							
$m=2$	0.35	0.33	0.30	0.68	0.67	0.65	0.81	0.80
$m=4$	0.17	0.19	0.16	0.50	0.48	0.46	0.67	0.66
$m=10$	0.05	0.08	0.07	0.22	0.22	0.20	0.41	0.39

Convergence factors as a function of the number of levels K, the polynomial approximation degree p, and the number of smoothing steps m. Richardson, Gauss-Seidel, and symmetric Gauss-Seidel smoothers

the h-multigrid schemes are completely analogous to the one observed in the two-dimensional test case.

References

1. P.F. Antonietti, P. Houston, A class of domain decomposition preconditioners for *hp*-discontinuous Galerkin finite element methods. J. Sci. Comput. **46**(1), 124–149 (2011)
2. P.F. Antonietti, M. Sarti, M. Verani, Multigrid algorithms for *hp*-discontinuous Galerkin discretizations of elliptic problems. SIAM J. Numer. Anal. **53**(1), 598–618 (2015)
3. P.F. Antonietti, B. Ayuso, S. Bertoluzza, M. Penacchio, Substructuring preconditioners for an *hp* domain decomposition method with interior penalty mortaring. Calcolo **52**(3), 289–316 (2015)
4. D.N. Arnold, An interior penalty finite element method with discontinuous elements. SIAM J. Numer. Anal. **19**(4), 742–760 (1982)
5. D.N. Arnold, F. Brezzi, B. Cockburn, L.D. Marini, Unified analysis of discontinuous Galerkin methods for elliptic problems. SIAM J. Numer. Anal. **39**(5), 1749–1779 (2001/2002)
6. F. Bassi, A. Ghidoni, S. Rebay, P. Tesini, High-order accurate p-multigrid discontinuous Galerkin solution of the Euler equations. Int. J. Numer. Methods Fluids **60**(8), 847–865 (2009)
7. J. Bramble, *Multigrid Methods*. Number 294 in Pitman Research Notes in Mathematics Series (Longman Scientific & Technical, London, 1993)
8. S.C. Brenner, Convergence of the multigrid V-cycle algorithm for second-order boundary value problems without full elliptic regularity. Math. Comput. **71**(238), 507–525 (2002)
9. S.C. Brenner, Convergence of nonconforming V-cycle and F-cycle multigrid algorithms for second order elliptic boundary value problems. Math. Comput. **73**(247), 1041–1066 (2004)
10. S.C. Brenner, J. Zhao, Convergence of multigrid algorithms for interior penalty methods. Appl. Numer. Anal. Comput. Math. **2**(1), 3–18 (2005)

11. K. Brix, M. Campos Pinto, C. Canuto, W. Dahmen, Multilevel preconditioning of discontinuous Galerkin spectral element methods. Part I: geometrically conforming meshes. IMA J. Numer. Anal. (2014). doi:10.1093/imanum/dru053
12. C. Canuto, L.F. Pavarino, A.B. Pieri, BDDC preconditioners for continuous and discontinuous Galerkin methods using spectral/*hp* elements with variable local polynomial degree. IMA J. Numer. Anal. **34**(3), 879–903 (2014)
13. B. Cockburn, C.-W. Shu, The local discontinuous Galerkin method for time-dependent convection-diffusion systems. SIAM J. Numer. Anal. **35**(6), 2440–2463 (electronic) (1998)
14. K.J. Fidkowski, T.A. Oliver, J. Lu, D.L. Darmofal, *p*-multigrid solution of high-order discontinuous Galerkin discretizations of the compressible Navier-Stokes equations. J. Comput. Phys. **207**(1), 92–113 (2005)
15. E.H. Georgoulis, E. Süli, Optimal error estimates for the *hp*-version interior penalty discontinuous Galerkin finite element method. IMA J. Numer. Anal. **25**(1), 205–220 (2005)
16. J. Gopalakrishnan, G. Kanschat, A multilevel discontinuous Galerkin method. Numer. Math. **95**(3), 527–550 (2003)
17. W. Hackbusch, *Multi-Grid Methods and Applications*. Springer Series in Computational Mathematics, vol. 4 (Springer, Berlin, 1985)
18. P. Houston, C. Schwab, E. Süli, Discontinuous *hp*-finite element methods for advection-diffusion-reaction problems. SIAM J. Numer. Anal. **39**(6), 2133–2163 (2002)
19. J. Kraus, P. Vassilevski, L. Zikatanov, Polynomial of best uniform approximation to $1/x$ and smoothing in two-level methods. Comput. Methods Appl. Math. **12**(4), 448–468 (2012)
20. H. Luo, J.D. Baum, R. Löhner, A *p*-multigrid discontinuous Galerkin method for the Euler equations on unstructured grids. J. Comput. Phys. **211**(2), 767–783 (2006)
21. B.S. Mascarenhas, B.T. Helenbrook, H.L. Atkins, Coupling *p*-multigrid to geometric multigrid for discontinuous Galerkin formulations of the convection-diffusion equation. J. Comput. Phys. **229**(10), 3664–3674 (2010)
22. C.R. Nastase, D.J. Mavriplis, High-order discontinuous Galerkin methods using an *hp*-multigrid approach. J. Comput. Phys. **213**(1), 330–357 (2006)
23. I. Perugia, D. Schötzau, An *hp*-analysis of the local discontinuous Galerkin method for diffusion problems. J. Sci. Comput. **17**(1–4), 561–571 (2002)
24. K. Shahbazi, D.J. Mavriplis, N.K. Burgess, Multigrid algorithms for high-order discontinuous Galerkin discretizations of the compressible Navier-Stokes equations. J. Comput. Phys. **228**(21), 7917–7940 (2009)
25. B. Stamm, T.P. Wihler, *hp*-optimal discontinuous Galerkin methods for linear elliptic problems. Math. Comput. **79**(272), 2117–2133 (2010)

BDDC Deluxe for Isogeometric Analysis

L. Beirão da Veiga, L.F. Pavarino, S. Scacchi, O.B. Widlund, and S. Zampini

1 Introduction

The main goal of this paper is to design, analyze, and test a BDDC (Balancing Domain Decomposition by Constraints, see [12, 23]) preconditioner for Isogeometric Analysis (IGA), based on a novel type of interface averaging, which we will denote by *deluxe scaling*, with either full or reduced set of primal constraints. IGA is an innovative numerical methodology, introduced in [17] and first analyzed in [1], where the geometry description of the PDE domain is adopted from a Computer Aided Design (CAD) parametrization usually based on Non-Uniform Rational B-Splines (NURBS) and the same NURBS basis functions are also used as the PDEs discrete basis, following an isoparametric paradigm; see the monograph [10]. Recent works on IGA preconditioners have focused on overlapping Schwarz preconditioners [3, 5, 7, 9], multigrid methods [16], and non-overlapping preconditioners [4, 8, 20].

L. Beirão da Veiga • L.F. Pavarino (✉) • S. Scacchi
Dipartimento di Matematica, Università di Milano, Via Saldini 50, 20133 Milano, Italy
e-mail: lourenco.beirao@unimi.it; luca.pavarino@unimi.it; simone.scacchi@unimi.it

O.B. Widlund
Courant Institute of Mathematical Sciences, 251 Mercer Street, New York, NY 10012, USA
e-mail: widlund@cims.nyu.edu

S. Zampini
Extreme Computing Research Center, King Abdullah University of Science and Technology, Thuwal 23955, Saudi Arabia
e-mail: stefano.zampini@kaust.edu.sa

T. Dickopf et al. (eds.), *Domain Decomposition Methods in Science and Engineering XXII*, Lecture Notes in Computational Science and Engineering 104, DOI 10.1007/978-3-319-18827-0_2

Deluxe scaling was recently introduced by Dohrmann and Widlund in a study of $H(curl)$ problems; see [14, 15, 29] and also [25] for its application to problems in $H(div)$ and [21] for Reissner–Mindlin plates. In our previous work on isogeometric BDDC [4], standard BDDC scalings were employed with averaging weights built directly from sone representative values of the elliptic coefficients in each subdomain (ρ-scaling) or from the values of the diagonal elements of local and global stiffness matrices (stiffness scaling). The novel deluxe scaling, originally developed to deal with elliptic problems with more than one variable coefficient, is instead based on solving local problems built from local Schur complements associated with sets of what are known as the dual variables. This new scaling turns out to be much more powerful than the standard ρ- and stiffness scalings in the present context, even for scalar elliptic problems with one variable coefficient. A novel adaptive strategy to select a reduced set of vertex primal constraints is also studied. The main result of our h-analysis shows that the condition number of the resulting deluxe BDDC preconditioner satisfies the same quasi-optimal polylogarithmic bound in the ratio H/h of subdomain to element diameters, as in [4], and that this bound is independent of the number of subdomains and jumps of the coefficients of the elliptic problem across subdomain interfaces. Moreover, our preliminary 2D numerical experiments with deluxe scaling show a remarkable improvement, in particular for increasing polynomial degree p of the isogeometric elements. Numerical tests in 3D can be found in [6].

2 Isogeometric Discretization of Scalar Elliptic Problems

We consider the model elliptic problem on a bounded and connected CAD domain $\Omega \subset \mathbb{R}^d$, $d = 2, 3$,

$$-\nabla \cdot (\rho \nabla u) = f \text{ in } \Omega, \quad u = 0 \text{ on } \partial\Omega, \tag{1}$$

where ρ is a scalar field satisfying $0 < \rho_{min} \leq \rho(x) \leq \rho_{max}$, $\forall x \in \Omega$. For simplicity, we describe our problem and preconditioner in the 2D single-patch case. Comments on the 3D extension can be found at the end of Sect. 3, and comments on the multi-patch extension can be found in [6]. We discretize (1) with IGA based on B-splines and NURBS basis functions. The bivariate B-spline discrete space is defined by

$$\hat{\mathcal{S}}_h := \text{span}\{B_{i,j}^{p,q}(\xi, \eta),\ i = 1, \ldots, n, j = 1, \ldots, m\}, \tag{2}$$

where the bivariate B-spline basis functions $B_{i,j}^{p,q}(\xi,\eta) = N_i^p(\xi)\, M_j^q(\eta)$ are defined by tensor products of 1D B-splines functions $N_i^p(\xi)$ and $M_j^q(\eta)$ of degree p and q, respectively (in our numerical experiments, we will only consider the case $p = q$). Analogously, the NURBS space is the span of NURBS basis functions defined in 1D by

$$R_i^p(\xi) := \frac{N_i^p(\xi)\omega_i}{\sum_{\hat{\imath}=1}^n N_{\hat{\imath}}^p(\xi)\omega_{\hat{\imath}}} = \frac{N_i^p(\xi)\omega_i}{w(\xi)}, \tag{3}$$

with the weight function $w(\xi) := \sum_{\hat{\imath}=1}^n N_{\hat{\imath}}^p(\xi)\omega_{\hat{\imath}} \in \hat{\mathcal{S}}_h$, and in 2D by

$$R_{i,j}^{p,q}(\xi,\eta) := \frac{B_{i,j}^{p,q}(\xi,\eta)\omega_{i,j}}{\sum_{\hat{\imath}=1}^n \sum_{\hat{\jmath}=1}^m B_{\hat{\imath},\hat{\jmath}}^{p,q}(\xi,\eta)\omega_{\hat{\imath},\hat{\jmath}}} = \frac{B_{i,j}^{p,q}(\xi,\eta)\omega_{i,j}}{w(\xi,\eta)}, \tag{4}$$

where $w(\xi,\eta)$ is the weight function and $\omega_{i,j} = (\mathbf{C}_{i,j}^{\omega})_3$ the positive weights associated with a $n \times m$ net of control points $\mathbf{C}_{i,j}$. The discrete space of NURBS functions on the domain Ω is defined as the span of the *push-forward* of the NURBS basis functions (4) (see, e.g., [17])

$$\mathcal{N}_h := \text{span}\{R_{i,j}^{p,q} \circ \mathbf{F}^{-1}, \text{ with } i = 1,\ldots,n; j = 1,\ldots,m\}, \tag{5}$$

with $\mathbf{F} : \hat{\Omega} \to \Omega$ the geometrical map between parameter and physical spaces defined by $\mathbf{F}(\xi,\eta) = \sum_{i=1}^n \sum_{j=1}^m R_{i,j}^{p,q}(\xi,\eta)\mathbf{C}_{i,j}$.

For simplicity, we will consider the case with a Dirichlet boundary condition imposed on all of $\partial\Omega$; we can then define the spline space in the parameter space and the NURBS space in physical space, respectively, as

$$\hat{V}_h := [\hat{\mathcal{S}}_h \cap H_0^1(\hat{\Omega})]^2 = [\text{span}\{B_{i,j}^{p,q}(\xi,\eta),\ i = 2,\ldots,n-1,\ j = 2,\ldots,m-1\}]^2,$$

$$V_h := [\mathcal{N}_h \cap H_0^1(\Omega)]^2 = [\text{span}\{R_{i,j}^{p,q} \circ \mathbf{F}^{-1}, \text{ with } i = 2,\ldots,n-1;\ j = 2,\ldots,m-1\}]^2.$$

The IGA formulation of problem (1) then reads:

$$\begin{cases} \text{Find } u_h \in V_h \text{ such that:} \\ a(u_h, v_h) = < f, v_h > \qquad \forall v \in V_h, \end{cases} \tag{6}$$

with the bilinear form $a(u_h, v_h) = \int_{\Omega} \rho \nabla u_h \nabla v_h dx$.

3 BDDC Preconditioners

When using iterative substructuring methods, such as BDDC, we first reduce the problem to one on the interface by implicitly eliminating the interior degrees of freedom, a process known as static condensation; see, e.g., Toselli and Widlund [28, Ch. 4].

Knots and Subdomain Decomposition A decomposition is first built for the underlying space of spline functions in the parametric space, and is then easily extended to the NURBS space in the physical domain. From the full set of knots, $\{\xi_1 = 0, \ldots, \xi_{n+p+1} = 1\}$, we select a subset $\{\xi_{i_k}, k = 1, \ldots, N+1\}$ of non-repeated knots with $\xi_{i_1} = 0, \xi_{i_{N+1}} = 1$. The interface knots are given by ξ_{i_k} for $k = 2, .., N$ and they define a decomposition of the closure of the reference interval into subdomains

$$\overline{(\hat{I})} = [0, 1] = \overline{\left(\bigcup_{k=1,..,N} \hat{I}_k \right)}, \text{ with } \hat{I}_k = (\xi_{i_k}, \xi_{i_{k+1}}),$$

that we assume to have similar lengths $H_k := \operatorname{diam}(\hat{I}_k) \approx H$. In more dimensions, we just use tensor products. Thus, in two dimension, we define the subdomains by

$$\hat{I}_k = (\xi_{i_k}, \xi_{i_{k+1}}), \quad \hat{I}_l = (\eta_{j_l}, \eta_{j_{l+1}}), \quad \hat{\Omega}_{kl} = \hat{I}_k \times \hat{I}_l, \quad 1 \le k \le N_1, \ 1 \le l \le N_2. \tag{7}$$

For simplicity, we reindex the subdomains using only one index to obtain the decomposition of our domain $\overline{\hat{\Omega}} = \bigcup_{k=1,..,K} \overline{\hat{\Omega}}_k$, into $K = N_1 N_2$ subdomains. Throughout this paper, we assume that both the subdomains and elements defined by the coarse and full sets of knot vectors are *shape regular* and with quasi-uniform characteristic diameters H and h, respectively.

The Schur Complement System As in classical iterative substructuring, we reduce the problem to one on the interface $\Gamma := \left(\bigcup_{k=1}^{K} \partial\hat{\Omega}_k \right) \setminus \partial\hat{\Omega}$ by static condensation, i.e., by eliminating the interior degrees of freedom associated with the basis functions with support in each subdomain. The resulting Schur complement for $\hat{\Omega}_k$ and its local interface $\Gamma_k := \partial\hat{\Omega}_k \setminus \partial\hat{\Omega}$ will be denoted by $S^{(k)}$. In the sequel, we will use the following sets of indices:

$$\Theta_\Omega = \{(i,j) \in \mathbb{N}^2 : 2 \le i \le n-1, 2 \le j \le m-1\},$$
$$\Theta_\Gamma = \{(i,j) \in \Theta_\Omega : \operatorname{supp}(B_{i,j}^{p,q}) \cap \Gamma \neq \emptyset\}.$$

We note that Θ_Γ consists of indices associated with a "fat" interface that typically consists of several layers of knots associated with the basis functions with support

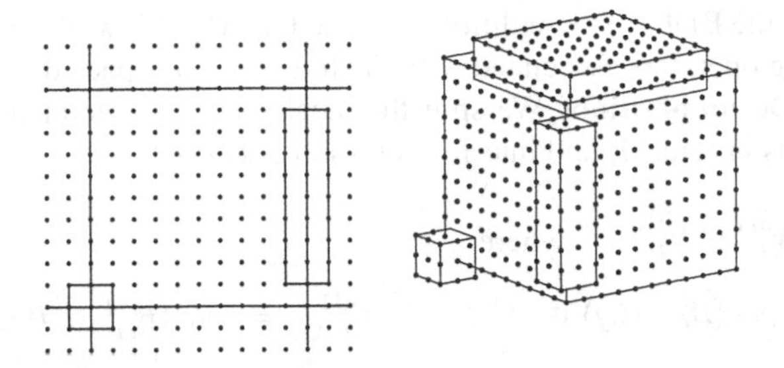

Fig. 1 Schematic illustration in index space of interface equivalence classes in 2D (*left*) and 3D (*right*) parametric space with $p = 3, \kappa = 2$: fat vertices, consisting of $(\kappa + 1)^2$ knots in 2D and $(\kappa + 1)^3$ in 3D; fat edges (without vertices), consisting of $(\kappa + 1)$ "slim" edges in 2D and $(\kappa + 1)^2$ in 3D; fat faces (without vertices and edges), consisting of $\kappa + 1$ slim faces in 3D

intersecting two or more subdomains, see e.g. Fig. 1. The discrete interface and local spaces are defined as

$$\hat{V}_\Gamma := \text{span}\{B_{i,j}^{p,q}, (i,j) \in \Theta_\Gamma\}, \quad V_I^{(k)} := \hat{V}_h \cap H_0^1(\hat{\Omega}_k). \tag{8}$$

The space $\hat{V}_h$ can be decomposed as $\oplus_{k=1}^K V_I^{(k)} + \mathcal{H}(\hat{V}_\Gamma)$, where $\mathcal{H} : \hat{V}_\Gamma \rightarrow \hat{V}_h$, is the piece-wise discrete spline harmonic extension operator, which provides the minimal energy extension of values given in $\hat{V}_\Gamma$. The interface component of the discrete solution satisfies the Schur complement reduced system

$$s(u_\Gamma, v_\Gamma) = <\hat{f}, v_\Gamma>, \quad \forall v_\Gamma \in \hat{V}_\Gamma, \tag{9}$$

with a suitable right-hand side $\hat{f}$ and a Schur complement bilinear form defined by $s(w_\Gamma, v_\Gamma) := a(\mathcal{H}(w_\Gamma), \mathcal{H}(v_\Gamma))$. For simplicity, in the sequel, we will drop the subscript Γ for functions in $\hat{V}_\Gamma$. In matrix form, (9) is the Schur complement system

$$\hat{S}_\Gamma w = \hat{f}, \tag{10}$$

where $\hat{S}_\Gamma = A_{\Gamma\Gamma} - A_{\Gamma I} A_{II}^{-1} A_{\Gamma I}^T$, $\hat{f} = f_\Gamma - A_{\Gamma I} A_{II}^{-1} f_I$, are obtained from the original discrete problem by Gaussian elimination after reordering the spline basis functions into sets of interior (subscript I) and interface (subscript Γ) basis functions. The Schur complement system (10) is solved by a Preconditioned Conjugate Gradient (PCG) iteration, where $\hat{S}_\Gamma$ is never explicitly formed since the action of $\hat{S}_\Gamma$ on a vector is computed by solving Dirichlet problems for individual subdomains and some sparse matrix-vector multiplies, which are also needed when working with the local Schur complements required by the application of the BDDC preconditioner defined below. The preconditioned Schur complement system solved by PCG is then

$$M_{\text{BDDC}}^{-1} \hat{S}_\Gamma w = M_{\text{BDDC}}^{-1} \hat{f}, \tag{11}$$

where M_{BDDC}^{-1} is the BDDC preconditioner, defined in (18) below using some restriction and scaling operators associated with the following subspace decompositions.

Subspace Decompositions We split the local space $V^{(k)}$ defined in (8) into a direct sum of its interior (I) and interface (Γ) subspaces, i.e.

$$V^{(k)} := V_I^{(k)} \oplus V_\Gamma^{(k)}, \quad \text{where}$$
$$V_I^{(k)} := \text{span}\{B_{i,j}^{p,q}, (i,j) \in \Theta_I^{(k)}\}, \qquad V_\Gamma^{(k)} := \text{span}\{B_{i,j}^{p,q}, (i,j) \in \Theta_\Gamma^{(k)}\},$$

which translate in the index sets

$$\Theta_I^{(k)} := \{(i,j) \in \Theta_\Omega : \text{supp}(B_{i,j}^{p,q}) \subset \hat{\Omega}_k\},$$
$$\Theta_\Gamma^{(k)} := \{(i,j) \in \Theta_\Gamma : \text{supp}(B_{i,j}^{p,q}) \cap (\partial\hat{\Omega}_k \cap \Gamma_k) \neq \emptyset\},$$

and we define the associated product spaces by

$$V_I := \prod_{k=1}^{K} V_I^{(k)}, \quad V_\Gamma := \prod_{k=1}^{K} V_\Gamma^{(k)}.$$

The functions in V_Γ are generally discontinuous (multi-valued) across Γ, while our isogeometric approximations belong to $\hat{V}_\Gamma$, the subspace of V_Γ of functions continuous (single-valued) across Γ. We will select some interface basis functions as *primal* (subscript Π), that will be made continuous across the interface and will be subassembled between their supporting elements, and we will call *dual* (subscript Δ) the remaining interface degrees of freedom that can be discontinuous across the interface and which vanish at the primal degrees of freedom. This splitting allows us to decompose each local interface space into primal and dual subspaces $V_\Gamma^{(k)} = V_\Pi^{(k)} \bigoplus V_\Delta^{(k)}$, and we can define the associated product spaces by

$$V_\Delta := \prod_{k=1}^{K} V_\Delta^{(k)}, \quad V_\Pi := \prod_{k=1}^{K} V_\Pi^{(k)}.$$

We also need an intermediate subspace $\tilde{V}_\Gamma \subset V_\Gamma$ of partially continuous basis functions

$$\tilde{V}_\Gamma := V_\Delta \bigoplus \hat{V}_\Pi,$$

where the product space V_Δ has been defined above and $\hat{V}_\Pi$ is a global subspace of the selected primal variables.

For two-dimensional problems, we will consider the primal space $\hat{V}_{\Pi}^{C}$ consisting of vertex basis functions with indices belonging to

$$\Theta_C = \{(i,j,k) \in \Theta_\Gamma : \text{supp}(B_{i,j,k}^{p,q,r}) \cap C \neq \emptyset\}. \tag{12}$$

In order to define our preconditioners, we will need the following restriction and interpolation operators represented by matrices with $\{0,1\}$ elements

$$\begin{array}{lll} \tilde{R}_{\Gamma\Delta} : \tilde{V}_\Gamma \longrightarrow V_\Delta, & \tilde{R}_{\Gamma\Pi} : \tilde{V}_\Gamma \longrightarrow \hat{V}_\Pi, & \hat{R}_\Pi : \hat{V}_\Gamma \longrightarrow \hat{V}_\Pi \\ R_\Delta^{(k)} : V_\Delta \longrightarrow V_\Delta^{(k)}, & R_\Pi^{(k)} : \hat{V}_\Pi \longrightarrow \hat{V}_\Pi^{(k)} & \hat{R}_\Delta^{(k)} : \hat{V}_\Gamma \longrightarrow V_\Delta^{(k)}. \end{array} \tag{13}$$

For any edge/face $\mathcal{F}$, we will use the symbol $R_\mathcal{F}$ to denote a restriction matrix to the ("fat") set of degrees of freedom associated with $\mathcal{F}$.

Deluxe Scaling We now apply to our isogeometric context the deluxe scaling proposed in [14]. Let Ω_k be any subdomain in the partition, $k = 1, 2, \dots, K$. We will indicate by Ξ_k the index set of all the $\Omega_j, j \neq k$, that share an edge $\mathcal{F}$ with Ω_k. For regular quadrilateral subdomain partitions in two dimensions, the cardinality of Ξ_k is 4 (or less for boundary subdomains).

In BDDC, the average $\bar{w} := E_D w$ of an element in $w \in \tilde{V}_\Gamma$, is computed separately for the sets of interface degrees of freedom of edge and face equivalence classes. We define the deluxe scaling for the class of $\mathcal{F}$ with only two elements, k, j, as for an edge in two dimensions. We define two principal minors, $S_\mathcal{F}^{(k)}$ and $S_\mathcal{F}^{(j)}$, obtained from $S^{(k)}$ and $S^{(j)}$ by removing all rows and columns which do not belong to the degrees of freedom which are common to the (fat) boundaries of Ω_k and Ω_j.

Let $w_\mathcal{F}^{(k)} := R_\mathcal{F} w^{(k)}$; the deluxe average across $\mathcal{F}$ is then defined as

$$\bar{w}_\mathcal{F} = \left(S_\mathcal{F}^{(k)} + S_\mathcal{F}^{(j)}\right)^{-1} \left(S_\mathcal{F}^{(k)} w_\mathcal{F}^{(k)} + S_\mathcal{F}^{(j)} w_\mathcal{F}^{(j)}\right). \tag{14}$$

If the Schur complements of an equivalence class have small dimensions, they can be computed explicitly, otherwise the action of $\left(S_\mathcal{F}^{(k)} + S_\mathcal{F}^{(j)}\right)^{-1}$ can be computed by solving a Dirichlet problem on the union of the relevant subdomains with a zero right hand side in the interiors of the subdomains.

Each of the relevant equivalence classes, which involve the subdomain Ω_k, will contribute to the values of $\bar{w}$. Each of these contributions will belong to $\hat{V}_\Gamma$, after being extended by zero to $\Gamma \setminus \mathcal{F}$; the resulting element is given by $R_\mathcal{F}^T \bar{w}_\mathcal{F}$. We then add the contributions from the different equivalence classes to obtain

$$\bar{w} = E_D w = w_\Pi + \sum_\mathcal{F} R_\mathcal{F}^T \bar{w}_\mathcal{F}. \tag{15}$$

E_D is a projection and its complementary projection is given by

$$P_D w := (I - E_D)w = w_\Delta - \sum_{\mathcal{F}} R_{\mathcal{F}}^T \bar{w}_{\mathcal{F}}. \tag{16}$$

With a small abuse of notation in what follows, we will consider $E_D w \in \hat{V}_\Gamma$ also as an element of $\tilde{V}_\Gamma$, by the obvious embedding $\hat{V}_\Gamma \subset \tilde{V}_\Gamma$. In order to rewrite E_D in matrix form, for each subdomain Ω_k, we define the block-diagonal scaling matrix

$$D^{(k)} = diag(D^{(k)}_{\mathcal{F}j_1}, D^{(k)}_{\mathcal{F}j_2}, \dots, D^{(k)}_{\mathcal{F}j_k}),$$

where $j_1, j_2, \dots, j_k \in \Xi_k$ and the diagonal blocks are given by the deluxe scaling $D^{(k)}_{\mathcal{F}} := \left(S^{(k)}_{\mathcal{F}} + S^{(j)}_{\mathcal{F}}\right)^{-1} S^{(k)}_{\mathcal{F}}$. We can now extend the operators defined in (13) and define the scaled local operators by $R^{(k)}_{D,\Gamma} := D^{(k)} R^{(k)}_\Gamma$, $\tilde{R}^{(k)}_{D,\Delta} := R^{(k)}_{\Gamma,\Delta} R^{(k)}_{D,\Gamma}$ and the global scaled operator

$$\tilde{R}_{D,\Gamma} := \text{ the direct sum } \hat{R}_\Pi \oplus_{k=1}^K \tilde{R}^{(k)}_{D,\Delta}, \tag{17}$$

so that the averaging operator is $E_D = \tilde{R}_\Gamma \tilde{R}^T_{D,\Gamma}$, where $\tilde{R}_\Gamma := \hat{R}_\Pi \oplus_{k=1}^K \tilde{R}^{(k)}_\Delta$.

The BDDC Preconditioner We denote by $A^{(k)}$ the local stiffness matrix restricted to subdomain $\bar{\Omega}_k$. By partitioning the local degrees of freedom into those in the interior (I) and those on the interface (Γ), as before, and by further partitioning the latter into dual (Δ) and primal (Π) degrees of freedom, then $A^{(k)}$ can be written as

$$A^{(k)} = \begin{bmatrix} A^{(k)}_{II} & A^{(k)^T}_{\Gamma I} \\ A^{(k)}_{\Gamma I} & A^{(k)}_{\Gamma\Gamma} \end{bmatrix} = \begin{bmatrix} A^{(k)}_{II} & A^{(k)^T}_{\Delta I} & A^{(k)^T}_{\Pi I} \\ A^{(k)}_{\Delta I} & A^{(k)}_{\Delta\Delta} & A^{(k)^T}_{\Pi\Delta} \\ A^{(k)}_{\Pi I} & A^{(k)}_{\Pi\Delta} & A^{(k)}_{\Pi\Pi} \end{bmatrix}.$$

Using the scaled restriction matrices defined in (13) and (17), the BDDC preconditioner can be written as

$$M^{-1}_{\text{BDDC}} = \tilde{R}^T_{D,\Gamma} \tilde{S}^{-1}_\Gamma \tilde{R}_{D,\Gamma}, \quad \text{where} \tag{18}$$

$$\tilde{S}^{-1}_\Gamma = \tilde{R}^T_{\Gamma\Delta} \left(\sum_{k=1}^K \begin{bmatrix} 0 & R^{(k)^T}_\Delta \end{bmatrix} \begin{bmatrix} A^{(k)}_{II} & A^{(k)^T}_{\Delta I} \\ A^{(k)}_{\Delta I} & A^{(k)}_{\Delta\Delta} \end{bmatrix}^{-1} \begin{bmatrix} 0 \\ R^{(k)}_\Delta \end{bmatrix} \right) \tilde{R}_{\Gamma\Delta} + \Phi S^{-1}_{\Pi\Pi} \Phi^T. \tag{19}$$

Here $S_{\Pi\Pi}$ is the BDDC coarse matrix and Φ is a matrix mapping primal degrees of freedom to interface variables, see e.g. [2, 22]. Our main theorem is the following (see [6] for a proof and more complete details).

Theorem 1 *Consider the model problem (1) in two dimensions and let the primal set be given by the subdomain corner set $\hat{V}_{\Pi}^{C}$ defined in (12). Then the condition number of the preconditioned operator is bounded by*

$$\mathrm{cond}\left(M_{BDDC}^{-1}\hat{S}_{\Gamma}\right) \leq C(1+\log(H/h))^2,$$

with $C > 0$ independent of h, H and the jumps of the coefficient ρ.

Comments on the Three-Dimensional Case The choice of primal degrees of freedom is fundamental for the construction of efficient BDDC preconditioners. The space $\hat{V}_{\Pi}^{C}$ is not sufficient to obtain scalable and fast preconditioners in three dimensions. In three dimensions, we can define an additional index set associated with fat edges

$$\Theta_{\mathrm{E}} = \{(i,j,k) \in \Theta_{\Gamma}/\Theta_{C} : \ \mathrm{supp}(B_{i,j,k}^{p,q,r}) \cap \mathrm{E} \neq \emptyset\},$$

and enrich the primal space with averages computed for each slim edge parallel to the subdomain edge (see Fig. 1). Three-dimensional numerical results (see [6]) show faster rates of convergence when considering such an enriched coarse space: in particular, the addition of edge slim averages is sufficient to obtain quasi-optimality and scalability as is the case with standard FEM discretizations. The deluxe convergence rate for increasing p seems to be orders of magnitude better than that of BDDC with stiffness scaling, but not as insensitive to p as in the 2D results of Table 1 in the next section.

Adaptive Choice of Reduced Sets of Primal Constraints In recent years, a number of people have investigated different adaptive choices of primal constraints in BDDC and FETI-DP methods, see e.g. [13, 18, 19, 24, 26, 27]. Most of these works focus on the adaptive selection of 2D edge or 3D face constraints, i.e. constraints associated with the interface between two subdomains, by solving some generalized eigenproblems. It is less clear how to extend such techniques to equivalence classes shared by more than two subdomains, such as 2D or 3D vertices and 3D edges. Here, inspired by the techniques of [13], we propose an adaptive selection of 2D primal vertices, driven by the desire to reduce the expensive fat vertex primal constrains used in the standard or deluxe BDDC method.

Let Ω_k be any subdomain in the partition, $k = 1, 2, \ldots, K$ and consider the associated local Schur complement $S^{(k)}$. Denote by $\mathcal{F}$ one of the equivalence classes (a vertex, edge, or face) and partition the degrees of freedom local to Ω_k into $\mathcal{F}$ and its complement $\mathcal{F}'$. Then $S^{(k)}$ can be partitioned as

$$S^{(k)} = \begin{pmatrix} S_{\mathcal{F}\mathcal{F}}^{(k)} & S_{\mathcal{F}\mathcal{F}'}^{(k)} \\ S_{\mathcal{F}'\mathcal{F}}^{(k)} & S_{\mathcal{F}'\mathcal{F}'}^{(k)} \end{pmatrix}. \quad (20)$$

For each equivalence class $\mathcal{F}$, define the new Schur complement

$$\tilde{S}^{(k)}_{\mathcal{F}\mathcal{F}} = S^{(k)}_{\mathcal{F}\mathcal{F}} - S^{(k)}_{\mathcal{F}\mathcal{F}'} S^{(k)^{-1}}_{\mathcal{F}'\mathcal{F}'} S^{(k)}_{\mathcal{F}'\mathcal{F}} \tag{21}$$

and define the generalize eigenvalue problem

$$S^{(k)}_{\mathcal{F}\mathcal{F}} v = \lambda \tilde{S}^{(k)}_{\mathcal{F}\mathcal{F}} v. \tag{22}$$

Given a threshold $\theta \geq 1$, we select the eigenvectors $\{v_1, v_2, \ldots, v_{N_c}\}$ associated to the eigenvalues of (22) greater than θ and we perform a BDDC change of basis in order to make these selected eigenvectors the primal variables.

4 Numerical Results

In this section, we report on numerical experiments with the isogeometric BDDC deluxe preconditioner for two-dimensional elliptic model problems (1), discretized with isogeometric NURBS spaces with a mesh size h, polynomial degree p and regularity κ. The domain is decomposed into K nonoverlapping subdomains of characteristic size H, as described in Sec. 3. The discrete Schur-complement problems are solved by the PCG method with the isogeometric BDDC preconditioner, with a zero initial guess and a stopping criterion of a 10^{-6} reduction of the Euclidean norm of the PCG residual. In the tests, we study how the convergence rate of the BDDC preconditioner depends on h, K, p, κ. The 2D tests have been performed with a MATLAB code based on the GeoPDEs library by De Falco et al. [11].

Scalability in K and Quasi-Optimality in H/h The condition number cond and iteration counts n_{it} of the BDDC deluxe preconditioner are reported in the table of Fig. 2 for a quarter-ring domain (shown on the left of the table), as a function of the number of subdomains K and mesh size h, for fixed $p = 3, \kappa = 2$ (top) or $p = 5, \kappa = 4$ (bottom). The results show that the proposed preconditioner is scalable, since moving along the diagonals of each table the condition number appears to be bounded from above by a constant independent of K. The results for higher degree $p = 5$ and regularity $\kappa = 4$ are even better than those for the lower degree case. The BDDC deluxe preconditioner appears to retain a very good performance in spite of the increase of the polynomial degree p, a property that was not always satisfied in [4]. To better understand this issue, we next study the BDDC performance for increasing values of p.

Dependence on p In this test, we compare the BDDC deluxe performance as a function of the polynomial degree p and the regularity κ. We recall that our theoretical work is only an h-analysis and does not cover the dependence of the convergence rate on p and κ. The domain considered is the quarter-ring discretized with a mesh size $h = 1/64$ and $K = 4 \times 4$ subdomains. The spline degree p varies from 2 to 10 and the regularity is always maximal ($\kappa = p - 1$) inside the

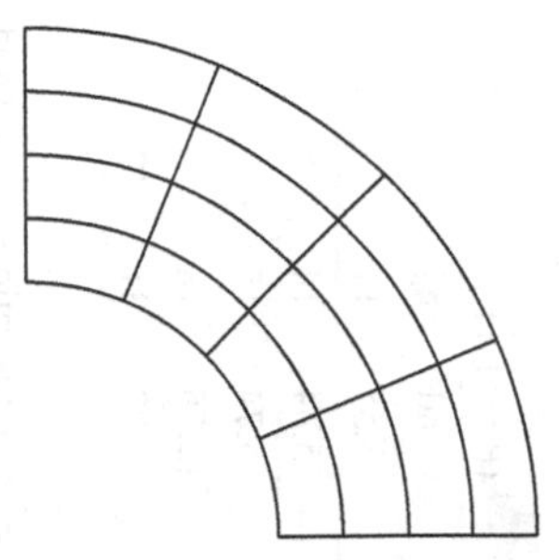

	K	$h = 1/16$		$h = 1/32$		$h = 1/64$		$h = 1/128$	
		cond	n_{it}	cond	n_{it}	cond	n_{it}	cond	n_{it}
	2×2	1.24	5	1.42	6	1.65	6	1.92	6
p D 3	4×4			2.02	8	2.68	10	3.46	11
κ D 2	8×8					2.39	10	3.29	12
	16×16							2.64	11
	2×2	1.19	5	1.35	6	1.55	6	1.78	6
p D 5	4×4			1.62	8	2.19	9	2.86	10
κ D 4	8×8					1.77	8	2.55	10
	16×16							1.87	8

Fig. 2 BDDC deluxe preconditioner for a 2D quarter-ring domain (*left*): condition number cond and iteration counts n_{it} as a function of the number of subdomains K and mesh size h. Fixed $p = 3, \kappa = 2$ (*top*), $p = 5, \kappa = 4$ (*bottom*)

Table 1 BDDC deluxe dependence on p in the 2D quarter-ring domain: condition number cond and iteration counts n_{it} as a function of the NURBS polynomial degree p. Fixed $h = 1/64,\ K = 4 \times 4,\ \kappa = p - 1$ (top), $\kappa = 2$ (bottom)

	p	2	3	4	5	6	7	8	9	10
$\kappa = p - 1$	Cond	3.22	2.68	2.41	2.19	2.04	1.91	1.80	1.72	1.62
	n_{it}	10	10	9	9	9	8	8	8	9
$\kappa = 2$	Cond	-	2.47	2.84	3.16	3.45	3.71	3.94	4.17	4.36
	n_{it}	-	10	11	11	11	12	12	12	12

subdomains, while at the subdomain interface is either maximal ($\kappa = p - 1$, top) or low ($\kappa = 2$, bottom). The results in Table 1 show that for $\kappa = p - 1$ the condition numbers and iteration counts are bounded independently of the degree p and actually improve slightly for increasing p, while for $\kappa = 2$ the condition numbers show a very modest sublinear growth with p, with associated iteration counts that are practically constant. This is a remarkable property that is not shared by any other nonoverlapping IGA preconditioner in the (current) literature.

Adaptive Choice of Vertex Primal Constraints Table 2 reports the results with the proposed adaptive choice of primal constraints applied only to the vertex constraints (the edge variables remain dual). We consider both an eigenvalue threshold $\theta = 2$ leading to the minimal choice of $N_c = 1$ primal vertex constraint (that turns out to be the average of the fat vertex values) and a lower threshold

Table 2 Adaptive choice of primal vertex constraints for the BDDC deluxe preconditioner on a square domain, with eigenvalue threshold θ and associated number N_c of selected primal constraints

	$N_c = 1$ ($\theta = 2$)		$N_c = 4$ ($\theta = 1.5$)	
K	Cond	n_{it}	Cond	n_{it}
2×2	1.81	7	1.66	8
4×4	12.74	14	6.74	13
8×8	14.74	24	7.48	18
16×16	15.67	26	7.78	18
32×32	16.13	24	7.87	17

a)

	$N_c = 1$ ($\theta = 2$)		$N_c = 4$ ($\theta = 1.5$)	
H/h	Cond	n_{it}	Cond	n_{it}
4	8.75	12	4.84	12
8	12.74	14	6.74	13
16	17.40	17	8.91	14
32	22.31	18	11.16	15
64	27.49	20	13.50	17

b)

	$N_c = 1$ ($\theta = 2$)		($\theta = 1.1$)		
p	Cond	n_{it}	Cond	n_{it}	N_c
2	6.09	13	3.55	11	3
3	17.40	17	5.34	14	5
4	230.9	21	5.74	15	8
5	7545.9	39	12.25	18	10
6	-	-	73.08	31	12

c)

Condition number cond and iteration counts n_{it} as a function of: a) the number of subdomains K for fixed $p = 3, \kappa = 2, H/h = 8$; b) the ratio H/h for fixed $p = 3, \kappa = 2, K = 4 \times 4$; c) the polynomial degree p for fixed $K = 4 \times 4, H/h = 16$ $(\kappa = p - 1)$

$\theta = 1.5$ leading to a richer choice of $N_c = 4$ primal vertex constraints for each subdomain vertex. In case of variable polynomial degree p, we also consider a very low threshold $\theta = 1.1$ that leads to a richer choice of approximately $N_c = 2p$ primal constraints for each subdomain vertex. The results in a) show that the BDDC deluxe preconditioner is scalable, since cond and n_{it} appears to be bounded from above by a constant independent of K, and the results in b) indicate that the preconditioner is quasi-optimal, since cond and n_{it} appears to grow polylogarithmically in H/h. The results in c) show that the minimal choice $N_c = 1$ does not perform well for increasing p (there is no convergence for $p = 6$), while with the richer choice corresponding to $\theta = 1.1$ we obtained only a mild performance degradation up to $p = 6$.

References

1. Y. Bazilevs, L. Beirão da Veiga, J.A. Cottrell, T.J.R. Hughes, G. Sangalli, Isogeometric analysis: approximation, stability and error estimates for h-refined meshes. Math. Models Methods Appl. Sci. **16**, 1–60 (2006)
2. L. Beirão da Veiga, C. Chinosi, C. Lovadina, L.F. Pavarino, Robust BDDC preconditioners for Reissner-Mindlin plate bending problems and MITC elements. SIAM J. Numer. Anal. **47**, 4214–4238 (2010)
3. L. Beirão da Veiga, D. Cho, L.F. Pavarino, S. Scacchi, Overlapping Schwarz methods for isogeometric analysis. SIAM J. Numer. Anal. **50**, 1394–1416 (2012)
4. L. Beirão da Veiga, D. Cho, L.F. Pavarino, S. Scacchi, BDDC preconditioners for isogeometric analysis. Math. Models Methods Appl. Sci. **23**, 1099–1142 (2013)
5. L. Beirão da Veiga, D. Cho, L.F. Pavarino, S. Scacchi, Isogeometric Schwarz preconditioners for linear elasticity systems. Comput. Methods Appl. Mech. Eng. **253**, 439–454 (2013)
6. L. Beirão da Veiga, L.F. Pavarino, S. Scacchi, O.B. Widlund, S. Zampini, Isogeometric BDDC preconditioners with deluxe scaling. SIAM J. Sci. Comput. **36**, A1118–A1139 (2014)
7. M. Bercovier, I. Soloveichik, *Overlapping non Matching Meshes Domain Decomposition Method in Isogeometric Analysis*. arXiv:1502.03756 [math.NA]
8. A. Buffa, H. Harbrecht, A. Kunoth, G. Sangalli, BPX-preconditioning for isogeometric analysis. Comput. Methods Appl. Mech. Eng. **265**, 63–70 (2013)
9. L. Charawi, Isogeometric overlapping additive Schwarz preconditioners for the Bidomain system, in *DD22 Proceedings*, 2014
10. J.A. Cottrell, T.J.R. Hughes, Y. Bazilevs, *Isogeometric Analysis. Towards integration of CAD and FEA* (Wiley, New York, 2009)
11. C. De Falco, A. Reali, R. Vazquez, GeoPDEs: a research tool for isogeometric analysis of PDEs. Adv. Eng. Softw. **42**, 1020–1034 (2011)
12. C.R. Dohrmann, A preconditioner for substructuring based on constrained energy minimization. SIAM J. Sci. Comput. **25**, 246–258 (2003)
13. C.R. Dohrmann, C. Pechstein, Constraints and weight selection algorithms for BDDC, in *Domain Decomposition Methods in Science and Engineering XXI*, Rennes, France, 2012. vol 98 (Springer LNCSE, Berlin, 2014)
14. C.R. Dohrmann, O.B. Widlund, Some recent tools and a BDDC algorithm for 3D problems in H(curl). In *Domain Decomposition Methods in Science and Engineering. XX*, San Diego, CA, 2011, vol. 91 (Springer LNCSE, Berlin, 2013), pp. 15–26
15. C.R. Dohrmann, O.B. Widlund, A BDDC algorithm with deluxe scaling for three-dimensional H(curl) problems. Comm. Pure Appl. Math. Appeared electronically in April 2015.

16. K. Gahalaut, J. Kraus, S. Tomar, Multigrid methods for isogeometric discretization. Comput. Methods Appl. Mech. Eng. **253**, 413–425 (2013)
17. T.J.R. Hughes, J.A. Cottrell, Y. Bazilevs, Isogeometric analysis: CAD, finite elements, NURBS, exact geometry, and mesh refinement. Comput. Methods Appl. Mech. Eng. **194**, 4135–4195 (2005)
18. H.H. Kim, E.T. Chung, A BDDC algorithm with enriched coarse spaces for two-dimensional elliptic problems with oscillatory and high contrast coefficients. Multiscale Model. Simul. **13**(2), 571–593 (2015)
19. A. Klawonn, M. Lanser, P. Radtke, O. Rheinbach, On an adaptive coarse space and on nonlinear domain decomposition. in *Domain Decomposition Methods in Science and Engineering*. XXI, Rennes, France, 2012, vol. 98 (Springer LNCSE, Berlin, 2014)
20. S.K. Kleiss, C. Pechstein, B. Jüttler, S. Tomar, IETI - isogeometric tearing and interconnecting. Comput. Methods Appl. Mech. Eng. **247–248**, 201–215 (2012)
21. J.H. Lee, A balancing domain decomposition by constraints deluxe method for numerically thin Reissner-Mindlin plates approximated with Falk–Tu elements. TR2013-951, Courant Institute, NYU, 2013
22. J. Li, O.B. Widlund, FETI-DP, BDDC, and block Cholesky methods. Int. J. Numer. Methods Eng. **66**, 250–271 (2006)
23. J. Mandel, C.R. Dohrmann, Convergence of a balancing domain decomposition by constraints and energy minimization. Numer. Linear Algebra Appl. **10**, 639–659 (2003)
24. J. Mandel, B. Sousedik, J. Sistek, Adaptive BDDC in three dimensions. Math. Comput. Simul. **82**(10), 1812–1831 (2012)
25. D.-S. Oh, O.B. Widlund, C.R. Dohrmann, A BDDC algorithm for Raviart-Thomas vector fields. TR2013-951, Courant Institute, NYU, 2013
26. C. Pechstein, C.R. Dohrmann, Modern domain decomposition methods - BDDC, deluxe scaling, and an algebraic approach. 2013. Seminar talk, Linz, December 2013. http://people.ricam.oeaw.ac.at/c.pechstein/pechstein-bddc2013.pdf
27. N. Spillane, V. Dolean, P. Hauret, P. Nataf, J. Rixen, Solving generalized eigenvalue problems on the interface to build a robust two-level FETI method. C. R. Math. Acad. Sci. Paris **351**(5–6), 197–201 (2013)
28. A. Toselli, O.B. Widlund, *Domain Decomposition Methods: Algorithms and Theory* (Springer, Berlin, 2004)
29. O.B. Widlund, C.R. Dohrmann, BDDC deluxe Domain Decomposition, in *DD22 Proceedings*, 2015

Auxiliary Space Multigrid Method for Elliptic Problems with Highly Varying Coefficients

Johannes Kraus and Maria Lymbery

1 Introduction

The robust preconditioning of linear systems of algebraic equations arising from discretizations of partial differential equations (PDE) is a fastly developing area of scientific research. In many applications these systems are very large, sparse and therefore it is vital to construct (quasi-)optimal iterative methods that converge independently of problem parameters.

The most established techniques to accomplish this objective are domain decomposition (DD), see, e.g., [23, 28], and multigrid (MG)/algebraic multilevel iteration (AMLI) methods, see, e.g., [10, 29, 30].

As demonstrated by Klawonn et al. [12], Toselli and Widlund [28], Graham et al. [9], two-level DD methods can be proven to be robust for scalar elliptic PDE with varying coefficient if the variations of the coefficient inside the coarse grid cells are assumed to be bounded. A key tool in the classical analysis of overlapping DD methods is the Poincaré inequality or its weighted analog as for problems with highly varying coefficients. It is well-known that the weighted Poincaré inequality holds only under certain conditions, e.g., in case of quasi-monotonic coefficients, see [26]. The concept of quasi-monotonic coefficients has been further developed in [25] for the convergence analysis of finite element tearing and interconnecting (FETI) methods.

J. Kraus (✉)
RICAM, Altenberger Str. 69, 4040 Linz, Austria
e-mail: johannes.kraus@oeaw.ac.at

M. Lymbery
IICT, Bulgarian Academy of Sciences, Acad. G. Bonchev Str., Bl. 25A, 1113 Sofia, Bulgaria
e-mail: mariq@parallel.bas.bg

T. Dickopf et al. (eds.), *Domain Decomposition Methods in Science and Engineering XXII*, Lecture Notes in Computational Science and Engineering 104, DOI 10.1007/978-3-319-18827-0_3

Recently the robustness of DD methods has also been achieved for problems with general coefficient variations using coarse spaces that are constructed by solving local generalized eigenvalue problems, see, e.g., [5, 8, 27].

In view of computational complexity, MG methods have been known to be most efficient as they have demonstrated optimality with respect to the problem size, see [10, 30]. Their design, however, needs careful adaptation for problems with large variations in the physical problem parameters. The AMLI framework contributes in achieving this goal, e.g. by providing more general polynomial acceleration techniques or Krylov cycles, see [1–3, 16].

The idea of integrating domain decomposition techniques into multigrid methods can be found as early as in [18]. The method that is presented in the following combines DD and MG techniques with those from auxiliary space preconditioning, see [31]. It is related to substructuring methods like FETI, see [6], and balancing domain decomposition (BDD) methods, see [19].

The most advanced of these methods, BDDC (BDD based on constraints), see [4], and FETI-DP (FETI dual-primal), see [7], can be formulated and analyzed in a common algebraic framework, see [20–22]. The BDDC method enforces continuity across substructure interfaces by a certain averaging operator. The additional constraints can be interpreted as subspace corrections where coarse basis functions are subject to energy minimization. From this point of view the BDDC method has a high degree of similarity with the present approach.

However, contrary to BDDC, the auxiliary space multigrid (ASMG) method considered here naturally allows overlapping of subdomains and coarse degrees of freedom (DOF) are associated in general not only with the interfaces of subdomains but also with their interior. Moreover, the aim is to define a full multilevel method by recursive application of a two-level method. In contrary to standard (variational) multigrid algorithms coarse-grid correction is replaced by an auxiliary space correction. The coarse-grid operator then appears as the exact Schur complement of the auxiliary matrix and defines an additive approximation of the Schur complement of the original system, see [14, 15].

The purpose of the present paper is to summarize the main steps of the construction of the ASMG method recently proposed in [17] on a less technical level (Sects. 2 and 4) and further to discuss its spectral properties and robustness with respect to highly varying coefficients (Sect. 3). The latter issue is also illustrated by numerical tests (Sect. 5).

2 Auxiliary Space Two-Grid Preconditioner

Consider the linear system of algebraic equations

$$A\mathbf{u} = \mathbf{f} \tag{1}$$

obtained after a finite element (FE) discretization of a partial differential equation (PDE) defined over a domain Ω, where A denotes the global stiffness matrix and $\mathbf{f}$ is a given right-hand side vector.

Consider a covering of Ω by n (overlapping) subdomains Ω_i, i.e., $\overline{\Omega} = \bigcup_{i=1}^{n} \overline{\Omega}_i$. Assume that for each subdomain Ω_i there is a symmetric positive semi-definite (SPSD) subdomain matrix A_i and that $A = \sum_{i=1}^{n} R_i^T A_i R_i$ where R_i restricts a global vector $\mathbf{v} \in V = \mathbb{R}^N$ to the local space $V_i = \mathbb{R}^{n_i}$ related to Ω_i. In practice the matrices A_i are assembled from scaled element matrices where the scaling factors account for the overlap of the subdomains. The DOF are split into two groups, coarse and fine, and the matrices A and A_i are partitioned accordingly into two-by-two blocks, where the lower right blocks (with index 22) are associated with coarse DOF, i.e.,

$$A = \begin{bmatrix} A_{11} & A_{12} \\ A_{21} & A_{22} \end{bmatrix}, \qquad A_i = \begin{bmatrix} A_{i:11} & A_{i:12} \\ A_{i:21} & A_{i:22} \end{bmatrix}, \qquad i = 1, \ldots, n.$$

Introduce the following auxiliary domain decomposition matrix

$$\tilde{A} = \begin{bmatrix} A_{1:11} & & & & A_{1:12}R_{1:2} \\ & A_{2:11} & & & A_{2:12}R_{2:2} \\ & & \ddots & & \vdots \\ & & & A_{n:11} & A_{n:12}R_{n:2} \\ R_{1:2}^T A_{1:21} & R_{2:2}^T A_{2:21} & \ldots & R_{n:2}^T A_{n:21} & \sum_{i=1}^{n} R_{i:2}^T A_{i:22} R_{i:2} \end{bmatrix}. \tag{2}$$

Denote $\tilde{A}_{11} = \operatorname{diag}\{A_{1:11}, \ldots, A_{n:11}\}$, $\tilde{A}_{22} = A_{22} = \sum_{i=1}^{n} R_{i:2}^T A_{i:22} R_{i:2}$, i.e., $\tilde{A} = \begin{bmatrix} \tilde{A}_{11} & \tilde{A}_{12} \\ \tilde{A}_{21} & \tilde{A}_{22} \end{bmatrix}$. The matrices $A \in \mathbb{R}^{N\times N}$ and $\tilde{A} \in \mathbb{R}^{\tilde{N}\times\tilde{N}}$ are related via

$$A = R\tilde{A}R^T \text{ where } R = \begin{bmatrix} R_1 & 0 \\ 0 & I_2 \end{bmatrix}, \; R_1 = \begin{bmatrix} R_{1:1}^T & \ldots & R_{n:1}^T \end{bmatrix}, \; A_{11} = R_1 \tilde{A}_{11} R_1^T.$$

Definition 1 ([15]) The additive Schur complement approximation (ASCA) of $S = A_{22} - A_{21}A_{11}^{-1}A_{12}$ is defined as the Schur complement Q of $\tilde{A}$:

$$Q := \tilde{A}_{22} - \tilde{A}_{21}\tilde{A}_{11}^{-1}\tilde{A}_{12} = \sum_{i=1}^{n} R_{i:2}^T(A_{i:22} - A_{i:21}A_{i:11}^{-1}A_{i:12})R_{i:2} \tag{3}$$

Next define a surjective mapping $\Pi_{\tilde{D}} : \tilde{V} \to V$ by

$$\Pi_{\tilde{D}} = (R\tilde{D}R^T)^{-1}R\tilde{D}, \tag{4}$$

where $\tilde{V} = \mathbb{R}^{\tilde{N}}$ and $\tilde{D}$ is a two-by-two block-diagonal SPD matrix.

The proposed auxiliary space two-grid preconditioner is defined by

$$B^{-1} := \overline{M}^{-1} + (I - M^{-T}A)C^{-1}(I - AM^{-1}) \tag{5}$$

where the operator M in (5) denotes an A-norm convergent smoother, i.e. $\|I - M^{-1}A\|_A \leq 1$, and $\overline{M} = M(M + M^T - A)^{-1}M^T$ is the corresponding symmetrized smoother. The matrix C defines a fictitious (auxiliary) space preconditioner approximating A and is given by

$$C^{-1} = \Pi_{\tilde{D}}\tilde{A}^{-1}\Pi_{\tilde{D}}^T. \tag{6}$$

Denote $\Pi = (I - M^{-T}A)\Pi_{\tilde{D}} = (I - M^{-T}A)(R\tilde{D}R^T)^{-1}R\tilde{D}$, then the preconditioner (5) can also be presented as

$$B^{-1} = \overline{M}^{-1} + \Pi\tilde{A}^{-1}\Pi^T. \tag{7}$$

The proposed auxiliary space two-grid method differs from the classical two-grid methods in the replacement of the coarse grid correction step by a subspace correction with iteration matrix $I - C^{-1}A$.

3 Spectral Properties and Robustness

As it has been shown in [17] the condition number of the two-grid preconditioner defined in (7) satisfies the estimate

$$\kappa(B^{-1}A) \leq (\bar{c} + c_\Pi)(\underline{c} + \eta)/\underline{c},$$

where $\rho_A = \lambda_{\max}(A)$, c_Π is the constant in the estimate $\|\Pi\tilde{\mathbf{v}}\|_A^2 \leq c_\Pi\|\tilde{\mathbf{v}}\|_{\tilde{A}}^2$ for all $\tilde{\mathbf{v}} \in \tilde{V}$, and the constants $\bar{c}$, $\underline{c}$ and η are such that the following properties hold:

$$\underline{c}\langle\mathbf{v},\mathbf{v}\rangle \leq \rho_A\langle\overline{M}^{-1}\mathbf{v},\mathbf{v}\rangle \leq \bar{c}\langle\mathbf{v},\mathbf{v}\rangle \quad \text{and} \quad \|M^{-T}A\mathbf{v}\|^2 \leq \frac{\eta}{\rho_A}\|\mathbf{v}\|_A^2.$$

Moreover, the ASCA defined in (3) is spectrally equivalent to S, i.e. $Q \simeq S$:

Theorem 1 ([17]) *Denote* $\pi_{\tilde{D}} = R^T \Pi_{\tilde{D}}$ *where* $\Pi_{\tilde{D}}$ *is defined as in* (4) *and* $\tilde{D}$ *is an arbitrary two-by-two block-diagonal SPD matrix for the same fine-coarse partitioning of DOF as used in the construction of* $\tilde{A}$.
Then $\langle A^{-1}\mathbf{u}, \mathbf{u}\rangle \leq \langle \Pi_{\tilde{D}} \tilde{A}^{-1} \Pi_{\tilde{D}}^T \mathbf{u}, \mathbf{u}\rangle \leq c \langle A^{-1}\mathbf{u}, \mathbf{u}\rangle \quad \forall \mathbf{u} \in V$ *where* $c := \|\pi_{\tilde{D}}\|_{\tilde{A}}^2$.
Hence,

$$\frac{1}{c}\langle S\mathbf{v}_2, \mathbf{v}_2\rangle \leq \langle Q\mathbf{v}_2, \mathbf{v}_2\rangle \leq \langle S\mathbf{v}_2, \mathbf{v}_2\rangle \quad \forall \mathbf{v}_2. \tag{8}$$

The upper bound in (8) *is sharp, the lower bound is sharp for* $\tilde{D} = \begin{bmatrix} \tilde{A}_{11} & 0 \\ 0 & I \end{bmatrix}$.

To verify that $\langle S\mathbf{v}_2, \mathbf{v}_2\rangle \leq c\langle Q\mathbf{v}_2, \mathbf{v}_2\rangle$ is robust with respect to an arbitrary variation of an elementwise constant coefficient $\alpha(\boldsymbol{x}) = \alpha_e$ for all $\boldsymbol{x} \in e$ and all elements e, see (15), one has to consider all possible distributions of $\{\alpha_e\}$ on the finest mesh. However, in the following we will show that the worst condition number (largest values of c) is obtained for a certain binary distribution of $\{\alpha_e\}$ so it suffices to study distributions of this type.

Let n_e denote the number of elements e and consider first an arbitrary distribution $\{\alpha_e\}$ of a piecewise constant coefficient where $\alpha_e \in (0, 1]$ for all e. Further, let A denote the global stiffness matrix corresponding to this distribution. Then there exists a set of binary distributions $\{\mathcal{C}_i : i = 1, 2, \ldots, n_e\}$ with $\mathcal{C}_i = \{\alpha_{e_j} : j = 1, 2, \ldots, n_e,\ \alpha_{e_j} = \beta_{e_i}$ if $j = i$ and $\alpha_{e_j} = \delta$ else$\}$ for some constants $0 < \delta \leq \beta_{e_i} \leq 1$ such that $A = \sum_{i=1}^{n_e} A_i$ where A_i is the global stiffness matrix corresponding to the distribution $\mathcal{C}_i$. It is easy to see that if A is SPD then A_i is SPD for all i. Now, let S_i denote the exact Schur complement of A_i and S be the Schur complement of A. Moreover, let Q_i denote the ASCA corresponding to A_i, i.e., $Q_i \simeq S_i$ where Q_i is the exact Schur complement of $\tilde{A}_i$, cf. (2).

Lemma 1 *Using the above notation, assume that*

$$\frac{1}{c_j}\langle S_j\mathbf{v}_2, \mathbf{v}_2\rangle \leq \langle Q_j\mathbf{v}_2, \mathbf{v}_2\rangle \leq \langle S_j\mathbf{v}_2, \mathbf{v}_2\rangle \quad \forall \mathbf{v}_2 \text{ and } j = 1, \ldots, n_e. \tag{9}$$

Further, denote $c_{\max} = \max_{i\in\{1,\ldots,n_e\}}\{c_i\}$. *Then the following relations hold:*

$$\frac{1}{c_{\max}}\langle S\mathbf{v}_2, \mathbf{v}_2\rangle \leq \langle Q\mathbf{v}_2, \mathbf{v}_2\rangle \leq \langle S\mathbf{v}_2, \mathbf{v}_2\rangle \quad \forall \mathbf{v}_2. \tag{10}$$

Proof The right inequality in (10) follows directly from the energy minimization property of Schur complements. In order to prove the left inequality we assume that (10) is wrong. Then there exists a vector $\mathbf{v}_2 \neq \mathbf{0}$ such that $\mathbf{v}_2^T S\mathbf{v}_2 \geq \bar{c}\,\mathbf{v}_2^T Q\mathbf{v}_2 > c_{\max}\mathbf{v}_2^T Q\mathbf{v}_2$, the left inequality of which can also be written in the form $\min_{\mathbf{v}_1} \begin{pmatrix} \mathbf{v}_1 \\ \mathbf{v}_2 \end{pmatrix}^T A \begin{pmatrix} \mathbf{v}_1 \\ \mathbf{v}_2 \end{pmatrix} \geq \bar{c}\, \mathbf{v}_2^T Q \mathbf{v}_2$, or, equivalently as $\min_{\mathbf{v}_1} \begin{pmatrix} \mathbf{v}_1 \\ \mathbf{v}_2 \end{pmatrix}^T \left(\sum_{j=1}^{n_e} A_j\right) \begin{pmatrix} \mathbf{v}_1 \\ \mathbf{v}_2 \end{pmatrix} \geq \bar{c} \min_{\tilde{\mathbf{v}}_1} \begin{pmatrix} \tilde{\mathbf{v}}_1 \\ \mathbf{v}_2 \end{pmatrix}^T \left(\sum_{j=1}^{n_e} \tilde{A}_j\right) \begin{pmatrix} \tilde{\mathbf{v}}_1 \\ \mathbf{v}_2 \end{pmatrix}$. From the latter inequality it follows that

$$\begin{pmatrix} \mathbf{v}_1 \\ \mathbf{v}_2 \end{pmatrix}^T \left(\sum_{j=1}^{n_e} A_j\right) \begin{pmatrix} \mathbf{v}_1 \\ \mathbf{v}_2 \end{pmatrix} \geq \bar{c} \sum_{j=1}^{n_e} \min_{\tilde{\mathbf{v}}_1} \begin{pmatrix} \tilde{\mathbf{v}}_1 \\ \mathbf{v}_2 \end{pmatrix}^T \tilde{A}_j \begin{pmatrix} \tilde{\mathbf{v}}_1 \\ \mathbf{v}_2 \end{pmatrix} \quad \forall \mathbf{v}_1,$$

which is equivalent to

$$\sum_{j=1}^{n_e} \begin{pmatrix} \mathbf{v}_1 \\ \mathbf{v}_2 \end{pmatrix}^T A_j \begin{pmatrix} \mathbf{v}_1 \\ \mathbf{v}_2 \end{pmatrix} \geq \bar{c} \sum_{j=1}^{n_e} \mathbf{v}_2^T Q_j \mathbf{v}_2 \quad \forall \mathbf{v}_1. \tag{11}$$

Then, since all matrices A_j and Q_j are SPSD, it follows from (11) that there exists at least one index $j_0 \in \{1, 2, \ldots, n_e\}$ such that

$$\begin{pmatrix} \mathbf{v}_1 \\ \mathbf{v}_2 \end{pmatrix}^T A_{j_0} \begin{pmatrix} \mathbf{v}_1 \\ \mathbf{v}_2 \end{pmatrix} \geq \bar{c}\, \mathbf{v}_2^T Q_{j_0} \mathbf{v}_2 \quad \forall \mathbf{v}_1.$$

Hence $\mathbf{v}_2^T S_{j_0} \mathbf{v}_2 = \min_{\mathbf{v}_1} \begin{pmatrix} \mathbf{v}_1 \\ \mathbf{v}_2 \end{pmatrix}^T A_{j_0} \begin{pmatrix} \mathbf{v}_1 \\ \mathbf{v}_2 \end{pmatrix} \geq \bar{c}\, \mathbf{v}_2^T Q_{j_0} \mathbf{v}_2$ which is in contradiction to (9) since $\bar{c} > c_{\max}$.

A crucial step in the application of the two-level preconditioner is the realization of the operator $\Pi_{\tilde{D}}$. We propose two different variants that correspond to the following choices of $\tilde{D}$:

[I] $\tilde{D} = \mathrm{diag}(\tilde{A})$;

[II] $\tilde{D} = \mathrm{blockdiag}(\tilde{A})$. The diagonal blocks are determined by the groups of fine DOF related to different macro structures whereas $\tilde{D} = \mathrm{diag}(\tilde{A})$ in rows corresponding to coarse DOF.

In variant [I] the matrix $R\tilde{D}R^T$ is diagonal, which makes the application of $\Pi_{\tilde{D}}$ notably simple and cost-efficient. In case of variant [II] the action of $(R\tilde{D}R^T)^{-1}$ can be implemented via an inner iterative method such as a preconditioned conjugate gradient (PCG) method, which then for reasons of efficiency requires a uniform preconditioner. A possible candidate is the one-level additive Schwarz (AS) preconditioner which however has to be adapted in order to be robust with respect to coefficient jumps. For this reason we study the scaled one-level AS preconditioner B_{AS} defined via

$$B_{\mathrm{AS}}^{-1} = SR\tilde{S}^{-1}(\tilde{S}\tilde{D}\tilde{S})^{-1}\tilde{S}^{-1}R^T S \tag{12}$$

which can be applied to the scaled system with the matrix

$$D_s = SDS = SR\tilde{D}R^T S,$$

where $S = [\text{diag}(A)]^{-1/2}$, if the result is then rescaled. Let us further denote

$$\tilde{D}_s = \tilde{S}\tilde{D}\tilde{S} \text{ and } R_s = SR\tilde{S}^{-1} \text{ where } \tilde{S} = [\text{diag}(\tilde{A})]^{-1/2}.$$

Then the following lemma holds:

Lemma 2 *The condition number of the preconditioned system using the scaled one-level AS preconditioner satisfies the estimate*

$$\kappa(B_{\text{AS}}^{-1}D_s) \leq \kappa(\tilde{D}_s). \tag{13}$$

Proof First we show that $\lambda_{\min}(B_{\text{AS}}^{-1}D_s) \geq 1$. Note that $D_s = R_s\tilde{D}_sR_s^T$ and

$$R_sR_s^T = SR\tilde{S}^{-1}\tilde{S}^{-1}R^TS = [\text{diag}(A)]^{-1/2}R\,[\text{diag}(\tilde{A})]R^T[\text{diag}(A)]^{-1/2} = I.$$

Consider next the matrix

$$\begin{bmatrix} R_s\tilde{D}_sR_s^T & I \\ I & R_s\tilde{D}_s^{-1}R_s^T \end{bmatrix} = \begin{bmatrix} R_s & 0 \\ 0 & R_s \end{bmatrix} \begin{bmatrix} \tilde{D}_s & I \\ I & \tilde{D}_s^{-1} \end{bmatrix} \begin{bmatrix} R_s^T & 0 \\ 0 & R_s^T \end{bmatrix}$$

which is SPSD with an SPD pivot block $D_s = R_s\tilde{D}_sR_s^T$. Consequently, its Schur complement is an SPSD matrix, i.e.

$$R_s\tilde{D}_s^{-1}R_s^T - (R_s\tilde{D}_sR_s^T)^{-1} \geq 0$$

which proves that $\lambda_{\min}(B_{\text{AS}}^{-1}D_s) \geq 1$.

On the other hand we have

$$\begin{aligned}
\lambda_{\max}(B_{\text{AS}}^{-1}D_s) &= \lambda_{\max}(R_s\tilde{D}_s^{-1}R_s^TD_s) \\
&= \lambda_{\max}(D_s^{1/2}R_s\tilde{D}_s^{-1}R_s^TD_s^{1/2}) \\
&= \lambda_{\max}(\tilde{D}_s^{-1/2}R_s^TD_sR_s\tilde{D}_s^{-1/2}) \\
&\leq \lambda_{\max}(\tilde{D}_s^{-1})\lambda_{\max}(R_s^TR_s\tilde{D}_sR_s^TR_s) \\
&\leq \lambda_{\max}(\tilde{D}_s^{-1})\lambda_{\max}(\tilde{D}_s)\lambda_{\max}(R_s^TR_s) = \kappa(\tilde{D}_s)
\end{aligned}$$

which completes the proof.

Remark 1 For conforming FEM discretization of the second order scalar elliptic PDE it is not difficult to show that $\kappa(\tilde{D}_s)$ is uniformly bounded with respect to jumps of an elementwise constant coefficient. Furthermore, $\tilde{D}_s$ is block-diagonal with small-sized blocks and thus $\kappa(\tilde{D}_s)$ is easily computable.

4 Auxiliary Space Multigrid Method

Consider the exact block factorization of the sequence of auxiliary stiffness matrices $\tilde{A}^k$, where the superscript $k = 0, 1, \dots, \ell - 1$ indicates the coarsening level:

$$\tilde{A}^{(k)\,-1} = \tilde{L}^{(k)\,T}\,\tilde{D}^{(k)}\tilde{L}^{(k)}, \quad A^{(k+1)} := Q^{(k)},$$

$$\tilde{L}^{(k)} = \begin{bmatrix} I & \\ -\tilde{A}_{21}^{(k)}\,\tilde{A}_{11}^{(k)\,-1} & I \end{bmatrix}, \qquad \tilde{D}^{(k)} = \begin{bmatrix} \tilde{A}_{11}^{(k)\,-1} & \\ & Q^{(k)-1} \end{bmatrix}.$$

Let the algebraic multilevel iteration (AMLI)-cycle auxiliary space multigrid (ASMG) preconditioner $B^{(k)}$ be defined by (see [17]):

$$\begin{aligned} B^{(k)-1} &:= \overline{M}^{(k)\,-1} \\ &\quad + (I - M^{(k)-T} A^{(k)})\Pi^{(k)}\,\tilde{L}^{(k)\,T}\overline{D}^{(k)}\tilde{L}^{(k)}\Pi^{(k)T}(I - A^{(k)}M^{(k)-1}), \\ \overline{D}^{(k)} &:= \begin{bmatrix} \tilde{A}_{11}^{(k)\,-1} & \\ & B_\nu^{(k+1)} \end{bmatrix}, \qquad B_\nu^{(\ell)} := A^{(\ell)-1}. \end{aligned}$$

In the nonlinear AMLI-cycle $B_\nu^{(k+1)} = B_\nu^{(k+1)}[\cdot]$ is a nonlinear mapping realized by ν iterations of a Krylov subspace method (e.g. the generalized conjugate gradient (GCG) method), thus employing the coarse level preconditioner $B^{(k+1)}$. In [13] the convergence of the multiplicative nonlinear AMLI has been first analyzed, while Notay and Vassilevski [24], Vassilevski [30], and Hu et al. [11] have provided the multigrid framework along with a comparative analysis.

We want to stress the fact that the presented construction provides a framework for both linear and nonlinear AMLI cycle multigrid as well as classical multigrid methods.

5 Numerical Results

Subject to numerical testing is the scalar elliptic boundary-value problem

$$-\nabla \cdot (\boldsymbol{k}(\boldsymbol{x})\nabla u(\boldsymbol{x})) = f(\boldsymbol{x}) \;\; in \;\; \Omega, \tag{14a}$$

$$u = 0 \;\; on \;\; \Gamma. \tag{14b}$$

Here Ω is a polygonal domain in $\mathbb{R}^2$, f is a given function in $L_2(\Omega)$ and

$$\boldsymbol{k}(\boldsymbol{x}) = \alpha(\boldsymbol{x})I = \alpha_e I. \tag{15}$$

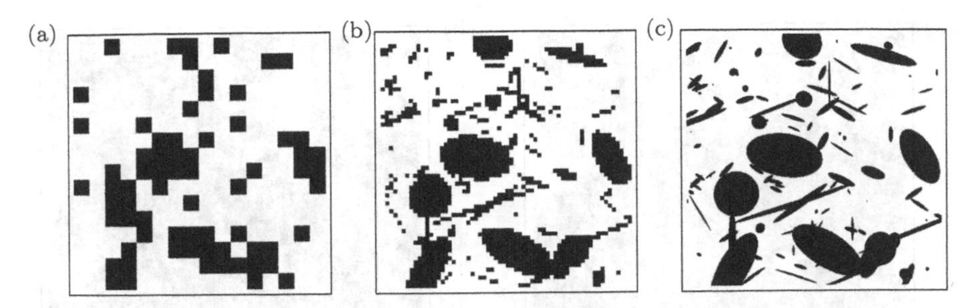

Fig. 1 Inclusions resolved on different fine scales (meshes). (**a**) 16 × 16 mesh. (**b**) 64 × 64 mesh. (**c**) 512 × 512 mesh

Upon the entire boundary of the domain Dirichlet boundary conditions have been imposed as other boundary conditions would not qualitatively affect the numerical results.

Piecewise bilinear functions have been used in the process of discretization of (14) leading to the linear system of algebraic equations (1). A uniform mesh consisting of $N \times N$ elements (squares) is considered where $N = 2^{\ell+2}$, $\ell = 1, \ldots, 7$, and the covering is assumed to consist of subdomains composed of 8×8 elements that overlap with half of their width or height. The mesh hierarchy is such that the coarsest mesh corresponds to $\ell = 1$ and is composed of $2^{1+2} \times 2^{1+2} = 64$ elements whereas the finest mesh is obtained by performing $\ell - 1 = 1, \ldots, 6$ steps of uniform mesh refinement.

The vector of all zeros was chosen to be the right hand side $\mathbf{f}$ in (1) while the outer iteration was initialized with a random vector. Three representative coefficient configurations are considered (on the respective finest mesh, see Fig. 1):

[0] log-uniformly distributed coefficient $\alpha_e = 10^{p_{rand}}$ where α_e is constant on each element e and $p_{rand} \in (0, q]$;
[1] inclusions with coefficient $\alpha_\iota = 10^{p_{rand}}$ against a background as in [0] where α_ι is constant on every inclusion ι and $p_{rand} \in (0, q]$, see Fig. 2a;
[2] stiff inclusions with coefficient $\alpha_\iota = 10^q$ against a background as in [0], see Fig. 2b.

In Table 1 we compare the condition numbers

$$\kappa(\tilde{D}_s) = \kappa(S\tilde{D}S), \quad \kappa(B_{\mathrm{AS}}^{-1}D_s) = \kappa(SR\tilde{S}^{-2}\tilde{D}^{-1}\tilde{S}^{-2}R^T S(SR\tilde{D}R^T S)),$$

with that of the corresponding unscaled preconditioned system

$$\kappa(R\tilde{D}^{-1}R^T(R\tilde{D}R^T))$$

for the coefficient distribution [0] on three different meshes with mesh size $h \in \{1/16, 1/32, 1/64\}$ and varying contrast q. The obtained numerical results are in accordance with Lemma 2; They further show that the scaled one-level additive

(a) 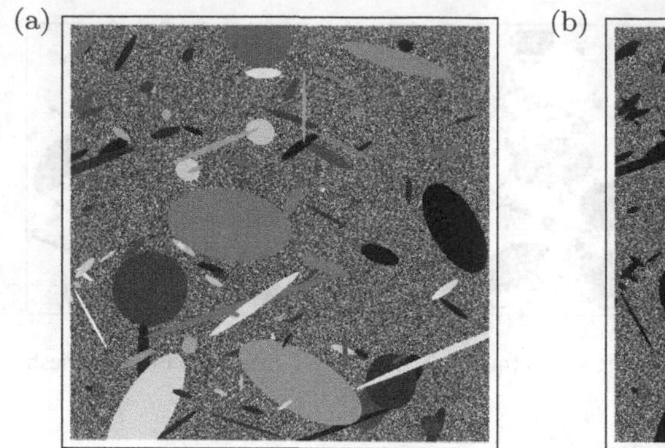 (b) 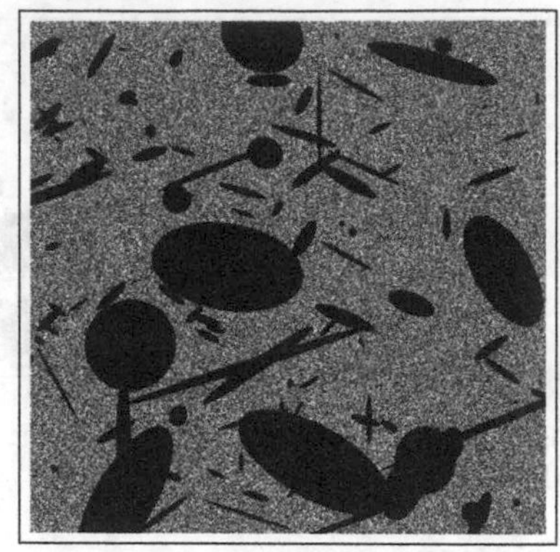

Fig. 2 Random and stiff inclusions against random background $\alpha_e = 10^{P_{rand}}$. (**a**) Coefficient for Problem (P1) on 512 × 512 mesh. (**b**) Coefficient for Problem (P2) on 512 × 512 mesh

Table 1 Condition numbers of AS-preconditioned systems versus $\kappa(\tilde{D}_s)$

	Unscaled AS method			Scaled AS method			$\kappa(\tilde{D}_s)$		
q \ h	1/16	1/32	1/64	1/16	1/32	1/64	1/16	1/32	1/64
1	9.76×10^1	9.47×10^1	9.35×10^1	1.25	1.26	1.26	4.73	4.73	4.73
2	2.25×10^2	3.69×10^2	5.89×10^2	1.28	1.27	1.29	4.73	4.73	4.73
3	6.93×10^2	2.42×10^3	3.70×10^3	1.29	1.32	1.33	4.73	4.73	4.73
4	1.93×10^4	1.97×10^4	3.77×10^4	1.33	1.33	1.33	4.73	4.73	4.73
5	1.78×10^5	1.87×10^5	2.16×10^5	1.32	1.33	1.33	4.73	4.73	4.73
6	3.07×10^5	1.34×10^6	2.15×10^6	1.33	1.33	1.33	4.73	4.73	4.73

Schwarz method yields a uniform preconditioner whereas its unscaled analog suffers from high-contrast coefficients.

Next, the numerical performance of the nonlinear (AMLI)-cycle ASMG method (V-cycle and W-cycle) utilizing the preconditioner B_{AS} is tested for:

(P1) Problem (14) with coefficient distributions [1] and variants [I] and [II] of $\Pi_{\tilde{D}}$. Variant [II] is realized by ten inner PCG iterations with the scaled one-level AS preconditioner.

(P2) Same as Problem (P1) but for coefficient distribution 2.

A comparison between variant [I] and variant [II] of the ℓ-level V-cycle and W-cycle is presented in Tables 2 and 3. Pre- and post-smoothing is performed by one symmetric point Gauss-Seidel iteration on each level except the coarsest one where all linear systems are solved directly.

The obtained results demonstrate that the choice of $\tilde{D}$ and consequently of the surjective mapping $\Pi_{\tilde{D}}$ affect the performance of the nonlinear AMLI-cycle ASMG method crucially. As for variant [I] the number of ASMG iterations required to

Table 2 Number of iterations for residual reduction by 10^6

Problem (P1)																								
	Nonlinear AMLI V-cycle												Nonlinear AMLI W-cycle											
	[I]						[II]						[I]						[II]					
q \ ℓ	2	3	4	5	6	7	2	3	4	5	6	7	2	3	4	5	6	7	2	3	4	5	6	7
1	4	5	6	6	7	8	5	5	6	6	7	8	4	5	5	5	5	5	5	5	5	5	5	5
2	5	5	6	6	7	8	5	5	6	6	7	8	5	5	5	5	5	5	5	5	5	5	5	5
3	5	6	6	7	7	8	5	6	6	7	7	8	5	6	6	6	6	6	5	5	5	5	5	5
4	5	6	7	8	8	9	5	6	7	8	8	8	5	6	6	6	6	6	5	6	6	6	6	6
5	5	7	7	8	9	9	5	6	7	8	8	8	5	6	6	6	7	7	5	6	6	6	6	6
6	5	7	8	9	13	15	5	7	8	8	8	9	5	6	6	7	9	10	5	6	6	6	6	6

Table 3 Number of iterations for residual reduction by 10^6

Problem (P2)																								
	Nonlinear AMLI V-cycle												Nonlinear AMLI W-cycle											
	[I]						[II]						[I]						[II]					
q \ ℓ	2	3	4	5	6	7	2	3	4	5	6	7	2	3	4	5	6	7	2	3	4	5	6	7
1	5	5	6	6	7	8	5	5	6	6	7	8	5	5	5	5	5	5	5	5	5	5	5	5
2	5	5	6	6	7	8	5	5	6	6	7	8	5	5	5	5	5	5	5	5	5	5	5	5
3	5	5	6	6	7	8	5	5	6	6	7	8	5	5	5	6	5	6	5	5	5	5	5	5
4	5	6	6	7	7	8	5	5	6	7	8	8	5	5	6	6	6	6	5	6	5	5	5	6
5	5	6	7	7	9	9	5	6	7	7	8	8	5	6	6	6	6	6	5	6	6	6	6	6
6	5	6	8	8	12	13	5	6	7	8	9	9	5	6	6	6	8	9	5	6	6	6	6	6

achieve the prescribed accuracy increases with the contrast, variant [II] shows full robustness.

References

1. O. Axelsson, P. Vassilevski, Algebraic multilevel preconditioning methods I. Numer. Math. **56**(2–3), 157–177 (1989)
2. O. Axelsson, P. Vassilevski, Algebraic multilevel preconditioning methods II. SIAM J. Numer. Anal. **27**(6), 1569–1590 (1990)
3. O. Axelsson, P. Vassilevski, Variable-step multilevel preconditioning methods, I: Self-adjoint and positive definite elliptic problems. Numer. Linear Algebra Appl. **1**, 75–101 (1994)
4. C.R. Dohrmann, A preconditioner for substructuring based on constrained energy minimization. SIAM J. Sci. Comput. **25**(1), 246–258 (2003)
5. Y. Efendiev, J. Galvis, R. Lazarov, J. Willems, Robust domain decomposition preconditioners for abstract symmetric positive definite bilinear forms. ESAIM Math. Model. Numer. Anal. **46**(05), 1175–1199 (2012)
6. C. Farhat, F.X. Roux, A method of finite element tearing and interconnecting and its parallel solution algorithm. Int. J. Numer. Methods Eng. **32**, 1205–1227 (1991)

7. C. Farhat, M. Lesoinne, P. LeTallec, K. Pierson, D. Rixen, FETI-DP: A dual-primal unified FETI method. I. A faster alternative to the two-level FETI method. Int. J. Numer. Methods Eng. **50**(7), 1523–1544 (2001)
8. J. Galvis, Y. Efendiev, Domain decomposition preconditioners for multiscale flows in high-contrast media. Multiscale Model. Simul. **8**(4), 1461–1483 (2010)
9. I.G. Graham, P.O. Lechner, R. Scheichl, Domain decomposition for multiscale PDEs. Numer. Math. **106**(4), 489–626 (2007)
10. W. Hackbusch, *Multi-Grid Methods and Applications* (Springer, Berlin, 2003)
11. X. Hu, P. Vassilevski, J. Xu, Comparative convergence analysis of nonlinear AMLI-cycle multigrid. SIAM J. Numer. Anal. **51**(2), 1349–1369 (2013)
12. A. Klawonn, O. Widlund, M. Dryja, Dual-primal FETI methods for three-dimensional elliptic problems with heterogeneous coefficients. SIAM J. Numer. Anal. **40**(1), 159–179 (2002)
13. J. Kraus, An algebraic preconditioning method for M-matrices: linear versus non-linear multilevel iteration. Numer. Linear Algebra Appl. **9**, 599–618 (2002)
14. J. Kraus, Algebraic multilevel preconditioning of finite element matrices using local Schur complements. Numer. Linear Algebra Appl. **13**, 49–70 (2006)
15. J. Kraus, Additive Schur complement approximation and application to multilevel preconditioning. SIAM J. Sci. Comput. **34**, A2872–A2895 (2012)
16. J. Kraus, P. Vassilevski, L. Zikatanov, Polynomial of best uniform approximation to 1/x and smoothing for two-level methods. Comput. Methods Appl. Math. **12**, 448–468 (2012)
17. J. Kraus, M. Lymbery, S. Margenov, Auxiliary space multigrid method based on additive Schur complement approximation. Numer. Linear Algebra Appl. (2014). doi:10.1002/nla.1959. Online (wileyonlinelibrary.com)
18. Y. Kuznetsov, Algebraic multigrid domain decomposition methods. Sov. J. Numer. Anal. Math. Model. **4**(5), 351–379 (1989)
19. J. Mandel, Balancing domain decomposition. Commun. Numer. Methods Eng. **9**(3), 233–241 (1993)
20. J. Mandel, C.R. Dohrmann, Convergence of a balancing domain decomposition by constraints and energy minimization. Numer. Linear Algebra Appl. **10**(7), 639–659 (2003)
21. J. Mandel, B. Sousedík, Adaptive selection of face coarse degrees of freedom in the BDDC and FETI-DP iterative substructuring methods. Comput. Methods Appl. Mech. Eng. **196**(8), 1389–1399 (2007)
22. J. Mandel, C.R. Dohrmann, R. Tezaur, An algebraic theory for primal and dual substructuring methods by constraints. Appl. Numer. Math. **54**(2), 167–193 (2005)
23. T.P.A. Mathew, *Domain Decomposition Methods for the Numerical Solution of Partial Differential Equations* (Springer, Berlin, 2008)
24. Y. Notay, P. Vassilevski, Recursive Krylov-based multigrid cycles. Numer. Linear Algebra Appl. **15**, 473–487 (2008)
25. C. Pechstein, R. Scheichl, Analysis of FETI methods for multiscale PDEs. Numer. Math. **111**(2), 293–333 (2008)
26. M. Sarkis, Schwarz preconditioners for elliptic problems with discontinuous coefficients using conforming and non-conforming elements. PhD thesis, Courant Institute, New York University (1994)
27. N. Spillane, V. Dolean, P. Hauret, F. Nataf, C. Pechstein, R. Scheichl, Abstract robust coarse spaces for systems of PDEs via generalized eigenproblems in the overlaps. Numer. Math. **126**(4), 741–770 (2014)
28. A. Toselli, O. Widlund, *Domain Decomposition Methods–Algorithms and Theory* (Springer, Berlin, 2005)
29. U. Trottenberg, C.W. Oosterlee, A. Schüller, *Multigrid* (Academic, San Diego, 2001)
30. P. Vassilevski, *Multilevel Block Factorization Preconditioners: Matrix-based Analysis and Algorithms for Solving Finite Element Equations* (Springer, New York, 2008)
31. J. Xu, The auxiliary space method and optimal multigrid preconditioning techniques for unstructured grids. Computing **56**, 215–235 (1996)

A Nonlinear FETI-DP Method with an Inexact Coarse Problem

Axel Klawonn, Martin Lanser, and Oliver Rheinbach

1 Introduction

We present a new nonoverlapping, nonlinear domain decomposition method with an inexact solution of the coarse problem. The method can be seen as an inexact reduced version of a recent nonlinear FETI-DP method [33].

In this method, the nonlinear problem is decomposed before linearization. This is opposed to standard Newton-Krylov-Domain-Decomposition methods where the decomposition is performed after linearization. Nonlinear FETI-DP methods were introduced in [32, 33] as nonlinear versions of the well known family of FETI-DP domain decomposition methods.

In domain decomposition methods of the FETI-DP [16, 17, 27, 29–31] and BDDC type [9, 13, 34–36] the coarse spaces are constructed from partial assembly of the finite elements. This has facilitated the extension of the scalability of these methods, see, e.g., [26, 28, 37, 41, 43, 44]. Inexact FETI-DP methods were introduced in [26] and their parallel scalability has been demonstrated in [29, 40].

A. Klawonn • M. Lanser
Mathematisches Institut, Universität zu Köln, Weyertal 86-90, 50931 Köln, Germany
e-mail: axel.klawonn@uni-koeln.de; martin.lanser@uni-koeln.de

O. Rheinbach (✉)
Fakultät für Mathematik und Informatik, Institut für Numerische Mathematik und Optimierung, TU Bergakademie Freiberg, Akademiestr. 6, 09599 Freiberg, Germany
e-mail: oliver.rheinbach@math.tu-freiberg.de

T. Dickopf et al. (eds.), *Domain Decomposition Methods in Science and Engineering XXII*, Lecture Notes in Computational Science and Engineering 104, DOI 10.1007/978-3-319-18827-0_4

Nonlinear approaches to domain decomposition are not new but have attracted recent interest as a strategy to localize computational work. Reduction of communication and synchronization is expected to be crucial to obtain good performance on future supercomputers.

The nonlinear, overlapping ASPIN (Additive Schwarz Preconditioned Inexact Newton) approach was introduced in [6]. See also [6, 7, 21, 22, 24, 25]. Nonlinear domain decomposition as a coupling method has been used, e.g., in fluid-structure interaction; see [10–12], or [18]; it has also been used for the coupling of multiphase flow, see, e.g., [19, 20]. Nonlinear FETI-1 methods were introduced in [39], nonlinear Neumann-Neumann methods, as a scalable solver approach, in [4]. Nonlinear Schwarz methods as a solver, i.e., not as a preconditioner, have already been considered much earlier, see, e.g., [5, 14]. The solution of local nonlinear problems can also be embedded into standard methods and has been denoted nonlinear localization; see [8].

2 Nonlinear FETI-DP Formulation

Let $\Omega_i, i = 1, \ldots, N$, be a decomposition of the domain $\Omega \subset \mathbb{R}^d$, $d = 2, 3$, into nonoverlapping subdomains. Each subdomain is a union of finite elements. We denote the associated local finite element spaces by W_i and the product space by $W = W_1 \times \ldots \times W_N$. We consider the minimization of a nonlinear energy $J : V^h \rightarrow \mathbb{R}$,

$$J(u) = \sum_{i=1}^{N} J_i(u_i), \tag{1}$$

where the $J_i : W_i \rightarrow \mathbb{R}, i = 1, \ldots, N$ are local energy functionals on the subdomains Ω_i. For standard problems, such as nonlinear elasticity, discretized by finite elements the global energy can be written as a sum of the local nonlinear energies on the nonoverlapping subdomains; for details, see [33].

Let $\varphi_{i,j}$, $i = 1, \ldots, N$, $j = 1, \ldots, N_i$ the nodal finite element basis functions for the local finite element space W_i. We write $J_i'(u_i)(\varphi_{i,j})$ in the form

$$J_i'(u_i)(\varphi_{i,j}) = (K_i(u_i) - f_i)_j$$

where $K_i(u_i)$ depends on u_i and f_i is independent of u_i.

Let us define the nonlinear, discrete block operator $K(u)$ and the corresponding block vectors u and f, i.e.,

$$K(u) := \begin{pmatrix} K_1(u_1) \\ \vdots \\ K_N(u_N) \end{pmatrix}, \quad f := \begin{pmatrix} f_1 \\ \vdots \\ f_N \end{pmatrix}, \quad \text{and} \quad u := \begin{pmatrix} u_1 \\ \vdots \\ u_N \end{pmatrix}. \tag{2}$$

We then define the nonlinear, partially assembled operator $\tilde{K}(\tilde{u}) := R_{\Pi}^T K(R_{\Pi}\tilde{u})$,and the corresponding partially assembled right hand side $\tilde{f} := R_{\Pi}^T f$. Here we use the FETI-DP partial assembly operator R_{Π}^T that is also used to define the coarse problem of standard (linear) FETI-DP methods; see, e.g., [27, 42] for the notation. Let B be the standard FETI-DP jump operator, we can then introduce the nonlinear FETI-DP master system [32, 33]

$$\begin{aligned} \tilde{K}(\tilde{u}) + B^T\lambda - \tilde{f} &= 0 \\ B\tilde{u} &= 0. \end{aligned} \tag{3}$$

The nonlinear FETI-DP methods Nonlinear-FETI-DP-1 (NL-1) and Nonlinear-FETI-DP-2 (NL-2), see [32, 33], are also based on the master system (3).

We assume that, as a result of a sufficient number of primal constraints, the operator $\tilde{K}$ is continuously differentiable and locally invertible. We use Newton's method applied to (3) to obtain fast local convergence and a line search as globalization strategy.

3 An Inexact Reduced Nonlinear FETI-DP Method

Newton's method applied to (3) results in the linearized system

$$\begin{bmatrix} D\tilde{K}(\tilde{u}) & B^T \\ B & 0 \end{bmatrix} \begin{bmatrix} \Delta\tilde{u} \\ \Delta\lambda \end{bmatrix} = \begin{bmatrix} \tilde{K}(\tilde{u}) + B^T\lambda - \tilde{f} \\ B\tilde{u} \end{bmatrix}. \tag{4}$$

Following the standard FETI-DP approach, we partition $\Delta\tilde{u}$ into the primal variables $\Delta\tilde{u}_{\Pi}$ and the dual variables $\Delta\tilde{u}_B$, i.e., $\Delta\tilde{u}^T = \left[\Delta u_B^T \; \Delta\tilde{u}_{\Pi}^T \right]$. We then obtain from (4) the system

$$\begin{bmatrix} (D\tilde{K}(\tilde{u}))_{BB} & (D\tilde{K}(\tilde{u}))_{\Pi B}^T & B_B^T \\ (D\tilde{K}(\tilde{u}))_{\Pi B} & (D\tilde{K}(\tilde{u}))_{\Pi\Pi} & 0 \\ B_B & 0 & 0 \end{bmatrix} \begin{bmatrix} \Delta u_B \\ \Delta\tilde{u}_{\Pi} \\ \Delta\lambda \end{bmatrix} = \begin{bmatrix} (\tilde{K}(\tilde{u}))_B + B_B^T\lambda - f_B \\ (\tilde{K}(\tilde{u}))_{\Pi} - \tilde{f}_{\Pi} \\ B_B u_B \end{bmatrix}. \tag{5}$$

Assuming enough primal constraints such that $(D\tilde{K}(\tilde{u}))_{BB}$ is invertible, we then eliminate of u_B and obtain a reduced system

$$\begin{bmatrix} \tilde{S}_{\Pi\Pi} & -(D\tilde{K}(\tilde{u}))_{\Pi B}(D\tilde{K}(\tilde{u}))_{BB}^{-1}B_B^T \\ -B_B(D\tilde{K}(\tilde{u}))_{BB}^{-1}(D\tilde{K}(\tilde{u}))_{\Pi B}^T & -B_B(D\tilde{K}(\tilde{u}))_{BB}^{-1}B_B^T \end{bmatrix} \begin{bmatrix} \Delta\tilde{u}_\Pi \\ \Delta\lambda \end{bmatrix}$$

$$= \begin{bmatrix} (\tilde{K}(\tilde{u}))_\Pi - \tilde{f}_\Pi - (D\tilde{K}(\tilde{u})_{\Pi B}(D\tilde{K}(\tilde{u}))_{BB}^{-1}((\tilde{K}(\tilde{u}))_B + B_B^T\lambda - f_B) \\ B_B u_B - B_B(D\tilde{K}(\tilde{u}))_{BB}^{-1}((\tilde{K}(\tilde{u}))_B + B_B^T\lambda - f_B) \end{bmatrix} \tag{6}$$

which we write as $\mathcal{A}_r x_r = \mathcal{F}_r$ using the same notation as in [26] for linear problems. The Schur complement

$$\tilde{S}_{\Pi\Pi} = (D\tilde{K}(\tilde{u}))_{\Pi\Pi} - (D\tilde{K}(\tilde{u}))_{\Pi B}(D\tilde{K}(\tilde{u}))_{BB}^{-1}(D\tilde{K}(\tilde{u}))_{\Pi B}^T \tag{7}$$

is the coarse problem of the FETI-DP method. In this paper, we will apply a preconditioned Krylov method to the block system (6), using the block-triangular preconditioner

$$\hat{\mathcal{B}}_r^{-1} = \begin{bmatrix} \hat{S}_{\Pi\Pi}^{-1} & 0 \\ -M^{-1}B_B(D\tilde{K}(\tilde{u})_{BB}^{-1}(D\tilde{K}(\tilde{u}))_{\Pi B}^T\hat{S}_{\Pi\Pi}^{-1} & -M^{-1} \end{bmatrix} \tag{8}$$

cf. [26, 29], where the irFETI-DP method (inexact reduced FETI-DP) for linear problems was introduced.

Here, M^{-1} is one of the standard FETI-DP preconditioners. In this paper, we always use the Dirichlet preconditioner [42]. Moreover, $\hat{S}_{\Pi\Pi}^{-1}$ is assumed to be a good preconditioner for the coarse problem $\tilde{S}_{\Pi\Pi}$. Since the preconditioner (8) is unsymmetric we have to use a Krylov space method suitable for unsymmetric systems. In this paper we will use GMRES. The use of conjugate gradients requires a symmetric reformulation.

In this nonlinear FETI-DP method the continuity of the solution is, in general, not reached until convergence of the Newton method. This is different from FETI-DP methods applied after Newton linearization where each Newton iterate is continuous. This method is thus not identical to a standard Newton-Krylov FETI-DP approach.

Note that the elimination of $\tilde{u}_\Pi$ from (6) leads to the Nonlinear-FETI-DP-1 (NL1) method $F_{NL1}\Delta\lambda = d$, introduced in [32, 33]. But this requires an exact solver for $\tilde{S}_{\Pi\Pi}$.

4 Initial Values for the Nonlinear FETI-DP Method

The convergence of Newton-type methods depends on a good initial value. We are interested to find a suitable initial value $\tilde{u}^{(0)}$ for the Newton iteration presented in Sect. 3. This initial value has to be continuous in all primal variables $\tilde{u}_\Pi^{(0)}$ but may be discontinuous in the dual variables $u_B^{(0)}$. Of course, it should provide a good local approximation of the problem. We can obtain such an initial value $\tilde{u}^{(0)}$ from solving the nonlinear problem

$$\tilde{K}(\tilde{u}^{(0)}) = \tilde{f} - B^T \lambda^{(0)} \tag{9}$$

by some Newton type iteration for some given initial value $\lambda^{(0)}$. In this paper we set $\lambda^{(0)} = 0$. The solution of (9) requires the solution of local nonlinear subdomain problems which are only coupled in the primal unknowns. This step thus requires only communication in the primal variables and is otherwise completely parallel. It may be seen as a nonlinear localization step.

Linearization of (9) results in

$$\begin{bmatrix} (D\tilde{K}(\tilde{u}))_{BB} & (D\tilde{K}(\tilde{u}))_{\Pi B}^T \\ (D\tilde{K}(\tilde{u}))_{\Pi B} & (D\tilde{K}(\tilde{u}))_{\Pi\Pi} \end{bmatrix} \begin{bmatrix} u_B \\ \tilde{u}_\Pi \end{bmatrix} = \begin{bmatrix} (\tilde{K}(\tilde{u}))_B + B_B^T\lambda - f_B \\ (D\tilde{K}(\tilde{u}))_\Pi - \tilde{f}_\Pi \end{bmatrix}.$$

A block elimination of u_B yields the symmetric system

$$\tilde{S}_{\Pi\Pi}\tilde{u} = \tilde{d}_\Pi \tag{10}$$

where $\tilde{S}_{\Pi\Pi}$ is defined as in (7). We solve (10) by a Krylov method using the preconditioner $\hat{S}_{\Pi\Pi}^{-1}$; see (8).

5 Numerical Results

In this section, we compare the standard Newton-Krylov approach, using either the standard FETI-DP method or the irFETI-DP [26, 29] method as a solver, and the new nonlinear domain decomposition approach, i.e., the irNonlinear-FETI-DP-1 approach. We have implemented the algorithm presented here using PETSc [1–3]. For all inexact algorithms, the preconditioner $\hat{S}_{\Pi\Pi}$ for the coarse problem $\tilde{S}_{\Pi\Pi}$ is formed by applying one iteration of BoomerAMG [23]. BoomerAMG is part of the Hypre library [15]. In all experiments we have used GMRES as a Krylov method. The Newton method is always combined with a line search using the strong Wolfe conditions; see [38]. For a minimization problem $\min_{x\in\mathbb{R}^n} J(x)$ and a descent direction Δx the strong Wolfe conditions read $J(x + t\,\Delta x) \leq J(x) + c_1 t\,\nabla^T J(x)\,\Delta x$ and $|\nabla^T J(x + t\,\Delta x)\,\Delta x| \leq c_2|\nabla^T J(x)\,\Delta x|$ with constants $0 < c_1 < c_2 < 1$, and where t is the step length.

First, we apply all algorithms to a standard linear diffusion problem, see Table 1, as a sanity check. For this linear problem, the initialization phase, see Sect. 4, is omitted as it is not necessary. The test runs on 16–1024 cores of a Cray XT6 show almost identical numerical and parallel performance of the different algorithms and

Table 1 Sanity check (irNonlinear-FETI-DP-1); Cray XT6: $H/h = 256$, standard linear Laplace, Alg. A

N (=Cores)	Solver	Krylov-It.	Factor.	Max. cond.	Max. It.	Krylov-time (s)	Runtime (s)
16	Newton-Krylov FETI-DP	11	1	7.3	11	0.74	5.3
	Newton-Krylov irFETI-DP	11	1	7.3	11	0.91	5.5
	irNonlinear-FETI-DP-1	11	1	7.3	11	0.92	5.4
64	Newton-Krylov FETI-DP	22	1	8.1	22	1.5	6.3
	Newton-Krylov irFETI-DP	22	1	8.0	22	2.0	6.7
	irNonlinear-FETI-DP-1	21	1	8.2	21	2.1	6.9
256	Newton-Krylov FETI-DP	32	1	8.3	32	2.3	7.4
	Newton-Krylov irFETI-DP	30	1	8.1	30	3.2	8.3
	irNonlinear-FETI-DP-1	30	1	8.3	30	4.7	9.9
1024	Newton-Krylov FETI-DP	32	1	8.4	32	2.5	8.8
	Newton-Krylov irFETI-DP	30	1	8.3	30	4.2	10.8
	irNonlinear-FETI-DP-1	28	1	8.4	28	4.3	11.0

Table 2 Comparison a standard Newton-Krylov irFETI-DP approach with the nonlinear method; Cray XT6: $H/h = 80$, $\Delta + 4\Delta_p, p = 4$, Alg. A

	NK-irFETI-DP			irNL-FETI-DP-1		
N (Cores)	Runtime (s)	Krylov-It.	Krylov-time (s)	Runtime (s)	Krylov-It.	Krylov-time (s)
64 (1)	92.3	92	23.6	90.5	19	5.5
256 (4)	126.5	88	31.4	107.0	20	7.2
1024 (16)	91.2	68	27.9	97.4	20	8.2
4096 (64)	111.8	67	30.2	100.9	20	9.1
16,384 (256)	113.7	67	28.5	102.5	20	8.5
65,536 (1024)	130.9	65	32.0	110.5	20	9.9

implementations. This is expected since, for a linear problem, the Newton-Krylov-irFETI-DP method and the irNonlinear-FETI-DP-1 method are equivalent. We do see some increase in the total runtime, mainly due to an increase in the Krylov iteration time. This increase is due to an inefficient parallel distribution of the coarse problem. A redistribution would be necessary on this architecture but was not performed here. In Table 2, we then perform a weak scaling test for a nonlinear problem on the Cray XT6 at Universität Duisburg-Essen using up to 1024 cores. We have considered a nonlinear diffusion problem $\Delta u + 4\Delta_p u = f$ for $p = 4$, where Δ is the standard Laplacian and Δ_p is the p-Laplacian. The step length is chosen according to a Wolfe rule. We have considered subdomains of quite small size, i.e., $H/h = 80$, but up to 65,536 subdomains.

Table 3 irNonlinear-FETI-DP-1 on the MIRA Supercomputer (BG/P) Argonne National Laboratory; $\Delta + 4\Delta_p, p = 4$, $H/h = 128$

Inexact-reduced-nonlinear-FETI-DP (irNL-FETI-DP-1)			
N (=Cores)	Step	Time (s)	Krylov-It.
16	Newton init 1:	5.2	0
	Newton init 2:	5.2	0
	Newton init 3:	5.2	0
	Newton init 4:	5.2	0
	Newton full 1:	7.3	9
64	Newton init 1:	5.3	0
	Newton init 2:	5.2	0
	Newton init 3:	5.2	0
	Newton init 4:	5.2	0
	Newton full 1:	8.2	17
256	Newton init 1:	5.4	0
	Newton init 2:	5.4	0
	Newton init 3:	5.4	0
	Newton init 4:	5.4	0
	Newton full 1:	9.5	21
1024	Newton init 1:	5.8	0
	Newton init 2:	5.9	0
	Newton init 3:	5.8	0
	Newton init 4:	5.9	0
	Newton full 1:	10.4	20
4096	Newton init 1:	7.6	0
	Newton init 2:	7.5	0
	Newton init 3:	7.5	0
	Newton init 4:	7.5	0
	Newton full 1:	13.1	20

Alg. A.; joint work with **B. Smith and S. Balay (Argonne National Laboratory)**; uses only 4 out of 16 BG/Q cores. "Newton Init" refers to a Newton step for solving (9) whereas "Newton Full" refers to a Newton step for solving (3). A single full Newton step is sufficient for this problem after four steps to compute the initial value

We can see that the new method is competitive and significantly reduces the number of Krylov iterations. As a result, the inexact reduced Nonlinear-FETI-DP-1 (irNL-1) method is slightly faster.

We then have performed a weak scalability test using 16–4096 processor cores of the MIRA supercomputer at the Argonne National Laboratory, see Table 3. We can see that, for this problem, up to four Newton steps are performed in the initialization phase, i.e., to solve (9). No Krylov iteration is necessary in this phase. A single Newton iteration, using between 9 and 21 Krylov iterations, is sufficient to solve the nonlinear problem (3) to the desired relative tolerance of 1e–9. The parallel efficiency drops to 56 % from 16 to 4096 processor cores. This was an unexpected result on the BG/Q architecture. Indeed, a performance bug in a parallel norm computation that limited scalability was identified as a result of these experiments.

After eliminating the performance bug we finally have performed a similar weak scalability test using 32–32,768 processor cores of the SuperMUC supercomputer at the Leibniz-Rechenzentrum in Munich. The results are presented in Table 4. To solve this problem eight Newton steps are performed in the initialization phase and then a single full Newton step is sufficient to reach a tolerance of 1e–10. Overall, the algorithm needs only between 26 and 34 Krylov iterations. The parallel scalability seems satisfactory and we reach an efficiency of 74 % using 32,768 cores compared to the baseline of 32 cores. Let us remark, that a non negligible amount of time is spent in the MPI initialization called by PETSc in the first Newton step and we expect to obtain even better results in the future.

Finally, in Table 5, we report on weak scalability for a problem of nonlinear hyperelasticity on the SuperMUC supercomputer.

Table 4 irNonlinear-FETI-DP-1 on the SuperMUC supercomputer at Leibniz-Rechenzentrum in Munich; $\Delta + 4\Delta_p, p = 4, Hx/hx = 768, Hy/hy = 384$; the algorithm uses all 16 cores of the node; "Newton Init" refers to a Newton step for solving (9) whereas "Newton Full" refers to a Newton step for solving (3)

Inexact-reduced-nonlinear-FETI-DP (irNL-FETI-DP-1)						
Nx x Ny = N (=Cores)	d.o.f.	Krylov-It.	Newton steps init/full	Krylov-time (s)	Runtime (s)	Eff. (%)
32	9,443,329	26	8/1	4.19	112.5	100
128	37,761,025	31	8/1	5.07	117.8	96
512	151,019,521	33	8/1	5.49	119.1	95
2048	604,028,929	33	8/1	5.65	119.1	95
8192	2,416,017,409	34	8/1	6.01	127.9	88
32,768	9,663,873,025	34	8/1	9.13	151.4	74

Table 5 irNonlinear-FETI-DP-1 on the SuperMUC supercomputer at Leibniz-Rechenzentrum (LRZ) in Munich; Neo-Hooke material; $E = 210{,}000$ in off-centered circular inclusions in the subdomains and $E = 210$ in the surrounding matrix material; Poisson ratio $\nu = 0.3$; a fixed displacement of 1 % in x-direction is prescribed on the boundary

irNL-FETI-DP-1 for hyperelasticity			
Cores	Problem size (M)	Total time (s)	Total effic. (%)
64	4	127	100
256	16	139	91
1024	67	128	99
4096	268	142	89

6 Summary

The new nonlinear FETI-DP method combines the approaches from [26, 33] and thus can be denoted inexact reduced Nonlinear-FETI-DP-1 (irNL1). An important building block of this method is the solution of nonlinear problems on the subdomains. Algorithmically, the same building blocks as standard FETI-DP methods are used. If exact solvers are used as building blocks the new method shows the same performance as the Nonlinear-FETI-DP-1 method [33]. If an efficient preconditioner is used for the coarse problem then the scalability can be extended substantially.

Acknowledgement The authors would like to thank **Satish Balay** and **Barry Smith**, Argonne National Laboratory (ANL), USA, for the fruitful cooperation and the assistance on running the authors' code on the MIRA Supercomputer (BG/P) at ANL. This work was supported in part by the German Research Foundation (DFG) through the Priority Programme 1648 "Software for Exascale Computing" (SPPEXA). The authors acknowledge the use of the Cray XT6 computer at Universität Duisburg-Essen. The authors also gratefully acknowledge the SuperMUC Supercomputer at Leibniz-Rechenzentrum (www.lrz.de).

References

1. S. Balay, W.D. Gropp, L.C. McInnes, B.F. Smith, Efficient management of parallelism in object oriented numerical software libraries, in *Modern Software Tools in Scientific Computing*, ed. by E. Arge, A.M. Bruaset, H.P. Langtangen (Birkhäuser, Basel, 1997), pp. 163–202
2. S. Balay, J. Brown, K. Buschelman, V. Eijkhout, W.D. Gropp, D. Kaushik, M.G. Knepley, L.C. McInnes, B.F. Smith, H. Zhang, PETSc users manual. Technical report ANL-95/11 - Revision 3.4, Argonne National Laboratory (2013)
3. S. Balay, J. Brown, K. Buschelman, W.D. Gropp, D. Kaushik, M.G. Knepley, L.C. McInnes, B.F. Smith, H. Zhang, PETSc Web page (2013). http://www.mcs.anl.gov/petsc
4. F. Bordeu, P.-A. Boucard, P. Gosselet, Balancing domain decomposition with nonlinear relocalization: parallel implementation for laminates, in *Proceedings of the First International Conference on Parallel, Distributed and Grid Computing for Engineering*, ed. by B.H.V. Topping, P. Ivnyi (Civil-Comp Press, Stirlingshire, 2009)
5. X.-C. Cai, M. Dryja, Domain decomposition methods for monotone nonlinear elliptic problems, in *Domain Decomposition Methods in Scientific and Engineering Computing (University Park, PA, 1993)*. Contemporary Mathematics, vol. 180 (American Mathematical Society, Providence, 1994), pp. 21–27
6. X.-C. Cai, D.E. Keyes, Nonlinearly preconditioned inexact Newton algorithms. SIAM J. Sci. Comput. **24**(1), 183–200 (electronic) (2002)
7. X.-C. Cai, D.E. Keyes, L. Marcinkowski, Non-linear additive Schwarz preconditioners and application in computational fluid dynamics. Int. J. Numer. Methods Fluids **40**(12), 1463–1470 (2002) [LMS Workshop on Domain Decomposition Methods in Fluid Mechanics (London, 2001)]
8. P. Cresta, O. Allix, C. Rey, S. Guinard, Nonlinear localization strategies for domain decomposition methods: application to post-buckling analyses. Comput. Methods Appl. Mech. Eng. **196**(8), 1436–1446 (2007)
9. J.-M. Cros, A preconditioner for the Schur complement domain decomposition method, in *Domain Decomposition Methods in Science and Engineering*, ed. by O. Widlund I. Herrera,

D. Keyes, R. Yates. Proceedings of the 14th International Conference on Domain Decomposition Methods, National Autonomous University of Mexico (UNAM), Mexico City, 2003, pp. 373–380. http://www.ddm.org/DD14 [ISBN 970-32-0859-2]
10. S. Deparis, Numerical analysis of axisymmetric flows and methods for fluid-structure interaction arising in blood flow simulation. Ph.D. thesis, EPFL, 2004
11. S. Deparis, M. Discacciati, G. Fourestey, A. Quarteroni, *Heterogeneus Domain Decomposition Methods for Fluid-Structure Interaction Problems*. Domain Decomposition Methods in Science and Engineering XVI, vol. 55 (Springer, Berlin, 2005)
12. S. Deparis, M. Discacciati, G. Fourestey, A. Quarteroni, Fluid-structure algorithms based on Steklov-Poincaré operators. Comput. Methods Appl. Mech. Eng. **195**, 5797–5812 (2006)
13. C.R. Dohrmann, A preconditioner for substructuring based on constrained energy minimization. SIAM J. Sci. Comput. **25**(1), 246–258 (2003)
14. M. Dryja, W. Hackbusch, On the nonlinear domain decomposition method. BIT **37**(2), 296–311 (1997)
15. R.D. Falgout, J.E. Jones, U.M. Yang, The design and implementation of hypre, a library of parallel high performance preconditioners, in *Numerical Solution of Partial Differential Equations on Parallel Computers*, ed. by A.M. Bruaset, P. Bjorstad, A. Tveito. Lecture Notes in Computational Science and Engineering, vol. 51 (Springer, Berlin, 2006), pp. 267–294. Also available as LLNL Technical report UCRL-JRNL-205459. doi:10.1007/3-540-31619-1_8
16. C. Farhat, M. Lesoinne, P. LeTallec, K. Pierson, D. Rixen, FETI-DP: a dual-primal unified FETI method - Part i: a faster alternative to the two-level FETI method. Int. J. Numer. Methods Eng. **50**, 1523–1544 (2001)
17. C. Farhat, M. Lesoinne, K. Pierson, A scalable dual-primal domain decomposition method. Numer. Linear Algebra Appl. **7**, 687–714 (2000)
18. M.Á. Fernández, J.-F. Gerbeau, A. Gloria, M. Vidrascu, Domain decomposition based Newton methods for fluid-structure interaction problems, in *CANUM 2006—Congrès National d'Analyse Numérique*. ESAIM Proceedings, vol. 22 (EDP Sciences, Les Ulis, 2008), pp. 67–82
19. B. Ganis, G. Pencheva, M.F. Wheeler, T. Wildey, I. Yotov, A frozen Jacobian multiscale mortar preconditioner for nonlinear interface operators. Multiscale Model. Simul. **10**(3), 853–873 (2012)
20. B. Ganis, K. Kumar, G. Pencheva, M.F. Wheeler, I. Yotov, A global Jacobian method for Mortar discretizations of a fully-implicit two-phase flow model. Technical report 2014-01, The Institute for Computational Engineering and Sciences, The University of Texas at Austin, January 2014
21. C. Groß, R. Krause, A generalized recursive trust-region approach - nonlinear multiplicatively preconditioned trust-region methods and applications. Technical report 2010-09, Institute of Computational Science, Universita della Svizzera italiana, March 2010
22. C. Groß, R. Krause, On the globalization of ASPIN employing trust-region control strategies - convergence analysis and numerical examples. Technical report 2011-03, Institute of Computational Science, Universita della Svizzera italiana, January 2011
23. V.E. Henson, U.M. Yang, Boomeramg: a parallel algebraic multigrid solver and preconditioner. Appl. Numer. Math. **41**, 155–177 (2002)
24. F.-N. Hwang, X.-C. Cai, Improving robustness and parallel scalability of Newton method through nonlinear preconditioning, in *Domain Decomposition Methods in Science and Engineering*. Lecture Notes in Computer Science and Engineering, vol. 40 (Springer, Berlin, 2005), pp. 201–208
25. F.-N. Hwang, X.-C. Cai, A class of parallel two-level nonlinear Schwarz preconditioned inexact Newton algorithms. Comput. Methods Appl. Mech. Eng. **196**(8), 1603–1611 (2007)
26. A. Klawonn, O. Rheinbach, Inexact FETI-DP methods. Int. J. Numer. Methods Eng. **69**(2), 284–307 (2007)
27. A. Klawonn, O. Rheinbach, Robust FETI-DP methods for heterogeneous three dimensional elasticity problems. Comput. Methods Appl. Mech. Eng. **196**(8), 1400–1414 (2007)

28. A. Klawonn, O. Rheinbach, A hybrid approach to 3-level FETI. PAMM **8**(1), 10841–10843 (2008). doi:10.1002/pamm.200810841 [Special Issue: 79th Annual Meeting of the International Association of Applied Mathematics and Mechanics (GAMM)]
29. A. Klawonn, O. Rheinbach, Highly scalable parallel domain decomposition methods with an application to biomechanics. ZAMM Z. Angew. Math. Mech. **90**(1), 5–32 (2010)
30. A. Klawonn, O.B. Widlund, Dual-primal FETI methods for linear elasticity. Commun. Pure Appl. Math. **59**(11), 1523–1572 (2006)
31. A. Klawonn, O.B. Widlund, M. Dryja, Dual-primal FETI methods for three-dimensional elliptic problems with heterogeneous coefficients. SIAM J. Numer. Anal. **40**(1), 159–179 (2002)
32. A. Klawonn, M. Lanser, P. Radtke, O. Rheinbach, On an adaptive coarse space and on nonlinear domain decomposition, in *Domain Decomposition Methods in Science and Engineering. Proceedings of the 21st International Conference on Domain Decomposition Methods*, Rennes, 25–29 June 2012. Lecture Notes on Computational Science and Engineering (Springer, Berlin, 2014). Accepted for publication, 12 pp.
33. A. Klawonn, M. Lanser, O. Rheinbach, Nonlinear FETI-DP and BDDC methods. SIAM J. Sci. Comput. **36**(2), A737–A765 (2014)
34. J. Li, O.B. Widlund, FETI–DP, BDDC, and Block Cholesky methods. Int. J. Numer. Methods Eng. **66**(2), 250–271 (2006)
35. J. Mandel, C.R. Dohrmann, Convergence of a balancing domain decomposition by constraints and energy minimization. Numer. Linear Algebra Appl. **10**, 639–659 (2003)
36. J. Mandel, C.R. Dohrmann, R. Tezaur, An algebraic theory for primal and dual substructuring methods by constraints. Appl. Numer. Math. **54**, 167–193 (2005)
37. J. Mandel, B. Sousedík, C.R. Dohrmann, On multilevel BDDC, in *Domain Decomposition Methods in Science and Engineering XVII*, ed. by U. Langer, M. Discacciati, D.E. Keyes, O.B. Widlund, W. Zulehner. Lecture Notes in Computational Science and Engineering, vol. 60 (Springer, Berlin, 2008), pp. 287–294
38. J. Nocedal, S.J. Wright, *Numerical Optimization* (Springer, New York, 2000)
39. J. Pebrel, C. Rey, P. Gosselet, A nonlinear dual-domain decomposition method: application to structural problems with damage. Int. J. Multiscale Comput. Eng. **6**(3), 251–262 (2008)
40. O. Rheinbach, Parallel iterative substructuring in structural mechanics. Arch. Comput. Meth. Eng. **16**(4), 425–463 (2009)
41. B. Sousedík, J. Mandel, On adaptive-multilevel BDDC, in *Domain Decomposition Methods in Science and Engineering XIX*. Lecture Notes in Computational Science and Engineering, vol. 78 (Springer, Heidelberg, 2011), pp. 39–50
42. A. Toselli, O. Widlund, *Domain Decomposition Methods - Algorithms and Theory*. Springer Series in Computational Mathematics, vol. 34 (Springer, Berlin, 2004)
43. X. Tu, Three-level BDDC in three dimensions. SIAM J. Sci. Comput. **29**(4), 1759–1780 (electronic) (2007)
44. X. Tu, Three-level BDDC in two dimensions. Int. J. Numer. Methods Eng. **69**, 33–59 (2007)

Substructuring Methods in Nonlinear Function Spaces

Oliver Sander

1 Spaces of Manifold-Valued Functions

Let Ω be a domain in $\mathbb{R}^d$, and M a smooth, connected, finite-dimensional manifold with positive injectivity radius. We assume M to be equipped with a metric g, which induces an exponential map $\exp : TM \to M$, where TM is the tangent bundle of M [7].

In this article we consider spaces[1] of functions $v : \Omega \to M$. We first define functions of Sobolev smoothness.

Definition 1 Let $\iota : M \to \mathbb{R}^m$ be an isometric embedding for some $m \in \mathbb{N}$, and let $k \in \mathbb{N}_0$ and $p \in \mathbb{N}$. Define

$$W^{k,p}(\Omega, M) := \{v \in W^{k,p}(\Omega, \mathbb{R}^m) : v(x) \in \iota(M) \text{ a.e.}\},$$

where $W^{k,p}(\Omega, \mathbb{R}^m)$ is the usual Sobolev space of m-component vector-valued functions on Ω.

Note that $W^{k,p}(\Omega, M)$ does not have a linear structure. By the Sobolev embedding theorem, it is a Banach manifold if $k > d/p$ [10].

[1] We use the word *space* in a topologist's sense here, without implying the existence of a linear structure.

O. Sander (✉)
Institut für Geometrie und Praktische Mathematik, RWTH Aachen, Templergraben 55, 52056 Aachen, Germany
e-mail: sander@igpm.rwth-aachen.de

T. Dickopf et al. (eds.), *Domain Decomposition Methods in Science and Engineering XXII*, Lecture Notes in Computational Science and Engineering 104, DOI 10.1007/978-3-319-18827-0_5

To formulate variational problems in such spaces we need to construct test functions. Unlike in linear spaces, test function spaces for a function $u : \Omega \to M$ depend on u.

Definition 2 Let $u \in W^{k,p}(\Omega, M)$. A vector field along u is a map $\eta : \Omega \to TM$, such that $\eta(x) \in T_{u(x)}M$ for almost all $x \in \Omega$.

More abstractly, vector fields along u are sections in a certain vector bundle. While the concept of a vector bundle is standard (see, e.g. [7]), we state it here for completeness.

Definition 3 Let E and B be two differentiable manifolds, and $\pi : E \to B$ a surjective continuous map. The triple (E, π, B) is called a (continuous) vector bundle if each fiber $E_x := \pi^{-1}(x)$, $x \in B$ has an n-dimensional real vector space structure, and the following triviality condition holds: For each $x \in B$, there exists a neighborhood U and a homeomorphism

$$\varphi : \pi^{-1}(U) \to U \times \mathbb{R}^n$$

with the property that for every $y \in U \subset B$

$$\varphi|_{E_y} : E_y \to \{y\} \times \mathbb{R}^n$$

is a bijective linear map. Such a pair (φ, U) is called a bundle chart. A family (φ_i, U_i) of bundle charts such that the U_i cover B is called a bundle atlas.

In other words, vector bundles are spaces that locally look like products $U \times \mathbb{R}^n$. We call E the total space, B the base space, and π the bundle projection of the vector bundle. The prototypical vector bundle is the tangent bundle (TM, π, M) of a smooth manifold M. In this case, the bundle projection π maps tangent vectors to their base points.

Vector bundles allow to generalize the concept of a map between spaces. A vector bundle section is an object s that locally is a map $s|_U : U \to \mathbb{R}^n$.

Definition 4 Let (E, π, B) be a vector bundle. A (global) section of E is a map $s : B \to E$ with $\pi \circ s = \mathrm{Id}_B$.

In particular, a map $w : \Omega \to \mathbb{R}^n$ can be interpreted as a section in the trivial bundle $(\Omega \times \mathbb{R}^n, \pi, \Omega)$. A section in the tangent bundle TM of a smooth manifold M is a vector field on M.

Let now N be another smooth manifold, $f : B \to N$ a continuous map, and (E, π, N) a vector bundle over N. We pull back the bundle via f, to obtain a bundle f^*E over B, for which the fiber over $x \in B$ is $E_{f(x)}$, the fiber over the image of x. The following formal definition is given in [6, Def. 2.5.3].

Definition 5 Let $f : B \to N$ be a continuous map, and (E, π, N) a vector bundle over N. The pulled back bundle f^*E has as base space B, as total space E_1, which is the subspace of all pairs $(b, x) \in B \times E$ with $f(b) = \pi(x)$, and as projection the map $(b, x) \mapsto b$.

With these preliminaries we can interpret vector fields along a continuous function as vector bundle sections. The proof of the following lemma follows directly from the definitions.

Lemma 1 *Let $f : \Omega \to M$ be continuous. A vector field η in the sense of Definition 2 is a section in the bundle f^*TM.*

So far, we have not mentioned the regularity of sections of vector bundles. The following definition is given in [7].

Definition 6 Let (E, π, B) be a vector bundle, and $s : B \to E$ a section of E with compact support. We say that s is contained in the Sobolev space $W^{k,p}(E)$, if for any bundle atlas with the property that on compact sets all coordinate changes and all their derivatives are bounded, and for any bundle chart $\varphi : E|_U \to U \times \mathbb{R}^n$ from such an atlas, we have that $\varphi \circ s|_U$ is contained in $W^{k,p}(U, \mathbb{R}^n)$.

As a special case of this we can define vector fields of Sobolev smoothness along a given continuous function $f : \Omega \to M$.

Definition 7 Let $f : \Omega \to M$ be continuous, and η a vector field along f. We say that η is of k,p-Sobolev smoothness, and we write $\eta \in \Xi_f^{k,p}$, if it is a k,p-section in the sense of Definition 6.

Finally, we need a trace theorem for vector fields along a function. We restrict our attention to $k = 1, p = 2$. The following is a special case of a result proved in [5]. We denote by $\mathcal{D}(\Omega, E)$ the smooth sections in (E, π, Ω) and by $\mathcal{D}(\Omega, E|_\Gamma)$ the smooth sections of the bundle restriction on Γ.

Lemma 2 *Let Ω have a C^∞ boundary, and let (E, π, Ω) be a vector bundle over Ω. Let Γ be a part of the boundary of Ω, and suppose it is a submanifold of $\overline{\Omega}$. Then the pointwise restriction $\mathrm{tr}_\Gamma : \mathcal{D}(\Omega, E) \to \mathcal{D}(\Gamma, E|_\Gamma)$ extends to a linear and bounded operator from $W^{1,2}(E)$ onto $W^{\frac{1}{2},2}(E|_\Gamma)$, i.e.,*

$$\mathrm{tr}_\Gamma\, W^{1,2}(E) = W^{\frac{1}{2},2}(E|_\Gamma).$$

Moreover, tr_Γ has a linear and bounded right inverse, an extension operator $\mathrm{Ex}_\Omega : W^{\frac{1}{2},2}(E|_\Gamma) \to W^{1,2}(E)$.

For $p \neq 2, p \geq 1$ the trace operator still exists, but the traces are only contained in certain spaces of Besov type [5]. Trace theorems for functions in $W^{1,p}(\Omega, M)$ also exist (see, e.g. [8, Chap. 1.12]), but in the following we only look at continuous functions anyway.

2 Substructuring Formulation of Variational Problems

We now consider variational problems in the space $W^{1,p}(\Omega, M)$. Let α be a form on $W^{1,p}(\Omega, M) \cap C(\Omega, M)$, i.e., for each continuous $u \in W^{1,p}(\Omega, M)$, $\alpha[u]$ is a linear map $\Xi_u^{1,p} \to \mathbb{R}$. We look for zeros of such a form, subject to Dirichlet boundary conditions on part of the boundary of Ω. Since for that case we need the trace theorem (Lemma 2) we restrict ourselves to $p = 2$ again. Let Γ_D be a subset of positive $d-1$-dimensional measure of $\partial\Omega$. For a function $u_0 : \Gamma_D \to M$ sufficiently smooth define the space $H_D^1 := \{v \in W^{1,2}(\Omega, M) \cap C(\Omega, M) \; : \; \operatorname{tr}_{\Gamma_D} v = u_0\}$, and for each $u \in H_D^1$ define $\Xi_{u,0}^{1,2} = \{\eta \in \Xi_u^{1,2} \; : \; \operatorname{tr}_{\Gamma_D} \eta = 0\}$. We then look for a function $u \in H_D^1$ such that

$$\alpha[u](\eta) = 0 \qquad \text{for all } \eta \in \Xi_{u,0}^{1,2}. \tag{1}$$

Such problems occur, for example, as the optimality condition for minimization problems for functionals $J : W^{1,p}(\Omega, M) \to \mathbb{R}$. In that case, $\alpha[u]$ is the differential of J at u.

The weak problem (1) can be written as a coupled problem, consisting of two subdomain problems and suitable coupling conditions. This is well-known for linear problems in linear spaces ([11, Chap. 1.2]). We show that the argument used there also holds for nonlinear function spaces.

Assume that Ω is partitioned in two nonoverlapping subdomains Ω_1 and Ω_2, and that the interface $\Gamma := \overline{\Omega}_1 \cap \overline{\Omega}_2$ is a $d-1$-dimensional Lipschitz manifold. We note the following technical results, which follow directly from the corresponding results for scalar-valued Sobolev spaces and Definition 1 (see also [8, Thm. 1.12.3]).

Lemma 3

1. *If $u \in W^{1,p}(\Omega, M)$, then $u|_{\Omega_i} \in W^{1,p}(\Omega_i, M)$ for $i = 1, 2$.*
2. *Let $u_i \in W^{1,p}(\Omega_i, M)$ for $i = 1, 2$ and $\operatorname{tr}_\Gamma u_1 = \operatorname{tr}_\Gamma u_2$. Then the function $u : \Omega \to M$ defined by*

$$u(x) := \begin{cases} u_1(x) & \text{if } x \in \Omega_1 \\ u_2(x) & \text{if } x \in \Omega_2 \end{cases}$$

is contained in $W^{1,p}(\Omega, M)$.

Suppose that α is a linear form on $W^{1,p}(\Omega, M)$. We assume that α is separable in the sense that there are linear forms α_i on $W^{1,p}(\Omega_i, M)$, $i = 1, 2$, such that

$$\alpha[u](\eta) = \sum_{i=1}^{2} \alpha_i[u|_{\Omega_i}](\eta|_{\Omega_i}) \qquad \text{for all } u \in W^{1,p}(\Omega, M),\ \eta \in \Xi_u^{1,p}. \tag{2}$$

This holds in particular if α is defined as an integral over a local density.

For a formal statement of our substructuring result we need the following spaces.

Definition 8 Let $u_0 : \Gamma_D \to M$ be a function of prescribed Dirichlet values, of sufficient smoothness. For $i = 1, 2$ set

$$H_i := \{v_i \in W^{1,2}(\Omega_i, M) \cap C(\Omega_i, M) \ : \ v_i|_{\Gamma_D \cap \partial\Omega_i} = u_0\}.$$

For $i = 1, 2$ and each $v_i : \Omega_i \to M$ continuous set

$$V_{i,v_i} := \{\eta_i \in \Xi^{1,2}_{v_i} \ : \ \eta_i(x) = 0 \in T_{v_i(x)}M \text{ for almost all } x \in \Gamma_D\},$$

$$V^0_{i,v_i} := \{\eta_i \in \Xi^{1,2}_{v_i} \ : \ \eta_i(x) = 0 \in T_{v_i(x)}M \text{ for almost all } x \in \Gamma_D \cup \Gamma\}.$$

Also, we define the interface space

$$\Lambda := \{w : \Gamma \to M \text{ such that } \operatorname{tr}_\Gamma v = w \text{ for some } v \in H_1\},$$

and the corresponding spaces of test functions on Γ

$$\Xi^{1/2}_w := \Xi^{\frac{1}{2},2}_w$$

for each continuous $w \in \Lambda$.

Note that the V^0_{i,v_i} and $\Xi^{1/2}_w$ are linear spaces, whereas the H_i and Λ are not. Unlike in the linear case, the test function spaces are replaced by entire families of spaces, parametrized by functions $v_i \in H_i$ and $w \in \Lambda$, respectively.

Lemma 4 *The weak problem* (1) *is equivalent to: Find* $u_i \in H_i$, $i = 1, 2$, *such that*

$$\alpha_i[u_i](\eta_i) = 0 \qquad \forall\, \eta_i \in V^0_{i,u_i},\ i = 1, 2 \tag{3}$$

$$\operatorname{tr}_\Gamma u_1 = \operatorname{tr}_\Gamma u_2 \tag{4}$$

$$\alpha_1[u_1](\operatorname{Ex}_{\Omega_1} \mu) = -\alpha_2[u_2](\operatorname{Ex}_{\Omega_2} \mu) \qquad \textit{for all } \mu \in \Xi^{1/2}_{\operatorname{tr}_\Gamma u_1}, \tag{5}$$

where $\operatorname{Ex}_{\Omega_i}$, $i = 1, 2$ *is an extension operator from* $\Xi^{1/2}_{\operatorname{tr}_\Gamma u_1}$ *to* V_{i,u_i}.

Note that the existence of the extension operators $\operatorname{Ex}_{\Omega_i}$ is ensured by Lemma 2.

Proof We follow the argument in [11, Chap. 1.2], and show first that the substructuring formulation is a consequence of (1). Let u be a solution of (1). Consequently, it is an element of $W^{1,2}(\Omega, M) \cap C(\Omega, M)$, and by Lemma 3, the subdomain functions $u_i := u|_{\Omega_i}, i = 1, 2$ are in H_1 and H_2, respectively. Equation (4) follows because u is continuous. Also, (3) holds, because any test function $v_i \in V^0_{i,u_i}$ can be extended by zero to a test function in $\Xi^{1,2}_{u,0}$. Finally, for every $\mu \in \Xi^{1/2}_{\operatorname{tr}_\Gamma u_1}$ define

$$\operatorname{Ex} \mu := \begin{cases} \operatorname{Ex}_{\Omega_1} \mu & \text{in } \Omega_1, \\ \operatorname{Ex}_{\Omega_2} \mu & \text{in } \Omega_2, \end{cases}$$

and note that $\operatorname{Ex}\mu \in \Xi^1_{u,0}$. Therefore, $\operatorname{Ex}\mu$ is a valid test function for (1). Together with the separability (2) of α we get

$$0 = \alpha[u](\operatorname{Ex}\mu) = \alpha_1[u_1](\operatorname{Ex}_{\Omega_1}\mu) + \alpha_2[u_2](\operatorname{Ex}_{\Omega_2}\mu),$$

which is (5).

To show the other direction let u_i, $i = 1, 2$, be a solution of (3)–(5), and define

$$u := \begin{cases} u_1 & \text{in } \Omega_1 \\ u_2 & \text{in } \Omega_2. \end{cases}$$

Since $u_1 = u_2$ on Γ we can invoke Lemma 3 to obtain that $u \in W^{1,2}(\Omega, M)$; additionally, u is continuous.

Let $\eta \in \Xi^{1,2}_u$ be a test function at u. By Lemma 2 it has a trace $\mu := \operatorname{tr}\eta$ with $\mu \in \Xi^{1/2}_{\operatorname{tr}_\Gamma u}$. Then $(\eta|_{\Omega_i} - \operatorname{Ex}_{\Omega_i}\mu) \in V^0_{i,u_i}$. With this we can compute

$$\begin{aligned}
\alpha[u](\eta) &= \sum_{i=1}^{2} \alpha_i[u_i](\eta|_{\Omega_i}) && \text{(by separability (2))} \\
&= \sum_{i=1}^{2} \Big[\alpha_i[u_i](\underbrace{\eta|_{\Omega_i} - \operatorname{Ex}_{\Omega_i}\mu}_{\in V^0_{i,u_i}}) + \alpha_i[u_i](\operatorname{Ex}_{\Omega_i}\mu)\Big] && \text{(by lin. of } \alpha_i[u_i](\cdot)) \\
&= \sum_{i=1}^{2} \alpha_i[u_i]([\operatorname{Ex}_{\Omega_i}\mu) && \text{(by (3))} \\
&= 0 && \text{(by (5))}.
\end{aligned}$$

Hence u solves (1).

3 Steklov–Poincaré Formulation

Following the standard substructuring approach we now write the coupled problem (3)–(5) as a single equation on an interface space. In our setting this interface space is the nonlinear space Λ.

We first introduce the Steklov–Poincaré operators for the subdomain problems. For each subdomain, these map Dirichlet values on Γ to the Neumann traces of the corresponding subdomain solutions on Γ. These Neumann traces are sections in a certain dual bundle.

Definition 9 Let $u : \Omega \to M$ be continuous. For any Sobolev space $\Xi_u^{k,p}$ of sections in u^*TM we call $(\Xi_u^{k,p})^*$ its dual, i.e., the set of all linear functionals $L : \Xi_u^{k,p} \to \mathbb{R}$ such that $L(\eta)$ is finite for all $\eta \in \Xi_u^{k,p}$.

We denote by $(\Xi^{k,p})^*$ the disjoint union of all spaces $(\Xi_u^{k,p})^*$ for all continuous u. This concept allows to generalize the space $(H^{\frac{1}{2}}(\Gamma))^*$ used for the Neumann traces of linear problems.

Definition 10 We call S_i the Dirichlet-to-Neumann map associated to the i-th subdomain. That is, for any $\lambda \in \Lambda$ we set $S_i\lambda \in (\Xi_\lambda^{1/2})^*$ to be such that

$$S_i\lambda[\mu] = \alpha_i[u_i](\mathrm{Ex}_{\Omega_i}\,\mu) \qquad \text{for all } \mu \in \Xi_\lambda^{1/2}, \tag{6}$$

where u_i fulfills $\mathrm{tr}_\Gamma\, u_i = \lambda$ and solves

$$\alpha_i[u_i](\eta) \qquad \text{for all } \eta \in V_{i,u_i}^0.$$

Remark 1 We assume here for simplicity that the S_i are single-valued, i.e., that for given Dirichlet data λ the corresponding subdomain problems have unique solutions.

Using the Steklov–Poincaré operators we can write the coupled problem (3)–(5) as a problem on the interface space alone.

Lemma 5 *The coupled problem* (3)–(5) *is equivalent to the Steklov–Poincaré equation*

$$S_1\lambda + S_2\lambda = 0. \tag{7}$$

Note that $S_1\lambda$ and $S_2\lambda$ are from the same linear space $(\Xi_\lambda^{1/2})^*$. Hence the addition is justified.

Proof Let $\lambda \in \Lambda$. Then the subdomain solutions u_1, u_2 used in the definition of S_1 and S_2 solve the subdomain problems (3) by construction. Also, since they both assume the same value λ on Γ they are continuous on the interface. Finally, inserting (6) into (7) yields (5). Conversely, if u_1, u_2 solve (3)–(5), then $\lambda := \mathrm{tr}_\Gamma\, u_1 = \mathrm{tr}_\Gamma\, u_2$ solves (7).

4 Nonlinear Preconditioned Richardson Iteration

The natural algorithm for the Steklov–Poincaré interface equation (7) is the preconditioned Richardson iteration. Depending on the preconditioner, various different domain decomposition algorithms result, which we will describe below.

Let $k \in \mathbb{N}$ and $\lambda^k \in \Lambda$ be an iterate of the interface variable. Following [3], we write one iteration of the preconditioned Richardson iteration in three steps:

1. Compute residual $\sigma^k \in (\Xi^{1/2}_{\lambda^k})^*$ by
$$\sigma^k = S_1\lambda^k + S_2\lambda^k.$$

2. Get correction $v^k \in \Xi^{1/2}_{\lambda^k}$ by preconditioning the negative residual
$$v^k = P^{-1}_{\lambda^k}(-\sigma^k).$$

3. Do a damped geodesic update
$$\lambda^{k+1} = \exp_{\lambda^k} \omega v^k,$$
where $\omega \in (0, \infty)$ is a parameter, and the map $\exp_{\lambda^k}$ is to be understood pointwise.

The preconditioner P is a vector bundle morphism from $\Xi^{1/2}$ to $(\Xi^{1/2})^*$, that is, a mapping from $\Xi^{1/2}$ to $(\Xi^{1/2})^*$ such that $\pi(Pv) = \pi v$ for all $v \in \Xi^{1/2}$, and such that for each $\lambda \in \Lambda$ the induced map from $\Xi^{1/2}_\lambda$ to $(\Xi^{1/2}_\lambda)^*$ is linear. It maps infinitesimal corrections to generalized stresses. We additionally require that each P_{λ^k} be invertible. Consequently, its inverse $P^{-1}_{\lambda^k}$ maps generalized stresses to corrections.

The update step 3 needs to use the exponential map to apply the correction v^k (which is a vector field along λ^k) to the current iterate λ^k. The correction is multiplied with a positive damping factor ω. More generally, this factor can be replaced by a linear map ω^k from the tangent space $\Xi^{1/2}_{\lambda^k}$ onto itself. If M is a linear space the exponential map degenerates to the addition of its argument to λ^k.

Remark 2 The two subdomain solves needed for Step 1 of the Richardson iteration can be performed in parallel. Since Step 1 is by far the most costly part this parallelization leads to considerable performance gains.

To construct preconditioners we introduce the derivatives of the Steklov–Poincaré operators. For $S_i : \Lambda \to (\Xi^{1/2})^*$ we interpret the derivative at a $\lambda \in \Lambda$ as a linear map $S_i'(\lambda)$ from $\Xi^{1/2}_\lambda$ to $(\Xi^{1/2}_\lambda)^*$.

Remark 3 This interpretation is most easily understood if we assume for a second that the space Λ is smooth enough to form a Banach manifold. We can then write vector fields as elements of the tangent bundle $T\Lambda$. The Steklov–Poincaré operator S_i becomes a map $S_i : \Lambda \to T^*\Lambda$, and its derivative at $\lambda \in \Lambda$ is the linear map $S_i' : T_\lambda\Lambda \to T_{S_i\lambda}T^*_\lambda\Lambda$. Since $T^*_\lambda\Lambda$ is a linear space we can identify $T_{S_i\lambda}T^*_\lambda\Lambda$ with $T^*_\lambda\Lambda$, and therefore interpret $S_i'(\lambda)$ as a linear map from $T_\lambda\Lambda$ to $T^*_\lambda\Lambda$. This corresponds to a map from $\Xi^{1/2}_\lambda$ to $(\Xi^{1/2}_\lambda)^*$ if Λ is not sufficiently smooth.

We now describe various preconditioners and the algorithms that result from them.

- *Dirichlet–Neumann Preconditioner:* The simplest choice for a preconditioner is the inverse of the linearized Steklov–Poincaré operator of one of the subproblems. We define the Dirichlet–Neumann preconditioner as

$$P_{\mathrm{DN},k} := S_1'[\lambda^k].$$

With this choice, the damped preconditioned Richardson iteration reads

$$\lambda^{k+1} = \exp_{\lambda^k}(\omega P_{\mathrm{DN},k}^{-1}(-\sigma^k)) = \exp_{\lambda^k}\left[\omega(S_1'[\lambda^k])^{-1}(-S_1\lambda^k - S_2\lambda^k)\right].$$

Using instead the second subdomain for preconditioning we define the Neumann–Dirichlet preconditioner

$$P_{\mathrm{ND},k} := S_2'[\lambda^k].$$

- *Neumann–Neumann Preconditioner:* We can generalize the above construction by allowing arbitrary convex combinations of the Dirichlet–Neumann and Neumann–Dirichlet preconditioners. Let γ_1, γ_2 be two non-negative real numbers with $\gamma_1 + \gamma_2 > 0$. Then

$$P_{\mathrm{NN},k}^{-1} := \gamma_1(S_1'[\lambda^k])^{-1} + \gamma_2(S_2'[\lambda^k])^{-1} \tag{8}$$

is the Neumann–Neumann preconditioner. When M is a linear space and the equation to be solved is linear, then the Richardson iteration together with the preconditioner (8) reduces to the usual Neumann–Neumann iterative scheme.

- *Robin Preconditioner:* Finally, we generalize the Robin–Robin method. Let again γ_1 and γ_2 be two non-negative coefficients such that $\gamma_1 + \gamma_2 > 0$. Further, let F be a vector bundle morphism from $\Xi^{1/2}$ to $(\Xi^{1/2})^*$ that is invertible on each fiber. For each $\lambda^k \in \Lambda$, F_{λ^k} is a linear map from $\Xi_{\lambda^k}^{1/2}$ to $(\Xi_{\lambda^k}^{1/2})^*$. We then define the Robin–Robin preconditioner

$$P_{\mathrm{RR},k} := \frac{1}{\gamma_1 + \gamma_2}\left[\gamma_1 F_{\lambda^k} + S_1'(\lambda^k)\right]F_{\lambda^k}^{-1}\left[\gamma_2 F_{\lambda^k} + S_2'(\lambda^k)\right].$$

For the linear finite-dimensional case, the identity map can be chosen for F. In that case the equivalence of this preconditioner to the Robin–Robin iterative method has been shown in [4].

5 Numerical Results

We demonstrate the performance of the Richardson iteration with a numerical example. Consider a hyperelastic Cosserat shell. Configurations of such a shell are pairs of functions $(\varphi, R) : \Omega \to \mathbb{R}^3 \times SO(3)$, where Ω is a two-dimensional domain, and SO(3) is the set of orthogonal 3×3-matrices R with $\det R = 1$. For $x \in \Omega$ we interpret $\varphi(x) \in \mathbb{R}^3$ as the position of a point of the shell midsurface, and $R_3(x) \in \mathbb{R}^3$ (the third column of $R(x) \in SO(3)$) as a transverse direction. The remaining two orthonormal vectors R_1 and R_2 describe an in-plane rotation (Fig. 1). This choice of kinematics allows to model size-effects and microstructure. We use a hyperelastic material with the energy functional proposed by Neff [9, Chap. 7]. For this energy, existence and partial regularity of minimizers have been shown [9], but no further analytical results are available.

As an example problem we use a rectangular strip of dimensions 10 mm × 1 mm. The thickness parameter is set to 0.05 mm. Both the displacement φ and the orientation R are clamped at one of the short ends. At the other short end we prescribe a time-dependent Dirichlet boundary condition to the midsurface position φ and rotations R, which describes a uniform rotation from 0 to 4π about the long central axis of the strip. The positions and rotations at the long sides are left free. This makes the strip coil up. Note that we need the hyperelastic shell energy with the nonlinear membrane term proposed in Chap. 7 of [9] for this to work, because it is a finite strain example. The resulting model is quasi-static, i.e., it does not contain inertia terms. Time enters only through the time-dependence of the boundary conditions, which is necessary to obtain the coiling behavior.

For the material parameters we choose the Lamé constants $\mu = 3.8462 \cdot 10^5\,\text{N/mm}^2$, $\lambda = 2.7149 \cdot 10^5\,\text{N/mm}^2$, and the Cosserat couple modulus $\mu_c = 0\,\text{N/mm}^2$. The internal length scale is set to $L_c = 0.1$ mm, and the curvature exponent is $p = 1$ (see [9] for details on these parameters).

We divide the domain into two subdomains of dimensions 5 mm × 1 mm, and the time interval in 20 uniform time steps. For each time step we solve the spatial problem with a nonlinear Richardson iteration and the Neumann–Neumann preconditioner of Sect. 4, with $\gamma_1 = \gamma_2 = \frac{1}{2}$. The subdomain problems

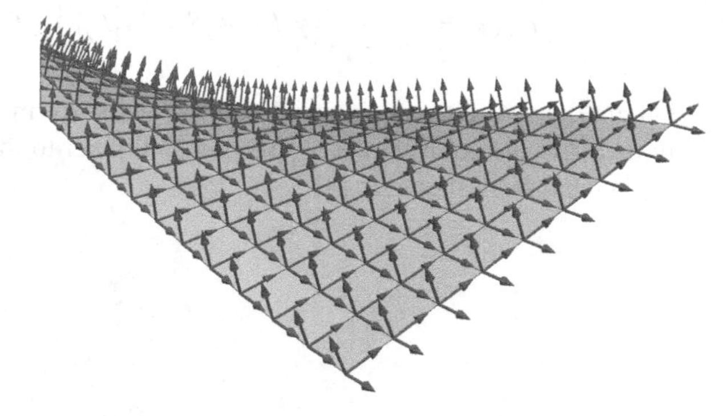

Fig. 1 Cosserat shell configurations consist of the deformation field φ of the mid-surface, and an orientation field R which can be interpreted as a field of three orthogonal director vectors

Fig. 2 Twisted elastic strip at rotation angles 0, $\frac{4}{5}\pi$, $\frac{8}{5}\pi$, $\frac{12}{5}\pi$, $\frac{16}{5}\pi$, and 4π

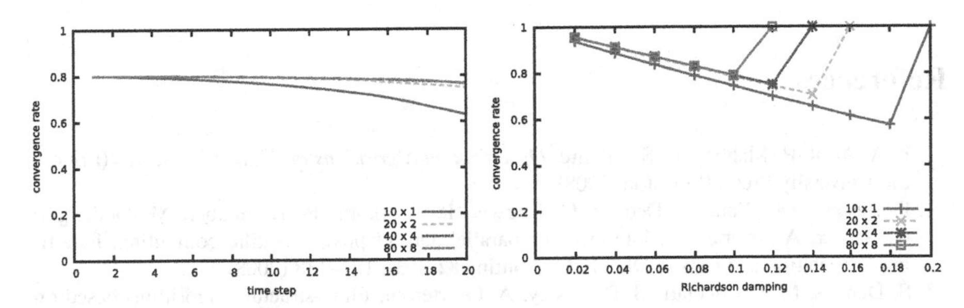

Fig. 3 *Left*: Convergence rates as a function of time for the Richardson damping parameter $\omega = 0.1$, and different grid resolutions. *Right*: Convergence rates averaged over time, for several grid resolutions and values of ω

are discretized using first-order geodesic finite elements [12] on a uniform grid with quadrilateral elements, and the resulting nonlinear algebraic minimization problems are solved using a Riemannian trust-region algorithm [1, 12]. The linear preconditioner problems are solved using a CG method. The code was implemented on top of the DUNE libraries [2].

Figure 2 shows several snapshots from the evolution of the strip. One can see how the strip coils up following the rotation prescribed to the boundary.

To assess the convergence speed of the substructuring method we monitor the traces λ^k defined on the interface $\Gamma = \{5\} \times [0, 1]$. We estimate the convergence rate of the Neumann–Neumann solver at iteration k by $\rho^k := \|v^k\| / \|v^{k-1}\|$, where v^{k-1} and v^k are two consecutive corrections produced by the Richardson iteration. For the norm $\|\cdot\|$ we use the Sobolev norm $H^1(\Gamma, \mathbb{R}^3 \times \mathbb{R}^4)$, using the canonical embedding of SO(3) into the quaternions to embed tangent vectors of SO(3) into $\mathbb{R}^4$. This norm is well-defined for discrete functions. We let the domain decomposition algorithm iterate until the H^1-norm of the correction drops below 10^{-3}. The overall convergence rate for one time step is then determined by taking the geometric average over the ρ^k.

We measure the rates as a function of the grid resolution and of the Richardson damping parameter ω. One observes immediately that a rather small value for ω is needed to make the algorithm converge. Figure 3, left, shows the convergence rates for $\omega = 0.1$ and four different grids as a function of time. Grid resolutions range from 10×1 to 80×8, created by uniform refinement. We see that the convergence rate is rather independent of the time step and of the grid resolution, with the exception of the coarsest grid, for which convergence rates ameliorate over time.

To get a better idea of the dependence of the convergence speed on the damping parameter ω we therefore average the rates over time and plot the results in Fig. 3, right. We observe that the optimal ω decreases and the optimal convergence rate increases as the grid is refined. This matches what is known for the linear case. A more detailed study of the behavior at vanishing mesh sizes, along with a proof of convergence, however, has to be left for future work.

References

1. P.-A. Absil, R. Mahony, R. Sepulchre, *Optimization Algorithms on Matrix Manifolds* (Princeton University Press, Princeton, 2008)
2. P. Bastian, M. Blatt, A. Dedner, C. Engwer, R. Klöfkorn, R. Kornhuber, M. Ohlberger, O. Sander, A generic grid interface for parallel and adaptive scientific computing. Part II: implementation and tests in DUNE. Computing **82**(2–3), 121–138 (2008)
3. S. Deparis, M. Discacciati, G. Fourestey, A. Quarteroni, Fluid-structure algorithms based on Steklov–Poincaré operators. Comput. Methods Appl. Mech. Eng. **195**, 5797–5812 (2006)
4. M. Discacciati, An operator-splitting approach to non-overlapping domain decomposition methods. Technical Report 14.2004, Ecole Polytechnique Fédérale de Lausanne, 2004
5. N. Große, C. Schneider, Sobolev spaces on Riemannian manifolds with bounded geometry: general coordinates and traces. Math. Nachr. **286**(16), 1586–1613 (2013). arXiv 1301.2539v1
6. D. Husemoller, *Fibre Bundles* (McGraw-Hill, New York, 1966)
7. J. Jost, *Riemannian Geometry and Geometric Analysis* (Springer, Berlin, 2011)
8. N.J. Korevaar, R.M. Schoen, Sobolev spaces and harmonic maps for metric space targets. Commun. Anal. Geom. **1**(4), 561–659 (1993)
9. P. Neff, A geometrically exact planar Cosserat shell-model with microstructure: existence of minimizers for zero Cosserat couple modulus. Math. Models Methods Appl. Sci. **17**(3), 363–392 (2007)
10. R.S. Palais, *Foundations of Global Non-linear Analysis*, vol. 196 (Benjamin, New York, 1968)
11. A. Quarteroni, A. Valli, *Domain Decomposition Methods for Partial Differential Equations* (Oxford Science Publications, Oxford, 1999)
12. O. Sander, Geodesic finite elements on simplicial grids. Int. J. Numer. Methods Eng. **92**(12), 999–1025 (2012)

Robin-Neumann Schemes for Incompressible Fluid-Structure Interaction

Miguel A. Fernández, Mikel Landajuela, Jimmy Mullaert, and Marina Vidrascu

1 Introduction

Mathematical problems involving the coupling of an incompressible viscous flow with an elastic structure appear in a large variety of engineering fields (see, e.g., [14, 17, 19–21]). This problem is considered here within a heterogenous domain decomposition framework, with the aim of using independent well-suited solvers for the fluid and the solid. One of the main difficulties that have to be faced under this approach is that the coupling can be very stiff. In particular, traditional Dirichlet-Neumann explicit coupling methods, which solve for the fluid (Dirichlet) and for the solid (Neumann) only once per time-step, are unconditionally unstable whenever the amount of added-mass effect in the system is large (see, e.g., [5, 12]). Typically this happens when the fluid and solid densities are close and the fluid domain is slender, as in hemodynamical applications. This explains, in part, the tremendous amount of work devoted over the last decade to the development of alternative coupling paradigms (see, e.g., [7] for a review).

In this paper we will review several explicit coupling procedures recently reported in the literature and present some new developments (Sect. 3.2). The common feature of these methods is that they are based on Robin-Neumann transmission conditions, whose nature depends on the thin- or thick-walled character of the structure (see Fig. 1).

M.A. Fernández • M. Landajuela • J. Mullaert • M. Vidrascu (✉)
Inria Paris-Rocquencourt, BP 105, 78153 Le Chesnay Cedex, France

Sorbonne Universités, UPMC University Paris 6, Laboratoire Jacques-Louis Lions, 4 Place Jussieu, 75252 Paris Cedex 05, France
e-mail: Marina.Vidrascu@inria.fr

T. Dickopf et al. (eds.), *Domain Decomposition Methods in Science and Engineering XXII*, Lecture Notes in Computational Science and Engineering 104, DOI 10.1007/978-3-319-18827-0_6

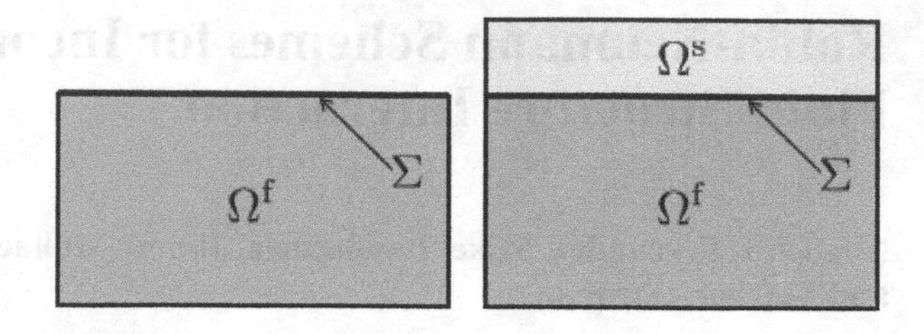

Fig. 1 Fluid-structure configurations for a thin- (*left*) and a thick-walled structure (*right*)

2 Problem Formulation

For the sake of simplicity we consider a low Reynolds regime and assume that the interface undergoes infinitesimal displacements. The fluid is described by the Stokes equations, in a fixed domain $\Omega^f \subset \mathbb{R}^d$ ($d = 2, 3$), and the structure by the linear (possibly damped) membrane equations written in the $(d-1)$-manifold $\Omega^s \equiv \Sigma$, which is also the fluid-structure interface (see Fig. 1 (left)).

The coupled model problem reads therefore as follows: find the fluid velocity $\boldsymbol{u} : \Omega^f \times \mathbb{R}^+ \rightarrow \mathbb{R}^d$, the fluid pressure $p : \Omega^f \times \mathbb{R}^+ \rightarrow \mathbb{R}$, the solid displacement $\boldsymbol{d} : \Sigma \times \mathbb{R}^+ \rightarrow \mathbb{R}^d$ and the solid velocity $\dot{\boldsymbol{d}} : \Sigma \times \mathbb{R}^+ \rightarrow \mathbb{R}^d$ such that

$$\begin{cases} \rho^f \partial_t \boldsymbol{u} - \boldsymbol{\nabla} \cdot \boldsymbol{\sigma}(\boldsymbol{u}, p) = \boldsymbol{0} & \text{in} \quad \Omega^f, \\ \boldsymbol{\nabla} \cdot \boldsymbol{u} = 0 & \text{in} \quad \Omega^f, \end{cases} \tag{1}$$

$$\begin{cases} \boldsymbol{u} = \dot{\boldsymbol{d}} & \text{on} \quad \Sigma, \\ \rho^s \epsilon \partial_t \dot{\boldsymbol{d}} + \boldsymbol{L}^e \boldsymbol{d} + \boldsymbol{L}^v \dot{\boldsymbol{d}} = -\boldsymbol{\sigma}(\boldsymbol{u}, p)\boldsymbol{n} & \text{on} \quad \Sigma, \\ \dot{\boldsymbol{d}} = \partial_t \boldsymbol{d} & \text{on} \quad \Sigma. \end{cases} \tag{2}$$

This system has to be complemented with appropriate initial and (external) boundary conditions, which will be omitted in the following since they are not relevant for the discussion. The symbols ρ^f and ρ^s denote, respectively, the fluid and solid densities, ϵ is the solid thickness and $\boldsymbol{n}$ stands for the unit normal vector on $\partial\Omega^f$. The fluid Cauchy-stress tensor is given by $\boldsymbol{\sigma}(\boldsymbol{u}, p) \stackrel{\text{def}}{=} -p\boldsymbol{I} + 2\mu\boldsymbol{\varepsilon}(\boldsymbol{u})$, with $\boldsymbol{\varepsilon}(\boldsymbol{u}) \stackrel{\text{def}}{=} \frac{1}{2}\left(\boldsymbol{\nabla}\boldsymbol{u} + \boldsymbol{\nabla}\boldsymbol{u}^{\mathrm{T}}\right)$ and where μ denotes the fluid dynamic viscosity. Finally, the surface differential operators $\boldsymbol{L}^e$ and $\boldsymbol{L}^v$ describe the membrane elastic and viscous behavior, respectively.

Remark 1 In two spatial dimensions and for the geometrical configuration of Fig. 1 (left) an example of solid elastic operator is given by where $\boldsymbol{d} = [0, \boldsymbol{d}_y]^T$ and $c_0, c_1 > 0$ are material dependent parameters. A widely used form of the solid viscous operator is $\boldsymbol{L}^v \dot{\boldsymbol{d}} = \alpha\rho^s\epsilon\dot{\boldsymbol{d}} + \beta\boldsymbol{L}^e\dot{\boldsymbol{d}}$, where $\alpha, \beta > 0$ are given parameters. In artery wall modeling, the zeroth-order term, $\alpha\rho^s\epsilon\dot{\boldsymbol{d}}$, describes the dissipative behavior of external tissues (see [19]), whereas the differential term, $\beta\boldsymbol{L}^e\dot{\boldsymbol{d}}$, corresponds to the Kelvin-Voigt model (see, e.g., [15, 22]).

Remark 2 Though simplified, problem (1)–(2) preserves some of the major numerical difficulties that arise in incompressible fluid-structure interaction.

3 Explicit Coupling Schemes

This section is devoted to the numerical approximation of the coupled problem (1)–(2). In the succeeding text, the symbol $\tau > 0$ denotes the time-step size, $t_n \stackrel{\text{def}}{=} n\tau$, for $n \in \mathbb{N}$, and $\partial_\tau x^n \stackrel{\text{def}}{=} (x^n - x^{n-1})/\tau$ the first order backward difference in time. In addition, the superscript $\star$ is used to indicate zeroth- (i.e., without), first-order or second-order extrapolation from the previous time-steps, namely, $x^\star = 0$ if $r = 0$, $x^\star = x^{n-1}$ if $r = 1$ and $x^\star = 2x^{n-1} - x^{n-2}$ if $r = 2$, where r denotes the extrapolation order.

The methods discussed in this review paper are explicit coupling schemes, in the sense that they enable a decoupled time-marching of the fluid and the solid. Traditional Dirichlet-Neumann explicit coupling procedures, as reported in Algorithm 1, are known to be unconditionally unstable, whenever the amount of added-mass effect in the system is large (see, e.g., [5]). Stability in explicit coupling for incompressible fluid-structure interaction demands a different treatment of the interface coupling conditions $(2)_{1,2}$.

A stable explicit coupling alternative is given by the Robin-Robin methods introduced in [3, 4], which build on a Nitsche treatment of the interface coupling. A salient feature of these methods is that they do not depend on the thin- or thick-walled nature of the solid. Unfortunately, the explicit treatment of the Nitsche's penalty induces a deterioration of the accuracy, which demands restrictive CFL constraints, unless correction iterations with suitable extrapolations are performed (see [4]). Numerical evidence suggests that optimal first-order accuracy can be

Algorithm 1 Dirichet-Neumann Explicit Coupling Scheme

For $n \geq 1$:

1. Fluid step: find $\boldsymbol{u}^n : \Omega^{\mathrm{f}} \times \mathbb{R}^+ \rightarrow \mathbb{R}^d$ and $p^n : \Omega^{\mathrm{f}} \times \mathbb{R}^+ \rightarrow \mathbb{R}$ such that

$$\begin{cases} \rho^{\mathrm{f}} \partial_\tau \boldsymbol{u}^n - \boldsymbol{\nabla} \cdot \boldsymbol{\sigma}(\boldsymbol{u}^n, p^n) = \boldsymbol{0} & \text{in} \quad \Omega^{\mathrm{f}}, \\ \boldsymbol{\nabla} \cdot \boldsymbol{u}^n = 0 & \text{in} \quad \Omega^{\mathrm{f}}, \\ \boldsymbol{u}^n = \dot{\boldsymbol{d}}^{n-1} & \text{on} \quad \Sigma. \end{cases}$$

2. Solid step: find $\boldsymbol{d}^n : \Sigma \times \mathbb{R}^+ \rightarrow \mathbb{R}^d$ such that

$$\begin{cases} \rho^{\mathrm{s}} \epsilon \partial_\tau \dot{\boldsymbol{d}}^n + \boldsymbol{L}^{\mathrm{e}} \boldsymbol{d}^n + \boldsymbol{L}^{\mathrm{v}} \dot{\boldsymbol{d}}^n = -\boldsymbol{\sigma}(\boldsymbol{u}^n, p^n)\boldsymbol{n} & \text{on} \quad \Sigma, \\ \dot{\boldsymbol{d}}^n = \partial_\tau \boldsymbol{d}^n & \text{on} \quad \Sigma. \end{cases}$$

achieved by using a non-symmetric penalty-free formulation (see [4, Section 4.3]). The rigorous stability analysis of the resulting schemes remains, however, an open problem.

3.1 Robin-Neumann Schemes

The key difficulty is hence the derivation of alternative splitting methods which guarantee stability without compromising accuracy. The Robin-Neumann methods proposed in [8, 10] achieve this purpose. The fundamental ingredient in the derivation of these schemes is the interface *Robin consistency* featured by the continuous problem (1)–(2). Indeed, from $(2)_{1,2}$ it follows that

$$\boldsymbol{\sigma}(\boldsymbol{u},p)\boldsymbol{n} + \rho^{\mathrm{s}}\epsilon\partial_t\boldsymbol{u} = -\boldsymbol{L}^{\mathrm{e}}\boldsymbol{d} - \boldsymbol{L}^{\mathrm{v}}\dot{\boldsymbol{d}} \quad \text{on} \quad \Sigma, \tag{3}$$

which can be viewed as a Robin-like boundary condition for the fluid. Hence, instead of performing the fluid solid time splitting in terms of $(2)_{1,2}$ as in Algorithm 1, we consider (3) and $(2)_2$. The resulting schemes are detailed in Algorithm 2.

Algorithm 2 completely uncouples the fluid and solid time-marchings. This is achieved via the explicit Robin condition $(4)_3$ derived from (3). Note that only the solid inertial effects are implicitly treated in $(4)_3$, this is enough to guarantee added-mass free stability. It is also worth noting that, from $(5)_1$, the explicit Robin condition $(4)_3$ can be reformulated as

$$\boldsymbol{\sigma}(\boldsymbol{u}^n,p^n)\boldsymbol{n} + \frac{\rho^{\mathrm{s}}\epsilon}{\tau}\boldsymbol{u}^n = \frac{\rho^{\mathrm{s}}\epsilon}{\tau}\left(\dot{\boldsymbol{d}}^{n-1} + \tau\partial_\tau\dot{\boldsymbol{d}}^\star\right) + \boldsymbol{\sigma}(\boldsymbol{u}^\star,p^\star)\boldsymbol{n} \quad \text{on} \quad \Sigma.$$

Algorithm 2 Robin-Neumann Explicit Coupling Schemes (from [10]).

For $n \geq r+1$:

1. Fluid step: find $\boldsymbol{u}^n : \Omega^{\mathrm{f}} \times \mathbb{R}^+ \to \mathbb{R}^d$ and $p^n : \Omega^{\mathrm{f}} \times \mathbb{R}^+ \to \mathbb{R}$ such that

$$\begin{cases} \rho^{\mathrm{f}}\partial_\tau\boldsymbol{u}^n - \boldsymbol{\nabla}\cdot\boldsymbol{\sigma}(\boldsymbol{u}^n,p^n) = \boldsymbol{0} & \text{in} \quad \Omega^{\mathrm{f}}, \\ \boldsymbol{\nabla}\cdot\boldsymbol{u}^n = 0 & \text{in} \quad \Omega^{\mathrm{f}}, \\ \boldsymbol{\sigma}(\boldsymbol{u}^n,p^n)\boldsymbol{n} + \dfrac{\rho^{\mathrm{s}}\epsilon}{\tau}\boldsymbol{u}^n = \dfrac{\rho^{\mathrm{s}}\epsilon}{\tau}\dot{\boldsymbol{d}}^{n-1} - \boldsymbol{L}^{\mathrm{e}}\boldsymbol{d}^\star - \boldsymbol{L}^{\mathrm{v}}\dot{\boldsymbol{d}}^\star & \text{on} \quad \Sigma. \end{cases} \tag{4}$$

2. Solid step: find $\boldsymbol{d}^n : \Sigma \times \mathbb{R}^+ \to \mathbb{R}^d$ such that

$$\begin{cases} \rho^{\mathrm{s}}\epsilon\partial_\tau\dot{\boldsymbol{d}}^n + \boldsymbol{L}^{\mathrm{e}}\boldsymbol{d}^n + \boldsymbol{L}^{\mathrm{v}}\dot{\boldsymbol{d}}^n = -\boldsymbol{\sigma}(\boldsymbol{u}^n,p^n)\boldsymbol{n} & \text{on} \quad \Sigma, \\ \dot{\boldsymbol{d}} = \partial_\tau\boldsymbol{d}^n & \text{on} \quad \Sigma. \end{cases} \tag{5}$$

The advantage of this new expression is its intrinsic character, in the sense that it avoids extrapolations of the solid viscoelastic terms within the fluid solver.

Remark 3 It should be noted that the implicit treatment of the solid-damping term $\boldsymbol{L}^{\mathrm{v}}$ in (4), as advocated in [2, 13, 18], yields a coupling scheme which is not explicit: it is semi-implicit. Moreover, the resulting solution procedure is not partitioned either, since the solid viscous contribution $\boldsymbol{L}^{\mathrm{v}}$ has to be integrated within the fluid solver.

Theoretical results on the stability and accuracy of Algorithm 2 have been reported in [8, 10]. A fundamental ingredient in the analysis is the fact that Algorithm 2 can be viewed as a fully implicit scheme with the following perturbed kinematic constraint

$$\boldsymbol{u}^n = \dot{\boldsymbol{d}}^n + \frac{\tau}{\rho^{\mathrm{s}}\epsilon}\left[\boldsymbol{L}^{\mathrm{e}}(\boldsymbol{d}^n - \boldsymbol{d}^\star) + \boldsymbol{L}^{\mathrm{v}}(\dot{\boldsymbol{d}}^n - \dot{\boldsymbol{d}}^\star)\right] \quad \text{on} \quad \Sigma. \tag{6}$$

The stability and the accuracy of Algorithm 2 are hence driven by the impact of this perturbation (i.e., the last term of (6)) on the stability and accuracy of the underlying implicit coupling scheme. Unconditional energy stability can be proved for $r = 0$ and $r = 1$. The scheme with $r = 2$ is energy stable under a CFL-like condition. As regards accuracy, the error analysis shows that the splitting error induced by the kinematic perturbation (6) scales as $\mathcal{O}(\tau^{2^{r-1}})$. Thus, Algorithm 2 with $r = 1$ or $r = 2$ yields an overall optimal first-order time-accuracy $\mathcal{O}(\tau)$ in the energy-norm, while a sub-optimal time convergence rate $\mathcal{O}(\tau^{\frac{1}{2}})$ is expected for the scheme with $r = 0$.

Remark 4 In the particular case of an undamped thin-walled solid (i.e., $\boldsymbol{L}^{\mathrm{v}} = \boldsymbol{0}$), Algorithm 2 with $r = 0$ yields the splitting scheme reported in [13], which is known to deliver very poor accuracy (see [8, 10] and the example below).

We conclude this section with a numerical illustration based on the balloon-like example proposed in [16, Section 7.1] and using a non-linear version of (1)–(2). This type of problems involving fully enclosed fluids cannot be solved using Algorithm 1 (or iterative variants) due to the constraint enforced by the fluid incompressibility on the interface solid velocity (unless it is directly prescribed in the solid solver, see [16]). Figure 2 (left) presents some snapshots of the fluid velocity magnitude in the deformed configuration obtained with a non-linear version of Algorithm 2 ($r = 1$ and $\tau = 0.05$). The fluid equations are discretized in space with $\mathbb{Q}_1/\mathbb{Q}_1$ finite elements and a SUPG/PSPG stabilized formulation. Quadrilateral MITC4 (locking-free) shell elements are considered for the structure (see, e.g., [6]). For comparison purposes, Fig. 2 (right) shows the maximal displacement magnitude on the interface obtained with Algorithm 2 and the implicit coupling scheme. Algorithm 2 with $r =$ 1 or $r = 2$ provides numerical solutions close to the implicit scheme. The superior accuracy of the variant with $r = 2$, induced by the second-order extrapolation in (4), is clearly noticeable. On the contrary, Algorithm 2 with $r = 0$ (see Remark 4) yields

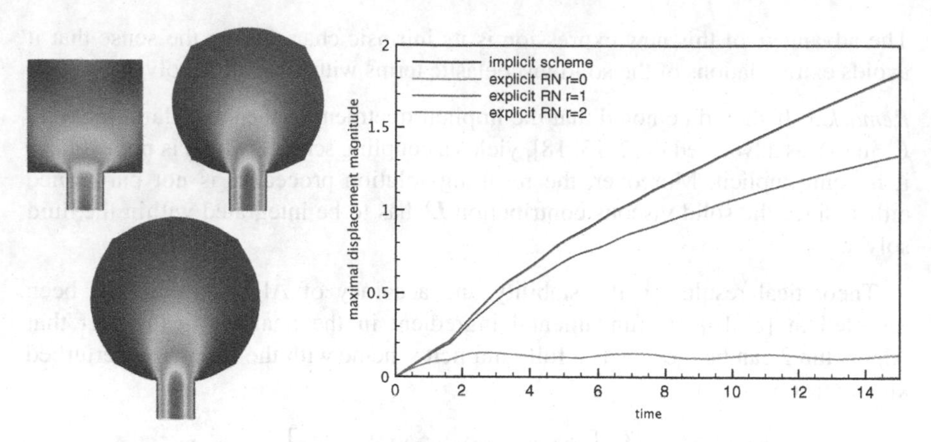

Fig. 2 *Left*: Snapshots of the fluid velocity magnitude in the deformed configuration at $t = 0.15$, 7.5, 15 (Algorithm 2 with $r = 1$ and $\tau = 0.05$). *Right*: Comparison of the solid displacements *vs.* time obtained with Algorithm 2 and the implicit scheme ($\tau = 0.05$)

an extremely poor approximation. This is a clear indication of the $\mathcal{O}(\tau^{\frac{1}{2}})$-loss in the accuracy of the scheme predicted by the error analysis.

3.2 Second-Order Accuracy

So far no explicit stable second-order time-accurate scheme is known for general fluid-structure interaction. For purely elastic thin-structures, some attempts have been presented in [18] by combining a Strang operator splitting approach with the ideas reported in [13]. Though the accuracy of the splitting is improved, second-order time-accuracy is still not achieved.

In this section we show how the Robin-Neumann explicit coupling paradigm of Sect. 3.1 can be adapted to deliver second-order time-accuracy. This is achieved by combining a Crank-Nicholson time-stepping in both the fluid and the solid subproblems, with an enhanced time-discretization of (3) based on either second-order extrapolation or defect-correction iterations. It is worth noting that this strategy for enhancing accuracy might lead to stability problems when applied to other explicit coupling paradigms (see, e.g., [4]).

The resulting schemes are displayed in Algorithm 3, where $K \geq 0$ denotes the number of correction iterations and $x^{n-\frac{1}{2},k} \stackrel{\text{def}}{=} \left(x^{n,k}+x^{n-1}\right)/2$ stands for the midpoint between the previous value x^{n-1} and the k-stage corrected one $x^{n,k}$.

Algorithm 3 Second-Order Robin-Neumann Schemes

For $n \geq 0$ if $r = 0, 1$ *or* for $n \geq 1$ if $r = 2$:

1. Extrapolation: $d^{n,0} = d^\star$, $\dot{d}^{n,0} = \dot{d}^\star$.
2. For $k = 1, \ldots, K+1$:
   a. Fluid step: Find $u^{n,k} : \Omega^{\mathrm{f}} \times \mathbb{R}^+ \to \mathbb{R}^d$ and $p^{n-\frac{1}{2},k} : \Omega^{\mathrm{f}} \times \mathbb{R}^+ \to \mathbb{R}$ such that

$$\begin{cases} \dfrac{\rho^{\mathrm{f}}}{\tau}(u^{n,k} - u^{n-1}) - \nabla \cdot \sigma(u^{n-\frac{1}{2},k}, p^{n-\frac{1}{2},k}) = \mathbf{0} & \text{in} \quad \Omega^{\mathrm{f}}, \\ \nabla \cdot u^{n-\frac{1}{2},k} = 0 & \text{in} \quad \Omega^{\mathrm{f}}, \\ \sigma(u^{n-\frac{1}{2},k}, p^{n-\frac{1}{2},k})n + \dfrac{\rho^{\mathrm{s}}\epsilon}{\tau}u^{n,k} = \dfrac{\rho^{\mathrm{s}}\epsilon}{\tau}\dot{d}^{n-1} + L^{\mathrm{e}}d^{n-\frac{1}{2},k-1} + L^{\mathrm{v}}\dot{d}^{n-\frac{1}{2},k-1} & \text{on} \quad \Sigma. \end{cases}$$

   b. Solid step: Find $d^{n,k} : \Sigma \times \mathbb{R}^+ \to \mathbb{R}^d$ such that

$$\begin{cases} \dfrac{\rho^{\mathrm{s}}\epsilon}{\tau}(\dot{d}^{n,k} - \dot{d}^{n-1}) + L^{\mathrm{e}}d^{n-\frac{1}{2},k} + L^{\mathrm{v}}\dot{d}^{n-\frac{1}{2},k} = -\sigma(u^{n-\frac{1}{2},k}, p^{n-\frac{1}{2},k})n & \text{on} \quad \Sigma, \\ \dot{d}^{n-\frac{1}{2},k} = \dfrac{1}{\tau}(d^{n,k} - d^{n-1}) & \text{on} \quad \Sigma. \end{cases}$$

3. Solution update: $u^n = u^{n,K+1}$, $p^{n-\frac{1}{2}} = p^{n-\frac{1}{2},K+1}$, $d^n = d^{n,K+1}$, $\dot{d}^n = \dot{d}^{n,K+1}$.

Similarly to Algorithm 2, Algorithm 3 with $K = 0$ can be regarded as interface kinematic perturbations of an underlying second-order implicit scheme. Hence, in order to achieve overall second-order time-accuracy, two approaches are investigated:

1. $r = 1$ and $K > 0$: Recall that the consistency errors induced by the kinematic perturbations with $r = 1$ scale as $\mathcal{O}(\tau)$. Thus, after $K > 0$ defect-corrections the perturbation of the kinematic constraint scales as $\mathcal{O}(\tau^{K+1})$. Hence, in order to retrieve second-order time-accuracy $K = 1$ will be enough.
2. $r = 2$ and $K = 0$ (genuine explicit scheme): Since the consistency error induced with $r = 2$ scales as $\mathcal{O}(\tau^2)$, no defect-correction is needed.

To give some insight into the stability properties of Algorithm 3, we consider a simplification of the model problem (1)–(2) at hand (see, e.g., [1, 5]). Specifically, we take $\Omega^{\mathrm{f}} = [0, L] \times [0, R] \subset \mathbb{R}^2$, $\Sigma = \{y = R\}$, the solid operators of Remark 1 and $\mu = 0$ (potential fluid). In this framework the following proposition holds.

Proposition 1 *Take* $K = 0$ *(no defect-correction) in Algorithm 3 and write* $d^n_y = \sum_{i=1}^{\infty} d^n_{y,i}\phi_i$ *where we consider the orthonormal basis on* $L^2_0(\Sigma)$ *given by* $\left\{\phi_i(x) = \sqrt{2/L}\sin(i\pi x/L)\right\}_{i=1}^{\infty}$. *Under the problem setting described in the previous paragraph, we have:*

1. If $r = 0$ *or* $r = 1$, $\quad |d^n_{y,i}| \underset{n\to+\infty}{\longrightarrow} 0 \quad \forall\, i \in \{1, \ldots, \infty\}$.

2. If $r = 2$, $|\boldsymbol{d}_{y,i}^n| \underset{n \to +\infty}{\longrightarrow} 0$ *with* $i \in \{1, \ldots, \infty\}$ *provided*

$$\begin{cases} 4a_i b_i + 4b_i^2 \tau + 4d_i b_i \tau^2 + \dfrac{d_i^2 \left(-2\mu_i \rho^{\mathrm{f}} + \epsilon \rho^{\mathrm{s}}\right)}{\epsilon \rho^{\mathrm{s}}} \tau^3 \geq 0, \\ 4e_i \epsilon \rho^{\mathrm{s}} (b_i + d_i \tau)(4a_i + d_i \tau^2 + 4\tau b_i) - 4d_i \tau e_i^2 \\ \qquad - 16\epsilon \rho^{\mathrm{s}} \tau (b_i + d_i \tau)^2 (a_i \epsilon \rho^{\mathrm{s}} + 2b_i \mu_i \rho^{\mathrm{f}} \tau) \geq 0, \end{cases} \tag{7}$$

where $a_i = \mu_i \rho^{\mathrm{f}} + \epsilon \rho^{\mathrm{s}}, b_i = \beta c_1 \lambda_i + \alpha \epsilon \rho^{\mathrm{s}}, d_i = c_0 + c_1 \lambda_i, e_i = 4a_i \epsilon \rho^{\mathrm{s}} + \tau(b_i \epsilon \rho^{\mathrm{s}} + 2d_i \mu_i \rho^{\mathrm{f}} \tau)$ *and* $\mu_i = \frac{L}{i\pi \tanh\left(\frac{i\pi R}{L}\right)}$, $\lambda_i = \frac{i^2 \pi^2}{L^2}$ *are the eigenvalues with respect to* ϕ_i *of the Neumann-to-Dirichlet map and* ∂_{xx} *operators.*

Proposition 1 establishes that whenever the Fourier series expansion of $\boldsymbol{d}_y^n$ is truncated (i.e., whenever the spatial discretization is fixed) the solution of Algorithm 3 with $K = 0$, under the above assumptions, is unconditionally stable with zeroth- and first-order extrapolations. For $r = 2$, the conditions (7) might be too restrictive since they do not explicitly take into account the effect of the spatial discretization step h.

In order to numerically illustrate the accuracy and stability of Algorithm 3, we consider the two-dimensional example of [10]. To provide evidence on the $\mathcal{O}(h + \tau^2)$ convergence behavior for the first and second order extrapolated variants, Fig. 3 (left) reports the time-convergence history, with $h = 10^{-1}/4$ fixed, of the solid displacement at time $t = 0.015$, in the relative elastic energy-norm, obtained with Algorithm 3 and a fully implicit second-order scheme. The reference solution has been generated using the implicit scheme with $\tau = 10^{-6}$ and the same h. The h-uniformity is guaranteed by Fig. 3 (right) were we have refined both in time and space according $h = \mathcal{O}(\tau^2)$. The reference solution has been now obtained with $\tau = 10^{-6}$ and $h = 3 \times 10^{-3}$.

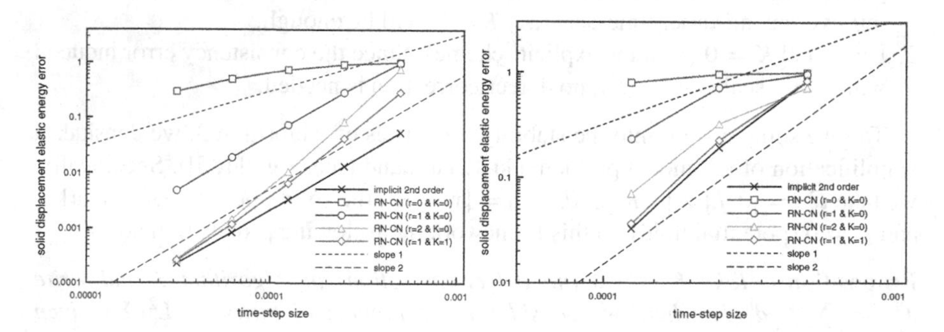

Fig. 3 *Left*: displacement convergence history in time with $h = 10^{-1}/4$ fixed. *Right*: displacement convergence history in time with $h = \mathcal{O}(\tau^2)$

3.3 Coupling with Thick-Walled Structures

In this section we briefly describe the extension of the Robin-Neumann explicit coupling paradigm of Algorithm 2 to the case of the coupling with thick-walled structures (see Fig. 1 (right)). Thus, in the coupled problem (1)–(2), the relations (2) are replaced by the linear elastodynamics equations

$$\begin{cases} \rho^s \partial_t \dot{\boldsymbol{d}} + \alpha \rho^s \dot{\boldsymbol{d}} - \nabla \cdot \boldsymbol{\Pi}(\boldsymbol{d}, \dot{\boldsymbol{d}}) = \boldsymbol{0} & \text{in} \quad \Omega^s, \\ \dot{\boldsymbol{d}} = \partial_t \boldsymbol{d} & \text{in} \quad \Omega^s, \end{cases} \tag{8}$$

together with the kinematic and kinetic coupling conditions

$$\begin{cases} \boldsymbol{u} = \dot{\boldsymbol{d}} & \text{on} \quad \Sigma, \\ \boldsymbol{\Pi}(\boldsymbol{d}, \dot{\boldsymbol{d}})\boldsymbol{n}^s = -\boldsymbol{\sigma}(\boldsymbol{u}, p)\boldsymbol{n} & \text{on} \quad \Sigma. \end{cases} \tag{9}$$

Here, the symbol $\boldsymbol{n}^s$ stands for the unit normal vector on $\partial\Omega^s$, the solid stress tensor is given by $\boldsymbol{\Pi}(\boldsymbol{d}, \dot{\boldsymbol{d}}) \stackrel{\text{def}}{=} \boldsymbol{\pi}(\boldsymbol{d}) + \beta \boldsymbol{\pi}(\dot{\boldsymbol{d}})$, where $\boldsymbol{\pi}(\boldsymbol{d}) \stackrel{\text{def}}{=} 2\lambda_1 \boldsymbol{\varepsilon}(\boldsymbol{d}) + \lambda_2 (\nabla \cdot \boldsymbol{d})\boldsymbol{I}$ and λ_1, λ_2 denote the Lamé coefficients. Damping effects in the solid are thus modeled via the Rayleigh-like term $\alpha \rho^s \dot{\boldsymbol{d}} - \beta \nabla \cdot \boldsymbol{\pi}(\dot{\boldsymbol{d}})$.

The fundamental ingredient in the derivation of the schemes described in the previous sections is the interface Robin consistency (3) featured by the continuous problem (1)–(2). Unfortunately, this property is not shared by the coupled problem (1), (8) and (9), since the inertial term in (8) is distributed on the whole solid domain Ω^s and $\Sigma \neq \Omega^s$. The following *generalized* interface Robin consistency can however be recovered after discretization in space, using a lumped-mass approximation in the structure (see [11]):

$$\boldsymbol{\sigma}(\boldsymbol{u}, p)\boldsymbol{n} + \rho^s \boldsymbol{B}_h \partial_t \boldsymbol{u} = \rho^s \boldsymbol{B}_h \partial_t \dot{\boldsymbol{d}} - \boldsymbol{\Pi}(\boldsymbol{d}, \dot{\boldsymbol{d}})\boldsymbol{n}^s \quad \text{on} \quad \Sigma. \tag{10}$$

Note that, instead of the usual identity operator, the interface condition (10) involves the discrete interface operator $\boldsymbol{B}_h$, which consistently accounts for the solid inertial effects within the fluid. In fact, at the algebraic level, this operator is given by the interface entries of the solid lumped-mass matrix. Instead of formulating the time splitting in terms of (9), we consider (10) and $(9)_2$. This yields the following Robin-Neumann splitting of (9):

$$\begin{cases} \boldsymbol{\sigma}(\boldsymbol{u}^n, p^n)\boldsymbol{n} + \dfrac{\rho^s}{\tau} \boldsymbol{B}_h \boldsymbol{u}^n = \dfrac{\rho^s}{\tau} \boldsymbol{B}_h \left(\dot{\boldsymbol{d}}^{n-1} + \tau \partial_\tau \dot{\boldsymbol{d}}^\star \right) - \boldsymbol{\Pi}(\boldsymbol{d}^\star, \dot{\boldsymbol{d}}^\star)\boldsymbol{n}^s, \\ \boldsymbol{\Pi}(\boldsymbol{d}^n, \dot{\boldsymbol{d}}^n)\boldsymbol{n}^s = -\boldsymbol{\sigma}(\boldsymbol{u}^n, p^n)\boldsymbol{n}. \end{cases} \tag{11}$$

The analysis reported in [11] shows that the splitting (11) preserves the energy stability of the original Robin-Neumann explicit coupling paradigm (Algorithm 2).

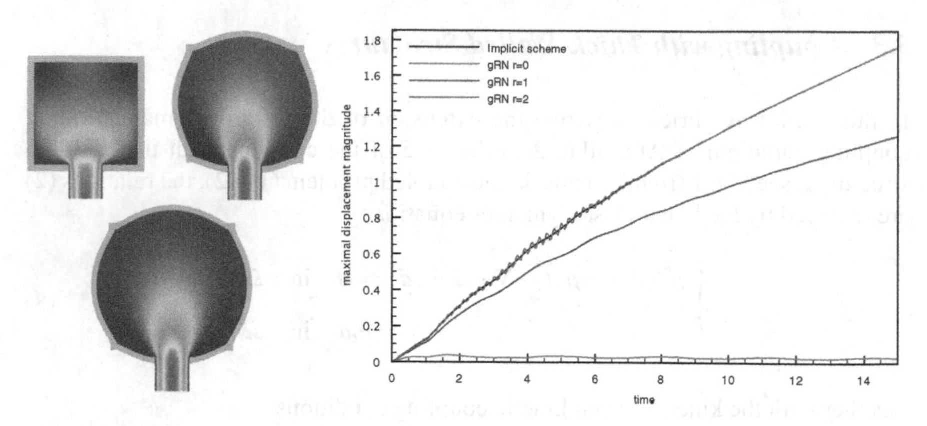

Fig. 4 *Left*: Snapshots of the fluid velocity magnitude in the deformed configurations at $t = 0.15$, 7.5, 15 (generalized Robin-Neumann explicit coupling (11) with $r = 1$ and $\tau = 0.025$). *Right*: Comparison of the solid displacements *vs.* time obtained with the generalized Robin-Neumann explicit coupling (11) and the implicit scheme ($\tau = 0.025$)

Numerical evidence indicates, however, that their optimal (first-order) accuracy is not preserved. Indeed, the order of the kinematic perturbation induced by the splitting (11) is expected to be $\mathcal{O}(\tau^{2^{r-1}}/h^{\frac{1}{2}})$. Interestingly, the factor $h^{-\frac{1}{2}}$ is intrinsically related to the thick-walled character of the structure, through the non-uniformity of the discrete viscoelastic operator, and not to the mass lumping approximation (see [9]).

We conclude this section by considering the balloon-like example of Sect. 3.1 but, this time, involving a thick-walled structure. In Fig. 4 (left) we have reported some snapshots of the fluid velocity magnitude and of the deformed configurations obtained with the generalized Robin-Neumann splitting (11) with $r = 1$ and $\tau = 0.025$. A comparison of the different variants with the implicit schemes is given in Fig. 4 (right). Note that spurious oscillations are visible for the explicit coupling with $r = 2$. This is consistent with the fact that stability conditions are expected to be more restrictive in the case of the coupling with thick-walled structures. Considering that the value of τ is twice smaller than in Sect. 3.1, the poor accuracy of the explicit scheme with $r = 0$ is even more striking. For $r = 1$ and $r = 2$ we obtain practically the same results as in Sect. 3.1. This is a clear indication of the $h^{-\frac{1}{2}}$ perturbation introduced by the splitting: the time-step length must be reduced to achieve a similar level of accuracy as in the thin-walled case.

4 Conclusion

We have discussed a class of explicit coupling schemes for incompressible fluid-structure interaction. The key ingredient in the derivation of the methods is the notion of interface Robin consistency which depends on the thin- or thick-walled character of the structure. In the case of the coupling with a thin-walled structure, energy stability and optimal first-order accuracy are retrieved without any restriction on the discretization parameters. Besides, under this structure regime, two promising extensions which deliver second-order time-accuracy have been presented. The main issue regarding thick-walled structures is accuracy, since the perturbation induced by the splitting is not uniform with respect to the spatial discretization step h. It is worth noting, however, that the scheme with first-order extrapolation yields convergence under a standard hyperbolic-condition without the need of correction iterations.

Acknowledgements This work was supported by the French National Research Agency (ANR) through the EXIFSI project (ANR-12-JS01-0004)

References

1. S. Badia, F. Nobile, C. Vergara, Fluid-structure partitioned procedures based on Robin transmission conditions. J. Comput. Phys. **227**(14), 7027–7051 (2008)
2. M. Bukač, S. Čanić, R. Glowinski, J. Tambača, A. Quaini, Fluid-structure interaction in blood flow capturing non-zero longitudinal structure displacement. J. Comput. Phys. **235**, 515–541 (2013)
3. E. Burman, M.A. Fernández, Stabilization of explicit coupling in fluid-structure interaction involving fluid incompressibility. Comput. Methods Appl. Mech. Eng. **198**(5–8), 766–784 (2009)
4. E. Burman, M.A. Fernández, Explicit strategies for incompressible fluid-structure interaction problems: Nitsche type mortaring versus Robin-Robin coupling. Int. J. Numer. Methods Eng. **97**(10), 739–758 (2014)
5. P. Causin, J.F. Gerbeau, F. Nobile, Added-mass effect in the design of partitioned algorithms for fluid-structure problems. Comput. Methods Appl. Mech. Eng. **194**(42–44), 4506–4527 (2005)
6. D. Chapelle, K.J. Bathe, *The Finite Element Analysis of Shells—Fundamentals*. Computational Fluid and Solid Mechanics (Springer, Berlin, 2011)
7. M.A. Fernández, Coupling schemes for incompressible fluid-structure interaction: implicit, semi-implicit and explicit. *SeMA J.* **55**(55), 59–108 (2011)
8. M.A. Fernández, Incremental displacement-correction schemes for incompressible fluid-structure interaction: stability and convergence analysis. Numer. Math. **123**(1), 21–65 (2013)
9. M.A. Fernández, J. Mullaert, Convergence analysis for a class of explicit coupling schemes in incompressible fluid-structure interaction (2014). Research report RR-8670, Inria, 2015
10. M.A. Fernández, J. Mullaert, M. Vidrascu, Explicit Robin-Neumann schemes for the coupling of incompressible fluids with thin-walled structures. Comput. Methods Appl. Mech. Eng. **267**, 566–593 (2013)
11. M.A. Fernández, J. Mullaert, M. Vidrascu, Generalized Robin-Neumann explicit coupling schemes for incompressible fluid-structure interaction: stability analysis and numerics. Int. J. Numer. Methods Eng. **101**(3), 199–229 (2015). doi:10.1002/nme.4785. https://hal.inria.fr/hal-00875819
12. C. Förster, W.A. Wall, E. Ramm, Artificial added mass instabilities in sequential staggered coupling of nonlinear structures and incompressible viscous flows. Comput. Methods Appl. Mech. Eng. **196**(7), 1278–1293 (2007)
13. G. Guidoboni, R. Glowinski, N. Cavallini, S. Čanić, Stable loosely-coupled-type algorithm for fluid-structure interaction in blood flow. J. Comput. Phys. **228**(18), 6916–6937 (2009)

14. M. Heil, A.L. Hazel, Fluid-structure interaction in internal physiological flows. Annu. Rev. Fluid Mech. **43**, 141–162 (2011)
15. P. Kalita, R. Schaefer. Mechanical models of artery walls. Arch. Comput. Methods Eng.**15**(1), 1–36 (2008)
16. U. Küttler, C. Förster, W.A. Wall, A solution for the incompressibility dilemma in partitioned fluid-structure interaction with pure Dirichlet fluid domains. Comput. Mech. **38**, 417–429 (2006)
17. M. Lombardi, N. Parolini, A. Quarteroni, G. Rozza, Numerical simulation of sailing boats: dynamics, FSI, and shape optimization, in *Variational Analysis and Aerospace Engineering: Mathematical Challenges for Aerospace Design*, ed. by G. Buttazzo, A. Frediani. Springer Optimization and Its Applications (Springer, Berlin, 2012), pp. 339–377
18. M. Lukáčová-Medvid'ová, G. Rusnáková, A. Hundertmark-Zaušková, Kinematic splitting algorithm for fluid-structure interaction in hemodynamics. Comput. Methods Appl. Mech. Eng. **265**(1), 83–106 (2013)
19. P. Moireau, N. Xiao, M. Astorino, C.A. Figueroa, D. Chapelle, C.A. Taylor, J.-F. Gerbeau, External tissue support and fluid-structure simulation in blood flows. Biomech. Model. Mechanobiol. **11**, 1–18 (2012)
20. M.P. Païdoussis, S.J. Price, E. de Langre, *Fluid-Structure Interactions: Cross-Flow-Induced Instabilities* (Cambridge University Press, Cambridge, 2011)
21. K. Takizawa, T.E. Tezduyar, Computational methods for parachute fluid-structure interactions. Arch. Comput. Methods Eng. **19**, 125–169 (2012)
22. D. Valdez-Jasso, H.T. Banks, M.A. Haider, D. Bia, Y. Zocalo, R.L. Armentano, M.S. Olufsen, Viscoelastic models for passive arterial wall dynamics. Adv. Appl. Math. Mech. **1**(2), 151–165 (2009)

Optimal Finite Element Methods for Interface Problems

Jinchao Xu and Shuo Zhang

1 Introduction

There are many physical problems such as multiphase flows and fluid-structure interactions whose solutions are piecewise smooth but may have discontinuity across some curved interfaces. The direct application of standard finite element method may not perform well. In this paper, we study some special finite element methods for this type of problems. For simplicity of exposition, we consider the case that there is only one interface which is smooth. Let $\Omega, \Omega_1 \subset \mathbb{R}^2$ be two bounded domains with $\Omega_1 \subset \Omega$. We assume that $\Gamma = \partial\Omega_1$ is sufficiently smooth, and $\Gamma \cap \partial\Omega = \emptyset$. To be focused on the influence of Γ, we assume $\Omega = (-1, 1)^2$.

Both authors are supported by the Department of Energy (DOE) Grants DE-SC0006903 and DE-SC0009249 (through the Applied Mathematics Program within the DOE Office of Advanced Scientific Computing Research (ASCR) as part of the Collaboratory on Mathematics for Mesoscopic Modeling of Materials (CM4)) and the Center for Computational Mathematics and Applications of the Pennsylvania State University. The second author is also supported by the NSFC Grant 11101415 and SRF for ROCS by SEM of P. R. China.

J. Xu (✉)
Department of Mathematics, Center for Computational Mathematics and Applications, The Pennsylvania State University, University Park, PA 16802, USA
e-mail: xu@math.psu.edu

S. Zhang
LSEC, ICMSEC, NCMIS, Academy of Mathematics and System Sciences, Chinese Academy of Sciences, Beijing 100190, China
e-mail: szhang@lsec.cc.ac.cn

T. Dickopf et al. (eds.), *Domain Decomposition Methods in Science and Engineering XXII*, Lecture Notes in Computational Science and Engineering 104, DOI 10.1007/978-3-319-18827-0_7

To be specific, we consider the homogeneous boundary value problems of the diffusion equation $-\mathrm{div}(\alpha\nabla u) = f$, and the Stokes equation $-\mathrm{div}(\alpha\nabla \underset{\sim}{u} - p\underset{\sim}{I}) = \underset{\sim}{f}$ with the incompressibility condition $\mathrm{div}\underset{\sim}{u} = 0$. In both of the equations, α represents a piecewise smooth function, namely $\alpha \in (\mathcal{C}^\infty(\Omega_1) \oplus \mathcal{C}^\infty(\Omega_2)) \setminus \mathcal{C}(\Omega)$, such that $0 < \alpha_1 \leqslant \alpha \leqslant \alpha_2$ for two constants α_1 and α_2, and $\Omega_2 = \Omega \setminus \Omega_1$.

Because of the discontinuity of the coefficient α, the solutions lose their smoothness near the interface. Accuracy would be lost if we use general uniform grids for discretisation. A way to remedy the accuracy of approximation is to use interface-fitted/resolved grids. This way, the non-smoothness of the solution can be restricted to a "narrow" subdomain with respect to the grid near the interface, and the approximation error due to the non-smoothness can thus be dominated.

In [10] (English translation: [12]), the following error estimate was obtained:

$$\|u - u_I\|_{0,\Omega} + h|u - u_I|_{1,\Omega} \leqslant C|\log h|^{1/2} h^2 |u|_{2,\Omega_1\cup\Omega_2}, \tag{1}$$

where u_I is the nodal interpolation of u to the linear element space. Here and after, we use $|w|_{m,\Omega_1\cup\Omega_2}$ or $\|w\|_{m,\Omega_1\cup\Omega_2}$ to denote $|w|_{m,\Omega_1} + |w|_{m,\Omega_2}$ or $\|w\|_{m,\Omega_1} + \|w\|_{m,\Omega_2}$, respectively, for $w \in H^m(\Omega_1 \cup \Omega_2) := \{v \in L^2(\Omega) : v|_{\Omega_i} \in H^m(\Omega_i),\ i = 1, 2\}$, with $m \in \{0, 1, 2\}$. See [6] for a same result. A sharper estimate was given in [3]:

$$\|u - u_I\|_{0,\Omega} + h|u - u_I|_{1,\Omega} \leqslant Ch^2 |u|_{2,\Omega_1\cup\Omega_2}. \tag{2}$$

The interface-fitting assumption in the works above can be loosened slightly to that the interface Γ is "$\mathcal{O}(h^2)$-resolved by the mesh", see [8], and the shape-regularity restriction of the grid can be loosened to maximal-angle-bounded grids, see [7]. The optimal approximation accuracy of linear element space can also be proved on these grids.

We refer to [7] for an algorithm to generate an interface-fitted grid from a shape-regular grid which is not interface-fitted. (c.f. Fig. 1.) The algorithm is easy to implement and the generated grid is maximal-angle-bounded. With the linear element functions constructed thereon, the piecewise smooth functions can be approximated optimally and economically.

In this paper, we discuss the linear element schemes for the diffusion equation and the Stokes equation with discontinuous coefficients on interface-fitted maximal-angle-bounded grids. We will consider the conforming (c.f. also [7]) and nonconforming linear element schemes for the diffusion equation, and the $P_1 - P_0$ element pair for the Stokes equation. Thanks to the above approximation results, the optimal accuracy of conforming linear element discretisation for the diffusion equation is straightforwardly obtained. When the nonconforming element dicretization is considered, the issue of consistency error needs to be addressed. Because of the irregularity of the grid, the traditional technique by trace theorem and scaling argument cannot be applied easily. In this paper, we use the relationship between the nonconforming linear element space and the lowest-order Raviart-

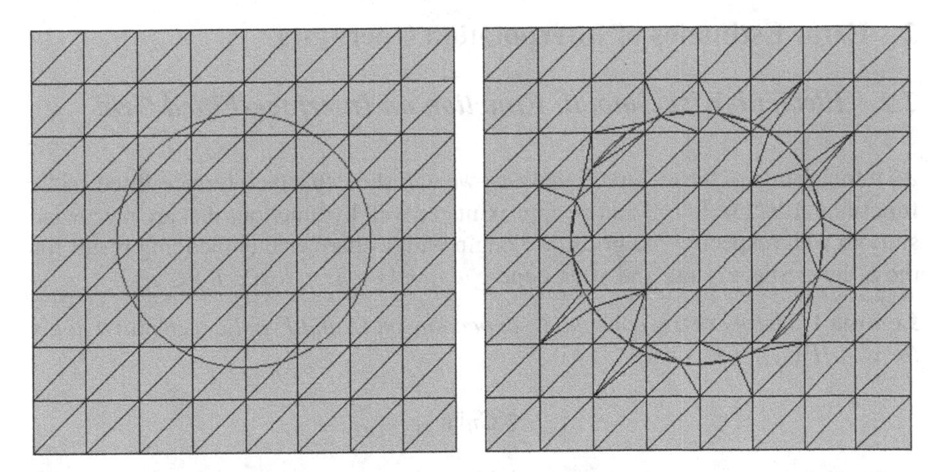

Fig. 1 *Left*: interface-unfitted mesh; *Right*: interface-fitted mesh

Thomas (R-T for short) element space suggested by Acosta and Durán [1], and obtain the optimal accuracy of the consistency error. As to the incompressible Stokes problem, we have that the $P_1 - P_0$ pair satisfies the inf-sup condition, and prove that it has optimal accuracy.

Then we discuss the optimal multigrid solver for the generated linear system. Particularly, we consider the special grid that is generated from a uniform grid with the algorithm of [7]. As the underlying grid is obtained by refining an original uniform structured grid, the finite element space thereon is different from the one on the original grid only near the interface. We use the original grid (finite element space thereon) as a coarse grid (subspace, respectively), with some smoothing operations added near the interface, to formulate a nested geometrical multigrid method. We take the conforming linear element system, which is less complicated, for a demonstration, and show the optimality of the formulated multigrid method.

Through the paper, we make use of this notation. Without bringing in ambiguity, we use $|\cdot|$ for the measure of subdomains, especially the area of a 2D manifold or the length of a 1D manifold. We use "$\underset{\sim}{\ }$" for a tensor, and a bold letter for a unit vector (direction). In the paper, "K" will always denote a triangular cell, unless special indication. When the triangulation $\mathcal{T}_h$ is considered, we denote $H^m(\mathcal{T}_h) := \{w \in L^2(\Omega) : w|_K \in H^m(K),\ \forall K \in \mathcal{T}_h\}, m = 0, 1, 2.$

The remaining of the paper is organised as follows. In Sect. 2, we collect some existing and new estimation results on interpolation operators, especially for piecewise smooth functions on interface-fitted and maximal-angle-bounded grids. In Sects. 3 and 4, we discuss the optimal finite element methods for the interface problems of the diffusion equation and of the Stokes equation, respectively. In Sect. 5, we give an optimal multigrid method for the conforming linear element scheme for diffusion equation. Finally, in Sect. 6, some concluding remarks are given.

2 Error Estimates of Interpolation Operators

2.1 Element-Wise Smooth Function on Interface-Fitted Grid

As a foundation of the technical analysis, we will show that on interface-fitted grids, functions in $H^m(\Omega_1 \cup \Omega_2)$ can be approximated well by functions that are piecewise smooth with respect to the grids. We begin with a sharpened embedding result for the Sobolev space. Here and after, denote $\omega_\eta := \{x \in \Omega : dist(x, \Gamma) \leqslant \eta\}$.

Lemma 1 *There exists a constant C, depending on Ω and Γ only, such that it holds for $w \in H^1(\Omega_1 \cup \Omega_2)$ that*

$$\|w\|_{0,\omega_\eta}^2 \leqslant C\eta \|w\|_{1,\Omega_1 \cup \Omega_2}^2.$$

The proof of Lemma 1 follows from Theorem 1.1 of [2] directly, and we omit it here. We also refer to [3, 8] for similar results.

Lemma 2 *Let $\mathcal{T}_h$ be an interface-fitted grid of Ω, with h the biggest diameter of $K \in \mathcal{T}_h$. Then there exists a constant C depending on Ω and Γ only, such that these inequalities hold:*

1. *given $w \in H^1(\Omega_1 \cup \Omega_2)$, there exists a $\tilde{w} \in H^1(\mathcal{T}_h)$, such that*

$$\sum_{K \in \mathcal{T}_h} \|\tilde{w}\|_{1,K}^2 \leqslant C(\|w\|_{1,\Omega_1}^2 + \|w\|_{1,\Omega_2}^2), \ \|w - \tilde{w}\|_{0,\Omega}^2 \leqslant Ch^2(\|w\|_{1,\Omega_1}^2 + \|w\|_{1,\Omega_2}^2);$$

2. *given $w \in H^2(\Omega_1 \cup \Omega_2)$, there exists $\tilde{w} \in H^2(\mathcal{T}_h)$, such that*

$$\sum_{K \in \mathcal{T}_h} \|\tilde{w}\|_{2,K}^2 \leqslant C(\|w\|_{2,\Omega_1}^2 + \|w\|_{2,\Omega_2}^2), \ \sum_{K \in \mathcal{T}_h} (\|w - \tilde{w}\|_{1,K\cap\Omega_1}^2 + \|w - \tilde{w}\|_{1,K\cap\Omega_2}^2) \leqslant Ch^2(\|w\|_{2,\Omega_1}^2 + \|w\|_{2,\Omega_2}^2);$$

   moreover, if $w \in H^1(\Omega) \cap H^2(\Omega_1 \cup \Omega_2)$, then $\tilde{w} = w$ on Γ;

3. *given $\underset{\sim}{w} \in (H^1(\Omega_1 \cup \Omega_2))^2 \cap H(\mathrm{div}; \Omega)$, there exists $\tilde{\underset{\sim}{w}} \in H^1(\mathcal{T}_h) \cap H(\mathrm{div}; \Omega)$, such that $\tilde{\underset{\sim}{w}} \cdot \mathbf{n} = \underset{\sim}{w} \cdot \mathbf{n}$ on Γ, and*

$$\sum_{K \in \mathcal{T}_h} \|\tilde{\underset{\sim}{w}}\|_{1,K}^2 \leqslant C(\|\underset{\sim}{w}\|_{1,\Omega_1}^2 + \|\underset{\sim}{w}\|_{1,\Omega_2}^2), \ \sum_{K \in \mathcal{T}_h} (\|\underset{\sim}{w} - \tilde{\underset{\sim}{w}}\|_{0,K}^2 \leqslant Ch^2(\|w\|_{1,\Omega_1}^2 + \|w\|_{1,\Omega_2}^2).$$

Proof We only prove the third item. The others can be found in [3].

First of all, given $K \in \mathcal{T}_h$, since $\mathcal{T}_h$ is interface fitted, K does not has vertices in different subdomains simultaneously. Besides, by approximation theory, there exists a constant C_0, depending on Γ and Ω, such that if K has a vertex in Ω_i, then $(K \cap \Omega_{3-i}) \subset \omega_{C_0h^2}$.

Now given $\underset{\sim}{w} \in (H^1(\Omega_1 \cup \Omega_2))^2$, by extension theorem, there exist $\underset{\sim}{w}_1, \underset{\sim}{w}_2 \in H^1(\Omega)^2$, such that $(\underset{\sim}{w}_i - \underset{\sim}{w})|_{\Omega_i} = 0$, and $\|\underset{\sim}{w}_i\|_{1,\Omega} \leqslant C\|\underset{\sim}{w}\|_{1,\Omega_i}$, C depending on Ω and Γ only. Then we define $\tilde{\underset{\sim}{w}}$ by $\tilde{\underset{\sim}{w}}|_K = \underset{\sim}{w}_i|_K$, if K has a vertex in Ω_i. Here, without loss of generality, we assume the vertices of K are not all on Γ. By the analysis above, $\underset{\sim}{w} - \tilde{\underset{\sim}{w}} = 0$, on $\Omega \setminus \omega_{Ch^2}$. Therefore, $\|\underset{\sim}{w} - \tilde{\underset{\sim}{w}}\|^2_{0,\Omega} = \|\underset{\sim}{w} - \tilde{\underset{\sim}{w}}\|^2_{0,\omega_{Ch^2}} \leqslant 3(\|\underset{\sim}{w}\|^2_{0,\omega_{Ch^2}} + \|\underset{\sim}{w}_1\|^2_{0,\omega_{Ch^2}} + \|\underset{\sim}{w}_2\|^2_{0,\omega_{Ch^2}})$. Further, by Lemma 1, $\|\underset{\sim}{w} - \tilde{\underset{\sim}{w}}\|^2_{0,\Omega} \leqslant Ch^2\|\underset{\sim}{w}\|^2_{1,\Omega_1\cup\Omega_2}$, with C depending on Γ and Ω.

Besides, that $\underset{\sim}{w} \in (H^1(\Omega_1 \cup \Omega))^2 \cap H(\mathrm{div};\Omega)$ implies $[\![\underset{\sim}{w} \cdot \mathbf{n}]\!]$ vanishes along Γ, this further implies that $\underset{\sim}{w} \cdot \mathbf{n}$, $\underset{\sim}{w}_1 \cdot \mathbf{n}$ and $\underset{\sim}{w}_2 \cdot \mathbf{n}$ are the same along the interface, thus $\tilde{\underset{\sim}{w}} \cdot \mathbf{n} = \underset{\sim}{w} \cdot \mathbf{n}$ along Γ. Here and after, we use $[\![\cdot]\!]$ to denote the jump between different sides. This finishes the proof.

2.2 *Interpolation Error for Piecewise Smooth Functions*

Let $\mathcal{T}_h$ be a grid on Ω. Denote Q_h the piecewise constant space on $\mathcal{T}_h$, V_h^{CR} the linear Crouzeix-Raviart element space, namely $V_h^{\mathrm{CR}} := \{w_h \in L^2(\Omega) : w_h|_K \in P_1(K),\ \forall K \in \mathcal{T}_h,\ \int_e [\![w_h]\!] = 0, \text{ on any interior edge } e\}$, V_h the continuous piecewise linear function space, and $\mathbb{V}_h^{\mathrm{RT}}$ the lowest order Raviart-Thomas element space, namely $\mathbb{V}_h^{\mathrm{RT}} := \{\underset{\sim}{w}_h \in (L^2(\Omega))^2 : \underset{\sim}{w}_h|_K \in (P_0)^2 \oplus \underset{\sim}{x}P_0,\ \int_e [\![\underset{\sim}{w}_h]\!] \cdot \mathbf{n} = 0, \text{ on any interior edge } e\}$. Associated with the local interpolations, we have these globally defined interpolations. Denote by P_h^0 the L^2 projection operator to Q_h, by I_h the interpolation operator to V_h, by Π_h^{CR} the interpolation operator to V_h^{CR}, and by Π_h^{RT} the interpolation operator to $\mathbb{V}_h^{\mathrm{RT}}$. It is evident that $\nabla_h \Pi_h^{\mathrm{CR}} w = P_h^0 \nabla w$, and $\mathrm{div}\Pi_h^{\mathrm{RT}} \underset{\sim}{w} = P_h^0 \mathrm{div}\underset{\sim}{w}$.

Lemma 3 *Let $\mathcal{T}_h$ be an interface-fitted grid of Ω, with h the biggest size of the elements. With constants C_2 and C_3 depending on the maximal angle of the grid, while C_1 not, we have:*

1. *let $w \in H^1(\Omega_1 \cup \Omega_2)$, then $\|w - P_h^0 w\|_{0,\Omega} \leqslant C_1 h\|w\|_{1,\Omega_1\cup\Omega_2}$;*
2. *let $u \in H^1(\Omega) \cap H^2(\Omega_1 \cup \Omega_2)$, then $\inf_{v_h \in V_h^{\mathrm{CR}}} |u - v_h|_{1,h} \leqslant |u - I_h u|_{1,\Omega} \leqslant C_2 h\|u\|_{2,\Omega_1\cup\Omega_2}$;*
3. *let $\underset{\sim}{w} \in (H^1(\Omega_1 \cup \Omega_2))^2 \cap H(\mathrm{div};\Omega)$, then $\|\underset{\sim}{w} - \Pi_h^{\mathrm{RT}}\underset{\sim}{w}\|_{0,\Omega} \leqslant C_3 h\|\underset{\sim}{w}\|_{1,\Omega_1\cup\Omega_2}$.*

Proof We only prove the third item, and the first one is similar. We refer to, e.g., [7] for the second item.

We begin with a stability result. Let K be a triangle, with e_1, e_2 and e_3 its edges. On K, it holds for $i = 1, 2, 3$ that $\int_{e_i} \Pi_h^{\rm RT} \underset{\sim}{w} \cdot \mathbf{n}_{e_i} = \int_{e_i} \underset{\sim}{w} \cdot \mathbf{n}_{e_i}$. Direct calculation leads to that $\|\Pi_h^{\rm RT} \underset{\sim}{w}\|_{0,K}^2 \leqslant \frac{3}{16|K|} \sum_i \left[(\int_{e_i} \underset{\sim}{w} \cdot \mathbf{n}_{e_i})^2 \sum_{j\neq i} |e_j|^2 \right]$. Now, given $\underset{\sim}{w} \in (H^1(\Omega_1 \cup \Omega_2))^2 \cap H(\text{div}; \Omega)$, by Lemma 2, there exists $\tilde{\underset{\sim}{w}} \in H^1(\mathcal{T}_h)^2 \cap H(\text{div}; \Omega)$, such that $\tilde{\underset{\sim}{w}} \cdot \mathbf{n} = \underset{\sim}{w} \cdot \mathbf{n}$ on Γ, $\sum_{K\in\mathcal{T}_h} \|\tilde{\underset{\sim}{w}}\|_{1,K}^2 \leqslant C(\|\underset{\sim}{w}\|_{1,\Omega_1}^2 + \|\underset{\sim}{w}\|_{1,\Omega_2}^2)$, and $\|\underset{\sim}{w} - \tilde{\underset{\sim}{w}}\|_{0,\Omega}^2 \leqslant Ch^2(\|\underset{\sim}{w}\|_{1,\Omega_1}^2 + \|\underset{\sim}{w}\|_{1,\Omega_2}^2)$. By triangle inequality,

$$\|\underset{\sim}{w} - \Pi_h^{\rm RT} \underset{\sim}{w}\|_{0,\Omega} \leqslant \|\underset{\sim}{w} - \tilde{\underset{\sim}{w}}\|_{0,\Omega} + \|\tilde{\underset{\sim}{w}} - \Pi_h^{\rm RT} \tilde{\underset{\sim}{w}}\|_{0,\Omega} + \|\Pi_h^{\rm RT} \tilde{\underset{\sim}{w}} - \Pi_h^{\rm RT} \underset{\sim}{w}\|_{0,\Omega} := I_1 + I_2 + I_3. \tag{3}$$

For I_3, we only have to estimate $\Pi_h^{\rm RT} \tilde{\underset{\sim}{w}} - \Pi_h^{\rm RT} \underset{\sim}{w}$ on such K that $K \cap \Gamma \neq \emptyset$. Without loss of generality, we choose $K = [P_0, P_1, P_2]$, such that $P_0 \in \Omega_1$, and $K \cap \Omega_2 \neq \emptyset$; particularly, Γ goes through K from P_1 to P_2, c.f. Fig. 2. Denote $e = [P_1, P_2]$ and $K' = K \setminus \Omega_1$. Then $\int_e (\tilde{\underset{\sim}{w}} - \underset{\sim}{w}) \cdot \mathbf{n}_e = \int_{K'} \nabla \cdot (\tilde{\underset{\sim}{w}} - \underset{\sim}{w}) - \int_{e'} (\tilde{\underset{\sim}{w}} - \underset{\sim}{w}) \cdot \mathbf{n}_{e'}$, where $e' = \partial K' \setminus e$ thus $e' \subset \Gamma$. Note that $(\tilde{\underset{\sim}{w}} - \underset{\sim}{w}) \cdot \mathbf{n}_{e'} = 0$ on e', and thus $\int_e (\tilde{\underset{\sim}{w}} - \underset{\sim}{w}) \cdot \mathbf{n}_e = \int_{K'} \nabla \cdot (\tilde{\underset{\sim}{w}} - \underset{\sim}{w}) \leqslant |K'|^{1/2} \|\nabla \cdot (\tilde{\underset{\sim}{w}} - \underset{\sim}{w})\|_{0,K'}$. Thus,

$$\|\Pi_h^{\rm RT} (\tilde{\underset{\sim}{w}} - \underset{\sim}{w})\|_{0,K}^2 \leqslant \frac{3}{8} h_K^2 \frac{|K'|}{|K|} \|\nabla \cdot (\tilde{\underset{\sim}{w}} - \underset{\sim}{w})\|_{0,K'}^2 \leqslant h_K^2 \|\nabla \cdot (\tilde{\underset{\sim}{w}} - \underset{\sim}{w})\|_{0,K'}^2$$
$$\leqslant h_K^2 \|\nabla \cdot (\tilde{\underset{\sim}{w}} - \underset{\sim}{w})\|_{0,K\cap\omega_{Ch^2}}^2.$$

Further, $\|\Pi_h^{\rm RT} (\tilde{\underset{\sim}{w}} - \underset{\sim}{w})\|_{0,\Omega}^2 \leqslant \sum_{K\in\mathcal{T}_h} Ch_K^2 \|\nabla \cdot (\tilde{\underset{\sim}{w}} - \underset{\sim}{w})\|_{0,K\cap\omega_{Ch^2}}^2 \leqslant Ch^2 \|\nabla \cdot (\tilde{\underset{\sim}{w}} - \underset{\sim}{w})\|_{0,\omega_{Ch^2}}^2 \leqslant Ch^2(\|\underset{\sim}{w}\|_{1,\Omega_1\cup\Omega_2}^2)$. Then by (3), we have $\|\underset{\sim}{w} - \Pi_h^{\rm RT} \underset{\sim}{w}\|_{0,\Omega} \leqslant$

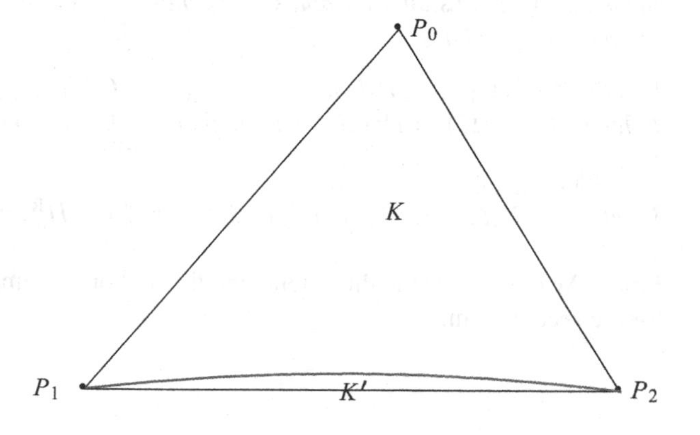

Fig. 2 Illustration of a cell K, and the edge $e = P_1 P_2$

$C_1h\|\underset{\sim}{w}\|_{1,\Omega_1\cup\Omega_2} + C_2h\|\underset{\sim}{w}\|_{1,\Omega_1\cup\Omega_2} + C_3h\|\underset{\sim}{w}\|_{1,\Omega_1\cup\Omega_2} \leqslant Ch\|\underset{\sim}{w}\|_{1,\Omega_1\cup\Omega_2}$, where C_2 depends on the maximal angle of the triangulation (c.f. [1]). This finishes the proof.

3 Optimal Linear Element Methods for Diffusion Equation

We consider the boundary-interface value problem:

$$\begin{cases} -\nabla\cdot(\alpha(x)\nabla u) = f, \text{ in } \Omega, \\ u = 0, \text{ on } \partial\Omega, \\ [\![u]\!] = 0,\ [\![\alpha\nabla u\cdot\mathbf{n}]\!] = 0, \text{ on } \Gamma, \end{cases} \tag{4}$$

where $\mathbf{n}$ is the normal direction of Γ. The variational formulation of the above problem is: Find $u \in H_0^1(\Omega)$ such that

$$a(u, v) = (f, v), \forall v \in H_0^1(\Omega), \tag{5}$$

where $a(u, v) = \int_\Omega \alpha(x)\nabla u\cdot\nabla v$, and $(f, v) = \int_\Omega fv$.

Evidently, given the coefficient α, the energy norm of the boundary value problem is equivalent to the H^1 norm (or piecewise H^1 norm for nonconforming element space). In the sequel, we focus ourselves on the analysis of the H^1 norm.

In this section and Sect. 4, we assume $\mathcal{T}_h$ is an interface-fitted triangulation of Ω, with h the biggest diameter of all $K \in \mathcal{T}_h$. We consider the case $\mathcal{T}_h$ is one in a maximal-angle-bounded family.

3.1 A Conforming Linear Element Method

Let $V_{h0} = V_h \cap H_0^1(\Omega)$. The finite element problem is to find $u_h \in V_{h0}$, such that

$$a(u_h, v_h) = (f, v_h), \quad \forall\, v_h \in V_{h0}. \tag{6}$$

Let u be the solution of (5), then by Cea lemma, it is straightforward that

$$|u - u_h|_{1,\Omega} \leqslant C \inf_{v_h\in V_{h0}} |u - v_h|_{1,\Omega} \leqslant Ch\|u\|_{2,\Omega_1\cup\Omega_2}.$$

We also refer to [3, 7, 8] for related discussions.

3.2 A Nonconforming Linear Element Method

Let $V_{h0}^{\mathrm{CR}} \subset V_h^{\mathrm{CR}}$ consist of the C-R element functions that vanish at the midpoints of the boundary edges. Then the C-R element scheme of the boundary value problem is to find $u_h \in V_{h0}^{\mathrm{CR}}$, such that

$$(\alpha\nabla_h u_h, \nabla_h v_h) = (f, v_h), \quad \forall\, v_h \in V_{h0}^{\mathrm{CR}}. \tag{7}$$

Here ∇_h denotes the piecewise gradient.

Theorem 1 *Let u and u_h be the solutions of (5) and (7), respectively. Assume $u \in H^2(\Omega_1 \cup \Omega_2) \cap H^1(\Omega)$. Then it holds with a constant C independent of h that*

$$\|\nabla_h(u - u_h)\|_{0,\Omega} \leqslant Ch(\|u\|_{2,\Omega_1\cup\Omega_2} + \|f\|_{0,\Omega}). \tag{8}$$

Proof Firstly, recall the Strang lemma and we have, with $|\cdot|_{1,h} = \|\nabla_h \cdot\|_{0,\Omega}$,

$$|u - u_h|_{1,h} \lesssim \inf_{v_h\in V_{h0}^{\mathrm{CR}}} |u - v_h|_{1,h} + \sup_{w_h\in V_{h0}^{\mathrm{CR}}} \frac{(\alpha\nabla u, \nabla w_h) - (f, w_h)}{|w_h|_{1,h}}. \tag{9}$$

By Lemma 3, we have to estimate the consistency error, which is (c.f. also [1])

$$(\alpha\nabla u, \nabla_h w_h) - (f, w_h) = (\alpha\nabla u - \Pi_h^{\mathrm{RT}}\alpha\nabla u, \nabla w_h) - (-\mathrm{div}\alpha\nabla u + \mathrm{div}\Pi_h^{\mathrm{RT}}\alpha\nabla u, w_h) := I - II. \tag{10}$$

By Lemma 3, $|II| = |(f + P_h^0\mathrm{div}\alpha\nabla u, w_h)| = |(f - P_h^0 f, w_h)| = |(f, w_h - P_h^0 w_h)| \leqslant Ch\|f\|_{0,\Omega}|w_h|_{1,h}$. Besides, as $\alpha\nabla u \in (H^1(\Omega_1 \cup \Omega_2))^2 \cap H(\mathrm{div};\Omega)$, $|I| \leqslant \|\alpha\nabla u - \Pi_h^{\mathrm{RT}}(\alpha\nabla u)\|_{0,\Omega}\|\nabla_h w_h\|_{0,\Omega} \leqslant Ch\|\alpha\nabla u\|_{1,\Omega_1\cup\Omega_2}|w_h|_{1,h}$. Substituting all above into (9) finishes the proof.

4 The $P^1 - P^0$ Element Method for Stokes Interface Problem

4.1 Model Problem and Finite Element Discretization

Now we consider the system of Stokes equation,

$$\begin{cases} -\mathrm{div}(\alpha\nabla \underset{\sim}{u} - p\underset{\sim}{\mathrm{Id}}) = \underset{\sim}{f}, \text{ in } \Omega, \\ \mathrm{div}\underset{\sim}{u} = 0, \text{ in } \Omega, \\ \underset{\sim}{u} = \underset{\sim}{0}, \text{ on } \partial\Omega, \\ [\![\underset{\sim}{u}]\!] = 0,\ [\![(\alpha\nabla\underset{\sim}{u} - p\underset{\sim}{\mathrm{Id}})\cdot\mathbf{n}]\!] = 0, \text{ on } \Gamma. \end{cases} \tag{11}$$

Here $\underset{\sim}{\mathrm{Id}} \in \mathbb{R}^{2\times 2}$ is the identity. The variational formulation is to find $(\underset{\sim}{u}, p) \in (H_0^1(\Omega))^2 \times L_0^2(\Omega)$, such that

$$\begin{cases} (\alpha \nabla \underset{\sim}{u}, \nabla \underset{\sim}{v}) - (p\underset{\sim}{\mathrm{Id}}, \nabla \underset{\sim}{v}) = (\underset{\sim}{f}, \underset{\sim}{v}), \ \forall \ \underset{\sim}{v} \in (H_0^1(\Omega))^2, \\ \qquad (q, \mathrm{div}\, \underset{\sim}{u}) = 0, \qquad \forall \, q \in L_0^2(\Omega). \end{cases} \tag{12}$$

Let $\tilde{Q}_h$ be the space of piecewise constant with zero average, then the finite element problem is to find $(\underset{\sim}{u}_h, p_h) \in (V_{h0}^{\mathrm{CR}})^2 \times \tilde{Q}_h$, such that

$$\begin{cases} (\alpha \nabla_h \underset{\sim}{u}_h, \nabla_h \underset{\sim}{v}_h) - (p_h \underset{\sim}{\mathrm{Id}}, \nabla_h \underset{\sim}{v}_h) = (\underset{\sim}{f}, \underset{\sim}{v}_h), \ \forall \ \underset{\sim}{v}_h \in (V_{h0}^{\mathrm{CR}})^2, \\ \qquad (q_h, \nabla_h \cdot \underset{\sim}{u}_h) = 0, \qquad \forall \, q_h \in \tilde{Q}_h. \end{cases} \tag{13}$$

It is well known that, by the commutative property and the inf-sup condition of the model problem (12), the discrete inf-sup condition follows as $\sup_{\underset{\sim}{v}_h \in (V_{h0}^{\mathrm{CR}})^2} \dfrac{(q_h, \mathrm{div}_h \underset{\sim}{v}_h)}{\|q_h\|_{0,\Omega} \|\underset{\sim}{v}_h\|_{1,h}} \geqslant C$, for $q_h \in \tilde{Q}_h$. Note that the constant does not depend on the triangulation.

4.2 Accuracy Analysis

Theorem 2 *Let $(\underset{\sim}{u}, p)$ and $(\underset{\sim}{u}_h, p_h)$ be the solutions of (12) and (13), respectively. Assume $\underset{\sim}{u} \in (H^2(\Omega_1 \cup \Omega_2) \cap H_0^1(\Omega))^2$, and $p \in H^1(\Omega_1 \cup \Omega_2) \cap L_0^2(\Omega)$. Then it holds with a constant C independent of h that*

$$|\underset{\sim}{u} - \underset{\sim}{u}_h|_{1,h} + \|p - p_h\|_{0,\Omega} \leqslant Ch(\|\underset{\sim}{u}\|_{2,\Omega_1 \cup \Omega_2} + \|p\|_{1,\Omega_1 \cup \Omega_2} + \|\underset{\sim}{f}\|_{0,\Omega}). \tag{14}$$

Proof We start with this fundamental estimate [4]:

$$|\underset{\sim}{u} - \underset{\sim}{u}_h|_{1,h} + \|p - p_h\|_{0,\Omega} \lesssim \inf_{\underset{\sim}{v}_h \in (V_{h0}^{\mathrm{CR}})^2} |\underset{\sim}{u} - \underset{\sim}{v}_h|_{1,h} + \inf_{q_h \in \tilde{Q}_h} \|p - q_h\|_{0,\Omega}$$
$$+ \sup_{\underset{\sim}{w}_h \in (V_{h0}^{\mathrm{CR}})^2} \frac{(\alpha \nabla \underset{\sim}{u} - p\underset{\sim}{\mathrm{Id}}, \nabla \underset{\sim}{w}_h) - (\underset{\sim}{f}, \underset{\sim}{w}_h)}{|\underset{\sim}{w}_h|_{1,h}}.$$

By Lemma 3, we only have to estimate the consistency error. Since $\alpha\nabla \underset{\sim}{u} - p\underset{\sim}{\mathrm{Id}} \in (H(\mathrm{div};\Omega)\cap(H^1(\Omega_1\cup\Omega_2))^2)^2$, we can use the same technique as that of Theorem 1 and obtain

$$|(\alpha\nabla \underset{\sim}{u} - p\underset{\sim}{\mathrm{Id}}, \nabla \underset{\sim}{w}_h) - (\underset{\sim}{f}, \underset{\sim}{w}_h)| \leqslant Ch(\|\alpha\nabla \underset{\sim}{u} - p\underset{\sim}{\mathrm{Id}}\|_{1,\Omega_1\cup\Omega_2} + \|\underset{\sim}{f}\|_{0,\Omega})|\underset{\sim}{w}_h|_{1,h}. \quad (15)$$

Summing all above finishes the proof.

5 A Two-Level Geometric Multigrid Method

In this section, we consider the optimal solver of the finite element problem (6). Define $A_h : V_{h0} \to V_{h0}$ by $(A_h w_h, v_h) = a_h(w_h, v_h)$, for any $w_h, v_h \in V_{h0}$. In this section, $\tilde{\mathcal{T}}_h$ is a uniform grid with multilevel structure, and $\mathcal{T}_h$ is an interface-fitted grid generated from $\tilde{\mathcal{T}}_h$ by local operations near the interface by the algorithm in [7]. (See Fig. 1 for $\tilde{\mathcal{T}}_h$(left) and $\mathcal{T}_h$(right).) Particularly, $\tilde{\mathcal{T}}_h$ is shape regular, and $\mathcal{T}_h$ is maximal-angle-bounded. Let $\mathcal{N}_h$ and $\tilde{\mathcal{N}}_h$ be the sets of vertices of $\mathcal{T}_h$ and $\tilde{\mathcal{T}}_h$, respectively. Denote $\tilde{\mathcal{N}}_h^c := \mathcal{N}_h \setminus \tilde{\mathcal{N}}_h$.

5.1 Theory of Successive Subspace Correction Method

In this section we give some general result of the successive subspace correction method of solving on a linear vector space V with inner product $(\cdot,\cdot)$ the equation $(Au, v) = (f, v)$, where $A : V \to V$ is a symmetric positive definite operator. The presentation follows closely to [5, 11, 14, 15].

We decompose the space $V = \sum_{i=0}^{J} V_i$ as the summation of subspaces $V_i \subset V$. We do not assume the summation is a direct sum. The original problem associates sub-problems in each V_i with smaller size which are relatively easier to solve. We use the following operators, for $i = 0, 1, \ldots, J$:

- $Q_i : V \to V_i$ the projection in the inner product $(\cdot,\cdot)$;
- $I_i : V_i \to V$ the natural inclusion which is often called prolongation;
- $P_i : V \to V_i$ the projection in the inner product $(\cdot,\cdot)_A = (A\cdot,\cdot)$;
- $A_i : V_i \to V_i$ the restriction of A to the subspace V_i;
- $R_i : V_i \to V_i$ an approximation of A_i^{-1} (often known as smoother);
- $T_i : V \to V_i$, $T_i = R_i Q_i A = R_i A_i P_i$.

It is easy to verify $Q_i A = A_i P_i$ and $Q_i = I_i^t$ with $(I_i^t u, v_i) := (u, I_i v_i)$. The operator I_i^t is often called restriction. If $R_i = A_i^{-1}$, then we have an exact local solver and $T_i = P_i$. With slightly abused notation, we still use T_i to denote the restriction $T_i|_{V_i} : V_i \to V_i$ and $T_i^{-1} = (T_i|_{V_i})^{-1} : V_i \to V_i$.

The Successive Subspace Correction (SSC) method performs the correction in every subspace in a successive way. In operator form, it reads, given some approximation solution u^k,

$$v^0 = u^k, v^{i+1} = v^i + I_i R_i I_i^t (f - Av^i), i = 0, \ldots, J, u^{k+1} = v^{J+1}, \tag{16}$$

and the corresponding error equation is

$$u - u^{k+1} = \left[\prod_{i=0}^{J} (I - I_i R_i I_i^t A) \right] (u - u^k) = \left[\prod_{i=0}^{J} (I - T_i) \right] (u - u^k). \tag{17}$$

Here we assume there is a built-in ordering from $i = 0$ to J. The multiplicative multigrid method for finite element systems is a special SSC method with subspaces constructed by finite element functions on multilevel grids. For the convergence, we have this fundamental estimate.

Lemma 4 (X-Z identity for SSC) *If there is a $\rho < 1$, such that $\|I - T_i\|_{A_i} \leqslant \rho,\ i = 0, \ldots, J$, then it holds that*

$$\left\| \prod_{i=0}^{J} (I - T_i) \right\|_A^2 = 1 - \frac{1}{c_1}, \tag{18}$$

where

$$c_1 = \sup_{\|v\|_A = 1} \inf_{\sum_{i=0}^{J} v_i = v} \sum_{i=0}^{J} (\bar{T}_i^{-1} (v_i + T_i^* w_i), v_i + T_i^* w_i)_A, \tag{19}$$

with $w_i = \sum_{j>i} v_j$, and $\bar{T}_i = T_i + T_i^* - T_i^* T_i$, T_i^* the adjoint operator of T_i with respect to $(\cdot, \cdot)_A$.

Remark 1 If we perform a two-level method, and particularly, we perform an exact solver on a subspace V_0, then we have $c_1 = \sup_{\|v\|_A = 1} (\|P_0 v\|_A^2 + \|v - \Pi_h v\|_{\bar{R}_1^{-1}})$ where $P_0 : V \to V_0$ and $\Pi_h : V \to V_1$ are the projection operators with respect to $(\cdot, \cdot)_A$ and $(\cdot, \cdot)_{\bar{R}_1^{-1}}$, respectively, and $\bar{R}_1 = R_1^t + R_i - R_i^t A R_i$.

5.2 *An Optimal Multigrid Method for (6)*

Let $\tilde{V}_h^c \subset V_h$ be space of nodal basis functions that vanish on $\tilde{\mathcal{N}}_h$. Then $V_h = \tilde{V}_h \oplus \tilde{V}_h^c$, where $\tilde{V}_h$ is the linear element space on $\tilde{\mathcal{T}}_h$. Let $\tilde{I}_h$ be the nodal interpolation on $\tilde{V}_h$. Then $(I - \tilde{I}_h) V_h = \tilde{V}_h^c$ and $\tilde{I}_h V_h = \tilde{V}_h$. Let $\tilde{A}_h$ and $\tilde{A}_h^c$ be the restrictions of A_h on $\tilde{V}_{h0} := \tilde{V}_h \cap V_{h0}$ and $\tilde{V}_h^c$, respectively.

Lemma 5 *It holds for $w_h \in V_{h0}$ that $\|\tilde{I}_h w_h\|_{\tilde{A}_h} \leqslant \Lambda \|w_h\|_{A_h}$, with Λ a constant independent of h.*

Proof When h is sufficiently small, for any $p \in \tilde{N}_h$, there exists a segment e with p being one of its ends, such that e is an edge of $\tilde{\mathcal{T}}_h$ and $\mathcal{T}_h$ simultaneously, and thus $\tilde{I}_h w_h = w_h$ on e. Therefore, by the standard technique alike to the stability of Scott-Zhang operator [9] and a Scott-Zhang type operator [5], we have $|\tilde{I}_h w_h|_{1,\Omega} \leqslant C|w_h|_{1,\Omega}$ with C depending on the shape regularity of $\tilde{\mathcal{T}}_h$ only. This finishes the proof.

Let $\tilde{R}_h : \tilde{V}_{h0} \to \tilde{V}_{h0}$ be approximately an inverse of $\tilde{A}_h$. We have this two-level successive subspace correction method (Algorithm 1).

Algorithm 1 *Implement this iterative procedure until converge:*

1. *do subspace correction on $\tilde{V}_{h0}$ with an inexact solver $\tilde{R}_h$;*
2. *do subspace correction on $\tilde{V}_h^c$ with an exact solver $(\tilde{A}_h^c)^{-1}$.*

Obviously, Algorithm 1 defines an iterative method for solving $A_h u_h = f_h$. Let $\tilde{P}_h^c$ and $\tilde{Q}_h$ be the projection operator onto $\tilde{V}_h^c$ and $\tilde{V}_{h0}$ with respect to $a_h(\cdot,\cdot)$ and $(\cdot,\cdot)$, respectively. Denote by B_h the iterator of the method. Then the error contract operator on V_{h0} is $I - B_h A_h = (I - \tilde{P}_h^c)(I - \tilde{R}_h \tilde{Q}_h A_h)$.

Theorem 3 *Assume that $\|I - \tilde{R}_h \tilde{A}_h\|_{\tilde{A}_h} \leqslant \rho < 1$. Then Algorithm 1 is uniformly convergent with respect to the mesh size with*

$$\|I - B_h A_h\|_{A_h}^2 \leqslant \frac{\Lambda}{1 - \rho^2 + \Lambda}.$$

Proof By the X-Z identity for the successive subspace correction method, (c.f., e.g., [14]) we have

$$\|I - B_h A_h\|_{A_h}^2 = 1 - \frac{1}{c_1},$$

with

$$c_1 = \sup_{v_h \in V_{h0}, \|v_h\|_{A_h} = 1} \Big(\|\tilde{P}_h^c v_h\|_{A_h}^2 + \inf_{\tilde{v}_h \in \tilde{V}_h^c, v_h - \tilde{v}_h \in \tilde{V}_h} ((\tilde{R}_h^t + \tilde{R}_h - \tilde{R}_h^t \tilde{A}_h \tilde{R}_h)^{-1}(v - \tilde{v}_h), (v - \tilde{v}_h)) \Big).$$

Since $\|I-\tilde{R}_h\tilde{A}_h\|_{\tilde{A}_h} \leqslant \rho < 1$, we have $\|I-(\tilde{R}_h^t+\tilde{R}_h-\tilde{R}_h^t\tilde{A}_h\tilde{R}_h)\|_{\tilde{A}_h} \leqslant \|I-\tilde{R}^t\tilde{A}_h\|_{\tilde{A}_h}\|I-\tilde{R}\tilde{A}_h\|_{\tilde{A}_h} \leqslant \rho^2$, and thus $\lambda_{\max}((\tilde{R}_h^t\tilde{A}_h+\tilde{R}_h\tilde{A}_h-\tilde{R}_h^t\tilde{A}_h\tilde{R}_h\tilde{A}_h)^{-1}) \leqslant \frac{1}{1-\rho^2}$. Therefore

$$
\begin{aligned}
&((\tilde{R}_h^t+\tilde{R}_h-\tilde{R}_h^t\tilde{A}_h\tilde{R}_h)^{-1}(v_h-\tilde{v}_h),(v_h-\tilde{v}_h))\\
&=((\tilde{R}_h^t\tilde{A}_h+\tilde{R}_h\tilde{A}_h-\tilde{R}_h^t\tilde{A}_h\tilde{R}_h\tilde{A}_h)^{-1}(v_h-\tilde{v}_h),\tilde{A}_h(v_h-\tilde{v}_h))\\
&\leqslant \frac{1}{1-\rho^2}(v_h-\tilde{v}_h,\tilde{A}_h(v_h-\tilde{v}_h)).
\end{aligned}
$$

Since evidently $\|\tilde{P}_h^c v_h\|_{A_h} \leqslant \|v_h\|_{A_h}$, we have $c_1 \leqslant \sup\limits_{v_h\in V_h,\|v_h\|_{A_h}=1}(1+\frac{1}{1-\rho^2}\inf\limits_{\tilde{v}_h\in\tilde{V}_h^c,v_h-\tilde{v}_h\in\tilde{V}_h}\|v_h-\tilde{v}_h\|_{\tilde{A}_h}^2)$. Then by Lemma 5, we have $c_1 \leqslant 1+\frac{\Lambda}{1-\rho^2}$, and finally obtain $\|I-B_hA_h\|_{A_h}^2 \leqslant \frac{\Lambda}{1-\rho^2+\Lambda}$.

When $\tilde{\mathcal{T}}_h$ is a shape-regular grid with a geometrical multilevel structure, then a geometric multigrid process can be implemented on $\tilde{V}_{h0}$, and the approximate inverse $\tilde{R}_h$ of $\tilde{A}_h$ can be chosen to be the iterator of V-cycle multigrid method. The assumption of Theorem 3 holds (see [11, 13, 14]).

5.3 Numerical Examples

To test the numerical methods, we consider the following example. Let the interface Γ be a circle centered at the origin with radius r_0. Let the exact solution be $u(x) = u(\mathbf{r}) = 2\mathbf{r}^4 + |\mathbf{r}^4 - r_0^4|$, where $\mathbf{r} = dist(x, \mathbf{0})$. Moreover, we choose $\alpha(x) = 1$ if $\mathbf{r} > r_0$ and $\alpha(x) = 3$ if $\mathbf{r} < r_0$, and the right hand side can be computed accordingly. Hereafter we set $r_0 = 0.6$.

We implement Algorithm 1, with $V(1,1)$ cycle geometric multigrid based on the original unfitted grid playing as the coarse grid corrector. We record the numerical results in Table 1. In these examples, the initial guess is $\mathbf{0}$, and the stopping criterion is the l^2 norm of the relative residual being smaller than 10^{-10}. From Table 1, we can see that the multigrid method converges uniformly with respect to the mesh size, which confirms our theoretical results.

Table 1 Numerical performance of Algorithm 1

h	2^{-4}	2^{-5}	2^{-6}	2^{-7}	2^{-8}	2^{-9}	2^{-10}
#iter	14	13	13	13	13	13	13

6 Concluding Remarks

In this paper, we discussed the optimal finite element method for the interface boundary value problem of the diffusion equation and the Stokes equation. We proved that the linear Crouzeix-Raviart element schemes provide optimal accuracy with respect to the mesh size for the two interface boundary value problems on grids that are interface-fitted and maximal-angle-bounded.

Given a uniform grid, an interface-fitted and maximal-angle-bounded grid can be generated by some local operations close to the interface. On the grids generated that way, we discussed the optimal multigrid method of the discrete linear systems. We took the conforming linear element system, the theory of which is less complicated, for a demonstration, and show that by the methodology of using the original grid as a coarse grid and reinforcing the smoothing effect near the interface, we obtain an optimal multigrid method.

Some other optimal finite element methods and their optimal multigrid solvers for interface boundary value problems will be discussed in the future works.

Acknowledgement The authors would like to thank Dr. Xiaozhe Hu for his help on the numerical examples.

References

1. G. Acosta, R.G. Durán, The maximum angle condition for mixed and nonconforming elements: application to the Stokes equations. SIAM J. Numer. Anal. **37**(1), 18–36 (1999) (electronic). ISSN 0036-1429. doi:10.1137/S0036142997331293. http://dx.doi.org/10.1137/S0036142997331293
2. J.M. Arrieta, A. Rodríguez-Bernal, J.D. Rossi, The best Sobolev trace constant as limit of the usual Sobolev constant for small strips near the boundary. Proc. R. Soc. Edinb. Sect. A **138**(2), 223–237 (2008). ISSN 0308-2105. doi:10.1017/S0308210506000813. http://dx.doi.org/10.1017/S0308210506000813
3. J.H. Bramble, J.T. King, A finite element method for interface problems in domains with smooth boundaries and interfaces. Adv. Comput. Math. **6**(2), 109–138 (1996). ISSN 1019-7168. doi:10.1007/BF02127700. http://dx.doi.org/10.1007/BF02127700
4. F. Brezzi, M. Fortin, *Mixed and Hybrid Finite Element Methods*. Springer Series in Computational Mathematics, vol. 15 (Springer, New York, 1991). ISBN 0-387-97582-9. doi:10.1007/978-1-4612-3172-1. http://dx.doi.org/10.1007/978-1-4612-3172-1
5. L. Chen, R.H. Nochetto, J. Xu, Optimal multilevel methods for graded bisection grids. Numer. Math. **120**(1), 1–34 (2012). ISSN 0029-599X. doi:10.1007/s00211-011-0401-4. http://dx.doi.org/10.1007/s00211-011-0401-4
6. Z. Chen, J. Zou, Finite element methods and their convergence for elliptic and parabolic interface problems. Numer. Math. **79**(2), 175–202 (1998). ISSN 0029-599X. doi:10.1007/s002110050336. http://dx.doi.org/10.1007/s002110050336.
7. Z. Chen, Z. Wu, Y. Xiao, An adaptive immersed finite element method with arbitrary lagrangian-eulerian scheme for parabolic equations in variable domains. Int J. Numer. Anal. Model. **12**(3), 567–591 (2015)

8. J. Li, J.M. Melenk, B. Wohlmuth, J. Zou, Optimal a priori estimates for higher order finite elements for elliptic interface problems. Appl. Numer. Math. **60**(1–2), 19–37 (2010). ISSN 0168-9274. doi:10.1016/j.apnum.2009.08.005. http://dx.doi.org/10.1016/j.apnum.2009.08.005
9. L.R. Scott, S. Zhang, Finite element interpolation of nonsmooth functions satisfying boundary conditions. Math. Comput. **54**(190), 483–493 (1990). ISSN 0025-5718. doi:10.2307/2008497. http://dx.doi.org/10.2307/2008497
10. J. Xu, Estimate of the convergence rate of finite element solutions to elliptic equations of second order with discontinuous coefficients (in chinese). *Natural Science Journal of Xiangtan University* (1982), pp. 1–5
11. J. Xu, Iterative methods by space decomposition and subspace correction. SIAM Rev. **34**(4), 581–613 (1992) ISSN 0036-1445. doi:10.1137/1034116. http://dx.doi.org/10.1137/1034116
12. J. Xu, Estimate of the convergence rate of finite element solutions to elliptic equations of second order with discontinuous coefficients. arXiv preprint arXiv:1311.4178 (2013)
13. J. Xu, Y. Zhu, Uniform convergent multigrid methods for elliptic problems with strongly discontinuous coefficients. Math. Models Methods Appl. Sci. **18**(1), 77–105 (2008) ISSN 0218-2025. doi:10.1142/S0218202508002619. http://dx.doi.org/10.1142/S0218202508002619
14. J. Xu, L. Zikatanov, The method of alternating projections and the method of subspace corrections in Hilbert space. J. Am. Math. Soc. **15**(3), 573–597 (2002). ISSN 0894-0347. doi:10.1090/S0894-0347-02-00398-3. http://dx.doi.org/10.1090/S0894-0347-02-00398-3
15. J. Xu, L. Chen, R.H. Nochetto, Optimal multilevel methods for H(grad), H(curl), and H(div) systems on graded and unstructured grids, in *Multiscale, Nonlinear and Adaptive Approximation* (Springer, Berlin, 2009), pp. 599–659. doi:10.1007/978-3-642-03413-8_14. http://dx.doi.org/10.1007/978-3-642-03413-8_14

BDDC Deluxe Domain Decomposition

Olof B. Widlund and Clark R. Dohrmann

1 Introduction

We will consider BDDC domain decomposition algorithms for finite element approximations of a variety of elliptic problems. The BDDC (Balancing Domain Decomposition by Constraints) algorithms were introduced by Dohrmann [5], following the introduction of FETI-DP by Farhat et al. [9]. These two families are closely related algorithmically and have a common theory. The design of a BDDC algorithm involves the choice of a set of *primal degrees of freedom* and the choice of an *averaging operator*, which restores the continuity of the approximate solution across the interface between the subdomains into which the domain of the given problem has been partitioned. We will also refer to these operators as *scalings*.

This paper principally address the latter choice. All our efforts aim at developing effective *preconditioners* for the stiffness matrices. These approximate inverses are then combined with the conjugate gradient method. We are primarily interested in hard problems with very many subdomains and to obtain convergence rates independent of that number and with rates that deteriorate slowly with the size of the subdomain problems. Our bounds can often be made independent of jumps in the coefficients between subdomains and our numerical results indicate that our new BDDC deluxe algorithm is quite promising and robust.

Among our applications are problems formulated in $H(\mathbf{curl})$, $H(\mathrm{div})$, and for Reissner-Mindlin plates. We have worked mostly with the lowest order finite

O.B. Widlund (✉)
Courant Institute, 251 Mercer Street, New York, NY 10012, USA
e-mail: widlund@cims.nyu.edu

C.R. Dohrmann
Sandia National Laboratories, Albuquerque, NM, USA
e-mail: crdohrm@sandia.gov

T. Dickopf et al. (eds.), *Domain Decomposition Methods in Science and Engineering XXII*, Lecture Notes in Computational Science and Engineering 104, DOI 10.1007/978-3-319-18827-0_8

element methods for self-adjoint elliptic problems but we have also helped develop solvers for isogeometric analysis. After introducing the general ideas, we will focus on our recent work on three-dimensional problems in $H(\mathbf{curl})$, see [7], since other applications are discussed in other papers of this volume or elsewhere; cf. [1–4, 8, 12–14].

2 BDDC, Finite Element Meshes, and Equivalence Classes

The BDDC algorithms work with decompositions of the domain Ω of the elliptic problem into non-overlapping subdomains Ω_i, each often with tens of thousands of degrees of freedom. In-between the subdomains is the interface Γ, which does not cut through any elements. The local interface of Ω_i is defined by $\Gamma_i := \partial\Omega_i \setminus \partial\Omega$.

For nodal finite element methods, most nodes are typically interior to individual subdomains while the others belong to several subdomain interfaces or to the boundary of the given region. We partition the nodes on Γ into equivalence classes determined by the set of indices of the local interfaces Γ_j to which they belong. In three dimensions, we have equivalence classes of face nodes, associated with two local interfaces, and classes of edge nodes and subdomain vertex nodes typically associated with more than two. For $H(\mathbf{curl})$ and Nédélec (edge) elements, there are only equivalence classes of element edges for subdomain faces and for subdomain edges. For $H(\text{div})$ and Raviart-Thomas elements, we only have degrees of freedom for element faces and the only equivalence classes are associated with the subdomain faces. These equivalence classes play a central role in the design, analysis, and programming of domain decomposition methods.

These preconditioners are constructed using partially subassembled stiffness matrices built from the subdomain stiffness matrices $A^{(i)}$ of the subdomains Ω_i, $i = 1, \ldots, N$. We will first consider nodal finite element problems. The nodes of $\Omega_i \cup \Gamma_i$ are divided into those in the interior (I) and those on the interface (Γ). The interface set is further divided into a primal set (Π) and a dual set (Δ).

We can then represent the subdomain stiffness matrix, of Ω_i, as

$$A^{(i)} = \begin{pmatrix} A_{II}^{(i)} & A_{I\Delta}^{(i)} & A_{I\Pi}^{(i)} \\ A_{\Delta I}^{(i)} & A_{\Delta\Delta}^{(i)} & A_{\Delta\Pi}^{(i)} \\ A_{\Pi I}^{(i)} & A_{\Pi\Delta}^{(i)} & A_{\Pi\Pi}^{(i)} \end{pmatrix}.$$

This matrix represents the stiffness contributed by Ω_i. Throughout the iteration, we enforce continuity of the primal variables, as in the given finite element model, but allow multiple values of the dual variables when working with a *partially subassembled* model as in Fig. 1. Partially subassembling the subdomain matrices

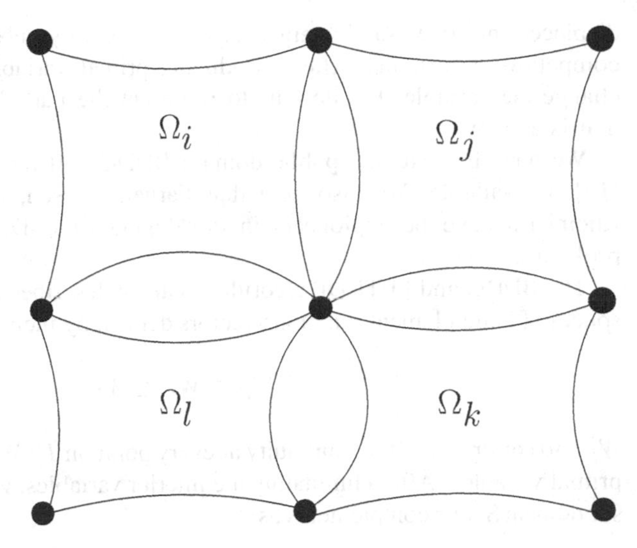

Fig. 1 Torn 2D scalar elliptic problem; primal variables at subdomain vertices only

and noting that the matrix $\tilde{A}_{\Pi\Pi}$ is assembled from the submatrices $A^{(i)}_{\Pi\Pi}$, we obtain

$$\tilde{A} = \begin{pmatrix} A^{(1)}_{II} & A^{(1)}_{I\Delta} & & & & A^{(1)}_{I\Pi} \\ A^{(1)}_{\Delta I} & A^{(1)}_{\Delta\Delta} & & & & A^{(1)}_{\Delta\Pi} \\ & & \ddots & & & \vdots \\ & & & A^{(N)}_{II} & A^{(N)}_{I\Delta} & A^{(N)}_{I\Pi} \\ & & & A^{(N)}_{\Delta I} & A^{(N)}_{\Delta\Delta} & A^{(N)}_{\Delta\Pi} \\ A^{(1)}_{\Pi I} & A^{(1)}_{\Pi\Delta} & \cdots & A^{(N)}_{\Pi I} & A^{(N)}_{\Pi\Delta} & \tilde{A}_{\Pi\Pi} \end{pmatrix}.$$

This partially subassembled stiffness matrix of this alternative finite element model is an important component of the BDDC preconditioners. The primal variables provide a necessary, global component of the preconditioners and they make the partially assembled matrix invertible.

Solving a linear system with the matrix $\tilde{A}$ is much cheaper than when using the fully assembled model but results in multiple values of the dual interface variables. When using BDDC, we therefore restore the continuity of the original finite element problem by averaging across the interface. When using FETI-DP, we instead employ Lagrange multipliers.

For scalar second order elliptic equations in the plane, as in Fig. 1, the approach outlined yields a condition number bound of $C(1 + \log(H/h))^2$, where $H/h := \max_i(H_i/h_i)$ with H_i the diameter of Ω_i and h_i that of the smallest of the elements of Ω_i. These results can be made independent of jumps in the coefficients, if the interface averages are chosen carefully, but for three dimensions the primal set of variables should include averages (and possibly first moments) of the displacements over subdomain edges (and possibly subdomain faces) to obtain competitive algorithms. After introducing primal variables of this type, we can change the variables to allow us to represent the partially subassembled system matrix as above.

We note that parallel, public domain BDDC software, developed by Zampini [17], is available. We also note that Farhat, Pierson, et al. and Klawonn and Rheinbach have been pioneers in developing FETI-DP software for elasticity problems.

The BDDC and FETI-DP algorithms can be described in terms of three product spaces of finite element functions/vectors defined by their interface nodal values:

$$\hat{W}_\Gamma \subset \tilde{W}_\Gamma \subset W_\Gamma.$$

W_Γ: no constraints; $\hat{W}_\Gamma$: continuity at every point on Γ; $\tilde{W}_\Gamma$: common values of the primal variables. After eliminating the interior variables, we can write the resulting subdomain Schur complements as

$$S^{(i)} := \begin{pmatrix} S^{(i)}_{\Delta\Delta} & S^{(i)}_{\Delta\Pi} \\ S^{(i)}_{\Pi\Delta} & S^{(i)}_{\Pi\Pi} \end{pmatrix} := \begin{pmatrix} A^{(i)}_{\Delta\Delta} & A^{(i)}_{\Delta\Pi} \\ A^{(i)}_{\Pi\Delta} & A^{(i)}_{\Pi\Pi} \end{pmatrix} - \begin{pmatrix} A^{(i)}_{\Delta I} \\ A^{(i)}_{\Pi I} \end{pmatrix} \left(A^{(i)}_{II}\right)^{-1} \left(A^{(i)}_{I\Delta}\, A^{(i)}_{I\Pi}\right).$$

By partially subassembling the $S^{(i)}$, we obtain $\tilde{S}$.

Let us denote the BDDC averaging operator, which maps $\tilde{W}_\Gamma$ into $\hat{W}_\Gamma$, by E_D. In each BDDC iteration, we first compute the residual of the fully assembled Schur complement equation. We then apply E_D^T to obtain a right-hand side for the partially subassembled linear system. We solve this system and then apply E_D. In the conventional BDDC algorithms the averaging across the interface is done pointwise and that leads to non-zero residuals at the nodes next to Γ. In each iteration, subdomain Dirichlet solves are then used to eliminate them, but in the deluxe variant this step is not needed. The iteration is accelerated by using a preconditioned conjugate gradient method.

The core of any theory for BDDC algorithms is the norm of the average operator E_D. By an algebraic argument known, for FETI-DP, since the publication of [11, Proof of Theorem 1], we have

$$\kappa(M^{-1}A) \leq \|E_D\|_{\tilde{S}},$$

which then provides an upper bound for the number of iterations required of the preconditioned conjugate gradient method; for details on the BDDC case, see, e.g., [1]. Here M^{-1} represents the action of the preconditioner.

3 The New Algorithmic Idea

When designing a BDDC algorithm, we have to choose an effective set of primal constraints and also a good recipe for the averaging across the interface. Traditional averaging recipes were found not to work uniformly well for three dimensional problems in $H(\mathbf{curl})$; see [6]. This is directly related to the fact that there are two material parameters. An alternative was found and will be outlined in this section. It has also proven to be very robust for $H(\text{div})$ problems, see [14], and for Reissner-Mindlin plates, see [12].

We note that experimentally, the condition numbers are often quite small. For Reissner-Mindlin plates, in Lee's experiments, they are ≤ 4 while without preconditioning the condition numbers can exceed 10^{11} for very thin plates with the parameter $t = 10^{-5}$. The results in the $H(\text{div})$-study are quite similar and experiments with the deluxe version of BDDC for isogeometric analysis show considerable improvement over older variants.

Across a subdomain face $F \subset \Gamma$, common to two subdomains Ω_i and Ω_j, the deluxe E_D is defined in terms of two Schur complements, which are principal minors of $S^{(i)}$ and $S^{(j)}$, respectively:

$$S_F^{(k)} := A_{FF}^{(k)} - A_{FI}^{(k)} A_{II}^{(k)^{-1}} A_{IF}^{(k)}, \; k = i, j.$$

The face contribution of the deluxe averaging operator is then defined by

$$\bar{w}_F := (E_D w)_F := (S_F^{(i)} + S_F^{(j)})^{-1} (S_F^{(i)} w_F^{(i)} + S_F^{(j)} w_F^{(j)}).$$

This action of this component of E_D can be implemented by solving a Dirichlet problem on $\Omega_i \cup F \cup \Omega_j$. This local problem is larger than those of the conventional BDDC algorithms, and we are currently exploring the effects of using cheaper, inexact solvers for these subproblems.

Similar formulas can also be written down for subdomain edges and other equivalence classes of interface variables. The operator E_D is assembled from these components.

We will now show that the analysis of BDDC deluxe can be reduced to bounds for individual subdomains. Arbitrary jumps in two coefficients can then often be well accommodated. We also note that the analysis of traditional BDDC algorithms requires an extension theorem; the deluxe version does not.

Instead of estimating $(R_F^T \bar{w}_F)^T S^{(i)} R_F^T \bar{w}_F$, where the restriction operator R_F maps the values on Γ onto those on F, we will work with the norm of $R_F^T(w_F^{(i)} - \bar{w}_F)$. Thus, instead of estimating the norm of E_D, we will estimate the norm of $I - E_D$; a bound on the norm of E_D will, as we previously have noted, give a bound on the condition number.

We easily find that

$$w_F^{(i)} - \bar{w}_F = (S_F^{(i)} + S_F^{(j)})^{-1} S_F^{(j)} (w_F^{(i)} - w_F^{(j)}).$$

By some more algebra and noting that $R_F S^{(i)} R_F^T = S_F^{(i)}$, we find that

$$(R_F^T(w_F^{(i)} - \bar{w}_F))^T S^{(i)} (R_F^T(w_F^{(i)} - \bar{w}_F)) =$$
$$(w_F^{(i)} - w_F^{(j)})^T S_F^{(j)} (S_F^{(i)} + S_F^{(j)})^{-1} S_F^{(i)} (S_F^{(i)} + S_F^{(j)})^{-1} S_F^{(j)} (w_F^{(i)} - w_F^{(j)}).$$

We now add the corresponding expression for the subdomain Ω_j and, after a simplification, find that this sum can be written as

$$(w_F^{(i)} - w_F^{(j)})^T (S_F^{(i)-1} + S_F^{(j)-1})^{-1} (w_F^{(i)} - w_F^{(j)}).$$

We then find that, for any element w_Π in the primal space,

$$(R_F^T(w_F^{(i)} - \bar{w}_F))^T S^{(i)} R_F^T(w_F^{(i)} - \bar{w}_F) + (R_F^T(w_F^{(j)} - \bar{w}_F))^T S^{(j)} R_F^T(w_F^{(j)} - \bar{w}_F)$$
$$\leq 2(w_F^{(i)} - R_F w_\Pi)^T S_F^{(i)} (w_F^{(i)} - R_F w_\Pi) + 2(w_F^{(j)} - R_F w_\Pi)^T S_F^{(j)} (w_F^{(j)} - R_F w_\Pi).$$

Each of the two terms on the right hand side are local to only one subdomain.

For the subdomain faces, what now remains is to estimate, after a suitable shift w_Π, $(w_F^{(i)} - R_F w_\Pi)^T S_F^{(i)} (w_F^{(i)} - R_F w_\Pi)$ by $w^{(i)^T} S^{(i)} w^{(i)}$. This is routine for $H^1(\Omega_i)$ using standard estimates in the domain decomposition literature such as a *face lemma* [16, Lemma 4.24] in which we estimate the energy of the extension of the face values by zero to the rest of Γ_i with that of the minimal energy extension. A factor of $C(1 + \log(H/h))^2$ results. For $H^1(\Omega_i)$, all these estimates have been available for 20 years. But for $H(\mathrm{div})$ and $H(\mathbf{curl})$, new tools have been required.

Similar estimates are required for subdomain edges. Let R_E be the restriction matrix which maps the values on Γ onto those on a subdomain edge E. If this edge is common to three subdomains Ω_i, Ω_j, and Ω_k, the edge average $\bar{w}_E$ is defined by

$$\bar{w}_E := (S_E^{(i)} + S_E^{(j)} + S_E^{(k)})^{-1} (S_E^{(i)} w_E^{(i)} + S_E^{(j)} w_E^{(j)} + S_E^{(k)} w_E^{(k)}).$$

Here $S_E^{(i)} := R_E S^{(i)} R_E^T$, etc. We can show that,

$$(R_E^T(w_E^{(i)} - \bar{w}_E))^T \, S^{(i)} \, R_E^T(w_E^{(i)} - \bar{w}_E) \leq$$
$$3(w_E^{(i)})^T S_E^{(j)} w_E^{(i)} + 3/4 (w_E^{(j)})^T \, S_E^{(j)} \, w_E^{(j)} + 3/4 (w_E^{(k)})^T S_E^{(k)} w_E^{(k)}.$$

Other bounds, e.g., with a shift with an element of the primal space, can also be developed, but the one given here has proven of use in our work on problems in $H(\mathbf{curl})$. We can also develop similar bounds for any edge, common to more than three subdomains, using the same kinds of arguments.

4 H(curl) Problems in Three Dimensions

Consider the variational problem: Find $\mathbf{u} \in H_0(\mathbf{curl}; \Omega)$ such that

$$a(\mathbf{u}, \mathbf{v})_\Omega = (\mathbf{f}, \mathbf{v})_\Omega \quad \forall \mathbf{v} \in H_0(\mathbf{curl}; \Omega),$$

where $\mathbf{u} \times \mathbf{n} = 0$ on $\partial\Omega$ and where

$$a(\mathbf{u}, \mathbf{v})_\Omega := \int_\Omega [\alpha \nabla \times \mathbf{u} \cdot \nabla \times \mathbf{v} + \beta \mathbf{u} \cdot \mathbf{v}]\, dx, \quad (\mathbf{f}, \mathbf{v})_\Omega := \int_\Omega \mathbf{f} \cdot \mathbf{v}\, dx.$$

Here, $\alpha(x) \geq 0$ and $\beta(x)$ strictly positive. For coefficients constant in each subdomain, we have

$$a(\mathbf{u}, \mathbf{v})_\Omega = \sum_{i=1}^{N} (\alpha_i (\nabla \times \mathbf{u}, \nabla \times \mathbf{v})_{\Omega_i} + \beta_i (\mathbf{u}, \mathbf{v})_{\Omega_i}).$$

In our work, there are two relevant finite element spaces, namely $W^{h_i}_{\text{curl}}$ of the lowest order triangular Nédeléc elements and $W^{h_i}_{\text{grad}}$ of the standard piecewise linear, continuous elements, on the same triangulation.

The space of Nédélec finite element functions, $W^{h_i}_{\text{curl}}$, can be represented as the range of an interpolation operator Π^h which is well defined, for sufficiently smooth elements of $\mathbf{w} \in H(\mathbf{curl}, \Omega)$, by

$$\Pi^h(\mathbf{w}) := \sum_e \lambda_e(\mathbf{w}) \mathbf{N}_e \ \textit{where}\ \lambda_e(\mathbf{w}) := \frac{1}{|e|} \int_e \mathbf{w} \cdot \mathbf{t}_e ds.$$

Here $\mathbf{t}_e$ is a unit vector in the direction of the element edge e and $\mathbf{N}_e$ the standard Nédélec basis function.

We have been able to build on the work by Toselli [15]. Thus, for Nédélec elements, the use of the basis based on $\{\mathbf{N}_e\}$ results in a poor result since the coupling between the subdomain faces and edges is far too strong. Following Toselli, we change the variables associated with the subdomain edges using a constant along each subdomain edge and the gradient of the standard $W^{h_i}_{\text{grad}}$ basis functions for all the interior nodes of the subdomain edges. After this change of variables, a quite stable decomposition can be found.

Domain decomposition theory always involves establishing the stability of a decomposition; in our context, a new auxiliary result is then needed:

Lemma 1 *Let F be a face of a polyhedral subdomain Ω_i. Further, let $f_{\partial F} \in W^{h_i}_{grad}(\Omega_i)$ have vanishing nodal values everywhere in Ω_i except on ∂F. There then exists a $\mathbf{g}_{iF} \in W^{h_i}_{curl}(\Omega_i)$ such that $\lambda_e(\mathbf{g}_{iF}) = \lambda_e(\nabla f_{\partial F})$ for all element edges in the* *interior of the face F, $\lambda_e(\mathbf{g}_{iF}) = 0$ for all other element edges on $\partial\Omega_i$, and*

$$\|\mathbf{g}_{iF}\|^2_{L^2(\Omega_i)} \le C((1+\log(H_i/h_i))\|f_{\partial F}\|^2_{L^2(\partial F)} + H_i^2\|\nabla f_{\partial F}\cdot \mathbf{t}_{\partial F}\|^2_{L^2(\partial F)}),$$
$$\|\nabla \times \mathbf{g}_{iF}\|^2_{L^2(\Omega_i)} \le C(1+\log(H_i/h_i))\|\nabla f_{\partial F}\cdot \mathbf{t}_{\partial F}\|^2_{L^2(\partial F)}.$$

For a proof of this result, see [7, Lemma 3.5]. We also use several standard auxiliary results for $W^{h_i}_{\text{grad}}$ as collected in [16, Subsection 4.6].

A key to our work is also a result by Hiptmair and Xu [10, Lemma 5.1]:

Lemma 2 *For any polyhedral subdomain Ω_i and any $\mathbf{u}_h \in W^{h_i}_{curl}(\Omega_i)$, there exist $\mathbf{\Psi}_h \in (W^{h_i}_{grad}(\Omega_i))^3$, $p_h \in W^{h_i}_{grad}(\Omega_i)$, and $\mathbf{q}_h \in W^{h_i}_{curl}(\Omega_i)$, such that*

$$\mathbf{w}_i = \mathbf{q}_i + \Pi^{h_i}(\mathbf{\Psi}_i) + \nabla p_i,$$
$$\|\nabla p_i\|^2_{L^2(\Omega_i)} \le C(\|\mathbf{w}_i\|^2_{L^2(\Omega_i)} + H_i^2\|\nabla\times\mathbf{w}_i\|^2_{L^2(\Omega_i)}),$$
$$\|h_i^{-1}\mathbf{q}_i\|^2_{L^2(\Omega_i)} + \|\mathbf{\Psi}_i\|^2_{H^1(\Omega_i)} \le C\|\nabla\times\mathbf{w}_i\|^2_{L^2(\Omega_i)}.$$

We note that these bounds are local and that the result has been established for polyhedra which are not necessarily convex.

This result is essential to Hiptmair and Xu's work on algebraic multigrid algorithms for $H(\mathbf{curl})$ in which AMG Poisson solvers are used.

In contrast to earlier results on domain decomposition algorithms for $H(\mathbf{curl})$, we do not have to rely on any trace theorem in our proof.

Toselli primarily advocates the use of two primal variables for each subdomain edge: the average and first moment and so do we. We have improved Toselli's condition number bound from

$$C\max_i(1+\frac{\beta_i H_i^2}{\alpha_i})(1+\log(H_i/h_i))^4$$

to an estimate, with the best possible power of $(1+\log(H_i/h_i))$:

$$C\max_i \min((H_i/h_i)^2, (1+\frac{\beta_i H_i^2}{\alpha_i}))(1+\log(H_i/h_i))^2.$$

We have fewer restrictions on the coefficients than Toselli; our constant C is independent of the α_i and β_i.

So far, we have not mastered the case where $\frac{\beta_i H_i^2}{\alpha_i}$ is large. We note that for $H(\text{div})$, one simple primal space works well in all cases; not so for $H(\mathbf{curl})$.

5 Numerical Results

Numerical results are presented in this section, which confirm the theory and demonstrate that in certain cases, the deluxe BDDC algorithm is much more robust than older BDDC variants. In our tables *iter* and *cond* denote the number of iterations and the estimated condition numbers obtained using a relative tolerance of 10^{-8} of the ℓ_2-norm of the residual as a stopping criterion. The subdomain problems are discretized using the lowest order hexahedral edge elements, for which our theory is equally valid.

In the first example, a unit cube is subdivided into N_d^3 smaller cubes, which are each subdivided into $64 = 4^3$ elements. Table 1 illustrates that the rate of convergence is independent of the number of subdomains.

In the next set of experiments, we study the behavior of our algorithm for increasing values of H/h, the number of elements across each subdomain. We note a much more rapid growth of the condition number for the mass-dominated cases, i.e., with $\beta_i H_i^2 >> \alpha_i$, represented by the first column of Table 2.

In our final table, Table 3, we consider a case of a three-dimensional checkerboard arrangement of the material parameters with $\alpha_i = 10^4$, $\beta_i = 10^{-2}$ for the red subdomains and $\alpha_i = 10^2$, $\beta_i = 1$ for the black. We indeed find a considerable improvement for the deluxe variant over two standard scalings. In the final columns,

Table 1 Results for unit cube decomposed into smaller cubical subdomains with $H/h = 4$

	$\alpha = 10^{-4}$		$\alpha = 10^{-2}$		$\alpha = 1$		$\alpha = 10^2$		$\alpha = 10^4$	
N_d	Iter	Cond	Iter	Cond	Iter	Cond	Iter	Cond	Iter	Cond
2	9	2.49	8	1.59	10	1.99	10	2.03	10	2.03
4	12	2.36	10	1.79	14	2.63	15	2.70	16	2.70
6	11	2.12	12	2.07	15	2.81	16	2.88	17	2.88
8	11	2.02	13	2.25	15	2.87	16	2.95	17	2.95
10	11	1.97	13	2.35	16	2.91	17	2.98	18	2.98
12	11	1.92	14	2.44	16	2.93	17	2.99	18	2.99

The material properties are constant with $\alpha_i = \alpha$ and $\beta_i = 1$

Table 2 Results for unit cube decomposed into 27 smaller cubical subdomains

	$\alpha = 10^{-7}$		$\alpha = 10^{-2}$		$\alpha = 1$		$\alpha = 10^2$		$\alpha = 10^4$	
H/h	Iter	Cond	Iter	Cond	Iter	Cond	Iter	Cond	Iter	Cond
4	12	2.74	9	1.63	13	2.41	13	2.47	14	2.47
6	15	4.51	12	2.15	14	2.93	15	3.01	16	3.01
8	19	6.89	14	2.70	16	3.34	17	3.44	18	3.44
10	22	9.98	15	3.22	17	3.69	18	3.79	19	3.79
12	24	13.8	16	3.69	17	3.98	19	4.09	20	4.10
14	28	18.3	17	4.13	18	4.24	19	4.36	21	4.36
16	30	23.5	18	4.55	19	4.47	20	4.60	22	4.60

The material properties are constant with $\alpha_i = \alpha$ and $\beta_i = 1$

Table 3 Results for unit cube decomposed into 27 smaller cubical subdomains with a checkerboard arrangement of material properties

	Stiffness scaling		Cardinality scaling		Deluxe scaling		e-deluxe scaling	
H/h	Iter	Cond	Iter	Cond	Iter	Cond	Iter	Cond
4	50	272	80	156	6	1.06	6	1.06
6	67	342	100	207	7	1.20	7	1.20
8	78	398	117	247	8	1.33	8	1.33
10	87	445	128	281	9	1.45	9	1.45
12	95	486	140	310	10	1.55	10	1.55
14	102	522	151	336	10	1.63	10	1.63
16	109	554	160	360	11	1.71	11	1.71

marked e-deluxe, results of replacing the solver over pairs of subdomains with a common face, by a solver over only a thin neighborhood of the face, which just includes the element edges next to the face, are given.

Acknowledgement The first author was supported in part by the National Science Foundation Grant DMS-1216564 and in part by the U.S. Department of Energy under contract DE-FE02-06ER25718. The second author is a member of the Computational Solid Mechanics and Structural Dynamics Department, Sandia National Laboratories, Albuquerque, New Mexico. Sandia National Laboratories is a multi-program laboratory managed and operated by Sandia Corporation, a wholly owned subsidiary of Lockheed Martin Corporation, for the U.S. Department of Energy's National Nuclear Security Administration under contract DE-AC04-94AL85000.

References

1. L. Beirão da Veiga, L.F. Pavarino, S. Scacchi, O.B. Widlund, S. Zampini, Isogeometric BDDC preconditioners with deluxe scaling. SIAM J. Sci. Comput. **36**(3), A1118–A1139 (2014a)
2. L. Beirão da Veiga, L.F. Pavarino, S. Scacchi, O.B. Widlund, S. Zampini, BDDC deluxe for isogeometric analysis, in *These Proceedings* (2014b)
3. J.G. Calvo, A BDDC algorithm with deluxe scaling for H(curl) in two dimensions with irregular domains. Technical report TR2014-965, Courant Institute of Mathematical Sciences, 2014. Math. Comput. (to appear)
4. E.T. Chung, H.H. Kim, A deluxe FETI-DP for a hybrid staggered discontinuous Galerkin method for H(curl)-elliptic problems. Int. J. Numer. Methods Eng. **98**(1), 1–23 (2014)
5. C.R. Dohrmann, A preconditioner for substructuring based on constrained energy minimization. SIAM J. Sci. Comput. **25**(1), 246–258 (2003)
6. C.R. Dohrmann, O.B. Widlund, Some recent tools and a BDDC algorithm for 3D problems in H(curl), in *Domain Decomposition Methods in Science and Engineering XX*, ed. by R. Bank, M. Holst, O. Widlund, J. Xu. Proceedings of the Twentieth International Conference on Domain Decomposition Methods held at the University of California at San Diego, 9–13 February 2011. Lecture Notes in Computational Science and Engineering, vol. 91 (Springer, Heidelberg–Berlin, 2013), pp. 15–26
7. C.R. Dohrmann, O.B. Widlund, A BDDC algorithm with deluxe scaling for three-dimensional H(curl) problems. Technical report TR2014-964, Courant Institute, New York University, March 2014. Appeared electronically in Comm. Pure Appl. Math. (April 2015)

8. M. Dryja, J. Galvis, M.V. Sarkis, A deluxe FETI-DP preconditioner for full DG discretization of elliptic problems, in *These Proceedings* (2014)
9. C. Farhat, M. Lesoinne, P.L. Tallec, K. Pierson, D. Rixen, FETI-DP: a dual-primal unified FETI method – Part I: a faster alternative to the two-level FETI method. Int. J. Numer. Methods Eng. **50**, 1523–1544 (2001)
10. R. Hiptmair, J. Xu, Nodal auxiliary space preconditioning in H(curl) and H(div) spaces. SIAM J. Numer. Anal. **45**(6), 2483–2509 (electronic) (2007)
11. A. Klawonn, O.B. Widlund, M. Dryja, Dual-primal FETI methods for three-dimensional elliptic problems with heterogeneous coefficients. SIAM J. Numer. Anal. **40**(1), 159–179 (2002)
12. J.H. Lee, A balancing domain decomposition by constraints deluxe method for numerically thin Reissner-Mindlin plates approximated with Falk-Xu finite elements. SIAM J. Numer. Anal. **53**(1), 63–81 (2014)
13. D.-S. Oh, A BDDC preconditioner for problems posed in H(div) with deluxe scaling, in *These Proceedings*, 2014
14. D.-S. Oh, O.B. Widlund, C.R. Dohrmann, A BDDC algorithm for Raviart-Thomas vector fields. Technical report TR2013-951, Courant Institute, New York University, February 2013
15. A. Toselli, Dual–primal FETI algorithms for edge finite–element approximations in 3D. IMA J. Numer. Anal. **26**, 96–130 (2006)
16. A. Toselli, O. Widlund, *Domain Decomposition Methods - Algorithms and Theory*. Springer Series in Computational Mathematics, vol. 34 (Springer, Berlin 2005)
17. S. Zampini (2013). http://www.mcs.anl.gov/petsc/petsc-dev/docs/manualpages/PC/PCBDDC.html

Part II
Talks in Minisymposia (MT)

A Stochastic Domain Decomposition Method for Time Dependent Mesh Generation

Alexander Bihlo and Ronald D. Haynes

1 Introduction

We are interested in PDE based mesh generation. The mesh is computed as the solution of a mesh PDE which is coupled to the physical PDE of interest. In [3] we proposed a stochastic domain decomposition (SDD) method to find adaptive meshes for steady state problems by solving a linear elliptic mesh generator. The SDD approach, as originally formulated in [1], relies on a numerical evaluation of the probabilistic form of the exact solution of the linear elliptic boundary value problem. Monte-Carlo simulations are used to evaluate this probabilistic form only at the sub-domain interfaces. These interface approximations can be computed independently and are then used as Dirichlet boundary conditions for the deterministic sub-domain solves. It is generally not necessary to solve the mesh PDEs with high accuracy. Only a good quality mesh, one that allows an accurate representation of the physical PDE, is required. This lower accuracy requirement makes the proposed SDD method computationally more attractive, reducing the number of Monte-Carlo simulations required.

Grid adaptation by an SDD approach does generate interesting issues in its own right. Grid quality should be monitored during the interface solves to give a suitable stopping criteria for the stochastic portion of the algorithm. Such a stopping criteria can be readily implemented by checking the mesh quality (as measured e.g. through mesh smoothness, alignment or equidistribution, see [5]) after every nth Monte-Carlo simulation. If the mesh quality is below a threshold, the additive nature of expected value computations allows one to resume the Monte-Carlo simulations and hence improve the mesh generation.

A. Bihlo • R.D. Haynes (✉)
Memorial University of Newfoundland, St. John's, Newfoundland and Labrador, Canada
e-mail: abihlo@mun.ca; rhaynes@mun.ca

T. Dickopf et al. (eds.), *Domain Decomposition Methods in Science and Engineering XXII*, Lecture Notes in Computational Science and Engineering 104, DOI 10.1007/978-3-319-18827-0_9

As mentioned, in [3] only the steady grid generation problem was considered. Of course, in practice, the problem of grid generation is coupled with the process of solving the system of physical, often time dependent, PDEs. It is this latter issue that we begin to explore in this paper.

We are interested in time dependent PDEs whose solutions evolve on disparate space and time scales. The solution behaviour lends itself to the use of time dependent meshes which automatically adapt and evolve to efficiently resolve the solution features. The generation of these time dependent grids can be done either by statically applying an elliptic mesh generator using the physical solution obtained at the previous time step or by employing a time relaxation of the static mesh PDE resulting in a parabolic moving mesh PDE, as in [5]. The extension of the SDD approach to (linear) parabolic mesh generators is possible due to the existence of a stochastic representation of the exact solution of such linear parabolic problems. For the sake of illustration, we will work with the time-relaxed form of the Winslow-Crowley variable diffusion mesh generation method, first described in [9].

2 Winslow's Method

The *equipotential* method of mesh generation in 2D, as described in [4], found the mesh lines in the physical co-ordinates x and y as the level curves of the potentials ξ and η satisfying Laplace's equations

$$\nabla^2 \xi = 0, \qquad \nabla^2 \eta = 0, \tag{1}$$

and appropriate boundary conditions which ensure grid lines lie along the boundary of the domain. Here derivatives are with respect to the physical co-ordinates. The mesh transformation, $x(\xi, \eta)$ and $y(\xi, \eta)$, in the physical domain Ω_p, can be found by (inverse) interpolation of the solution of (1) onto a (say) uniform (ξ, η) grid. In practice, the inversion to the physical co-ordinates is not necessary. Instead one could transform the physical PDE of interest to the computational co-ordinate system.

Winslow [10] generalized (1) by adding a diffusion coefficient $w(x, y)$ depending on the gradient or other aspects of the solution. This gives the linear elliptic mesh generator

$$\nabla \cdot (w \nabla \xi) = 0 \quad \text{and} \quad \nabla \cdot (w \nabla \eta) = 0. \tag{2}$$

The function $w(x, y)$, known as a mesh density function, characterizes regions where additional mesh resolution is needed and in general depends on the solution of the physical PDE. We assume w and $1/w$ are strictly positive, bounded C^2-functions.

Here we assume the solution of the physical PDE is time dependent and hence the mesh density function is changing with time, $w = w(t, x, y)$. One could still use (2) to solve the mesh transformation at each time t. For time dependent PDEs this would result a system of differential-algebraic equations for the physical solution and the mesh. Instead, we choose to relax (2) to obtain a parabolic linear mesh generator of the form

$$\xi_t = \frac{1}{T}(\nabla w \cdot \nabla \xi + w\nabla^2 \xi) \quad \text{and} \quad \eta_t = \frac{1}{T}(\nabla w \cdot \nabla \eta + w\nabla^2 \eta). \tag{3}$$

This gives a mesh PDE which depends explicitly on the mesh speed and provides a degree of temporal smoothing for the mesh. In fact one can show the difference between the solution of (3) and the solution of (2) goes to zero as $T \to 0$, see [5]. In the following, we set $T = 1$.

Below we only work with prescribed functions for w. In practice, however, the monitor function would be linked to the solution of a physical PDE. For example, one could use an arc-length type function $\rho = \sqrt{1 + \alpha(u_x^2 + u_y^2)}$ and choose $w = 1/\rho$. We also note that our algorithm uses an interpolated form of w instead of the analytic expression. In practice, this is necessary since u is only known on the current grid as we alternately solve the mesh and physical PDEs.

3 Linear Parabolic Differential Equations and Stochastic Domain Decomposition

The system of mesh PDEs (3) is of the form

$$\xi_t = \mathrm{L}\xi, \quad \eta_t = \mathrm{L}\eta, \tag{4}$$

where $\xi(t, x, y)$ and $\eta(t, x, y)$ are the computational coordinates defined over $[0, T] \times \Omega_\mathrm{p}$. In system (4), L is a linear elliptic operator of the form

$$\mathrm{L} = a_{ij}\frac{\partial^2}{\partial x_i \partial x_j} + b_i \frac{\partial}{\partial x_i},$$

with continuous coefficient matrix $a(t, x, y) = (a_{ij})(t, x, y)$, $i, j \in \{1, 2\}$, and drift vector $\mathbf{b} = (b_1, b_2)^\mathrm{T}(t, x, y)$. Here we employ the summation convention over repeated indices. System (4) is accompanied by smooth boundary and initial conditions $\xi|_{\partial\Omega_\mathrm{p}} = f(t, x, y)$, $\eta|_{\partial\Omega_\mathrm{p}} = g(t, x, y)$, $\xi(0, x, y) = \xi_0(x, y)$, and $\eta(0, x, y) = \eta_0(x, y)$.

The solution of such linear parabolic problems can be described using the tools of stochastic calculus [2, 7]. Provided that ξ and η are C^1-functions in t and C^2 in (x, y), the point-wise solution of system (4) at $(t, x, y) \in [0, T] \times \Omega_p$ is given probabilistically as

$$\xi(t,x,y) = \mathrm{E}\left[\xi_0(\mathbf{X}(t))\mathbf{1}_{[\tau_{\partial\Omega_\mathrm{p}}>t]}\right] + \mathrm{E}\left[f(t-\tau_{\partial\Omega_\mathrm{p}}, \mathbf{X}(\tau_{\partial\Omega_\mathrm{p}}))\mathbf{1}_{[\tau_{\partial\Omega_\mathrm{p}}<t]}\right], \tag{5}$$

where the process $\mathbf{X}(t) = (x(t), y(t))^\mathrm{T}$ satisfies, in the Îto sense, the stochastic differential equation (SDE)

$$\mathrm{d}\mathbf{X}(t) = \mathbf{b}(t, \mathbf{X}(t))\mathrm{d}t + \sigma(t, \mathbf{X}(t))\mathrm{d}\mathbf{W}(t).$$

The relation between σ and (a_{ij}) is given through

$$\frac{1}{2}\sigma(t,x,y)\sigma(t,x,y)^\mathrm{T} = a(t,x,y)$$

for all $(t,x,y) \in [0,T] \times \Omega_\mathrm{p}$. The solution for $\eta(t,x,y)$ is completely analogous.

In (4), the $\mathrm{E}[\cdot]$ denotes the expected value, $\tau_{\partial\Omega_\mathrm{p}}$ is the time when the stochastic path starting at (x,y) first hits the boundary of the physical domain Ω_p, $\mathbf{W}$ is two-dimensional Brownian motion and $\mathbf{1}$ is the indicator function. See [7] for a proper definition of the required probability space.

The time dependent mesh generator (3) is a special case of the general form (4) with

$$a(t,x,y) = wI_2, \quad b_1(t,x,y) = w_x, \quad b_2(t,x,y) = w_y, \tag{6}$$

where I_2 is the 2×2 identity matrix.

For our two dimensional mesh generator we choose the initial conditions $\xi(t = 0,x,y) = \xi_0(x,y) = x$ and $\eta(t = 0,x,y) = \eta_0(x,y) = y$, corresponding to an initial uniform mesh, and the static boundary conditions $\xi(t,x_\mathrm{l},y) = 0$, $\xi(t,x_\mathrm{r},y) = 1$, $\eta(t,x,y_\mathrm{l}) = 0$ and $\eta(t,x,y_\mathrm{u}) = 1$. This ensures we use the standard computational domain $\Omega_\mathrm{c} = [0,1] \times [0,1]$ and the rectangular physical domain $\Omega_\mathrm{p} = [x_\mathrm{l}, x_\mathrm{r}] \times [y_\mathrm{l}, y_\mathrm{u}]$. The remaining boundary conditions for $\xi(t,x,y_\mathrm{l}), \xi(t,x,y_\mathrm{u}), \eta(t,x_\mathrm{l},y)$ and $\eta(t,x_\mathrm{r},y)$ are determined by solving the 1D version of (2) along the boundaries. Collectively, we use f and g to denote these boundary conditions for ξ and η as in Eq. (5).

Hence we have to solve the SDE

$$\mathrm{d}\mathbf{X}(t) = \nabla w\,\mathrm{d}t + \sqrt{2}w\,\mathrm{d}\mathbf{W}(t), \tag{7a}$$

for the single path $\mathbf{X}(t)$. The stochastic form of the exact solution of Eq. (3) for ξ is then obtained by evaluating

$$\xi(t,x,y) = \mathrm{E}\left[\xi_0(\mathbf{X}(t))\mathbf{1}_{[\tau_{\partial\Omega_\mathrm{p}}>t]}\right] + \mathrm{E}\left[f(\mathbf{X}(\tau_{\partial\Omega_\mathrm{p}}))\mathbf{1}_{[\tau_{\partial\Omega_\mathrm{p}}<t]}\right]. \tag{7b}$$

The point-wise solution for $\eta(t,x,y)$ is obtained in an analogous fashion.

In principle, the probabilistic solution (7) allows one to determine the computational coordinates ξ and η at each point in the space-time domain $[0, T] \times \Omega_{\mathrm{p}}$. However, this is prohibitively expensive (unless a sufficiently large number of compute cores is available). A more efficient approach is to evaluate the solution (7) only at points along artificially imposed interfaces. These solutions serve as boundary values for the DD implementation. Moreover, one can reduce the number of stochastic solves along the interfaces even further as described at the end of the next section, cf. [2].

In the mesh generation context it is not possible to obtain the solution of (5) at all times, as the solution of the mesh PDE is coupled to the physical solution. That is, rather than solving (5) for a time $t \in [0, T]$, it is generally only possible to use this stochastic solution to advance the solution of (4) over one single time step from t^n to t^{n+1}. In this case, ξ_0 and η_0 should be interpreted as the values of ξ and η at time t^n and the monitor function, w, is given at either t^n or t^{n+1} and remains constant over the time step.

4 The Numerical Method

Stochastic Solver and Domain Decomposition The use of the stochastic solution (5) for the time-relaxed Winslow mesh generator with parameters (6) is straightforward. We solve (7a) using the classical Euler-Maruyama scheme, i.e. we employ linear time-stepping. An alternative would be to use exponential time-stepping as advocated e.g. in [1, 3, 6]. In our tests, linear time-stepping gives sufficient accuracy. The components of the Brownian motion $\mathbf{dW}(t)$ are computed as $\sqrt{\Delta t}\mathcal{N}(0, 1)$, where $\mathcal{N}(0, 1)$ is a normally distributed random number with mean zero and variance one [7].

The time dependent weight only becomes available at each time step (due to a possible coupling with a physical PDE). Hence we are only able to employ formula (7b) to integrate over a single time step, i.e. from t^n to t^{n+1}. Over this time step, the weight function is evaluated at t^n and held constant, i.e. we have $w^n(x, y) = w(t^n, x, y)$ in (7a). Accordingly, ξ_0 in Eq. (7b) is to be interpreted as $\xi_0^n = \xi(t^n, x, y)$, i.e. the values of the computational coordinates at the current time t^n. Moreover, the boundary functions f and g are updated at each time step to reflect changes in the physical solution.

We then numerically integrate the SDE (7a) from t^n to t^{n+1}. The drift vector $\mathbf{b} = \nabla w$ is required everywhere along the path of the stochastic process $\mathbf{X}(t)$ but is only available directly at the grid points of the domain. Bilinear interpolation is used to obtain the values of $\mathbf{b}$ in between these grid points. The quantity ∇w is approximated using finite differences.

In the DD context, the stochastic solution is only required at a selection of points, (x_k^i, y_k^i), which live on the interfaces between sub-domains. One time step Δt is split into several smaller sub-time steps in order to numerically integrate the SDE (7a) from t^n to t^{n+1}. We found this splitting of Δt into sub-time steps necessary to determine, with sufficient accuracy, whether the stochastic processes has left the domain Ω_p during Δt. This is not unlike the M^k approach for mesh generation discussed in [5]. At each sub-time step, a boundary test is performed to determine whether the stochastic process has left the domain Ω_p. If this is the case, the process contributes via the second term in Eq. (7b) to the approximation of $\xi(t^{n+1}, x_k^i, y_k^i)$. If the stochastic process did not leave the domain until t^{n+1} is reached, it contributes to the first term in the approximation of $\xi(t^{n+1}, x_k^i, y_k^i)$ in Eq. (7b). The computation of $\eta(t^{n+1}, x_k^i, y_k^i)$ is handled analogously. The expected values are then replaced by arithmetic means. Note, it is not desirable to make Δt itself smaller, as this would degrade the efficiency of the (deterministic) implicit sub-domain solver, which is described below.

Deterministic Sub-domain Solver The values of ξ and η along the sub-domain interfaces serve as boundary conditions for the sub-domain solver. The sub-domain solver we employ is an implicit finite-difference discretization of Eq. (3). The matrix system is solved using an LU-factorization.

Parallelization and Further Speed-up It is well-known that Monte-Carlo techniques converge rather slowly [8] and are usually most competitive for problems in high dimensions. The use of the stochastic solution to obtain the interface values for a DD solution, however, is considerably more efficient and provides a fully parallel grid generation algorithm. Moreover, the DD method requires no iteration. The stochastic solutions on the interfaces can be determined at each point separately and each Monte-Carlo simulation is independent. Additionally, each sub-domain solution could potentially be assigned to a single processor once the interface solutions are obtained, yielding excellent scalability. Due to the fully parallel nature of the algorithm, the method is also fault tolerant. This renders the method suitable for an implementation on massively parallel computing architectures, cf. [1–3].

A further source of improvement stems from the fact that ξ and η do not have to be computed at all grid points along the interfaces. As proposed in [1] it may be sufficient to use the stochastic solution only at few points on the interface and recover the solution at the remaining interface points using interpolation. In [3] we have used a relatively simple optimal placement strategy to determine the most important locations on the interface where the stochastic solution should be computed. We use the same strategy in the present algorithm, i.e. the stochastic solution is computed near the maxima and minima of w_x and w_{xx} along the horizontal interfaces and w_y and w_{yy} along the vertical interfaces.

5 Numerical Results

We present an example for our SDD method to generate an adaptive (moving) mesh for the weight function $w = 1/\rho$, where

$$\rho = 1 + \alpha \exp\left(\beta \left| \left(x - \frac{1}{2} - \frac{1}{4}\cos(2\pi t)\right)^2 + \left(y - \frac{1}{2} - \frac{1}{4}\sin(2\pi t)\right)^2 - \frac{1}{100} \right| \right).$$

We choose the parameters $\alpha = 10$ and $\beta = -50$ used in [5]. Both the physical and computational domains are the unit square. The grid we generate has 41×41 nodes and is divided into four sub-domains. On the interfaces we determine the stochastic solution at the key points using the optimal placement strategy mentioned in the previous section. Piecewise cubic Hermite interpolation is used to determine the remaining interface points. We integrate (3) up to $t = 0.75$ using $\Delta t = 0.001$. Each time step is split into 20 sub-time steps while solving the SDE (7a) and $N = 10000$ Monte-Carlo simulations are used to estimate the expected value in (7b). The resulting meshes at $t = 0.25$, $t = 0.5$, and $t = 0.75$ are depicted in Fig. 1.

The method is able to produce smooth meshes over the physical domain that adapt well to the time-dependent monitor function. No explicit smoothing was applied to the final meshes in this example. In general we have found sub-domain smoothing to be a way to further reduce the number of Monte-Carlo simulations needed in the probabilistic expression (7b), see [3].

6 Conclusion

In this paper we have proposed a new stochastic domain decomposition method for the construction of adaptive moving meshes suitable for time-dependent problems. The method is fully parallelizable as the values of the computational coordinates ξ and η on the single sub-domains can be determined without information exchange from neighboring sub-domains and all the interface values can be computed independently.

Future refinements include the use of exponential time-stepping to solve the SDE (7a). More generally, more sophisticated boundary tests could better determine the first exit time of a stochastic process. This will allow using larger time steps in the solution of (7a) thus making the method more efficient. An alternate approach to generate time dependent meshes is to apply the SDD method from [3] to the sequence of elliptic problems which result from discretizing (2) in time.

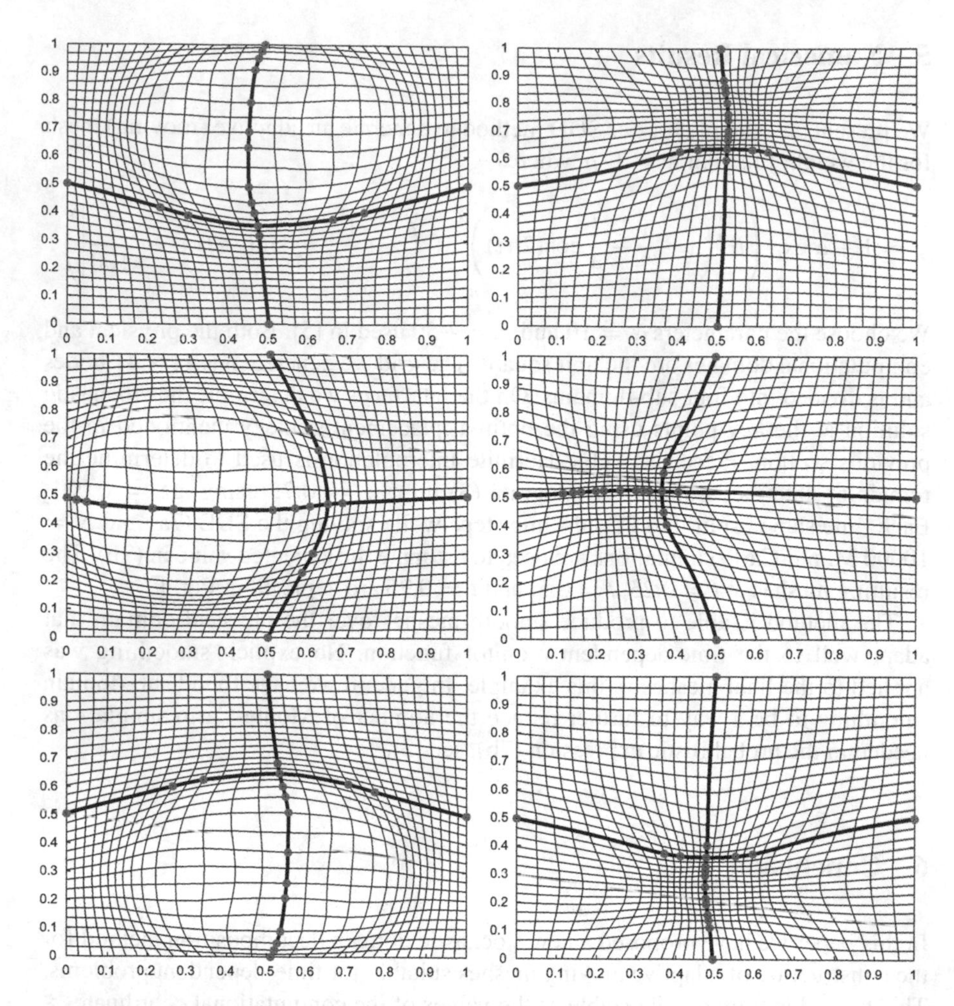

Fig. 1 *Top to bottom*: Meshes obtained from the parabolic mesh generator (3) using the SDD method at $t = 0.25$, $t = 0.5$, and $t = 0.75$. *Left*: Meshes over the physical domain. *Right*: Meshes over the computational domain obtained from the former using natural neighbor interpolation. *Thick line*: Sub-domain interfaces. *Circles*: Points where the mesh is obtained using the stochastic solution (7)

Acknowledgements This research was supported by NSERC (Canada). AB is a recipient of an APART Fellowship of the Austrian Academy of Sciences. The authors thank Professor Weizhang Huang (Kansas) and the two anonymous referees for helpful remarks.

References

1. J.A. Acebrón, M.P. Busico, P. Lanucara, R. Spigler, Domain decomposition solution of elliptic boundary-value problems via Monte Carlo and quasi-Monte Carlo methods. SIAM J. Sci. Comput. **27**(2), 440–457 (2005)
2. J.A. Acebrón, Á. Rodríguez-Rozas, R. Spigler, Efficient parallel solution of nonlinear parabolic partial differential equations by a probabilistic domain decomposition. J. Sci. Comput. **43**(2), 135–157 (2010)
3. A. Bihlo, R.D. Haynes, Parallel stochastic methods for PDE based grid generation. Comput. Math. Appl. **68**(8), 804–820 (2014)
4. W.P. Crowley, An "equipotential" zoner on a quadrilateral mesh. Technical Report, 1962
5. W. Huang, R.D. Russell, *Adaptive Moving Mesh Methods* (Springer, New York, 2010)
6. K.M. Jansons, G.D. Lythe, Exponential timestepping with boundary test for stochastic differential equations. SIAM J. Sci. Comput. **24**(5), 1809–1822 (2003)
7. I. Karatzas, S.E. Shreve, Brownian Motion and Stochastic Calculus, in *Graduate Texts in Mathematics*, vol. 113 (Springer, New York, 1991)
8. W.H. Press, S.A. Teukolsky, W.T. Vetterling, B.P. Flannery, *Numerical Recipes 3rd Edition: The Art of Scientific Computing* (Cambridge University Press, Cambridge, 2007)
9. A.M. Winslow, Numerical solution of the quasilinear Poisson equation in a nonuniform triangle mesh. J. Comput. Phys. **1**(2), 149–172 (1966)
10. A.M. Winslow, Adaptive-mesh zoning by the equipotential method. Technical Report UCID-19062, Lawrence Livermore National Laboratory, CA, 1981

Some Geometric and Algebraic Aspects of Domain Decomposition Methods

D.S. Butyugin, Y.L. Gurieva, V.P. Ilin, and D.V. Perevozkin

1 Introduction

The DDMs include a variety of geometric, algebraic, and functional aspects which are aimed at a high performance solution of large-size problems on post-petaflop computers.

Numerous works and Internet sites are devoted to this problem: monographs, papers, conference proceedings, programs, etc. [2, 10]. The issues that are of most interest from the practical point of view are the requirements on high resolution of the numerical approaches to solving multi-dimensional interdisciplinary boundary value problems described by systems of partial differential equations (PDEs) or the corresponding variational statements in the computational domains with complicated piecewise smooth boundaries and contrasting material properties of their subdomains. Approximation of such problems by finite-volume or finite-element methods on nonstructured grids results in very large systems of linear algebraic equations (SLAEs) with 10^8–10^{10} unknowns with ill-conditioned or nondefinite sparse matrices with complicated portrait structures.

The solution of the SLAEs is a weak point of modern computing, and the DDMs are the main tool providing scalable parallelism on multi-processor and multi-core

D.S. Butyugin • Y.L. Gurieva • D.V. Perevozkin (✉)
Institute of Computational Mathematics and Mathematical Geophysics SB RAS, Novosibirsk, Russia
e-mail: foxillys@gmail.com

V.P. Ilin
Institute of Computational Mathematics and Mathematical Geophysics SB RAS, Novosibirsk, Russia

Novosibirsk State University, Novosibirsk, Russia

T. Dickopf et al. (eds.), *Domain Decomposition Methods in Science and Engineering XXII*, Lecture Notes in Computational Science and Engineering 104, DOI 10.1007/978-3-319-18827-0_10

systems. The goal of this paper is to experimentally investigate several approaches to automatic construction of balancing grid subdomains and to parallel solution of the resulting SLAE using the parametrized width of subdomain overlapping, different internal boundary conditions, aggregation techniques, see, for example, [9]. The results of a comparative analysis of the efficiency of various approaches for the model problems are presented. The computations were carried out with the Krylov library [1].

2 Grid Domain Decomposition Without Separator Nodes

Let the matrix of the SLAE $Au = f$ be split into P subsystems:

$$(Au)_p = A_{p,p}u_p + \sum_{\substack{q=1\\q\neq p}}^{P} A_{p,q}u_q = f_p, \;\; p = 1,\ldots,P, \;\; A = \{a_{i,j}\} \in \mathscr{R}^{N,N}, \tag{1}$$

$$A = \{A_{p,q} \in \mathscr{R}^{N_p,N_q}\}, \;\; u = \{u_p \in \mathscr{R}^{N_p}\}, \;\; f = \{f_p \in \mathscr{R}^{N_p}\}, \;\; p,q = 1,\ldots,P.$$

Assume that SLAE (1) is a system of grid equations approximating a multi-dimensional boundary value problem for a differential equation, so that the components of the vectors u, f correspond to a grid point, the total number of nodes in the grid computational domain $\Omega^h = \bigcup_{p=1}^{P} \Omega^h_p$ being equal to N. The block decomposition of the matrix and vectors corresponds to the partitioning of Ω^h into P non-overlapping subdomains Ω^h_p, each consisting of N_p nodes, $N_1 + \ldots + N_p = N$. The decomposition of Ω^h does not use separator nodes, i.e., the boundaries of the subdomains do not pass through the grid nodes.

The process of system (1) solving can be parallelized by the additive Schwarz method:

$$A_{p,p}u^n_p = f_p - \sum_{\substack{q=1\\q\neq p}}^{P} A_{p,q}u^{n-1}_q \equiv g^{n-1}_p. \tag{2}$$

The above matrix-algebraic representation of the structure of SLAE (1) can be extended by introducing a graph describing the same problem. Each ith grid node (or the ith row of the matrix A) can be associated with a vertex v_i of a graph G, and the mesh edge connecting the nodes i and $j \in \Omega^h$, can be associated with the edge of the graph $G = (V, E)$, $V = \{v_i;\; i = 1,\ldots,N\}$, $E = \{(v_i, v_j)\; |a_{i,j} \neq 0, i,j = 1,\ldots,N\}$.

Define an extended subdomain $\bar{\Omega}_p^h \supset \Omega_p^h$ with overlapping, whose breadth is defined in terms of the number of layers, or fronts, of the grid nodes.

Let $\Gamma_p^0 \in \Omega_p^h$ denote a set of internal near-boundary nodes, i.e., nodes $P_i \in \Omega_p^h$, in which one of the neighbors does not lie in Ω_p^h ($P_j \notin \Omega_p^h$, $j \in \omega_i$, $j \neq i$). In Γ_p^0, define a subset of nodes $\Gamma_{p,q}^0$ in which the neighboring nodes belong to the adjacent subdomain Ω_q^h, $q \in \bar{\omega}_p$, where $\bar{\omega}_p$ is a set of numbers of the subdomains adjacent to Ω_p^h. Thus, $\Gamma_p^0 = \bigcup\limits_{q \in \bar{\omega}_p} \Gamma_{p,q}^0$, and the subsets $\Gamma_{p,q}^0$ may intersect, i.e. they can contain near-boundary nodes with neighbors from different subdomains.

Let Γ_p^1 denote a set of nodes adjacent to the nodes from Γ_p^0 but not belonging to Ω_p^h and Γ_p^0; and let Γ_p^2 be a set of nodes adjacent to the nodes from Γ_p^1 but not belonging to the union $\Gamma_p^1 \bigcup \Omega_p^h$, etc. These sets will be called the first external layer (front) of nodes, the second layer, etc., respectively. The resulting collection of nodes $\Omega_p^\Delta = \Omega_p^h \bigcup \Gamma_p^1 \ldots \bigcup \Gamma_p^\Delta$ will be called the extended pth grid subdomain, and Δ, the extension breadth. The case $\Delta = 0$ actually means the decomposition of the domain Ω^h into subdomains without intersections ($\Omega_p^0 = \Omega_p^h$).

The set $\Gamma_p^\Delta \in \Omega_p^\Delta$ presents internal near-boundary nodes of the extended subdomain Ω_p^Δ, and $\Gamma_p^{\Delta+1}$, a set of external near-boundary nodes. Thus the geometric boundary of Ω_p^Δ runs between Γ_p^Δ and $\Gamma_p^{\Delta+1}$. Similarly to Γ_p^0, the set Γ_p^Δ can be partitioned into subsets of near-boundary nodes $\Gamma_p^\Delta = \Gamma_{p,q_1}^\Delta \bigcup \Gamma_{p,q_2}^\Delta \ldots \bigcup \Gamma_{p,q_{m_p}}^\Delta$ whose neighboring nodes are located, respectively, in the subdomains $\Omega_{q_1}^h, \Omega_{q_2}^h, \ldots, \Omega_{q_{m_p}}^h$ (here m_p denotes the number of subdomains that intersect Ω_p^Δ, and $q_1, q_2, \ldots, q_{m_p}$ are the numbers of these subdomains).

Consider iterative process (2) for the equation corresponding to the ith near-boundary node in $\bar{\Omega}_p^h$. Some of the neighbors belong to other subdomains $\bar{\Omega}_q^h$, $q \neq p$ but do not belong to $\bar{\Omega}_p^h$:

$$\Big(a_{i,i} + \theta_i \sum_{j \notin \bar{\Omega}_p} a_{i,j}\Big) u_i^n + \sum_{j \in \bar{\Omega}_p} a_{i,j} u_j^n = f_i + \sum_{j \notin \bar{\Omega}_p} a_{i,j} (\theta_i u_i^{n-1} - u_j^{n-1}). \qquad (3)$$

Here $\theta_i \in [0, 1]$ are parameters, corresponding for $\theta_i = 0$ or $\theta_i = 1$ to the Dirichlet or Neumann boundary conditions, and for $0 < \theta_i < 1$, to the Robin condition.

Introduce matrices $\bar{A}_{p,p} \in \mathscr{R}^{\bar{N}_p, \bar{N}_p}$, $\bar{A}_{p,q} \in \mathscr{R}^{\bar{N}_p, \bar{N}_q}$ for Eq. (3). Then the iterative process can be transformed to the form

$$\bar{A}_{p,p} \bar{u}_p^n = \bar{f}_p - \sum_{\substack{q=1 \\ q \neq p}}^{P} \bar{A}_{p,q} \bar{u}_q^{n-1} \equiv \bar{g}_p^{n-1}. \qquad (4)$$

In the above discussion, we have considered the extension of the subdomain Ω_p^h towards its outer side. The same procedures are performed for the neighboring subdomains, which results in the construction of fronts inside Ω_p^h. These procedures can be implemented at the grid layers (fronts) in the extension of the neighboring subdomains $\Omega_q^h, q \neq p$.

Formula (4) does not describe the iterative process exactly since $\bar{N}_1 + \ldots + \bar{N}_p \geq N$. The vector u^n can be determined by partitioning the unit:

$$u_i^n = \sum_{q_i \in \bar{\omega}_i} \eta_{q_i} (\bar{u}_{q_i}^n)_i, \quad \sum_{q_i \in \bar{\omega}_i} \eta_{q_i} = 1, \tag{5}$$

where $\bar{\omega}_i$ is a set of the extended subdomains $\bar{\Omega}_{q_i}^h$ including the node P_i. Particular, but important, cases in (5) are $\eta_{q_i} = 1$ for $P_i \in \Omega_{q_i}^h$ and $\eta_{q_i} = 0$ for $P_i \neq \Omega_{q_i}^h$.

An alternative approach is to use iterations "in traces". Let $\bar{\Gamma}_p = \Gamma_p^\Delta \bigcup \Gamma_p^{\Delta+1}$ define the trace of the extended subdomain $\bar{\Omega}_p^h$ for $\theta_i \neq 0, P_i \in \Gamma_p^\Delta$, and $\bar{\Gamma}_p = \Gamma_p^{\Delta+1}$, for $\theta_i = 0$. We can write $\bar{\Gamma}_q = \bigcup \bar{\Gamma}_{q,p}$, where $\bar{\Gamma}_{p,q} = \bar{\Gamma}_p \bigcap \Omega_q^h$. From (2) we have

$$\bar{u}_p^n = \bar{A}_{p,p}^{-1} (\bar{f}_p - \sum_{\substack{q=1 \\ q \neq p}}^{P} \hat{A}_{p,q} \hat{u}_p^{n-1}). \tag{6}$$

Here the matrices $\bar{A}_{p,p}$ are assumed to be non-singular, $\hat{A}_{p,q} \in \mathcal{R}^{\bar{N}_p, \hat{N}_{p,q}}$ and $\hat{u}_p = \{u_i; P_i \in \bar{\Gamma}_{p,q}\} \in \mathcal{R}^{\hat{N}_{p,q}}$, $\hat{N}_{p,q}$ being the number of nodes in $\bar{\Gamma}_{p,q}$.

If $\bar{u}_p^n \rightarrow \bar{u}_p$ for $n \rightarrow \infty$, iterations (6) provide the solution of the preconditioned SLAE

$$\bar{A}u = \bar{f}, \quad \bar{f} \in \mathcal{R}^N, \quad \bar{A} \in \mathcal{R}^{N,N}. \tag{7}$$

Multiplying Eq. (6) by $\bar{A}_{q,p}$ and denoting $\bar{A}_{p,q}\bar{u}_q^n = \hat{A}_{p,q}\hat{u}_q^n = v_{p,q}^n \in \mathcal{R}^{\bar{N}_p}$, we obtain the algebraic system "in traces":

$$v_{q,p} + \bar{A}_{q,p}\bar{A}_{p,p}^{-1} \sum_{\substack{q=1 \\ q \neq p}}^{P} v_{p,q} = \bar{A}_{q,p}\bar{A}_{p,p}^{-1}\bar{f}_p, \quad p = 1, \ldots, P; \; q \in \bar{\omega}_p. \tag{8}$$

The degree of freedom of this SLAE is $\hat{N} = \sum_{p=1}^{P} \hat{N}_p = \sum_{p=1}^{P} \sum_{q \in \bar{\omega}_p} \hat{N}_{p,q} \ll N$. Iterative solution of Eq. (8) can be implemented by a Krylov method. To speed up the iterative DDM process, various approaches, for example, deflation, coarse grid correction, and smoothed aggregation can be used. We consider the SLAE reduction procedure based on an interpolation principle, under the assumption of smooth behavior of the solution to be sought for in each subdomain.

Define a prolongation matrix $W^T = \{w_p\} \in \mathscr{R}^{N,P}$, where the vectors (columns) w_p have nonzero (unit) entries corresponding to the subdomain Ω_p only. Then $\hat{A} = W^T A W \in \mathscr{R}^{P,P}$ presents the global aggregation matrix, and $B = W\hat{A}^{-1}W^T$ is, in a sense, an aggregating preconditioning matrix. For simplicity, we consider non-overlapping subdomains. In this case, the matrix in (7) has a simple form $\bar{A} = B_J A$, where B_J is the block Jacobi preconditioner [9].

DDM-exploiting iterative processes can be constructed in various ways. We use a simple one, namely, the FGMRES [9] with dynamic preconditioner B_n: $B_n = B_A$ for $n = km + 1$, $k = 0, 1, \ldots$, and $B_n = B_J$ otherwise. The stopping criteria of this process are

$$||\hat{r}^n|| \leq ||\hat{f} - \hat{A}u^n|| \leq \varepsilon^e ||\hat{f}||, \quad \varepsilon^e \ll 1, \text{ or } n \leq n^e_{max}. \tag{9}$$

Subdomain SLAEs are solved by either the direct solver PARDISO [5] or the iterative BiCGStab method [9].

In the latter case of a two-level iterative algorithm, various approaches can be chosen for defining the internal stopping criteria $\varepsilon^i \leq \varepsilon^e$ and n^i_{max}, similarly to (9).

3 Parallel Implementation of Algorithms

The major question in high-performance implementation of DDMs is automatic construction of balancing grid subdomains, based, for instance, on CSR format of the original SLAE. This problem is solved by the graph partitioning approach in two stages. First, we define P subdomains Ω^h_p, $p = 1, \ldots, P$, without intersections. Then extended subdomains $\bar{\Omega}^h_p$ with a given breadth Δ of overlapping are constructed on the basis of the following algorithm.

The non-overlapping grid subdomains Ω^h_p are formulated as subgraphs $G_p(V_p, E_p)$ with possibly small diameters containing approximately equal numbers of vertices $N_p \approx N/P$. In practice, the task consists in transforming the original CSR format to the CSR_p formats for P subdomains, which should be distributed among the corresponding MPI processes.

The graph partitioning is a multi-level aggregation procedure of the sequential macrographs $G^{(l)}(V^{(l)}, E^{(l)}) = \{G^{(l)}_p(V^{(l)}_p, E^{(l)}_p)\}$, $l = 0, 1 \ldots, L$, $p = 1, \ldots, P_l$. Here L and P_l are the number of levels and the number of macrovertices at the lth level, respectively, whose macro-vertices include several vertices of a lower level. If $G^{(0)}(V^{(0)}, E^{(0)})$ denotes the original grid graph, the first aggregation step can be described by the following pseudocode (breadth-first search [7]):

$i = 1$, *while* $\{u \in V \mid C(u) = 0\} \neq \emptyset$
  pick any v *from* $\{u \in V \mid C(u) = 0\}$
  $Q := \{v\}, \quad n = 0$
  while $(n < n_{max}$ *and* $Q \neq \emptyset)$
    $v \leftarrow Q, \qquad C(v) := i$
    $Q \leftarrow (Adj(v) \cap \{u \in V \mid C(u) = 0\}) \setminus Q$
    $n = n + W(v)$
  end while
  $i = i + 1$
end while

Here $C(u)$ and $W(u)$ are the color and weight (integers) of the vertex u, respectively, with the initial values $C(u) = 0$, $W(u) = 1$, Adj(v) is a set of vertices adjacent to u, and Q is the queue type data structure. Later, $C(u)$ presents the number of a subdomain (macrovertex) containing the vertex (grid point) u, and $W(u)$ is the resulting number of nodes in the subdomain ($W(u) \leq n_{max}$). This algorithm is repeated for the levels $l = 1, \ldots, L$.

Parallel implementation of DDM–FGMRES is performed using hybrid programming with MPI processes on distributed memory for subdomains and OpenMP tools for each of the multi-core processors with shared memory.

4 Numerical Experiments

We present the results of some numerical experiments on solving a model Dirichlet boundary value problem for the 2D and the 3D Laplace equation in the unit computational domain $\Omega = [0, 1]^d$, $d = 2, 3$, which is approximated by a standard (2d+1)-point finite difference scheme on a square mesh (which is cubic in 3D) with the degree of freedom $N = N_x^d$, for different values of N_x. The stopping criteria for FGMRES without restarts were $\varepsilon^e = 10^{-7}$ and $n_{max}^e = \infty$. The exact solution and initial guess for the iterations were taken equal to unit and zero, respectively. All the experiments were carried out on the NKS-30T cluster [6] with standard double-precision arithmetic.

Table 1 shows the efficiency of the proposed algorithm for automatic construction of 3D balancing grid subdomains for $P = 1, 8, 16, 32, 64$. The subdomain SLAEs were solved either by the direct method PARDISO from Intel MKL or by the preconditioned BiCGStab method (Eisenstadt modification of incomplete factorization [4]) with the parameters $\varepsilon^i = 0.1$, $n_{max}^i = 5$ (these values are nearly optimal for the given problem data). Note that the PARDISO was run with 12 threads, whereas the BiCGStab was implemented without any parallelization. In Table 1, the upper and lower figures in each line correspond to grids with 128^3 and 256^3 unknowns, respectively, and the left and right figures in each column present the numbers of

Table 1 Comparative analysis of DDM without overlapping for direct and iterative subdomain solvers, $\theta = 0, N = 128^3, 256^3$

Method	P: 1	8	16	32	64
Direct	1 885	53 30.1	75 20.4	108 12.6	130 18.1
	– –	72 332	102 212	142 138	169 189
Iterative	18 64.9	68 20.5	92 12.5	103 13.0	197 11.9
	18 606	99 296	132 203	197 139	262 115

Table 2 Numbers of external iterations for solving SLAEs with aggregation preconditioning

N	m: 0	1	5	10	15	20
	82	50	46	41	42	46
128^3	132	62	53	52	57	58
	143	70	54	51	53	53
256^3	193	60	72	61	62	68

external iterations and execution time in seconds. In this case, the DDM parameters $\Delta = \theta = 0$ were used.

Table 2 presents the number of iterations for the aggregation approach for the same model SLAEs with the exact solution $u = 1000 + x + y$ and initial guess $u^0 = 0$. The aggregation preconditioner was used once every m steps (with $m = 10$ as an optimal value). Note that the behavior is also observed for different numbers of subdomains, whereas the results are given here for $P = 16$ and 32 (upper and lower cell values, respectively). The case $m = 0$ means solving without aggregation.

In the other experiments, 2D problems were solved on square meshes with $N = 128^2, 256^2$, and $P = 4, 16, 64$ equal square subdomains. The systems in the subdomains were solved by the PARDISO, and the external iterations were carried out by the iterative BiCGStab method "in traces".

Table 3 presents the iterative process versus the overlapping value Δ. The cells present the same data as in Table 1 for $\theta = 0$, and $N = 128^2, 256^2$ (upper and lower lines in each row, respectively). We see that the number of iterations decreases monotonically with increasing Δ, but for the run time there is some minimum for a sufficiently small value $\Delta \leq 4$.

Table 4 contains the number of iterations versus θ values. The left and right cell values correspond to $N = 128^2$ and $N = 256^2$, respectively. No overlapping takes place, i.e. $\Delta = 0$.

These results demonstrate that the constant parameter θ is appropriate only for a sufficiently small P. The experiments have also shown that for the overlapping decomposition ($\Delta > 0$) it is better to take $\theta = 0$.

Table 3 Numerical results for different overlapping values Δ, $\theta = 0$, $N = 128^2, 256^2$

	Δ					
P	0	1	2	3	4	5
4	18 1.75	11 1.45	9 1.37	7 1.26	7 1.26	6 1.20
	27 6.85	16 4.37	12 3.51	10 3.02	9 2.82	8 2.49
16	32 1.42	18 1.18	14 1.19	12 1.09	11 0.89	9 0.79
	41 3.85	24 2.83	20 2.20	17 1.80	14 1.38	14 1.66
64	43 1.56	26 1.66	19 1.39	16 1.50	14 1.56	12 0.86
	60 4.75	36 4.16	27 3.35	22 3.11	20 3.00	18 4.66

Table 4 Number of iterations for non-overlapping DDMs ($\Delta = 0$) with different θ, $N = 128^2, 256^2$

	θ				
P	0	0.5	0.6	0.7	0.9975
4	18 27	16 26	16 24	14 23	10 12
16	32 41	28 40	27 39	27 40	31 75
64	43 60	42 56	40 55	41 55	93 86

5 Conclusions

Our preliminary numerical results show that the DDMs considered have reasonable efficiency. However, there are too many approaches needing systematic experimental investigation to construct high-performance code. This concerns, in particular, the application of various optimized Schwarz methods [3, 8] with different values of parameter θ and coarse grid correction for overlapping or non-overlapping DDM. Of course, the problem of creating an adapted environment for robust SLAE solvers on modern supercomputers requires coordinated efforts of algebraists and programmers.

References

1. D.S. Butyugin, V.P. Il'in, D.V. Perevozkin, Parallel methods for solving the linear systems with distributed memory in the library Krylov (in Russian). Her. SUSU Ser. Comput. Math. Inf. **47**(306), 5–19 (2012)
2. Domain Decomposition Methods (2013). http://ddm.org
3. M.J. Gander, Optimized Schwarz methods. SIAM J. Num. Anal. **44**, 699–731 (2006)
4. V.P. Ilin, *Finite Element Methods and Technologies (in Russian)* (ICM&MG SBRAS Publishers, Novosibirsk, 2007)
5. Intel (R) Math Kernel Library (2013). http://software.intel.com/en-us/articles/intel-mkl/
6. NKS-30T Cluster (2009). http://www2.sscc.ru/HKC-30T/HKC-30T.htm
7. S. Pissanetsky, *Sparse Matrix Technology* (Academic, New York, 1984)
8. L. Qin, X. Xu, On a parallel Robin-type nonoverlapping domain decomposition method. SIAM J. Numer. Anal. **44**(6), 2539–2558 (2006). ISSN 0036-1429

9. Y. Saad, *Iterative Methods for Sparse Linear Systems*, 2nd edn. (Society for Industrial and Applied Mathematics, Philadelphia, PA, 2003). doi:10.1137/1.9780898718003. ISBN:0-89871-534-2. http://dx.doi.org/10.1137/1.9780898718003
10. 22nd International Conference on Domain Decomposition Methods (DD22) (2013). http://dd22.ics.usi.ch/

Isogeometric Overlapping Additive Schwarz Solvers for the Bidomain System

Lara Antonella Charawi

1 Introduction

The electrical activity of the heart is a complex phenomenon strictly related to its physiology, fiber structure and anatomy.

At the cellular level the cell membrane separates both the intra- and extracellular environments consisting of a dilute aqueous solution of dissolved salts dissociated into ions. Differences in ion concentrations on opposite sides of the membrane lead to a voltage called the *transmembrane potential*, v_M, defined as the difference between the intra- and extracellular potentials, (u_I and u_E). The bioelectric activity of a cardiac cell is described by the time course of v_M, the so called *action potential*. At the tissue level the most complete mathematical model of cardiac electrophysiology is the Bidomain model, consisting of a degenerate reaction-diffusion system of a parabolic and an elliptic partial differential equation modelling v_M and u_E of the anisotropic cardiac tissue, coupled nonlinearly with a membrane model. The multiscale nature of the Bidomain models yields very high computational costs for its numerical resolution. The starting point for a spatial discretization is a geometrical representation that encompasses the required anatomical and structural details, and that is also suitable for computational studies. Detailed models were proposed based on structured grids with cubic Hermite interpolation functions, which enable a smooth representation of ventricular geometry with relatively few elements, see e.g. [14]. In this study we used an alternative approach based on Isogeometric Analysis (IGA), a novel method for the discretization of partial differential equations introduced in [7]. This method adopts the same spline or Non-Uniform Rational B-spline (NURBS) basis functions used to design domain

L.A. Charawi (✉)
Department of Mathematics, Università di Pavia, Via Ferrata 1, 27100 Pavia, Italy
e-mail: lara.charawi@unipv.it

T. Dickopf et al. (eds.), *Domain Decomposition Methods in Science and Engineering XXII*, Lecture Notes in Computational Science and Engineering 104, DOI 10.1007/978-3-319-18827-0_11

geometries in CAD to construct both trial and test spaces in the discrete variational formulation of the differential problem, and provides a higher control on the regularity of the discrete space. The IGA discretization of the Bidomain model in space and semi-implicit (IMEX) finite differences in time lead to the resolution at each time step of a large and very ill-conditioned linear system. Since the iteration matrix is symmetric semidefinite, it is natural to use the preconditioned conjugate gradient method.

We have developed and analyzed an overlapping additive Schwarz preconditioner for the isogeometric discretization of the cardiac Bidomain model. We have proved that the resulting solver is scalable and optimal in the ratio of subdomain/overlap size. Several tests confirm the theoretical bound on three-dimensional NURBS domains. We note that Isogeometric overlapping Schwarz preconditioners were first introduced in [2] for scalar elliptic problems, while multilevel Schwarz preconditioners for FEM discretizations of the Bidomain system were studied in [10].

2 The Bidomain Model

The macroscopic Bidomain representation of cardiac tissue volume is obtained by considering the superposition of two anisotropic continuous media the intra- (I) and extra- (E) cellular media, coexisting at every point of the tissue and separated by a distributed continuous cellular membrane; see [12] for a derivation of the Bidomain model from homogenization of cellular model. The cardiac tissue consists of an arrangement of fibers that rotate counterclockwise from epi- to endocardium, and that have a laminar organization modeled as a set of muscle sheets running radially from epi- to endocardium, see [8]. The anisotropy of the intra- and extracellular media is described by the orthotropic conductivity tensors $D_I(x)$ and $D_E(x)$, see e.g. [4]. We denote by Ω the bounded physical region occupied by the cardiac tissue and introduce a parabolic-elliptic formulation of the Bidomain system. Given an extracellular applied stimulus per unit volume I^E_{app}, we seek the transmembrane and the extracellular potentials, v_M and u_E, respectively, and the gating variable w satisfying the system

$$\begin{cases} c_m \frac{\partial v_M}{\partial t} - \mathrm{div}(D_I \nabla(v_M + u_E)) + I_{ion}(v_M, w) = 0 & \text{on } \Omega \times (0,T) \\ -\mathrm{div}((D_I + D_E)\nabla u_E) - \mathrm{div}(D_I \nabla v_M) = I^E_{app} & \text{on } \Omega \times (0,T) \\ \frac{\partial w}{\partial t} - R(v_M, w) = 0 & \text{on } \Omega \times (0,T) \end{cases} \quad (1)$$

with insulating boundary conditions, suitable initial conditions on v_M, u_E and w, while c_m is the membrane capacitance per unit volume. The non-linear reaction term I_{ion}, the ionic current of the membrane per unit volume, and the ODE system for the gating variables are given by the chosen ionic membrane model. Here we will consider the (LR1) membrane model by Luo and Rudy [9]. The system uniquely determines v_M, while u_E is defined only up to a same additive time-dependent constant, chosen by imposing $\int_{\Omega} u_E \, d\mathbf{x} = 0$.

3 Discretization and Numerical Methods

Isogeometric Space Discretization In the three-dimensional case, our domain Ω, representing the left ventricle, is modeled by a family of truncated ellipsoids. According to the isoparametric approach we discretized the Bidomain system (1) with IGA based on NURBS basis functions, see e.g. [5]. NURBS functions are built from B-spline functions.

In what follows, let $d \geq 2$ be the dimension of the physical domain of interest. For any $\alpha = 1, \ldots, d$, we introduce the open knot vector, a set of non decreasing real numbers $\Xi_{\alpha} = \{0 = \xi_{1,\alpha}, \xi_{2,\alpha}, \ldots, \xi_{n_\alpha+p+1,\alpha} = 1\}$, where p is the order of the B-spline and n_α is the number of basis functions necessary to describe it. Given the knot vector, it is possible to define univariate B-spline basis functions, $B^p_{i,\alpha}(\xi)$, and by tensor product the multivariate B-spline basis functions, $B^p_{i_1,\ldots,i_d}$. Therefore the tensor product spline space living in the parametric domain is

$$\hat{V} := \text{span}\{B^p_{i_1 \ldots i_d}, \; i_\alpha = 1, \ldots, n_\alpha, 1 \leq \alpha \leq d\}.$$

Given $\omega_{i_1 \ldots i_d}$ the *weights* associated to $\mathbf{C}_{i_1 \ldots i_d}$, a mesh of *control points*, we can define the NURBS basis function on the parametric domain

$$R^p_{i_1 \ldots i_d}(\boldsymbol{\xi}) = \frac{B^p_{i_1 \ldots i_d}(\boldsymbol{\xi})\omega_{i_1 \ldots i_d}}{w(\boldsymbol{\xi})},$$

with $w(\boldsymbol{\xi}) := \sum_{i_i \ldots i_d}^{n_1 \ldots n_d} B^p_{i_1 \ldots i_d}(\boldsymbol{\xi})\omega_{i_1 \ldots i_d}$.

Since the single patch domain Ω is a NURBS region, we define a geometrical map $\mathbf{F} : (0, 1)^d \to \Omega$ as

$$\mathbf{F}(\xi) = \sum_{i_1=1}^{n_1} \cdots \sum_{i_d=1}^{n_d} R^p_{i_1 \ldots i_d}(\boldsymbol{\xi})\mathbf{C}_{i_1 \ldots i_d},$$

and the physical space V as the span of the pushforward of the NURBS basis functions

$$V := \text{span}\{R^p_{i_1 \ldots i_d} \circ \mathbf{F}^{-1}, i_\alpha = 1, \ldots, n_\alpha, 1 \leq \alpha \leq d\}.$$

A semidiscrete problem of (1) is obtained by applying a standard Galerkin procedure. We denote by M the mass matrix, by $A_{I,E}$ the symmetric stiffness matrices associated to the intra- and extra anisotropic conductivity tensors, respectively.

Time Discretization The time discretization is performed by a decoupled semi-implicit method consisting of the two following steps:

- Given $\mathbf{v}_M^n$, $\mathbf{u}_E^n$ and $\mathbf{w}^n$ at the previous step n, we first solve the ODEs system for the gating and ionic concentration variables. Since the membrane model employed is the LR1, the ODE integration approach is based on the Rush-Larsen method, see [13].
- Once computed $\mathbf{w}^{n+1}$, a semi-implicit scheme is applied to the reaction-diffusion part, see [1], i.e., by using the implicit Euler method for the diffusion term, while the nonlinear reaction term I_{ion} is treated explicitly. As a consequence at each time step we need to solve the linear system

$$\begin{bmatrix} \frac{c_m}{\Delta t}M + A_I & A_I \\ A_I & A_I + A_E \end{bmatrix} \begin{pmatrix} \mathbf{v}_M^{n+1} \\ \mathbf{u}_E^{n+1} \end{pmatrix} = \begin{pmatrix} \frac{c_m}{\Delta t}M\mathbf{v}_M^n - \mathbf{i}_{ion}(\mathbf{v}_M^n, \mathbf{w}^{n+1}) \\ \mathbf{I}_{app}^E \end{pmatrix} \tag{2}$$

imposing $\mathbf{1}^T M \mathbf{u_E}^{n+1} = 0$. Due to the ill-conditioning of the iteration matrix and the large number of unknowns required by realistic simulations of cardiac excitation in three-dimensional domains, a scalable and efficient preconditioner is required.

We recall that the linear system (2) is equivalent to the elliptic variational problem: given $f \in L^2(\Omega)$,

$$\text{find } u \in U \text{ such that } \quad a_{bido}(u, z) = (f, z_M) \quad \forall z = [z_M, z_E] \in U,$$

where $U := V \times \tilde{V}$, with $\tilde{V} := \{u_E \in V \, : \, \int_\Omega u_E = 0\}$, while for the definition and the properties of the bilinear form a_{bido} see [11].

4 Overlapping Schwarz Preconditioners

In this section, we construct an isogeometric overlapping additive Schwarz preconditioner for the Bidomain system, using the general framework developed in [2] for a model elliptic problem, and in [10] for the Bidomain system discretized using FEM.

For $\alpha = 1, \ldots, d$, we define a decomposition of the reference interval $\hat{I}$ selecting from the open knot vector Ξ_α a subset of $N_\alpha + 1$ nonrepeated interface knots $\{\xi_{i_{m_\alpha},\alpha}, m_\alpha = 1, \ldots, N_\alpha + 1 | \xi_{i_1,\alpha} = 0, \ \xi_{i_{N_\alpha+1},\alpha} = 1\}$. Thus, the closure of $\hat{I}$ can be decomposed into N_α intervals $\hat{I}_{m_\alpha,\alpha} := (\xi_{i_{m_\alpha},\alpha}, \xi_{i_{m_\alpha+1},\alpha})$, assuming that they have a similar diameter on order H. For each of the interface knots there exists at least one index $s_{m_\alpha,\alpha}$ such that $2 \leq s_{m_\alpha,\alpha} \leq n_\alpha - 1$ and so that the support of the basis function $B^p_{s_{m_\alpha,\alpha}}$ intersects both $\hat{I}_{m_\alpha-1,\alpha}$ and $\hat{I}_{m_\alpha,\alpha}$.

Let r be an integer counting the basis functions shared by adjacent subdomains. We are able to define N_α subspaces $\{\hat{V}_{m_\alpha,\alpha}\}_{m_\alpha=1}^{N_\alpha}$ forming an overlapping decomposition of the B-spline univariate space, $\hat{V}$, as

$$\hat{V}_{m_\alpha,\alpha} := \text{span}\{B^p_{j,\alpha}(\xi) | s_{m_\alpha,\alpha} - r \leq j \leq s_{m_\alpha+1,\alpha} + r\} \quad m_\alpha = 1, \dots, N_\alpha.$$

We build the coarse space $\hat{V}_{0,\alpha}$ from the partition of $\hat{I}$. Let

$$\Xi_{0,\alpha} = \{\xi_{1,\alpha}, \dots, \xi_{p,\alpha}, \xi_{i_1,\alpha}, \xi_{i_2,\alpha}, \dots, \xi_{i_{N_\alpha-1},\alpha}, \xi_{i_{N_\alpha},\alpha}, \xi_{i_{N_\alpha+1},\alpha}, \dots, \xi_{i_{N_\alpha+p+1},\alpha}\}$$

an open knot vector and let $\{\mathring{B}^p{}_{i,\alpha}\}_{i=1}^{N_{0,\alpha}}$ be the corresponding $N_{0,\alpha}$ basis functions, then the coarse space is

$$\hat{V}_{0,\alpha} = \text{span}\{\mathring{B}^p{}_{i,\alpha}, \quad i = 1 \dots N_{0,\alpha}\}.$$

In more than one dimension, we proceed by using tensor product. Let $N := \prod_1^d N_\alpha$, for $m = 1, \dots, N$ the local and the coarse subspaces are then

$$\hat{V}_m \equiv \hat{V}_{m_1,\dots,m_d} := \text{span}\{B^p_{i_1,\dots,i_d}, \ s_{m_\alpha} - r \leq i_\alpha \leq s_{m_\alpha+1} + r, \ \alpha = 1, \dots, d\};$$

$$\hat{V}_0 := \text{span}\{\mathring{B}^p{}_{i_1,\dots,i_d}, \quad i_\alpha = 1 \dots N_{0,\alpha}, \ \alpha = 1, \dots, d\}.$$

The decomposition of the NURBS space V and therefore of U in the physical domain is trivial:
$U_m := V_m \times V_m$ and $U_0 := V_0 \times \widetilde{V_0}$ with

$$V_m \equiv V_{m_1,\dots,m_d} := \text{span}\{R^p_{i_1,\dots,i_d} \circ \mathbf{F}^{-1}, \ s_{m_\alpha} - r \leq i_\alpha \leq s_{m_\alpha+1} + r, \ \alpha = 1, \dots, d\};$$
$$V_0 := \text{span}\{\mathring{R}^p{}_{i_1,\dots,i_d} \circ \mathbf{F}^{-1}, \ i_\alpha = 1 \dots N_{0,\alpha}, \ \alpha = 1, \dots, d\} \quad \text{and} \quad \widetilde{V_0} := V_0 \cap \tilde{V}.$$

We are now able to construct a two-level overlapping Additive Schwarz method for the Bidomain system (2). We remark that $U_0 \subset U$, whereas U_m is not a subset of U, $m = 1, \dots, N$. We define therefore the interpolation operators $\mathbf{I}_m : U_m \to U$ as

$$\text{given } u = (v_M, \ u_E) \in U_m, \ \mathbf{I}_m u = (\mathbf{I}_{m,M} u, \mathbf{I}_{m,E} u) := (v_M, u_E - \frac{1}{|\Omega|} \int_\Omega u_E),$$

whereas $\mathbf{I}_0 : U_0 \to U$ is simply the embedding operator. We define the local projectors operators $\tilde{\mathbf{T}}_m : U \to U_m$ for $m = 0, \dots, N$ by

$$a_{bido}(\tilde{\mathbf{T}}_m u, v) = a_{bido}(u, \mathbf{I}_m v) \quad \forall v \in U_m.$$

Defining $\mathbf{T}_m = \mathbf{I}_m \tilde{\mathbf{T}}_m$, the 2-level Overlapping Additive Schwarz (OAS) operator is then

$$\mathbf{T}_{OAS} := T_0 + \sum_{m=1}^{N} \mathbf{T}_m.$$

We have the following result about the convergence rate bound, see [3].

Theorem 1 *Under the assumptions that the parametric mesh is quasi-uniform and the overlap index r is bounded from above by a fixed constant, the condition number of the preconditioner operator T_{OAS} is bounded by*

$$\kappa(\mathbf{T}_{OAS}) \leq C\left(1 + \frac{H}{\delta}\right), \tag{3}$$

where $\delta := h(2r+2)$ is the overlap parameter and C is a constant independent of h, H, N and δ but not of p and the regularity k.

5 Numerical Results

Numerical results presented in this section refer to the 3D Bidomain problem on a portion of the truncated ellipsoid, representing a simplified ventricular geometry. The IGA discretization with mesh size h and polynomial degree p and regularity k is carried out by in MATLAB, using the library GeoPDEs [6]. The domain is decomposed in N overlapping subdomains of characteristic size H and overlap index r.

Table 1 shows the scalability of the 2-level OAS preconditioner for a 3D NURBS domain decomposed into an increasing number of subdomains, such that their size

Table 1 **OAS preconditioner in 3D ellipsoidal domain. Scalability test**: iteration counts (it.), condition number κ and extreme eigenvalues (λ_{max} and λ_{min}) as a function of the number of subdomains N for fixed $H/h = 4$ for unpreconditioned (Unpc.), 1-level and 2-level OAS preconditioners. $p = 3$, $k = 2$ and $r = 0, 1$

					2-level OAS			
	Unpc.		1-level OAS		$r = 0$		$r = 1$	
N	it.	κ	it.	$\kappa = \lambda_{max}/\lambda_{min}$	it.	$\kappa = \lambda_{max}/\lambda_{min}$	it.	$\kappa = \lambda_{max}/\lambda_{min}$
$2 \times 2 \times 1$	175	4.98e3	21	65=4.0/6.09e−2	12	11.07=4.74/4.12e−1	6	5.24 = 5.00/0.95
$3 \times 3 \times 2$	185	4.44e3	44	331=8.0/2.41e−2	22	32.13=8.60/2.72e−1	9	10.87 = 9.21/0.85
$4 \times 4 \times 3$	206	6.32e3	61	627=8.0/1.27e−2	23	31.90=8.63/2.73e−1	8	9.00 = 9.31/1.03
$5 \times 5 \times 4$	247	8.89e3	78	1020=8.0/7.84e−3	23	32.09=8.64/2.69e−1	8	10.39 = 9.20/0.89
$6 \times 6 \times 5$	297	1.20e4	94	1507=8.0/5.31e−3	23	31.60=8.64/2.27e−1	7	9.16 = 6.95/1.32

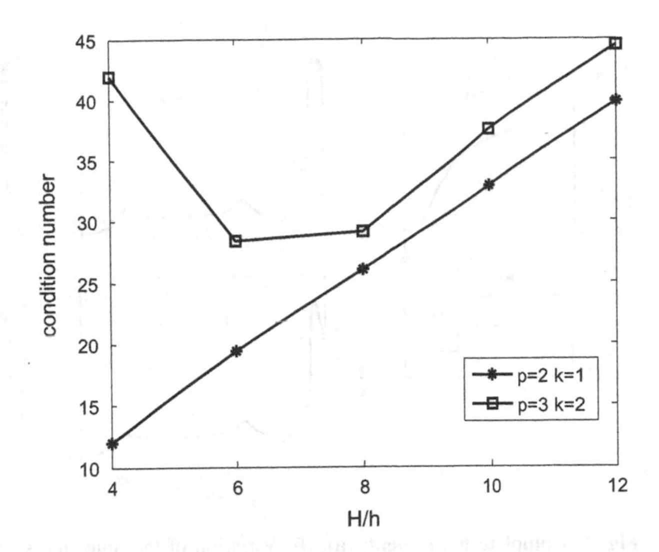

Fig. 1 2-Level OAS dependence on $\frac{H}{h}$: plot of κ as a function of $\frac{H}{h}$, for $p = 2, 3$ and $k = p - 1$

are fixed $\frac{H}{h} = 4, p = 3, k = 2$ and $r = 0, 1$. The simulation is run for 30 time steps, 1.5 ms, and the condition number is estimated using the usual Lanczos' method. As expected the 1-level preconditioner (without coarse problem) has a condition number growing with N, and the performances of the 2-level OAS improve when increasing the overlap size. Additional results, for $p = 3, 2$ and $k = p-1$, are plotted in Fig. 1, and confirm that the condition number, κ, of the 2-level preconditioned problem grows linearly with the increasing ratio $\frac{H}{h}$, as predicted by (3) using minimal overlap ($r = 0$).

Finally, Fig. 2 compares the variation of the condition number and iteration count during a complete heartbeat (300 ms) by using 1- and 2-level OAS solvers or unpreconditioned Conjugate Gradient. These variations are strictly related to the time step size (Δt), that changes according to the adaptive strategy described in [4], following the different phases of a ventricular action potential. In this test the number of the subdomains is $6 \times 6 \times 5$ and the ratio $\frac{H}{h} = 4$. We can note that the depolarization is the most intense computationally phase, nevertheless OAS solvers keep the condition number quite uniform for all the duration of the cycle. As expected, the 2-level greatly improves the conditioning of the problem.

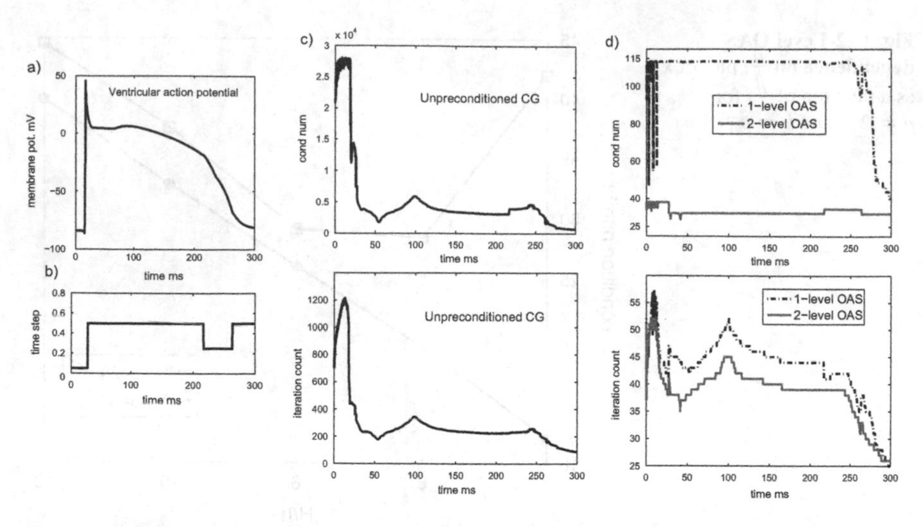

Fig. 2 **Complete heart beat**. (**a**), (**b**) Variation of the time step size following the phases of a ventricular action potential. (**c**), (**d**) Time course of κ (*upper panels*) and iteration count (*lower panels*) during a heartbeat: comparison between unpreconditioned operator (**c**) and 1- and 2-level OAS (**d**). $N = 6 \times 6 \times 5$, $\frac{H}{h} = 4$, $p = 3$, $k = 2$ and $r = 0$

References

1. U.M. Ascher, S.J. Ruuth, B.T.R. Wetton, Implicit-explicit methods for time-dependent partial differential equations. SIAM J. Numer. Anal. **32**(3), 797–823 (1995)
2. L. Beirão da Veiga, D. Cho, L.F. Pavarino, S. Scacchi, Overlapping Schwarz methods for isogeometric analysis. SIAM J. Numer. Anal. **50**(3), 1394–1416 (2012)
3. L.A. Charawi, Isogeometric overlapping Schwarz preconditioners in computational electrocardiology. Ph.D. thesis, Università degli Studi di Milano, 2014
4. P. Colli Franzone, L.F. Pavarino, B. Taccardi, Simulating patterns of excitation, repolarization and action potential duration with cardiac bidomain and monodomain models. Math. Biosci. **197**(1), 35–66 (2005)
5. J.A. Cottrell, T.J.R. Hughes, Y. Bazilevs, *Isogeometric Analysis: Toward Integration of CAD and FEA* (Wiley, Chichester, 2009)
6. C. De Falco, A. Reali, R. Vázquez, GeoPDEs: a research tool for isogeometric analysis of pdes. Adv. Eng. Softw. **42**(12), 1020–1034 (2011)
7. T.J.R. Hughes, J.A. Cottrell, Y. Bazilevs, Isogeometric analysis: CAD, finite elements, NURBS, exact geometry and mesh refinement. Comput. Methods Appl. Mech. Eng. **194** (39), 4135–4195 (2005)
8. I.J. LeGrice, B.H. Smaill, L.Z. Chai, S.G. Edgar, J.B. Gavin, P.J. Hunter, Laminar structure of the heart: ventricular myocyte arrangement and connective tissue architecture in the dog. Am. J. Physiol-Heart C **269**(2), H571–H582 (1995)
9. C.-H. Luo, Y. Rudy, A model of the ventricular cardiac action potential, depolarization, repolarization, and their interaction. Circ. Res. **68**(6), 1501–1526 (1991)
10. L.F. Pavarino, S. Scacchi, Multilevel additive Schwarz preconditioners for the bidomain reaction-diffusion system. SIAM J. Sci. Comput. **31**(1), 420–445 (2008)

11. L.F. Pavarino, S. Scacchi, Parallel multilevel Schwarz and block preconditioners for the Bidomain parabolic-parabolic and parabolic-elliptic formulations. SIAM J. Sci. Comput. **33**(4), 1897–1919 (2011)
12. M. Pennacchio, G. Savaré, P. Colli Franzone, Multiscale modeling for the bioelectric activity of the heart. SIAM J. Math. Anal. **37**(4), 1333–1370 (2005)
13. S. Rush, H. Larsen, A practical algorithm for solving dynamic membrane equations. IEEE Trans. Biomed. Eng. **25**(4), 389–392 (1978)
14. N.P. Smith, D.P. Nickerson, E.J. Crampin, P.J. Hunter, Multiscale computational modelling of the heart. Acta Numer. **13**(1), 371–431 (2004)

On the Minimal Shift in the Shifted Laplacian Preconditioner for Multigrid to Work

Pierre-Henri Cocquet and Martin J. Gander

1 Introduction

Multigrid is an excellent iterative solver for discretized elliptic problems with diffusive nature, see [12] and the references therein. It is natural that substantial research was devoted to extend the multigrid method for solving the Helmholtz equation

$$-\Delta u - k^2 u = f \quad \text{in } \Omega \tag{1}$$

with the same efficiency, but it turned out that this is a very difficult task. Textbooks mention that there are substantial difficulties, see [3, p. 72], [11, p. 212], [12, p. 400], and also the review [7] for why in general iterative methods have difficulties when applied to the Helmholtz equation (1).

Motivated by the early proposition in [2] to use the Laplacian to precondition the Helmholtz equation, the shifted Laplacian has been advocated over the past decade as a way of making multigrid work for the indefinite Helmholtz equation, see [1, 4–6, 10] and references therein. The idea is to shift the wave number into the complex plane to obtain a good preconditioner for a Krylov method when solving (1). The hope is that due to the shift, it becomes possible to use standard multigrid to invert the preconditioner, and if the shift is not too big, it is still an effective preconditioner for the Helmholtz equation with a real wave number. This implies however two conflicting requirements: the shift should be not too large for the shifted preconditioner to be a good preconditioner, and it should be large enough for multigrid to work. It was already indicated in [7] that it is not possible to satisfy

P.-H. Cocquet • M.J. Gander (✉)
University of Geneva, 2-4 rue du Lièvre, CP 64, 1211 Genève, Switzerland
e-mail: martin.gander@unige.ch; pierre-henri.cocquet@unige.ch

T. Dickopf et al. (eds.), *Domain Decomposition Methods in Science and Engineering XXII*, Lecture Notes in Computational Science and Engineering 104, DOI 10.1007/978-3-319-18827-0_12

both these requirements, see also [4]. It was then rigorously proved in [9] that the preconditioner is effective, i.e. iteration numbers stay bounded independently of the wave number k if the shift is at most of the size of the wavenumber. We prove here rigorously for a one dimensional model problem that if the complex shift is less than the size of the wavenumber squared, multigrid will not work. It is therefore not possible to solve the shifted Laplace preconditioner with multigrid in the regime where it is a good preconditioner. We also show that if the complex shift is of the size of the wave number squared and the constant is large enough, then multigrid will solve the preconditioner independently of the wave number k. For a different shift idea as a dispersion correction, where the shift is real and one obtains in one dimension a multigrid solver with standard components that solves the original Helmholtz problem (1) independently of the wave number, see [8].

2 Model Problem and Discretization

To study the shifted Laplacian preconditioner for the Helmholtz equation (1) in 1d, we consider the 1d shifted Helmholtz equation

$$-u''(x) - (k^2 + i\varepsilon)u(x) = f(x) \quad x \text{ in } (0, 1) \tag{2}$$

with homogeneous Dirichlet boundary conditions $u(0) = u(1) = 0$. We discretize (2) using a standard 3-point centered finite difference approximation on a uniform mesh with n interior grid points and mesh size $h = 1/(n+1)$ to get a linear system $A_h\mathbf{u} = \mathbf{f}$ with system matrix

$$A_h = \frac{1}{h^2}\text{tridiag}(-1, 2 - (k^2 + i\varepsilon)h^2, -1). \tag{3}$$

It is this system matrix which is used as a preconditioner for solving (1), and therefore following the idea of the shifted Laplacian preconditioner, systems with this matrix have to be solved effectively using multigrid. We analyze here in detail a two grid method: we use a Jacobi smoother,

$$\mathbf{u}_{m+1} = \mathbf{u}_m + \omega D^{-1}(\mathbf{f} - A_h\mathbf{u}_m),$$

where $D = \text{diag}(A)$, and ω is a relaxation parameter, which we choose here based on the optimal choice of the case without relaxation, see [8], to be

$$\omega := \frac{2 - (k^2 + i\varepsilon)h^2}{3 - (k^2 + i\varepsilon)h^2}.$$

For the coarse correction, we assume n to be a power of two minus one, and use the extension operator based on interpolation,

$$I_H^h = \frac{1}{2}\begin{bmatrix} 2 & & & \\ 1 & 1 & & \\ & 2 & & \\ & 1 & & \\ & & \ddots & 1 \\ & & & 2 \\ & & & 1\ 1 \\ & & & \ \ 2 \end{bmatrix} \in \mathbb{R}^{(n+2)\times(N+2)}, \quad N := (n+1)/2 - 1,$$

and for the restriction $I_h^H = \frac{1}{2}(I_H^h)^T$, with the coarse grid matrix obtained by Galerkin projection, $A_H := I_h^H A_h I_H^h$. The resulting two grid operator with ν_1 pre-smoothing and ν_2 post-smoothing steps is then of the form

$$T := (I - \omega D^{-1}A_h)^{\nu_1}(I - I_H^h A_H^{-1} I_h^H A_h)(I - \omega D^{-1}A_h)^{\nu_2}. \tag{4}$$

Using the subspaces

$$\operatorname{span}\{\mathbf{v}_1^h, \mathbf{v}_n^h\},\ \operatorname{span}\{\mathbf{v}_2^h, \mathbf{v}_{n-1}^h\},\ \ldots,\ \operatorname{span}\{\mathbf{v}_N^h, \mathbf{v}_{N+2}^h\},\ \operatorname{span}\{\mathbf{v}_{N+1}^h\} \tag{5}$$

defined by the eigenfunctions of A_h given by $\mathbf{v}_j^h := [\sin j\ell\pi h]_{\ell=1}^n, j = 1, \ldots, n$, one can block diagonalize the two grid operator (4), see [8]. The action of T on these one- and two-dimensional subspaces is represented by the block diagonal matrix $\operatorname{diag}(T_1, \ldots, T_N, T_{N+1})$ with

$$T_j = \begin{bmatrix} \sigma_j & 0 \\ 0 & \sigma_{j'} \end{bmatrix}^{\nu_2} \begin{bmatrix} 1 - c_j^4 \frac{\lambda_j^h}{\lambda_j^H} & c_j^2 s_j^2 \frac{\lambda_{j'}^h}{\lambda_j^H} \\ c_j^2 s_j^2 \frac{\lambda_j^h}{\lambda_j^H} & 1 - s_j^4 \frac{\lambda_{j'}^h}{\lambda_j^H} \end{bmatrix} \begin{bmatrix} \sigma_j & 0 \\ 0 & \sigma_{j'} \end{bmatrix}^{\nu_1}, \quad T_{N+1} = \sigma_{N+1}^{\nu_1+\nu_2}, \tag{6}$$

where $c_j := \cos\frac{j\pi h}{2}$, $s_j := \sin\frac{j\pi h}{2}$, $j = 1, \ldots, N$, $\sigma_j := 1 - \omega(1 - \frac{2\cos(j\pi h)}{2-(k^2+i\varepsilon)h^2})$, $j = 1, \ldots, n$, and

$$\lambda_j^h := \frac{4}{h^2}\sin^2\frac{j\pi h}{2} - (k^2 + i\varepsilon), \qquad j = 1, \ldots, n, \tag{7}$$

$$\lambda_j^H := \frac{4}{H^2}\sin^2\frac{j\pi H}{2} - (k^2 + i\varepsilon), \qquad j = 1, \ldots, N, \tag{8}$$

are the eigenvalues of A_h and A_H, with $j' := N + 1 - j$ denoting the complementary mode index. To prove convergence of the two grid method, one has to prove that the spectral radius of T_j is smaller than one for all $j = 1, \ldots, N + 1$, since this implies that the spectral radius of the two grid operator T is less than one. We will show in the next section that if the shift is not large enough, the spectral radius of T will actually be bigger than one, and hence the two grid method can not be convergent.

3 Analysis

We first study the case of a shift $\varepsilon = Ck^{2-\delta}$, $0 \leq \delta < 2$. The following theorem shows that with such a shift, it is not possible to obtain robust multigrid convergence, because for any small mesh parameter h, there exists a wavenumber of the Helmholtz equation for which the two grid method diverges.

Theorem 1 (Divergence with Too Small Shift) *Assume that we are performing $\nu = \nu_1 + \nu_2$ smoothing steps and that $\varepsilon = Ck^{2-\delta}$ for $2 > \delta \geq 0$. Then, for h small enough, there exists a wavenumber $k(h)$ such that the spectral radius of the two grid method satisfies*

$$\rho(T) \geq \left(\frac{3^{\delta/2}}{3Ch^{\delta}}\right)^{\nu} + o\left(\frac{1}{h^{\delta\nu}}\right),$$

and hence the two grid method diverges for this mesh size and wavenumber.

Proof Denoting by μ_j the eigenvalues of the iteration operator T we have

$$\rho(T) \geq |\mu_j|,\ j = 1, \cdots, n.$$

Using the block diagonal form of the two-grid iteration matrix we have obtained in (6), we have in particular

$$\rho(T) \geq |\sigma_{N+1}^{\nu_1+\nu_2}| = |1-\omega|^{\nu} = |\mu_{N+1}|^{\nu},$$

with

$$|\mu_{N+1}| := \frac{1}{\sqrt{(3-k^2h^2)^2 + C^2h^4k^{4-2\delta}}}. \tag{9}$$

We now wish to find the maximum of $|\mu_{N+1}|$ as a function of the wavenumber k. Taking a derivative with respect to k, we obtain

$$\partial_k(|\mu_{N+1}(k)|^2) = k^{2\delta+1}\frac{2h^2(C^2k^2\delta h^2 - 2C^2k^2h^2 - 2k^{2\delta+2}h^2 + 6k^{2\delta})}{(C^2k^4h^4 + h^4k^{4+2\delta} - 6k^{2+2\delta}h^2 + 9k^{2\delta})^2},$$

and hence the maximum is reached at $k(h)$ satisfying

$$C^2k^2(\delta-2)h^2 - 2k^{2\delta+2}h^2 + 6k^{2\delta} = 0. \tag{10}$$

Since this equation can not be solved in closed form, we compute an asymptotic expansion of $k(h)$ for small mesh size h. We make the Ansatz

$$k(h) = \frac{\alpha_0}{h} + o\left(\frac{1}{h}\right)$$

and obtain for h small enough the expansions

$$k(h)^2 = \frac{\alpha_0^2}{h^2} + o\left(\frac{1}{h^2}\right), \quad k(h)^{2\delta} = \frac{\alpha_0^{2\delta}}{h^{2\delta}} + o\left(\frac{1}{h^{2\delta}}\right).$$

Substituting the above expressions into the equation (10) satisfied by $k(h)$ and considering only the leading order terms, we find

$$\frac{1}{h^{2\delta}}\left(6\alpha_0^{2\delta+2} - 2\alpha_0^{2\delta}\right) + o\left(\frac{1}{h^{2\delta}}\right) = 0,$$

and therefore

$$\alpha_0 = \sqrt{3},$$

and one can check that this is indeed asymptotically a maximum. We now replace the asymptotic expansion of $k(h)$ into the expression for $|\mu_{N+1}(k(h))|$ given in (9). Since $k(h)h = \sqrt{3} + o(1)$, a Taylor expansion shows that

$$\rho(T) \geq |\mu_{N+1}(k(h))| = \frac{1}{\sqrt{(3 - k(h)^2h^2)^2 + C^2h^4k(h)^{4-2\delta}}} = \frac{3^{\delta/2}}{3Ch^{\delta}} + o\left(\frac{1}{h^{\delta}}\right),$$

which gives the result.

Remark 1 In our proof, we only gave the first term of the asymptotic expansion of $k(h)$, since this was sufficient to obtain divergence. One could however compute the asymptotic expansion also to any order without additional difficulties.

Now we study the case $\varepsilon = Ck^2$. Substituting this value into the blocks (6) of the block diagonal representation, we notice that the matrices become homogeneous functions of the product kh. One can therefore study the spectral radius directly as a function of $kh > 0$ and $c_j \in (0, 1)$, using trigonometric formulas to replace the dependency on s_j. We show in Fig. 1 on the left for $\nu_1 = 1$, $\nu_2 = 0$ the maximum over all kh of the spectral radius of the matrix T as a function of C for $\varepsilon = Ck^2$. We clearly see that for C small, multigrid does not converge. For C larger however, we get convergence. The value C^* where the spectral radius equals one can be computed, it is $C^* = 0.3850$. We show on the right in Fig. 1 the spectrum of the blocks T_j, represented as a continuous function of $c_j \in (0, 1)$ and kh for $C = C^*$, and one can clearly see where the maximum value one is reached.

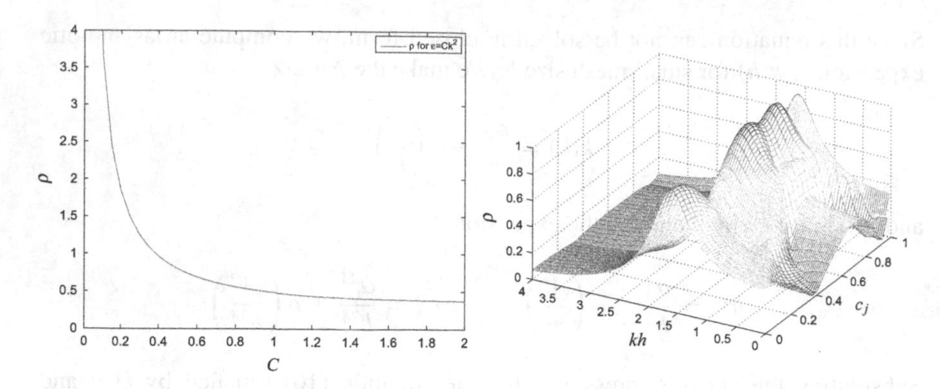

Fig. 1 Maximum over kh of the spectral radius of the two grid operator for shift $\varepsilon = Ck^2$ as a function of C on the left, and for $C = 0.3850$ the spectrum as a function of kh and c_j on the right

Remark 2 The value C^* is larger than the limiting value $C = 1/3$ found from the limiting case as δ goes to zero in Theorem 1 for which divergence can be guaranteed. This is because Theorem 1 only provides a lower bound for which divergence can be guaranteed. As we see from the sharper analysis above, divergence even set in a bit earlier.

Remark 3 From Fig. 1 on the left, we also see that making C very large will eventually not lead to further improvement, the curve has an asymptote which one can compute to be at $1/3$. Hence, the best contraction factor one can achieve with the two grid method applied to the shifted Helmholtz equation with shift $\varepsilon = Ck^2$ for C large in our example is $1/3$. Note also that the two grid convergence is uniform in k as soon as $C > C^*$.

4 Numerical Experiments

We present in this section several numerical illustrations of Theorem 1 and our additional estimate for the shift $\varepsilon = Ck^2$. We assume that the source term in the shifted Helmholtz equation (2) is $f = 0$ giving $u = 0$ as the unique solution. We use for our simulations the parameters

$$n = 511, \quad h = \frac{1}{512}, \quad k = \frac{\sqrt{3}}{h}, \quad \nu = 1,$$

so that we are in the regime of Theorem 1 where divergence should be observed if the shift is not sufficient. We perform twenty iterations of the two grid method applied to the shifted problem, starting with a random initial guess.

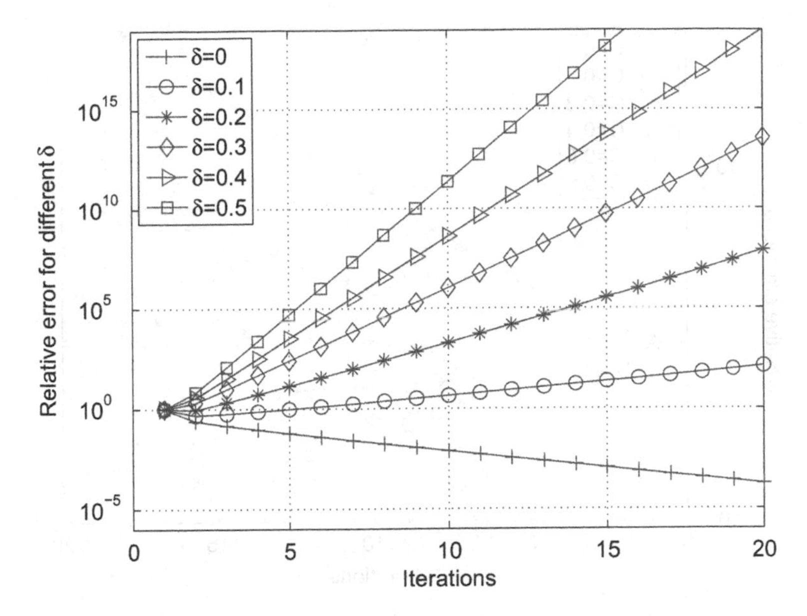

Fig. 2 Relative error versus iteration index for $C = 0.45$ and various values of $\delta = 0.2$

We first illustrate the result of Theorem 1. We choose $C = 0.45$ in the shift $\varepsilon = Ck^{2-\delta}$. Figure 2 shows the relative error of the two-grid scheme versus the number of iterations for various values of δ. We see that the two grid method converges for $\delta = 0$, but diverges for all other values $\delta > 0$. For the value of $h = 1/512$ in our experiment, and the constant $C = 0.45$, we see that the two grid method would still converge for a very small, but positive value of δ. This is not in disagreement with Theorem 1, which only makes a statement for h small enough.

We next show an experiment to illustrate that even with the shift $\varepsilon = Ck^2$, the constant still needs to be bigger than $C^* = 0.3850$ for the two grid method to converge, see also Remark 2. In Fig. 3 we show the relative error versus the iteration index for various values of C in this case. We observe that for $C < C^*$ the multigrid method does not converge, the shift is not enough. For $C > C^*$ however the multigrid method converges, and convergence gets faster as C increases, as expected. There is however a limit, as we have seen in Remark 3, the contraction factor of the two grid method will not be better than $1/3$.

5 Conclusions

We have analyzed for the shifted Helmholtz operator how large a shift of the form $\varepsilon = Ck^{2-\delta}$ has to be to obtain a uniformly convergent two grid method. We have proved for a one dimensional model problem that uniform convergence in the wavenumber k is not possible if $\delta > 0$. For $\delta = 0$, we have shown that if the constant

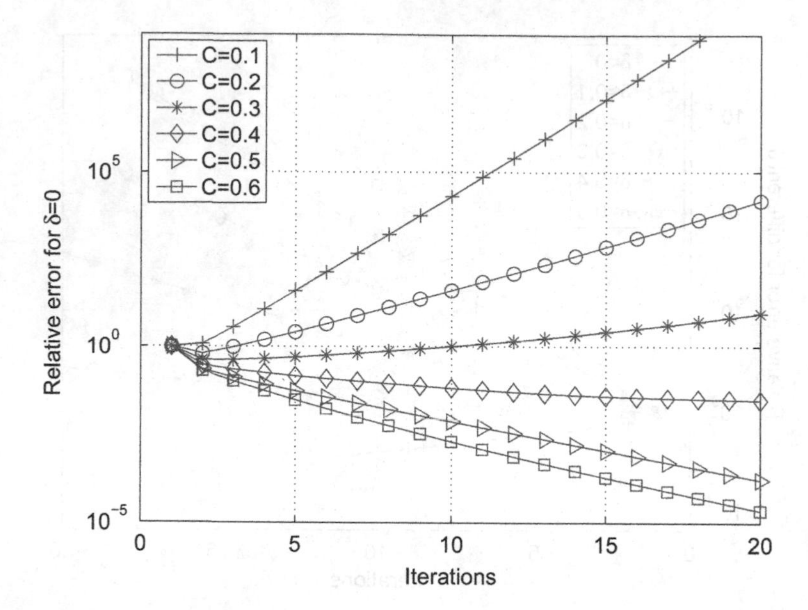

Fig. 3 Relative error versus iteration index for $\delta = 0$ and various values of C

$C > C^* = 0.3850$, then uniform convergence in the wavenumber k can be achieved. Our results are for the particular case of a one dimensional problem with a second order finite difference discretization, using a Galerkin coarse grid correction with full weighting and a Jacobi smoother with particular relaxation parameter. Using a different relaxation parameter, for example $\omega = 2/3$, leads to slightly worse results in this case, e.g. C^* becomes approximately 0.75 instead of 0.3850. Our analysis can be generalized, for example to higher dimensions, or other discretizations.

There is therefore indeed a big gap in the requirements for using the shifted Laplacian as a preconditioner when solving discretized Helmholtz problems: for multigrid to invert the preconditioner efficiently, the shift needs to be $O(k^2)$, but to prove that the preconditioner is effective, the shift needed to be at most $O(k)$, see [9], where numerical experiments also indicate that this estimate is sharp. Any compromise with the shift, i.e. using a shift of $O(k^\alpha)$ with $\alpha \in (1,2)$, will therefore lead to a preconditioner which is outside the requirements one would like to impose.

References

1. T. Airaksinen, E. Heikkola, A. Pennanen, J. Toivanen, An algebraic multigrid based shifted-laplacian preconditioner for the Helmholtz equation. J. Comput. Phys. **226**(1), 1196–1210 (2007)
2. A. Bayliss, C. Goldstein, E. Turkel, An iterative method for the Helmholtz equation. J. Comput. Phys. **49**, 443–457 (1983)

3. A. Brandt, O.E. Livne, in *Multigrid Techniques, 1984 Guide with Applications to Fluid Dynamics*, Revised Edition. Classics in Applied Mathematics, vol. 67 (SIAM, Philadelphia, 2011)
4. S. Cools, W. Vanroose, Local Fourier analysis of the complex shifted Laplacian preconditioner for Helmholtz problems. Numerical Linear Algebra with Applications **19**(2), 232–252 (2013)
5. Y. Erlangga, Advances in iterative methods and preconditioners for the Helmholtz equation. Arch. Comput. Meth. Eng. **15**, 37–66 (2008)
6. Y. Erlangga, C. Vuik, C. Oosterlee, On a class of preconditioners for solving the Helmholtz equation. Appl. Numer. Math. **50**, 409–425 (2004)
7. O. Ernst, M. Gander, Why it is Difficult to Solve Helmholtz Problems with Classical Iterative Methods, in *Numerical Analysis of Multiscale Problems*, ed. by I. Graham, T. Hou, O. Lakkis, R. Scheichl (Springer, Berlin, 2012), pp. 325–363
8. M. Gander, O. Ernst, Multigrid Methods for Helmholtz Problems: A Convergent Scheme in 1d Using Standard Components, in *Direct and Inverse Problems in Wave Propagation and Applications* (De Gruyter, Boston, 2013), pp. 135–186
9. M. Gander, I.G. Graham, E.A. Spence, How should one choose the shift for the shifted laplacian to be a good preconditioner for the Helmholtz equation? Numer. Math. (2015). doi:10.1007/s00211-015-0700-2
10. M.V. Gijzen, Y. Erlangga, C. Vuik, Spectral analysis of the discrete Helmholtz operator preconditioned with a shifted Laplacian. SIAM J. Sci. Comput. **29**(5), 1942–1958 (2007)
11. W. Hackbusch, *Multi-Grid Methods and Applications* (Springer, Berlin, 1985)
12. U. Trottenberg, C.C.W. Oosterlee, A. Schüller, *Multigrid* (Academic Press, New York, 2001)

Multitrace Formulations and Dirichlet-Neumann Algorithms

Victorita Dolean and Martin J. Gander

1 Introduction

Multitrace formulations (MTF) for boundary integral equations (BIE) were developed over the last few years in [1, 2, 4] for the simulation of electromagnetic problems in piecewise constant media, see also [3] for associated boundary integral methods. The MTFs are naturally adapted to the developments of new block preconditioners, as indicated in [5], but very little is known so far about such associated iterative solvers. The goal of our presentation is to give an elementary introduction to MTFs, and also to establish a natural connection with the more classical Dirichlet-Neumann algorithms that are well understood in the domain decomposition literature, see for example [6, 7]. We present for a model problem a convergence analysis for a naturally arising block iterative method associated with the MTF, and also first numerical results to illustrate what performance one can expect from such an iterative solver.

V. Dolean (✉)
University of Strathclyde, Glasgow, UK
e-mail: Victorita.Dolean@strath.ac.uk

M.J. Gander
University of Geneva, Geneva, Switzerland
e-mail: martin.gander@unige.ch

T. Dickopf et al. (eds.), *Domain Decomposition Methods in Science and Engineering XXII*, Lecture Notes in Computational Science and Engineering 104, DOI 10.1007/978-3-319-18827-0_13

2 One-Dimensional Example

In this section we introduce the Calderon projectors and the multitrace formulation for the one dimensional model problem

$$Au := u''(x) - a^2 u(x) = 0, \quad a > 0. \tag{1}$$

The family of bounded solutions of (1) on the domains $\Omega^\pm = \mathbb{R}^\pm$ is given by $u(x) = Ce^{\mp ax}$, where $C = u(0)$. We say that the solution spaces of the operator A on $\mathbb{R}^\pm$ are given by

$$Z^\pm = \{u \in L^2(\Omega) | u(x) = Ce^{\mp ax},\ C \in \mathbb{R}\} = \mathbb{R}e^{\mp ax}.$$

Note that any $u_\pm \in Z^\pm$ satisfies the relation $u'_\pm(0) = \pm a u_\pm(0)$ and thus the space of all possible Cauchy data of the solutions of (1) on $\mathbb{R}^\pm$ is given by

$$V^\pm = \{(g_0, g_1) = C(1, \pm a),\ C \in \mathbb{R}\} = \mathbb{R}\begin{pmatrix} 1 \\ \pm a \end{pmatrix}.$$

Definition 1 (Calderon Projectors) Let $\rho^\pm : Z^\pm \to V^\pm$ be the operator that associates to any solution of $Au = 0$ on $\mathbb{R}^\pm$ its pair of traces $(u(0), u'(0))$. Let $K^\pm : \mathbb{R}^2 \to Z^\pm$ be the operator that associates to any pair $(h_0, h_1) \in \mathbb{R}^2$ the quantity $K^\pm(h_0, h_1) = c_\mp e^{\mp ax}$, where $u(x) = c_+ e^{ax} + c_- e^{-ax}$ is the unique solution of (1) with Cauchy data (h_0, h_1),

$$Au = 0,\ u(0) = h_0 \text{ and } u'(0) = h_1. \tag{2}$$

Calderon projectors are defined as the projections $P^\pm : \mathbb{R}^2 \to V^\pm$, such that

$$P^\pm = \rho^\pm \circ K^\pm.$$

The expressions of $P^\pm$ for our model problem can be computed explicitly. The solution of (2) is

$$u(x) = \frac{1}{2a}(ah_0 + h_1)e^{ax} + \frac{1}{2a}(ah_0 - h_1)e^{-ax},$$

and thus $K^\pm(h_0, h_1) = \frac{1}{2a}(ah_0 \mp h_1)e^{\mp ax}$ and

$$P^\pm(h_0, h_1) := (\rho^\pm \circ K^\pm)(h_0, h_1) = \begin{pmatrix} \frac{1}{2a}(ah_0 \mp h_1) \\ \mp\frac{1}{2}(ah_0 \mp h_1) \end{pmatrix} \Rightarrow P^\pm = \begin{bmatrix} \frac{1}{2} & \mp\frac{1}{2a} \\ \mp\frac{a}{2} & \frac{1}{2} \end{bmatrix}.$$

Remark 1 From the previous construction we see that the Calderon projector is unique. When working with subdomains, it is however more convenient to introduce normal derivatives at interfaces, instead of $u'(0)$, and we thus define the Calderon projectors for normal derivatives with the modified sign

$$\mathbb{P}^{\pm}(h_0, h_1) := P^{\pm}(h_0, \mp h_1) \Rightarrow \mathbb{P}^{+} = \mathbb{P}^{-} = \begin{bmatrix} \frac{1}{2} & \frac{1}{2a} \\ \frac{a}{2} & \frac{1}{2} \end{bmatrix}, \tag{3}$$

and we will use $\mathbb{P}^{\pm}$ in what follows.

Definition 2 (Cauchy Traces) Following the notations in [4], we denote by

$$\mathbb{T}^{\pm}u := \begin{pmatrix} u(0) \\ \mp u'(0) \end{pmatrix} \tag{4}$$

the Cauchy trace (Dirichlet and Neumann) on the boundary $\{x = 0\}$ of a solution u of the equation $Au = 0$ posed on the half space $\mathbb{R}^{\pm}$.

Suppose now we have a decomposition of $\mathbb{R}$ into two subdomains $\Omega_1 = \Omega^-$ and $\Omega_2 = \Omega^+$ and we want to solve Eq. (1) by an iterative algorithm involving Dirichlet and Neumann traces on the interface $\{x = 0\}$. Let $\mathbb{T}_{1,2}$ be the trace operators as defined in (4) ($\mathbb{T}_1 = \mathbb{T}^-$ and $\mathbb{T}_2 = \mathbb{T}^+$) for the subdomains $\Omega_{1,2}$, and $\mathbb{P}_{1,2}$ the corresponding Calderon projectors as defined in (3) ($\mathbb{P}_1 = \mathbb{P}^-$ and $\mathbb{P}_2 = \mathbb{P}^+$) .

Definition 3 (Multitrace Formulation) The *multitrace formulation* from [4] states that the pairs $(\mathbb{T}_i u_i)_{i=1,2}$ are traces of the solution defined on Ω_i if they verify the relations

$$\begin{cases} (\mathbb{P}_1 - I)\mathbb{T}_1 u_1 - \sigma_1 \left(\mathbb{T}_1 u_1 - \begin{pmatrix} 1 & 0 \\ 0 & -1 \end{pmatrix} \mathbb{T}_2 u_2 \right) = 0, \\ (\mathbb{P}_2 - I)\mathbb{T}_2 u_2 - \sigma_2 \left(\mathbb{T}_2 u_2 - \begin{pmatrix} 1 & 0 \\ 0 & -1 \end{pmatrix} \mathbb{T}_1 u_1 \right) = 0, \end{cases} \tag{5}$$

where $\sigma_{1,2}$ are some relaxation parameters.

We see that a natural iterative method (also introduced in [5]) for (5) starts with some initial guesses $(u_i^0, v_i^0)_{i=1,2}$ for the traces, and computes for $n = 1, 2, \ldots$ the new trace pairs from the relations

$$\begin{cases} (\mathbb{P}_1 - I) \begin{pmatrix} u_1^n \\ v_1^n \end{pmatrix} - \sigma_1 \begin{pmatrix} u_1^n \\ v_1^n \end{pmatrix} = -\sigma_1 \begin{pmatrix} u_2^{n-1} \\ -v_2^{n-1} \end{pmatrix}, \\ (\mathbb{P}_2 - I) \begin{pmatrix} u_2^n \\ v_2^n \end{pmatrix} - \sigma_2 \begin{pmatrix} u_2^n \\ v_2^n \end{pmatrix} = -\sigma_2 \begin{pmatrix} u_1^{n-1} \\ -v_1^{n-1} \end{pmatrix}. \end{cases} \tag{6}$$

By introducing the expressions of $\mathbb{P}_i$, we can rewrite the iteration in the form

$$\begin{bmatrix} -(\sigma_1+\frac{1}{2}) & \frac{1}{2a} & & \\ \frac{a}{2} & -(\sigma_1+\frac{1}{2}) & & \\ & & -(\sigma_2+\frac{1}{2}) & \frac{1}{2a} \\ & & \frac{1}{2} & -(\sigma_2+\frac{1}{2}) \end{bmatrix} \begin{pmatrix} u_1^n \\ v_1^n \\ u_2^n \\ v_2^n \end{pmatrix} = \begin{pmatrix} -\sigma_1 u_2^{n-1} \\ \sigma_1 v_2^{n-1} \\ -\sigma_2 u_1^{n-1} \\ \sigma_2 v_1^{n-1} \end{pmatrix}, \tag{7}$$

or when solving for the new iterates

$$\begin{pmatrix} u_1^n \\ v_1^n \\ u_2^n \\ v_2^n \end{pmatrix} = \begin{bmatrix} 0 & A_1 \\ A_2 & 0 \end{bmatrix} \begin{pmatrix} u_1^{n-1} \\ v_1^{n-1} \\ u_2^{n-1} \\ v_2^{n-1} \end{pmatrix} =: A \begin{pmatrix} u_1^{n-1} \\ v_1^{n-1} \\ u_2^{n-1} \\ v_2^{n-1} \end{pmatrix}, \tag{8}$$

where

$$A_i = \frac{1}{2(\sigma_i+1)} \begin{bmatrix} 2\sigma_i+1 & -\frac{1}{a} \\ a & -(1+2\sigma_i) \end{bmatrix}, \quad i = 1,2.$$

The convergence factor of (6) is therefore given by the spectral radius of the iteration matrix A, whose eigenvalues are

$$\lambda(A) := \left\{ -\sqrt{\frac{\sigma_1}{\sigma_1+1}}, \sqrt{\frac{\sigma_1}{\sigma_1+1}}, -\sqrt{\frac{\sigma_2}{\sigma_2+1}}, \sqrt{\frac{\sigma_2}{\sigma_2+1}} \right\}. \tag{9}$$

We see that the convergence factor is independent of a and thus only depends on the relaxation parameters σ_i. If we suppose by symmetry that $\sigma_1 = \sigma_2 =: \sigma$, the convergence factor becomes $\rho(A) = \sqrt{\frac{\sigma}{\sigma+1}}$, and we show a plot of $\rho(A)$ as a function of σ in Fig. 1. We see that the algorithm diverges for $\sigma < -\frac{1}{2}$, stagnates for $\sigma = -\frac{1}{2}$ and converges for $\sigma > -\frac{1}{2}$. For $\sigma = 0$, the convergence factor vanishes, but a closer look at the iteration formula (7) shows that the matrix is then singular and thus the algorithm is no longer well defined for this value. On the other hand, the associated iteration (8) is still well defined, the latter being equivalent to (7) only for $\sigma \neq 0$. Overall, we see that algorithm (7) converges rapidly when the relaxation parameter is chosen close to 0.

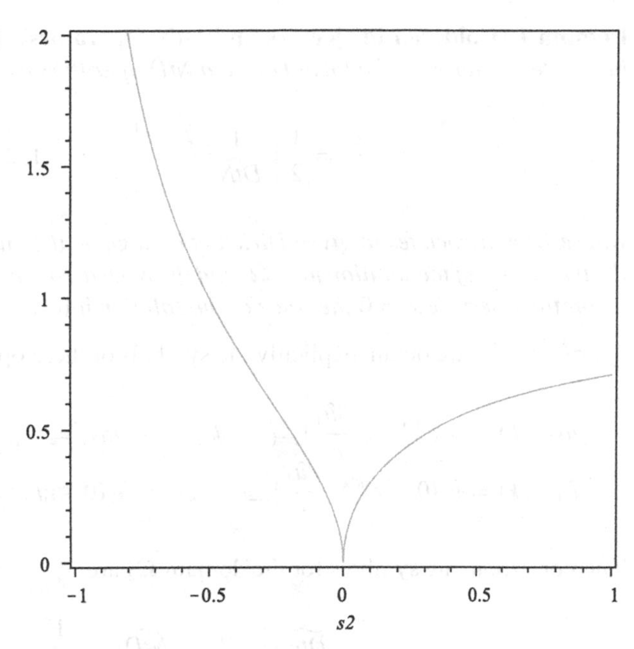

Fig. 1 Convergence factor of the iterative multitrace formulation in 1d as function of the relaxation parameter σ

3 Two-Dimensional Example

Suppose we want to solve the Laplace equation

$$u_{xx} + u_{yy} = 0, \quad \text{in } \Omega = \mathbb{R}^2, \tag{10}$$

using the two subdomains $\Omega_1 := \mathbb{R}^- \times \mathbb{R}$ and $\Omega_2 := \mathbb{R}^+ \times \mathbb{R}$ and a multitrace formulation. To use our results from the previous section we take a Fourier transform in the y variable,

$$\hat{u}_{xx} - k^2 \hat{u} = 0.$$

We can now follow the reasoning of the previous section in Fourier space, replacing a by $|k|$. Thus any given pair of boundary functions $(\hat{h}_0(k), \hat{h}_1(k))$ can be projected to become compatible boundary traces using the symbol of the *Calderon projectors*

$$\hat{\mathbb{P}}_i \begin{pmatrix} \hat{h}_0 \\ \hat{h}_1 \end{pmatrix} = \begin{bmatrix} \frac{1}{2} & \frac{1}{2|k|} \\ \frac{|k|}{2} & \frac{1}{2} \end{bmatrix} \begin{pmatrix} \hat{h}_0 \\ \hat{h}_1 \end{pmatrix}, \quad i = 1, 2.$$

We next express the Calderon projectors in terms of Dirichlet-to-Neumann (DtN) and Neumann-to-Dirichlet (NtD) operators.

Lemma 1 (Calderon Projectors and DtN Operators) *Calderon projectors can be written in terms of the local DtN and NtD operators as*

$$\hat{\mathbb{P}}_i = \frac{1}{2}\begin{bmatrix} 1 & \widehat{NtD}_i \\ \widehat{DtN}_i & 1 \end{bmatrix}, \quad i = 1, 2, \tag{11}$$

where DtN_i associates to given Dirichlet data $\hat{g}_0$ on the interface $x = 0$ the normal derivative $\frac{\partial u_i}{\partial n_i}$ of the solution u_i in Ω_i and the NtD_i associates to given Neumann data $\hat{g}_1$ on the interface $x = 0$ the trace of the solution $\hat{u}_i(0, k)$ on the same boundary.

Proof On Ω_1, we obtain explicitly the symbols of these operators from

$$\hat{u}_1(x,k) = \hat{g}_0 e^{|k|x} \Rightarrow \frac{\partial \hat{u}_1}{\partial x}|_{x=0} = |k|\hat{g}_0 \Rightarrow \widehat{DtN}_1 = |k|,$$
$$\hat{u}_1(x,k) = \hat{u}_1(0,k)e^{|k|x}, \ \frac{\partial \hat{u}_1}{\partial x}|_{x=0} = \hat{g}_1 \Rightarrow \hat{u}_1(0,k)|k| = \hat{g}_1 \Rightarrow \widehat{NtD}_1 = \frac{1}{|k|}.$$

The corresponding symbols for the domain Ω_2 are

$$\widehat{DtN}_2 = |k|, \quad \widehat{NtD}_2 = \frac{1}{|k|}.$$

Inserting these expressions into (11) concludes the proof. □

We are ready now to establish the link between these algorithms and the classical DtN iterations.

Theorem 1 (Link with the DtN Iterations) *The iterative multitrace formulation for the special choice $\sigma_1 = \sigma_2 = -\frac{1}{2}$ computes simultaneously a Dirichlet-Neumann iteration (u_1^n, v_2^n) and a Neumann-Dirichlet iteration (v_1^n, u_2^n) without a relaxation parameter.*

Proof According to the results of Lemma 1, in two dimensions, iteration (6) becomes

$$\begin{aligned} \frac{1}{2}\begin{bmatrix} -1-2\sigma_1 & \widehat{NtD}_1 \\ \widehat{DtN}_1 & -1-2\sigma_1 \end{bmatrix}\begin{pmatrix} \hat{u}_1^n \\ \hat{v}_1^n \end{pmatrix} &= -\sigma_1\begin{pmatrix} \hat{u}_2^{n-1} \\ -\hat{v}_2^{n-1} \end{pmatrix}, \\ \frac{1}{2}\begin{bmatrix} -1-2\sigma_2 & \widehat{NtD}_2 \\ \widehat{DtN}_2 & -1-2\sigma_2 \end{bmatrix}\begin{pmatrix} \hat{u}_2^n \\ \hat{v}_2^n \end{pmatrix} &= -\sigma_2\begin{pmatrix} \hat{u}_1^{n-1} \\ -\hat{v}_1^{n-1} \end{pmatrix}. \end{aligned} \tag{12}$$

We see that for the special choice $\sigma_1 = \sigma_2 = -\frac{1}{2}$, iteration (12) simplifies to

$$\begin{cases} \widehat{NtD}_1 \hat{v}_1^n = \hat{u}_2^{n-1}, \\ \widehat{DtN}_1 \hat{u}_1^n = -\hat{v}_2^{n-1}, \end{cases} \quad \begin{cases} \widehat{NtD}_2 \hat{v}_2^n = \hat{u}_1^{n-1}, \\ \widehat{DtN}_2 \hat{u}_2^n = -\hat{v}_1^{n-1}. \end{cases} \tag{13}$$

From the symbols, we see that $\widehat{NtD}_i^{-1} = \widehat{DtN}_i$, and hence iteration (13) becomes

$$\begin{cases} \hat{v}_1^n = \widehat{DtN}_1 \hat{u}_2^{n-1}, \\ \hat{u}_1^n = -\widehat{NtD}_1 \hat{v}_2^{n-1}, \end{cases} \quad \begin{cases} \hat{v}_2^n = \widehat{DtN}_2 \hat{u}_1^{n-1}, \\ \hat{u}_2^n = -\widehat{NtD}_2 \hat{v}_1^{n-1}, \end{cases}$$

which leads to the conclusion. □

In order to study the role of the relaxation parameters σ_i, we check first under which conditions iteration (12), written explicitly as

$$\begin{aligned} B_1 \begin{pmatrix} \hat{u}_1^n \\ \hat{v}_1^n \end{pmatrix} &:= \frac{1}{2} \begin{bmatrix} -1-2\sigma_1 & \frac{1}{|k|} \\ |k| & -1-2\sigma_1 \end{bmatrix} \begin{pmatrix} \hat{u}_1^n \\ \hat{v}_1^n \end{pmatrix} = -\sigma_1 \begin{pmatrix} \hat{u}_2^{n-1} \\ -\hat{v}_2^{n-1} \end{pmatrix}, \\ B_2 \begin{pmatrix} \hat{u}_2^n \\ \hat{v}_2^n \end{pmatrix} &:= \frac{1}{2} \begin{bmatrix} -1-2\sigma_2 & \frac{1}{|k|} \\ |k| & -1-2\sigma_2 \end{bmatrix} \begin{pmatrix} \hat{u}_2^n \\ \hat{v}_2^n \end{pmatrix} = -\sigma_2 \begin{pmatrix} \hat{u}_1^{n-1} \\ -\hat{v}_1^{n-1} \end{pmatrix}, \end{aligned} \tag{14}$$

is well defined. This is the case if the matrices B_i are invertible. Since $\det(B_i) = 4\sigma_i(\sigma_i + 1)$,the multitrace iteration is well defined if $\sigma_i \neq \{0, -1\}$. In this case (14) is equivalent to

$$\begin{aligned} \begin{pmatrix} \hat{u}_1^n \\ \hat{v}_1^n \end{pmatrix} &= B_1^{-1} \begin{pmatrix} \hat{u}_2^{n-1} \\ \hat{v}_2^{n-1} \end{pmatrix} = \begin{pmatrix} \frac{1+2\sigma_1}{2(\sigma_1+1)} \hat{u}_2^{n-1} - \frac{1}{2(\sigma_1+1)} \widehat{NtD}_1 \hat{v}_2^{n-1} \\ \frac{1}{2(\sigma_1+1)} \widehat{DtN}_1 \hat{u}_2^{n-1} - \frac{1+2\sigma_1}{2(\sigma_1+1)} \hat{v}_2^{n-1} \end{pmatrix}, \\ \begin{pmatrix} \hat{u}_2^n \\ \hat{v}_2^n \end{pmatrix} &= B_2^{-1} \begin{pmatrix} \hat{u}_1^{n-1} \\ \hat{v}_1^{n-1} \end{pmatrix} = \begin{pmatrix} \frac{1+2\sigma_2}{2(\sigma_2+1)} \hat{u}_1^{n-1} - \frac{1}{2(\sigma_2+1)} \widehat{NtD}_2 \hat{v}_1^{n-1} \\ \frac{1}{2(\sigma_2+1)} \widehat{DtN}_2 \hat{u}_1^{n-1} - \frac{1+2\sigma_2}{2(\sigma_2+1)} \hat{v}_1^{n-1} \end{pmatrix}. \end{aligned} \tag{15}$$

Algorithm (15) has the same convergence properties as (8), since we obtain the same convergence factor independent of the Fourier variable k, which means convergence is going to be mesh independent.

4 Numerical Results

We now show some numerical experiments for illustration purposes on our two-dimensional model problem (10) on the domain $\Omega = (-1, 1) \times (0, 1)$ decomposed into the two subdomains $\Omega_1 = (-1, 0) \times (0, 1)$ and $\Omega_2 = (0, 1) \times (0, 1)$. We use standard five point finite differences for the discretization and simulate directly the error equations corresponding to the algorithm (15) for different values of the parameter σ_i. For $\sigma_i = -0.6$, our analysis shows that the algorithm does not converge, and we see how the error grows in the iteration in Fig. 2. For $\sigma_i = -0.5$, our analysis predicts stagnation, and this is also observed in Fig. 3. For $\sigma_i = 0.1$, we obtain the predicted rapid convergence seen in Fig. 4. We finally show in Fig. 5 on the left how the error evolves in the maximum norm as the iteration progresses for different values of σ, and on the right the numerically estimated contraction factor, which looks very similar to the predicted behavior shown in Fig. 1.

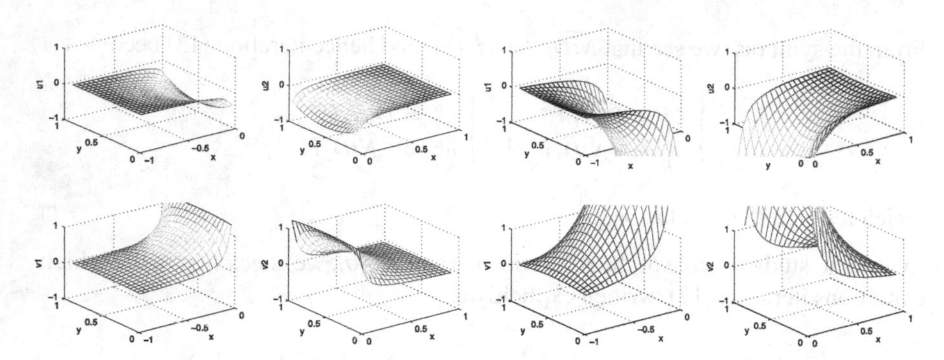

Fig. 2 Evolution of the error for $\sigma = -0.6$ after 2 Iterations (*left*), 10 iterations (*right*)

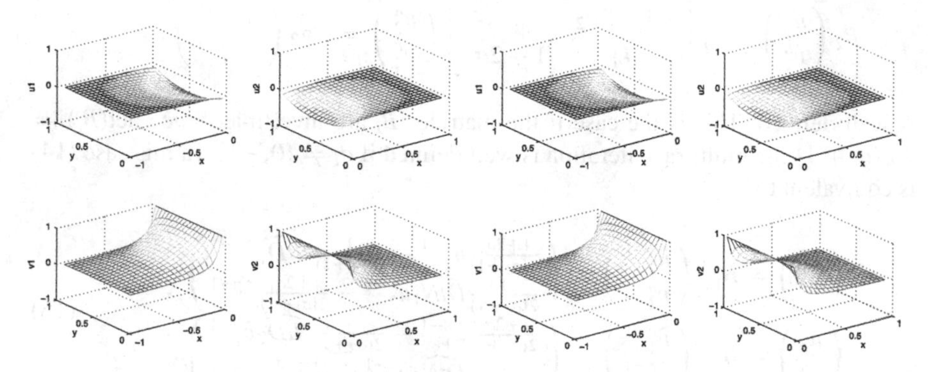

Fig. 3 Evolution of the error for $\sigma = -0.5$ after 2 Iterations (*left*), 10 iterations (*right*)

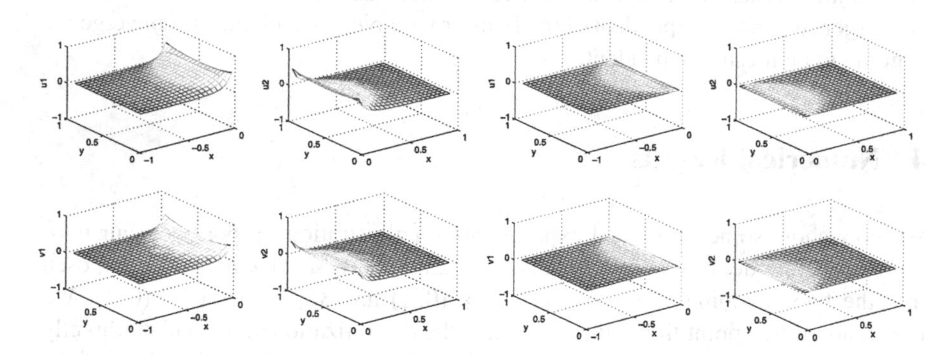

Fig. 4 Evolution of the error for $\sigma = 0.1$ after 2 Iterations (*left*), 10 iterations (*right*)

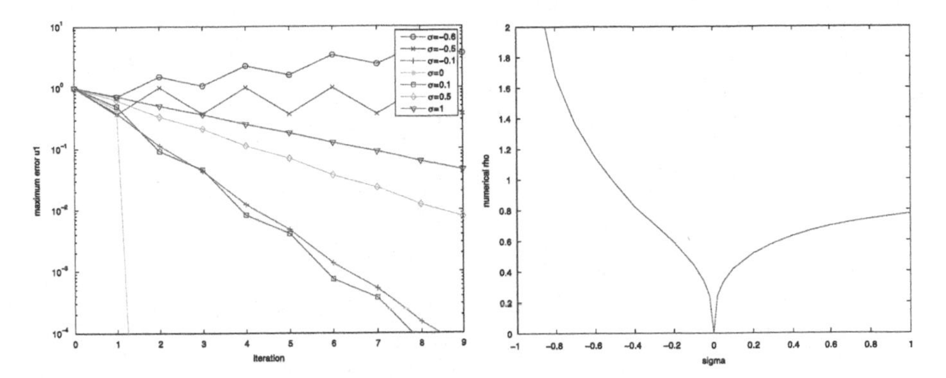

Fig. 5 Error in the maximum norm as a function of the iteration number for different values of σ (*left*), and numerically measured contraction factor of the multitrace iteration as function of σ (*right*)

5 Conclusion

Using a simple model problem and two subdomains, we explained multitrace formulations and a naturally associated iterative method of domain decomposition type. Using the formalism of Dirichlet to Neumann operators, we showed that for a particular choice of the relaxation parameter in the multitrace iteration, a combined sequence of an unrelaxed Dirichlet-Neumann and Neumann-Dirichlet algorithm is obtained. Our analysis also indicates good choices for the relaxation parameter in the multitrace iteration, which was confirmed by numerical experiments.

References

1. X. Claeys, R. Hiptmair, Electromagnetic scattering at composite objects: a novel multi-trace boundary integral formulation. ESAIM Math. Model. Numer. Anal. **46**(6), 1421–1445 (2012)
2. X. Claeys, R. Hiptmair, Multi-trace boundary integral formulation for acoustic scattering by composite structures. Commun. Pure Appl. Math. **66**(8), 1163–1201 (2013)
3. X. Claeys, R. Hiptmair, E. Spindler, A second-kind Galerkin boundary element method for scattering at composite objects, in *Technical Report 2013-13 (revised), Seminar for Applied Mathematics*, ETH Zürich (2013)
4. R. Hiptmair, C. Jerez-Hanckes, Multiple traces boundary integral formulation for Helmholtz transmission problems. Adv. Comput. Math. **37**(1), 39–91 (2012)
5. R. Hiptmair, C. Jerez-Hanckes, J. Lee, Z. Peng, Domain decomposition for boundary integral equations via local multi-trace formulations, in *Technical Report 2013-08 (revised), Seminar for Applied Mathematics*, ETH Zürich (2013)
6. A. Quarteroni, A. Valli, *Domain Decomposition Methods for Partial Differential Equations* (Oxford Science Publications, Oxford, 1999)
7. A. Toselli, O. Widlund, *Domain Decomposition Methods: Algorithms and Theory*. Springer Series in Computational Mathematics, vol. 34 (Springer, Berlin, 2004)

A Deluxe FETI-DP Method for Full DG Discretization of Elliptic Problems

Maksymilian Dryja, Juan Galvis, and Marcus Sarkis

1 Introduction, Differential and Discrete Problems

In this paper we consider a boundary value problem for elliptic second order partial differential equations with highly discontinuous coefficients in a 2D polygonal region Ω. The problem is discretized by a (full) DG method on triangular elements using the space of piecewise linear functions. The goal of this paper is to study a special version of FETI-DP preconditioner, called *deluxe*, for the resulting discrete system of this discretization. The deluxe version for continuous FE discretization is considered in [1], for standard FETI-DP methods for composite DG method, see [4], for full DG, see [4], and for conforming FEM, see the book [5].

Now we discuss the continuous and discrete problems we take into consideration for preconditioning.

M. Dryja
Department of Mathematics, Warsaw University, Banacha 2, 00-097 Warsaw, Poland
e-mail: dryja@mimuw.edu.pl

J. Galvis (✉)
Departamento de Matemáticas, Universidad Nacional de Colombia, Bogotá, Colombia
e-mail: jcgalvisa@unal.edu.co

M. Sarkis
Instituto Nacional de Matemática Pura e Aplicada (IMPA), Estrada Dona Castorina 110, CEP 22460-320, Rio de Janeiro, Brazil

Department of Mathematical Sciences at Worcester Polytechnic Institute, 100 Institute Road, Worcester, MA 01609, USA
e-mail: msarkis@wpi.edu

T. Dickopf et al. (eds.), *Domain Decomposition Methods in Science and Engineering XXII*, Lecture Notes in Computational Science and Engineering 104, DOI 10.1007/978-3-319-18827-0_14

Differential Problem *Find $u_{ex}^* \in H_0^1(\Omega)$ such that*

$$a(u_{ex}^*, v) = f(v) \quad \text{for all } v \in H_0^1(\Omega), \tag{1}$$

$$a(u, v) := \sum_{i=1}^N \int_{\Omega_i} \rho_i \nabla u \cdot \nabla v \, dx \quad \text{and} \quad f(v) := \int_\Omega f v \, dx,$$

where the ρ_i are positive constants and $f \in L^2(\Omega)$.

We assume that $\overline{\Omega} = \cup_{i=1}^N \overline{\Omega}_i$ and the substructures Ω_i are disjoint shaped regular polygonal subregions of diameter $O(H_i)$. We assume that the partition $\{\Omega_i\}_{i=1}^N$ is geometrically conforming, i.e., for all i and j with $i \neq j$, the intersection $\partial\Omega_i \cap \partial\Omega_j$ is either empty, a common corner or a common edge of Ω_i and Ω_j. For clarity we stress that here and below the identifier *edge* means a curve of continuous intervals and its two endpoints are called corners. The collection of these corners on $\partial\Omega_i$ are referred as the set of corners of Ω_i. Let us denote $\bar{E}_{ij} := \partial\Omega_i \cap \partial\Omega_j$ as an edge of $\partial\Omega_i$ and $\bar{E}_{ji} := \partial\Omega_j \cap \partial\Omega_i$ as an edge of $\partial\Omega_j$. Let us denote by $\mathcal{J}_H^{i,0}$ the set of indices j such that Ω_j has a common edge E_{ji} with Ω_i. To take into account edges of Ω_i which belong to the global boundary $\partial\Omega$, let us introduce a set of indices $\mathcal{J}_H^{i,\partial}$ to refer these edges. The set of indices of all edges of Ω_i is denoted by $\mathcal{J}_H^i = \mathcal{J}_H^{i,0} \cup \mathcal{J}_H^{i,\partial}$.

Discrete Problem Let us introduce a shape regular and quasiuniform triangulation (with triangular elements) $\mathcal{T}_h^i$ on Ω_i and let h_i represent its mesh size. The resulting triangulation on Ω is matching across $\partial\Omega_i$. Let $X_i(\Omega_i) := \prod_{\tau \in \mathcal{T}_h^i} X_\tau$ be the product space of finite element (FE) spaces X_τ which consists of linear functions on the element τ belonging to $\mathcal{T}_h^i$. We note that a function $u_i \in X_i(\Omega_i)$ allows discontinuities across elements of $\mathcal{T}_h^i$. We also note that we do not assume that functions in $X_i(\Omega_i)$ vanish on $\partial\Omega$. The global DG finite element space we consider is defined by $X(\Omega) = \prod_{i=1}^N X_i(\Omega_i) \equiv X_1(\Omega_1) \times X_2(\Omega_2) \times \cdots \times X_N(\Omega_N)$.

We define $\mathcal{E}_h^{i,0}$ as the set of edges of the triangulation $\mathcal{T}_h^i$ which are inside Ω_i, and by $\mathcal{E}_h^{i,j}$, for $j \in \mathcal{J}_H^i$, the set of edges of the triangulation $\mathcal{T}_h^i$ which are on E_{ij}. An edge $e \in \mathcal{E}_h^{i,0}$ is shared by two elements denoted by τ_+ and τ_- of $\mathcal{T}_h^i$ with outward unit normal vectors $\boldsymbol{n}^+$ and $\boldsymbol{n}^-$, respectively, and denote $\{\nabla u\} = \frac{1}{2}(\nabla u_{\tau_+} + \nabla u_{\tau_-})$ and $[u] = u_{\tau_+}\boldsymbol{n}^+ + u_{\tau_-}\boldsymbol{n}^-$.

The discrete problem we consider by the DG method is of the form: *Find $u^* = \{u_i^*\}_{i=1}^N \in X(\Omega)$ where $u_i^* \in X_i(\Omega_i)$, such that*

$$a_h(u^*, v) = f(v) \quad \text{for all } v = \{v_i\}_{i=1}^N \in X(\Omega), \tag{2}$$

where the global bilinear from a_h and the right hand side f are assembled as

$$a_h(u, v) := \sum_{i=1}^N a_i'(u, v) \ \text{ and } \ f(v) := \sum_{i=1}^N \int_{\Omega_i} f v_i \, dx.$$

Here, the local bilinear forms a_i', $i = 1, \ldots, N$, are defined as

$$a_i'(u, v) := a_i(u_i, v_i) + s_{0,i}(u_i, v_i) + p_{0,i}(u, v) + s_{\partial,i}(u, v) + p_{\partial,i}(u, v) \tag{3}$$

where a_i, $s_{0,i}$ and $p_{0,i}$ are defined by,
$a_i(u_i, v_i) := \sum_{\tau \in \mathcal{T}_h^i} \int_\tau \rho_i \nabla u_i \cdot \nabla v_i \, dx$,
$s_{0,i}(u_i, v_i) := -\sum_{e \in \mathcal{E}_h^{i,0}} \int_e (\rho_i \{\nabla u_i\} \cdot [v_i] + \rho_i \{\nabla v_i\} \cdot [u_i]) \, ds$, and
$p_{0,i}(u, v) := \sum_{e \in \mathcal{E}_h^{i,0}} \int_e \delta \frac{\rho_i}{h_e} [u_i].[v_i] \, ds$. The corresponding forms over the local interface edges are given by

$$s_{\partial,i}(u, v) := \sum_{j \in \mathcal{J}_H^i} \sum_{e \in \mathcal{E}_h^{i,j}} \int_e \frac{1}{l_{ij}} \left(\rho_{ij} \frac{\partial u_i}{\partial n}(v_j - v_i) + \rho_{ij} \frac{\partial v_i}{\partial n}(u_j - u_i) \right) ds,$$

$$p_{\partial,i}(u, v) := \sum_{j \in \mathcal{J}_H^i} \sum_{e \in \mathcal{E}_h^{i,j}} \int_e \frac{\delta}{l_{ij}} \frac{\rho_{ij}}{h_e} (u_i - u_j)(v_i - v_j) \, ds,$$

respectively. Here $\rho_{ij} = 2\rho_i\rho_j/(\rho_i + \rho_j)$, h_e denotes the length of the edge e. When $j \in \mathcal{J}_H^{i,0}$ we set $l_{ij} = 2$, when $j \in \mathcal{J}_H^{i,\partial}$ we denote the boundary edges $E_{ij} \subset \partial\Omega_i$ by $E_{i\partial}$ and set $l_{i\partial} = 1$, and on the artificial edge $E_{ji} \equiv E_{\partial i}$ we set $u_\partial = 0$ and $v_\partial = 0$. The partial derivative $\frac{\partial}{\partial n}$ denotes the outward normal derivative on $\partial\Omega_i$ and δ is the penalty positive parameter.

The discrete formulation used here is convenient for our FETI-DP method. We also mention that problem (2) has a unique solution for sufficiently large δ and its error bound is known, see for example, [3, 4].

2 Schur Complement Matrices and Harmonic Extensions

In this section, we describe the elimination of unknowns interior to the subdomains required on the FETI-DP formulation for DG discretizations.

Let the set of degrees of freedom associated to subdomain Ω_i be defined by

$$\Omega_i' := \overline{\Omega}_i \bigcup \{\cup_{j \in \mathcal{J}_H^{i,0}} \bar{E}_{ji}\}$$

i.e., it is the union of $\overline{\Omega}_i$ and the $\bar{E}_{ji} \subset \partial\Omega_j$ such that $j \in \mathcal{J}_H^{i,0}$. Define $\Gamma_i := \overline{\partial\Omega_i \backslash \partial\Omega}$ and $\Gamma_i' := \Gamma_i \bigcup \{\cup_{j \in \mathcal{J}_H^{i,0}} \bar{E}_{ji}\}$. We also introduce the sets

$$\Gamma := \bigcup_{i=1}^N \Gamma_i, \ \Gamma' := \prod_{i=1}^N \Gamma_i', \ I_i := \Omega_i' \backslash \Gamma_i' \text{ and } I := \prod_{i=1}^N I_i. \tag{4}$$

Let $W_i(\Omega_i')$ be the FE space of functions defined by nodal values on Ω_i'

$$W_i(\Omega_i') = W_i(\overline{\Omega}_i) \times \prod_{j \in \mathcal{J}_H^{0,i}} W_i(\bar{E}_{ji}), \tag{5}$$

where $W_i(\overline{\Omega}_i) := X_i(\Omega_i)$ and $W_i(\bar{E}_{ji})$ is the trace of the DG space $X_j(\Omega_j)$ on $\bar{E}_{ji} \subset \partial\Omega_j$ for all $j \in \mathcal{J}_H^{i,0}$. A function $u_i' \in W_i(\Omega_i')$ is defined by the nodal values on Ω_i', i.e., by the nodal values on $\overline{\Omega}_i$ and the nodal values on all adjacent faces $\bar{E}_{ji}$ for all $j \in \mathcal{J}_H^{i,0}$. Below, we denote u_i' by u_i if it is not confused with functions of $X_i(\Omega_i)$. A function $u_i \in W_i(\Omega_i')$ is represented as $u_i = \{(u_i)_i, \{(u_i)_j\}_{j \in \mathcal{J}_H^{i,0}}\}$, where $(u_i)_i := u_{i|\overline{\Omega}_i}$ (u_i restricted to $\overline{\Omega}_i$) and $(u_i)_j := u_{i|\bar{E}_{ji}}$ (u_i restricted to $\bar{E}_{ji}$). Here and below we use the same notation to identify both DG functions and their vector representations. Note that $a_i'(\cdot,\cdot)$, see (3), is defined on $W_i(\Omega_i') \times W_i(\Omega_i')$ with corresponding stiffness matrix A_i' defined by

$$a_i'(u_i, v_i) = \langle A_i' u_i, v_i \rangle \quad u_i, v_i \in W_i(\Omega_i'), \tag{6}$$

where $\langle u_i, v_i \rangle$ denotes the ℓ_2 inner product of nodal values associated to the vector space in consideration. We also represent $u_i \in W_i(\Omega_i')$ as $u_i = (u_{i,I}, u_{i,\Gamma'})$ where $u_{i,\Gamma'}$ represents values of u_i at nodal points on Γ_i' and $u_{i,I}$ at the interior nodal points in I_i, see (4). Hence, let us represent $W_i(\Omega_i')$ as the vector spaces $W_i(I_i) \times W_i(\Gamma_i')$. Using the representation $u_i = (u_{i,I}, u_{i,\Gamma'})$, the matrix A_i' can be represented as

$$A_i' = \begin{pmatrix} A_{i,II}' & A_{i,I\Gamma'}' \\ A_{i,\Gamma'I}' & A_{i,\Gamma'\Gamma'}' \end{pmatrix}. \tag{7}$$

The Schur complement of A_i' with respect to $u_{i,\Gamma'}$ is of the form

$$S_i' := A_{i,\Gamma'\Gamma'}' - A_{i,\Gamma'I}' (A_{i,II}')^{-1} A_{i,I\Gamma'}' \tag{8}$$

and introduce the block diagonal matrix $S' = \text{diag}\{S_i'\}_{i=1}^N$.

Let us introduce the product space

$$W(\Omega') := \prod_{i=1}^N W_i(\Omega_i'),$$

i.e., $u \in W(\Omega')$ means that $u = \{u_i\}_{i=1}^N$ where $u_i \in W_i(\Omega_i')$; see (5) for the definition of $W_i(\Omega_i')$. Recall that we write $(u_i)_i = u_{i|\overline{\Omega}_i}$ (u_i restricted to $\overline{\Omega}_i$) and $(u_i)_j = u_{i|\bar{E}_{ji}}$ (u_i restricted to $\bar{E}_{ji}$). Using the representation $u_i = (u_{i,I}, u_{i,\Gamma'})$ where $u_{i,I} \in W_i(I_i)$ and $u_{i,\Gamma'} \in W_i(\Gamma_i')$ were used in (7), let us introduce the product space

$$W(\Gamma') := \prod_{i=1}^N W_i(\Gamma_i'),$$

i.e., $u_{\Gamma'} \in W(\Gamma')$ means that $u_{\Gamma'} = \{u_{i,\Gamma'}\}_{i=1}^N$ where $u_{i,\Gamma'} \in W_i(\Gamma_i')$. The space $W(\Gamma')$ which was defined on Γ' only, is also interpreted below as the subspace of $W(\Omega')$ of functions which are discrete $\mathcal{H}_i'$-harmonic in the sense of $a_i'(.,.)$ in each i.

3 FETI-DP with Corner Constraints

We now design a FETI-DP method for solving (2). We follow the abstract approach described in pages 160–167 in [5].

We introduce the nodal points associated to the corner unknowns by

$$\mathcal{V}_i' := \mathcal{V}_i \bigcup \{\cup_{j\in\mathcal{J}_H^{i,0}} \partial E_{ji}\} \qquad \text{where} \qquad \mathcal{V}_i := \{\cup_{j\in\mathcal{J}_H^{i,0}} \partial E_{ij}\}.$$

We now consider the subspace $\tilde{W}(\Omega') \subset W(\Omega')$ (and $\tilde{W}(\Gamma') \subset W(\Gamma')$) as the space of functions which are continuous on all the $\mathcal{V}_i'$ as follows.

Definition 1 (Subspaces $\tilde{W}(\Omega')$ and $\tilde{W}(\Gamma')$) We say that $u = \{u_i\}_{i=1}^N \in \tilde{W}(\Omega')$ if it is continuous at the corners $\mathcal{V}_i'$, that is, if for $1 \le i \le N$ we have

$$(u_i)_i(x) = (u_j)_i(x) \quad \text{at } x \in \partial E_{ij} \text{ for all } j \in \mathcal{J}_H^{i,0}, \quad \text{and} \tag{9}$$

$$(u_i)_j(x) = (u_j)_j(x) \quad \text{at } x \in \partial E_{ji} \text{ for all } j \in \mathcal{J}_H^{i,0}. \tag{10}$$

Analogously we define $\tilde{W}(\Gamma')$.

Note that $\tilde{W}(\Gamma') \subset W(\Gamma'.)$ Let $\tilde{A}$ be the stiffness matrix which is obtained by assembling the matrices A_i' for $1 \le i \le N$, from $W(\Omega')$ to $\tilde{W}(\Omega')$. Note that the matrix $\tilde{A}$ is no longer block diagonal since there are couplings between variables at the corners $\mathcal{V}_i'$ for $1 \le i \le N$. We represent $u \in \tilde{W}(\Omega')$ as $u = (u_I, u_\Pi, u_\Delta)$ where the subscript I refers to the interior degrees of freedom at nodal points $I = \prod_{i=1}^N I_i$, the Π refers to the corners $\mathcal{V}_i'$ for all $1 \le i \le N$, and the Δ refers to the remaining nodal points, i.e., the nodal points of $\Gamma_i' \backslash \mathcal{V}_i'$, for all $1 \le i \le N$. The vector $u = (u_I, u_\Pi, u_\Delta) \in \tilde{W}(\Omega')$ is obtained from the vector $u = \{u_i\}_{i=1}^N \in W(\Omega')$ using Eqs. (9) and (10), i.e., the continuity of u on $\mathcal{V}_i'$ for all $1 \le i \le N$. Using the decomposition $u = (u_I, u_\Pi, u_\Delta) \in \tilde{W}(\Omega')$ we can partition $\tilde{A}$ as

$$\tilde{A} = \begin{pmatrix} A'_{II} & A'_{I\Pi} & A'_{I\Delta} \\ A'_{\Pi I} & \tilde{A}_{\Pi\Pi} & A'_{\Pi\Delta} \\ A'_{\Delta I} & A'_{\Delta\Pi} & A'_{\Delta\Delta} \end{pmatrix}.$$

We note that the only couplings across subdomains are through the variables Π where the matrix $\tilde{A}$ is subassembled.

A Schur complement of $\tilde{A}$ with respect to the Δ-unknowns (eliminating the I- and the Π-unknowns) is of the form

$$\tilde{S} := A'_{\Delta\Delta} - (A'_{\Delta I}\, A'_{\Delta\Pi}) \begin{pmatrix} A'_{II} & A'_{I\Pi} \\ A'_{\Pi I} & \tilde{A}_{\Pi\Pi} \end{pmatrix}^{-1} \begin{pmatrix} A'_{I\Delta} \\ A'_{\Pi\Delta} \end{pmatrix}. \tag{11}$$

A vector $u \in \tilde{W}(\Gamma')$ can uniquely be represented by $u = (u_\Pi, u_\Delta)$, therefore, we can represent $\tilde{W}(\Gamma') = \hat{W}_\Pi(\Gamma') \times W_\Delta(\Gamma')$, where $\hat{W}_\Pi(\Gamma')$ refers to the Π-degrees of freedom of $\tilde{W}(\Gamma')$ while $W_\Delta(\Gamma')$ to the Δ-degrees of freedom of $\tilde{W}(\Gamma')$. The vector space $W_\Delta(\Gamma')$ can be decomposed as

$$W_\Delta(\Gamma') = \prod_{i=1}^{N} W_{i,\Delta}(\Gamma'_i) \tag{12}$$

where the local space $W_{i,\Delta}(\Gamma'_i)$ refers to the degrees of freedom associated to the nodes of $\Gamma'_i \backslash \mathcal{V}'_i$ for $1 \le i \le N$. Hence, a vector $u \in \tilde{W}(\Gamma')$ can be represented as $u = (u_\Pi, u_\Delta)$ with $u_\Pi \in \hat{W}_\Pi(\Gamma')$ and $u_\Delta = \{u_{i,\Delta}\}_{i=1}^N \in W_\Delta(\Gamma')$ where $u_{i,\Delta} \in W_{i,\Delta}(\Gamma'_i)$. Note that $\tilde{S}$, see (11), is defined on the vector space $W_\Delta(\Gamma')$.

In order to measure the jump of $u_\Delta \in W_\Delta(\Gamma')$ across the Δ-nodes let us introduce the space $\hat{W}_\Delta(\Gamma)$ defined by

$$\hat{W}_\Delta(\Gamma) = \prod_{i=1}^{N} X_i(\Gamma_i \backslash \mathcal{V}_i),$$

where $X_i(\Gamma_i \backslash \mathcal{V}_i)$ is the restriction of $X_i(\Omega_i)$ to $\Gamma_i \backslash \mathcal{V}_i$. To define the jumping matrix $B_\Delta : W_\Delta(\Gamma') \to \hat{W}_\Delta(\Gamma)$, let $u_\Delta = \{u_{i,\Delta}\}_{i=1}^N \in W_\Delta(\Gamma')$ and let $v := B_\Delta u$ where $v = \{v_i\}_{i=1}^N \in \hat{W}_\Delta(\Gamma)$ is defined by

$$v_i = (u_{i,\Delta})_i - (u_{j,\Delta})_i \ \text{ on } E_{ijh} \ \text{ for all } \ j \in \mathcal{J}_H^{i,0}, \tag{13}$$

where E_{ijh} is the set of interior nodal points on E_{ij}. The jumping matrix B_Δ can be written as

$$B_\Delta = (B_\Delta^{(1)}, B_\Delta^{(2)}, \cdots, B_\Delta^{(N)}), \tag{14}$$

where the rectangular matrix $B_\Delta^{(i)}$ consists of columns of B_Δ attributed to the (i) components of functions of $W_{i,\Delta}(\Gamma'_i)$ of the product space $W_\Delta(\Gamma')$, see (12). The entries of the rectangular matrix consist of values of $\{0, 1, -1\}$. It is easy to see that the Range $B_\Delta = \hat{W}_\Delta(\Gamma)$, so B_Δ is full rank.

We can reformulate the problem (2) as the variational problem with constraints in $W_\Delta(\Gamma')$ space: *Find* $u^*_\Delta \in W_\Delta(\Gamma')$ *such that*

$$J(u^*_\Delta) = \min J(v_\Delta) \tag{15}$$

subject to $v_\Delta \in W_\Delta(\Gamma')$ with constraints $B_\Delta v_\Delta = 0$. Here $J(v_\Delta) := \frac{1}{2}\langle \tilde{S} v_\Delta, v_\Delta\rangle - \langle \tilde{g}_\Delta, v_\Delta\rangle$ with $\tilde{S}$ given in (11) and $\tilde{g}_\Delta$ is easily obtained using the fact that it can be represented as $f = (f_I, f_\Pi, f_{\Gamma\setminus\Pi})$. Note that $\tilde{S}$ is symmetric and positive definite since $\tilde{A}$ has these properties. Introducing Lagrange multipliers $\lambda \in \hat{W}_\Delta(\Gamma)$, the problem (15) reduces to the saddle point problem of the form: *Find* $u^*_\Delta \in W_\Delta(\Gamma')$ *and* $\lambda^* \in \hat{W}_\Delta(\Gamma)$ *such that*

$$\begin{cases} \tilde{S}u^*_\Delta + B^T_\Delta \lambda^* = \tilde{g}_\Delta \\ B_\Delta u^*_\Delta = 0. \end{cases} \tag{16}$$

Hence, (16) reduces to

$$F\lambda^* = g \tag{17}$$

where $F := B_\Delta \tilde{S}^{-1} B^T_\Delta$ and $g := B_\Delta \tilde{S}^{-1}\tilde{g}_\Delta$.

3.1 Dirichlet Preconditioner

We now define the FETI-DP preconditioner for F, see (17). Let $S'_{i,\Delta}$ be the Schur complement of S'_i, see (8), restricted to $W_{i,\Delta}(\Gamma'_i) \subset W_i(\Gamma'_i)$, i.e., taken S'_i on functions in $W_i(\Gamma'_i)$ which vanish on $\mathcal{V}'_i$. Let

$$S'_\Delta := \mathrm{diag}\{S'_{i,\Delta}\}_{i=1}^N.$$

In other words, $S'_{i,\Delta}$ is obtained from S'_i by deleting rows and columns corresponding to nodal values at nodal points of $\mathcal{V}'_i \subset \Gamma'_i$.

Let us introduce diagonal scaling operators $D^{(i)}_\Delta : W_{i,\Delta}(\Gamma'_i) \to W_{i,\Delta}(\Gamma'_i)$, for $1 \le i \le N$. They are based on partial Schur complements of $S'_{i,\Delta}$ used in [1] for continuous FE discretization and this is know in the literature as the deluxe version of FETI-DP preconditioner. We first introduce $W_{i,\Delta,E_{ij}}(\Gamma'_i)$ as the space of $u_i \in W_{i,\Delta}(\Gamma'_i)$ which vanish on $\partial\Omega_i \setminus E_{ij}$ and $E_{ki} \subset \partial\Omega_k$ for $k \neq j$. Let $S'_{i,\Delta,E_{ij}}$ denote the Schur complement of $S'_{i,\Delta}$ restricted to $W_{i,\Delta,E_{ij}}$. In a similar way it is defined the restricted Schur complement $S'_{j,\Delta,E_{ji}}$. The operator $D^{(i)}_\Delta$ on $E_{ij} \subset \partial\Omega_i$ is defined as

$$D^{(i)}_{\Delta,E_{ij}} = (S'_{i,\Delta,E_{ij}} + S'_{j,\Delta,E_{ji}})^{-1} S'_{j,\Delta,E_{ji}}. \tag{18}$$

Let $B_{D,\Delta} = (B_\Delta^{(1)} D_\Delta^{(1)}, \cdots, B_\Delta^{(N)} D_\Delta^{(N)})$ and $P_\Delta := B_{D,\Delta}^T B_\Delta$, which maps $W_\Delta(\Gamma')$ into itself. It can be checked straightforwardly that P_Δ preserves jumps in the sense that $B_\Delta P_\Delta = B_\Delta$ and $P_\Delta^2 = P_\Delta$.

In the FETI-DP method, the preconditioner M^{-1} is defined as follows:

$$M^{-1} = B_{D,\Delta} S'_\Delta B_{D,\Delta}^T = \sum_{i=1}^{N} B_\Delta^{(i)} D_\Delta^{(i)} S'_{i,\Delta} (D_\Delta^{(i)})^T (B_\Delta^{(i)})^T.$$

Note that M^{-1} is a block-diagonal matrix, and each block is invertible since $S'_{i,\Delta}$ and $D_\Delta^{(i)}$ are invertible and $B_\Delta^{(i)}$ is a full rank matrix. The following theorem holds.

Theorem 1 *For any $\lambda \in \hat{W}_\Delta(\Gamma)$ it holds that*

$$\langle M\lambda, \lambda \rangle \le \langle F\lambda, \lambda \rangle \le C \left(1 + \log \frac{H}{h}\right)^2 \langle M\lambda, \lambda \rangle$$

where $\log(\frac{H}{h}) := \max_{i=1}^{N} \log(\frac{H_i}{h_i})$, C is a positive constant independent of h_i, h_i/h_j, H_i, λ and the jumps of ρ_i.

The complete proof of Theorem 1 will be presented elsewhere.

Remark 2 The FETI-DP method is introduced for a composite DG discretization in the 3-D case in [2]. In order to extend the deluxe scaling FETI-DP method for 3-D DG discretizations, we need to introduce the averaging of the deluxe operators for faces and edges. The face operators are introduced similarly as described as in (18) by replacing edges E_{ij} by faces F_{ij}. For the edge operators, consider for instance that E_{ijk} is an edge of Ω_i common to Ω_j and Ω_k. Let E_{jik} and E_{kij} be edges equal to E_{ijk} but belonging to Ω_j and Ω_k, respectively. Let $W_{i,\Delta,E_{ijk}}(\Gamma'_i)$ be a subspace of $W_{i,\Delta}(\Gamma'_i)$ with nonzero data on E_{ijk}, E_{jik} and E_{kij} only. Let $S'_{i,\Delta,E_{ijk}}$ be the restriction of $S'_{i,\Delta}$ to the space $W_{i,\Delta,E_{ijk}}$. In the same way we introduce $S'_{j,\Delta,E_{jik}}$ and $S'_{k,\Delta,E_{kij}}$. For the deluxe FETI-DP method with non-redundant Lagrange multipliers on edges, see [5], it is enough to define the edge averaging operators as follows:

$$D_{\Delta,E_{ijk},1}^{(i)} = (S'_{i,\Delta,E_{ijk}} + S'_{j,\Delta,E_{jik}} + S'_{k,\Delta,E_{kij}})^{-1} S'_{j,\Delta,E_{jik}}, \text{ and}$$

$$D_{\Delta,E_{ijk},2}^{(i)} = (S'_{i,\Delta,E_{ijk}} + S'_{j,\Delta,E_{jik}} + S'_{k,\Delta,E_{kij}})^{-1} S'_{k,\Delta,E_{kij}}.$$

In the 3-D case $B_{D,\Delta}$ is modified by setting $B_{D,\Delta} = (B_\Delta D_\Delta B_\Delta^T)^{-1} B_\Delta D_\Delta$ and $M^{-1} = B_{D,\Delta} S'_\Delta B_{D,\Delta}^T$ where $D_\Delta = \text{diag}\{D_\Delta^{(i)}\}$ and $D_\Delta^{(i)}$ is a block diagonal containing the averaging operators corresponding to faces and edges defined above. The operator $P_\Delta = B_{D,\Delta}^T B_\Delta$ preserves the jumps and is a projection.

Acknowledgement The authors thank the anonymous referee for his suggestions that helped to improve the paper.

This research was supported in part by the Polish Sciences Foundation under grant 2011/01/B/ST1/01179 (Maksymilian Dryja).

References

1. C.R. Dohrmann, O.B. Widlund, Some recent tools and a BDDC algorithm for 3D problems in H(curl), in *Domain Decomposition Methods in Science and Engineering XX*. Lecture Notes in Computer Science Engineering, vol. 91 (Springer, New York, 2013), pp. 15–25
2. M. Dryja, M. Sarkis, 3-D FETI-DP preconditioners for composite finite element-discontinuous Galerkin methods, in *Domain Decomposition Methods in Science and Engineering XXI*. Lecture Notes in Computer Science Engineering, vol. 99 (Springer, Heidelberg, 2014), pp. 127–140
3. M. Dryja, J. Galvis, M. Sarkis, A FETI-DP preconditioner for a composite finite element and discontinuous Galerkin method. SIAM J. Numer. Anal. **51**(1), 400–422 (2013)
4. M. Dryja, J. Galvis, M. Sarkis, *The analysis of FETI-DP preconditioner for full DG discretization of elliptic problems in two dimensions*, (Springer Berlin Heidelberg, 2015), pp. 1–34 [ISSN 0029-599X]. doi:10.1007/s00211-015-0705-x. http://dx.doi.org/10.1007/s00211-015-0705-x
5. A. Toselli, O. Widlund, *Domain Decomposition Methods—Algorithms and Theory*. Springer Series in Computational Mathematics, vol. 34 (Springer, Berlin, 2005) [ISBN 3-540-20696-5]

Additive Schwarz Methods for DG Discretization of Elliptic Problems with Discontinuous Coefficient

Maksymilian Dryja and Piotr Krzyżanowski

1 Introduction

In this paper we consider a second order elliptic problem defined on a polygonal region Ω, where the diffusion coefficient is a discontinuous function. The problem is discretized by a symmetric interior penalty discontinuous Galerkin (DG) finite element method with triangular elements and piecewise linear functions. Our goal is to design and analyze an additive Schwarz method (ASM), see the book by Toselli and Widlund [11], for solving the resulting discrete problem with rate of convergence independent of the jumps of the coefficient. The method is two-level and without overlap of the substructures into which the original region Ω is partitioned.

Usually, two level ASMs for discretizations on fine mesh of size h are being built by introducing a partitioning of the domain into subdomains of size $H > h$, where local solvers are applied in parallel. A global coarse problem is then typically based on the same partitioning. This approach has been generalized for nonoverlapping domain decomposition methods for DG discretizations by Feng and Karakashian [10] and further extended by Antonietti and Ayuso [1] by allowing the coarse grid with mesh size H to be a refinement of the original partitioning into subdomains where the local solvers are applied.

The ASM discussed here is a generalization to non-constant diffusion coefficient and very small subdomains of methods mentioned above and of those presented in [7, 8]. Other recent works towards domain decomposition preconditioning of DG discretizations of problems with strongly varying coefficients include [2, 4, 5]. In this paper, local solvers act on subdomains which are equal to single elements of

M. Dryja • P. Krzyżanowski (✉)
University of Warsaw, Warsaw, Poland
e-mail: m.dryja@mimuw.edu.pl; p.krzyzanowski@mimuw.edu.pl; przykry@mimuw.edu.pl

T. Dickopf et al. (eds.), *Domain Decomposition Methods in Science and Engineering XXII*, Lecture Notes in Computational Science and Engineering 104, DOI 10.1007/978-3-319-18827-0_15

the fine mesh. By allowing single element subdomains we substantially increase the level of parallelism of the method. Very small and cheap to solve local systems come in huge quantities, which possibly can be an advantage on new multithreaded processors. Moreover, small subdomains give more flexibility in assigning them to processors in coarse grain parallel processing. The price to be paid for this in some sense extreme parallelism is worse condition number of the preconditioned system, which is of order $O(H^2/h^2)$, where H and h are the coarse and the fine mesh parameters, respectively. This bound is independent of the jumps of diffusion coefficient if its variation inside substructures is bounded. Numerical experiments confirm theoretical results.

The paper is organized as follows. In Sect. 2, differential and discrete DG problems are formulated. In Sect. 3, ASM for solving the discrete problem is designed and analyzed. Numerical experiments are presented in Sect. 4.

In the paper, for nonnegative scalars x, y, we shall write $x \lesssim y$ if there exits a positive constant C, independent of x, y and the mesh parameters h, H, and of the jumps of the diffusion coefficient ρ as well, such that $x \leq Cy$. If both $x \lesssim y$ and $y \lesssim x$, we shall write $x \simeq y$.

2 Differential and Discrete DG Problems

Let us consider the following variational problem in a polygonal region Ω:

Find $u^* \in H_0^1(\Omega)$ such that

$$a(u^*, v) = (f, v)_\Omega, \qquad v \in H_0^1(\Omega), \tag{1}$$

where

$$a(u, v) = \int_\Omega \rho \nabla u \cdot \nabla v dx, \qquad (f, v)_\Omega = \int_\Omega f\, v.$$

We assume that $\rho \in L^\infty(\Omega)$ and that there exist constants α_0 and α_1 such that $0 < \alpha_0 \leq \rho \leq \alpha_1$ in Ω. In addition we assume that $f \in L^2(\Omega)$.

2.1 Discrete Problem

Let $\mathcal{T}_H$ be a subdivision of Ω into N_H disjoint open polygonal regions Ω_i, $i = 1, \ldots, N_H$, such that $\bar{\Omega} = \bigcup_{i=1,\ldots,N_H} \bar{\Omega}_i$ and that the number of neighboring regions is uniformly bounded. We set $H_i = \operatorname{diam}(\Omega_i)$ and $H = \max_{i=1,\ldots,N_H} H_i$. Further, let $\mathcal{T}_h$ denote an affine, shape regular conforming triangulation (with triangles) of Ω, $\bar{\Omega} = \bigcup_{\kappa \in \mathcal{T}_h} \bar{\kappa}$, which is derived from $\mathcal{T}_H$ by some refinement procedure. Thus, each

Ω_i is a union of certain elements from $\mathcal{T}_h$. The diameter of a triangle $\kappa \in \mathcal{T}_h$ will be denoted by h_κ and the mesh parameter is $h = \max_{\kappa \in \mathcal{T}_h} h_\kappa$.

In what follows we shall assume that ρ is piecewise constant (possibly with large discontinuities) on $\mathcal{T}_h$, so that $\rho_{|\kappa}$ is constant on each $\kappa \in \mathcal{T}_h$.

By $\mathcal{E}_h^0$ we denote the set of all common (internal) faces of elements in $\mathcal{T}_h$, so that $e_{ij} \in \mathcal{E}_h$ iff $e_{ij} = \kappa_i \cap \kappa_j$ is of positive measure. We will use symbol $\mathcal{E}_h$ to denote the set of all faces, that is those either in $\mathcal{E}_h^0$ or on the boundary $\partial\Omega$; for $e \in \mathcal{E}_h$, we also set $|e| = \mathrm{diam}(e)$. We shall assume local quasi-uniformity of the grid, i.e. if $e_{ij} \in \mathcal{E}_h^0$ is such that $e_{ij} = \kappa_i \cap \kappa_j$, then $h_i \simeq h_j$.

For $p \in \{0, 1\}$, we denote by $\mathcal{P}_p(\kappa)$ the set of polynomials of degree not greater than p on $\bar{\kappa}$. Then we define the finite element space V_h, in which we will approximate (1),

$$V_h = \{v \in L^2(\Omega) : v_{|\kappa} \in \mathcal{P}_1(\kappa), \forall \kappa \in \mathcal{T}_h\}. \tag{2}$$

Note that the traces of the functions from V_h are multi-valued on the interface $\mathcal{E}_h^0$.

We define the discrete problem as the symmetric interior penalty discontinuous Galerkin method, see for example [9] or [6]:

Find $u \in V_h$ such that

$$A_h(u, v) = (f, v)_\Omega, \qquad v \in V_h, \tag{3}$$

where

$$A_h(u, v) \equiv \sum_{\kappa \in \mathcal{T}_h} (\rho\, \nabla u, \nabla v)_\kappa + \sum_{e \in \mathcal{E}_h} \langle \gamma [u], [v] \rangle_e$$

$$- \sum_{e \in \mathcal{E}_h} \left(\langle [u], \{\rho \nabla v\}_\omega \rangle_e + \langle \{\rho \nabla u\}_\omega, [v] \rangle_e \right),$$

and $\delta > 0$ is sufficiently large to ensure positive definiteness of $A_h(\cdot, \cdot)$, and on $e_{ij} = \kappa_i \cap \kappa_j$

$$\gamma = \frac{\delta}{|e_{ij}|} \frac{\rho_i \rho_j}{\rho_i + \rho_j}, \qquad \{\rho \nabla u\}_\omega = \omega_j \rho_i \nabla u_i + \omega_i \rho_j \nabla u_j, \qquad [u] = u_i\, n_i + u_j\, n_j,$$

with $\omega_j = \rho_j / (\rho_i + \rho_j)$. Here, for any function φ we use the convention that φ_i (resp.φ_j) refers to the value of $\varphi_{|\kappa_i}$ (resp. $\varphi_{|\kappa_j}$) on e_{ij}. The unit normal vector pointing outward κ_i is denoted by n_i. On the boundary of Ω, we set $\{\rho \nabla u\}_\omega = \rho \nabla u$ and $[u] = u\, n$.

Let us introduce a simplified form

$$D_h(u, v) = \sum_{\kappa \in \mathcal{T}_h} (\rho\, \nabla u, \nabla v)_\kappa + \sum_{e \in \mathcal{E}_h} \langle \gamma [u], [v] \rangle_e.$$

Then it is well known that $D_h(\cdot, \cdot)$ is spectrally equivalent to $A_h(\cdot, \cdot)$, i.e.

$$A_h(u, u) \simeq D_h(u, u) \quad \forall u \in V_h.$$

3 Additive Schwarz Methods

3.1 Additive Schwarz Method, Version I

Let N_h be the number of elements in $\mathcal{T}_h$. We decompose V_h as follows:

$$V_h = V_0 + \sum_{i=1}^{N_h} V_i$$

where

$$V_0 = \{v \in V_h : v_{|_\kappa} \in \mathcal{P}_0(\kappa) \text{ on } \kappa \in \mathcal{T}_h\}$$

and

$$V_i = \{v \in V_h : v_{|_\kappa} = 0 \text{ for all } \kappa \in \mathcal{T}_h \text{ such that } \kappa \neq \kappa_i\}. \tag{4}$$

Using the above decomposition we define local operators $T_i : V_h \rightarrow V_i$, $i = 1, \ldots, N_h$, with inexact solver

$$D_h(T_i u, v) = A_h(u, v) \qquad \forall v \in V_i,$$

so that we solve for $u_i = T_i u$ defined on $\kappa_i \in \mathcal{T}_h$ such that

$$(\rho_i \nabla u_i, \nabla v_i)_{\kappa_i} + \sum_{e \subset \partial \kappa_i} \int_e \gamma u_i v_i = A_h(u, v_i) \qquad \forall v_i \in V_i,$$

and set $(T_i u)_{|_{\kappa_j}} = 0$ for $j \neq i$. The coarse solve operator is $T_0 : V_h \rightarrow V_0$ defined analogously as

$$D_h(T_0 u, v_0) = A_h(u, v_0) \qquad \forall v_0 \in V_0.$$

Note that on V_0, the approximate form $D_h(\cdot, \cdot)$ coincides with $A_h(\cdot, \cdot)$ and simplifies to

$$D_h(u_0, v_0) = \sum_{e \in \mathcal{E}_h} \langle \gamma [u_0], [v_0] \rangle_e \qquad \forall u_0, v_0 \in V_0.$$

Theorem 1 *Let* $T = T_0 + \sum_{i=1}^{N_h} T_i$. *Then*

$$A_h(Tu, u) \simeq A_h(u, u) \qquad \forall u \in V_h.$$

This means that the condition number of the resulting system is uniformly bounded independently of h, H and ρ. However, the method is not robust, because $\dim V_0 = N_h$ is very large. The proof of Theorem 1 will appear elsewhere.

3.2 Additive Schwarz Method, Version II

Since version I described above suffers from the very large size of the coarse space V_0 (based on edges of the fine triangulation $\mathcal{T}_h$, with averaged coefficients on them), here we consider a coarse space which is set up on the edges of $\mathcal{T}_H$, the coarse partition. In this way the method regains high level of parallelism, as the coarse problem now can in principle be solved on a single processor. Note that this approach is similar to that of [10].

We decompose V_h as follows:

$$V_h = \bar{V}_0 + \sum_{i=1}^{N_h} V_i$$

where

$$\bar{V}_0 = \{v \in V_h : v_{|\Omega_i} \in \mathcal{P}_0(\Omega_i),\ i = 1, \ldots, N_H\}$$

and the local spaces V_i, $i = 1, \ldots, N_h$, remain as defined in (4). Now, the coarse operator $\bar{T}_0 : V_h \to \bar{V}_0$ is defined such that $\bar{T}_0 u = \bar{u}_0$ where

$$D_h(\bar{u}_0, v) = A_h(u, v) \qquad \forall v \in \bar{V}_0.$$

In order to formulate the condition number result, we shall assume uniformly bounded level of variation of the coefficient within subdomain: there exist positive constants c and C such that

$$c\,\bar{\rho}_i \le \rho_{|\Omega_i} \le C\,\bar{\rho}_i, \qquad i = 1, \ldots, N_H, \tag{5}$$

where

$$\bar{\rho}_i := \frac{1}{|\Omega_i|} \int_{\Omega_i} \rho.$$

Theorem 2 *Let* $H_i = \text{diam}(\Omega_i)$ *and let* $T = \bar{T}_0 + \sum_{i=1}^{N_h} T_i$. *Under the above assumptions,*

$$\beta^{-1} A_h(u, u) \lesssim A_h(Tu, u) \lesssim A_h(u, u)$$

where $\beta = \max_{i=1,\dots,N_H}\{\frac{H_i^2}{\min_{\kappa \in \mathcal{T}_h, \kappa \subset \Omega_i} h_\kappa^2}\}$.

Remark 1 Detailed proofs of Theorems 1 and 2 will be provided elsewhere due to the page limits. Here we only briefly sketch the idea of the proof of Theorem 2. We follow the abstract theory from the book by Toselli and Widlund [11]. Since the local stability and strengthened Schwarz inequality assumptions are straightforward, it remains to prove the existence of stable decomposition for any $v \in V_h$. To this end, we make use of the coarse space which makes it possible to extract subdomain average from v and deal only with functions with zero average on each subdomain. Applying Friedrichs inequality for discontinuous functions, [3], and making use of (5) we prove the stability constant of the decomposition is of order β.

4 Numerical Experiments

Let us choose the unit square as the domain Ω and for some prescribed integer M divide it into $N_H = 2^M \times 2^M$ smaller squares Ω_i $(i = 1, \dots, N_H)$ of equal size. This decomposition of Ω is then further refined into a uniform triangulation $\mathcal{T}_h$ based on a square $2^m \times 2^m$ grid $(m \geq M)$ with each square split into two triangles of identical shape. Hence, the fine mesh parameter is $h = 2^{-m}$, while the coarse grid parameter is $H = 2^{-M}$. We discretize the problem (1) on the fine triangulation using the method (3) with $\delta = 7$.

In the following tables we report the number of Preconditioned Conjugate Gradient iterations for operator T (defined in Sect. 3.2) which are required to reduce the initial Euclidean norm of the residual by a factor of 10^6 and (in parentheses) the condition number estimate for T. We consider two sets of test problems: with either continuous or discontinuous coefficient ρ. We always choose a random vector for the right hand side and a zero as the initial guess.

4.1 ASM Version II vs. "Standard" ASM

First let us consider the performance of ASM version II against a more "standard" ASM, see [8, Section 3.3], where the local solve is restricted not to a single element of size h, but to a single subdomain Ω_i of size H. For the diffusion coefficient we

Table 1 Dependence of the number of iterations and the condition number (in parentheses) on $H = 2^{-M}$ and $h = 2^{-m}$ for the method of Sect. 3.2

Fine (m) → ↓ Coarse (M)	4	5	6	7
4	29 (22)	39 (40)	59 ($1.1 \cdot 10^2$)	96 ($3.8 \cdot 10^2$)
5		30 (23)	39 (40)	59 ($1.1 \cdot 10^2$)
6			30 (23)	38 (40)
7				30 (23)

Table 2 Dependence of the number of iterations and the condition number (in parentheses) on $H = 2^{-M}$ and $h = 2^{-m}$ for the method of [8, Sect. 3.3]

Fine (m) → ↓ Coarse (M)	4	5	6	7
4	27 (20)	35 (34)	46 (67)	62 ($1.3 \cdot 10^2$)
5		28 (20)	35 (34)	46 (67)
6			28 (20)	35 (34)
7				28 (20)

take a continuous function, $\rho(x) = x_1^2 + x_2^2 + 1$. As it turns out from Tables 1 and 2, the condition number of the method considered in Sect. 3.2 indeed shows an $O((H/h)^2)$ behavior, as predicted by Theorem 2, while methods which use local solves on subdomains of diameter at least H (e.g. [8] or, similarly, [1, 10]) exhibit more favorable $O(H/h)$ dependence.

4.2 *Discontinuous Coefficient*

Next, let us consider ρ with discontinuities aligned with an auxiliary partitioning of Ω into 4×4 squares. Precisely, we introduce a red–black checkerboard coloring of this partitioning and set $\rho = 1$ in red regions, and the value of ρ_1 reported in Table 3 in black ones. In this way, our fine and coarse triangulations, with $m = 7$ and $M = 4$, will always be aligned with the discontinuities. Table 3 shows the independence of the condition number on ρ_1 in this case.

Finally, we consider elementwise discontinuous coefficient, with $\rho = 1$ on odd and $\rho = \rho_1$ on even-numbered triangles. Table 4 shows that in this case the coarse space fails (a dash means the method did not converge in 600 iterations). This confirms the importance of the assumption of mild variation of the coefficient (5).

Table 3 Dependence of the number of iterations and the condition number (in parentheses) on the discontinuity when the coefficient is constant inside subdomains

ρ_1	10^0	10^{-2}	10^{-4}	10^{-6}
Iter (cond)	134 $(3.8 \cdot 10^2)$	141 $(3.7 \cdot 10^2)$	161 $(3.7 \cdot 10^2)$	179 $(3.8 \cdot 10^2)$

Red–black 4×4 distribution of ρ, aligned with domain decomposition. Fixed $H/h = 8$

Table 4 Dependence of the number of iterations and the condition number (in parentheses) on the discontinuity when the coefficient elementwise discontinuous

ρ_1	10^0	10^{-2}	10^{-4}	10^{-6}
Iter (cond)	134 $(3.8 \cdot 10^2)$	435 $(3.8 \cdot 10^3)$	– $(3.1 \cdot 10^5)$	– $(2.5 \cdot 10^7)$

Fixed $H/h = 8$

5 Conclusions

A nonoverlapping ASM for symmetric interior penalty DG discretization of second order elliptic PDE with discontinuous coefficient has been presented, in which a very large number of very small local problems is solved in parallel, together with one coarse problem of moderate size. Under mild assumptions, the condition number of the resulting system is $O((H/h)^2)$, independently of the jumps of the coefficient.

Acknowledgement We would like to thank an anonymous referee whose comments and remarks helped to improve the paper. This research has been supported by the Polish National Science Centre grant 2011/01/B/ST1/01179.

References

1. P.F. Antonietti, B. Ayuso, Schwarz domain decomposition preconditioners for discontinuous Galerkin approximations of elliptic problems: non-overlapping case. Math. Model. Numer. Anal. **41**(1), 21–54 (2007). doi:10.1051/m2an:2007006. ISSN:0764-583X. http://www.dx.doi.org/10.1051/m2an:2007006
2. B. Ayuso de Dios, M. Holst, Y. Zhu, L. Zikatanov, Multilevel preconditioners for discontinuous Galerkin approximations of elliptic problems with jump coefficients. Math. Comput. **83**, 1083–1120 (2014)
3. S.C. Brenner, Poincaré-Friedrichs inequalities for piecewise H^1 functions. SIAM J. Numer. Anal. **41**(1), 306–324 (2003). doi:10.1137/S0036142902401311. ISSN:0036-1429. http://www.dx.doi.org/10.1137/S0036142902401311
4. K. Brix, C. Pinto, C. Canuto, W. Dahmen, Multilevel preconditioning of Discontinuous-Galerkin spectral element methods part I: geometrically conforming meshes (IGPM Preprint, RWTH Aachen, Aachen, 2013)
5. C. Canuto, L.F. Pavarino, A.B. Pieri, BDDC preconditioners for continuous and discontinuous Galerkin methods using spectral/*hp* elements with variable local polynomial degree. IMA J. Numer. Anal. **34**(3), 879–903 (2014). doi:10.1093/imanum/drt037. ISSN:0272-4979. http://www.dx.doi.org/10.1093/imanum/drt037

6. M. Dryja, On discontinuous Galerkin methods for elliptic problems with discontinuous coefficients. Comput. Methods Appl. Math. **3**(1), 76–85 (2003). ISSN:1609-4840
7. M. Dryja, M. Sarkis, Additive average Schwarz methods for discretization of elliptic problems with highly discontinuous coefficients. Comput. Methods Appl. Math. **10**(2), 164–176 (2010). doi:10.2478/cmam-2010-0009. ISSN:1609-4840. http://www.dx.doi.org/10.2478/cmam-2010-0009
8. M. Dryja, P. Krzyżanowski, M. Sarkis, Additive Schwarz method for DG discretization of anisotropic elliptic problems, in *Domain Decomposition Methods in Science and Engineering XXI*, ed. by J. Erhel, M.J. Gander, L. Halpern, G. Pichot, T. Sassi, O. Widlund. Lecture Notes in Computational Science and Engineering, vol. 98 (Springer, New York, 2014), pp. 407–415
9. A. Ern, A.F. Stephansen, P. Zunino, A discontinuous Galerkin method with weighted averages for advection-diffusion equations with locally small and anisotropic diffusivity. IMA J. Numer. Anal. **29**(2), 235–256 (2009). doi:10.1093/imanum/drm050. ISSN:0272-4979. http://www.dx.doi.org/10.1093/imanum/drm050
10. X. Feng, O.A. Karakashian, Two-level additive Schwarz methods for a discontinuous Galerkin approximation of second order elliptic problems. SIAM J. Numer. Anal. **39**(4), 1343–1365 (2001). ISSN:1095-7170
11. A. Toselli, O. Widlund, *Domain Decomposition Methods—Algorithms and Theory*. Springer Series in Computational Mathematics, vol. 34 (Springer, Berlin, 2005). ISBN:3-540-20696-5

Algebraic Multigrid for Discontinuous Galerkin Methods Using Local Transformations

Christian Engwer, Klaus Johannsen, and Andreas Nüßing

1 Introduction

Discontinuous Galerkin methods are popular discretization methods for partial differential equations for over a decade. For the resulting linear system, the need arises for robust and efficient solvers. A geometric multigrid algorithm which maintains the properties of the discretization along the grid hierarchy has been presented in [6]. The grid transfer is based on an L^2-projection and an overlapping element block smoother is applied on each level. For cases where the construction of a geometric grid hierarchy is not feasible, certain classes of algebraic multigrid methods have been developed. In [2], an iterative method has been proposed, based on the splitting of the function space into two non-overlapping subspaces. On those spaces, the problem can be solved more efficiently. Another approach has been followed in [7]. There, an algebraic multigrid method has been presented which uses a smoothed aggregation method to form the coarser grid levels. A combination of both approaches has been developed in [4]. The algebraic multigrid being described there uses a projection of the discontinuous space onto the conforming subspace of linear elements. An agglomeration strategy is employed to create the smoother and the coarse grid levels. This strategy drops the block structure of the linear system and loses the information of the discontinuous Galerkin discretization on coarser grid levels. In addition, it is not applicable to the Stokes equation since the inf-

C. Engwer • A. Nüßing (✉)
Institute for Computational and Applied Mathematics, University of Münster, Münster, Germany
e-mail: christian.engwer@uni-muenster.de; andreas.nuessing@uni-muenster.de

K. Johannsen
Uni Computing, Uni Research AS, Bergen, Norway

T. Dickopf et al. (eds.), *Domain Decomposition Methods in Science and Engineering XXII*, Lecture Notes in Computational Science and Engineering 104, DOI 10.1007/978-3-319-18827-0_16

sup stability is lost on the first order conforming subspace due to the equal order discretization of velocity and pressure.

The aim of this paper is to develop and evaluate an algebraic multigrid method for discontinuous Galerkin discretizations, which preserves and uses the block structure on each grid level and can be applied to different problems, including the Stokes equation. We follow the general structure of the geometric multigrid of [6] but also take ideas from [4] into account. The derivation of the method uses the Poisson equation and includes comments on the differences for the Stokes equation when applicable. The paper is structured as follows: Sect. 2 provides a short introduction to the discretization of the Poisson equation and the resulting linear system. In Sect. 3 the algebraic multigrid algorithm is presented, including the transfer between different grid levels and the smoothing operator. The algorithm is evaluated in Sect. 4 and finally a short conclusion is given.

2 Preliminaries

We describe our method using the discontinuous Galerkin discretization for the Poisson equation, cf. [1]. Let $T_h(\Omega) := \{\Omega_0, \ldots, \Omega_{N-1}\}$ define a triangulation of the domain Ω with the size parameter $h \in \mathbb{R}$. The broken Sobolev space is defined as $V_h := \{u \in L^2(\Omega)|u_{|\Omega_i} \in P(\Omega_i)\}$ for some polynomial spaces $P(\Omega_i)$. The discontinuous Galerkin formulation of the Poisson equation with homogeneous Dirichlet boundary conditions reads: find $u_h \in V_h$ such that $a_\epsilon(u_h, v) = f(v)$ holds for all $v \in V_h$ (cf. [1] for a derivation and definition of a_ϵ). The method parameter is denoted by ϵ and the penalty parameter by $\eta \in \mathbb{R}$. For each grid element Ω_i, we assume there is a diffeomorphism $\mu_i : \mathbb{R}^n \to \mathbb{R}^n$ with $\mu_i(\hat{\Omega}) = \Omega_i$, mapping local coordinates on a reference element $\hat{\Omega}$ to global coordinates on Ω. Next we introduce local polynomial basis functions on the reference element:

$$\phi_i : \mathbb{R}^n \to \mathbb{R}, \quad i \in \{0, \ldots, N_b - 1\} \tag{1}$$

In order to simplify the description, we assume the same local basis on all elements. Note that this restriction is not essential. Using the local to global transformations, we define the basis function in global coordinates as $\phi_{i,k} = \phi_i \circ \mu_k^{-1}$. Introducing a representation of u_h and v with respect to the global basis functions in the discontinuous Galerkin formulation yields the linear system for the Poisson equation:

$$\begin{aligned} Ax = b, \quad A &= (A_{kl})_{k,l} \\ A_{kl} = (a_{ij}^{kl})_{ij} \in \mathbb{R}^{N_b \times N_b}, \quad a_{ij}^{kl} &= a_\epsilon(\phi_{j,l}, \phi_{i,k}) \end{aligned} \tag{2}$$

Besides for the Poisson equation, we also construct the multigrid method for the Stokes equation. We will not present its discontinuous Galerkin formulation here, but refer to [8]. We block the degrees of freedom for pressure and velocity element wise, which again yields a sparse block linear system.

3 Algorithm

3.1 General

The proposed algebraic multigrid method is a method to solve a linear sparse block system $Ax = b$ resulting from a discontinuous Galerkin discretization using only geometric information on the finest grid level. The grid levels are numbered from coarse to fine with $0, \ldots, L$, such that 0 denotes the coarsest grid level. By $N_l \in \mathbb{N}$ we denote the number of elements on level l. We will mark matrices and vectors with the level they are associated with. If the level index is missing, the matrix or vector refers to the finest level, if not stated otherwise. On each grid level but L, we assume there is a prolongation operator P^l mapping a coefficient vector from grid level l to the next finer level $l+1$. The restriction R^l of a vector from level $l+1$ to level l is accomplished using the transposed of the prolongation $R^l := (P^l)^t$. We compute the coarse grid matrices recursively from the finest matrix by applying the Galerkin product $A^{l-1} = R^{l-1}A^lP^{l-1}$. To reduce oscillating error frequencies, we apply the smoother S^l on level l. Both, the prolongation and the smoother are described in the remainder of this section.

3.2 Grid Transfer

The spaces on coarse levels are constructed recursively as subspaces of the space at the next finer level using a semi coarsening approach. A semi coarsening can be constructed based on a matching in the block matrix graph of the block matrix A. The graph $G(A) = (V(A), E(A))$ consists of the nodes and the edges:

$$\begin{aligned} V(A) &= \{0, \ldots, N_L - 1\} \\ E(A) &= \big\{(i,j) \in V(A) \times V(A) : i < j \wedge A_{ij} \neq 0\big\} \end{aligned} \tag{3}$$

Since the sparsity pattern of A is symmetric, $G(A)$ is undirected. In the following, when selecting edges for coarsening, we only consider strong edges $E_s(A) \subset E(A)$.

We divide the edges into weak and strong ones, using the same criterion as in [4]. The strength of an edge $(i,j) \in E$ is defined as

$$\rho((i,j)) := \frac{\|A_{ij}\| \|A_{ji}\|}{\|A_{ii}\| \|A_{jj}\|} \tag{4}$$

An edge is called strong if its strength is greater than β times the maximum strength among its neighbors, for a constant $\beta \in [0, 1]$. The selection of disjoint strong edges corresponds to finding a graph matching. A graph matching of strong edges is a subset of $E_s(A)$ such that every node is part of at most one edge.

The transfer between two grid levels is constructed using so called *shift matrices*, consisting of local basis transformations. For a pair of elements, we select the polynomial basis of the first element to be the basis of the combined element and embed the basis of the second element into the one of the first. The shift from l to k for two neighboring elements l and k is defined as

$$\begin{aligned} S_{kl} &:= M_k^{-1}\tilde{S}_{kl} \\ M_k &:= (m_{ij}^k)_{ij} \quad m_{ij}^k := \langle \phi_{i,k}, \phi_{j,k} \rangle_{L^2(\Omega_k)} \\ \tilde{S}_{kl} &:= (\tilde{s}_{ij}^{kl})_{ij} \quad \tilde{s}_{ij}^{kl} := \langle \phi_{i,k}, \phi_{j,l} \rangle_{L^2(\Omega_k)} \end{aligned} \tag{5}$$

These local shift matrices can be combined into a global sparse block matrix. Due to the coupling of neighboring elements in the discontinuous Galerkin discretization, the global shift matrix has the same sparsity pattern as the matrix A. The shift matrices on coarser grid levels can be obtained from the next finer level by successive shifting into neighboring elements.

Having selected a set of pairs to be coarsened, we can construct the prolongation matrix which transfers a block coefficient vector from the coarse to the fine level. For an element which has not been selected for coarsening, we keep its basis on the coarse grid and therefore set the associated prolongation block to the identity matrix. For each pair, we keep the basis of the first element and again set the block to the identity matrix. The basis of the other element gets transferred into the basis of the first using the local shift matrices described above. This approach yields the prolongation as a sparse block matrix P^l which can be defined as

$$P^l(\ldots, x_e, \ldots) := (\ldots, [x_e, S_{fe}x_e], \ldots) \tag{6}$$

for each selected pair (e,f). We define the corresponding restriction matrix as $R^l := (P^l)^T$. The domain Ω_i^l associated with an element i on level l is defined as the union of all elements on the finest grid level which have been aggregated in element i. Accordingly, the function space V^l on level l is spanned by the bases of the elements of level l.

3.3 Smoother

The smoother S^l should reduce oscillating components of the error on the current level. As presented in [6], we use an additive and multiplicative Schwarz method as a smoother. Let $V_i^l \subset V^l$ define a subspace of V^l for each $i \in I^l$ with an index set I^l to be defined later. For each subspace $V_i^l, i \in I^l$, we solve $a^l(u_k^l + c_{k,i}^l, v) = f(v) \quad \forall v \in V_i^l$ for $c_{k,i}^l \in V_i^l$. The *additive Schwarz method* is then given by

$$u_{k+1}^l = u_k^l + \theta^l \sum_{i \in I^l} c_{k,i}^l \tag{7}$$

with a damping parameter $\theta^l \in \mathbb{R}$. In a similar way, the *multiplicative Schwarz method* can be introduced, where the updates are computed and applied successively.

In [6] different types of patches have been evaluated in a geometric multigrid setting. The results indicate that non-overlapping element block patches do not yield a robust smoother. Overlapping vertex based patches, depicted in Fig. 1, show robust smoothing behavior and are therefore used by the smoother in our method. It should be pointed out, that the geometric information about vertices and their connection to elements is only available at the finest level. We need to adopt this information along the coarsening process. This is done, by keeping only those vertices from level $l+1$, which have not become internal vertices between two elements. The connectivity information between the remaining vertices and their adjoining elements on the coarse level can be transferred from the fine level: a vertex on level l is connected to an element i on level l if it was connected to an element on level $l+1$ which has been aggregated into i. The smoother is said to fulfill the *smoothing property*, if $\|A^l(S^l)^\nu\| \leq C\eta(\nu)$ with a function $\eta(\nu) \to 0$ for $\nu \to \infty$.

For the Poisson equation, we set I^l to be the index set of grid vertices on level l in the algebraic sense. V_i^l is the linear subspace spanned by the degrees of freedom associated with an element which is connected to the grid vertex i. The numerical results in [6] indicate for the one dimensional problem using the NIPG method, that the additive smoother fulfills the smoothing property with $1/\nu$.

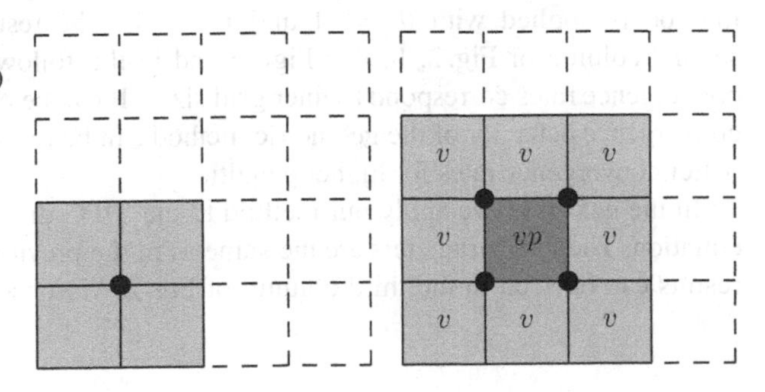

Fig. 1 Overlapping patches for the Poisson equation (*left*) and Stokes equation (*right*)

For the Stokes equation, in addition to the vertex based patches, we need to take into account the saddle point structure of the problem. We adopt the idea of the Vanka type smoother from [10], where, in a staggered grid context, a pressure degree of freedom is combined with all coupling velocity degrees of freedom. In addition we include the vertex based approach in order to construct a robust smoother. Combining both approaches in the context of the discontinuous Galerkin formulations, we set I^l to be the index set of elements on level l. V_i^l is the linear subspace spanned by the degrees of freedom associated with an element which shares a grid vertex with element i (see Fig. 1). Based on experimental results, we apply a different damping factor depending on the position of an element inside the patch. Theoretic results from [9] and numerical experiments indicate that for Stokes SIPG, the additive smoother fulfills the smoothing property with at least $1/\sqrt{\nu}$.

4 Evaluation

We implemented the algebraic multigrid method using the *Distributed and Unified Numerics Environment (DUNE)* (see [3]), using the PDELab toolbox (see [5]) for the PDE discretization. First, we apply our method to a two dimensional Poisson problem with $\Omega = [0, 1]^2$ on a structured grid with rectangular elements, in order to reproduce the results given in [6] for a geometric multigrid method. As local basis functions we use an orthogonalized Q_k basis, with $Q_k := \{(x, y) \mapsto x^{\alpha_x} y^{\alpha_y} : \alpha_x, \alpha_y \in \mathbb{N}, \alpha_x, \alpha_y \leq k\}$. For the following tests, we set $k := 2$. We use a NIPG discretization with different penalties and different sizes of the finest grid level. The penalty ranges from 10^{-3} to 10^6 and the fine grid size is increased by successive uniform refinement starting with a size of 5×5. The convergence rate is measured as

$$\rho := \left(\frac{\|d_{20}\|_2}{\|d_{10}\|_2} \right)^{\frac{1}{10}}, \tag{8}$$

where d_i denotes the defect in iteration i. We apply the additive method with damping $\theta = \frac{1}{2}$ and $\nu = 4$ pre- and post smoothing steps. The multiplicative method is applied with $\theta = 1$ and $\nu = 1$. The results can be seen in the second column of Fig. 2. In this Figure and in the following, graphs with higher convergence rates correspond to finer grid sizes. It can be observed, that the general convergence behavior of the geometric method can be reproduced, while producing better convergence rates for higher penalties.

In the next test, we apply our method to the SIPG discretization of the Poisson equation. The test parameters are the same as in the previous test. The convergence results can be seen in the third column of Fig. 2. We observe similar convergence

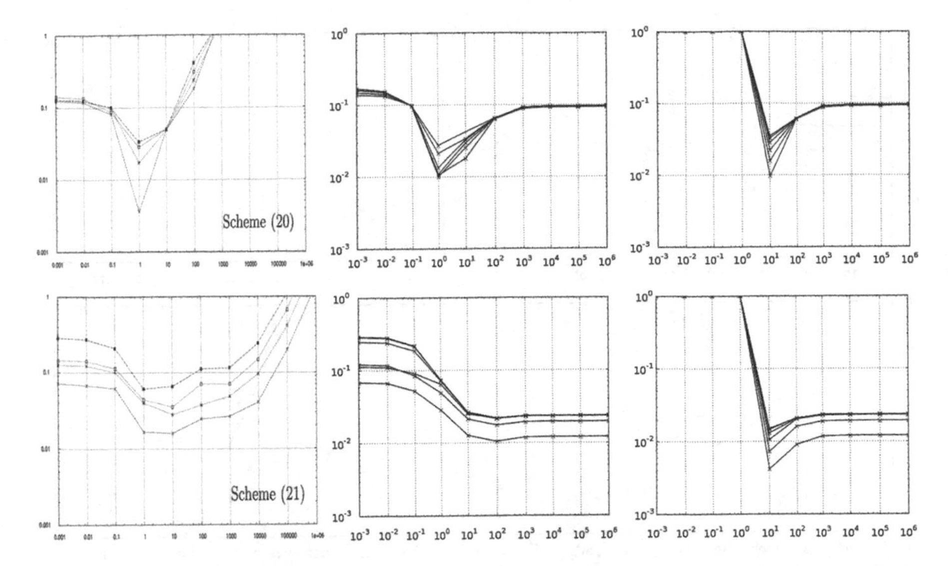

Fig. 2 Convergence rates for the Poisson equation, *left*: NIPG method using the geometric multigrid from [6], *center*: NIPG method using our multigrid, *right*: SIPG method using our multigrid, *top*: additive smoother, *bottom*: multiplicative smoother

Table 1 Results for the Poisson equation using the NIPG method on unstructured tetrahedral grids with the multigrid as a preconditioner for the BiCGSTAB algorithm

	Unit sphere					Unit cube				
Elements	2104	8270	33,418	139,572	547,038	2406	9386	38,202	154,194	635,216
Levels	11	11	10	10	10	11	12	13	12	11
Iterations	3	3	3	3	4	3	3	3	3	3

behavior as in the NIPG case. The method does not converge for a penalty less than $\sigma_0 < 10$, which corresponds to the theoretic findings in [1].

Next, we use the method as a preconditioner in a BiCGSTAB solver for a second order NIPG discretization on different unstructured grids. For different values of h, we create triangulations of the unit sphere and unit cube using tetrahedral elements. We use the multiplicative smoother and stop the iteration at a relative defect reduction of 10^{-10}. The results can be seen in Table 1.

Finally, we test for NIPG and SIPG discretizations of the Stokes equation on the unit square. We choose the orthogonalized Q_k basis for the velocity components and an orthogonalized P_{k-1} basis for the pressure, where $P_{k-1} := \{(x, y) \mapsto x^{\alpha_x} y^{\alpha_y} : \alpha_x, \alpha_y \in \mathbb{N}, \alpha_x + \alpha_y \leq k - 1\}$. Again we use a structured grid with rectangular elements, choose $k = 2$ and apply the method with different penalties and grid sizes. We use the same damping parameters as before, but weight the velocity degrees of freedom differently depending on their local patch position when applying an update. The weight for the central element of a patch is set to $\frac{1}{2}$ and the weight for the outer elements is set to $\frac{1}{2m}$, where m denotes the number of outer elements in

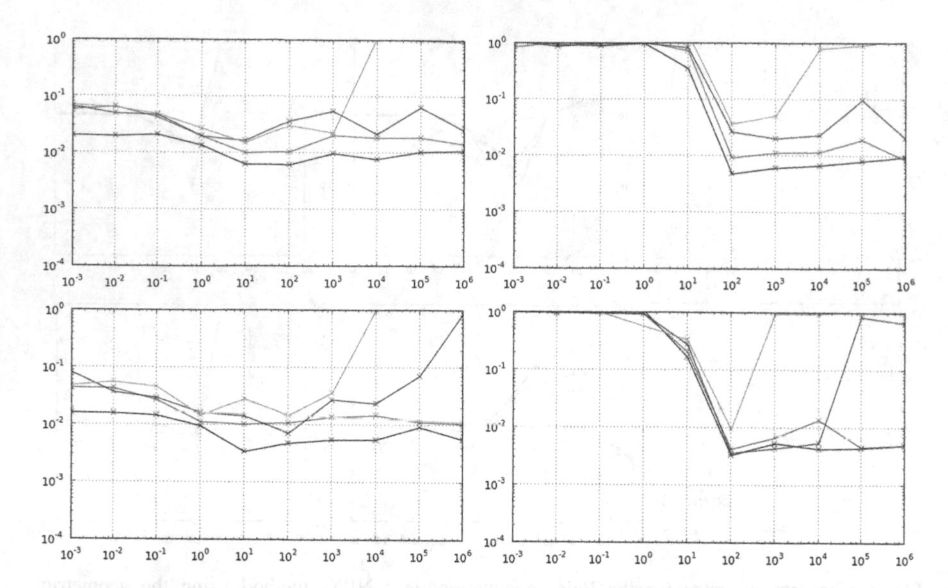

Fig. 3 Convergence rates for the Stokes equation, *left*: NIPG method, *right*: SIPG method, *top*: additive smoother, *bottom*: multiplicative smoother

the patch. Our method is used as a preconditioner for the BiCGSTAB algorithm. The results can be seen in Fig. 3. We observe increased convergence rates when compared to the Poisson equation. In addition, we observe larger convergence rates for finer grids and larger penalties.

5 Conclusion

We proposed an algebraic multigrid method for the discontinuous Galerkin discretization of the Poisson and Stokes problem. It shows good convergence rates and is flexible enough to be applied to different types of problems, which are not covered in this paper. Currently, one drawback of the method is its large computational cost. This effort is dominated by the application of the overlapping block smoother on the finest grid level. Reducing this effort by applying different smoothing strategies has not yielded the desired convergence behavior so far. To avoid increasing convergence rates for finer grids and higher penalties, one can develop different local shift strategies. Instead of projecting into the local basis of a single element, one can investigate the possibility to project into a common basis on all aggregated elements. In order to get a better understanding of the smoother, an investigation of the smoothing property might be worthwhile.

Acknowledgement This work was supported by the "German Academic Exchange Service" (DAAD) with the project 54570350 of the "German-Norwegian collaborative research support scheme".

References

1. D. Arnold, F. Brezzi, B. Cockburn, L. Marini, Unified analysis of discontinuous Galerkin methods for elliptic problems. SIAM J. Numer. Anal. **39**(5), 1749–1779 (2001/2002). ISSN 0036-1429
2. B. Ayuso de Dios, L. Zikatanov, Uniformly convergent iterative methods for discontinuous Galerkin discretizations. J. Sci. Comput. **40**(1–3), 4–36 (2009). ISSN 0885-7474
3. P. Bastian, M. Blatt, A. Dedner, C. Engwer, R. Klöfkorn, R. Kornhuber, M. Ohlberger, O. Sander, A generic grid interface for parallel and adaptive scientific computing. Part II: implementation and tests in DUNE. Computing **82**(2–3), 121–138 (2008). ISSN 0010-485X
4. P. Bastian, M. Blatt, R. Scheichl, Algebraic multigrid for discontinuous Galerkin discretizations of heterogeneous elliptic problems. Numer. Linear Algebra Appl. **19**(2), 367–388 (2012). ISSN 1070-5325
5. P. Bastian, F. Heimann, S. Marnach, Generic implementation of finite element methods in the distributed and unified numerics environment (DUNE). Kybernetika (Prague) **46**(2), 294–315 (2010). ISSN 0023-5954
6. K. Johannsen, Multigrid methods for NIPG. Technical Report, University of Texas at Austin, 2005
7. F. Prill, M. Lukáčová-Medviďová, R. Hartmann, Smoothed aggregation multigrid for the discontinuous Galerkin method. SIAM J. Sci. Comput. **31**(5), 3503–3528 (2009). ISSN 1064-8275
8. B. Rivière, V. Girault, Discontinuous finite element methods for incompressible flows on subdomains with non-matching interfaces. Comput. Methods Appl. Mech. Eng. **195**(25–28), 3274–3292 (2006). ISSN 0045-7825
9. J. Schöberl, W. Zulehner, On Schwarz-type smoothers for saddle point problems. Numer. Math. **95**(2), 377–399 (2003). ISSN 0029-599X
10. S.P. Vanka, Block-implicit multigrid solution of Navier-Stokes equations in primitive variables. J. Comput. Phys. **65**(1), 138–158 (1986). ISSN 0021-9991

Concepts for Flexible Parallel Multi-domain Simulations

Christian Engwer and Steffen Müthing

1 Introduction

Domain Decomposition methods provide a flexible tool for developing multi-physics simulations and coupling different discretization methods. In general, multi-physics simulations will require the handling of non-matching grids. Domain Decomposition methods like the Mortar method [3] enable us to simulate complex applications like contact problems, mechanics of moving parts, or heterogeneous coupling like surface-/groundwater flow.

As we will discuss, coupling unrelated parallel meshes poses significant practical problems. To our knowledge only very few implementations exist: both the well-known MpCCI library [7] and the SIERRA framework implement a parallel rendezvous algorithm [8] based on intersection algorithms, but neither of them is publicly available. An alternative approach can be based on radial basis functions, see [5].

The DUNE framework [1] offers different strategies for Domain Decomposition methods, which are available as DUNE extensions. One approach is to construct individual meshes for each sub-domain and relate them afterwards, the alternative is to create one mesh for the whole domain and define sub-domain meshes as appropriate sub-meshes. In this paper we only discuss the first approach. In [6, 6] describe a new algorithm that improves the complexity of matching unrelated meshes from $O(n^2)$ to $O(n)$, where n is the number of coupling elements. This algorithm is implemented in

C. Engwer (✉)
Institute for Computational and Applied Mathematics, University of Münster, Münster, Germany
e-mail: christian.engwer@wwu.de

S. Müthing
Interdisciplinary Center for Scientific Computing, Heidelberg University, Heidelberg, Germany
e-mail: steffen.muething@iwr.uni-heidelberg.de

T. Dickopf et al. (eds.), *Domain Decomposition Methods in Science and Engineering XXII*, Lecture Notes in Computational Science and Engineering 104, DOI 10.1007/978-3-319-18827-0_17

the DUNE GRID-GLUE [2] library. We discuss extensions of this library for handling distributed meshes.

When using methods like Dirichlet-Neumann coupling in the parallel context the user is forced to manage distributed data, as the necessary coupling information is not available locally. We present an abstraction that hides this non-locality and allows the user to implement his Domain Decomposition strategy in a clear mathematical setting. By introducing two auxiliary Finite Element spaces on the coupling interface we can reformulate the original domain decomposition algorithm and hide all parallel data handling from the user. In a proof of concept we implement these auxiliary spaces for the DUNE PDELAB library, where they are created in a completely automatic fashion.

2 Relating Unrelated Meshes

In the following we only describe a non-overlapping scenario, although the presented techniques are applicable to more general settings.

We consider a domain $\Omega \subset \mathbb{R}^d$. Ω is partitioned into two sub-domains Ω_0 and Ω_1 which meet at an interface Γ. The domains are triangulated into meshes $\mathcal{T}_0$ and $\mathcal{T}_1$ which are independent and in general do not match at the interface. Each mesh describes a set of entities, e.g. cells, faces, etc. We select a subset of entities which covers the interface Γ, i.e. the patches $\mathcal{P}_0$, $\mathcal{P}_1$; on these we impose the coupling conditions.

In order to relate information on Ω_0 and Ω_1 one has to transfer data like approximate solutions and evaluations of local residuals. We follow a mesh intersection approach, requiring us to compute the intersections of all entities in $\mathcal{P}_0$ with those in $\mathcal{P}_1$ (see Fig. 1). Based on the algorithm presented in [6] we identify pairs of overlapping entities from both sides, for which we then compute entity clippings, yielding a set of polyhedral intersections.

This algorithm is available as `Dune::GridGlue::Merger` within DUNE GRID-GLUE; it is provided as a native implementation and as an interface to legacy codes. Using a predicate m_i the coupling patches are defined as $\mathcal{P}_i = \{\gamma | \gamma \in$

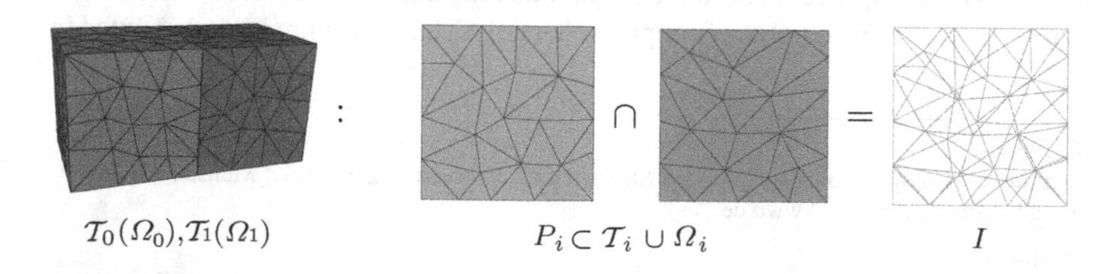

Fig. 1 Intersecting the coupling patches $\mathcal{P}_0$ and $\mathcal{P}_1$ yields a set I of intersections, which can be used to evaluate the coupling conditions

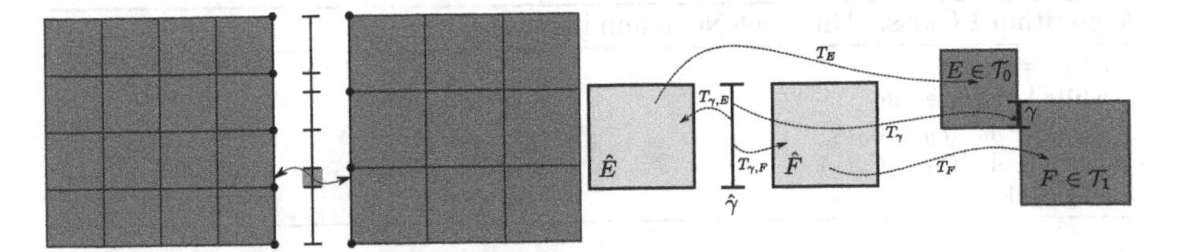

Fig. 2 *Left*: intersections relate adjacent cells of unrelated grids. *Right*: geometric mappings provided by an intersection

$\mathcal{T}_i \cap \partial\Omega_i \wedge m_i(\gamma)\}$. The computed intersections are modelled as the intersections in the DUNE grid interface and exposed as `Dune::GridGlue::Intersection`, which provides topological and geometrical information. In the sequential case it gives access to the adjacent cells in the two grids $\mathcal{T}_0$ and $\mathcal{T}_1$. To compute coupling conditions, intersections provide a mapping from local coordinates to global coordinates as well as mappings to the local coordinate systems of the adjacent cells (see Fig. 2).

2.1 Coupling via Intersections

As a short example, let us consider a two-domain Poisson problem with Dirichlet-Neumann coupling condition: Find u_0 and u_1 such that

$$\begin{aligned} -\Delta u_i &= 0 && \text{on } \Omega_i,\ i \in 0,1 \\ u_i &= g && \text{at } \partial\Omega_i \setminus \Gamma,\ i \in 0,1 \\ u_0 &= u_1 && \text{at } \Gamma \\ \nabla u_1 \cdot \mathbf{n} &= \nabla u_0 \cdot \mathbf{n} && \text{at } \Gamma\,. \end{aligned} \tag{1}$$

We follow the usual approach and introduce discrete trial and test spaces V_0, V_1 on Ω_0 and Ω_1. In the simplest case this might be a conforming Lagrange discretization. Testing with functions $v_i \in V_i$ and integration by parts yields the problem in its weak formulation. On Ω_0 we impose Dirichlet boundary conditions along Γ, whereas Neumann boundary conditions are imposed along Γ on Ω_1. As the interface Γ is in general non-conforming, we can employ a Clément interpolation to interpolate the solution u_1 onto Ω_0. For given bases Φ_0, Φ_1, we obtain a system matrix of the following form, where C_0 and C_1 correspond to Dirichlet and Neumann coupling blocks:

$$\begin{pmatrix} A_0 & C_0 \\ C_1 & A_1 \end{pmatrix} \cdot \begin{pmatrix} u_0 \\ u_1 \end{pmatrix} = \begin{pmatrix} b_0 \\ b_1 \end{pmatrix} \tag{2}$$

Algorithm 1 Classic Dirichlet-Neumann iteration

u^0, u^1 = initial
while ! converged **do**
  $u_0 \leftarrow A_0^{-1}(b_0 - C_\sigma u_1)$
  $u_1 \leftarrow A_1^{-1}(b_1 - C_\sigma u_0)$
end while

The matrix entries in the off-diagonal blocks are given by

$$C_0^{i,j} = -\langle \nabla \phi_0^i \mathbf{n}, \phi_1^j \rangle_\Gamma , \qquad C_1^{i,j} = -\omega_{\phi_0^j} \langle \phi_1^i, \phi_0^j \rangle_\Gamma ,$$

with $\omega_{\phi_0^j} = 1/\langle 1, \phi_0^j \rangle_\Gamma$ the weights of the Clément operator and $\phi_*^i \in \Phi_*$.

A straightforward approach to solving this problem iteratively is a fix point iteration on the split problem. In order to better illustrate the differences to the following parallel setting, we sketch this iteration in Algorithm 1.

2.2 *Concepts of Parallel Mesh Coupling*

Based on the previously introduced local grid matching algorithm we derive a parallel grid matching algorithm, see Algorithm 2. We extract the local part of the coupling patches $\mathcal{P}_0$, $\mathcal{P}_1$, merge these and communicate the data in a ring. We retrieve the neighboring patches and intersect them with our local patches. This yields the set of all intersections of local entities, either in Ω_0 or Ω_1, with any other entity, including remote entities. This provides all topological and geometric information required to evaluate the coupling conditions, but in general, as illustrated in Fig. 3, we lack access to the data in the adjacent domain. We therefore assign a globally unique ID to each intersection to provide parallel communication on the interfaces. This communication is built upon the *parallel IndexSets* [4] of DUNE and allows a gather/scatter mechanism to send and receive data across domain intersection patches. In analogy to the parallel communication in the DUNE grid interface, the user has to provide a `DataHandle` object which implements the gather and scatter operations. The communicated data depends on the chosen Domain Decomposition method, thus the user is usually required to implement the data communication himself. For high level frameworks this a very unsatisfactory situation.

For methods like Mortar or FETI-DP the problems are less immanent as we have no direct coupling along the sub-domain faces. These methods introduce additional degrees of freedom on the interface, the sub-domains couple only to the interface and then the arising Schur-Complement system for the interface is solved.

Other methods like classic non-overlapping Schwarz methods or Dirichlet-Neumann coupling directly couple the sub-domains and require explicit communication of remote data. The main difference is that in the latter case we cannot

Algorithm 2 Parallel grid matching algorithm

parallel GridGlue
  P ▷ P: # of parallel processes
  $\mathcal{T}(\Omega_0), \mathcal{T}(\Omega_1)$ ▷ Sub domain meshes
  m_0, m_1 ▷ Predicates for Ω_1 and Ω_2
  process Π $[p \in \{0, \dots, P-1\}]$
    $\mathcal{P}_0 = \{\gamma | \gamma \in \mathcal{T}_0|_p \cap \partial\Omega_0 \wedge m_0(\gamma)\}$ ▷ Local coupling patches
    $\mathcal{P}_1 = \{\gamma | \gamma \in \mathcal{T}_1|_p \cap \partial\Omega_1 \wedge m_1(\gamma)\}$
    $I^p \leftarrow \text{merge}(\mathcal{P}_0, \mathcal{P}_1)$ ▷ Set of intersection
    $(\widehat{\mathcal{P}}_0, \widehat{\mathcal{P}}_1) \leftarrow (\mathcal{P}_0, \mathcal{P}_1)$
    for $i \in [0, P-2)$ **do**
      asend: $(\widehat{\mathcal{P}}_0, \widehat{\mathcal{P}}_1) \longrightarrow (p+1)\%P$ ▷ send to right neighbor
      arecv: $(\widehat{\mathcal{P}}_0, \widehat{\mathcal{P}}_1) \longrightarrow (p-1+P)\%P$ ▷ receive from left neighbor
      $I^p \leftarrow I^p \cup \text{merge}(\widehat{\mathcal{P}}_0, \mathcal{P}_1)$ ▷ merge remote patches
      $I^p \leftarrow I^p \cup \text{merge}(\mathcal{P}_0, \widehat{\mathcal{P}}_1)$ ... with local patches
    end for
  end process
end parallel

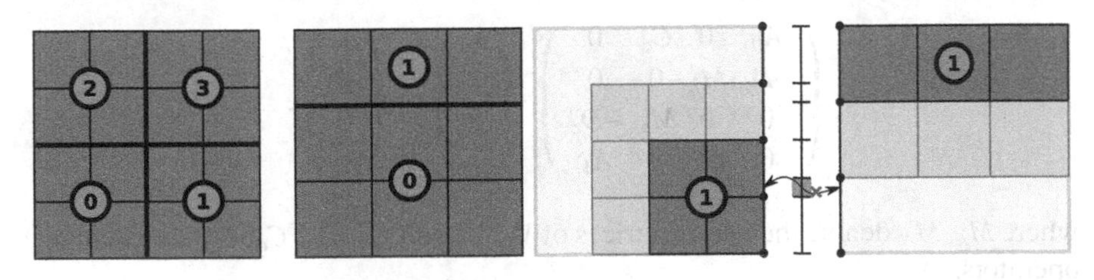

Fig. 3 When coupling distributed grids, neighboring cells of the remote mesh might not be accessible locally, making it impossible to evaluate coupling conditions (*numbers in circles* denote the process rank)

fully represent the local part of the Poincaré-Steklov operator on a single processor, but only the local contributions.

3 Hiding Parallel Communication Using Auxiliary Spaces

We now describe a mathematical abstraction which allows implementations to hide all communications from the user. We introduce additional function spaces V_λ and V_σ on the coupling interface Γ, see Fig. 4. The definition of these function spaces is general; they can thus be constructed automatically as

$$V_\lambda = \left\{ v \in L^2(\Gamma) \,\middle|\, v|_\gamma \in P_k(\gamma), \gamma \in I, k = \text{order}(V_0) \right\} \supseteq \text{tr}(V_0)$$

$$V_\sigma = \left\{ v \in L^2(\Gamma) \,\middle|\, v|_\gamma \in P_k(\gamma), \gamma \in I, k = \text{order}(V_1) \right\} \supseteq \text{tr}(V_1)\,,$$

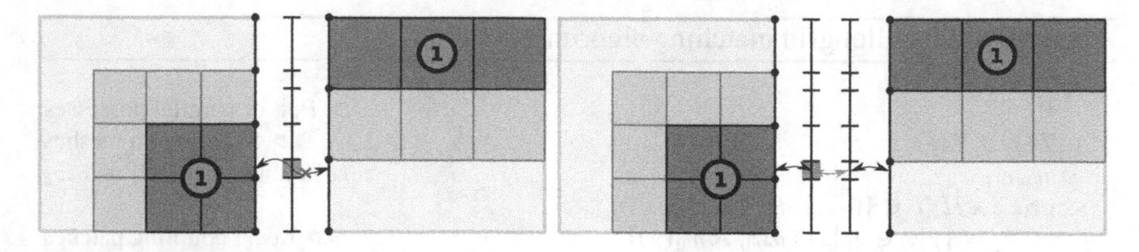

Fig. 4 Through the use of auxiliary spaces on the coupling interface Γ, direct access to non-local cells of the neighboring domain is avoided (*numbers in circles* denote the process rank)

where P_k denotes the space of polynomial functions up to degree k. V_λ and V_σ are defined as discontinuous polynomial spaces on the interface, where V_λ is the minimal DG space containing the trace spaces of V_0 and V_σ for V_1, respectively. For efficiency we choose L^2 orthonormal bases. Note that for $\text{order}(V_0) = \text{order}(V_1)$ it follows that $V_\lambda = V_\sigma$. The arising structure of the global system is as follows, although it is never assembled as a whole:

$$\begin{pmatrix} A_0 & 0 & C_\sigma & 0 \\ -D_\lambda & M_\lambda & 0 & 0 \\ 0 & 0 & M_\sigma & -D_\sigma \\ 0 & C_\lambda & 0 & A_1 \end{pmatrix} \cdot \begin{pmatrix} u_0 \\ \lambda \\ \sigma \\ u_1 \end{pmatrix} = \begin{pmatrix} b_0 \\ 0 \\ 0 \\ b_1 \end{pmatrix},$$

where M_λ, M_σ denote the mass matrices of V_λ, V_σ and $C_\lambda, D_\lambda, C_\sigma, D_\sigma$ are coupling operators.

The auxiliary spaces V_λ and V_σ eliminate the direct coupling between A_0 and A_1. We split the original coupling operator C_1 to obtain the pair C_λ, D_λ and proceed analogously for C_0. As we have chosen L^2 orthonormal basis functions for V_λ and V_σ, the mass matrices reduce to the identity $\mathbb{1}$. Therefore the coupling operators can be evaluated on the fly in an efficient fashion. All computations are completely local and can be handled by a generic gather/scatter implementation. The relation between C_1 and C_λ, D_λ becomes obvious when eliminating λ or σ, respectively. We use $M_* = \mathbb{1}$ and obtain the classical coupled system as in (2)

$$\begin{pmatrix} A_0 & C_\sigma D_\sigma \\ C_\lambda D_\lambda & A_1 \end{pmatrix} \cdot \begin{pmatrix} u_0 \\ u_1 \end{pmatrix} = \begin{pmatrix} b_0 \\ b_1 \end{pmatrix}$$

In analogy to Algorithm 1, we can solve the coupled parallel system using Algorithm 3. As we recover the original DD method, it is also possible to use it as a preconditioner in existing Krylov methods.

Algorithm 3 Auxiliary space iterative algorithm

u^0, u^1 = initial
while ! converged **do**
  $\sigma \leftarrow D_\sigma u_1$ ▷ implicit data communication
  $u_0 \leftarrow A_0^{-1}(b_0 - C_\sigma \sigma)$ ▷ parallel solver on Ω_0
  $\lambda \leftarrow D_\lambda u_0$ ▷ implicit data communication
  $u_1 \leftarrow A_1^{-1}(b_1 - C_\sigma \lambda)$ ▷ parallel solver on Ω_1
end while

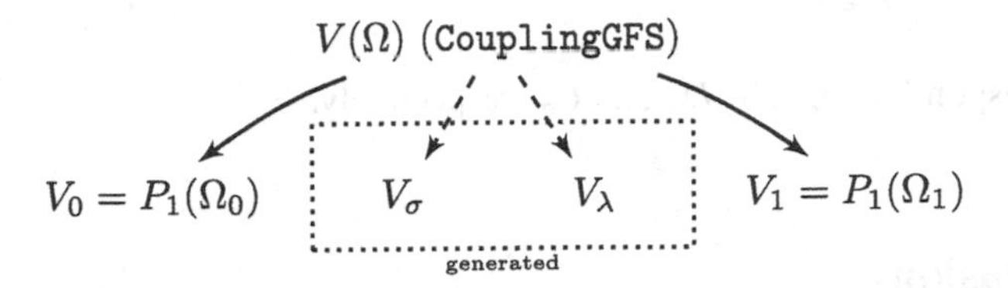

Fig. 5 Sketch of the hierarchic construction of the global function space for a coupled problem. DUNE PDELAB automatically generated the spaces V_λ and V_σ

3.1 Implementation in DUNE PDELab

When implementing the Poisson example from Sect. 2.1 with the auxiliary spaces approach, DUNE PDELAB transparently synthesizes the auxiliary spaces V_λ and V_σ and represents the overall solution space $V = V_0 \times V_1 \times V_\lambda \times V_\sigma$ as a tree of elementary function spaces (cf. Fig. 5). Given a weak problem of the form $u \in U : a(u, v) = b(v) \; \forall \; v \in V$, DUNE PDELAB splits the (bi)linear forms into sums of entity-local contributions α_e, α_s and α_b for cells, interior facets and boundary facets, respectively, isolating the user from mesh and DOF handling. $a(u, v)$ thus reads

$$a(u,v) = \sum_{e \in E_h} R_e^E(\alpha_v, u, v) + \sum_{f \in F_h^{(i)}} R_f^F(\alpha_s, u, v) + \sum_{f \in F_h^{(b)}} R_b^B(\alpha_b, u, v). \tag{3}$$

R^E, R^F and R^B map the global spaces U and V to the element-local restrictions on the cells adjacent to the current entity, leaving the user with the task of implementing the local contributions α_e, α_s and α_b.

The coupling operators D_λ, C_λ, D_σ and C_σ resemble the interior facet terms in that they involve restricted function spaces with different supports, but differ in that the restrictions do not belong to the same global space. Those terms consequently require an extension of Eq. (3) with additional coupling terms on the interface Γ and the two sub-domains.

D_λ and D_σ form projection operators onto V_λ and V_σ, whereas C_λ and C_σ mimic the operators C_0 and C_1. The first one behaves like a source on the interface, whereas the second one is a direct adoption of the Clément operator. Given local bases Φ_*^γ on γ (with $V_*|\gamma = \text{span}(\Phi_*^\gamma)$) the user has to implement the following local contributions to the global stiffness matrix:

$$\alpha_\gamma^{0,\lambda}(\Phi_0^\gamma,\Phi_\lambda^\gamma) = \sum_{\substack{\phi_\lambda\in\Phi_\lambda^\gamma\\ \phi_0\in\Phi_0^\gamma}} -\langle\nabla\phi_0\mathbf{n},\phi_\lambda\rangle_\gamma\,, \quad \alpha_\gamma^{\lambda,1}(\Phi_\lambda^\gamma,\Phi_1^\gamma) = \sum_{\substack{\phi_1\in\Phi_1^\gamma\\ \phi_\lambda\in\Phi_\lambda^\gamma}} \langle\phi_\lambda,\phi_1\rangle_\gamma\,,$$

$$\alpha_\gamma^{1,\sigma}(\Phi_1^\gamma,\Phi_\sigma^\gamma) = \sum_{\substack{\phi_\sigma\in\Phi_\sigma^\gamma\\ \phi_1\in\Phi_1^\gamma}} -\langle\phi_1,\phi_\sigma\rangle_\gamma\,, \qquad \alpha_\gamma^{\sigma,0}(\Phi_\sigma^\gamma,\Phi_0^\gamma) = \sum_{\substack{\phi_0\in\Phi_0^\gamma\\ \phi_\sigma\in\Phi_\sigma^\gamma}} -\omega_{\phi_0}\langle\phi_\sigma,\phi_0\rangle_\gamma\,,$$

which correspond to D_λ, C_λ, D_σ and C_σ, respectively.

4 Conclusions

The DUNE GRID-GLUE library offers software infrastructure for the coupling of unrelated grids. We presented recent extensions to DUNE GRID-GLUE to work in the context of distributed meshes. Reconstructed geometrical and topological relations between the grids are encapsulated as intersection objects. Although presented for non-overlapping intersections, the parallel implementation also handles overlapping and mixed-dimensional setups.

The coupling of distributed grids usually requires substantial changes to the user code and explicit use of parallel communication. We discussed a concept to reformulate the numerical scheme using auxiliary spaces on the coupling interface Γ, which allows the implementation of domain decomposition methods in a common framework that can hide the parallel communication from the user. This reformulated coupling problem integrates nicely with the hierarchic function space and operator concepts available in DUNE PDELAB.

The presented parallel mesh matching is available in the current version of the DUNE GRID-GLUE library. A prototype implementation for DUNE PDELAB is available, a more general implementation is under development. The code is available under an open source license from the DUNE website http://www.dune-project.org/.

Acknowledgement This work was supported by the German Research Foundation (DFG) through the Priority Programme 1648 "Software for Exascale Computing" (SPPEXA) and the DFG EXC 1003 Cells in Motion—Cluster of Excellence, Münster, Germany.

References

1. P. Bastian, M. Blatt, A. Dedner, C. Engwer, R. Klöfkorn, R. Kornhuber, M. Ohlberger, O. Sander, A generic grid interface for parallel and adaptive scientific computing. Part II: implementation and tests in DUNE. Computing **82**(2–3), 121–138 (2008)

2. P. Bastian, G. Buse, O. Sander, Infrastructure for the coupling of dune grids, in *Proceedings of ENUMATH 2009* (Springer, Berlin, 2010), pp. 107–114
3. C. Bernardi, Y. Maday, A.T. Patera, Domain decomposition by the mortar element method, in *Asymptotic and Numerical Methods for Partial Differential Equations with Critical Parameters*, ed. by H.G. Kaper, M. Garbey, G.W. Pieper (Springer, The Netherlands, 1993), pp. 269–286
4. M. Blatt, P. Bastian, On the generic parallelisation of iterative solvers for the finite element method. Int. J. Comput. Sci. Eng. **4**(1), 56–69 (2008)
5. S. Deparis, D. Forti, A. Quarteroni, A rescaled localized radial basis functions interpolation on non-cartesian and non-conforming grids. Technical Report 37.2013, EPFL (2013)
6. M. Gander, C. Japhet, Algorithm 932: PANG: software for non-matching grid projections in 2d and 3d with linear complexity. ACM Trans. Math. Softw. **40**(1), 6:1–6:25 (2013)
7. W. Joppich, M. Kürschner, MpCCI – a tool for the simulation of coupled applications. Concurr. Comput. Pract. Exp. **18**(2), 183–192 (2006)
8. S.J. Plimpton, B. Hendrickson, J.R. Stewart, A parallel rendezvous algorithm for interpolation between multiple grids. J. Parallel Distrib. Comput. **64**(2), 266–276 (2004)

Domain Decomposition and Parallel Direct Solvers as an Adaptive Multiscale Strategy for Damage Simulation in Quasi-Brittle Materials

Frank P.X. Everdij, Oriol Lloberas-Valls, Angelo Simone, Daniel J. Rixen, and Lambertus J. Sluys

1 Introduction

Understanding failure processes of heterogeneous materials is an active research field in computational mechanics. The failure analysis of quasi-brittle materials such as concrete is a topic of particular interest in civil engineering. Failure in quasi-brittle materials is characterized by the initial formation of cracks at a microscopic level followed by their coalescence into macroscopic cracks leading to weakening and fracture. Because the fracturing process of these materials occurs at different length scales, care must be taken to provide an accurate description which accounts for all the relevant mechanical processes while maintaining acceptable computation costs. With this in mind, we propose a multiscale approach capable of switching between different spatial discretizations and material representations depending on the local mechanical behaviour.

In this contribution, we present a non-local damage finite element analysis of a wedge-split test used to evaluate fracture properties in concrete-like materials. We apply the classical FETI framework [7] to a non-linear gradient-enhanced damage (GD) model [15] using both iterative and direct solvers to the interface problem as

F.P.X. Everdij (✉) • A. Simone • L.J. Sluys
Faculty of Civil Engineering and Geosciences, Delft University of Technology, P.O. Box 5048, 2600 GA Delft, The Netherlands
e-mail: F.P.X.Everdij@tudelft.nl

O. Lloberas-Valls
International Center for Numerical Methods in Engineering (CIMNE), Campus Nord UPC, Edifici C-1, C/Jordi Girona 1-3, 08034 Barcelona, Spain

D.J. Rixen
Faculty of Mechanical Engineering, Technische Universität München, Boltzmannstrasse 15, 85748 Garching, Germany

T. Dickopf et al. (eds.), *Domain Decomposition Methods in Science and Engineering XXII*, Lecture Notes in Computational Science and Engineering 104, DOI 10.1007/978-3-319-18827-0_18

well as using a direct solver for the entire set of equations of the fully dual assembled system.

2 Framework

2.1 Gradient-Enhanced Damage Model

The gradient-enhanced damage model by Peerlings et al. [15] is employed to model concrete failure. The GD model is non-local: it consists of a coupled set of differential equations involving the modified Helmholtz equation for the non-local equivalent strain and the classical quasistatic equilibrium equations. Damage evolution is highly non-linear, requiring the use of a loop control dividing the total load into small steps with an iterative Newton-Raphson (NR) scheme for each step to assure equilibrium.

The damage parameter ω, which modifies the stress–strain relation according to

$$\boldsymbol{\sigma} = (1 - \omega)\mathsf{D}^e : \boldsymbol{\varepsilon} \,, \tag{1}$$

varies from 0 for undamaged to 1 for fully damaged material. Its evolution,

$$\omega\left(\kappa\right) = \begin{cases} 0 & \kappa \leq \kappa_0 \\ 1 - \frac{\kappa_0}{\kappa}\left(1 - \alpha\left(1 - e^{-\beta(\kappa - \kappa_0)}\right)\right) & \kappa > \kappa_0 \end{cases} , \tag{2}$$

is a function of the history parameter κ which is defined as the maximum value ever attained by the nonlocal equivalent strain. In the above equations, D^e is the elasticity fourth-order tensor, $\boldsymbol{\sigma}$ is the second-order stress tensor, $\boldsymbol{\varepsilon}$ is the second order strain tensor, κ_0, α and β are parameters governing the shape of the damage evolution law.

The underlying damage formalism results in an asymmetric stiffness matrix. To solve the set of equations, a solver supporting asymmetry, both in direct and iterative approaches, is required.

2.2 Multiscale Domain Decomposition

The key to solving the discrete system of equations in a reasonable amount of time is to use two different representations of the problem under examination. One numerical model has a fine mesh with a detailed representation of the mesostructure of the material. The other numerical model has a coarse mesh with homogenized material properties which have been determined to approximate the response of the 'fine' model in the linear regime. Both numerical models have been decomposed into a fixed amount of domains. Each domain in the 'fine' model has a corresponding domain in the 'coarse' model matching its shape.

The calculation starts with the 'coarse' numerical model for all domains. In each step and for each domain, a check for the condition of onset of non-linearity is performed. For every node, the non-local equivalent strain difference is calculated from the displacement field of the current and two previous steps. Onset of non-linearity occurs if for a single node the strain difference exceeds a chosen damage initiation threshold value κ_0. The domains for which this condition is met are subsequently replaced by domains with the fine scale mesh. To preserve continuity of the displacements and forces, a boundary value problem is solved for each replaced domain followed by a global relaxation step.

Computing the strain difference for the onset of the non-linearity condition is a choice that should match the nature of the formation of non-linearities. For tensile test calculations and the gradient-enhanced damage model, our current choice yields satisfactory results [13].

2.3 Classical FETI Method

In order to solve the multiscale system with a mixture of coarse and fine meshes for each domain, the classical FETI method [7] is used. Lagrange multipliers ensure continuity of the solution field between interface nodes of adjacent domains. Linear multipoint constraints and full-collocation are used for fine mesh interface nodes which do not have a corresponding coarse mesh node on the adjoining domain [14].

Boundary conditions are also included by means of Lagrange multipliers, thus implying that all domains in this framework are floating. This method is known as the Total-FETI method [6]. Rigid body motion vectors are constructed to enforce compatibility between domains. To solve the local equations for each domain, we use QR factorization of the domain stiffness matrix which can be stored for later use in computing the Lagrange multipliers by means of either the iterative or direct solve of the global interface problem as shown in [12, 13].

3 Numerical Computation

3.1 Model

We use a two-dimensional model of a wedge split specimen for the quasistatic damage simulation of the heterogeneous sample of concrete shown in Fig. 1.

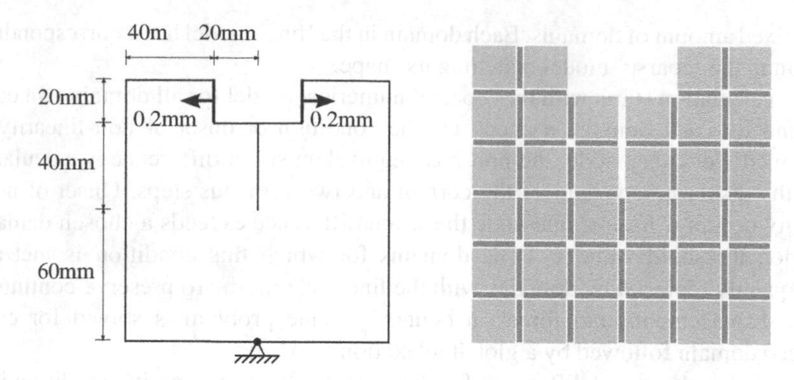

Fig. 1 Dimensions and domain decomposition of the wedge split model test. The interface is represented in *dark-grey*

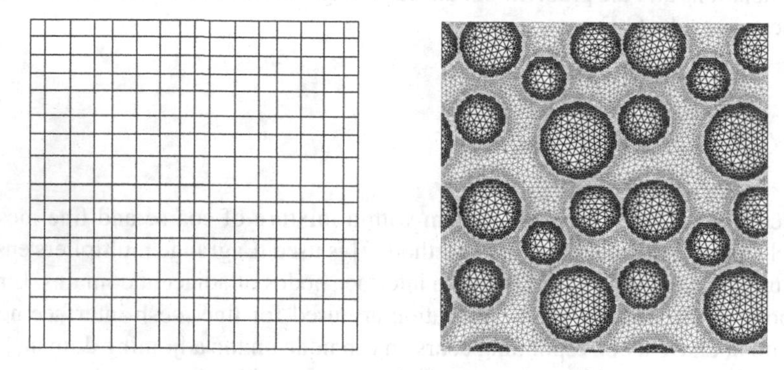

Fig. 2 Coarse (*left*) and fine (*right*) scale domain meshes. Coloring in the fine domain: aggregates in *black*, cement matrix in *grey* and ITZ in *light grey*

For the multiscale framework we use two different meshes: a homogeneous mesh consisting of quadrilateral elements with four integration points for the coarse domains, and a heterogeneous mesh with triangular elements and one integration point for fine scale domains. Both meshes are shown in Fig. 2. The fine-scale mesh is representative of a typical concrete mesostructure which consists of spherical aggregates, an interface transition zone (ITZ) surrounding the aggregates, and a cementitious matrix material in which the aggregates are embedded. Because of the independence of the individual domains, we are not restricted in mesh, element and material choice per domain provided that the solution field is continuous across the interface.

The parameters are listed in Table 1. Plane strain conditions are considered. The Young's modulus for the homogeneous coarse-scale mesh is an effective Young's modulus derived from the heterogeneous mesh. This is necessary for an accurate material-averaged linear response in the coarse description of the model.

Table 1 Material data

	Material parameters		Aggregates	Matrix	ITZ
E	Young's modulus	(GPa)	35.0	30.0	20.0
ν	Poisson's ratio	(–)	0.2	0.2	0.2
ε_{eq}	Non-local equivalent strain	(–)	Mazars	Mazars	Mazars
κ_0	Damage initiation threshold	(–)	dummy	8.5×10^{-5}	5×10^{-5}
c	Gradient parameter	(mm^2)	0.75	0.75	0.75
$\omega(\kappa)$	Damage evolution law	(–)	Exponential	Exponential	Exponential
α	Residual stress parameter	(–)	0.999	0.999	0.999
β	Softening rate parameter	(–)	150	150	150

3.2 *Software Framework and Solvers*

The non-linear quasistatic calculation is performed by dividing the total applied displacement into 200 load increments. In each load increment the non-linear GD model is evaluated iteratively using an NR scheme with a convergence threshold of 1.0×10^{-6} for the relative error in energy. Usually 3–4 NR iterations are sufficient for the solution to converge.

In the FETI calculations, all factorizations of the domain stiffness matrices are being performed by SuiteSparseQR [4]. Solving the flexibility problem iteratively requires projection to ensure positive semi-definiteness of the matrix, allowing the iterative solvers to converge. Because of the asymmetry of the flexibility matrix, only few iterative solvers like BiCGStab by van der Vorst [20] and GMRES by Saad and Schultz [17] are suitable. We chose BiCGStab with projection using openMP for the product of the projected stiffness matrix and solution vector (Eqs. (9)–(12) in [12]).

Superlumped (SL), lumped (L) and Dirichlet (D) type preconditioners from [16] are used to accelerate iterative convergence, as well as the multiplicity (m), stiffness (k) and Dirichlet (s) scaling to augment the preconditioners.

The flexibility interface problem can also be solved directly, using openMP for evaluating the flexibility matrix by distributing the domain contributions to the sum over all available parallel cores, followed by a dense matrix solver such as UMFPACK [3]. Even though this approach was discouraged in [7] because of the large amounts of solutions required, we have performed this direct calculation since it does provide an upper time limit for finding the Lagrange multipliers with an iterative approach.

An alternative approach is the solution of the set of equations from which the FETI method originates:

$$\begin{bmatrix} \mathbf{K} & \mathbf{B}^T \\ \mathbf{B} & \mathbf{0} \end{bmatrix} \begin{bmatrix} \mathbf{u} \\ \boldsymbol{\lambda} \end{bmatrix} = \begin{bmatrix} \mathbf{f} \\ \mathbf{0} \end{bmatrix}. \tag{3}$$

Because of the reduction in degrees of freedom, obtained by starting with all coarse domains and a simplified model description, and only substituting domains with fine, heterogeneous counterparts where it is needed, the full dual assembled matrix is much smaller than the full numerical solution (FNS) and can be solved using parallel direct solvers.

In this contribution we have selected a couple of solvers with the requirement of being able to handle asymmetric cases: MUMPS by Amestoy et al. [1, 2], Pardiso by Schenk et al. [18], PaSiX by Hénon et al. [9], WSMP by Gupta [8] and SuperLU by Li [10], Li et al. [11], and Demmel et al. [5]. These solvers can also be applied to obtain the FNS.

4 Results

The full numerical solution and the 34 domain FETI-direct calculations show identical damage patterns and displacements as shown in Fig. 3. However, none of the iterative FETI calculations, regardless of preconditioner and scaling combination, succeed in completing the calculation within the 1000 BiCGStab iteration limit.

Figure 4 shows a significant rise in BiCGStab iterations as the damage calculation progresses. This indicates the inability of the iterative preconditioners and scalings to deal with progressive damage evolution, possibly due to large differences in material stiffness. In order to ascertain this assumption we study the number of iterations for one linear elastic calculation with a domain decomposed mesh, consisting of the 26 zoomed-in domains, by choosing three different load increments i and their corresponding damage profiles ω_i from the FETI-direct calculation and substituting the Young's modulus E by $(1 - \omega_i)E$. This approach enables us to observe the dependency of the damage evolution versus the number of iterative steps needed for convergence.

From Table 2 we confirm that the iterations strongly depend on the damage profile: the iterations increase dramatically upon progressively growing differences in material stiffness. This is caused by the differences of orders of magnitudes in the

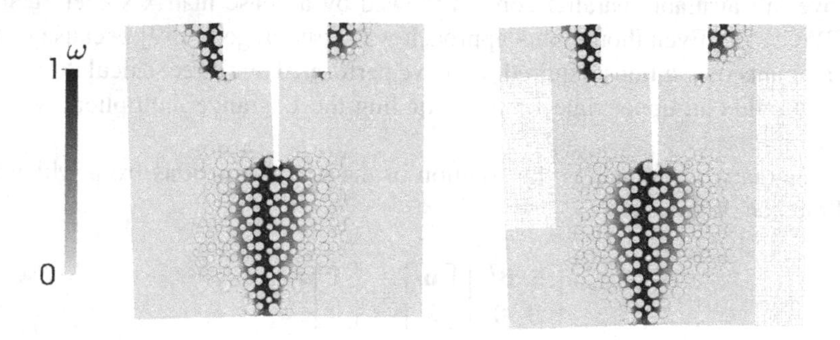

Fig. 3 Comparison of final damage profile of FNS (*right*) and FETI-direct 34 domain

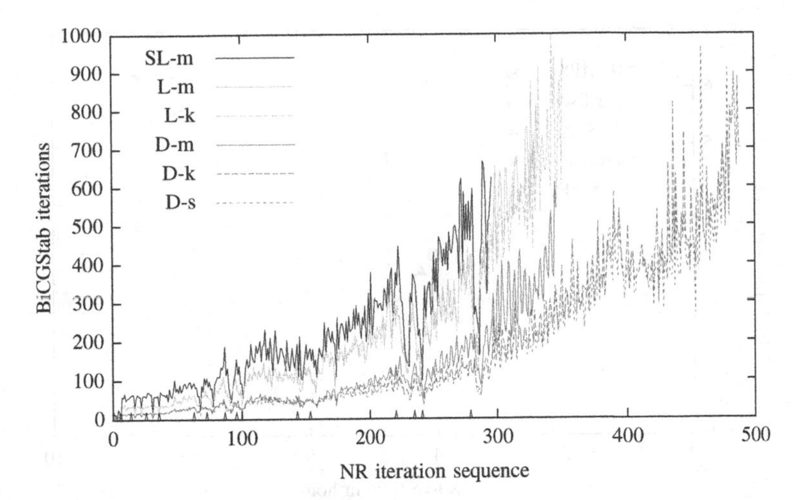

Fig. 4 BiCGStab iteration trend per NR-iteration number. Refer to Sect. 3.2 for an explanation of used preconditioner and scaling acronyms

Table 2 Linear elastic BiCGStab iteration count as a function of damage profile for a given load increment. Two different preconditioner/scaling results are shown

	Load increment		
Preconditioner + scaling	0	100	Final
Dirichlet + k scaling	16	233	1936
Lumped + k scaling	39	781	>5000

matrix entries. We therefore conclude that the standard preconditioners and scalings fail to accelerate the BiCGStab iterative solver in situations of substantial damage.

Improving the preconditioners for these type of systems involves adapting new techniques in combination with the damage model, for instance using eigenvalue analysis in FETI-GenEO [19]. This is a challenging research topic because of the asymmetric nature of the stiffness matrix in the GD model.

If we instead turn our attention to the parallel direct solvers for both the FNS and full assembly of the FETI system, we see a favourable reduction of time and used memory of the full assembly compared to the FNS for all solvers (Fig. 5). The reduction is not very large, as was expected since the used model system shows an extensive damage pattern affecting 75 % of the domains. We are confident that for larger 3D model systems undergoing damage the amount of zoomed in domains will be much smaller and therefore more economic in terms of computation time.

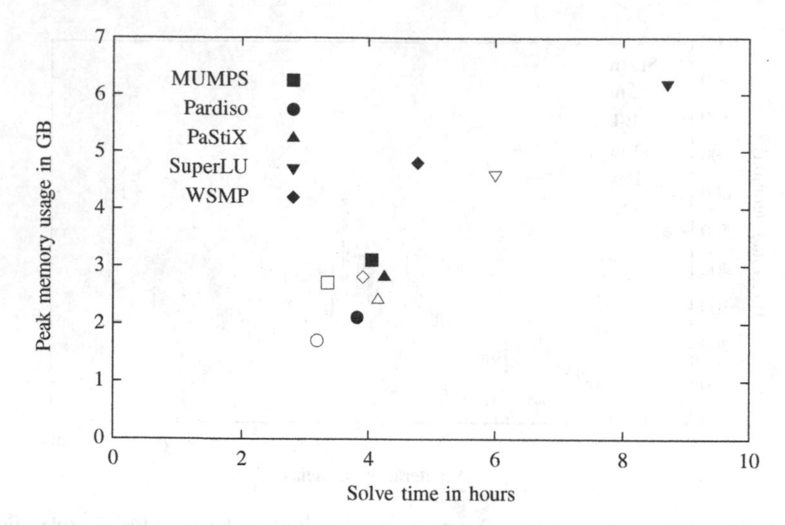

Fig. 5 Comparison of parallel direct solvers. *Solid symbols* denote FNS, *outlined symbols* denote multiscale DD

5 Conclusions

The multiscale framework proposed by Lloberas-Valls et al. [13] in combination with a classic FETI approach is shown to provide a reduction of degrees of freedom necessary to efficiently simulate damage evolution in multiscale models of concrete-like materials. By using parallel direct solvers the calculation can be done in less time and memory than the FNS.

In the iterative FETI approach, a high iteration count of the iterative solver is caused by the large differences in material stiffness along domain interface boundaries because of damage evolution. This poses a challenge for existing preconditioners and scalings. We nevertheless expect the iterative FETI to become the most efficient algorithm for very large problems once suitable preconditioners have been identified.

References

1. P.R. Amestoy, I.S. Duff, J. Koster, J.-Y. L'Excellent, A fully asynchronous multifrontal solver using distributed dynamic scheduling. SIAM J. Matrix Anal. A **23**(1), 15–41 (2001)
2. P.R. Amestoy, A. Guermouche, J.-Y. L'Excellent, S. Pralet, Hybrid scheduling for the parallel solution of linear systems. Parallel Comput. **32**(2), 136–156 (2006)
3. T.A. Davis, Algorithm 832: UMFPACK, an unsymmetric-pattern multifrontal method. ACM Trans. Math. Softw. **30**(2), 196–199 (2004)
4. T.A. Davis, Algorithm 915: SuiteSparseQR: multifrontal multithreaded rank-revealing sparse QR factorization. ACM Trans. Math. Softw. **38**(1), 8:1–8:22 (2011)

5. J.W. Demmel, J.R. Gilbert, X.S. Li, An asynchronous parallel supernodal algorithm for sparse gaussian elimination. SIAM J. Matrix Anal. A **20**(4), 915–952 (1999)
6. Z. Dostál, D. Horák, R. Kučera, Total FETI—an easier implementable variant of the FETI method for numerical solution of elliptic PDE. Commun. Numer. Methods Eng. **22**(12), 1155–1162 (2006)
7. C. Farhat, F.-X. Roux, A method of finite element tearing and interconnecting and its parallel solution algorithm. Int. J. Numer. Methods Eng. **32**(6), 1205–1227 (1991)
8. A. Gupta, A shared- and distributed-memory parallel sparse direct solver, in *Proceedings of the 7th International Conference on Applied Parallel Computing: State of the Art in Scientific Computing, PARA'04* (Springer, Berlin/Heidelberg, 2006), pp. 778–787
9. P. Hénon, P. Ramet, J. Roman, PaStiX: a high-performance parallel direct solver for sparse symmetric definite systems. Parallel Comput. **28**(2), 301–321 (2002)
10. X.S. Li, An overview of SuperLU: algorithms, implementation, and user interface. ACM Trans. Math. Softw. **31**(3), 302–325 (2005)
11. X.S. Li, J.W. Demmel, J.R. Gilbert, L. Grigori, M. Shao, I. Yamazaki. SuperLU users' guide. Technical Report LBNL-44289, Lawrence Berkeley National Laboratory (1999), http://www.crd.lbl.gov/~xiaoye/SuperLU/. Last update: August 2011
12. O. Lloberas-Valls, D.J. Rixen, A. Simone, L.J. Sluys, Domain decomposition techniques for the efficient modeling of brittle heterogeneous materials. Comput. Methods Appl. Mech. Eng. **200**(13–16), 1577–1590 (2011)
13. O. Lloberas-Valls, D.J. Rixen, A. Simone, L.J. Sluys, Multiscale domain decomposition analysis of quasi-brittle heterogeneous materials. Int. J. Numer. Methods Eng. **89**(11), 1337–1366 (2012a)
14. O. Lloberas-Valls, D.J. Rixen, A. Simone, L.J. Sluys, On micro-to-macro connections in domain decomposition multiscale methods. Comput. Methods Appl. Mech. Eng. **225–228**, 177–196 (2012b)
15. R.H.J. Peerlings, R. de Borst, W.A.M. Brekelmans, J.H.P. de Vree, Gradient enhanced damage for quasi-brittle materials. Int. J. Numer. Methods Eng. **39**(19), 3391–3403 (1996)
16. D.J. Rixen, C. Farhat, A simple and efficient extension of a class of substructure based preconditioners to heterogeneous structural mechanics problems. Int. J. Numer. Methods Eng. **44**(4), 489–516 (1999)
17. Y. Saad, M. Schultz, Gmres: a generalized minimal residual algorithm for solving nonsymmetric linear systems. SIAM J. Sci. Stat. Comput. **7**(3), 856–869 (1986)
18. O. Schenk, K. Gärtner, W. Fichtner, A. Stricker, PARDISO: a high-performance serial and parallel sparse linear solver in semiconductor device simulation. Futur. Gener. Comput. Syst. **18**(1), 69–78 (2001)
19. N. Spillane, D.J. Rixen, Automatic spectral coarse spaces for robust finite element tearing and interconnecting and balanced domain decomposition algorithms. Int. J. Numer. Methods Eng. **95**(11), 953–990 (2013)
20. H. van der Vorst, Bi-cgstab: a fast and smoothly converging variant of bi-cg for the solution of nonsymmetric linear systems. SIAM J. Sci. Stat. Comput. **13**(2), 631–644 (1992)

Schwarz Methods for the Time-Parallel Solution of Parabolic Control Problems

Martin J. Gander and Felix Kwok

1 Introduction

Suppose we are interested in the following distributed control problem: given a system governed by the parabolic PDE $\dot{y}+\mathcal{L}y = u$ on the time interval $[0,T]$ (where $\dot{y}$ denotes the time derivative of y), we wish to choose the forcing term $u = u(t)$ to minimize the discrepancy between the trajectory and the desired state $\hat{y} = \hat{y}(t)$. After semi-discretization in space, we obtain for a given choice of parameters $\gamma, \nu > 0$ the following minimization problem:

$$\min_{y,u} \frac{1}{2}\int_0^T \|y-\hat{y}\|^2\,dt + \frac{\gamma}{2}\|y(T)-\hat{y}(T)\|^2 + \frac{\nu}{2}\int_0^T \|u\|^2\,dt \qquad \text{subject to} \quad \dot{y}+Ay = u, \quad y(0) = y_0, \tag{1}$$

where A is the matrix obtained by semi-discretization of the operator $\mathcal{L}$. While the PDE in (1) may resemble an initial-value problem, the minimization problem is in fact a two-point boundary value problem in time, since the first-order optimality conditions couple the PDE to an adjoint equation that is backwards in time and contains a final condition, see Sect. 2. To solve such systems in parallel, one can use multiple shooting methods, see [8] and references therein, or parareal-type algorithms in a reduced Hessian formulation, see [4, 7]. A Schwarz preconditioner

M.J. Gander
Department of Mathematics, University of Geneva, Geneva, Switzerland
e-mail: Martin.Gander@unige.ch

F. Kwok (✉)
Department of Mathematics, Hong Kong Baptist University, Kowloon, Hong Kong
e-mail: felix_kwok@hkbu.edu.hk

T. Dickopf et al. (eds.), *Domain Decomposition Methods in Science and Engineering XXII*, Lecture Notes in Computational Science and Engineering 104, DOI 10.1007/978-3-319-18827-0_19

in time for such systems was presented in [1], where on each subinterval $I_j = [T_j, T_{j+1}]$, one uses an initial condition for y from I_{j-1} and a final condition for the adjoint state λ from I_{j+1}. To the authors' knowledge, no convergence analysis is available for this method.

We study in this paper Schwarz methods for the time-parallel solution of (1). We present a rigorous convergence analysis for the case of two subdomains, which shows that the classical Schwarz method converges, even without overlap! Reformulating the algorithm reveals that this is because imposing initial conditions for y and final conditions on λ is equivalent to using Robin transmission conditions between time subdomains for y. Using well chosen linear combinations of y and λ as transmission conditions allows us to optimize the Robin conditions for performance, and leads to much faster Schwarz methods, especially when the spatial operator has eigenvalues close to zero. We illustrate our results with numerical experiments.

2 Schwarz Methods in Time

Using the Lagrange multiplier approach (see e.g. the historical review [6]), one can derive the forward and adjoint problems to be

$$\begin{cases} \dot{y} + Ay = u & \text{on } (0,T), \\ y(0) = y_0, \end{cases} \qquad \begin{cases} \dot{\lambda} - A^T\lambda = y - \hat{y} & \text{on } (0,T), \\ \lambda(T) = -\gamma(y(T) - \hat{y}(T)), \end{cases}$$

where the control u and adjoint state λ are related by the algebraic equation $\lambda(t) = \nu u(t)$ for all $t \in (0,T)$. Eliminating u, the above system can thus also be written as

$$\begin{bmatrix} \dot{y} \\ \dot{\lambda} \end{bmatrix} + \begin{bmatrix} A & -\nu^{-1}I \\ -I & -A^T \end{bmatrix} \begin{bmatrix} y \\ \lambda \end{bmatrix} = \begin{bmatrix} 0 \\ -\hat{y} \end{bmatrix}. \tag{2}$$

Suppose we wish to divide the time interval $(0,T)$ into two subintervals $I_1 = (0,\beta)$ and $I_2 = (\alpha,T)$ with $\alpha \leq \beta$ in order to solve the two subdomain problems in parallel. Then for any choice of parameters $p, q \geq 0$, we propose the following parallel Schwarz algorithm: for $k = 1, 2, \ldots$, solve

$$\begin{cases} \begin{bmatrix} \dot{y}_1^k \\ \dot{\lambda}_1^k \end{bmatrix} + \begin{bmatrix} A & -\nu^{-1}I \\ -I & -A^T \end{bmatrix} \begin{bmatrix} y_1^k \\ \lambda_1^k \end{bmatrix} = \begin{bmatrix} 0 \\ -\hat{y} \end{bmatrix} \quad \text{on } I_1 = (0,\beta), \\ y_1^k(0) = y_0, \\ \lambda_1^k(\beta) + p y_1^k(\beta) = \lambda_2^{k-1}(\beta) + p y_2^{k-1}(\beta), \end{cases} \tag{3a}$$

$$\begin{cases} \begin{bmatrix} \dot{y}_2^k \\ \dot{\lambda}_2^k \end{bmatrix} + \begin{bmatrix} A & -\nu^{-1}I \\ -I & -A^T \end{bmatrix} \begin{bmatrix} y_2^k \\ \lambda_2^k \end{bmatrix} = \begin{bmatrix} 0 \\ -\hat{y} \end{bmatrix} & \text{on } I_2 = (\alpha, T), \\ y_2^k(\alpha) - q\lambda_2^k(\alpha) = y_1^{k-1}(\alpha) - q\lambda_1^{k-1}(\alpha), \\ \lambda_2^k(T) = -\gamma(y_2^k(T) - \hat{y}(T)). \end{cases} \tag{3b}$$

For $p = q = 0$, the transmission conditions reduce to the classical conditions from [1]. To understand why we consider transmission conditions of this form, suppose that $A = A^T \in \mathbb{R}^{m\times m}$, so that A can be diagonalized as $A = QDQ^T$, with $Q^TQ = I$ and $D = \text{diag}(d_1, \ldots, d_m)$. Then the ODE system in (3a) can be written as

$$\begin{cases} \begin{bmatrix} \dot{z}_1^k \\ \dot{\mu}_1^k \end{bmatrix} + \begin{bmatrix} D & -\nu^{-1}I \\ -I & -D \end{bmatrix} \begin{bmatrix} z_1^k \\ \mu_1^k \end{bmatrix} = \begin{bmatrix} 0 \\ -\hat{z} \end{bmatrix} & \text{on } I_1 = (0, \beta), \\ z_1^k(0) = z_0, \\ \mu_1^k(\beta) + pz_1^k(\beta) = \mu_2^{k-1}(\beta) + pz_2^{k-1}(\beta), \end{cases} \tag{4}$$

where $z_j^k = Q^Ty_j^k$, $\mu_j^k = Q^T\lambda_j^k$ for $j = 1, 2$ and $\hat{z} = Q^T\hat{y}$, $z_0 = Q^Ty_0$. Thus, we obtain m independent 2×2 systems of the form

$$\dot{z}_1^{(i),k} + d_i z_1^{(i),k} - \nu^{-1}\mu_1^{(i),k} = 0, \qquad \dot{\mu}_1^{(i),k} - d_i\mu^{(i),k} - z_1^{(i),k} = \hat{z}^{(i)}, \tag{5}$$

where $z_1^{(i),k}$ and $\mu_1^{(i),k}$ are the ith components of z_1^k and μ_1^k respectively, and d_i is the ith eigenvalue of A. By isolating μ from the first equation in (5) and substituting into the second, we obtain the second-order ODE

$$\ddot{z}_1^{(i),k} - (d_i^2 + \nu^{-1})z_1^{(i),k} = -\nu^{-1}\hat{z}^{(i)}, \tag{6}$$

whereas the boundary conditions become

$$z_1^{(i),k}(0) = z_0^{(i)}(0), \; \dot{z}_1^{(i),k} + (d_i + p\nu^{-1})z_1^{(i),k}\Big|_{t=\beta} = \dot{z}_2^{(i),k-1} + (d_i + p\nu^{-1})z_2^{(i),k-1}\Big|_{t=\beta}.$$

Hence, once we eliminate the adjoint state, it becomes apparent that we are in fact imposing a Robin transmission condition on the elliptic boundary value problem (6), even with the classical Schwarz method $p = q = 0$ from [1]. With the additional parameter p and q, one can now optimize the convergence, as in optimized Schwarz methods [5]. Boundary conditions of the form $y - q\lambda$ in (3b) can be explained similarly; here, the minus sign is chosen so that the subdomain problem is well-posed for $q \geq 0$ whenever A is symmetric semi-positive definite.

Remark on Implementation Since we are primarily interested in the behavior of the Schwarz method, we will regard solvers for the subdomain problems (3a) and (3b) as black boxes. We emphasize however that final conditions of the form $\lambda + py$ already appear when the objective function contains the target term $\frac{\gamma}{2}|y(T) - \hat{y}(T)|^2$, see (3b). Thus, existing solvers can be used as is or easily modified to handle the optimized conditions, see [2] or [3].

3 Convergence Analysis

In this section, we assume A to be symmetric and semi-positive definite, so that (3a)–(3b) can be diagonalized as in Sect. 2 with $d_i \geq 0$. Moreover, since the problem is linear, we can analyze the error equation, which means setting y_0 and $\hat{y}$ to zero and studying how y_j^k and λ_j^k converge to zero as $k \to \infty$. After diagonalization, the first subdomain solution satisfies (6) with homogeneous initial condition:

$$\ddot{z}_1^{(i),k} - (d_i^2 + \nu^{-1})z_1^{(i),k} = 0, \quad z_1^{(i),k}(0) = 0 \quad \Longrightarrow \quad z_1^{(i),k}(t) = A_i^k \sinh(\sigma_i t), \tag{7}$$

where $\sigma_i = \sqrt{d_i^2 + \nu^{-1}} > 0$, and A_i^k is a constant determined by the boundary condition $\nu \dot{z}_1^{(i),k} + (p + \nu d_i)z_1^{(i),k}|_{t=\beta} = g^{(i),k}$. Substituting the solution from (7) and isolating A_i^k yields $A_i^k = \frac{g_1^{(i),k}}{\nu[\sigma_i \cosh(\sigma_i\beta) + (d_i + p\nu^{-1})\sinh(\sigma_i\beta)]}$. Next, we consider the subdomain $I_2 = (\alpha, T)$ at iteration $k + 1$. The boundary data at $t = \alpha$ can be written as

$$\begin{aligned} h^{(i),k+1} := z_1^{(i),k} - q\mu_1^{(i),k}\Big|_{t=\alpha} &= -\nu q \dot{z}_1^{(i),k} + (1 - \nu q d_i)z_1^{(i),k}\Big|_{t=\alpha} \\ &= -g^{(i),k} \frac{\sigma_i q \cosh(\sigma_i\alpha) + (q d_i - \nu^{-1})\sinh(\sigma_i\alpha)}{\sigma_i \cosh(\sigma_i\beta) + (d_i + p\nu^{-1})\sinh(\sigma_i\beta)}. \end{aligned} \tag{8}$$

On the other hand, the ODE can be written as

$$\begin{aligned} \ddot{\mu}_2^{(i),k+1} - (d_i^2 + \nu^{-1})\mu_2^{(i),k+1} &= 0 \qquad \text{on } I_2 = (\alpha, T), \\ \mu_2^{(i),k+1}(T) + \gamma z_2^{(i),k+1}(T) = 0, \qquad z_2^{(i),k+1}(\alpha) &- q\mu_2^{(i),k+1}(\alpha) = h^{(i),k+1}. \end{aligned}$$

Since $z_2^{(i),k+1} = \dot{\mu}_2^{(i),k+1} - d_i\mu_2^{(i),k+1}$, the boundary conditions can be written as

$$\gamma\dot{\mu}_2^{(i),k+1}(T) + (1 - d_i\gamma)\mu_2^{(i),k+1}(T) = 0, \ \dot{\mu}_2^{(i),k+1}(\alpha) - (d_i + q)\mu_2^{(i),k+1}(\alpha) = h^{(i),k+1}.$$

The boundary condition at $t = T$ gives

$$\mu_2^{(i),k+1} = B_i^{k+1}\left[\sigma_i\gamma\cosh(\sigma_i(T-t)) + (1-d_i\gamma)\sinh(\sigma_i(T-t))\right],$$

where B_i^{k+1} is a constant. The boundary condition at $t = \alpha$ allows us to determine this constant (after some algebra) to be

$$B_i^{k+1} = \frac{-h^{(i),k+1}}{(\sigma_i(1+q\gamma))\cosh(\sigma_i(T-\alpha)) + (d_i(1-q\gamma) + q + \nu^{-1}\gamma)\sinh(\sigma_i(T-\alpha))}.$$

Note that the denominator does not vanish for any choice of $q, \gamma \geq 0$: if we define $\theta_i = \tanh^{-1}(d_i/\sigma_i)$, which is possible because $0 \leq d_i < \sigma_i$, then we can write the denominator as

$$\begin{aligned}
&\sigma_i\cosh(\cdot) + d_i\sinh(\cdot) + q\gamma(\sigma_i\cosh(\cdot) - d_i\sinh(\cdot)) + (q+\nu^{-1}\gamma)\sinh(\cdot)\\
&\quad = \nu^{-1/2}\left[\cosh(\cdot+\theta_i) + q\gamma\cosh(\cdot-\theta_i)\right] + (q+\nu^{-1}\gamma)\sinh(\cdot) > 0.
\end{aligned}$$

If we now let $g^{(i),k+2} = \mu_2^{(i),k+1}(\beta) + pz_2^{(i),k+1}(\beta)$, we get

$$\begin{aligned}
&g^{(i),k+2}\\
&= h^{(i),k+1}\frac{\nu^{-1/2}\left[p\cosh(\sigma_i(T-\beta)+\theta_i) - \gamma\cosh(\sigma_i(T-\beta)-\theta_i)\right] - (1-\nu^{-1}p\gamma)\sinh(\sigma_i(T-\beta))}{\nu^{-1/2}\left[\cosh(\sigma_i(T-\alpha)+\theta_i) + q\gamma\cosh(\sigma_i(T-\alpha)-\theta_i)\right] + (q+\nu^{-1}\gamma)\sinh(\sigma_i(T-\alpha))}.
\end{aligned}$$

Substituting (8) into the above equations and taking absolute values, we obtain

Theorem 1 *The parallel Schwarz method* (3a)–(3b) *converges whenever* $\rho < 1$, *where*

$$\begin{aligned}
\rho = \max_{d_i\in\lambda(A)} &\left|\frac{\sigma_i q\cosh(\sigma_i\alpha) + (qd_i - \nu^{-1})\sinh(\sigma_i\alpha)}{\sigma_i\cosh(\sigma_i\beta) + (d_i + p\nu^{-1})\sinh(\sigma_i\beta)}\right.\\
&\left.\cdot\frac{\nu^{-1/2}\left[p\cosh(\sigma_i(T-\beta)+\theta_i) - \gamma\cosh(\sigma_i(T-\beta)-\theta_i)\right] - (1-\nu^{-1}p\gamma)\sinh(\sigma_i(T-\beta))}{\nu^{-1/2}\left[\cosh(\sigma_i(T-\alpha)+\theta_i) + q\gamma\cosh(\sigma_i(T-\alpha)-\theta_i)\right] + (q+\nu^{-1}\gamma)\sinh(\sigma_i(T-\alpha))}\right|^{1/2},
\end{aligned}$$

where the maximum is taken over all the set of eigenvalues of A.

To gain a better understanding of the convergence, let us assume that $A = A^T$ is positive semi-definite (so that $d_i \geq 0$) and consider a few special cases.

Classical Transmission Conditions ($p = q = 0$) Here the expression simplifies to

$$\rho^2 = \max_i\left(\frac{\sinh(\sigma_i\alpha)}{\cosh(\sigma_i\beta+\theta_i)}\cdot\frac{\nu^{1/2}\sinh(\sigma_i(T-\beta)) + \gamma\cosh(\sigma_i(T-\beta)-\theta_i)}{\gamma\sinh(\sigma_i(T-\alpha)) + \nu^{1/2}\cosh(\sigma_i(T-\alpha)+\theta_i)}\right).$$

If $\gamma \le \sqrt{\nu}$, then $\rho < 1$ and the method converges; this is because

$$\sinh(\sigma_i\alpha) \le \cosh(\sigma_i\alpha) \le \cosh(\sigma_i\beta + \theta_i)$$

and, since $\sinh(\sigma_i(T-\beta)) \le \cosh(\sigma_i(T-\beta)+\theta_i)$, we have

$$\begin{aligned}\nu^{1/2}\sinh(\sigma_i(T-\beta)) &+ \gamma\cosh(\sigma_i(T-\beta)-\theta_i)\\ &\le \nu^{1/2}\sinh(\sigma_i(T-\alpha)) + \gamma\cosh(\sigma_i(T-\alpha)+\theta_i)\\ &\le \gamma\sinh(\sigma_i(T-\alpha)) + \nu^{1/2}\cosh(\sigma_i(T-\alpha)-\theta_i).\end{aligned}$$

However, it is possible for the method to diverge if $\gamma > \nu^{1/2}$, see Sect. 4. In the case when $\gamma = 0$, i.e., when the target state does not appear explicitly in the objective function, it is possible to estimate the convergence factor directly. Here we have

$$\rho^2 = \max_i \frac{\sinh(\sigma_i\alpha)\sinh(\sigma_i(T-\beta))}{\cosh(\sigma_i\beta+\theta_i)\cosh(\sigma_i(T-\alpha)+\theta_i)} < 1,$$

since $\alpha \le \beta$. The term inside the maximum is a function of the eigenvalues d_i via $\sigma_i = \sqrt{d_i^2 + \nu^{-1}}$ and $\theta_i = \operatorname{arctanh}(d_i/\sigma_i)$. It can be shown that this function is decreasing with respect to d_i on $[0,\infty)$, see also Fig. 1; thus, if $d_{\min} \ge 0$ is the minimum eigenvalue of A and $\sigma_{\min}$ and $\theta_{\min}$ are the corresponding values, then one can estimate ρ by

$$\rho \le \left(\frac{\exp(\sigma_{\min}(\alpha + T - \beta))}{\exp(\sigma_{\min}(\beta + T - \alpha) + 2\theta_{\min})}\right)^{1/2} = e^{-\sigma_{\min}(\beta-\alpha)-\theta_{\min}},$$

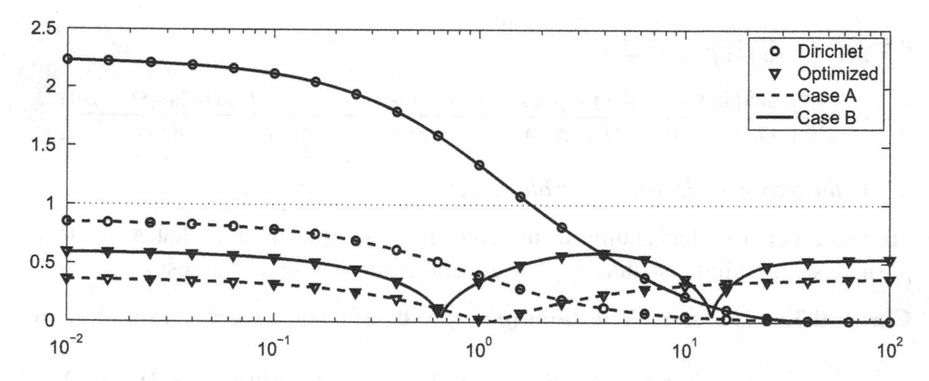

Fig. 1 A comparison of contraction factors as a function of eigenvalues d_i for classical ($p = q = 0$) and optimized transmission conditions (p and q obtained by equioscillation)

where we used the bounds $\sinh(x) \leq \frac{1}{2}\exp(x)$ and $\cosh(x) \geq \frac{1}{2}\exp(x)$, valid for all $x \geq 0$. Thus, when the subdomains overlap, i.e., when $\beta - \alpha > 0$, the convergence factor decreases exponentially with respect to the overlap size $\beta - \alpha$. When there is no overlap, i.e., when $\alpha = \beta$, it is still possible to bound ρ by estimating $e^{-\theta_{\min}}$ since $\tanh(\theta_{\min}) = d_{\min}/\sigma_{\min}$ by definition, we get

$$\frac{e^{\theta_{\min}} - e^{-\theta_{\min}}}{e^{\theta_{\min}} + e^{-\theta_{\min}}} = \frac{1 - e^{-2\theta_{\min}}}{1 + e^{-2\theta_{\min}}} = \frac{d_{\min}}{\sigma_{\min}} \implies 1 - \frac{d_{\min}}{\sigma_{\min}} = e^{-2\theta_{\min}}\left(1 + \frac{d_{\min}}{\sigma_{\min}}\right).$$

This implies

$$e^{-2\theta_{\min}} = \frac{\sigma_{\min} - d_{\min}}{\sigma_{\min} + d_{\min}} = \frac{\nu^{-1}}{\left(\sqrt{d_j^2 + \nu^{-1}} + d_j\right)^2}.$$

Taking square roots, we obtain the following estimate:

Theorem 2 *Suppose A is symmetric positive definite and* $\gamma = 0$. *Then the parallel Schwarz method* (3a)–(3b) *with classical transmission conditions* ($p = q = 0$) *converges for all initial guesses with the estimate*

$$\rho \leq \frac{e^{-(\beta-\alpha)\sqrt{d^2+\nu^{-1}}}}{\sqrt{1 + \nu d^2} + \nu^{1/2} d},$$

where $\beta - \alpha \geq 0$ *is the overlap size and* $d \geq 0$ *is the smallest eigenvalue of A.*

Note that if A arises from a spatial discretization of a differential operator, then the smallest eigenvalue of A typically does not vary much as the spatial grid is refined. Thus, *the convergence of the method is independent of the mesh parameter* h. However, if A is singular ($d = 0$) and there is no overlap, then convergence can be very slow, see the example in Sect. 4.

Optimized Transmission Conditions, No Target State ($\gamma = 0$) To accelerate the convergence of the method when A is singular, let us consider choosing the parameters p and q to be equal but non-zero. Then the convergence factor becomes

$$\rho = \max_{d_i \in \lambda(A)} \left| \frac{\sigma_i p \cosh(\sigma_i \alpha) + (p d_i - \nu^{-1}) \sinh(\sigma_i \alpha)}{\sigma_i \cosh(\sigma_i \beta) + (d_i + p\nu^{-1}) \sinh(\sigma_i \beta)} \cdot \frac{p \sigma_i \cosh(\sigma_i (T-\beta)) + (p d_i - 1) \sinh(\sigma_i (T-\beta))}{\sigma_i \cosh(\sigma_i (T-\alpha)) + (p + d_i) \sinh(\sigma_i (T-\alpha))} \right|^{1/2}.$$

A plot of the right-hand side as a function of d_i for fixed $p > 0$ is shown in Fig. 1. We see that as $d_i \to \infty$, we have

$$\rho \longrightarrow p \cdot \lim_{d_i \to \infty} \left(\frac{\cosh(\sigma_i \alpha + \theta_i) \cosh(\sigma_i (T - \beta) + \theta_i)}{\cosh(\sigma_i \beta + \theta_i) \cosh(\sigma_i (T - \alpha) + \theta_i)} \right)^{1/2}.$$

Thus, if no overlap is used, then the method converges only if $0 \le p < 1$. On the other hand, for $d_i = 0$, we have

$$\rho(d_i = 0) = \left| \frac{p\cosh(\sigma_i\alpha) - \nu^{-1/2}\sinh(\sigma_i\alpha)}{\cosh(\sigma_i\beta) + p\nu^{-1/2}\sinh(\sigma_i\beta)} \cdot \frac{\nu^{-1/2}p\cosh(\sigma_i(T-\beta)) - \sinh(\sigma_i(T-\beta))}{\nu^{-1/2}\cosh(\sigma_i(T-\alpha)) + p\sinh(\sigma_i(T-\alpha))} \right|^{1/2}.$$

Thus, if we assume the eigenvalues of A can be anywhere in the interval $[0, \infty)$, then the smallest convergence factor is obtained when $|\rho(d_i = 0)| = |\rho(d_i \to \infty)|$, i.e., by equioscillation.

4 Numerical Experiments

To understand how convergence depends on the different parameters, we consider for each ODE two different test cases:

Case A: The time interval $\Omega = [0, 3]$ is subdivided into $\Omega_1 = (0, 1)$, $\Omega_2 = (1, 3)$ (no overlap), and the objective function has no explicit target term ($\gamma = 0$). The regularization parameter is $\nu = 1$.

Case B: The subdomains are $\Omega_1 = (0, 2.9)$ and $\Omega_2 = (2.9, 3)$, and the objective function has a target term with $\gamma = 10$. The regularization parameter is still $\nu = 1$.

For each test case, we plot in Fig. 1 the convergence factor ρ as a function of the frequency d_i, both for classical ($p = q = 0$) and optimized transmission conditions. Based on the equioscillation criterion, we choose $p = q = 0.37$ for case A and $p = q = 0.55$ for case B. We see that when classical conditions are used, the method converges in case A for all frequencies, whereas in case B, the method only converges when the lowest eigenvalue of the spatial operator is larger than about 2. However, when optimized conditions are used, the parameters can be chosen so that the method converges for all frequencies, and the spectral radius can be made much smaller than in the classical case (0.37 versus about 0.9 for classical).

Next, we solve numerically the optimal control problem (1) with governing PDE $\partial_t u = \partial_{xx} u$ and regularization parameters $\nu = 1$, $\gamma = 0$. The problem is discretized using the second-order Crank–Nicolson method with spatial and temporal mesh size $h = 1/32$ and $1/64$. The problem is then solved in parallel using two time windows $\Omega_1 = (0, 1)$ and $\Omega_2 = (1, 3)$. Again we consider two cases: in the first case, we use Dirichlet boundary conditions in space, which means the operator A in (1) has lowest eigenvalue $\pi^2 \approx 9.87$. From Fig. 2, we see that the method converges very quickly with a rate that is indeed independent of h (see remark after Theorem 2). The fast convergence can be explained by Fig. 1: the spectral radius curve beyond the point $d_i = 9.87$ is very close to zero, so the convergence is very quick indeed.

In the second case, we consider the same PDE, but with Neumann boundary conditions in space. In this case, zero is an eigenvalue of the spatial operator, meaning

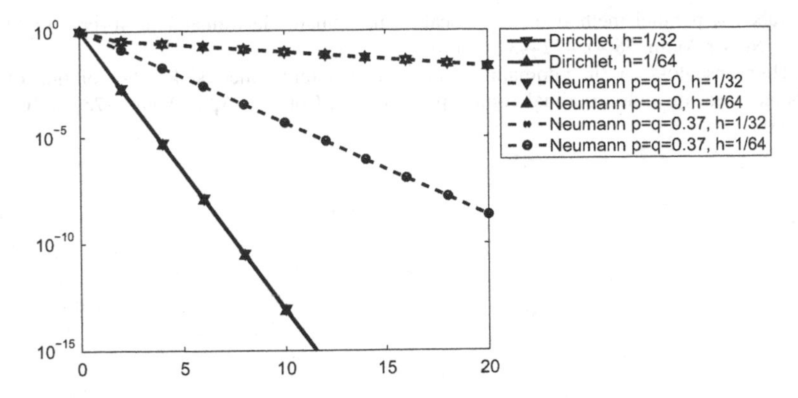

Fig. 2 Convergence of algorithm (3a)–(3b) for different parameters and boundary conditions

we have to minimize the convergence factor over the whole interval $d_i \in [0, \infty)$. Here, the method with classical transmission conditions ($p = q = 0$) converges very slowly, whereas convergence is much faster with optimized transmission conditions. Again the convergence is independent of the spatial mesh size, as expected.

5 Conclusions

We have presented a first analysis of Schwarz methods in time for parabolic control problems. We have shown that classical Schwarz methods already use Robin type transmission conditions, and introduced a parameter which can be chosen to obtain substantially faster convergence, especially when the spatial operator has eigenvalues close to zero. We are currently working on error estimates for the many-subdomain case and on higher order transmission conditions.

References

1. A.T. Barker, M. Stoll, Domain decomposition in time for PDE-constrained optimization (2014, submitted)
2. A. Borzì, V. Schulz, *Computational Optimization of Systems Governed by Partial Differential Equations*, vol. 8 (SIAM, Philadelphia, 2012)
3. H.S. Dollar, N.I. Gould, M. Stoll, A.J. Wathen, Preconditioning saddle-point systems with applications in optimization. SIAM J. Sci. Comput. **32**(1), 249–270 (2010)
4. X. Du, M. Sarkis, C.E. Schaerer, D.B. Szyld, Inexact and truncated parareal-in-time Krylov subspace methods for parabolic optimal control problems. Electron. Trans. Numer. Anal. **40**, 36–57 (2013)
5. M.J. Gander, Optimized Schwarz methods. SIAM J. Numer. Anal. **44**(2), 699–731 (2006)
6. M. Gander, F. Kwok, G. Wanner, Constrained Optimization: From Lagrangian Mechanics to Optimal Control and PDE Constraints, in *Optimization with PDE Constraints* (Springer, New York, 2014), pp. 151–202

7. E. Haber, A parallel method for large scale time domain electromagnetic inverse problems. Appl. Numer. Math. **58**(4), 422–434 (2008)
8. M. Heinkenschloss, A time-domain decomposition iterative method for the solution of distributed linear quadratic optimal control problems. J. Comput. Appl. Math. **173**(1), 169–198 (2005)

On the Relation Between Optimized Schwarz Methods and Source Transfer

Zhiming Chen, Martin J. Gander, and Hui Zhang

1 Introduction

Optimized Schwarz methods (OS) use Robin or higher order transmission conditions instead of the classical Dirichlet ones. An optimal Schwarz method for a general second-order elliptic problem and a decomposition into strips was presented in [13]. Here optimality means that the method converges in a finite number of steps, and this was achieved by replacing in the transmission conditions the higher order operator by the subdomain exterior Dirichlet-to-Neumann (DtN) maps. It is even possible to design an optimal Schwarz method that converges in two steps for an arbitrary decomposition and an arbitrary partial differential equation (PDE), see [6], but such algorithms are not practical, because the operators involved are highly non-local. Substantial research was therefore devoted to approximate these optimal transmission conditions, see for example the early reference [11], or the overview [5] which coined the term "optimized Schwarz method", and references therein.

Z. Chen
LSEC, Institute of Computational Mathematics, Academy of Mathematics and System Sciences, Chinese Academy of Sciences, Beijing 100190, China
e-mail: zmchen@lsec.cc.ac.cn

M.J. Gander
Section de Mathématiques, Université de Genève, 2-4 rue du Lièvre, Case postale 64, 1211 Genève 4, Switzerland
e-mail: martin.gander@unige.ch

H. Zhang (✉)
Department of Mathematics, Zhejiang Ocean University, Zhoushan, Zhejiang, China

Section de Mathématiques, Université de Genève, 2-4 rue du Lièvre, Case postale 64, 1211 Genève 4, Switzerland
e-mail: mike.hui.zhang@hotmail.com; hui.zhang@unige.ch

T. Dickopf et al. (eds.), *Domain Decomposition Methods in Science and Engineering XXII*, Lecture Notes in Computational Science and Engineering 104, DOI 10.1007/978-3-319-18827-0_20

In particular for the Helmholtz equation, Gander et al. [9] presents optimized second-order approximations of the DtN, Toselli [17] (improperly) and Schädle and Zschiedrich [14] (properly) tried for the first time using perfectly matched layers (PML, see [1]) to approximate the DtN in OS.

The DtN map arises also naturally in the analytic factorization of partial differential operators. This has been identified by Gander and Nataf [7] with the Schur complement occurring in the block LU factorization of block tridiagonal matrices, which led to analytic incomplete LU (AILU) preconditioners. The AILU preconditioners consist of one forward and one backward sweep corresponding to block "L" and "U" solves. In particular, second-order differential approximations of the DtN were studied by Gander and Nataf [8] for AILU for the Helmholtz equation. The connection between the DtN and the block LU factorization was rediscovered in [4], where a PML approximation of the DtN was used to improve the AILU preconditioners, and this has quickly inspired more research: Stolk [16] showed a "rapidly converging" domain decomposition method (DDM) based on sweeps, Chen and Xiang [2, 3] presented and analyzed the source transfer DDM (STDDM), and Geuzaine and Vion [10] proposed to use the sweeping process to accelerate Jacobi-type optimized Schwarz methods. All these new algorithms use PML but apparently in different formulations. In order to show their tight connection, we present here the relation between STDDM and OS. Such close connections also exist between OS and AILU, the sweeping preconditioner, and the method in [16], but these results, as well as the corresponding discrete formulations will appear elsewhere.

2 Algorithms and Equivalence

We consider a linear second order PDE of the form

$$\mathcal{L}u = f \text{ in } \Omega, \quad \mathcal{B}u = g \text{ on } \partial\Omega, \tag{1}$$

where Ω could either be $\mathbb{R}^d$, or a truncated domain padded with PML, in which case we consider the PML region as part of the domain. We decompose Ω into either overlapping or non-overlapping strips (or slices in higher dimensions) called subdomains Ω_j, $j = 1, \ldots, J$, which are in turn decomposed into boundary layers (overlaps) that are shared with neighboring subdomains, and non-shared interior, i.e. $\Omega_j = \Gamma_{j-1} \cup I_j \cup \Gamma_j$, see Fig. 1 for examples.

We start by introducing the optimized Schwarz method of symmetric Gauss-Seidel type (OS-SGS) for the strip decomposition we consider here, see also [12]. This method is based on subdomain solves that are performed first by sweeping forward across the subdomains, and then backward, a technique often used in the linear algebra community to render a Gauss-Seidel preconditioner symmetric. We then rewrite the OS-SGS method in residual correction form, in order to show how closely related it is to the STDDM from [2, 3]. All our formulations are at the continuous level, but one can also develop the corresponding discrete variants.

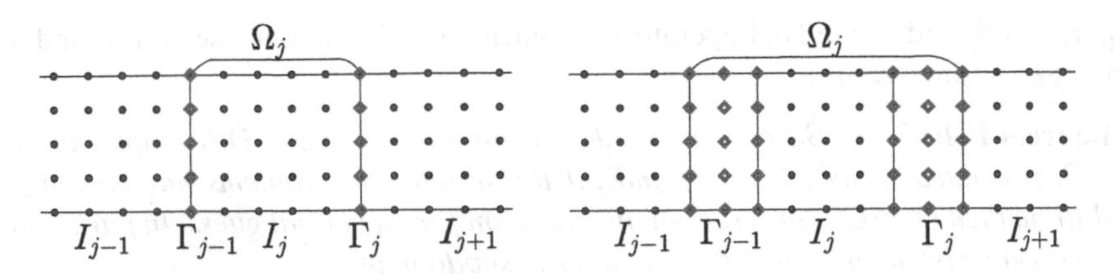

Fig. 1 Non-overlapping and overlapping domain decomposition into strips

In OS-SGS (see below), $\mathcal{S}_j$, $\tilde{\mathcal{S}}_j$ and $\mathcal{T}_j$, $\tilde{\mathcal{T}}_j$, are tangential operators on the left and right interfaces of Ω_j which need to ensure well-posedness of the subdomain problems. Note that on Ω_1 and Ω_J we did for simplicity not specify the modification due to the physical boundary there. If $\mathcal{T}_1 = \tilde{\mathcal{T}}_1$, then $v_1^{(n+1)} = u_1^{(n)}$, because the subdomain problems solved coincide, and so we need only to solve one of them. Even if $\mathcal{T}_1 \neq \tilde{\mathcal{T}}_1$, $u_1^{(n)}$ is not necessary for iteration $(n+1)$, only to complete iteration (n).

OS-SGS (interface transmission form)

Forward sweep: given $(u_j^{(n-1)})_{j=1}^J$ on $(\Omega_j)_{j=1}^J$ at iteration step $(n-1)$, solve successively for $j = 1, \ldots, J-1$ the subdomain problems

$$\begin{aligned} \mathcal{L}\, v_j^{(n)} &= f \ \text{ in } \Omega_j, \\ \mathcal{B}\, v_j^{(n)} &= g \ \text{ on } \partial\Omega \cap \partial\Omega_j, \\ (\tfrac{\partial}{\partial \mathbf{n}_j} + \mathcal{S}_j)(v_j^{(n)} - v_{j-1}^{(n)}) &= 0 \ \text{ on } \partial\Omega_j \cap \Omega_{j-1}, \\ (\tfrac{\partial}{\partial \mathbf{n}_j} + \mathcal{T}_j)(v_j^{(n)} - u_{j+1}^{(n-1)}) &= 0 \ \text{ on } \partial\Omega_j \cap \Omega_{j+1}. \end{aligned} \tag{2}$$

Backward sweep: solve successively for $j = J, \ldots, 1$ the subdomain problems

$$\begin{aligned} \mathcal{L}\, u_j^{(n)} &= f \ \text{ in } \Omega_j, \\ \mathcal{B}\, u_j^{(n)} &= g \ \text{ on } \partial\Omega \cap \partial\Omega_j, \\ (\tfrac{\partial}{\partial \mathbf{n}_j} + \tilde{\mathcal{S}}_j)(u_j^{(n)} - v_{j-1}^{(n)}) &= 0 \ \text{ on } \partial\Omega_j \cap \Omega_{j-1}, \\ (\tfrac{\partial}{\partial \mathbf{n}_j} + \tilde{\mathcal{T}}_j)(u_j^{(n)} - u_{j+1}^{(n)}) &= 0 \ \text{ on } \partial\Omega_j \cap \Omega_{j+1}. \end{aligned}$$

Definition 1 The Dirichlet to Neumann (DtN) map exterior to Ω_j is

$$\mathrm{DtN}_j^c : g_D \to g_N = \partial_n v, \ \text{s.t.} \quad \begin{aligned} \mathcal{L}\, v &= 0, \ \text{ in } \Omega \backslash \Omega_j, \\ \mathcal{B}\, v &= 0, \ \text{ on } \partial\Omega_j \cap \partial\Omega, \\ v &= g_D, \text{ on } \partial\Omega_j \backslash \partial\Omega. \end{aligned}$$

The optimal choice for the transmission conditions in the optimal Schwarz method is to use the DtN, see [13]. We show here that it suffices to choose for the tangential operators $\mathcal{S}_j$ and $\tilde{\mathcal{S}}_j$ the DtN operators, independent of what one uses for $\mathcal{T}_j$ and $\tilde{\mathcal{T}}_j$, to get an optimal result:

Theorem 1 *If $\mathcal{S}_j = \tilde{\mathcal{S}}_j$, $j = 1, \ldots, J-1$ correspond to the DtN maps exterior to Ω_j restricted to $\partial\Omega_j \cap \Omega_{j-1}$, and all the subdomain problems have a unique solution, then OS-SGS converges in one iteration for any initial guess. In particular, convergence is independent of the number of subdomains.*

This result can either be proved following the arguments in [13] using the error equations, or by the approach in [6] at the discrete level or in [2] at the continuous level to substitute exterior source terms with transmission data represented by subdomain solutions. We omit the details here.

In optimized Schwarz methods, one replaces the DtN with an approximation, for example an absorbing boundary condition, or a PML. For the latter, we define the approximation DtN_j^L by

$$\mathrm{DtN}_j^L : g_D \to g_N = \partial_n v, \text{ s.t. } \begin{aligned} \tilde{\mathcal{L}} v &= 0, && \text{in } \Omega_j^L, \\ \mathcal{B} v &= 0, && \text{on } \partial\Omega_j \cap \partial\Omega, \\ v &= g_D, && \text{on } \partial\Omega_j \backslash \partial\Omega, \end{aligned}$$

where Ω_j^L is the PML region exterior to Ω_j and $\tilde{\mathcal{L}}$ is chosen such that DtN_j^L closely approximates DtN_j^c from Definition 1. We notice that if $\mathcal{S}_j = \mathrm{DtN}_j^L$, the subdomain problem (2) is equivalent to solve one PDE in $\Omega_j \cup \Omega_j^L$ with the Dirichlet and Neumann traces $v_j^{(n)} - v_{j-1}^{(n)}$ continuous across $\partial\Omega_j^L \cap \partial\Omega_j$.

OS-SGS in the interface transmission form requires the evaluation of operators on data, such as $(\partial_{\mathbf{n_j}} + \mathcal{S}_j)v_{j-1}^{(n)}$, which can be inconvenient, especially if $\mathcal{S}_j$ is complicated. This can be avoided if we solve for the corrections. To this end, we introduce in the forward sweep $\delta v_j^{(n)} := v_j^{(n)} - \tilde{v}_j^{(n-1)}$ for some $\tilde{v}_j^{(n-1)}$ which has the same Dirichlet and Neumann traces as $v_{j-1}^{(n)}$ on $\partial\Omega_j \cap \Omega_{j-1}$ and $u_{j+1}^{(n-1)}$ on $\partial\Omega_j \cap \Omega_{j+1}$. For example, following [2] (see also [15] at the discrete level), we introduce the weighting functions α_j and β_j such that

$$\frac{\partial}{\partial \mathbf{n}_j}\alpha_j = 0, \alpha_j = 1 \text{ on } \partial\Omega_j \cap \Omega_{j-1}, \quad \frac{\partial}{\partial \mathbf{n}_j}\beta_j = 0, \beta_j = 1 \text{ on } \partial\Omega_j \cap \Omega_{j+1}. \tag{3}$$

Then, we can define the auxiliary function $\tilde{v}_j^{(n-1)}$ as

$$\tilde{v}_j^{(n-1)} = \begin{cases} \alpha_j v_{j-1}^{(n)} + (1-\alpha_j) w_j^{(n-1)}, & \text{on } \Gamma_{j-1}, \\ w_j^{(n-1)}, & \text{in } I_j, \\ \beta_j u_{j+1}^{(n-1)} + (1-\beta_j) w_j^{(n-1)}, & \text{on } \Gamma_j, \end{cases} \tag{4}$$

where $w_j^{(n-1)}$ is an arbitrary function. One can verify that

$$\begin{aligned}
\tfrac{\partial}{\partial \mathbf{n}_j}\tilde{v}_j^{(n-1)} &= \tfrac{\partial}{\partial \mathbf{n}_j}v_{j-1}^{(n)}, \quad \tilde{v}_j^{(n-1)} = v_{j-1}^{(n)} \text{ on } \partial\Omega_j \cap \Omega_{j-1},\\
\tfrac{\partial}{\partial \mathbf{n}_j}\tilde{v}_j^{(n-1)} &= \tfrac{\partial}{\partial \mathbf{n}_j}u_{j+1}^{(n-1)}, \ \tilde{v}_j^{(n-1)} = u_{j+1}^{(n-1)} \text{ on } \partial\Omega_j \cap \Omega_{j+1},
\end{aligned}$$

which together with (2) imply

$$\begin{aligned}
(\tfrac{\partial}{\partial \mathbf{n}_j} + \mathcal{S}_j)(v_j^{(n)} - \tilde{v}_j^{(n-1)}) &= 0 \text{ on } \partial\Omega_j \cap \Omega_{j-1},\\
(\tfrac{\partial}{\partial \mathbf{n}_j} + \mathcal{T}_j)(v_j^{(n)} - \tilde{v}_j^{(n-1)}) &= 0 \text{ on } \partial\Omega_j \cap \Omega_{j+1}.
\end{aligned}$$

Similar identities also hold for the backward sweep. Therefore, the OS-SGS algorithm in interface transmission form can equivalently be written in the residual-correction form (see below).

Remark 1 Usually one uses the subdomain iterates for defining $\tilde{v}_j^{(n-1)}$ and $\tilde{u}_j^{(n-1)}$, e.g. $w_j^{(n-1)} := u_j^{(n-1)}$ in (4), thus gluing the subdomain solutions together to obtain a global approximation. If the weighting functions $\{\beta_j\}$ for the gluing are the indicator functions of the corresponding non-overlapping partition, we obtain the so called *restricted* Schwarz methods; other choices give the same subdomain iterates but only different global iterates.

OS-SGS (residual-correction form)

Forward sweep: given $(u_j^{(n-1)})_{j=1}^J$ on $(\Omega_j)_{j=1}^J$ at iteration $(n-1)$, solve successively for $j = 1, \ldots, J-1$ the subdomain problems

$$\begin{aligned}
\mathcal{L}\,\delta v_j^{(n)} &= f - \mathcal{L}\,\tilde{v}_j^{(n-1)} && \text{in } \Omega_j,\\
\mathcal{B}\,\delta v_j^{(n)} &= g - \mathcal{B}\,\tilde{v}_j^{(n-1)} && \text{on } \partial\Omega \cap \partial\Omega_j,\\
(\tfrac{\partial}{\partial \mathbf{n}_j} + \mathcal{S}_j)\delta v_j^{(n)} &= 0 && \text{on } \partial\Omega_j \cap \Omega_{j-1},\\
(\tfrac{\partial}{\partial \mathbf{n}_j} + \mathcal{T}_j)\delta v_j^{(n)} &= 0 && \text{on } \partial\Omega_j \cap \Omega_{j+1},
\end{aligned}$$

each followed by letting $v_j^{(n)} \leftarrow \tilde{v}_j^{(n-1)} + \delta v_j^{(n)}$ and setting $\tilde{v}_{j+1}^{(n-1)}$ as in (4).

Backward sweep: solve successively for $j = J, \ldots, 1$ the subdomain problems

$$\begin{aligned}
\mathcal{L}\,\delta u_j^{(n)} &= f - \mathcal{L}\tilde{u}_j^{(n)} && \text{in } \Omega_j,\\
\mathcal{B}\,\delta u_j^{(n)} &= g - \mathcal{B}\tilde{u}_j^{(n)} && \text{on } \partial\Omega \cap \partial\Omega_j,\\
(\tfrac{\partial}{\partial \mathbf{n}_j} + \tilde{\mathcal{S}}_j)\delta u_j^{(n)} &= 0 && \text{on } \partial\Omega_j \cap \Omega_{j-1},\\
(\tfrac{\partial}{\partial \mathbf{n}_j} + \tilde{\mathcal{T}}_j)\delta u_j^{(n)} &= 0 && \text{on } \partial\Omega_j \cap \Omega_{j+1},
\end{aligned}$$

each followed by setting $u_j^{(n)} \leftarrow \tilde{u}_j^{(n)} + \delta u_j^{(n)}$ and setting $\tilde{u}_{j-1}^{(n-1)}$ as in (4).

Theorem 2 *The source transfer domain decomposition method defined in [2] is an overlapping optimized Schwarz method of symmetric Gauss-Seidel type, with the overlap covering half the subdomains, and using PML transmission conditions on the left and right interfaces in the forward sweep and Dirichlet instead of PML on the right interfaces in the backward sweep. In addition, the source terms are consistently modified in the forward sweep.*

Proof As we have seen for OS-SGS, the residual-correction form is equivalent to the interface transmission form. The only difference of STDDM from the residual-correction form of OS-SGS is that in the forward sweep the residual for $1 \leq j \leq J-1$ in the overlap with the right neighbor is set to zero, see ALGORITHM 3.1 in [2]. This modification can also be interpreted as taking the boundary layer Γ_j as part of the PML on the right of the subdomain so the physical subdomains become effectively non-overlapping.

To see the consistency of STDDM, we assume $u_j^{(n-1)}$ is equal to the exact solution of the original problem in Ω_j for $1 \leq j \leq J$ and check whether $u_j^{(n)} = u_j^{(n-1)}$ holds, i.e. the exact solution is a fixed point of the iteration. We note that STDDM uses $w_j^{(n-1)} = u_j^{(n-1)}$ in (4). In this case, by the assumption on $u_1^{(n-1)}$ and $u_2^{(n-1)}$, we can show $\tilde{v}_1^{(n-1)} = u_1^{(n-1)}$ and so the residual vanishes in Ω_1 both for OS-SGS and STDDM. Therefore, the correction $\delta v_1^{(n)} = v_1^{(n)} - \tilde{v}_1^{(n-1)}$ must be zero because the sub-problem has a unique solution, which gives $v_1^{(n)} = \tilde{v}_1^{(n-1)} = u_1^{(n-1)}$. By induction, we then show that $u_j^{(n)} = u_j^{(n-1)}$ for $1 \leq j \leq J$. □

3 Numerical Experiments

We solve the Helmholtz equation in rectangles discretized by Q1 finite elements. For the free space and open cavity problems, the wave speed is constant, $c = 1$, and the point source is at (0.5177, 0.6177) while the Marmousi model problem has a variable wave speed and the point source at (6100, 2200). PML are padded around all the domains except for the open cavity problem, where homogeneous Neumann conditions are imposed at the top and bottom. We use the same depth (counted with mesh elements) of PML for the original domain and the subdomains since already for a PML with two layers the dominating error is around the point source. The PML complex stretching function we use is given by $s(d) = \frac{1}{1-\mathrm{i}4\pi d^2/(L^3 k)}$ where $k = \omega/c$ is the wavenumber, d is the distance to the physical boundary and L is the geometric depth of the PML. We use the same mesh size and element-wise constant material coefficients in the physical and PML regions. We use a zero initial guess for GMRES with relative residual (preconditioned) tolerance 10^{-6}. The results are shown in Table 1 where "STDDM2" is the STDDM without changing transmission from PML to Dirichlet in the backward sweep, "PMLh" represents OS-SGS with two elements overlap and PML on all boundaries, "TO2h" ("TO0h") is

Table 1 Minimal depth m of the PML layer and the corresponding number nx of mesh elements for each subdomain in the x-direction to reach the given iteration number it (J is the number of subdomains. Two nx are presented for STDDM because in the backward sweep the right interfaces use Dirichlet instead of PML)

Free space problem on the unit square

$\frac{\omega}{2\pi}$	$\frac{1}{10h}$	$\frac{J}{10}$	STDDM			STDDM2			PMLh			TO2h	TO0h	O2h		
			it	m	nx	it	m	nx	it	m	nx	it	it	it	m	nx
20	20	2	4	4	26, 22	4	2	22	4	2	16	11	12	14	0	12
20	40	4	4	7	32, 25	4	2	22	4	3	18	21	25	17	0	12
20	80	8	6	7	32, 25	4	4	26	4	7	26	63	45	–	0	12
40	40	4	6	4	26, 22	4	2	22	5	2	16	15	25	49	0	12
80	80	8	6	9	36, 27	4	3	24	6	3	18	25	89	–	0	12
160	160	16	10	22	62, 40	6	4	26	6	5	22	49	>200	–	0	12

Open cavity problem on the unit square

$\frac{\omega}{2\pi}$	$\frac{1}{10h}$	$\frac{J}{10}$	STDDM			STDDM2			PMLh			TO2h	TO0h	O2h		
			it	m	nx	it	m	nx	it	m	nx	it	it	it	m	nx
20	20	2	11	2	22, 20	6	2	22	8	2	16	41	85	33	0	12
20	40	4	13	3	24, 22	10	2	22	12	5	22	76	216	55	0	12
20	80	8	19	5	28, 23	18	4	26	22	6	24	142	392	–	0	12
40	40	4	19	2	22, 20	10	2	22	14	2	16	123	292	119	0	12
80	80	8	–	–	–	22	2	22	30	2	16	429	–	–	0	12

Marmousi model

$\frac{\omega}{2\pi}$	h	J	STDDM			STDDM2			PMLh			TO2h	TO0h	O2h		
			it	m	nx	it	m	nx	it	m	nx	it	it	it	m	nx
2	74	10	5	2	26, 24	4	2	26	4	2	18	9	6	7	0	14
2	37	20	6	4	30, 26	4	2	26	4	3	20	16	10	11	0	14
2	18	40	6	5	32, 27	5	2	26	5	3	20	37	17	19	0	14
4	37	20	6	3	28, 25	4	2	26	5	2	18	10	9	8	0	14
8	18	40	8	4	30, 26	5	2	26	6	2	18	12	14	18	0	14
16	9	80	10	4	30, 26	6	2	26	6	2	18	16	25	–	0	14

the OS-SGS with the Taylor second- (zero-) order transmission conditions and two elements overlap. The optimized transmission conditions from [9] are also tested with overlap and the results for the optimized condition of second-order are listed (the original boundaries still use Taylor second-order conditions) under the name "O2h". The optimized condition of zero-order suffers from too many subdomains and can not converge to the correct solution in all cases. We implemented all the algorithms in the residual-correction form. We also tested the classical Schwarz method of symmetric Gauss-Seidel type with Dirichlet transmission conditions but the preconditioned system is very ill-conditioned so that the obtained solution comprises a significant error even if the preconditioned residual is reduced by the tolerance factor. The same failure happens in Table 1 indicated by middle bars. From the table, we find that, for our particular test problems with open boundaries on both left and right sides, STDDM2 which uses always PML on both sides works better than STDDM which changes to Dirichlet on the right side in the backward sweep.

Acknowledgements This work was supported by the Université de Genève. HZ thanks the International Science and Technology Cooperation Program of China (2010DFA14700).

References

1. J.-P. Berenger, A perfectly matched layer for absorption of electromagnetic waves. J. Comput. Phys. **114**, 185–200 (1994)
2. Z. Chen, X. Xiang, A source transfer domain decomposition method for Helmholtz equations in unbounded domain. SIAM J. Numer. Anal. **51**(4), 2331–2356 (2013a)
3. Z. Chen, X. Xiang, A source transfer domain decomposition method for Helmholtz equations in unbounded domain part II: extensions. Numer. Math. Theory Methods Appl. **6**(3), 538–555 (2013b)
4. B. Engquist, L. Ying, Sweeping preconditioner for the Helmholtz equation: moving perfectly matched layers. Multiscale Model. Simul. **9**(2), 686–710 (2011)
5. M.J. Gander, Optimized Schwarz methods. SIAM J. Numer. Anal. **44**, 699–731 (2006)
6. M.J. Gander, F. Kwok, Optimal interface conditions for an arbitrary decomposition into subdomains, in *Domain Decomposition Methods in Science and Engineering XIX*, ed. by Y. Huang, R. Kornhuber, O.B. Widlund, J. Xu (Springer, Heidelberg, 2011), pp. 101–108
7. M.J. Gander, F. Nataf, AILU: a preconditioner based on the analytic factorization of the elliptic operator. Numer. Linear Algebra Appl. **7**, 505–526 (2000)
8. M.J. Gander, F. Nataf, An incomplete preconditioner for problems in acoustics. J. Comput. Acoust. **13**, 455–476 (2005)
9. M.J. Gander, F. Magoulès, F. Nataf, Optimized Schwarz methods without overlap for the Helmholtz equation. SIAM J. Sci. Comput. **24**, 38–60 (2002)
10. C. Geuzaine, A. Vion, Double sweep preconditioner for Schwarz methods applied to the Helmholtz equation, in *Domain Decomposition Methods in Science and Engineering XXII* (Springer, Heidelberg, 2015)
11. C. Japhet, Optimized Krylov-Ventcell method. Application to convection-diffusion problems, in *Ninth International Conference on Domain Decomposition Methods*, ed. by P.E. Bjorstad, M.S. Espedal, D.E. Keyes (ddm.org, Bergen, 1998)
12. F. Nataf, F. Nier, Convergence rate of some domain decomposition methods for overlapping and nonoverlapping subdomains. Numer. Math. **75**, 357–377 (1997)

13. F. Nataf, F. Rogier, E. de Sturler, Optimal interface conditions for domain decomposition methods. Technical report, Polytechnique (1994)
14. A. Schädle, L. Zschiedrich, Additive Schwarz method for scattering problems using the PML method at interfaces, in *Domain Decomposition Methods in Science and Engineering XVI*, ed. by O.B. Widlund, D.E. Keyes (Springer, Heidelberg, 2007), pp. 205–212
15. A. St-Cyr, M.J. Gander, S.J. Thomas, Optimized multiplicative, additive, and restricted additive Schwarz preconditioning. SIAM J. Sci. Comput. **29**, 2402–2425 (2007)
16. C. Stolk, A rapidly converging domain decomposition method for the Helmholtz equation. J. Comput. Phys. **241**, 240–252 (2013)
17. A. Toselli, Overlapping methods with perfectly matched layers for the solution of the Helmholtz equation, in *Eleventh International Conference on Domain Decomposition Methods*, ed. by C.-H. Lai, P. Bjorstad, M. Cross, O.B. Widlund (1999), pp. 551–558

Domain Decomposition in Shallow Lake Modelling for Operational Forecasting of Flooding

Menno Genseberger, Edwin Spee, and Lykle Voort

1 Introduction

The Netherlands is a highly urbanized area. In addition to flooding from the sea due to storm surges and high water discharges from rivers, flooding from major lakes is also a threat. Since 2011 there is a new system in operational use (24 h per day, 7 days per week), for the prediction of flooding at Lake IJssel, Lake Marken, and the lakes bordering them. This system, RWsOS Meren [5] enables a real-time dynamic forecasting of wind driven waves, water flow, wave runup, and overtopping at dikes.

At the moment the time horizon of forecasts with RWsOS Meren is 2 days ahead. To enlarge this time horizon, medium-range global weather forecasts from ECMWF [4] up to 15 days (two forecasts per day) and short-to-medium range forecasts of extreme and localised weather events from COSMO-LEPS (limited area ensemble prediction system) [3] up to 5.5 days (one forecast per day) will be used as input for RWsOS Meren. In RWsOS Meren, only the two shallow-water models of the lakes will be run with this input (and not the models for waves, wave runup, and overtopping). ECMWF and COSMO-LEPS use ensembles (51 and 16 ensemble members, respectively). Therefore, also the two shallow-water models will be run

M. Genseberger (✉)
Deltares, P.O. Box 177, 2600 MH Delft, The Netherlands

CWI, P.O. Box 94079, 1090 GB, Amsterdam, The Netherlands
e-mail: Menno.Genseberger@deltares.nl

E. Spee
Deltares, P.O. Box 177, 2600 MH Delft, The Netherlands
e-mail: Edwin.Spee@deltares.nl

L. Voort
SURFsara, P.O. Box 94613, 1090 GP, Amsterdam, The Netherlands
e-mail: lykle.voort@surfsara.nl

T. Dickopf et al. (eds.), *Domain Decomposition Methods in Science and Engineering XXII*, Lecture Notes in Computational Science and Engineering 104, DOI 10.1007/978-3-319-18827-0_21

in ensemble mode. As a consequence, for these models 204 runs with a simulation period of 15 days and 32 runs with a simulation period of 5.5 days have to finish within a reasonable time on a daily basis. This asks for a balance between low computational times per ensemble member and the efficient use of the available hardware (and energy) resources. In this paper we investigate how to manage this on current hardware.

Here, the essential ingredient is the domain decomposition technique in the shallow-water solver Simona [6, 10, 2] that we apply. The implementation of this domain decomposition technique in Simona has the nice property that it enables (sub)structuring, distribution, and minimizing the exchange of data in a practical and efficient way. This is both on the high—modelling level (decomposition in physical subdomains with absorbing boundary conditions), intermediate—numerical level (parallel solver with minimized iteration count) and low—implementation level (data distribution with minimized data exchange between different memory blocks). A lower level inherits the gain in efficiency from a higher level. Therefore, most gain is on the high level and on the lower levels some fine-tuning remains. However, gain in efficiency on the high level will not always automatically be there and some effort is needed. This will be illustrated here for the practical example of the shallow lake models in RWsOS Meren.

The paper is organized as follows. First, the physical characteristics and the shallow-water models of the lakes are described in Sect. 2. Then, in Sect. 3 we apply domain decomposition in Simona for these models in two stages (automatic partitioning in Sect. 3.1 and fine-tuning in Sect. 3.2). For this purpose, we investigate the consequences for computational times and (parallel) efficiency by numerical experiments.

2 Shallow Lake Modelling

The operational system RWsOS Meren [5] covers eight major lakes of the Netherlands: Lake IJssel (IJsselmeer in Dutch), Lake Marken (Markermeer), and six smaller lakes at the borders (with Dutch names Ketelmeer, Vossemeer, Zwarte Meer, IJmeer, Gooimeer, and Eemmeer), see Fig. 1. All lakes are quite shallow: depths are in the order of several meters whereas horizontal dimensions are in the order of kilometers. Ketelmeer, Vossemeer, and Zwarte Meer are in open connection with Lake IJssel. IJmeer, Gooimeer, and Eemmeer are in open connection with Lake Marken. Lake Marken is separated from Lake IJssel by a dike ("Houtribdijk") with two sluices. On the north, Lake IJssel is separated from the Wadden Sea by a dike ("Afsluitdijk") with two sluices. Most important driving force of the water system is wind. However in specific situations, for instance after heavy rainfall, river discharges are also important. Here, the largest contribution is from the river IJssel that enters Ketelmeer. Furthermore, river Overijsselse Vecht enters Zwarte Meer (via river Zwarte Water) and river Eem enters Eemmeer. The water level of Lake IJssel is kept to a fixed level by draining off superfluous water via the two

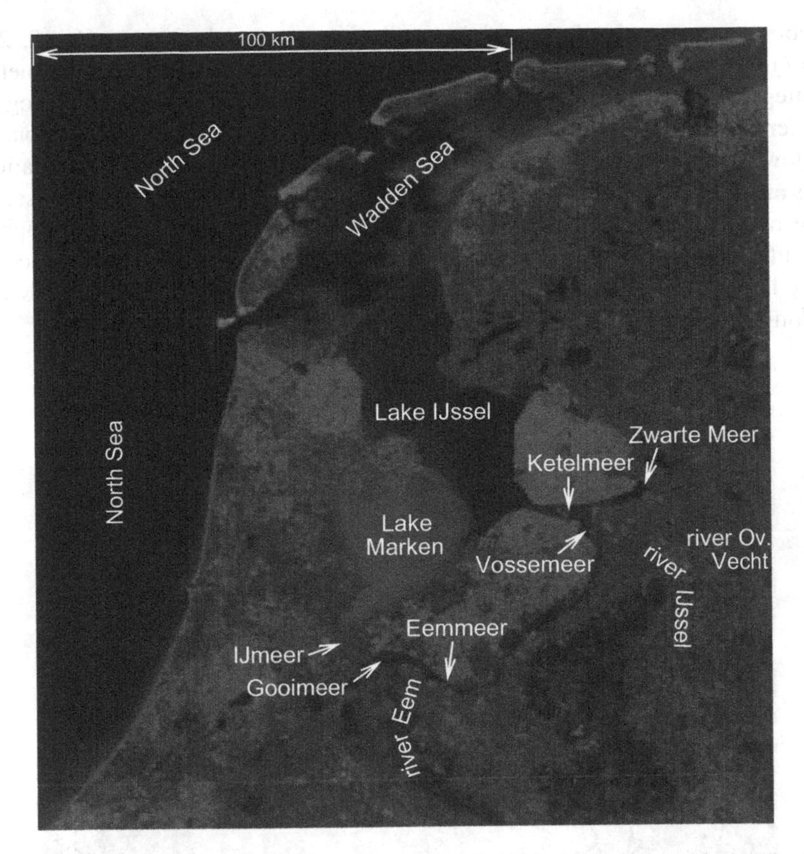

Fig. 1 Geographical domain with eight major lakes of the Netherlands. (Color figure online.)

sluices to the Wadden Sea. Lake Marken is also kept to a fixed water level, however discharges through the sluices are much smaller.

For computing flow of water based on medium-range global and short-to-medium range weather forecasts, the same two models will be used as in the current operational system of RWsOS Meren. One is the shallow-water model for Lake IJssel including the smaller lakes Ketelmeer, Vossemeer, and Zwarte Meer and parts of the rivers IJssel, Zwarte Water, and Overijsselse Vecht. The other is the shallow-water model for Lake Marken including the smaller lakes IJmeer, Gooimeer, and Eemmeer and the river Eem with its floodplain. For rivers IJssel and Overijsselse Vecht boundary conditions are imposed through discharges. Close to the sluices on the side of the Wadden Sea boundary conditions are imposed through water levels. Here, both discharges and water levels are a combination of observed values and predicted values (from neighbouring operational systems). Wind predictions (as computed externally) are downscaled to the required sizes for the models of the lakes.

For the numerical solution of the shallow-water models Simona [6, 10, 2] is being used. Simona applies a so-called alternating direction implicit (ADI) method to integrate the shallow-water equations numerically in time, using an orthogonal staggered grid with horizontal curvilinear coordinates. For this application, the shallow-water models are depth averaged. The sizes of the horizontal computational grids are 486×1983 and 430×614 for the shallow-water models of Lake IJssel and Lake Marken, respectively. See Figs. 2 and 4 for the corresponding geographical lay-out. The grids are relatively fine in (the floodplain areas of) the rivers and coarse in the larger lakes. For the shallow-water model of Lake IJssel this can be observed by comparing the geographical lay-out with the memory lay-out in Fig. 2.

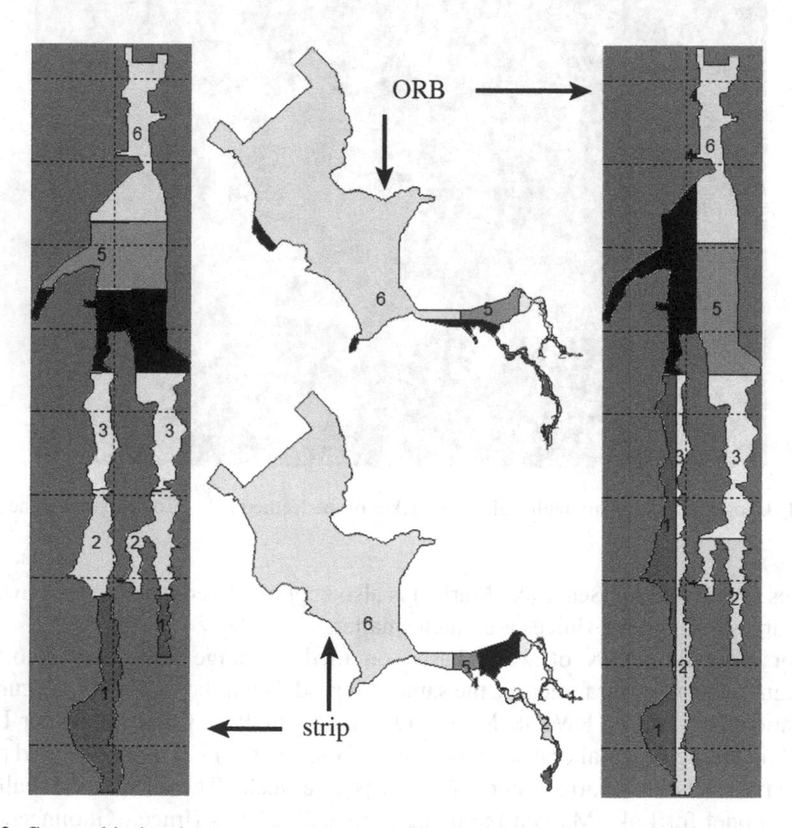

Fig. 2 Geographical and memory lay-out of computational grid of shallow-water model for Lake IJssel with automatic partitioning by domain decomposition. *Middle bottom*: geographical lay-out of domain decomposed in 6 subdomains (in different colours and numbered from 1 to 6) with stripwise partitioning, *left*: corresponding memory lay-out. *Middle top*: geographical lay-out of domain decomposed in 6 subdomains (in different colours and numbered from 1 to 6) with partitioning via orthogonal recursive bisection (ORB), *right*: corresponding memory lay-out. (Color figure online.)

Calibration and validation of the models was carried out for periods with historical storms (including typical wind behavior, some also with high river discharges). For this, measured values of discharges, waterlevels, rainfall, and evaporation were used with some small corrections due to missing terms in the water balance for the physical system.

Here we take a simulation period of 32 hours for both shallow-water models. Note that computational times of Simona are almost not influenced by the physical conditions in a given simulation period (storm or mild wind conditions and/or high or low river discharges). To get the computational times for an ECMWF (COSMO-LEPS) ensemble member with a simulation period of 15 (5.5) days the computational time has to be multiplied with a factor 11.25 (4.125).

Domain decomposition will be used to have a good balance between computational times and (parallel) efficiency for running ensembles with the two shallow-water models.

3 Domain Decomposition

The domain decomposition technique in the current versions of Simona is based on a nonoverlapping Schwarz method with optimized coupling at the subdomain interfaces [2]. This approach has shown to yield excellent parallel performance for practical flow problems from civil engineering. However, the two shallow-water models have a complicated geometry and a relatively small number of computational grid points. Because of this, obtaining a good balance is not straightforward: increasing the number of subdomains can lower computational times more but may result in less efficient use of the available hardware (and energy) resources.

As we can not investigate all possibilities, we proceed with a pragmatic approach. First, we analyse the parallel performance for two automatic partitioning methods as a function of the number of computational cores in Sect. 3.1. Then, for a nearly optimal number of subdomains from Sect. 3.1, we try to get efficient ensemble runs with the models on current hardware by fine-tuning in Sect. 3.2.

3.1 Automatic Partitioning

Here we analyse the parallel performance of both shallow-water models by a numerical experiment. For this we varied the number of subdomains from 1 to 16 for two automatic partitioning methods. Here, one subdomain is assigned to one computational core. Both methods are based on domain decomposition of the active computational grid points. One method makes a stripwise partitioning in one direction of the domain. The other method decomposes the domain based on orthogonal recursive bisection (ORB) [1].

The numerical experiment was performed on the H4+ linux-cluster at Deltares (nodes interconnected with Gigabit Ethernet, each node contains 1 Intel quad-core i7-2600 processor "Sandy Bridge" ([7, Sect. 2.8.5.3]; [8, Sect. 2.8.4.1]), 3.4 GHz/core, hyperthreading off) with the 2011 version of Simona (compiled with Intel Fortran 11 compiler and OpenMPI for Linux 64 bits platform). For the distribution of the memory blocks (each block contains the unknowns in one subdomain) over the nodes two options were considered: round-robin (memory blocks are distributed alternated over the nodes) and compact (option tries to position each memory block close to blocks of neighbouring subdomains).

Figure 2 (Fig. 4) shows the corresponding geographical lay-out of the computational grid of the shallow-water model for Lake IJssel (Lake Marken) in case of 6 subdomains. The wall-clock time as a function of the number of computational cores for this model is shown on the left (right) in Fig. 3. Reported wall-clock times are averages of three measurements. For all cases the corresponding standard deviation is less than 3 % of the average.

The speed up is not as ideal as linear (for that case lines will have a downward slope of 45° in the double logarithmic figures: doubling the number of computational cores will half the wall-clock time). But, in general, from Fig. 3 it can be observed that for both models the wall-clock time can be reduced substantially for decompositions in up to 6 subdomains. Based on this observation, we choose 6 as the nearly optimal number of subdomains for both models.

Furthermore, one of the automatic partitioning methods does not clearly seem to be more beneficial than the other (Fig. 3). This indicates the possibility to further optimize the decomposition by inspecting the configurations in 6 subdomains of both methods. That will be subject in Sect. 3.2. Overall, the memory option compact improves the results of round-robin for more than four computational cores (i.e. the cases that more nodes are used). This is as expected: for option compact more

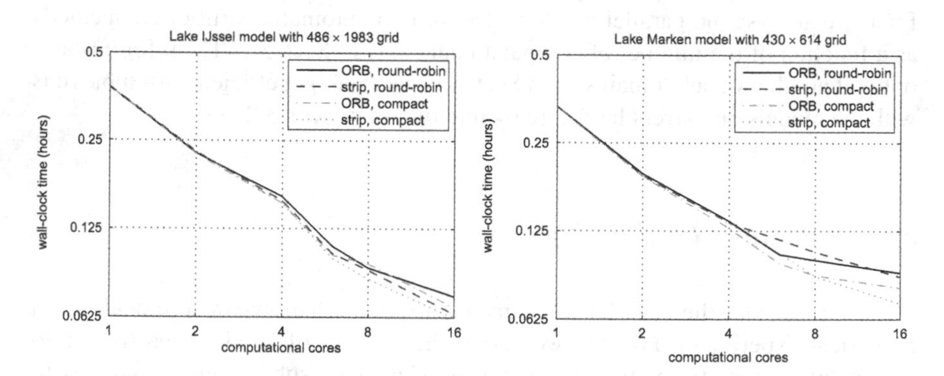

Fig. 3 Wall-clock time (in hours) for shallow-water model of Lake IJssel (*left*) and Lake Marken (*right*) as a function of the number of computational cores. Shown are results for two automatic partitioning methods: stripwise and ORB (orthogonal recursive bisection) and two options for memory distribution: round-robin and compact. (Color figure online.)

neighbouring subdomains are positioned inside the same node, therefore there is less communication between nodes resulting in lower computational times.

3.2 Fine-Tuning

For the nearly optimal number of 6 subdomains for both models from Sect. 3.1, we try to get efficient ensemble runs with the models on current hardware by fine-tuning.

We considered the following hardware at SURFsara:

- 2 socket L5640 node (2 Intel six-core Xeon L5640 processors "Westmere-EP" [7, Sect. 2.8.5.2], 2.26 GHz/core) (Lisa),
- 2 socket 2650L node (2 Intel eight-core Xeon E5-2650L processors "Sandy Bridge" ([7, Sect. 2.8.5.3]; [8, Sect 2.8.4.1]), 1.8 GHz/core) (Lisa),
- 2 socket 2695 v2 node (2 Intel twelve-core Xeon E5-2695 v2 processors "Ivy Bridge" [9, Sect. 2.8.4], 2.4 GHz/core) (Cartesius).

Note that, with 6 subdomains, multiple runs (2 runs for a 2 socket L5640 or 2650L node, 4 runs for a 2 socket 2695 v2 node) of the models fit in a single node. Instead of using more than one node for a single run to lower computational times more (like the numerical experiment in Sect. 3.1), for efficiency we will consider here the use of a single node for multiple runs simultaneously. A 2013 version of Simona compiled with Intel Fortran 13 and OpenMPI for Linux 64 bits platform was used.

First, we try to further optimize the decomposition in 6 subdomains by inspecting the configurations of the two automatic partitioning methods from Sect. 3.1. For that purpose we used the Visipart package of Simona. By comparing the geographical lay-out of subdomains for the shallow-water model of Lake Marken for the two automatic partitioning methods (left and middle picture) in Fig. 4 one can see that for the stripwise decomposition (left picture) there is a very long subdomain interface and a part of a subdomain is quite thin. This has a negative effect on the computational times. Relatively long subdomain interfaces require more data communication. Very thin subdomains with widths of less than a dozen grid cells affect the validity of the applied local optimized coupling in Simona. Therefore, we used the results of the other automatic partitioning method, by ORB (middle picture) as a basis for further optimization. The right picture of Fig. 4 illustrates the resulting geographical lay-out of subdomains for the shallow-water model of Lake Marken. In a similar way, the decomposition in 6 subdomains for the shallow-water model of Lake IJssel has been optimized. This strategy for further optimization is confirmed by the wall-clock times as shown in columns 2 (automatic stripwise

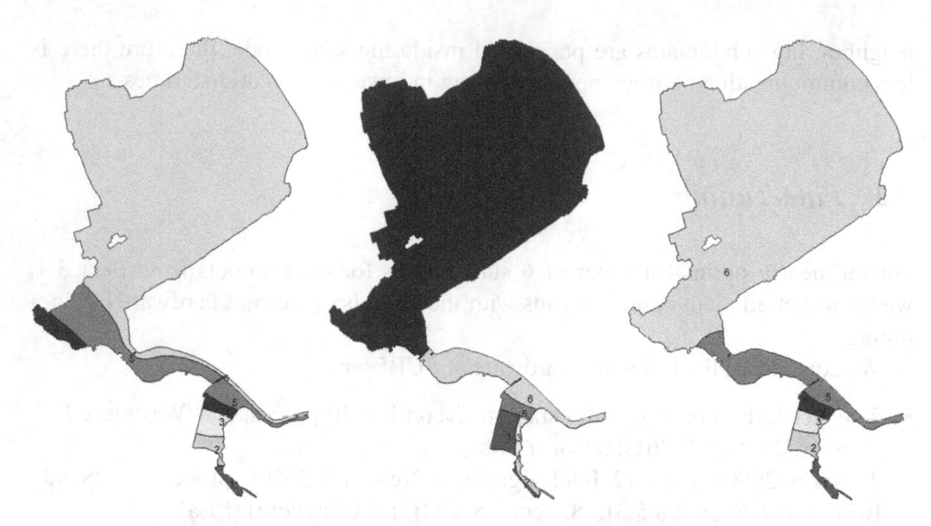

Fig. 4 Geographical lay-out of computational grid of shallow water-model for Lake Marken with partitioning by domain decomposition in 6 subdomains (in different colours and numbered from 1 to 6). *Left*: geographical lay-out of subdomains with automatic stripwise partitioning. *Middle*: geographical lay-out of subdomains with automatic partitioning via orthogonal recursive bisection (ORB). *Right*: geographical lay-out of subdomains with manual fine-tuning of the partitioning. (Color figure online.)

partitioning), 3 (automatic partitioning with ORB), and 4 (fine-tuning of one of the automatic partitionings) of Table 1 (Lake IJssel) and Table 2 (Lake Marken). Here, the reported wall-clock times are averages of three measurements and the corresponding standard deviation is given after the ± symbol.

Then, with the further optimized decomposition we ran two models simultaneously on one single 2 socket L5640 and 2650L node. Corresponding wall-clock times are shown in column 5 of Table 1 (Lake IJssel) and Table 2 (Lake Marken). By comparing these times with column 4 (same decomposition but only one model run on the node) one can see there is some price to pay. We can relieve a part of this pain by binding one of the runs to 6 successive cores of socket 1 and the other run to 6 successive cores of socket 2 as shown in column 6 of both tables. Here data of each model stays inside one socket and no communication is needed between the sockets (this is somehow similar to the situation—with nodes instead of sockets—for memory option compact from Sect. 3.1). On one single 2 socket 2695 v2 node we were not able to run multiple models without binding. For this type of node we observe from columns 4, 6, and 7 in the tables that they can be used efficiently for running 4 models simultaneously.

Table 1 Wall clock time (in minutes) for shallow-water model of Lake IJssel on one two socket node for different decompositions in 6 subdomains and memory distributions

Decomp. type of	Strip	ORB	Fine-tuned	Fine-tuned	Fine-tuned	Fine-tuned
# simultaneous runs on node	1	1	1	2	2	4
Binding	No	No	No	No	Yes	Yes
2 socket L5640 node	8.07 ± 0.25	8.64 ± 0.16	7.83 ± 0.11	10.98 ± 0.24	9.79 ± 0.02	
2 socket 2650L node	6.98 ± 0.10	7.28 ± 0.06	6.41 ± 0.10	8.06 ± 0.12	7.44 ± 0.06	
2 socket 2695 v2 node	6.56 ± 0.00	7.08 ± 0.01	6.07 ± 0.01		6.15 ± 0.01	7.96 ± 0.01

Table 2 Wall clock time (in minutes) for shallow-water model of Lake Marken on one two socket node for different decompositions in 6 subdomains and memory distributions

Decomp. type	Strip	ORB	Fine-tuned	Fine-tuned	Fine-tuned	Fine-tuned
# simultaneous runs on node	1	1	1	2	2	4
Binding	No	No	No	No	Yes	Yes
2 socket L5640 node	8.02 ± 0.14	6.84 ± 0.19	6.75 ± 0.26	9.02 ± 0.29	8.26 ± 0.04	
2 socket 2650L node	7.19 ± 0.20	6.15 ± 0.04	5.85 ± 0.02	7.14 ± 0.32	6.62 ± 0.02	
2 socket 2695 v2 node	6.71 ± 0.01	5.70 ± 0.02	5.31 ± 0.00		5.35 ± 0.01	6.15 ± 0.02

4 Conclusions

We studied how to run efficiently shallow-water models of an operational system for prediction of flooding at the borders of the major Dutch lakes. Aim is to combine the shallow-water models with short-to-medium weather ensemble forecasts to enlarge the time horizon. This asks for a balance between low computational times per ensemble member and the efficient use of the available resources on current hardware. Here, the essential ingredient is the domain decomposition technique in the applied shallow-water solver.

First, the parallel performance for two automatic partitioning methods of the shallow-water models was analyzed. Although the models have a complicated geometry and a relatively small number of computational grid points, the wall-clock time can be reduced substantially for decompositions in up to 6 subdomains. Then, for a nearly optimal partitioning, we tried to get efficient ensemble runs on current hardware by fine-tuning. The resulting optimized decompositions show relatively short internal interfaces between the subdomains (less communication needed) and subdomains that are not too thin (very thin ones affect the validity of the locally optimized domain decomposition coupling). Finally, multiple models can be run simultaneously in an efficient way on one 2 socket node of current hardware by binding the subdomains of each model to successive cores of one socket.

Acknowledgement We thank SURFsara (www.surfsara.nl) for their support in using the Lisa linux-cluster and the Dutch national supercomputer Cartesius. We acknowledge PRACE for awarding a part of this support.

References

1. M.J. Berger, S.H. Bokhari, A partitioning strategy for nonuniform problems on multiprocessors. IEEE Trans. Comput. **36**(5), 570–580 (1987)
2. M. Borsboom, M. Genseberger, B. van 't Hof, E. Spee, Domain decomposition in shallow-water modelling for practical flow applications, in *Domain Decomposition Methods in Science and Engineering XXI*, ed. by J. Erhel et al. (Springer, Berlin, 2014), pp. 557–565
3. COSMO: Consortium for small-scale modeling (2015), www.cosmo-model.org (online)
4. ECMWF: European centre for medium-range weather forecasts (2015), www.ecmwf.int (online)
5. M. Genseberger, A. Smale, H. Hartholt, Real-time forecasting of flood levels, wind driven waves, wave runup, and overtopping at dikes around Dutch lakes, in *2nd European Conference on FLOODrisk Management, Comprehensive Flood Risk Management*, ed. by F. Klijn, T. Schweckendiek (Taylor & Francis Group, London, 2013), pp. 1519–1525
6. Simona: WAQUA/TRIWAQ - two- and three-dimensional shallow-water flow model (2012), http://apps.helpdeskwater.nl/downloads/extra/simona/release/doc/techdoc/waquapublic/sim1999-01.pdf
7. A.J. van der Steen, Overview of recent supercomputers. Technical Report (2011), Online at http://www.euroben.nl/reports.php [An earlier version of this report appeared in J.J. Dongarra and A.J. van der Steen. High-performance computing systems: status and outlook. Acta Numerica **21**, 379–474 (2012)]

8. A.J. van der Steen, Overview of recent supercomputers. Technical Report (2012). [See A.J. van der Steen (2011)]
9. A.J. van der Steen, Overview of recent supercomputers. Technical report (2013). [See A.J. van der Steen (2011)]
10. E.A.H. Vollebregt, M.R.T. Roest, J.W.M. Lander, Large scale computing at rijkswaterstaat. Parallel Comput. **29**, 1–20 (2003)

Parallel Double Sweep Preconditioner for the Optimized Schwarz Algorithm Applied to High Frequency Helmholtz and Maxwell Equations

A. Vion and C. Geuzaine

1 Non-overlapping Optimized Schwarz Algorithm

We consider the optimized Schwarz algorithm for the Helmholtz and Maxwell equations. The algorithm makes use of impedance boundary conditions on the artificial interfaces; although overlapping variants of it exist, we focus on the non-overlapping version, with a partition of the domain into N_d subdomains $\Omega_{1\leq i\leq N_d}$, such that $\cup\bar{\Omega}_i = \bar{\Omega}$ and with $\Sigma_{ij} = \bar{\Omega}_i \cap \bar{\Omega}_j$ the common boundary between two adjacent domains. An iteration of the algorithm for Helmholtz (see e.g. [5] for the Maxwell formulation) is the solution of the subproblems:

$$\begin{aligned} -(\Delta + k^2)u_i^{(k+1)} &= 0 \quad \text{in } \Omega_i, \\ (\partial_n + \mathcal{S})u_i^{(k+1)} &= g_{ij}^{(k)} \quad \text{on } \Sigma_{ij}, \end{aligned} \tag{1}$$

with boundary conditions on the external boundaries inherited from the original problem. The iteration completes with the update relations:

$$\begin{aligned} g_{ij}^{(k+1)} &= -\partial_n u_j^{(k+1)} + \mathcal{S}u_j^{(k+1)} \quad \text{on } \Sigma_{ij}, \\ &= -g_{ji}^{(k)} + 2\mathcal{S}u_j^{(k+1)}. \end{aligned} \tag{2}$$

The algorithm can classically be accelerated by rewriting it in a compact form as a fixed point iteration involving an iteration operator $\mathcal{A}$:

$$g^{(k+1)} = \mathcal{A}g^{(k)} + b. \tag{3}$$

A. Vion • C. Geuzaine (✉)
University of Liège, Montefiore Institute, Grande Traverse, 10, 4000 Liège, Belgium
e-mail: cgeuzaine@ulg.ac.be

T. Dickopf et al. (eds.), *Domain Decomposition Methods in Science and Engineering XXII*, Lecture Notes in Computational Science and Engineering 104, DOI 10.1007/978-3-319-18827-0_22

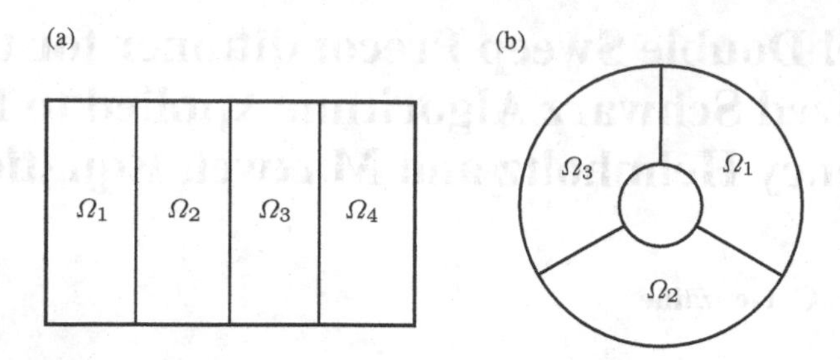

Fig. 1 Two topologies of a decomposed domain into non-overlapping subdomains, without crosspoint: (**a**) "layered" decomposition; (**b**) "cyclic" decomposition around an object

Its solution g satisfies the linear system $\mathcal{F}g = b$, with $\mathcal{F} = \mathcal{I} - \mathcal{A}$ and b the right-hand side containing the contribution of the physical sources. Operator $\mathcal{F}$ involves the solution of subproblems and the update of the interface quantities g_{ij}; as we will see in Sect. 2, it is non-symmetric, hence amenable to a GMRes iterative solver. The optimal choice for the operator $\mathcal{S}$ used in the transmission conditions is the Dirichlet-to-Neumann (DtN) map, as shown in [4]. It is a non-local operator, hence difficult to manipulate in local discretization methods like the Finite Element Method. The literature proposes different local approximations of it, among which we choose a truncated rational approximation of order $(2, 2)$ (see [1, 2]).

In order to circumvent the difficulties associated with the so-called crosspoints (points that are at the intersection of more than two subdomains), we will consider two kinds of decompositions that naturally avoid them: layered or 1d-like decompositions, and cyclic decompositions around an object. Figure 1 shows basic examples of such decompositions.

2 Study of the Iteration Operator

Because the unknown g of the system can be regarded as a composite vector of unknown functions g_{ij}, the iteration operator $\mathcal{F}$ can be written as a matrix F, whose coefficients are operators acting on the interface functions. They take as input a function defined on one side of a domain and transfer the information over the domain, to the opposite interface, where a homogeneous transmission condition is imposed. We will refer to them as transport operators. There are two transport operators defined on the ith subdomain, that we denote by $\mathcal{B}_i^f$ and $\mathcal{B}_i^b$, where the f and b indices refer to the forward or backward direction of the transfer. This distinction is important for what follows, as we will see in Sect. 3 that the convergence of the algorithm can be accelerated by propagating information over longer distances, simultaneously in the forward and backward directions.

We will first consider the case where the "true" DtN map $\mathcal{D}$ is used as transmission operator, leading to a perfectly non-reflecting boundary condition that lets the wave freely propagate outside the domain, without reflection. The effect of an imperfect transmission condition on the structure and the properties of the operator will be considered in a second step. The matrix writes, with the vector of unknown functions $g = [g_{12}, g_{21}, g_{23}, \dots]^T$ and a layered topology of the decomposition (Fig. 1a):

$$F_{\mathcal{D}}(N_d) = \begin{bmatrix} \mathcal{I} & \mathcal{B}_2^b & & & & \\ & \mathcal{I} & & & & \\ & & \mathcal{I} & \ddots & & \\ & \mathcal{B}_2^f & \mathcal{I} & & & \\ & & & & \mathcal{B}^b_{N_d-1} & \\ & & \ddots & \ddots & & \\ & & & & \mathcal{I} & \\ & & & \mathcal{B}^f_{N_d-1} & & \mathcal{I} \end{bmatrix}. \quad (4)$$

Even when the optimized Schwarz algorithm is used with the optimal choice of transmission operator, its convergence is strongly impacted by the number of subdomains N_d, and can become very slow for large numbers of domains. This is classically understood as being caused by the local interactions of the subdomains in the algorithm, that are able to exchange information only with their direct neighbours at each iteration. There are situations, like in waveguides, where the information needs to travel through all the domains before the algorithm is able to build an acceptable solution everywhere. The situation is even worse if the information is distorted while being passed through a non-ideal transmission condition. That intuitive explanation is supported by the spectral properties of the iteration operator, that is defective (lacks a full basis of eigenvectors, while still being invertible) in the case of exact DtN map, which is known to cause slow convergence of Krylov solvers. With an approximate DtN map $\tilde{\mathcal{D}}$ and large N_d, another source of poor convergence resides in the fact that some of the eigenvalues get close to 0 for large N_d, leading to large condition numbers, while the operator can still be considered close to defective.

3 Preconditioning Strategy for Convergence Acceleration

We start from the principle that a preconditioner should be a good approximation of the inverse of the system to be solved, and observe that the inverse of the matrix of the operator with exact DtN map can be easily obtained via a recurrence relation, for an arbitrary number of subdomains. Therefore, we design our preconditioner as having the same structure as the inverse of the ideal operator (4) $\mathcal{F}_{\mathcal{D}}^{-1}$, though using an approximate DtN map. Its product with a vector can be obtained as a matrix-free routine that performs a double sequence of subproblem solves, in the forward and backward directions, hence the name "double sweep" preconditioner [7, 8]. This is made possible by the fact that we can give an interpretation of the coefficients of the inverse matrix, that are products of transport operators $\mathcal{B}_i^{\{f,b\}}$, as the transport of information between distant subdomains. As the two sweeps are independent from each other, they can be performed in parallel, as can be seen on the left diagram of Fig. 3. Because we do not need to know the exact nature of the transport operators, the strategy is exactly the same for Helmholtz and Maxwell problems.

The effect of the preconditioner on the spectrum of the preconditioned non-ideal operator $\mathcal{F}_{\tilde{\mathcal{D}}}\mathcal{F}_{\mathcal{D}}^{-1}$ is a strong clustering of the eigenvalues around $(1, 0)$, which ensures a good conditioning of the operator. That being so, the eigenvectors are now well distinct from each other, which enables fast convergence of the modified algorithm.

4 Parallelization of the Double Sweep

An important shortcoming of the double sweep preconditioner is its sequential nature, that destroys the scalability of the algorithm on parallel computers: assigning each subdomain to a separate CPU makes the preprocessing and the application of the iteration operator fully parallel, but these CPUs will remain idle during most of the application of the sweeps. An alternative strategy is to perform shorter sweeps over smaller groups of subdomains, independently of the other groups, by cutting the long sequence into smaller ones (Fig. 2). This method still enables the sharing of information over longer distances than a single domain, yet not over the whole domain as before. The advantage is of course that the sweeps over each group can be performed simultaneously, therefore partially restoring scalability. Consequently, one can expect a degradation of the preconditioner performance compared to the original version, since it approximates the inverse of the Schwarz operator less accurately. The timeline of subdomains solves reported on Fig. 3 highlights the improved level of parallelism when using 2 cuts (right) instead of none (left).

A similar preconditioning strategy can be followed when the domain is decomposed as in Fig. 1b: introducing (at least) one cut in the cyclic decomposition allows to use the double sweep preconditioner as is.

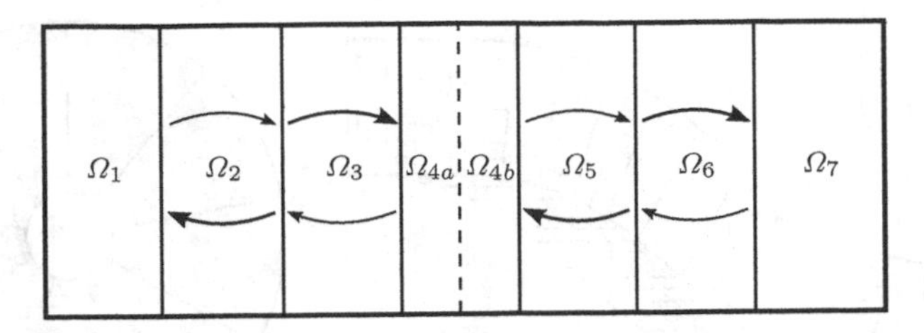

Fig. 2 Partial sweeps cover non-overlapping groups of domains, separated by the *dashed line*. The position of the cut inside the domain is not important as the first and last domains are not solved in our sweeps, as shown by the *arrows*

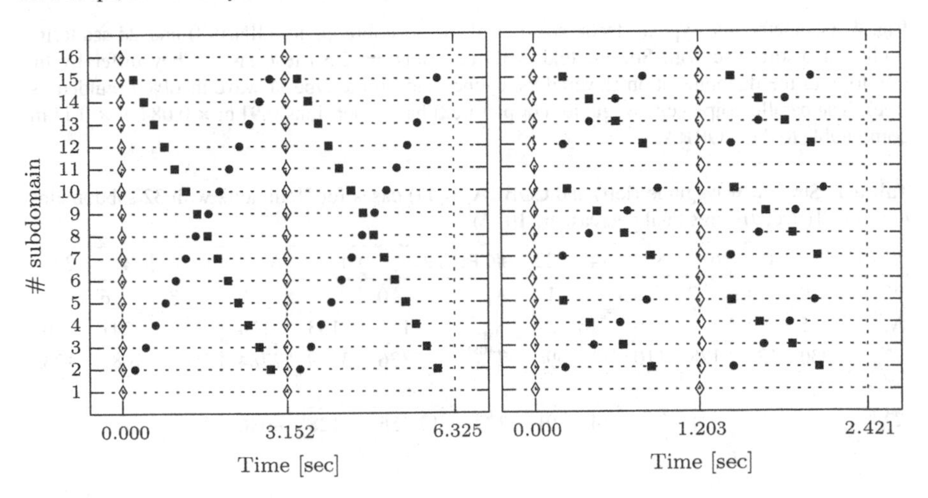

Fig. 3 Introducing 2 cuts in the double sweep preconditioner (*right*) enables parallel execution of the partial sweeps, reducing the application time of the preconditioner without cuts (*left*). The *white diamonds* indicate solves performed in the iteration operator; the *black circles* and *squares* indicate solves in the forward and backward sweeps, respectively. These time lines were obtained for the COBRA test case of Sect. 5, with 16 subdomains and cuts in subdomains 6 and 11

5 Numerical Results

We present results obtained on three different test geometries: a straight 3d (parallelepipedic) waveguide, a 3d S-shaped cavity (the COBRA benchmark defined by the JINA98 workgroup) and the open 2d scattering problem by a circular object (Fig. 4). The first two are solved using a layered decomposition while the third uses a cyclic decomposition. The COBRA is solved for both Helmholtz and Maxwell, while the other two are solved for Helmholtz only. Earlier work [1, 7, 8] has shown that without preconditioner, the iteration count for such problems typically grows linearly with the number of domains, and that with the use of the double sweep it becomes almost independent for layered decompositions, provided that the approximation of the DtN map is sufficiently accurate.

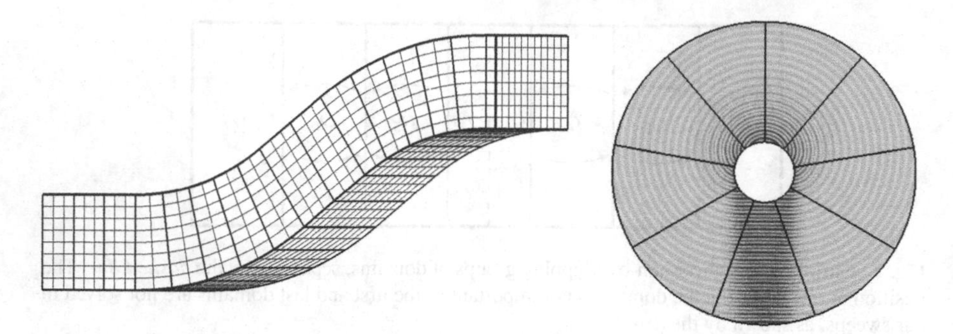

Fig. 4 Geometry and typical decomposition of the 3d cobra cavity (JINA98) and 2d scattering (unit sound-soft disc with Sommerfeld ABC at radius = 5 m) test cases. They differ by the topology of the decomposition (layered vs. cyclic) and by the type of wave involved (guided vs. free). The parallelepipedic waveguide (not pictured) has dimensions 0.91 m × 0.084 m × 0.11 m, comparable to the COBRA

Table 1 Straight waveguide (left) and COBRA (right) cases for Helmholtz with 32 subdomains, $k = 314.16$ (relative residual decrease by 10^{-4})

#CPU	2	4	6	8	14	22
N_c	0	1	2	3	6	10
$N_{\text{it}}^{(\text{ds})}$	5	6	8	10	16	24
$T_{\text{sol}}^{(\text{ds})}$	230	138	128	110	112	96
$N_{\text{it}}^{(\text{np})}$	62					
$T_{\text{sol}}^{(\text{np})}$	992	496	331	248	142	91

#CPU	2	4	6	8	14	22
N_c	0	1	2	3	6	10
$N_{\text{it}}^{(\text{ds})}$	116	153	174	188	241	308
$T_{\text{sol}}^{(\text{ds})}$	5336	3519	2784	2068	1687	1232
$N_{\text{it}}^{(\text{np})}$	766					
$T_{\text{sol}}^{(\text{np})}$	12,256	6128	4086	3064	1751	1115

Tables 1, 2, and 3 summarize the number of iterations required by each algorithm to converge to the prescribed tolerance, together with an estimation of the normalized times required for the completion of the algorithm. Provided that at least 2 CPUs are alloted per group of domains, the time required for the application of the standard Schwarz operator and the double sweep preconditioner with N_d subdomains, N_c cuts and C_{tot} CPUs (assumed evenly distributed between the groups of subdomains) are approximately given, in the case of a layered decomposition by:

$$T_{\text{Sch}} = \frac{N_d}{C_{\text{tot}}} T_p \quad \text{and} \quad T_{\text{sw}}(N_c) = \left\lceil \frac{N_d - N_c - 2}{N_c + 1} \right\rceil T_p,$$

with T_p the solution time for one subproblem (supposed identical for all subdomains). Note that T_{sw} would be doubled if only one CPU is available to perform the double sweep per group of domains. Slightly different estimations hold in the case of the cyclic decomposition. The total solution times for the unpreconditioned and double sweep algorithms are then $T_{\text{sol}}^{(np)} = T_{Sch} N_{it}^{(np)}$ and $T_{\text{sol}}^{(ds)}(N_c) = (T_{\text{Sch}} + T_{\text{sw}}(N_c)) N_{it}^{(ds)}$.

Table 2 COBRA test case for Maxwell with 32 subdomains, $k = 157.08$ (left) and $k = 314.16$ (right) (relative residual decrease by 10^{-4})

#CPU	2	4	6	8	14	22
N_c	0	1	2	3	6	10
$N_{\text{it}}^{(\text{ds})}$	21	34	48	62	104	160
$T_{\text{sol}}^{(\text{ds})}$	966	782	768	682	728	640
$N_{\text{it}}^{(\text{np})}$	448					
$T_{\text{sol}}^{(\text{np})}$	7168	3584	2390	1792	1024	652

#CPU	2	4	6	8	14	22
N_c	0	1	2	3	6	10
$N_{\text{it}}^{(\text{ds})}$	44	74	105	135	230	354
$T_{\text{sol}}^{(\text{ds})}$	2024	1702	1680	1485	1610	1416
$N_{\text{it}}^{(\text{np})}$	>1000					
$T_{\text{sol}}^{(\text{np})}$	>16,016	>8008	>5339	>4004	>2288	>1456

Table 3 Scattering test case for Helmholtz with 128 subdomains, k = 6.28 (left) and $k = 25.13$ (right) (relative residual decrease by 10^{-4})

#CPU	2	52	86	#CPU	2	52	86
N_c	1	26	43	N_c	1	26	43
$N_{\text{it}}^{(\text{ds})}$	24	27	31	$N_{\text{it}}^{(\text{ds})}$	20	29	37
$T_{\text{sol}}^{(\text{ds})}$	4584	189	124	$T_{\text{sol}}^{(\text{ds})}$	3820	203	148
$N_{\text{it}}^{(\text{np})}$	55			$N_{\text{it}}^{(\text{np})}$	85		
$T_{\text{sol}}^{(\text{np})}$	3520	136	82	$T_{\text{sol}}^{(\text{np})}$	5440	210	127

Tables 1, 2, and 3 show that in all cases the behaviour of the algorithm is similar. The preconditioner strongly reduces the number of iterations, and thus the number of overall linear system solves. Moreover, the parallel version of the preconditioner makes it also an appealing proposition with respect to the overall computational (wall-clock) time when the number of CPUs is smaller than the number of subdomains, especially in the high frequency regime. For example, in the challenging COBRA case for Maxwell, with 32 domains on 8 CPUs (3 cuts), with $k = 100\pi$, the preconditioned version requires $135\times(32+2\times(32-2-3)) = 11,610$ system solves instead of $>1000 \times 32$ and runs about 3 times faster than the standard algorithm.

6 Conclusion

We have presented a double sweep preconditioning strategy for the optimized Schwarz algorithm and a variant of it that performs the double sweeps in parallel on groups of subdomains, rather than over all subdomains. Numerical results highlight the potential of the approach for both Helmholtz and Maxwell in the high frequency regime.

Acknowledgements Work supported in part by the Belgian Science Policy (IAP P7/02). Computational resources provided by CÉCI, funded by F.R.S.-FNRS under Grant No. 2.5020.11.

References

1. Y. Boubendir, X. Antoine, C. Geuzaine, A quasi-optimal non-overlapping domain decomposition algorithm for the Helmholtz equation. J. Comput. Phys. **231**(2), 262–280 (2012). ISSN 0021-9991. doi: 10.1016/j.jcp.2011.08.007. http://dx.doi.org/10.1016/j.jcp.2011.08.007
2. M. El Bouajaji, X. Antoine, C. Geuzaine, Approximate local magnetic-to-electric surface operators for time-harmonic Maxwell's equations. J. Comput. Phys. **279**, 241–260 (2014). ISSN 0021-9991. doi: 10.1016/j.jcp.2014.09.011. http://dx.doi.org/10.1016/j.jcp.2014.09.011
3. B. Engquist, L. Ying, Sweeping preconditioner for the Helmholtz equation: moving perfectly matched layers. Multiscale Model. Simul. **9**(2), 686–710 (2011). ISSN 1540-3459. doi: 10.1137/100804644. http://dx.doi.org/10.1137/100804644

4. F. Nataf, Interface connections in domain decomposition methods, in *Modern Methods in Scientific Computing and Applications* (Montréal, QC, 2001). NATO Science Series II: Mathematics, Physics and Chemistry, vol. 75 (Kluwer Academic Publishers, Dordrecht, 2002), pp. 323–364
5. Z. Peng, V. Rawat, J.-F. Lee, One way domain decomposition method with second order transmission conditions for solving electromagnetic wave problems. J. Comput. Phys. **229**(4), 1181–1197 (2010). ISSN 0021-9991. doi: 10.1016/j.jcp.2009.10.024. http://dx.doi.org/10.1016/j.jcp.2009.10.024
6. C.C. Stolk, A rapidly converging domain decomposition method for the Helmholtz equation. J. Comput. Phys. **241**(0), 240–252 (2013). ISSN 0021-9991. doi: http://dx.doi.org/10.1016/j.jcp.2013.01.039
7. A. Vion, C. Geuzaine, Double sweep preconditioner for optimized Schwarz methods applied to the Helmholtz problem. J. Comput. Phys. **266**, 171–190 (2014). ISSN 0021-9991. doi: 10.1016/j.jcp.2014.02.015. http://dx.doi.org/10.1016/j.jcp.2014.02.015
8. A. Vion, R. Bélanger-Rioux, L. Demanet, C. Geuzaine, A DDM double sweep preconditioner for the Helmholtz equation with matrix probing of the DtN map, in *Mathematical and Numerical Aspects of Wave Propagation WAVES 2013*, June 2013

A Multiscale Domain Decomposition Method for Flow and Transport Problems

Victor Ginting and Bradley McCaskill

1 Background

It has been widely recognized that one of the major challenges in the simulation of flow and transport problems is finding the numerical solution of the pressure equation [2]. Typically we seek to find the pressure solution, p, such that

$$\begin{cases} -\nabla \cdot (k\nabla p) = f & \text{in } \Omega \\ p = p_{\mathrm{D}} & \text{on } \Gamma_{\mathrm{D}} \\ -k\nabla p \cdot \boldsymbol{n} = g_{\mathrm{N}} & \text{on } \Gamma_{\mathrm{N}}, \end{cases} \tag{1}$$

where k represents the positive elliptic coefficient, and f represents a forcing function. The associated Dirichlet, and Neumann boundary conditions are given by p_{D} and g_{N} respectively. The corresponding variational formulation is to find p, with $(p - p_{\mathrm{D}}) \in V$, that satisfies

$$a(p, v) = \ell(v) \ \forall v \in V, \tag{2}$$

where $V = \{v \in H^1(\Omega) : v = 0 \text{ on } \Gamma_{\mathrm{D}}\}$, and

$$a(p, v) = \int_{\Omega} k\nabla p \cdot \nabla v \, \mathrm{d}\boldsymbol{x}, \text{ and } \ell(v) = \int_{\Omega} f v \, \mathrm{d}\boldsymbol{x} - \int_{\Gamma_{\mathrm{N}}} g_{\mathrm{N}} v \, \mathrm{d}\boldsymbol{l}. \tag{3}$$

V. Ginting • B. McCaskill (✉)
Department of Mathematics, University of Wyoming, Laramie, WY 82071, USA
e-mail: vginting@uwyo.edu; bmccaski@uwyo.edu

T. Dickopf et al. (eds.), *Domain Decomposition Methods in Science and Engineering XXII*, Lecture Notes in Computational Science and Engineering 104, DOI 10.1007/978-3-319-18827-0_23

Assuming sufficient regularity of the data, the Lax-Milgram Theorem guarantees a unique solution of (3). The chief difficulty in approximating p stems from the heterogeneity of k, which can occur in multiple scales. This heterogeneity directly dictates the degree of the mesh resolution on which the approximate solution is found. In turn this results in a very high dimensional algebraic system which must be solved.

With the advances of parallel computing, domain decomposition as a general framework has gained a stronger role in efficiently finding accurate solutions to problems of this type. In this paper, we propose an iterative procedure for solving (3) that relies on a one-time preprocessing step where a set of independent subdomain problems are computed. This preprocessing step yields a set of so called multiscale basis functions with which the global solution is represented. Continuity of the solution at the interface is established by imposing Robin Transmission conditions on each subdomain interface. This imposition is accomplished in an iterative manner. In the following section we describe an iterative domain decomposition technique that serves as the backbone for our proposed procedure.

2 A Domain Decomposition with Robin Transmission Conditions

We decompose the domain Ω into a set of non-overlapping subdomains $\{\Omega_j\}_{j=1}^{N_{sd}}$, and construct a local problem on each subdomain. For ease of notation we define $\mathcal{N}_m$ to be the set of indices for subdomains that share an edge with Ω_m. For example, $\mathcal{N}_m = \{l, r, b, t\}$ is associated with the subdomain presented in Fig. 1. Each local problem is supplied with a boundary condition that allows for the continuity of the solution and its flux at each subdomain interface to be maintained. In particular, for each $n \in \mathcal{N}_m$, we impose

$$-k\nabla p_m \cdot \boldsymbol{e}_m - \gamma_{mn} p_m = g_{mn} \text{ on } \Gamma_{mn}, \tag{4}$$

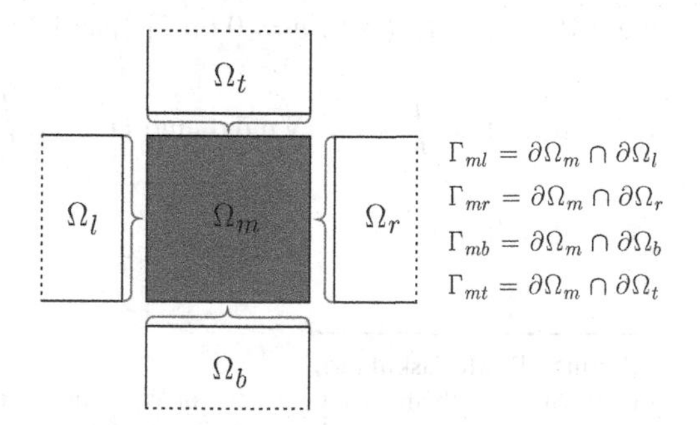

Fig. 1 An internal rectangular subdomain Ω_m, and its neighbouring subdomains $\{\Omega_n\}$

where γ_{mn} is a positive constant, $\boldsymbol{e}_m$ represents the exterior unit normal respective to subdomain Ω_m, $\Gamma_{mn} = \partial\Omega_m \cap \partial\Omega_n$, and the value of g_{mn} comes from the neighbouring subdomain Ω_n, expressed as

$$g_{mn} = k\nabla p_n \cdot \boldsymbol{e}_n - \gamma_{nm} p_n \text{ on } \Gamma_{mn}. \tag{5}$$

To establish the iterative procedure, it is assumed that g_{mn} is known, namely, from the previous iteration level. The resulting local variational formulation is to find $p_m \in H^1(\Omega_m)$ such that

$$a_m(p_m, w) + \sum_{n\in\mathcal{N}_m} b_{mn}(p_m, w) = \ell_m(w) + \sum_{n\in\mathcal{N}_m} r_{mn}(w) \; \forall w \in H^1(\Omega_m), \tag{6}$$

where

$$\begin{aligned} a_m(v, w) &= \int_{\Omega_m} k\nabla v \cdot \nabla w \, \mathrm{d}\boldsymbol{x}, \quad b_{mn}(v, w) = \int_{\Gamma_{mn}} \gamma_{mn} v w \, \mathrm{d}\boldsymbol{l}, \\ \ell_m(w) &= \int_{\Omega_m} f w \, \mathrm{d}\boldsymbol{x}, \quad r_{mn}(w) = -\int_{\Gamma_{mn}} g_{mn} w \, \mathrm{d}\boldsymbol{l}. \end{aligned} \tag{7}$$

We use (6) to develop an iterative technique for approximating (2) whose algorithm is presented in Algorithm 1. At the practical level, this iteration does converge to the true solution [3–5], but it requires that we calculate a new local solution on every subdomain for each step of the iteration. Depending on the initial guess, and the number of subdomains, this can greatly exceed the computational time required to solve the problem with traditional methods.

Algorithm 1

Set initial guess for $\{p_m^{(0)}\}_{m=0}^{N_{sd}}$
for $it = 1$ until convergence **do**
  Construct $g_{mn}^{(it-1)}$, for all $n \in \mathcal{N}_m$, $m = 1, \ldots, N_{sd}$
  Solve (6) to get $p_m^{(it)}$ for $m = 1, \ldots, N_{sd}$
end for

3 Incorporation of Multiscale Basis Functions

To alleviate the aforementioned burden of calculation, our strategy is to form a preprocessing step aimed at collecting the finescale heterogeneity information on each subdomain. This information is stored in the so called subdomain multiscale basis functions. Here our motivation is to find an approximate solution to (6) that is expressed as a linear combination of these multiscale basis functions.

For each $n \in \mathcal{N}_m$ we decompose Γ_{mn} into a union of nonoverlapping segments $\{I^i_{mn}\}_{i=1}^{k_{mn}}$, and denote by $\{z^i_{mn}\}_{i=0}^{k_{mn}}$ the associated vertices. For simplicity we assume uniformity of these segments as they relate to neighbouring subdomains.

We set

$$\tilde{g}_{mn} = \sum_{i=0}^{k_{mn}} g_{mn}(z^i_{mn})\phi^i_{mn}, \tag{8}$$

where $\{\phi^i_{mn}\}_{i=0}^{k_{mn}}$ is the usual "hat" nodal basis function corresponding to $\{z^i_{mn}\}_{i=0}^{k_{mn}}$ expressed in a parametric form associated with Γ_{mn}. Examples of these "hat" functions are presented in Fig. 2. For our approximate solution we construct a new variational formulation. Find $\tilde{p}_m \in H^1(\Omega_m)$, satisfying

$$a_m(\tilde{p}_m, w) + \sum_{n \in \mathcal{N}_m} b_{mn}(\tilde{p}_m, w) = \ell_m(w) + \sum_{n \in \mathcal{N}_m} \tilde{r}_{mn}(w) \ \forall w \in H^1(\Omega_m), \tag{9}$$

where

$$\tilde{r}_{mn}(w) = -\sum_{i=0}^{k_{mn}} g_{mn}(z^i_{mn}) \int_{\Gamma_{mn}} \phi^i_{mn} w \, \mathrm{d}l. \tag{10}$$

With this formulation, the same iteration as in Algorithm 1 could have been done. It is worth noting that there are two sources of error that are committed when conducting the iteration based on (9). The first error is shared by the iteration using (6), namely resulting from the fact that in practice only a finite number of iterations are used. The second error stems from the replacement of g_{mn} by $\tilde{g}_{mn}$, i.e., an approximation error. There is a nonlinear interaction between these two error components. We expect, however, that at the asymptotic level of systematic

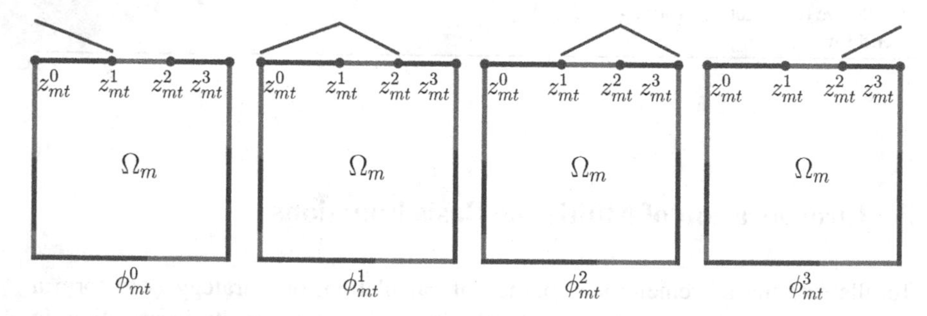

Fig. 2 Example of "hat" functions associated with an edge Γ_{mt}. On edges $\Gamma_{mb}, \Gamma_{mr}, \Gamma_{ml}$ the value of these functions is zero

refinement ($k_{mn} \to \infty$ and convergence is reached), $\tilde{p} = \sum_{m=1}^{N_{sd}} \tilde{p}_m 1_{\Omega_m}$ should converge to p. What is more important is that the formulation (9) provides a building block for the construction of subdomain multiscale basis functions as part of the preprocessing step. The approximate solution on each subdomain is then represented using these basis functions.

To each Γ_{mn} we associate a set of multiscale basis functions $\{\psi_{mn}^i\}_{i=0}^{k_{mn}}$, where $\psi_{mn}^i \in H^1(\Omega_m)$ is the solution to the variational formulation

$$a_m(\psi_{mn}^i, w) + \sum_{n \in \mathcal{N}_m} b_{mn}(\psi_{mn}^i, w) = r_{mn}^i(w) \ \forall w \in H^1(\Omega_m). \tag{11}$$

The linear functional in (11) is given by

$$r_{mn}^i(w) = -\int_{\Gamma_{mn}} \phi_{mn}^i w \, \mathrm{d}l. \tag{12}$$

When $f \neq 0$, we compute an extra multiscale basis function $\hat{\psi}_m \in H^1(\Omega_m)$ that satisfies

$$a_m(\hat{\psi}_m, w) + \sum_{n \in \mathcal{N}_m} b_{mn}(\hat{\psi}_m, w) = \ell_m(w) \ \forall w \in H^1(\Omega_m). \tag{13}$$

On each subdomain we set $V_m = \mathrm{span}\{\psi_{mn}^i, i = 1, \cdots, k_{mn}, n \in \mathcal{N}_m, \hat{\psi}_m\}$ and seek $\tilde{p}_m \in V_m$, i.e.,

$$\tilde{p}_m = \hat{\psi}_m + \sum_{n \in \mathcal{N}_m} \sum_{i=0}^{k_{mn}} \alpha_{mn}^i \psi_{mn}^i \approx p_m. \tag{14}$$

An approximation of the global solution is now recaptured by determining the values of each $\boldsymbol{\alpha}_{mn} = [\alpha_{mn}^0, \ldots, \alpha_{mn}^{k_{mn}}]$ that induce the continuity condition outlined in (4), and imposed in (6). Thus, for each ϕ_j associated with an interface edge Γ_{mn} we require

$$\sum_{j=0}^{k_{mn}} \alpha_{mn}^j \int_{\Gamma_{mn}} \phi_{mn}^j \phi_{mn}^i \, \mathrm{d}l = \int_{\Gamma_{mn}} \tilde{g}_{mn} \phi_{mn}^i \, \mathrm{d}l, \ \forall \, i = 0, \cdots, k_{mn} + 1. \tag{15}$$

Here we note that this continuity condition yields a linear system governing $\boldsymbol{\alpha}_{mn}$. The associated matrix is tridiagonal and of dimension $k_{mn} + 1$. At a practical level the calculation of $\tilde{g}_{mn}$ can be performed using $\tilde{p}_n$, the multiscale representation of p_n. The iterative procedure presented in Algorithm 1, is now modified to be an iteration governing each $\boldsymbol{\alpha}_{mn}$. The modified iteration is presented in Algorithm 2.

Algorithm 2

Calculate $\{\psi_{mn}^{i}\}_{i=1}^{k_{mn}}$, for all $n \in \mathcal{N}_m$, $m = 1, \ldots, N_{\text{sd}}$
Set initial guess $,_{mn}^{(0)}$, for all $n \in \mathcal{N}_m$, $m = 1, \ldots, N_{\text{sd}}$
for $it = 1$ until convergence **do**
  Calculate $\tilde{g}_{mn}^{(it-1)}$, for all $n \in \mathcal{N}_m$, $m = 1, \ldots, N_{\text{sd}}$
  Solve for $\boldsymbol{\alpha}_{mn}^{(it)}$, satisfying (15) for all $n \in \mathcal{N}_m$, $m = 1, \ldots, N_{\text{sd}}$
  Set $\tilde{p}_m^{(it)}$, $m = 1, \ldots, N_{\text{sd}}$
end for

4 Numerical Examples

In this section we present two studies. First, we present a convergence study of our method when applied to a problem with a known solution. We then apply our method to a single phase flow model, and compare the results with traditional methods. To calculate the multiscale basis functions, we use the traditional continuous Galerkin FEM to solve (11) and (13).

4.1 Convergence Study

We first explore the behaviour of the approximate solution in terms of the discretization parameters. In particular, it is interesting to study the interaction between the subdomain and the segment configuration. The former determines how many local problems are created while the latter determines the number of multiscale basis functions to represent a particular local problem. The subdomain size is denoted by H and the segment size is denoted by $\tilde{h}$. The interplay between the two parameters reflects a choice of balancing the accuracy and efficiency of the approximate solution.

For this purpose, we choose a problem with a known solution. The problem is posed in $(0, 1)^2$ with a zero Neumann condition on $x_2 = 0, 1$ and a Dirichlet condition on $x_1 = 0, 1$. We assume that $f = 0$ and $k(\boldsymbol{x}) = a_1(x_1)a_2(x_2)$, where a_1 and a_2 are

$$a_1(x_1) = [0.25 - 0.999(x_1 - {x_1}^2)\sin(11.2\pi x_1)]^{-1}$$

$$a_2(x_2) = [0.25 - 0.999(x_2 - {x_2}^2)\cos(5.2\pi x_2)]^{-1},$$

yielding $k_{\max}/k_{\min} \approx 2 \times 10^4$. Comparison of the effect that various segment and subdomain configurations have on the accuracy of the resulting approximate solution are presented in Table 1. In this example, the finescale solution is found on a grid of 256×256 rectangles (i.e., $h = 1/256$) and this finescale mesh is the base for the configuration of $\tilde{h}$ after H is determined.

Table 1 Comparison of the L_2-norm, H_1-norm of the approximate solution found using various segment lengths, and subdomain sizes

	L_2		H_1	
$\tilde{h}$	$H = 0.250$	$H = 0.125$	$H = 0.250$	$H = 0.125$
h	0.000217	0.000217	0.02070	0.02070
$2h$	0.000217	0.000218	0.02074	0.02079
$4h$	0.000223	0.000241	0.02137	0.02201
$8h$	0.000252	0.000485	0.03068	0.03559
$16h$	0.001159	0.002538	0.08002	0.10412

We note that when $\tilde{h} = h$ the resulting solution has exactly the same error estimates as solutions found with the traditional Galerkin FEM on the fine mesh. For a fixed H, the errors of the proposed method stay relatively unchanged as $\tilde{h}$ is increased. This can be taken as a potential advantage of the proposed method; lower dimensional V_m can still produce a relatively accurate numerical solution. This of course reduces the number of multiscale basis functions which must be calculated. Furthermore, results in Table 1 indicate that the errors seem to be less sensitive to H. Traditionally, it has been established (see for example [4, 5]) that an increase in subdomain interfaces (i.e., the finer H is) can potentially increase the number of iterations needed for convergence to a desired tolerance. Thus, this indication suggests that only fewer subdomains (i.e., less interfaces) are required to extract accurate solutions, which results in fewer iterations for convergence. On the other hand, this can potentially mean that the multiscale basis functions are governed by a higher dimensional problem, which correlates to a higher computational load in the preprocessing step. In the end, a problem dependent choice of H and $\tilde{h}$ leads to an optimized scenario of calculation.

4.2 *Applications to Single Phase Flow*

The mathematical model is

$$\partial_t S + \boldsymbol{u} \cdot \nabla \lambda(S) = 0, \text{ with specified I.C. + B.C. and } \boldsymbol{u} = -k(\boldsymbol{x})\nabla p,$$

where S represents the saturation and $\nabla \cdot \boldsymbol{u} = 0$, i.e., elliptic PDE governing the pressure p. The boundary condition for p is the same as the one in the previous subsection. The model is a typical one way coupling equation where the pressure is first solved and the velocity $\boldsymbol{u}$ is constructed from it, which in turn is used as an input in solving the transport equation. We applied the postprocessing technique [1] to recover a locally conservative flux $\boldsymbol{u}\cdot\boldsymbol{n}$ on the finescale grid. Then a first order upwinding scheme is used to determine the time evolved saturation value. The elliptic coefficient that is used for this model is shown in Fig. 3. This elliptic coefficient

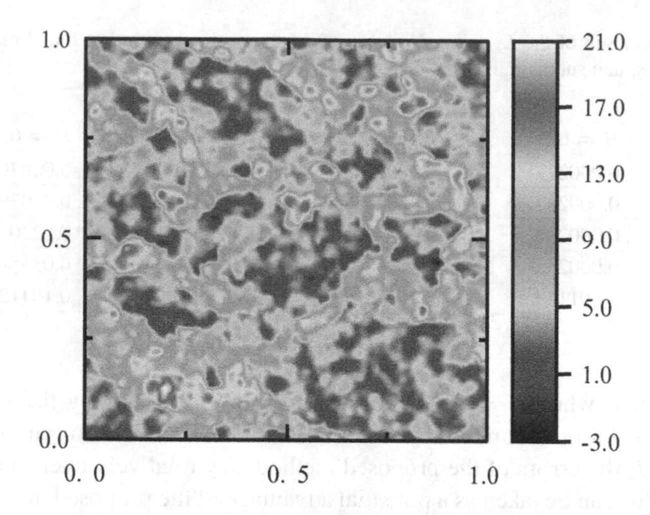

Fig. 3 A logarithmic plot of $k(x)$ used in single phase flow simulation

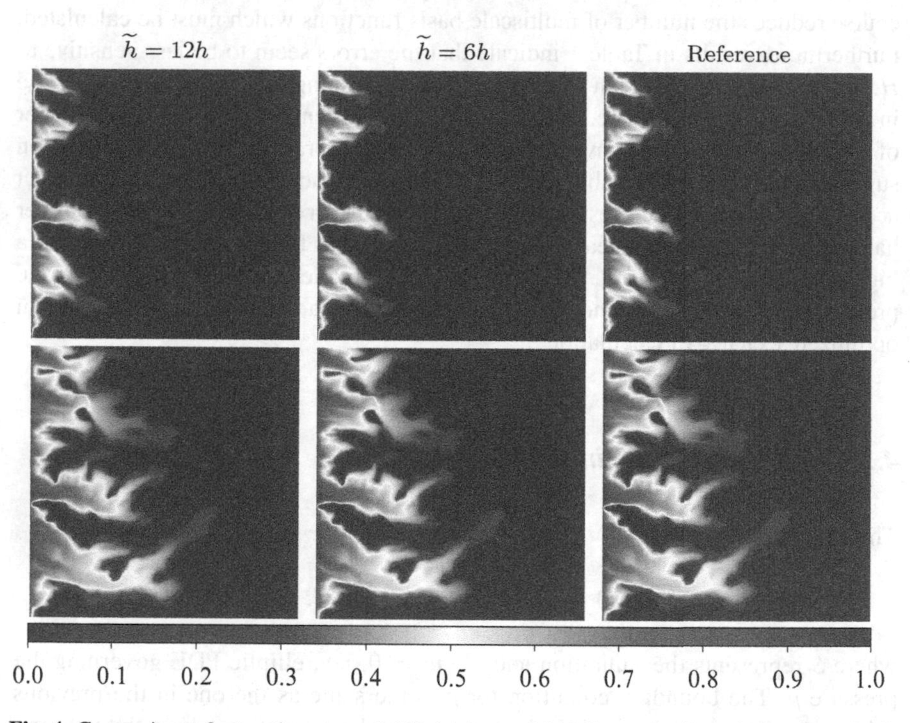

Fig. 4 Comparison of saturation at $t = 0.003$ (*top*), $t = 0.009$ (*bottom*), all results use $H = 0.25$

is posed on 240 × 240 grid and has a ratio $k_{\max}/k_{\min} \approx 6.4 \times 10^4$. In Fig. 4 we show a visual comparison of the saturation solution at various time steps, for our method and traditional methods. In Fig. 5 a plot of the relative difference between

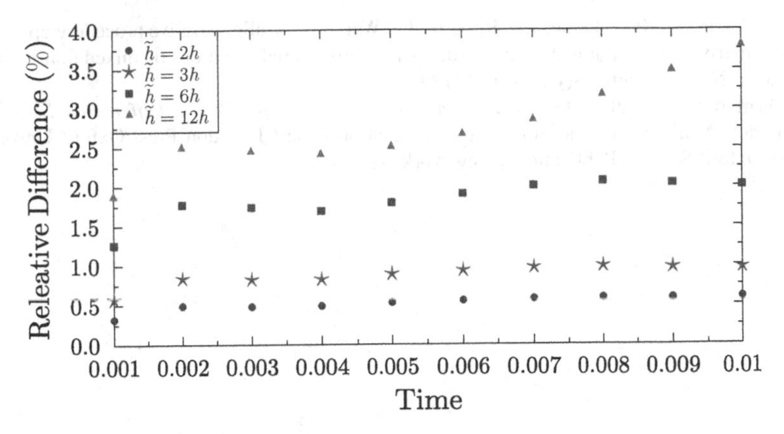

Fig. 5 Comparison of the L^2-error of the saturation difference between our method and traditional Galerkin FEM, for various choices of $\tilde{h}$. In all cases, $H = 0.25$

the solution found with the proposed method and the solution found with traditional methods is presented.

5 Conclusion

We have proposed an iterative multiscale domain decomposition method with certain favourable properties. By incorporating the multiscale basis functions into an iterative domain decomposition procedure we have reduced its computational demand. The numerical examples suggest that our method is capable of recapturing accurate solutions that are comparable to those found with traditional methods. In the future we will extend the capability of the method to multiphase flow models. We are also interested in conducting a rigorous convergence analysis of the proposed method.

References

1. L. Bush, V. Ginting, On the application of the continuous Galerkin finite element method for conservation problems. SIAM J. Sci. Comput. **35**(6), A2953–A2975 (2013)
2. Z. Chen, G. Huan, Y. Ma, *Computational Methods for Multiphase Flows in Porous Media*. Computational Science & Engineering (Society for Industrial and Applied Mathematics (SIAM), Philadelphia, 2006)
3. W. Chen, M. Gunzburger, F. Hua, X. Wang, A parallel Robin-Robin domain decomposition method for the Stokes-Darcy system. SIAM J. Numer. Anal. **49**(3), 1064–1084 (2011)

4. J. Douglas Jr., P.J. Paes-Leme, J.E. Roberts, J.P. Wang, A parallel iterative procedure applicable to the approximate solution of second order partial differential equations by mixed finite element methods. Numer. Math. **65**(1), 95–108 (1993)
5. A. Quarteroni, A. Valli, *Domain Decomposition Methods for Partial Differential Equations*. Numerical Mathematics and Scientific Computation (The Clarendon Press/Oxford University Press/Oxford Science Publications, New York, 1999)

An Optimized Schwarz Algorithm for a Discontinuous Galerkin Method

Soheil Hajian

1 Introduction

It has been shown in [4] that block Jacobi iterates of a discretization obtained from hybridizable discontinuous Galerkin methods (HDG) can be viewed as non-overlapping Schwarz methods with Robin transmission condition. The Robin parameter is exactly the penalty parameter μ of the HDG method. There is a stability constraint on the penalty parameter and the usual choice of μ results in slow convergence of the Schwarz method. In this paper we show how to overcome this problem without changing μ. To fix ideas, we consider the model problem

$$
\begin{aligned}
(\eta - \Delta)u &= f \text{ in } \Omega \subset \mathbb{R}^2, \\
u &= 0 \text{ on } \partial\Omega,
\end{aligned}
\tag{1}
$$

where Ω is a bounded polygon, $0 \leq \eta \leq \eta_0$ and $f \in L^2(\Omega)$. We then consider a hybridizable interior penalty (IPH) discretization and develop domain decomposition algorithms to solve the resulting linear system efficiently. For the sake of brevity we consider the two-subdomain case in this paper.

Our paper is organized as follows: in Sect. 2 we describe the IPH method. We introduce a Schur complement system for the IPH discretization and review some of its properties in Sect. 3. In Sect. 4 we introduce two iterative methods for the Schur complement and present their convergence behavior. Finally we present some numerical experiments in Sect. 5.

S. Hajian (✉)
Université de Genève, 2-4 rue du Lièvre, CP 64, CH-1211 Genève 4, Switzerland
e-mail: Soheil.Hajian@unige.ch

T. Dickopf et al. (eds.), *Domain Decomposition Methods in Science and Engineering XXII*, Lecture Notes in Computational Science and Engineering 104, DOI 10.1007/978-3-319-18827-0_24

2 Hybridizable Interior Penalty Method

IPH was first introduced in [2] and later studied as a member of the class of hybridizable DG methods in [1]. We first establish some notation and then define the IPH method in two different but equivalent forms. Let $\mathcal{T}_h = \{K\}$ be a shape-regular and quasi-uniform triangulation of the domain Ω. Let h_K be the diameter of an element of the triangulation and $h = \max_{K \in \mathcal{T}_h} h_K$. If e is an edge of an element, we denote by h_e the length of that edge.

We denote by $\mathcal{E}^0$ the set of interior edges, by $\mathcal{E}^\partial$ the set of boundary edges and all edges by $\mathcal{E} := \mathcal{E}^\partial \cup \mathcal{E}^0$. We introduce the broken Sobolev space $\mathrm{H}^l(\mathcal{T}_h) := \prod_{K \in \mathcal{T}_h} \mathrm{H}^l(K)$ where $\mathrm{H}^l(K)$ is the Sobolev space in $K \in \mathcal{T}_h$ and l is a positive integer. Therefore the element boundary traces of functions in $\mathrm{H}^l(\mathcal{T}_h)$ belong to $\mathrm{T}(\mathcal{E}) = \prod_{K \in \mathcal{T}_h} \mathrm{L}^2(\partial K)$, where $q \in \mathrm{T}(\mathcal{E})$ can be double-valued on $\mathcal{E}^0$, and is single-valued on $\mathcal{E}^\partial$.

We also define two trace operators: let $q \in \mathrm{T}(\mathcal{E})$ and $\boldsymbol{\sigma} \in [\mathrm{T}(\mathcal{E})]^2$. On $e = \partial K_1 \cap \partial K_2$ we then define average $\{\cdot\}$ and jump $[\![\cdot]\!]$ operators by

$$\begin{aligned} \{q\} &= \tfrac{1}{2}(q_1 + q_2), & [\![q]\!] &= q_1\,\mathbf{n}_1 + q_2\,\mathbf{n}_2, \\ \{\boldsymbol{\sigma}\} &= \tfrac{1}{2}(\boldsymbol{\sigma}_1 + \boldsymbol{\sigma}_2), & [\![\boldsymbol{\sigma}]\!] &= \boldsymbol{\sigma}_1 \cdot \mathbf{n}_1 + \boldsymbol{\sigma}_2 \cdot \mathbf{n}_2, \end{aligned} \tag{2}$$

where $\mathbf{n}_i$ is the unit outward normal of K_i on e, $q_i := q|_{\partial K_i \cap e}$ and $\boldsymbol{\sigma}_i := \boldsymbol{\sigma}|_{\partial K_i \cap e}$. On $\partial\Omega$ we set the average and jump operators to be $\{\boldsymbol{\sigma}\} = \boldsymbol{\sigma}$ and $[\![q]\!] = q\,\mathbf{n}$ respectively. Note that we do not need to specify $\{q\}$ and $[\![\boldsymbol{\sigma}]\!]$ on $e \in \mathcal{E}^\partial$ because it is not needed in the formulation.

We define a finite-dimensional broken space on $\mathcal{T}_h$ for the discrete approximation $V_h := \{v \in \mathrm{L}^2(\Omega) : v|_K \in \mathbb{P}_1(K), \forall K \in \mathcal{T}_h\}$, where $\mathbb{P}_k(K)$ is the space of polynomials of degree $\leq k$ in the simplex $K \in \mathcal{T}_h$.

For the sake of simplicity we denote the volume and surface integrals by $(a, b)_K = \int_K a\,b$ for $K \in \mathcal{T}_h$ and $\langle a, b\rangle_e = \int_e a\,b$ for $e \in \mathcal{E}$.

We now present IPH method in *primal* and *hybridizable* form. Let $u, v \in \mathrm{H}^2(\mathcal{T}_h)$, then the IPH bilinear form of the model problem (1) is defined as

$$\begin{aligned} a(u, v) := \eta(u, v)_{\mathcal{T}_h} &+ (\nabla u, \nabla v)_{\mathcal{T}_h} - \langle \{\nabla u\}, [\![v]\!]\rangle_{\mathcal{E}} - \langle \{\nabla v\}, [\![u]\!]\rangle_{\mathcal{E}} \\ &+ \langle \tfrac{\mu}{2}[\![u]\!], [\![v]\!]\rangle_{\mathcal{E}} - \left\langle \tfrac{1}{2\mu}[\![\nabla u]\!], [\![\nabla v]\!]\right\rangle_{\mathcal{E}^0}, \end{aligned} \tag{3}$$

where $\mu \in \mathrm{L}^2(\mathcal{E})$ is the penalty parameter. For a constant $\alpha > 0$ we set $\mu|_e = \alpha h_e^{-1}$. We should mention that this scaling cannot be weakened due to stability constraints. The IPH bilinear form is different from the classical IP only in the last term, i.e. the last term is not present in the IP bilinear form. For a formal derivation of the bilinear form (3) see [6], Section 1.2.2.

The IPH bilinear form is coercive over V_h provided $\alpha > 0$ and sufficiently large, that is we can show

$$a(v, v) \geq c\|v\|_{\mathrm{DG}}^2, \quad \forall v \in V_h,$$

where $0 < c < 1$ is a constant independent of h. Here the energy norm is defined as

$$\|v\|_{\mathrm{DG}}^2 := \eta\|v\|_{\mathcal{T}_h}^2 + \|\nabla v\|_{\mathcal{T}_h}^2 + \sum_{e\in\mathcal{E}} \mu_e \|[\![v]\!]\|_e^2, \quad \forall v \in V_h. \tag{4}$$

The discrete problem can be stated as: find $u_h \in V_h$ such that

$$a(u_h, v) = (f, v)_{\mathcal{T}_h}, \quad \forall v \in V_h. \tag{5}$$

Since $a(\cdot,\cdot)$ is coercive over V_h, we can conclude that there exists a unique discrete solution. Furthermore we can show that IPH has optimal approximation properties provided $\alpha > 0$ is sufficiently large; see [6].

We show now that one can write IPH in a hybridized form, such that static condensation with respect to a single-valued unknown is possible. This is not the case for most DG methods, e.g. classical IP. Let us decompose the domain into two non-overlapping subdomains Ω_1 and Ω_2. Denoting the interface by $\Gamma := \overline{\Omega}_1 \cap \overline{\Omega}_2$, we assume $\Gamma \subset \mathcal{E}^0$, i.e. the cut does not go through any element of the triangulation. This results in a natural partitioning of $\mathcal{T}_h$ into $\mathcal{T}_1$ and $\mathcal{T}_2$; for an example see Fig. 1 (right).

This naturally allows us to introduce local spaces on Ω_1 and Ω_2 by

$$V_{h,i} := \{v \in \mathrm{L}^2(\Omega_i) : v|_{K\in\mathcal{T}_i} \in \mathbb{P}_1(K)\}, \text{ for } i = 1, 2. \tag{6}$$

Note that this domain decomposition setting implies $V_h = V_{h,1} \oplus V_{h,2}$. We define on the interface the space of broken single-valued functions by

$$\Lambda_h := \{\varphi \in \mathrm{L}^2(\Gamma) : \varphi|_{e\in\Gamma} \in \mathbb{P}_1(e)\}. \tag{7}$$

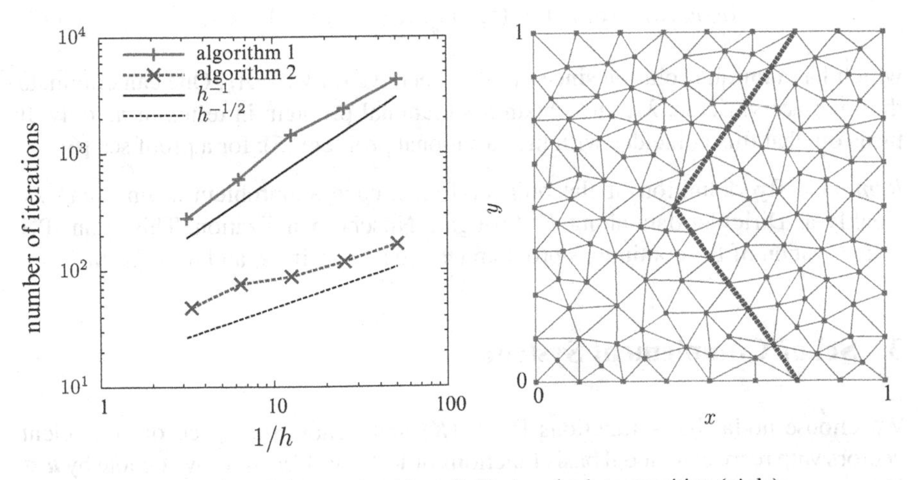

Fig. 1 Convergence of the Schwarz algorithms (*left*), domain decomposition (*right*)

For the sake of simplicity we denote the restriction of $v \in V_h$ on $V_{h,i}$ by v_i. Observe that the trace of $v_i \in V_{h,i}$ on Γ belongs to Λ_h.

Let $(u, \lambda), (v, \varphi) \in V_h \times \Lambda_h$ and consider the symmetric bilinear form

$$\tilde{a}((u, \lambda), (v, \varphi)) := \tilde{a}_\Gamma(\lambda, \varphi) + \sum_{i=1}^{2} \tilde{a}_i(u_i, v_i) + \tilde{a}_{i\Gamma}(v_i, \lambda) + \tilde{a}_{i\Gamma}(u_i, \varphi), \tag{8}$$

where $\tilde{a}_\Gamma(\lambda, \varphi) := 2\langle \mu\, \lambda, \varphi\rangle_\Gamma$, $\tilde{a}_{i\Gamma}(v_i, \varphi) := \left\langle \frac{\partial v_i}{\partial \mathbf{n}_i} - \mu v_i, \varphi \right\rangle_\Gamma$ and

$$\begin{aligned}\tilde{a}_i(u_i, v_i) := {} & \eta(u_i, v_i)_{\mathcal{T}_i} + (\nabla u_i, \nabla v_i)_{\mathcal{T}_i} - (\{\nabla u_i\}, [\![v_i]\!])_{\mathcal{E}_i^0} - (\{\nabla v_i\}, [\![u_i]\!])_{\mathcal{E}_i^0} \\ & + \left\langle \tfrac{\mu}{2}[\![u_i]\!], [\![v_i]\!]\right\rangle_{\mathcal{E}_i^0} - \left\langle \tfrac{1}{2\mu}[\![\nabla u_i]\!], [\![\nabla v_i]\!]\right\rangle_{\mathcal{E}_i^0} \\ & - \left\langle \tfrac{\partial u_i}{\partial \mathbf{n}_i}, v_i \right\rangle_{\partial\Omega_i} - \left\langle \tfrac{\partial v_i}{\partial \mathbf{n}_i}, u_i \right\rangle_{\partial\Omega_i} + \langle \mu\, u_i, v_i\rangle_{\partial\Omega_i}.\end{aligned} \tag{9}$$

The bilinear form $\tilde{a}(\cdot,\cdot)$ is also coercive at the discrete level if $\alpha > 0$, independent of h and sufficiently large:

$$\tilde{a}((v, \varphi), (v, \varphi)) \geq c\, \|(v, \varphi)\|_{\text{HDG}}^2 \quad \forall (v, \varphi) \in V_h \times \Lambda_h, \tag{10}$$

where c is independent of h and the HDG-norm is defined by

$$\|(v, \varphi)\|_{\text{HDG}}^2 := \sum_{i=1}^{2} \eta \|v_i\|_{\mathcal{T}_i}^2 + \|\nabla v_i\|_{\mathcal{T}_i}^2 + \mu \|[\![v_i]\!]\|_{\mathcal{E}_i \setminus \Gamma}^2 + \mu \|v_i - \varphi\|_\Gamma^2. \tag{11}$$

Consider the following discrete problem: find $(u_h, \lambda_h) \in V_h \times \Lambda_h$ such that

$$\tilde{a}((u_h, \lambda_h), (v, \varphi)) = (f, v)_{\mathcal{T}_h}, \quad \forall (v, \varphi) \in V_h \times \Lambda_h, \tag{12}$$

which has a unique solution since $\tilde{a}(\cdot,\cdot)$ is coercive on $V_h \times \Lambda_h$. One can eliminate the interface variable, λ_h, and obtain a variational problem in terms of u_h only. It turns out that this coincides with the variational problem (5); for a proof see [6].

Remark 1 By definition of the bilinear forms, each subproblem is imposing λ_h weakly as Dirichlet data along Γ through a Nitsche penalization. This is an IPH discretization of the continuous problem $(\eta - \Delta)w = f$ in Ω_i and $w = \lambda_h$ on Γ.

3 Schur Complement System

We choose nodal basis functions for $\mathbb{P}_1(K)$ and denote the space of coefficient vectors with respect to nodal basis functions of V_h by V. If $u_h \in V_h$ we denote by $\boldsymbol{u} \in V$ its corresponding coefficient vector. The variational problem in (5) is equivalent to the linear system $A\boldsymbol{u} = \boldsymbol{f}$. A is called the stiffness matrix. We decompose $\boldsymbol{u}$ into

$\{\boldsymbol{u}_1, \boldsymbol{u}_2\}$ where $\boldsymbol{u}_i$ corresponds to coefficients of nodal basis functions in Ω_i. Then we can arrange the entries of A and rewrite the linear system as

$$\begin{bmatrix} A_1 & A_{21} \\ A_{21} & A_2 \end{bmatrix} \begin{pmatrix} \boldsymbol{u}_1 \\ \boldsymbol{u}_2 \end{pmatrix} = \begin{pmatrix} \boldsymbol{f}_1 \\ \boldsymbol{f}_2 \end{pmatrix}. \tag{13}$$

We use nodal basis functions for Λ_h and denote by $\boldsymbol{\lambda}$ the corresponding coefficient vector for $\lambda_h \in \Lambda_h$. Then the variational form (12) can be written as

$$\begin{bmatrix} \tilde{A}_1 & & \tilde{A}_{1\Gamma} \\ & \tilde{A}_2 & \tilde{A}_{2\Gamma} \\ \tilde{A}_{\Gamma 1} & \tilde{A}_{\Gamma 2} & \tilde{A}_{\Gamma} \end{bmatrix} \begin{pmatrix} \boldsymbol{u}_1 \\ \boldsymbol{u}_2 \\ \boldsymbol{\lambda} \end{pmatrix} = \begin{pmatrix} \boldsymbol{f}_1 \\ \boldsymbol{f}_2 \\ \boldsymbol{0} \end{pmatrix}, \tag{14}$$

where $\tilde{A}_{\Gamma i} = \tilde{A}_{i\Gamma}^{\top}$. Note that the advantage of this formulation over (13) is that subdomains are communicating through $\boldsymbol{\lambda}$ and we can form a Schur complement for a *single-valued* function, λ_h. To do so we define $\tilde{B}_i := \tilde{A}_{\Gamma i}\tilde{A}_i^{-1}\tilde{A}_{i\Gamma}$ and $\boldsymbol{g}_\Gamma := \sum_{i=1}^{2} \tilde{A}_{\Gamma i}\tilde{A}_i^{-1}\boldsymbol{f}_i$. Then the Schur complement system reads

$$\tilde{S}_\Gamma \boldsymbol{\lambda} := \Big(\tilde{A}_\Gamma - \sum_{i=1}^{2} \tilde{B}_i\Big)\boldsymbol{\lambda} = \boldsymbol{g}_\Gamma. \tag{15}$$

We define $u_i := \mathcal{H}_i(\lambda_h)$ to be the discrete harmonic extension of $\lambda_h \in \Lambda_h$ into subdomain Ω_i, i.e. u_i satisfies $\tilde{A}_i \boldsymbol{u}_i + \tilde{A}_{i\Gamma}\boldsymbol{\lambda} = 0$; that is we impose λ_h as Dirichlet data (weakly) on Γ and solve inside Ω_i. The following result shows that an application of $\tilde{B}_i\boldsymbol{\lambda}$ can be viewed as finding the harmonic extension, $u_i := \mathcal{H}_i(\lambda_h)$, and then evaluating a "Robin-like trace" on the interface.

Proposition 1 *Let $\lambda_h \in \Lambda_h$ and define its harmonic extension by $u_i := \mathcal{H}_i(\lambda_h)$. Then $\boldsymbol{\varphi}^\top \tilde{B}_i \boldsymbol{\lambda} = \left\langle \mu u_i - \frac{\partial u_i}{\partial \mathbf{n}_i}, \varphi \right\rangle_\Gamma$ for all $\varphi \in \Lambda_h$.*

Proof Let $u_i := \mathcal{H}_i(\lambda_h)$. Then by definition of $\tilde{B}_i$ and $\tilde{a}_{i\Gamma}(\cdot,\cdot)$ we have

$$\boldsymbol{\varphi}^\top \tilde{B}_i \boldsymbol{\lambda} = \boldsymbol{\varphi}^\top \tilde{A}_{\Gamma i}\tilde{A}_i^{-1}\tilde{A}_{i\Gamma}\boldsymbol{\lambda} = -\boldsymbol{\varphi}^\top \tilde{A}_{\Gamma i}\boldsymbol{u}_i = \left\langle \mu u_i - \frac{\partial u_i}{\partial \mathbf{n}_i}, \varphi \right\rangle_\Gamma, \text{ for all } \varphi \in \Lambda_h,$$

which completes the proof, since $\tilde{A}_{\Gamma i} = \tilde{A}_{i\Gamma}^\top$. □

One can estimate the eigenvalues of $\{\tilde{B}_i\}$. They are useful in proving convergence of Schwarz methods later on. The proofs are technical and beyond the scope of this short paper. They can be found in [5].

Lemma 1 *$\tilde{B}_i$ is s.p.d. and there exists $\alpha > 0$, sufficiently large, such that*

$$c_B\, \mu\, \|\varphi\|_\Gamma^2 \le \boldsymbol{\varphi}^\top \tilde{B}_i \boldsymbol{\varphi} \le \Big(1 - C_B \frac{h}{\alpha}\Big)\, \mu\, \|\varphi\|_\Gamma^2,$$

where $0 < c_B < 1$ and $C_B > 0$. Both constants are independent of h. Moreover $\tilde{A}_\Gamma - 2\tilde{B}_i$ is s.p.d. for $i = 1, 2$.

4 Schwarz Methods for the Schur Complement System

One approach in solving the linear system (13) is to use the block Jacobi method:

$$M\mathbf{u}^{(n+1)} = N\mathbf{u}^{(n)} + \mathbf{f}, \quad M = \begin{bmatrix} A_1 & \\ & A_2 \end{bmatrix}, N = M - A. \tag{16}$$

Instead in this section we derive two Schwarz algorithms to solve the Schur complement system where the first one is equivalent to (16) and slow while the second one has much faster convergence.

Let us relax the constraint that λ_h is single-valued. Let $\lambda_{h,1}, \lambda_{h,2} \in \Lambda_h$. Assume $\lambda_{h,2}$ is known; that is we know $u_2 \in V_{h,2}$. Then we can split the Schur complement system (15) and solve for $\lambda_{h,1}$, through $(\tilde{A}_\Gamma - \tilde{B}_1)\boldsymbol{\lambda}_1 = \tilde{B}_2\boldsymbol{\lambda}_2 + \mathbf{g}_\Gamma$. Lemma 1 ensures that $(\tilde{A}_\Gamma - \tilde{B}_1)$ is invertible and we can obtain $\lambda_{h,1}$. This suggests an iterative method to find λ_h.

Algorithm 1 (Block Jacobi) *Let $\lambda_{h,1}^{(0)}, \lambda_{h,2}^{(0)} \in \Lambda_h$ be two arbitrary initial guesses. Then for $n = 1, 2, \ldots$ solve (17) for $\{\lambda_{h,i}^{(n)}\}$.*

$$\begin{aligned} (\tilde{A}_\Gamma - \tilde{B}_1)\boldsymbol{\lambda}_1^{(n)} &= \tilde{B}_2\boldsymbol{\lambda}_2^{(n-1)} + \mathbf{g}_\Gamma, \\ (\tilde{A}_\Gamma - \tilde{B}_2)\boldsymbol{\lambda}_2^{(n)} &= \tilde{B}_1\boldsymbol{\lambda}_1^{(n-1)} + \mathbf{g}_\Gamma. \end{aligned} \tag{17}$$

Note that at convergence we have $\tilde{A}_\Gamma(\boldsymbol{\lambda}_1 - \boldsymbol{\lambda}_2) = 0$ which implies $\boldsymbol{\lambda}_1 = \boldsymbol{\lambda}_2 = \tilde{S}_\Gamma^{-1}\mathbf{g}_\Gamma$ since $\tilde{A}_\Gamma$ is s.p.d. We show now that Algorithm 1 is equivalent to the block Jacobi iteration (16). It suffices to prove this for $f = 0$ ($\mathbf{g}_\Gamma = 0$).

Proposition 2 *Let $\lambda_{h,1}^{(0)}, \lambda_{h,2}^{(0)}$ be two random initial guesses. Set the initial guess of the block Jacobi iteration (16) to be $u_i^{(0)} = \mathcal{H}_i(\lambda_{h,i}^{(0)})$. Then $u_i^{(n)} = \mathcal{H}_i(\lambda_{h,i}^{(n)})$ for all $n > 0$, i.e. both methods produce the same iterates.*

Proof We start by subdomain Ω_1. Set $w_{h,i}^{(n)} = \mathcal{H}_i(\lambda_{h,i}^{(n)})$. By Proposition 1, we have $\boldsymbol{\varphi}^\top \tilde{B}_i\boldsymbol{\lambda}_i^{(n)} = \left\langle \mu w_{h,i}^{(n)} - \partial_{\mathbf{n}_i} w_{h,i}^{(n)}, \varphi \right\rangle_\Gamma$ for all $\varphi \in \Lambda_h$. Then the first equation in iteration (17) implies $\lambda_{h,1}^{(n)} = (\frac{1}{2} - \frac{1}{2\mu}\partial_{\mathbf{n}_1})w_{h,1}^{(n)} + (\frac{1}{2} - \frac{1}{2\mu}\partial_{\mathbf{n}_2})w_{h,2}^{(n-1)}$. Recall that $w_{h,1}^{(n)}$ is the harmonic extension, hence it satisfies $\tilde{a}_i(w_{h,1}^{(n)}, v_1) + \tilde{a}_{i\Gamma}(v_1, \lambda_{h,1}^{(n)}) = 0$ for all $v_1 \in V_{h,1}$. Now we substitute $\lambda_{h,1}^{(n)}$ in terms of $w_{h,1}^{(n)}$ and $w_{h,2}^{(n-1)}$. We arrive at exactly the first row of block Jacobi (16), i.e. $A_1\mathbf{w}_1^{(n)} + A_{12}\mathbf{w}_2^{(n-1)} = 0$. The proof for Ω_2 is similar. □

Convergence of the block Jacobi (16) or equivalently Algorithm 1 can be proved with the contraction factor $\rho_h \leq 1 - O(h)$. For details we refer the reader to [5].

The slow convergence of this algorithm is due to the fact that the transmission condition is of Robin type with Robin parameter $\mu = \alpha h^{-1}$; see [4]. According to optimized Schwarz theory the best choice is $\mu = O(h^{-1/2})$; see [3]. We would like to emphasize that for IPH, one cannot change the scaling of μ because of coercivity and approximation property constraints.

The remedy is to split the Schur complement differently. We know from Lemma 1 that $\tilde{A}_\Gamma - 2B_i$ is s.p.d. Therefore assuming $\boldsymbol{\lambda}_2$ is known we can multiply the Schur complement by $(1+\hat{p})$ where $\hat{p}$ is a constant and solve for $\boldsymbol{\lambda}_1$ such that

$$(\tilde{A}_\Gamma - (1+\hat{p})B_1)\boldsymbol{\lambda}_1 = -(\hat{p}\tilde{A}_\Gamma - (1+\hat{p})\tilde{B}_2)\boldsymbol{\lambda}_2 + (1+\hat{p})\mathbf{g}_\Gamma.$$

If $0 \le \hat{p} < 1$ then the left hand side is still s.p.d. We use $\hat{p}$ to obtain a fast converging solver. Note that for $\hat{p} = 0$ we have Algorithm 1.

Algorithm 2 (Optimized Schwarz) *Let $\lambda_{h,1}^{(0)}, \lambda_{h,2}^{(0)} \in \Lambda_h$ be two arbitrary initial guesses and $0 \le \hat{p} < 1$ be a constant. Then for $n = 1, 2, \ldots$ solve (18) for $\{\lambda_{h,i}^{(n)}\}$.*

$$\begin{aligned}(\tilde{A}_\Gamma - (1+\hat{p})\tilde{B}_1)\boldsymbol{\lambda}_1^{(n)} &= -(\hat{p}\tilde{A}_\Gamma - (1+\hat{p})\tilde{B}_2)\boldsymbol{\lambda}_2^{(n-1)} + (1+\hat{p})\mathbf{g}_\Gamma, \\ (\tilde{A}_\Gamma - (1+\hat{p})\tilde{B}_2)\boldsymbol{\lambda}_2^{(n)} &= -(\hat{p}\tilde{A}_\Gamma - (1+\hat{p})\tilde{B}_1)\boldsymbol{\lambda}_1^{(n-1)} + (1+\hat{p})\mathbf{g}_\Gamma.\end{aligned} \tag{18}$$

At convergence we have $(1-\hat{p})\tilde{A}_\Gamma(\boldsymbol{\lambda}_1 - \boldsymbol{\lambda}_2) = 0$ which implies $\boldsymbol{\lambda}_1 = \boldsymbol{\lambda}_2 = \tilde{S}_\Gamma^{-1}\mathbf{g}_\Gamma$ if $\hat{p} \neq 1$. An application of Proposition 1 and Remark 1 shows Algorithm 2 has a modified Robin parameter which we summarize in the next proposition.

Proposition 3 *Algorithm 2 is the discrete version of the non-overlapping optimized Schwarz method*

$$\begin{aligned}\mathcal{L}u_1^{(n+1)} &= f \quad \text{in } \Omega_1, & \mathcal{L}u_2^{(n+1)} &= f \quad \text{in } \Omega_2, \\ \mathcal{B}_1 u_1^{(n+1)} &= \mathcal{B}_1 u_2^{(n)} \text{ on } \Gamma, & \mathcal{B}_2 u_2^{(n+1)} &= \mathcal{B}_2 u_1^{(n)} \text{ on } \Gamma,\end{aligned}$$

where $\mathcal{L} := (\eta - \Delta)$, $\mathcal{B}_i := \hat{\mu} + \partial_{\mathbf{n}_i}$ and Robin parameter $\hat{\mu} := \frac{1-\hat{p}}{1+\hat{p}}\mu$.

A heuristic approach in obtaining optimal $\hat{p}$ is to set the *modified* Robin parameter to $\hat{\mu} = O(h^{-1/2})$ and solve for $\hat{p}$. This results in $\hat{p} = \frac{1-\sqrt{h}}{1+\sqrt{h}} < 1$. A rigorous proof at the discrete level in [5] gives same scaling and with this choice of $\hat{p}$ the contraction factor of Algorithm 2 is bounded by $\rho_h \le 1 - O(\sqrt{h})$.

5 Numerical Experiments

We consider $(\eta - \Delta)u = f$ in Ω and $u = 0$ on $\partial\Omega$ where we set $\eta = 1$, $\Omega = (0,1)^2$ and f such that the exact solution is $u(x,y) = \sin(\pi x)\sin(2\pi x + \frac{\pi}{4})\sin(2\pi y)$ in Ω. We set the penalty parameter to $\mu = 10h_e^{-1}$. We choose a non-straight interface as in Fig. 1 (right). We measure the number of iterations necessary to reduce the error $\|u_h - u_h^{(n)}\|_0$ to 10^{-10} on a sequence of (quasi-uniform) unstructured meshes while the interface is *fixed*. As for the initial guess, we set DOFs of the initial guess using a random number generator; in `Matlab` given by `rand(N_DOF)`.

In Fig. 1 (left) we observe for Algorithm 1 that the number of iterations grows like $O(h^{-1})$. This is equivalent to $\rho_h \leq 1 - O(h)$. For Algorithm 2 with the optimal value of $\hat{p}$ we see that it grows like $O(h^{-1/2})$ hence $\rho_h \leq 1 - O(\sqrt{h})$. This is in agreement with the results in Sect. 4. For more extensive numerical experiments see [5].

6 Conclusions

It has been shown in [4] that for some DG methods one can obtain a fast converging solver by just *modifying* the penalty parameter while for some other it is not possible, e.g. IPH. We showed that it is possible to define an iterative method, Algorithm 2, for IPH such that we obtain fast convergence without changing the penalty parameter. We are now studying a multi-subdomain version of Algorithm 2 and the case of higher polynomial degree, $k > 1$.

Acknowledgement The author thanks Martin J. Gander for his useful comments.

References

1. B. Cockburn, J. Gopalakrishnan, R. Lazarov, Unified hybridization of discontinuous Galerkin, mixed, and continuous Galerkin methods for second order elliptic problems. SIAM J. Numer. Anal. **47**(2), 1319–1365 (2009)
2. R.E. Ewing, J. Wang, Y. Yang, A stabilized discontinuous finite element method for elliptic problems. Numer. Linear Algebra Appl. **10**(1–2), 83–104 (2003)
3. M.J. Gander, Optimized Schwarz methods. SIAM J. Numer. Anal. **44**(2), 699–731 (2006)
4. M.J. Gander, S. Hajian, Block Jacobi for discontinuous Galerkin discretizations: no ordinary Schwarz methods, in *Domain Decomposition Methods in Science and Engineering XXI*, ed. by J. Erhel, M.J. Gander, L. Halpern, G. Pichot, T. Sassi, O. Widlund. Lecture Notes in Computational Science and Engineering (Springer, Berlin, 2013)
5. M.J. Gander, S. Hajian, Analysis of Schwarz methods for a hybridizable discontinuous Galerkin discretization. SIAM J. Numer. Anal. **53**(1), 573–597 (2015)
6. C. Lehrenfeld, Hybrid discontinuous Galerkin methods for incompressible flow problems. Master's thesis, RWTH Aachen, 2010

On Full Multigrid Schemes for Isogeometric Analysis

Clemens Hofreither and Walter Zulehner

1 Introduction

Isogeometric analysis (IGA), a numerical technique for the solution of partial differential equations first proposed in [11], has attracted considerable research attention in recent years. The use of spline spaces both for representation of the geometry and for approximation of the solution affords the method several very interesting features, such as the possibility to use exactly the geometry generated by CAD systems, refinement without further communication with the CAD system, the possibility of using high-continuity trial functions, the use of high-degree spaces with comparatively few degrees of freedom, and more. We refer to [1, 11] as well as the monograph [8] and the references therein for details on this method.

The efficient solution of the discretized systems arising in isogeometric analysis has been the topic of several publications, among these, [2, 3, 5, 7, 9, 12]. In the present paper, we investigate geometric full multigrid methods for IGA. It is known [9] that geometric multigrid solvers for IGA possess h-independent convergence rates for V-cycle iteration using standard smoothers. Our aim is to study more closely the performance of the full multigrid (FMG) iteration strategy, especially in dependence of the spline degree.

C. Hofreither (✉) • W. Zulehner
Institute of Computational Mathematics, Johannes Kepler University Linz, Altenbergerstr. 69, 4040 Linz, Austria
e-mail: chofreither@numa.uni-linz.ac.at; zulehner@numa.uni-linz.ac.at

T. Dickopf et al. (eds.), *Domain Decomposition Methods in Science and Engineering XXII*, Lecture Notes in Computational Science and Engineering 104, DOI 10.1007/978-3-319-18827-0_25

2 Isogeometric Analysis

We construct, in every direction $i = 1, \dots, d$, a B-spline space of degree p_i over an open knot vector which spans the parameter interval $(0, 1)$. *Open* means that the first and last knots are repeated $p_i + 1$ times. We restrict ourselves to maximum continuity, i.e., all knots in the interior are simple. For the definition of B-splines, see, e.g., [8, 14, 15]. Taking the tensor product of the B-splines bases over all directions i, we obtain a tensor product basis $\{B_j : (0,1)^d \to \mathbb{R}_0^+\}_j$. To each of its basis functions B_j, we associate a control point (coefficient) $C_j \in \mathbb{R}^d$ in such a way that we obtain an invertible geometry mapping $F = \sum_j C_j B_j : (0,1)^d \to \Omega$, where $\Omega \subset \mathbb{R}^d$ is the computational domain. The isogeometric basis functions on Ω are given by $B_j \circ F^{-1} : \Omega \to \mathbb{R}_0^+$, and their span is the isogeometric trial space on Ω.

In practice, NURBS, i.e., rational versions of the B-spline basis functions, are commonly used to represent the geometry. In this paper, we however restrict ourselves to the case of B-splines for the sake of simplicity.

In the following, let $\mathcal{V}_h \subset H_0^1(\Omega)$ denote a tensor product spline space over Ω as constructed above. An isogeometric method for the Poisson equation with Dirichlet boundary conditions is given by the discrete variational problem: find $u_h \in \mathcal{V}_h$ such that, for all $v_h \in \mathcal{V}_h$,

$$\int_\Omega \nabla u \cdot \nabla v \, dx =: a(u_h, v_h) = \langle F, v_h \rangle := \int_\Omega f v_h \, dx - a(\tilde{g}, v),$$

where $\tilde{g} \in H^1(\Omega)$ is a suitable extension of the Dirichlet data g. Here, $u_h + \tilde{g}$ is the approximation to the solution of the boundary value problem.

Essential boundary conditions require some care in isogeometric methods. In our setting, we construct an approximation g_h to g which lies in the spline space. Due to the use of open knot vectors, the degrees of freedom (DoFs) can be cleanly separated into boundary DoFs and interior DoFs. The values for the boundary DoFs of g_h are determined by solving a $(d-1)$-dimensional Lagrange interpolation problem on each face of the patch Ω, where the Gréville points of the spline basis are chosen as interpolation points. The interior DoFs of g_h are set to zero. In the variational setting, this corresponds to solving a problem with the approximate right-hand side

$$\langle F_h, v \rangle = \int_\Omega f v \, dx - a(g_h, v). \tag{1}$$

On the topic of essential boundary conditions in isogeometric analysis, we also refer to [6, 13, 17].

3 Geometric Multigrid Methods for IGA

In the following, we outline very briefly the construction of a geometric multigrid scheme for IGA. We refer to the multigrid literature [4, 10, 16] for further details.

Starting from a coarse isogeometric mesh, inserting a new knot at the midpoint of every non-empty knot span creates a "fine" spline space with a halved mesh size which contains all functions of the original "coarse" spline space, yielding the isogeometric analogue of uniform h-refinement.

Let $\hat{\mathcal{V}}_0$ denote a coarse parametric spline space over $(0, 1)^d$ which is rich enough to represent the geometry Ω exactly. With repeated uniform refinement steps, we obtain a sequence of h-refined spline spaces $\hat{\mathcal{V}}_1, \hat{\mathcal{V}}_2, \ldots$ The push-forward to the geometry yields isogeometric spline spaces $\mathcal{V}_0, \mathcal{V}_1, \mathcal{V}_2, \ldots$

Let $\mathcal{V}_H \subset \mathcal{V}_h$ denote two successive spline spaces with the canonical embedding $P : \mathcal{V}_H \to \mathcal{V}_h$. One step of the two-grid iteration process is given by a pre-smoothing step, the coarse-grid correction, and a post-smoothing step. Given a starting value $u_0 \in \mathcal{V}_h$, the next iterate u_1 is thus obtained from

$$\begin{aligned} u^{(1)} &:= u_0 + M^{-1}(f_h - A_h u_0), \\ u^{(2)} &:= u^{(1)} + P A_H^{-1} P^\top (f_h - A_h u^{(1)}), \\ u_1 := u^{(3)} &:= u^{(2)} + M^{-\top}(f_h - A_h u^{(2)}). \end{aligned}$$

Here, M is a suitable smoother for the fine-space stiffness matrix A_h. Common choices are the Richardson smoother (with M being a scalar multiple of identity), the damped Jacobi smoother (M being a scaled diagonal of A_h), and the Gauss-Seidel smoother (M being the lower triangular part of A_h). A multigrid scheme is obtained by considering a hierarchy of nested spline spaces and replacing the exact inverse A_H^{-1} in the above procedure recursively with the same procedure applied on the next coarser space, until $\mathcal{V}_0$ is reached, where an exact solver is used.

We set up a Poisson model problem, $-\Delta u = f$, with pure Dirichlet boundary conditions on the d-dimensional unit interval $\Omega = (0, 1)^d$. We choose tensor product B-spline basis functions defined on equidistant knot vectors with constant spline degrees $p_1 = \ldots = p_d = p$ and maximum continuity. The geometry mapping F is chosen as identity. The right-hand side f and the boundary conditions are chosen according to the prescribed analytical solution $u(x) = \prod_{i=1}^d \sin(\pi(x_i + 0.5))$.

As a comparison point, we test the V-cycle iteration numbers. For this, we choose a random starting vector u_0 and perform V-cycle iteration until the initial residual is reduced by a factor of 10^{-8} in the Euclidean norm. The resulting iteration numbers are shown in Table 1. We point out that very similar numbers have been obtained in [9]. In higher dimensions, in particular for $d = 3$, the number of iterations sees a dramatic increase as the spline degree is raised.

Table 1 V-cycle iteration numbers for the model Poisson problem

d	N	p				d	N	p				d	N	p			
		1	2	3	4			1	2	3	4			1	2	3	4
	~500	11	9	7	10		~4 k	9	12	37	140		~6 k	9	37	249	1935
1	~4.1 k	11	8	7	9	2	~66 k	9	11	37	127	3	~40 k	9	38	240	1682
	~262 k	11	8	7	9		~1.05 m	9	11	36	125		~290 k	9	38	236	1564

Columns, left to right: space dimension d, number of unknowns N, V-cycle iteration numbers for $p = 1$–4

4 Full Multigrid for IGA

We set up a full multigrid (FMG) method in the usual way. That is, we start from the exact coarse-grid solution $u_0 = A_0^{-1} f_0 \in \mathcal{V}_0$ and transfer it to the next higher level by means of a full interpolation operator, $u_1 = I_0^1(u_0)$. Here the solution is corrected by one multigrid V-cycle with a suitable coarse-space right-hand side f_i, and the result is again interpolated to the next higher level by means of I_1^2. This procedure is continued until the finest space $\mathcal{V}_\ell$ is reached, where one final V-cycle is applied. We found that two issues related to the treatment of Dirichlet boundary conditions need attention.

First, we need a sequence of full interpolation operators $I_i^{i+1} : \mathcal{V}_i \to \mathcal{V}_{i+1}$ which transfer solutions, as opposed to mere corrections, to the next finer level while maintaining a high order of accuracy. Dirichlet boundary conditions must be carefully taken into account here. Recall that the approximation to the solution of the boundary value problem on level i is given by $u_i + g_i$, where g_i is a spline function approximating the Dirichlet boundary data having non-zero coefficients only on the boundary DoFs, whereas u_i vanishes on the boundary DoFs since they were eliminated from the linear system. Prolonging both contributions separately, we see that $P_i^{i+1} u_i \in \mathcal{V}_{i+1}$ still vanishes on the boundary DoFs. On the other hand, the representation of g_i in $\mathcal{V}_{i+1}$ has non-zero contributions in some interior DoFs close to the boundary. This situation is illustrated in the 1D setting in Fig. 1. Therefore, the proper choice for the full interpolation operator is $I_i^{i+1}(u_i) := P_i^{i+1} u_i + \mathring{P}_i^{i+1} g_i$, where by $\mathring{P}_i^{i+1}$ we mean the operator which prolongs the boundary function and discards the boundary DoFs, keeping only the contributions to the interior DoFs.

The second issue is related to the choice of the coarse-space right-hand sides f_i, $i = 1, \ldots, \ell - 1$. The seemingly natural choice $f_i = (P_i^{i+1})^\top f_{i+1}$ does not take into account that the right-hand side vector f_ℓ stems from the approximated linear functional $\langle F_h, \cdot \rangle$ given in (1), where we have chosen a fine-grid spline approximation g_h for the Dirichlet data. This approximation by necessity depends on the mesh level: the fine-grid Dirichlet functions must have better approximation properties, but cannot be represented on coarser grids. We thus found it necessary to assemble f_i on every level separately.

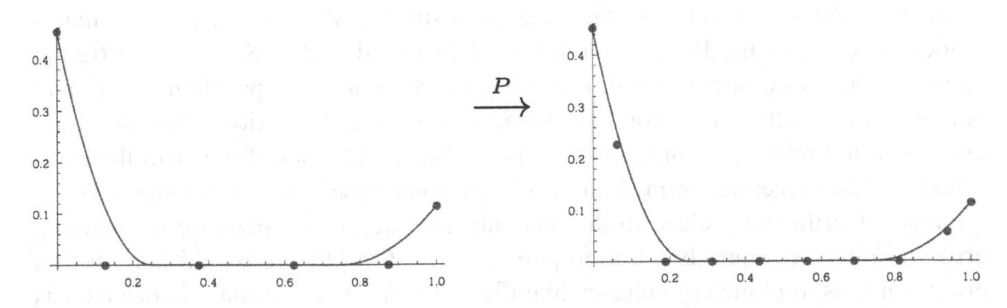

Fig. 1 Prolongation of boundary functions creates non-zero contributions to interior DoFs

Table 2 Errors after one full multigrid cycle in 1D

$p=1$			$p=2$		
N	L_2-error	Ratio	N	L_2-error	Ratio
33	$4.944858 \cdot 10^{-4}$	4.76	34	$3.722578 \cdot 10^{-6}$	8.09
65	$1.034253 \cdot 10^{-4}$	4.78	66	$4.646054 \cdot 10^{-7}$	8.01
129	$2.231043 \cdot 10^{-5}$	4.64	130	$5.808685 \cdot 10^{-8}$	8.00
257	$5.090763 \cdot 10^{-6}$	4.38	258	$7.263190 \cdot 10^{-9}$	8.00
513	$1.218115 \cdot 10^{-6}$	4.18	514	$9.081029 \cdot 10^{-10}$	8.00
$p=3$			$p=4$		
N	L_2-error	Ratio	N	L_2-error	Ratio
35	$6.536758 \cdot 10^{-8}$	16.71	36	$4.984064 \cdot 10^{-9}$	34.42
67	$3.966554 \cdot 10^{-9}$	16.48	68	$1.527129 \cdot 10^{-10}$	32.64
131	$2.439422 \cdot 10^{-10}$	16.26	132	$4.754326 \cdot 10^{-12}$	32.12
259	$1.512164 \cdot 10^{-11}$	16.13	260	$1.750396 \cdot 10^{-13}$	27.16
515	$9.883776 \cdot 10^{-13}$	15.30	516	$9.712425 \cdot 10^{-14}$	1.80

Table 3 Errors after one full multigrid cycle in 2D

$p=1$			$p=2$		
N	L_2-error	Ratio	N	L_2-error	Ratio
4225	$1.76004 \cdot 10^{-4}$	4.20	4356	$4.8427 \cdot 10^{-7}$	8.05
16,641	$4.28888 \cdot 10^{-5}$	4.10	16,900	$6.0407 \cdot 10^{-8}$	8.02
66,049	$1.05903 \cdot 10^{-5}$	4.05	66,564	$7.5446 \cdot 10^{-9}$	8.01
263,169	$2.63839 \cdot 10^{-6}$	4.01	264,196	$9.4271 \cdot 10^{-10}$	8.00
1,050,625	$6.59801 \cdot 10^{-7}$	4.00	1,052,676	$1.1782 \cdot 10^{-10}$	8.00
$p=3$			$p=4$		
N	L_2-error	Ratio	N	L_2-error	Ratio
4489	$6.8025 \cdot 10^{-9}$	17.47	4624	$1.2380 \cdot 10^{-9}$	28.05
17,161	$3.8527 \cdot 10^{-10}$	17.66	17,424	$6.1208 \cdot 10^{-11}$	20.23
67,081	$2.2387 \cdot 10^{-11}$	17.21	67,600	$3.6698 \cdot 10^{-12}$	16.68
265,225	$1.3527 \cdot 10^{-12}$	16.55	266,256	$2.4967 \cdot 10^{-13}$	14.70
1,054,729	$2.6654 \cdot 10^{-13}$	5.07			

With these issues taken care of, we apply a single FMG cycle using the Gauss-Seidel smoother to the Poisson model problem introduced in Sect. 3 for different values of the space dimension d, the spline degree p and the problem size N and compute the resulting L_2-error with respect to the exact solution. The errors are presented in Tables 2, 3, and 4 for the 1D, 2D and 3D cases along with the error ratio between successive refinement levels. (In some cases, the errors stagnate once a threshold sufficiently close to the machine accuracy is reached due to rounding errors.) From the approximation properties derived in [1], we would hope for an error which asymptotically behaves like $\mathcal{O}(h^{p+1})$. We observe that this behavior is achieved using a single FMG cycle for all tested spline degrees up to 4 in the 1D

Table 4 Errors after one full multigrid cycle in 3D

$p = 1$			$p = 2$		
N	L_2-error	Ratio	N	L_2-error	Ratio
125	$7.2738 \cdot 10^{-2}$	–	216	$3.0617 \cdot 10^{-3}$	–
729	$1.2829 \cdot 10^{-2}$	5.67	1000	$2.0613 \cdot 10^{-4}$	14.85
4913	$3.2797 \cdot 10^{-3}$	3.91	5832	$2.4836 \cdot 10^{-5}$	8.30
35,937	$7.9084 \cdot 10^{-4}$	4.15	39,304	$3.5042 \cdot 10^{-6}$	7.09
274,625	$1.9521 \cdot 10^{-4}$	4.05	287,496	$4.3712 \cdot 10^{-7}$	8.02
$p = 3$			$p = 4$		
N	L_2-error	Ratio	N	L_2-error	Ratio
343	$4.5383 \cdot 10^{-4}$	–	512	$1.1096 \cdot 10^{-4}$	–
1331	$5.4518 \cdot 10^{-5}$	8.32	1728	$2.5938 \cdot 10^{-5}$	4.28
6859	$3.4414 \cdot 10^{-6}$	15.84	8000	$3.1818 \cdot 10^{-6}$	8.15
42,875	$1.8704 \cdot 10^{-7}$	18.40	46,656	$1.9189 \cdot 10^{-7}$	16.58
300,763	$1.0657 \cdot 10^{-8}$	17.55	314,432	$9.5155 \cdot 10^{-9}$	20.17

Table 5 Errors for $p = 4$ after one FMG cycle with ν pre- and postsmoothing steps

$d = 2, \nu = 2$ N	L_2-error	Ratio
64	$7.809885 \cdot 10^{-5}$	–
144	$4.964622 \cdot 10^{-6}$	15.75
400	$1.426569 \cdot 10^{-7}$	34.80
1296	$3.616898 \cdot 10^{-9}$	39.44
4624	$8.554892 \cdot 10^{-11}$	42.28
17,424	$2.027914 \cdot 10^{-12}$	42.19
67,600	$5.222339 \cdot 10^{-14}$	38.83
266,256	$1.010133 \cdot 10^{-13}$	0.52
$d = 3, \nu = 3$ N	L_2-error	Ratio
512	$6.282724 \cdot 10^{-5}$	–
1728	$3.377428 \cdot 10^{-6}$	18.60
8000	$1.116225 \cdot 10^{-7}$	30.26
46,656	$2.372285 \cdot 10^{-9}$	47.05

case, and for degrees up to 3 in the 2D and 3D cases. One possible measure to restore the optimal convergence orders in the case $p = 4$ is to increase the number of pre- and postsmoothing steps. In Table 5, we display the resulting errors with two smoothing steps in 2D and with three smoothing steps in 3D.

We remark that the solution time using the FMG method was typically only a small fraction of the time used to assemble the problems.

Acknowledgement This work was supported by the National Research Network "Geometry + Simulation" (NFN S117, 2012–2016), funded by the Austrian Science Fund (FWF). The first author was also supported by the project AComIn "Advanced Computing for Innovation", grant 316087, funded by the FP7 Capacity Programme "Research Potential of Convergence Regions".

References

1. Y. Bazilevs, L. Beirão da Veiga, J.A. Cottrell, T.J.R. Hughes, G. Sangalli, Isogeometric analysis: approximation, stability and error estimates for h-refined meshes. Math. Models Methods Appl. Sci. **16**(07), 1031–1090 (2006)
2. L. Beirão da Veiga, D. Cho, L. Pavarino, S. Scacchi, Overlapping Schwarz methods for isogeometric analysis. SIAM J. Numer. Anal. **50**(3), 1394–1416 (2012). doi:10.1137/110833476. http://epubs.siam.org/doi/abs/10.1137/110833476
3. L. Beirão da Veiga, D. Cho, L. Pavarino, S. Scacchi, BDDC preconditioners for isogeometric analysis. Math. Models Methods Appl. Sci. **23**(6), 1099–1142 (2013). doi:10.1142/S0218202513500048
4. W.L. Briggs, V.E. Henson, S.F. McCormick, *A Multigrid Tutorial* (Society for Industrial and Applied Mathematics, Philadelphia, 2000) ISBN:9780898714623
5. A. Buffa, H. Harbrecht, A. Kunoth, G. Sangalli, BPX-preconditioning for isogeometric analysis. Comput. Meth. Appl. Mech. Eng. **265**, 63–70 (2013). doi:10.1016/j.cma.2013.05.014. ISSN:0045-7825
6. T. Chen, R. Mo, Z.W. Gong, Imposing essential boundary conditions in isogeometric analysis with Nitsche's method. Appl. Mech. Mater. **121–126**, 2779–2783 (2011). doi:10.4028/www.scientific.net/AMM.121-126.2779
7. N. Collier, D. Pardo, L. Dalcin, M. Paszynski, V.M. Calo, The cost of continuity: a study of the performance of isogeometric finite elements using direct solvers. Comput. Meth. Appl. Mech. Eng. **213–216**, 353–361 (2012) doi:10.1016/j.cma.2011.11.002. ISSN:0045-7825. http://www.sciencedirect.com/science/article/pii/S0045782511003392
8. J.A. Cottrell, T.J.R. Hughes, Y. Bazilevs, *Isogeometric Analysis: Toward Integration of CAD and FEA* (Wiley, Chichester, 2009)
9. K.P.S. Gahalaut, J.K. Kraus, S.K. Tomar, Multigrid methods for isogeometric discretization. Comput. Meth. Appl. Mech. Eng. **253**, 413–425 (2013). doi:10.1016/j.cma.2012.08.015. ISSN:0045-7825. http://www.sciencedirect.com/science/article/pii/S0045782512002678
10. W. Hackbusch, *Multi-Grid Methods and Applications*. Springer Series in Computational Mathematics. vol. 4 (Springer, Berlin, 2003)
11. T.J.R. Hughes, J.A. Cottrell, Y. Bazilevs, Isogeometric analysis: CAD, finite elements, NURBS, exact geometry and mesh refinement. Comput. Meth. Appl. Mech. Eng. **194**(39–41), 4135–4195 (2005). doi:10.1016/j.cma.2004.10.008. ISSN:00457825. http://dx.doi.org/10.1016/j.cma.2004.10.008
12. S.K. Kleiss, C. Pechstein, B. Jüttler, S. Tomar, IETI – isogeometric tearing and interconnecting. Comput. Meth. Appl. Mech. Eng. **247–248**, 201–215 (2012). doi:10.1016/j.cma.2012.08.007. ISSN:0045-7825
13. T.J. Mitchell, S. Govindjee, R.L. Taylor, A method for enforcement of Dirichlet boundary conditions in isogeometric analysis, in *Recent Developments and Innovative Applications in Computational Mechanics*, ed. by D.M.-Hoeppe, S. Loehnert, S. Reese (Springer, Berlin, 2011), pp. 283–293. doi:10.1007/978-3-642-17484-1_32. ISBN:978-3-642-17483-4. http://dx.doi.org/10.1007/978-3-642-17484-1_32
14. L. Piegl, W. Tiller, *The NURBS Book*. Monographs in Visual Communications (Springer, Berlin, 1997). ISBN:9783540615453
15. L. Schumaker, *Spline Functions: Basic Theory*. Cambridge Mathematical Library (Cambridge University Press, Cambridge, 2007). ISBN:9780521705127
16. U. Trottenberg, C.W. Oosterlee, A. Schuller, *Multigrid* (Elsevier, Amsterdam, 2000). ISBN:9780080479569
17. D. Wang, J. Xuan, An improved NURBS-based isogeometric analysis with enhanced treatment of essential boundary conditions. Comput. Meth. Appl. Mech. Eng. **199**(37–40), 2425–2436 (2010). doi:http://dx.doi.org/10.1016/j.cma.2010.03.032. ISSN:0045-7825. http://www.sciencedirect.com/science/article/pii/S004578251000109X

Simulation of Cavity Flows by an Implicit Domain Decomposition Algorithm for the Lattice Boltzmann Equations

Jizu Huang, Chao Yang, and Xiao-Chuan Cai

1 Introduction

The 2D steady state lid-driven cavity flow problem is a benchmark problem to test new numerical methods due to its simple geometry and interesting flow behaviors. There are several mathematical models available for simulating this flow, such as the Navier–Stokes (NS) equations and the Boltzmann equations among others. For problems satisfying the continuum assumption, the Boltzmann model and the NS model usually have the same solution in some sense, because the NS model can be derived from the Boltzmann model. But for problems that don't satisfy the continuum assumption, the NS model fails to provide a physically meaningful solution and the Boltzmann model can be viewed as a higher level model. In the past two decades, numerical methods based on the Boltzmann model, such as the lattice Boltzmann equations (LBEs) become increasingly popular [2, 10] in simulating the 2D lid-driven cavity flow. There are extensive numerical experiments carried

J. Huang
Institute of Software, Chinese Academy of Sciences, Beijing 100190, P.R. China

Institute of Computational Mathematics and Scientific/Engineering Computing, Academy of Mathematics and Systems Science, Chinese Academy of Sciences, Beijing 100190, China

C. Yang (✉)
Institute of Software, Chinese Academy of Sciences, Beijing 100190, P.R. China

State Key Laboratory of Computer Science, Chinese Academy of Sciences, Beijing 100190, P.R. China
e-mail: yangchao@iscas.ac.cn

X.-C. Cai
Department of Computer Science, University of Colorado Boulder, Boulder, CO 80309, USA

T. Dickopf et al. (eds.), *Domain Decomposition Methods in Science and Engineering XXII*, Lecture Notes in Computational Science and Engineering 104, DOI 10.1007/978-3-319-18827-0_26

out with the LBEs [10, 13, 14]. However, all existing approaches are explicit or semi-implicit and the time step size of these approaches is limited by the Courant–Friedrichs–Lewy (CFL) condition, and the numerical solutions obtained by using these methods are less accurate than those of the NS equations.

In this paper, we introduce a fully implicit and parallel Newton–Krylov–RAS algorithm for the LBEs, which is unconditionally stable and the time step size depends only on the accuracy requirement. The method is based on an inexact Newton method whose Jacobian systems are solved with an overlapping RAS preconditioned Krylov subspace method. To reduce the computational cost and improve the scalability of the RAS preconditioner, a first-order discretization is developed just for the preconditioner which is re-computed only once per time step. We report accuracy results and scalability studies on fine meshes and on a supercomputer with more than 10,000 processors.

2 Model Problem, Discretization, and Domain Decomposition Preconditioning

In this paper, the LBEs [2] are considered

$$\frac{\partial f_\alpha}{\partial t}(\mathbf{x},t) + \mathbf{e}_\alpha \cdot \nabla f_\alpha(\mathbf{x},t) = \Theta_\alpha, \quad \alpha = 0, 1, \cdots, 8, \quad \mathbf{x} \in \Omega, \quad t \in (0,T), \tag{1}$$

where f_α is the particle distribution function, $\mathbf{e}_\alpha = (e_{\alpha 1}, e_{\alpha 2})$ is the discrete particle velocity, Θ_α is the collision operator, $\Omega = (0,1)^2 \in R^2$ is the computational domain, and $(0,T)$ is the time interval. The macroscopic density ρ and the macroscopic velocity $\mathbf{u} = (u_1, u_2)$ of the fluid are respectively induced from the particle distribution function by

$$\rho = \sum_\alpha f_\alpha, \qquad \mathbf{u} = \frac{1}{\rho}\sum_\alpha f_\alpha \mathbf{e}_\alpha. \tag{2}$$

The collision operator is defined as $\Theta_\alpha = -\frac{1}{\tau}(f_\alpha(\mathbf{x},t) - f_\alpha^{(eq)}(\mathbf{x},t))$, where $\tau = c_s^{-2}\nu$ is the relaxation time of the fluid and $f_\alpha^{(eq)}$ is the local equilibrium distribution function (EDF) defined as

$$f_\alpha^{(eq)} = w_\alpha \rho\Big[1 + \frac{1}{c_s^2}\mathbf{e}_\alpha \cdot \mathbf{u} + \frac{1}{2c_s^4}(\mathbf{e}_\alpha \cdot \mathbf{u})^2 - \frac{1}{2c_s^2}|\mathbf{u}|^2\Big]. \tag{3}$$

Here ν is the shear viscosity, $c_s = 1/\sqrt{3}$ is the sound speed, $|\mathbf{u}| = (u_1^2 + u_2^2)^{1/2}$, and the discrete velocities are given by $\mathbf{e}_0 = (0,0)$, and $\mathbf{e}_\alpha = \lambda_\alpha(\cos\theta_\alpha, \sin\theta_\alpha)$, with $\lambda_\alpha = 1$, $\theta_\alpha = (\alpha-1)\pi/2$ for $\alpha = 1, 2, 3, 4$ and $\lambda_\alpha = \sqrt{2}$, $\theta_\alpha = (\alpha-5)\pi/2 + \pi/4$

for $\alpha = 5, 6, 7, 8$. The weighting factors are defined as $w_0 = 4/9$, $w_\alpha = 1/9$ for $\alpha = 1, 2, 3, 4$ and $w_\alpha = 1/36$ for $\alpha = 5, 6, 7, 8$.

Assume $(0, T)$ is divided into time intervals, where n is the time step index. A fully implicit backward Euler scheme is used to discretize the temporal derivative. Then we obtain a semi-discretized system for (1) as follows

$$\frac{f_\alpha^{n+1} - f_\alpha^n}{\Delta t_{n+1}} + \mathbf{e}_\alpha \cdot \nabla f_\alpha^{n+1} = \Theta_\alpha^{n+1}, \tag{4}$$

where the time step size is $\Delta t_{n+1} = t_{n+1} - t_n$, $f_\alpha^n(\mathbf{x}) \approx f_\alpha(\mathbf{x}, t^n)$, and $\Theta_\alpha^{n+1} \approx \Theta_\alpha(\mathbf{x}, t^{n+1})$. If $e_{\alpha k} \neq 0$, we implement a family of fully implicit finite difference schemes originally proposed in [10] for an explicit method to discretize the spatial derivative $\frac{\partial f_\alpha}{\partial x_k}$. We partition the domain Ω to a uniform $N \times N$ mesh with mesh size $h = 1/(N-1)$ and mesh points $(x_1^i, x_2^j), i, j = 0, 1, \ldots, N-1$. Let us define a scheme $\frac{\partial f_\alpha}{\partial x_k^i}|_m$ in the family as

$$\left.\frac{\partial f_\alpha}{\partial x_k^i}\right|_m = \epsilon \left.\frac{\partial f_\alpha}{\partial x_k^i}\right|_u + (1-\epsilon)\left.\frac{\partial f_\alpha}{\partial x_k^i}\right|_c, \qquad k = 1, 2, \quad 1 \leq i \leq N-2, \tag{5}$$

where $0 \leq \epsilon \leq 1$ is a control parameter that determines how much upwinding is added,

$$\left.\frac{\partial f_\alpha}{\partial x_k^i}\right|_c = \frac{1}{2h}\left[f_\alpha(x_k^{i+1}, \cdot) - f_\alpha(x_k^{i-1}, \cdot)\right],$$

and

$$\left.\frac{\partial f_\alpha}{\partial x_k^i}\right|_u = \begin{cases} \frac{e_{\alpha k}}{2h}\left[3 f_\alpha(x_k^i, \cdot) - 4 f_\alpha(x_k^{i-e_{\alpha k}}, \cdot) + f_\alpha(x_k^{i-2e_{\alpha k}}, \cdot)\right] & \text{if } \ 2 \leq i \leq N-3, \\ \frac{e_{\alpha k}}{h}\left[f_\alpha(x_k^i, \cdot) - f_\alpha(x_k^{i-e_{\alpha k}}, \cdot)\right] \ \text{ if } \ i = 1, \text{or } i = N-2. \end{cases}$$

Theoretically, the scheme is second-order in the interior of the domain and first-order near the boundary, but for our test cases, the scheme is effectively second-order. We also introduce a cheaper first-order upwinding scheme $\frac{\partial f_\alpha}{\partial x_k^i} = \frac{e_{\alpha k}}{h}\big[f_\alpha(x_k^i, \cdot)$ $-f_\alpha(x_k^{i-e_{\alpha k}}, \cdot)\big]$ to construct an efficient preconditioner for the scheme (5).

The initial condition is set to be the EDF, i.e. $f_\alpha(\mathbf{x}, 0) = f_\alpha^{(eq)}(\mathbf{x}, 0)$. The boundary conditions are obtained by a nonequilibrium extrapolation method [11]. Assume that $\mathbf{x}_b$ is a mesh point on the boundary of the domain, and $\mathbf{x}_{nb}$ is the nearest neighboring mesh point of $\mathbf{x}_b$ in the interior of the domain. According to the nonequilibrium extrapolation method, the particle distribution function at $\mathbf{x}_b$ is set to be

$$f_\alpha(\mathbf{x}_b) = f_\alpha^{(eq)}(\mathbf{x}_b) + \left[\, f_\alpha(\mathbf{x}_{nb}) - f_\alpha^{(eq)}(\mathbf{x}_{nb})\right]. \tag{6}$$

After the discretization, a system of nonlinear algebraic equations

$$\mathscr{F}^{n+1}(\mathbf{X}^{n+1}) := \frac{\mathbf{X}^{n+1} - \mathbf{X}^{n}}{\Delta t_{n+1}} + \mathscr{G}^{n+1}(\mathbf{X}^{n+1}) = 0, \quad n = 0, 1, \ldots \tag{7}$$

is obtained and needs to be solved at each time step. Here $\mathscr{G}^{n+1}$ is dependent on the spatial discretization and the collision term. We employ a Newton–Krylov–Schwarz (NKS) [6, 7] type algorithm to solve (7). At each Newton iteration, a Jacobian system is analytically computed and approximately solved by using a Krylov subspace method

$$J^{n+1}\mathbf{S}^{n+1} = -\mathscr{F}^{n+1}(\mathbf{X}^{n+1}), \tag{8}$$

where the Jacobian matrix $J^{n+1} = (\mathscr{F}^{n+1})'(\mathbf{X}^{n+1})$ and $\mathbf{S}^{n+1}$ is the search direction of the Newton method. A restarted GMRES (20) method is applied to approximately solve the right-preconditioned system

$$J^{n+1}(M^{n+1})^{-1}(M^{n+1}\mathbf{S}^{n+1}) = -\mathscr{F}^{n+1}(\mathbf{X}^{n+1}), \tag{9}$$

where M^{n+1} is the restricted additive Schwarz (RAS) preconditioner defined in [5]. The initial guess for the Newton iteration is chosen as the final solution from the previous time step.

3 Numerical Experiments

We implement the new algorithm described in the previous section based on PETSc [1]. A steady state driven cavity flow in 2D is carefully studied in this section. The numerical tests are carried out on a supercomputer Tianhe-2, which tops the Top-500 list as of June, 2013. The computing nodes of Tianhe-2 are interconnected via a proprietary high performance network. And there are two 12-core Intel Ivy Bridge Xeon CPUs and 24 GB local memory in each node. In the numerical experiments we use all 24 CPU cores in each node and assign one subdomain to each core.

In the 2D driven cavity flow problem, we assume the top boundary of the cavity moves from right to left with a constant velocity $U_0 = -0.1$ while the other three boundaries are fixed. The initial condition of macroscopic variables $\rho = 1.0$ and $\mathbf{u} = (0, 0)$ in the cavity. The Reynolds number is defined as $Re = U_0 H/\nu$ with $H = 1.0$. In our simulations, Re is chosen to be 100, 1000, 3200, 5000, 7500, and 10,000.

Simulating this flow by solving the NS equations is a popular approach [8, 9, 12], in which the presence of singularities at the corners is a well-known difficulty. At the corners (0,1) and (1,1), the pressure and the vorticity are unbounded, and at the corners (1,0) and (0,0) the second derivatives of the pressure and vorticity are unbounded. To improve the accuracy of the solution at the corners, Deng et al. [8]

Table 1 $Re = 100$, extrema of the velocity through the centerlines of the cavity

Reference	N	$u_{1,\max}$	$x_{2,\max}$	$u_{2,\max}$	$x_{1,\max}$	$u_{2,\min}$	$x_{1,\min}$
Present	65	0.2075	0.4531	0.1674	0.7656	−0.2471	0.1875
Present	97	0.2098	0.4583	0.1711	0.7604	−0.2492	0.1875
Present	129	0.2107	0.4609	0.1725	0.7656	−0.2500	0.1875
Present	161	0.2111	0.4562	0.1733	0.7625	−0.2504	0.1875
Present	257	0.2117	0.4609	0.1742	0.7617	−0.2510	0.1875
Botella and Peyret [3]	96	0.2140	0.4581	0.1796	0.7630	−0.2538	0.1896
Deng et al. [8]	64	0.2132	–	0.1790	–	−0.2534	–
Ghia et al. [9]	129	0.2109	0.4531	0.1753	0.7656	−0.2453	0.1953
Bruneau and Jouron [4]	129	0.2106	0.4531	0.1786	0.7656	−0.2521	0.1875

perform a Richardson extrapolation of solutions obtained by a finite volume method. In [3], a spectral method is developed to remove the pollution of the singularities. To check the accuracy of the discretization, we simulate the flow at $Re = 100$ with different mesh sizes. The maximum of u_1 on the vertical line $x_1 = 0.5$ is denoted as $u_{1,\max}$ and its location $x_{2,\max}$. The minimum and maximum of u_2 on the horizontal line $x_2 = 0.5$ are, respectively, denoted as $u_{2,\min}$ and $u_{2,\max}$; their locations are, respectively, denoted as $x_{1,\min}$ and $x_{1,\max}$. Table 1 shows the values of these extremum and previously published results obtained by the NS equations. Our results are in agreement with those of [4, 9], but less accurate than the results of [3, 8]. In [4, 9], second-order schemes are used to solve the NS equations. In [3, 8], higher order schemes are given to remove the pollution from the corner singularities.

The streamline contours for the cavity flow configurations with Re increasing from 100 to 10,000 are shown in Fig. 1. We observe that the flow structures are in good agreement with the benchmark results obtained by Ghia et al. [9]. These plots show clearly the effect of Re on the flow pattern. For flows with $Re \leq 1000$, only three vortices appear in the cavity; a primary one near the center and a pair of secondary ones in the corners of the cavity. At $Re = 3200$, a third secondary vortex is seen in the upper right corner. At $Re = 5000$, a tertiary vortex appears in the lower left corner. Furthermore, another tertiary vortex appears in the lower right corner as $Re \geq 7500$.

To show the parallel scalability of the implicit method, we consider a 4096×4096 uniform mesh. We use a fixed time step size $\Delta t = 0.0244$ and run the code for 10 time steps. We test two overlapping factors $\delta = h, 2h$ with different number of processors. We compare the point-block LU subdomain solver and the point-block ILU(l) solver. Here l is the fill-in level for the incomplete LU factorization. The point-block size is 9×9. We set the fill-in levels $l = 0, 1, 2, 3$. The numbers of linear and nonlinear iterations are reported in Table 2. The number of linear iterations grows slowly with the increase of the number of processors. Large overlap or larger fill-in helps reduce the total number of linear iterations. The compute time of both an explicit method [10] and the implicit method with different subdomain solvers is shown in Fig. 2. The optimal compute time can be obtained with fill-in level $l = 1$,

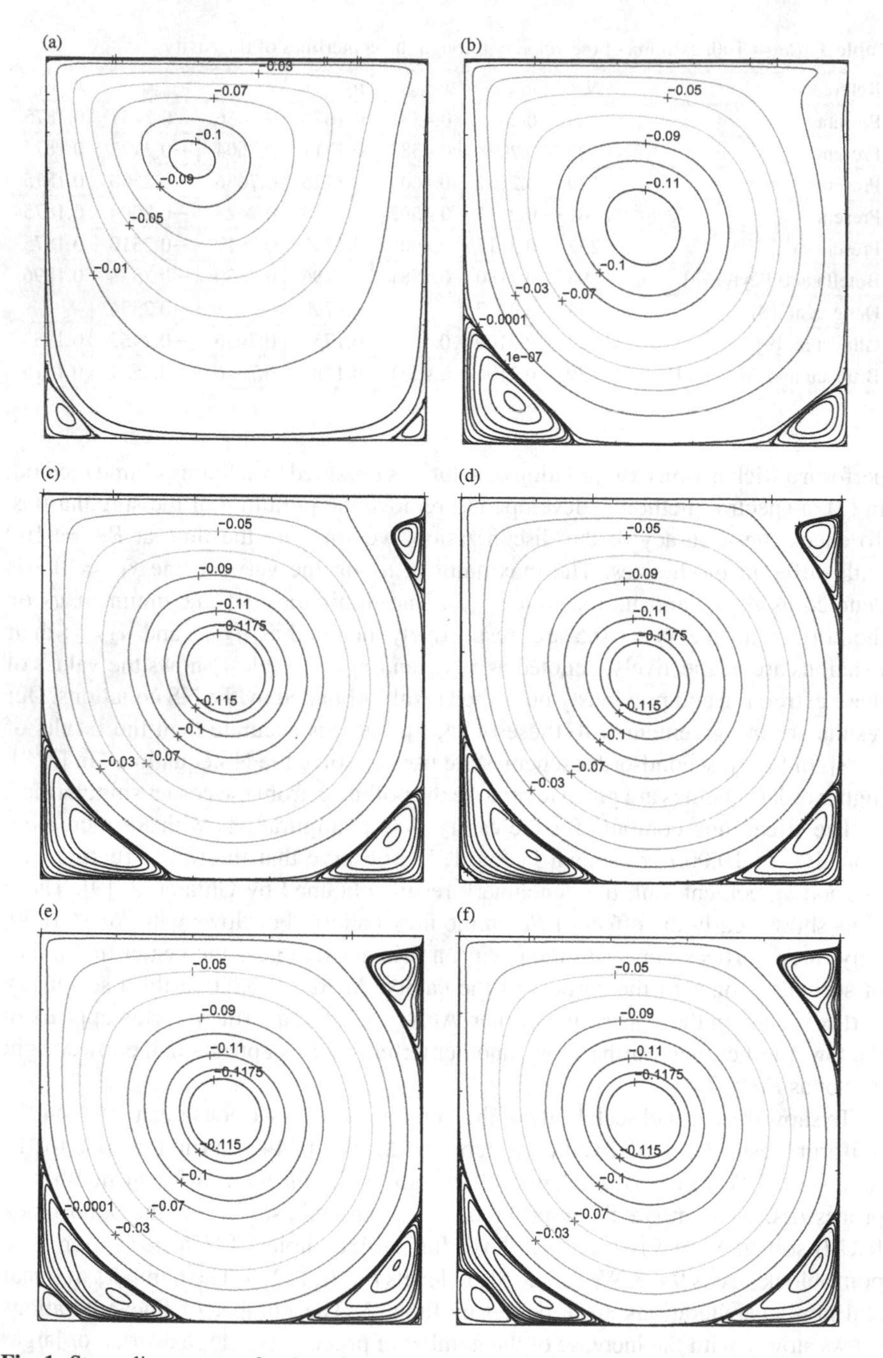

Fig. 1 Streamline patterns for the primary, secondary, and additional corner vortices. (**a**) $Re = 100$, 128×128 mesh. (**b**) $Re = 1000$, 128×128 mesh. (**c**) $Re = 3200$, 256×256 mesh. (**d**) $Re = 5000$, 512×512 mesh. (**e**) $Re = 7500$, 768×768 mesh. (**f**) $Re = 10{,}000$, 768×768 mesh

Table 2 Test results using different overlapping factors and number of processors, 4096 × 4096 mesh (# of unknowns = 150, 994, 944), $t_0 = 0$, time step size $\Delta t = 0.0244$, CFL = 100, $Re = 3200$, 10 time steps

		Newton(avg.)					GMRES/Newton				
δ	N_p	LU	ILU(0)	ILU(1)	ILU(2)	ILU(3)	LU	ILU(0)	ILU(1)	ILU(2)	ILU(3)
h	512	4.7	6.1	6	6	6	20.68	26.62	21.93	21.95	21.85
	1024	4.7	6.1	6	6	6	21.83	27.21	22.92	22.98	22.90
	2048	4.7	6.1	6	6	6	22.62	27.93	23.42	23.52	23.55
	4096	4.7	6.1	6	6	6	24.02	28.82	24.45	24.63	24.55
	8192	4.7	6.1	6	6	6	26.11	30.43	25.73	25.80	25.73
	16,384	4.7	6.1	6	6	6	27.60	32.08	26.98	27.07	26.98
$2h$	512	6	6.1	6	6	6	20.65	26.16	20.85	20.67	20.63
	1024	6	6.1	6	6	6	21.78	26.82	21.50	21.60	21.60
	2048	6	6.1	6	6	6	22.35	27.43	22.03	22.07	22.12
	4096	6	6.1	6	6	6	23.85	28.10	22.83	22.97	23.10
	8192	6	6.1	6	6	6	25.47	29.43	23.78	23.77	23.78
	16,384	6	6.1	6	6	6	26.85	30.97	24.90	24.68	25.20

which is less than that of the explicit method. Excellent speedup is obtained from 512 processors to 16,384 processors. From the figure we see that ILU is faster in terms of the total compute time than LU.

We also do some weak scaling tests of proposed implicit method with local solvers (LU or ILU(1)). It is observed that the method does not reach the ideal performance, because the number of GMRES iterations increases as more processor cores are used. We believe that coarse level corrections in the additive Schwarz preconditioner can improve the weak scaling performance of the fully implicit solver and plan to study this issue in the future. But, due to the page limit, the results are not given in the paper.

4 Conclusions

We developed a parallel, highly scalable fully implicit method for the LBEs. The accuracy of the method is comparable with that of the NS equations. The fully implicit method exhibits an excellent speedup with up to 150 million unknowns on a supercomputer with up to 16,384 processors. Without the CFL limit, the fully implicit method can be used with a suitable adaptive time stepping method that increases the time step size as the solution approaches steady state. Because of the page limit, the discussion related to adaptive time stepping and comparisons with other methods will be presented in a separate report.

Acknowledgements The work was supported in part by NSFC grants 61170075 and 973 grant 2011CB309701.

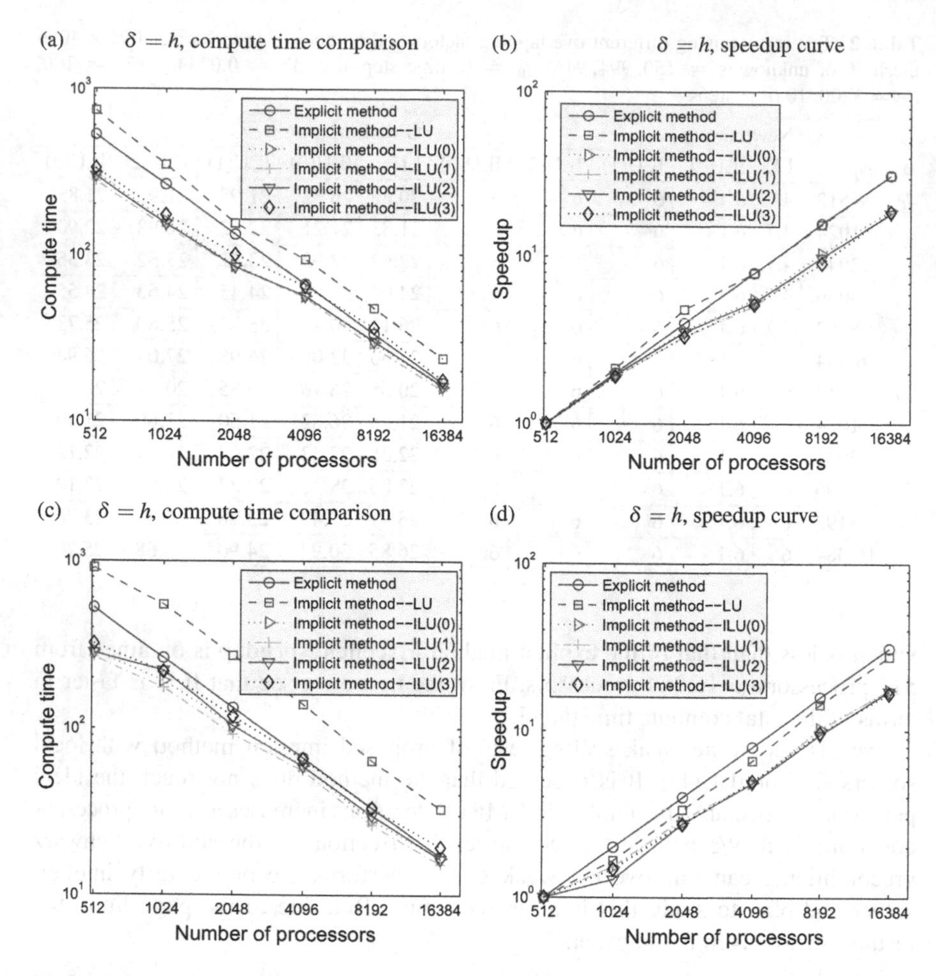

Fig. 2 Compute time and speedup comparison on a 4096×4096 mesh with 512, 1024, 2048, 4096, 8192, and 16,384 processors. Implicit method with different subdomain solvers: 10 time steps with a fixed time step size $\Delta t = 0.0244$. Explicit method [10]: 20,000 time steps with a fixed CFL = 0.05 (**a**, **c**) $\delta = h$, compute time comparison (**b**, **d**) $\delta = h$, speedup curve

References

1. S. Balay, K. Buschelman, W.D. Gropp, D. Kaushik, M. Knepley, L.C. Mcinnes, B.F. Smith, H. Zhang, *PETSc Users Manual* (Argonne National Laboratory, Argonne, 2013)
2. R. Benzi, S. Succi, M. Vergassola, The lattice Boltzmann equation: theory and applications. Phys. Rep. **222**, 145–197 (1992)
3. O. Botella, R. Peyret, Benchmark spectral results on the lid-driven cavity flow. Comput. Fluids **27**, 421–433 (1998)
4. C.H. Bruneau, C. Jouron, An efficient scheme for solving steady incompressible Navier-Stokes equations. J. Comput. Phys. **89**, 389–413 (1990)

5. X.-C. Cai, M. Sarkis, A restricted additive Schwarz preconditioner for general sparse linear systems. SIAM J. Sci. Comput. **21**, 792–797 (1999)
6. X.-C. Cai, W.D. Gropp, D.E. Keyes, M.D. Tidriri, Newton-Krylov-Schwarz methods in CFD, in *Notes in Numerical Fluid Mechanics: Proceedings of the International Workshop on the Navier-Stokes Equations*, ed. by R. Rannacher. (Vieweg Verlag, Braunschweig, 1994), pp. 123–135
7. X.-C. Cai, W.D. Gropp, D.E. Keyes, R.G. Melvin, D.P. Young, Parallel Newton-Krylov-Schwarz algorithms for the transonic full potential equation. SIAM J. Sci. Comput. **19**, 246–265 (1998)
8. G.B. Deng, J. Piquet, P. Queutey, M. Visonneau, Incompressible flow calculations with a consistent physical interpolation finite volume approach. Comput. Fluids **23**, 1029–1047 (1994)
9. U. Ghia, K.N. Ghia, C.T. Shin, High-*Re* solutions for incompressible flow using the Navier-Stokes equations and a multigrid method. J. Comput. Phys. **48**, 387–411 (1982)
10. Z.L. Guo, T.S. Zhao, Explicit finite-difference lattice Boltzmann method for curvilinear coordinates. Phys. Rev. E **67**, 066709(12p) (2003)
11. Z.L. Guo, C.G. Zheng, B.C. Shi, Non-equilibrium extrapolation method for velocity and boundary conditions in the lattice Boltzmann method. Chin. Phys. **11**(4), 0366–0374 (2002)
12. P. Luchini, Higher-order difference approximations of the Navier-Stokes equations. Int. J. Numer. Methods Fluids **12**, 491–506 (1991)
13. R. Mei, W. Shyy, On the finite difference-based lattice Boltzmann method in curvilinear coordinates. J. Comput. Phys. **143**, 426–448 (1998)
14. H.W. Xi, G. W. Peng, S.-H. Chou, Finite-volume lattice Boltzmann method. Phys. Rev. E **59**, 6202–6265 (1999)

Multiplicative Overlapping Schwarz Smoothers for H^{div}-Conforming Discontinuous Galerkin Methods for the Stokes Problem

Guido Kanschat and Youli Mao

1 Introduction

The efficient solution of the Stokes equations is an important step in the development of fast flow solvers. The saddle point structure due to the divergence constraint makes the solution process more complicated. Block preconditioners are often employed, but their performance is limited by the inf-sup constant of the problem and by the difficulty of finding a good preconditioner for the pressure Schur complement. This could be avoided, if the multigrid method operated on the divergence free subspace directly. Recently in [8], we introduced and analyzed a multigrid method with an additive overlapping Schwarz smoother. The main ingredients of our method are a smoother which implicitly operates on the divergence free subspace and a grid transfer operator from coarse to fine mesh which maps the coarse divergence free subspace into the fine one. In this contribution here, we now employ the multiplicative version of this Schwarz method and present numerical results for it.

G. Kanschat (✉)
Interdisziplinäres Zentrum für Wissenschaftliches Rechnen (IWR), Universität Heidelberg, Im Neuenheimer Feld 368, 69120 Heidelberg, Germany
e-mail: kanschat@uni-heidelberg.de

Y. Mao
Department of Mathematics, Texas A&M University, 3368 TAMU, College Station, TX 77843, USA
e-mail: youlimao@math.tamu.edu

T. Dickopf et al. (eds.), *Domain Decomposition Methods in Science and Engineering XXII*, Lecture Notes in Computational Science and Engineering 104, DOI 10.1007/978-3-319-18827-0_27

We consider discretizations of the Stokes equations with no-slip boundary conditions

$$\begin{aligned} -\Delta u + \nabla p &= f \quad \text{in } \Omega, \\ \nabla \cdot u &= 0 \quad \text{in } \Omega, \\ u &= u^B \quad \text{on } \partial\Omega, \end{aligned} \tag{1}$$

on a bounded domain $\Omega \subset \mathbb{R}^d$ of dimension $d = 2, 3$. The natural solution spaces for this problem are $V = H^1_0(\Omega; \mathbb{R}^d)$ for the velocity u and the space of mean value free square integrable functions $Q = L^2_0(\Omega)$ for the pressure p. We point out that other well-posed boundary conditions do not pose a problem.

In order to obtain a finite element discretization, we partition the domain Ω into a hierarchy of meshes $\{\mathbb{T}_\ell\}_{\ell=0,\dots,L}$ of parallelogram and parallelepiped cells in two and three dimensions, respectively. By $\mathbb{F}_\ell$ we denote the set of all faces of the mesh $\mathbb{T}_\ell$. The set $\mathbb{F}_\ell$ is composed of the set of interior faces $\mathbb{F}^i_\ell$ and the set of all boundary faces $\mathbb{F}^\partial_\ell$.

In order to discretize (1) on the mesh $\mathbb{T}_\ell$, we choose discrete subspaces $X_\ell = V_\ell \times Q_\ell$, where $Q_\ell \subset Q$. Following [6], we employ discrete subspaces V_ℓ of the space $H^{\text{div}}_0(\Omega)$, where

$$\begin{aligned} H^{\text{div}}(\Omega) &= \{ v \in L^2(\Omega; \mathbb{R}^d) \big| \nabla \cdot v \in L^2(\Omega) \}, \\ H^{\text{div}}_0(\Omega) &= \{ v \in H^{\text{div}}(\Omega) \big| v \cdot \mathbf{n} = 0 \quad \text{on } \partial\Omega \}. \end{aligned}$$

On each mesh cell T, we choose the Raviart–Thomas [9] space of degree k with $k \geq 1$, mapped by the Piola transformation if necessary and denoted by V_T. We point out that any pair of velocity spaces V_ℓ and pressure spaces Q_ℓ is admissible, if the key relation

$$\nabla \cdot V_\ell = Q_\ell \tag{2}$$

holds. We obtain the finite element spaces

$$\begin{aligned} V_\ell &= \{ v \in H^{\text{div}}_0(\Omega) \big| \forall T \in \mathbb{T}_\ell : v_{|T} \in V_T \}, \\ Q_\ell &= \{ q \in L^2_0(\Omega) \big| \forall T \in \mathbb{T}_\ell : q_{|T} \in Q_T \}. \end{aligned}$$

1.1 Discontinuous Galerkin Discretization

While the fact that V_ℓ is a subspace of $H^{\text{div}}_0(\Omega)$ implies continuity of the normal component of its functions across interfaces between cells, this is not true for tangential components. Thus, $V_\ell \not\subset H^1(\Omega; \mathbb{R}^d)$, and it cannot be used immediately to discretize (1). We follow the example in for instance [6] and apply a DG formulation to the discretization of the elliptic operator. Here, we focus on the interior penalty method [1]. Let T_1 and T_2 be two mesh cells with a joint face F, and let u_1 and u_2 be the traces of a function u on F from T_1 and T_2, respectively. On this face F, we introduce the averaging operator

$$\{\{u\}\} = \frac{u_1 + u_2}{2}. \tag{3}$$

Using the notation, that every integral form over a set of mesh cells or faces is the sum of the integrals over all objects in the set, the interior penalty bilinear form reads

$$\begin{aligned} a_\ell(u, v) = {} & (\nabla u, \nabla v)_{\mathbb{T}_\ell} + 4 \langle \sigma_L \{\{u \otimes \mathbf{n}\}\}, \{\{v \otimes \mathbf{n}\}\} \rangle_{\mathbb{F}_\ell^i} \\ & - 2 \langle \{\{\nabla u\}\}, \{\{\mathbf{n} \otimes v\}\} \rangle_{\mathbb{F}_\ell^i} - 2 \langle \{\{\nabla v\}\}, \{\{\mathbf{n} \otimes u\}\} \rangle_{\mathbb{F}_\ell^i} \\ & + 2 \langle \sigma_L u, v \rangle_{\mathbb{F}_\ell^\partial} - \langle \partial_n u, v \rangle_{\mathbb{F}_\ell^\partial} - \langle \partial_n v, u \rangle_{\mathbb{F}_\ell^\partial}. \end{aligned} \tag{4}$$

The operator "$\otimes$" denotes the Kronecker product of two vectors. We note that the term $4\{\{u \otimes \mathbf{n}\}\} : \{\{v \otimes \mathbf{n}\}\}$ actually denotes the product of the jumps of u and v.

The discrete weak formulation of (1) reads now: find $(u_\ell, p_\ell) \in V_\ell \times Q_\ell$, such that for all test functions $v_\ell \in V_\ell$ and $q_\ell \in Q_\ell$ there holds

$$\mathscr{A}_\ell\left(\begin{pmatrix} u_\ell \\ p_\ell \end{pmatrix}, \begin{pmatrix} v_\ell \\ q_\ell \end{pmatrix}\right) \equiv a_\ell(u_\ell, v_\ell) + (p_\ell, \nabla\cdot v_\ell) - (q_\ell, \nabla\cdot u_\ell) = \mathscr{F}(v_\ell, q_\ell) \equiv (f, v_\ell). \tag{5}$$

Discussion on the existence and uniqueness of such solutions can be found for instance in [5]. Here, we summarize, that $a_\ell(.,.)$ is symmetric and, if σ_L is sufficiently large, it is positive definite. Thus, we can define a norm on V_ℓ by

$$\|v_\ell\|_{V_\ell} = \sqrt{a_\ell(v_\ell, v_\ell)}. \tag{6}$$

In order to obtain optimal convergence results, σ_L is chosen as $\overline{\sigma}/h_L$, where h_L is mesh size on the finest level L and $\overline{\sigma}$ is a positive constant depending on the polynomial degree. A key result in the convergence analysis of this discretization as well as in the analysis of the additive Schwarz smoother is the inf-sup condition

$$\inf_{v \in V_\ell} \sup_{q \in Q_\ell} \frac{(q, \nabla\cdot v)}{\|v\|_{V_\ell} \|q\|_{Q_\ell}} \geq \gamma_\ell > 0 \tag{7}$$

where $\gamma_\ell = c\sqrt{\frac{h_L}{h_\ell}} = c\sqrt{2^{\ell-L}}$ and c is a constant independent of the grid level ℓ.

2 Multigrid Method

In this section we define a V-cycle multigrid preconditioner $\mathscr{B}_\ell$ for the operator $\mathscr{A}_\ell$. We define the action of the multigrid preconditioner $\mathscr{B}_\ell : X_\ell \to X_\ell$ recursively as the multigrid V-cycle with $m(\ell) \geq 1$ pre- and post-smoothing steps. Let $\mathscr{R}_\ell$ be a suitable smoother. Let $\mathscr{B}_0 = \mathscr{A}_0^{-1}$. For $\ell \geq 1$, define the action of $\mathscr{B}_\ell$ on a vector $\mathscr{L}_\ell = (f_\ell, g_\ell)$ by

1. Pre-smoothing: begin with $(u_0, p_0) = (0, 0)$ and let

$$\begin{pmatrix} u_i \\ p_i \end{pmatrix} = \begin{pmatrix} u_{i-1} \\ p_{i-1} \end{pmatrix} + \mathscr{R}_\ell \left(\mathscr{L}_\ell - \mathscr{A}_\ell \begin{pmatrix} u_{i-1} \\ p_{i-1} \end{pmatrix} \right) \qquad i = 1, \dots, m(\ell), \tag{8a}$$

2. Coarse grid correction:

$$\begin{pmatrix} u_{m(\ell)+1} \\ p_{m(\ell)+1} \end{pmatrix} = \begin{pmatrix} u_{m(\ell)} \\ p_{m(\ell)} \end{pmatrix} + \mathscr{B}_{\ell-1} \mathscr{I}_{\ell-1}^t \left(\mathscr{L}_\ell - \mathscr{A}_\ell \begin{pmatrix} u_{m(\ell)} \\ p_{m(\ell)} \end{pmatrix} \right), \tag{8b}$$

3. Post-smoothing:

$$\begin{pmatrix} u_i \\ p_i \end{pmatrix} = \begin{pmatrix} u_{i-1} \\ p_{i-1} \end{pmatrix} + \mathscr{R}_\ell \left(\mathscr{L}_\ell - \mathscr{A}_\ell \begin{pmatrix} u_{i-1} \\ p_{i-1} \end{pmatrix} \right), \qquad i = m(\ell) + 2, \dots, 2m(\ell) + 1 \tag{8c}$$

4. Assign:

$$\mathscr{B}_\ell \mathscr{L}_\ell = \begin{pmatrix} u_{2m(\ell)+1} \\ p_{2m(\ell)+1} \end{pmatrix} \tag{8d}$$

We distinguish between the standard V-cycle with $m(\ell) = m(L)$ and the variable V-cycle with $m(\ell) = m(L)2^{L-\ell}$, where the number $m(L)$ of smoothing steps on the finest level is a free parameter. We refer to $\mathscr{B}_L$ as the V-cycle preconditioner of $\mathscr{A}_L$. The iteration

$$\begin{pmatrix} u_{k+1} \\ p_{k+1} \end{pmatrix} = \begin{pmatrix} u_k \\ p_k \end{pmatrix} + \mathscr{B}_L \left(\mathscr{L}_L - \mathscr{A}_L \begin{pmatrix} u_k \\ p_k \end{pmatrix} \right) \tag{9}$$

is the V-cycle iteration.

2.1 *Overlapping Schwarz Smoothers*

In this subsection, we define a class of smoothing operators $\mathscr{R}_\ell$ based on a subspace decomposition of the space X_ℓ. Let $\mathscr{N}_\ell$ be the set of vertices in the triangulation $\mathbb{T}_\ell$, and let $\mathbb{T}_{\ell,\upsilon}$ be the set of cells in $\mathbb{T}_\ell$ sharing the vertex υ. They form a subdivision of Ω with N overlapping subdomains (also called patches) which we denote by $\{\Omega_{\ell,\upsilon}\}_{\upsilon=1}^N$.

The subspace $X_{\ell,\upsilon} = V_{\ell,\upsilon} \times Q_{\ell,\upsilon}$ consists of the functions in X_ℓ with support in $\Omega_{\ell,\upsilon}$. Note that this implies homogeneous slip boundary conditions on $\partial\Omega_{\ell,\upsilon}$ for the velocity subspace $V_{\ell,\upsilon}$ and zero mean value on $\Omega_{\ell,\upsilon}$ for the pressure subspace $Q_{\ell,\upsilon}$. The Ritz projection $\mathscr{P}_{\ell,\upsilon} : X_\ell \to X_{\ell,\upsilon}$ is defined by the equation

$$\mathscr{A}_\ell\left(\mathscr{P}_{\ell,\upsilon}\begin{pmatrix}u_\ell\\p_\ell\end{pmatrix}, \begin{pmatrix}v_{\ell,\upsilon}\\q_{\ell,\upsilon}\end{pmatrix}\right) = \mathscr{A}_\ell\left(\begin{pmatrix}u_\ell\\p_\ell\end{pmatrix}, \begin{pmatrix}v_{\ell,\upsilon}\\q_{\ell,\upsilon}\end{pmatrix}\right) \qquad \forall \begin{pmatrix}v_{\ell,\upsilon}\\q_{\ell,\upsilon}\end{pmatrix} \in X_{\ell,\upsilon}. \tag{10}$$

Note that each cell belongs to no more than four (eight in 3D) patches $\mathbb{T}_{\ell,\upsilon}$, one for each of its vertices.

We recall the additive Schwarz smoother

$$\mathscr{R}_{a,\ell} = \eta \sum_{\upsilon \in \mathscr{N}_\ell} \mathscr{P}_{\ell,\upsilon}\mathscr{A}_\ell^{-1}$$

where $\eta \in (0, 1]$ is a scaling factor, $\mathscr{R}_\ell$ is L^2 symmetric and positive definite. In [8], it was shown based on arguments from [2, 10], that this smoother yields a uniformly convergent multigrid method if η is chosen appropriately.

Here, we use the symmetric multiplicative Schwarz smoother $\mathscr{R}_{m,\ell}$ associated with the spaces $X_{\ell,\upsilon}$, defined by

$$\begin{aligned}\mathscr{R}_{m,\ell} &= (\mathscr{I} - \mathscr{E}_\ell)\mathscr{A}_\ell^{-1},\\ \mathscr{E}_\ell &= (\mathscr{I} - \mathscr{P}_{\ell,1})\dots(\mathscr{I} - \mathscr{P}_{\ell,N})\dots(\mathscr{I} - \mathscr{P}_{\ell,1}).\end{aligned}$$

We proved uniform convergence for the variable V-cycle iteration with the smoother $\mathscr{R}_{a,\ell}$ in [8] and showed its efficiency by numerical experiments. Since standard arguments from domain decomposition theory like stable decomposition and strengthened Cauchy-Schwarz inequalities are used, we conjecture that the analysis applies to the multiplicative version in the usual fashion. We note that the use of the variable V-cycle is induced by the level dependence of the inf-sup condition (7). Since optimality of this estimate has not been established, we study standard cycles as well.

3 Numerical Results

We present numerical results for the multiplicative Schwarz method in various V-cycle methods and different solvers in order to show that the contraction numbers are not only bounded away from one, but are actually small enough to make this method

Table 1 Number of iterations n_8 to reduce the residual by 10^{-8} with the variable V-cycle and the standard V-cycle iteration with one and two pre- and post-smoothing steps

	$m(\ell) = 2^{L-\ell}$			$m(\ell) = 1$			$m(\ell) = 2$		
L	RT_1	RT_2	RT_3	RT_1	RT_2	RT_3	RT_1	RT_2	RT_3
3	5	5	5	5	5	5	3	3	3
4	6	6	7	6	6	7	5	5	5
5	6	6	6	6	6	7	5	5	6
6	5	5	6	6	6	7	5	5	6
7	5	5	6	7	7	7	5	5	6
8	5	5	6	7	7	7	6	6	6

Penalty parameter dependent of the finest level mesh size 2^{1-L}

very efficient. The following results were produced using the deal.II library [3, 4] and its multigrid capabilities [7].

The experimental setup for most of the tables is as follows: the domain is $\Omega = [-1, 1]^2$, the coarsest mesh $\mathbb{T}_0$ consists of a single cell $T = \Omega$. The mesh $\mathbb{T}_\ell$ on level ℓ is obtained by dividing all cells in $\mathbb{T}_{\ell-1}$ into four quadrilaterals by connecting the edge midpoints. Thus, a mesh on level ℓ has 4^ℓ cells, and the length of their edges is $2^{1-\ell}$. The right hand side is $f = (1, 1)$.

In Table 1, we first study convergence of the linear multigrid method (preconditioned Richardson iteration) with the multiplicative Schwarz smoother using a variable V-cycle algorithm on a square domain with no-slip boundary condition. The penalty constant in the DG form (4) is chosen as $\bar{\sigma}/h_L$, where $\bar{\sigma} = (k+1)(k+2)$, on the finest level L and all lower levels ℓ. Results for pairs of RT_k/Q_k with orders k between one and three are reported in the table which show the fast and uniform convergence. On the right of this table, we keep the same experimental setup and present iteration counts for the standard V-cycle algorithm with one and two pre- and post-smoothing steps, respectively. Although not proven for this case, we still observe uniform convergence results. We also see that the variable V-cycle with a single smoothing step on the finest level is as fast as the standard V-cycle with two smoothing steps, and thus the variable V-cycle is more efficient.

In Table 2, we test the variable and standard V-cycles with penalty parameters depending on the mesh level ℓ, namely $\bar{\sigma}/h_\ell$ (where $\bar{\sigma}$ is a positive constant depending on the polynomial degree) in the DG form (4). This is the typical situation when the operators are assembled independently on each grid level.

In Table 3, we provide results with GMRES solver and $\mathscr{B}_L$ as preconditioner for experimental setups as in Tables 1 and 2, respectively. The second to fourth columns are results for the variable V-cycle with penalty parameter dependent of the finest level mesh size. The fifth and seventh columns are the results for the standard V-cycle with penalty parameter dependent of the finest level mesh size. The last three columns are the results for the standard V-cycle with penalty parameter depend on the mesh size of each level. From this table, we see that the GMRES method, as expected, is faster in every case.

Table 2 Penalty parameter dependent on the mesh size of each level

Level	Variable			Standard		
	RT_1	RT_2	RT_3	RT_1	RT_2	RT_3
3	6	6	6	6	6	6
4	6	6	6	6	6	7
5	6	6	6	6	6	7
6	5	5	6	6	6	7
7	5	5	6	6	6	7
8	5	5	6	6	6	7

Number of iterations n_8 to reduce the residual by 10^{-8} with variable and standard V-cycle iterations with $m(L) = 1$

Table 3 Number of iterations n_8 to reduce the residual by 10^{-8} with GMRES solver and preconditioner $\mathscr{B}_L$; variable and standard V-cycle with inherited forms, variable V-cycle with noninherited forms

Level	Variable			Standard			Noninherited		
	RT_1	RT_2	RT_3	RT_1	RT_2	RT_3	RT_1	RT_2	RT_3
3	2	2	2	2	2	2	3	3	3
4	3	3	4	4	4	4	5	5	5
5	5	5	5	5	5	5	5	5	5
6	4	4	5	5	5	5	5	5	5
7	4	4	5	5	5	5	5	5	5
8	5	4	5	5	5	5	5	5	5

One pre- and post-smoothing step on the finest level

Table 4 Three-dimensional domain

Level	Richardson		GMRES	
	RT_1	RT_2	RT_1	RT_2
2	1	1	1	1
3	5	5	4	4
4	6	5	4	4
5	6	5	4	4

Number of iterations n_8 to reduce the residual by 10^{-8} with the variable V-cycle algorithm with penalty parameter dependent of the finest level mesh size

In Table 4, we provide results in three dimensions for variable V-cycle methods with the same penalty parameter as we choose in Table 1. We keep the similar experimental setups: domain $\Omega = [-1, 1]^3$ and right hand side $f = (1, 1, 1)$. We observe the similar fast and uniform convergence performance as in two dimensions.

We finish our experiments by applying our method to a non-simply connected domain. We choose a square with a square hole, namely the domain $\Omega = [-1, 1] \setminus [-\frac{1}{3}, \frac{1}{3}]$. The coarse grid on level $\ell = 0$ consists of the squares of the form $[-1 + \frac{2i}{3}, -1 + \frac{2i+2}{3}] \times [-1 + \frac{2j}{3}, -1 + \frac{2j+2}{3}]$ with $0 \leq i, j \leq 2$, and with the index pair $(i, j) = (1, 1)$ missing. We note that the Hodge decomposition in this case is more

Table 5 Number of iterations n_8 to reduce the residual by 10^{-8}, different finite element orders and solvers on the domain with hole $[-1,1]^2 \setminus [-1/3,1/3]^2$

	Richardson		GMRES	
Level	RT_1	RT_2	RT_1	RT_2
2	6	6	4	4
3	6	6	4	4
4	6	6	4	4
5	5	5	4	4
6	5	5	4	4
7	5	5	4	4

complicated due to the presence of a harmonic form. Nevertheless, the results with the multiplicative Schwarz method in Table 5 exhibit the same performance we observed in the simply connected case.

References

1. D.N. Arnold, An interior penalty finite element method with discontinuous elements. SIAM J. Numer. Anal. **19**(4), 742–760 (1982)
2. D.N. Arnold, R.S. Falk, R. Winther, Preconditioning in $H(\text{div})$ and applications. Math. Comput. **66**(219), 957–984 (1997). ISSN 0025-5718. doi:10.1090/S0025-5718-97-00826-0
3. W. Bangerth, R. Hartmann, G. Kanschat, deal.II — a general purpose object oriented finite element library. ACM Trans. Math. Softw. **33**(4) (2007). doi:10.1145/1268776.1268779
4. W. Bangerth, T. Heister, L. Heltai, G. Kanschat, M. Kronbichler, M. Maier, B. Turcksin, T.D. Young, The `deal.II` library, version 8.2. Arch. Numer. Softw. **3**(100), 1–8 (2015)
5. B. Cockburn, G. Kanschat, D. Schötzau, C. Schwab, Local discontinuous Galerkin methods for the Stokes system. SIAM J. Numer. Anal. **40**(1), 319–343 (2002). doi:10.1137/S0036142900380121.
6. B. Cockburn, G. Kanschat, D. Schötzau, A note on discontinuous Galerkin divergence-free solutions of the Navier-Stokes equations. J. Sci. Comput. **31**(1–2), 61–73 (2007). doi:10.1007/s10915-006-9107-7
7. B. Janssen, G. Kanschat, Adaptive multilevel methods with local smoothing for H^1- and H^{curl}-conforming high order finite element methods. SIAM J. Sci. Comput. **33**(4), 2095–2114 (2011) doi:10.1137/090778523.
8. G. Kanschat, Y. Mao, Multigrid methods for $\mathbf{H}^{\text{div}}$-conforming discontinuous Galerkin methods for the Stokes equations. J. Numer. Math. (2014, to appear)
9. P.-A. Raviart, J.M. Thomas, A mixed method for second order elliptic problems, in *Mathematical Aspects of the Finite Element Method*, ed. by I. Galligani, E. Magenes (Springer, New York, 1977), pp. 292–315
10. J. Schöberl, Robust multigrid methods for parameter dependent problems. Dissertation, Johannes Kepler Universität Linz, 1999

A Newton-Krylov-FETI-DP Method with an Adaptive Coarse Space Applied to Elastoplasticity

Axel Klawonn, Patrick Radtke, and Oliver Rheinbach

1 Introduction

We consider a Newton-Krylov-FETI-DP algorithm to solve problems in elastoplasticity. First, the material model and its discretization will be described. The model contains a Prandtl-Reuss flow rule and a von Mises flow function. We restrict ourselves to the case of perfect elastoplasticity; thus, there is no hardening. For more information on elastoplasticity; see, e.g., [1, 4, 10]. In this material model we will have local nonlinearities introduced by plastic material behavior in activated zones of the domain. For the finite element discretization we follow the framework given in [1]. Second, we will briefly present the linearization and the FETI-DP method which is used to solve the linearized problems. For more details on the FETI-DP algorithm, see, e.g., [2, 7, 8, 11]. The convergence of the Newton-Krylov-FETI-DP method using a standard coarse space with vertices and edge averages can deteriorate when the plastically activated zone intersects the interface introduced by the domain decomposition. In this case, we use an adaptive coarse space which successfully decreases the number of cg iterations and the condition numbers of the preconditioned linearized systems. Only a small amount of adaptive constraints is needed if the plastically activated zone is restricted to a small part of the domain.

A. Klawonn • P. Radtke (✉)
Mathematisches Institut, Universität zu Köln, Weyertal 86-90, 50931 Köln, Germany
e-mail: axel.klawonn@uni-koeln.de; patrick.radtke@uni-koeln.de

O. Rheinbach
Fakultät für Mathematik und Informatus, Institut für Numerische Mathematik und Optimierung, Technical Universität Bergakademie Freiberg, Akademiestr. 6, 09596 Freiberg, Germany
e-mail: oliver.rheinbach@math.tu-freiberg.de

T. Dickopf et al. (eds.), *Domain Decomposition Methods in Science and Engineering XXII*, Lecture Notes in Computational Science and Engineering 104, DOI 10.1007/978-3-319-18827-0_28

Additional constraints are needed mainly in the final time and Newton steps. The additional constraints for the coarse space are chosen by a strategy proposed in [9] for linear elliptic problems. In contrast to their implementation, here, the additional constraints will be implemented using a deflation approach; see [6].

2 Elastoplastic Material Model and Discretization

The material model is derived from the quasi-static equation of equilibrium

$$\operatorname{div} \sigma(x,t) = f(x,t);$$

see, e.g., [1, 4, 10]. Let d be the dimension of the domain. Multiplying the equation with $v \in H^1_D(\Omega)^d := \{v \in H^1(\Omega)^d : v = 0 \text{ on } \Gamma_D\}$ and application of the Gauss theorem yields the weak formulation: Find $u \in H^1(\Omega)^d$ which satisfies $u = w$ on Γ_D, such that for all $v \in H^1_D(\Omega)^d$:

$$\int_\Omega \sigma(u) : \epsilon(v) dx = \int_\Omega f \cdot v dx + \int_{\Gamma_N} g \cdot v ds. \tag{1}$$

By discretization in time using the implicit Euler method we obtain:
Find $u_n \in H^1(\Omega)^d$ with $u_n = w$ on Γ_D, such that $\forall v \in H^1_D(\Omega)^d$

$$\int_\Omega \sigma_n : \varepsilon(v)\, dx = \int_\Omega f_n\, v\, dx + \int_{\Gamma_N} g_n \cdot v\, ds,$$

where σ_n is dependent on the displacement u_n. This dependency is determined by the von Mises flow function and the chosen type of hardening. In this article we consider perfect elastoplastic material behavior and hardening effects are absent. In this case the von Mises flow function is given by $\Phi(\sigma) = |\mathrm{dev}(\sigma)| - \sigma_y$, where σ_y is the yield point and $\mathrm{dev}\,(\sigma) = \sigma - \frac{1}{d}\mathrm{tr}\,(\sigma)\mathrm{I}_{d\times d}$. The tension tensor in the nth timestep is then linear elastic if $\Phi(\sigma_n) \leq 0$ and plastic otherwise. In the first case we have

$$\sigma_n = (\lambda + \mu)\mathrm{tr}(\varepsilon(u_n - u_{n-1}) + \mathbb{C}^{-1}\sigma_{n-1}) + 2\mu\mathrm{dev}(\varepsilon(u_n - u_{n-1}) + \mathbb{C}^{-1}\sigma_{n-1})$$

with the Lamé constants λ, μ and the fourth order elasticity tensor $\mathbb{C}$. In the second case the tension tensor in the nth timestep reads

$$\begin{aligned}\sigma_n = {} & (\lambda + \mu)\mathrm{tr}(\varepsilon(u_n - u_{n-1}) + \mathbb{C}^{-1}\sigma_{n-1}) \\ & + \sigma_y \frac{\mathrm{dev}(\varepsilon(u_n - u_{n-1}) + \mathbb{C}^{-1}\sigma_{n-1})}{|\mathrm{dev}(\varepsilon(u_n - u_{n-1}) + \mathbb{C}^{-1}\sigma_{n-1})|}.\end{aligned}$$

Note that in the first case, we have a linear relationship between the tension and the displacement, while in the second case, we have a nonlinearity introduced by normalizing the deviatoric term. For a more detailed description how to obtain the time discrete tension tensor explicitly for different types of hardening, see [1].

3 Linearization

We need to linearize the nonlinear discrete problem in every time step. For this we will represent the problem as a root finding problem. We define the pth component of the vector field F by

$$F_p(u_n) = \int_{\Omega} \sigma_n : \varepsilon(\varphi_p)\, dx - \int_{\Omega} f_n \cdot \varphi_p\, dx - \int_{\Gamma_N} g_n \cdot \varphi_p\, ds.$$

Then the nonlinear problem reads: Solve $F(u_n) = 0$. The Newton update in the $(k+1)$th Newton step is $u_n^{k+1} = u_n^k + \Delta u_n^{k+1}$ with Δu_n^{k+1} defined by

$$DF(u_n^k)\Delta u_n^{k+1} = -F(u_n^k),$$

where the tangential stiffness matrix DF is given by $(DF(u_n^k))_{pq} = \frac{\partial F_p(u_n^k)}{\partial u_{n,q}^k}$. In our numerical examples we iterate in each timestep until the residual satisfies $||F(u_n^k)||_2 \leq 10^{-10} + 10^{-6}||F(u_n^0)||_2$, where $u_n^0 := 0$; for the stopping criterion, see, e.g. [1], p. 171, l. 34 of the source code and [5], p. 73, (5.4). To guarantee the convergence we will use the Armijo rule, see, e.g., [5], as a line search algorithm. In each Newton iteration we will first set $\tau = 1$ as an initial step length and then assemble local stiffness matrices $K^{(i)} = DF(u_n^{k,(i)})$ and right-hand sides $f^{(i)} = F(u_n^{k,(i)})$, $i = 1, \ldots, N$. Then we will solve the linearized system

$$DF(u_n^k)\Delta u_n^{k+1} = -F(u_n^k)$$

with FETI-DP as described in the following section. Our trial update is given by $u_{n,\tau}^{k+1} = u_n^k + \tau \Delta u_n^{k+1}$. We test if the Armijo condition

$$||F(u_{n,\tau}^{k+1})||_2 < (1 - 10^{-4} \cdot \tau)||F(u_n^k)||_2$$

is satisfied. In this case we update $u_n^{k+1,(i)} \leftarrow u_{n,\tau}^{k+1,(i)}$. Otherwise we halve the step length $\tau \leftarrow \tau/2$.

4 FETI-DP and Deflation

We will now briefly describe the FETI-DP algorithm. For more details on FETI-DP, see, e.g., [7, 8, 11]. Let the primal variables, for example, vertices or edge averages in subdomain Ω_i be denoted by $u_{\Pi}^{(i)}$ and the remaining variables be denoted by $u_B^{(i)}$, and the corresponding stiffness matrices and right-hand sides be sorted accordingly. Then, we have for the local stiffness matrices $K^{(i)}$ and local load vectors $f^{(i)}$

$$K^{(i)} = \begin{bmatrix} K_{BB}^{(i)} & K_{\Pi B}^{(i)T} \\ K_{\Pi B}^{(i)} & K_{\Pi\Pi}^{(i)} \end{bmatrix} \quad \text{and} \quad f^{(i)} = \begin{bmatrix} f_B^{(i)} \\ f_{\Pi}^{(i)} \end{bmatrix},$$

respectively. We denote by $K_{BB} = \text{diag}_{i=1}^N K_{BB}^{(i)}$, $K_{\Pi\Pi} = \text{diag}_{i=1}^N K_{\Pi\Pi}^{(i)}$, and $K_{\Pi B} = [K_{\Pi B}^{(1)}, \dots, K_{\Pi B}^{(N)}]$. We introduce the following notation

$$\begin{bmatrix} K_{BB} & \tilde{K}_{\Pi B}^T \\ \tilde{K}_{\Pi B} & \tilde{K}_{\Pi\Pi} \end{bmatrix} = \begin{bmatrix} I_B & 0 \\ 0 & R_{\Pi}^T \end{bmatrix} \begin{bmatrix} K_{BB} & K_{\Pi B}^T \\ K_{\Pi B} & K_{\Pi\Pi} \end{bmatrix} \begin{bmatrix} I_B & 0 \\ 0 & R_{\Pi} \end{bmatrix},$$

where R_{Π}^T is the partial assembly operator in the primal variables. We define a jump operator B_B consisting of entries $0, 1$, and -1, which enforces continuity in the remaining unknowns by $B_B u_B = 0$. Then the FETI-DP system reads $F\lambda = d$, with

$$\begin{aligned} F &= B_B K_{BB}^{-1} B_B^T + B_B K_{BB}^{-1} \tilde{K}_{\Pi B}^T \tilde{S}_{\Pi\Pi}^{-1} \tilde{K}_{\Pi B} K_{BB}^{-1} B_B^T, \\ d &= B_B K_{BB}^{-1} f_B - B_B K_{BB}^{-1} \tilde{K}_{\Pi B}^T \tilde{S}_{\Pi\Pi}^{-1} \left(\tilde{f} - \tilde{K}_{\Pi B} K_{BB}^{-1} f_B\right), \end{aligned}$$

where $\tilde{S}_{\Pi\Pi} = \tilde{K}_{\Pi\Pi} - \tilde{K}_{\Pi B} K_{BB}^{-1} \tilde{K}_{\Pi B}^T$. We further partition the remaining variables $u_B^{(i)} = [u_I^{(i)T} \, u_{\Delta}^{(i)T}]^T$ into dual variables on the interface $u_{\Delta}^{(i)}$ and inner variables $u_I^{(i)}$ and the stiffness matrices and right-hand sides accordingly. Define $K_{\Delta\Delta} = \text{diag}_{i=1}^N K_{\Delta\Delta}^{(i)}$, $K_{II} = \text{diag}_{i=1}^N K_{II}^{(i)}$, and $K_{\Delta I} = [K_{\Delta I}^{(1)} \dots K_{\Delta I}^{(N)}]$. The FETI-DP algorithm is the preconditioned conjugate gradient algorithm applied to $F\lambda = d$ with the Dirichlet preconditioner

$$M^{-1} = B_{B,D} [0 \quad I_{\Delta}]^T \left(K_{\Delta\Delta} - K_{\Delta I} K_{II}^{-1} K_{\Delta I}^T\right) [0 \quad I_{\Delta}] B_{B,D}^T.$$

An additional coarse level in the FETI-DP method can be introduced by a deflation approach; see, e.g., [6] for more details. We will aggregate constraints as columns in a matrix U. The constraint $U^T B u = 0$ will be enforced by introducing projections $P = U(U^T F U)^{-1} U^T F$ and $Q = I - P$. Then the projected system $Q^T F \lambda = Q^T d$ will be solved iteratively, while $P^T F \lambda = P^T d$ will be solved directly. We can also solve the original system with the balancing preconditioner $M_{BP}^{-1} = Q M^{-1} Q^T + P F^{-1}$ where M^{-1} is the classical Dirichlet preconditioner instead; see, e.g., [6].

5 Adaptive Coarse Space

The presentation in this section follows the ideas proposed in [9] for linear elliptic problems. We will start our Newton-Krylov-FETI-DP algorithm with an initial coarse space consisting of vertex constraints as primal variables enforced by subassembly. It is well known that the condition number satisfies $\kappa(M^{-1}F) \leq \omega := \sup_{w\in\tilde{W}} \frac{|P_D w|_S^2}{|w|_S^2}$, where $P_D := B_{B,D}^T B_B$; see, e.g., [7, 11]. Consider a local edge between the subdomains Ω_i and Ω_j and define $S_{E_{ij}} := \text{diag}(S^{(i)}, S^{(j)})$. Let $B_{E_{ij}}$ be a local version of the Matrix B defined as the matrix with the rows of $[B^{(i)} \quad B^{(j)}]$, which consist of a 1 and a -1 and are zero elsewhere. Let $\tilde{W}_{E_{ij}}$ be the subspace of functions in $W^{(i)} \times W^{(j)}$ which are continuous in vertices which both subdomains have in common and define

$$\omega_{E_{ij}} := \sup_{w_{E_{ij}} \in \tilde{W}_{E_{ij}}} \frac{|P_{D,E_{ij}} w_{E_{ij}}|_{S_{E_{ij}}}^2}{|w_{E_{ij}}|_{S_{E_{ij}}}^2}$$

as the local *condition number estimator*, where $P_{D,E_{ij}} = B_{D,E_{ij}}^T B_{E_{ij}}$ and $B_{D,E_{ij}}$ is a scaled version of $B_{E_{ij}}$. Define $\tilde{\omega} := \max_{E_{ij} \subset \Gamma} \omega_{E_{ij}}$ as the maximum $\omega_{E_{ij}}$ of all edges on the interface. Then $\tilde{\omega}$ is expected to be a good estimator of the bound ω of the condition number $\kappa(M^{-1}F)$. We choose a prescribed tolerance TOL ≥ 1 for the condition number. With local orthogonal projections $\Pi_{E_{ij}}$ from $W^{(i)} \times W^{(j)}$ onto $\tilde{W}_{E_{ij}}$ and $\overline{\Pi}$ onto range $(\Pi_{E_{ij}} S_{E_{ij}} \Pi_{E_{ij}})$ we solve the following local generalized eigenvalue problem on each edge

$$\begin{aligned} &\overline{\Pi}\Pi_{E_{ij}} P_{D,E_{ij}}^T S_{E_{ij}} P_{D,E_{ij}} \Pi_{E_{ij}} \overline{\Pi} w_{E_{ij}} \\ &= \mu_{E_{ij}} \left(\overline{\Pi}\left(\Pi_{E_{ij}} S_{E_{ij}} \Pi_{E_{ij}} + \sigma(I - \Pi_{E_{ij}})\right)\overline{\Pi} + \sigma(I - \overline{\Pi})\right) w_{E_{ij}}, \end{aligned}$$

where $\sigma > 0$ is a shift parameter here chosen as $\max_i (S_{E_{ij}})_{ii}$; see also [9]. We are only interested in eigenvectors to eigenvalues which exceed the tolerance TOL. Let the eigenvalues $\mu_{E_{ij},l}$, $l = 1,..,n$ be sorted in a decreasing order. For each eigenvector $w_{E_{ij},l}$ to an eigenvalue $\mu_{E_{ij},l} \geq$ TOL, $l = 1,\ldots,k$ we set $\overline{u}_{E_{ij},l} = B_{D,E_{ij}} S_{E_{ij}} P_{D,E_{ij}} w_{E_{ij},l}$. Let $u_{E_{ij},l}$ be vectors representing functions in the Lagrange multiplier space that coincide with $\overline{u}_{E_{ij},l}$ on the edge E_{ij} and that are zero elsewhere. For each edge we collect the $u_{E_{ij},l}$ as columns of a matrix U and apply the modified Gram-Schmidt algorithm to detect and remove linearly dependent constraints.

6 Numerical Examples

In the following we will present numerical examples. Consider a square domain $\Omega = (0, 1)^2$ with zero Dirichlet boundary conditions imposed on the lower edge $\{(x, y) \in \partial\Omega | y = 0\}$ which is exposed to a surface force $g(x, y, t) = (150t, 0)^T$ if $x \in \{(x, y) \in \partial\Omega | y = 1\}$ and $g(x, y, t) = 0$ elsewhere. The material has a Young modulus of $E = 206, 900$, a Poisson ratio of $\nu = 0.29$ and $\sigma_y = 200$. We compute the solution in the time interval $T = [0, 0.45]$ in nine time steps of step length $\Delta t = 0.05$. The space is discretized with P2 finite elements in all our examples (Fig. 1).

In the first set of numerical experiments we consider a classical coarse space with only vertex and edge average constraints using different partitions into elements and subdomains. There are no problems with the classical coarse space if the plastically activated zone does not intersect the interface; see Table 1 for a decomposition

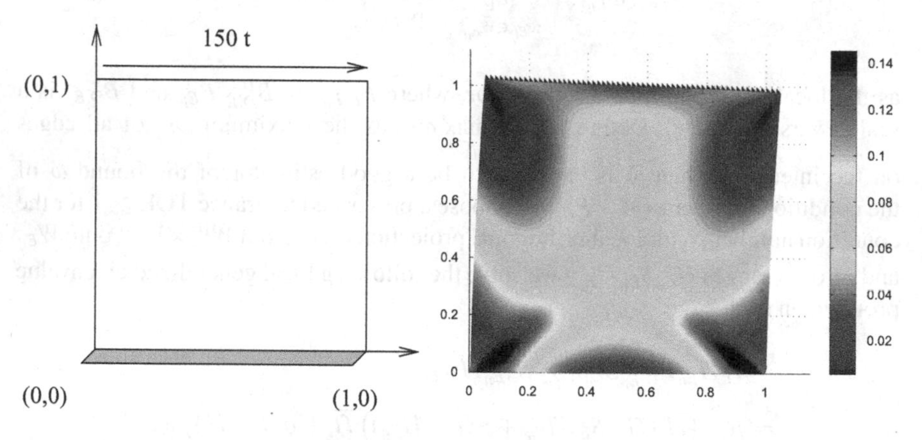

Fig. 1 Unit square with zero Dirichlet boundary conditions at the lower edge $y = 0$ exposed to a surface force $g(t) = (150t, 0)^T$ at the upper edge $y = 1$ (*left*). Displacement magnified by factor 20 and shear energy density in the last timestep (*right*). Material parameters $E = 206{,}900$, $\nu = 0.29$ and $\sigma_y = 200$

Table 1 FETI-DP maximal condition numbers and iteration counts in Newton's scheme with a coarse space consisting of vertices and edge averages

$n = H/h$	$N = 1/H$	Max. cond	Max. CG-It.	Newton its per timestep
20	2	4.06	13	1/1/1/4/4/6/7/9/11
30	2	4.53	14	1/1/3/5/5/7/8/10/11
40	2	4.87	14	1/1/3/4/5/7/9/13/13

We use P2 finite elements in all our examples

Table 2 Problems with the classical coarse space

$n = H/h$	$N = 1/H$	Max. cond	Max. CG-It.
4	15	900,837	371
6	15	$>10^6$	>1000
8	15	$>10^6$	>1000

FETI-DP maximal condition numbers and iteration counts

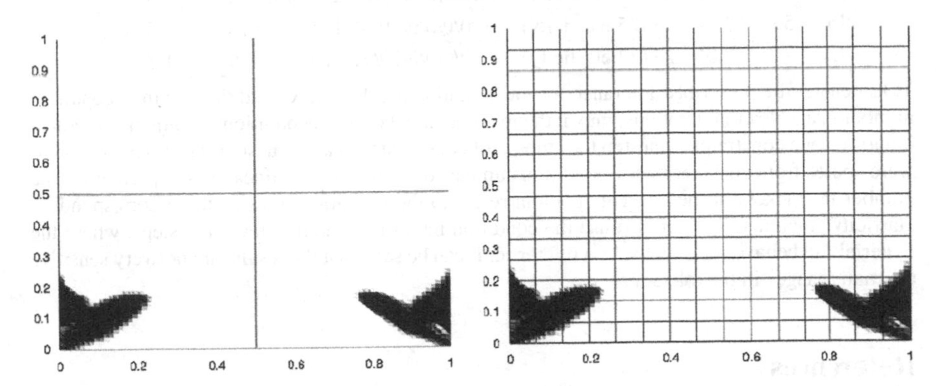

Fig. 2 Plastically activated zone in the last timestep. Decomposition into 2 × 2 subdomains. The plastically activated zones stay completely inside of subdomains (*left*). Decomposition into 15 × 15 subdomains. The plastically activated zones intersect the interface (*right*)

in 2 × 2 subdomains. In this case each linearized system can be analyzed as in [3] using a slab technique. However if the plastically activated zone intersects the interface, the condition numbers and iteration counts increase considerably; see Table 2 for the results with a decomposition in 15 × 15 subdomains (Fig. 2). For the results with the adaptive coarse space described in Sect. 5, see Table 3. The eigenpairs were computed using the MATLAB built-in function “eig”. The complexity thus is cubic with respect to the length of the subdomain edges. For constant H/h the length of the subdomain edges is constant. Moreover, the global number of subdomain edges, and thus also the number of eigenvalue problems, grows linearly with the number of subdomains. The solution of the eigenvalue problems can, of course, be performed in parallel. The condition numbers and iteration counts decrease for the cost of a few more primal constraints in the last time steps. The tolerance is currently determined heuristically; see Table 3.

Table 3 For each subdomain in each space direction, there are n finite elements and in each space direction there are N subdomains

$n = H/h$	$N = 1/H$	TOL	Max cond	Max it	Constraints/ timestep	Global dofs	Constraints/ global dofs (%)
4	15	6.0	5.84	25 (25 elasticity)	0/0/0/0/0/0/19/46/121	8316	1.5
6	15	7.0	7.06	28 (27 elasticity)	0/0/0/0/0/0/29/71/180	11,676	1.5
8	15	8.0	8.01	30 (29 elasticity)	0/0/0/0/0/0/35/81/225	15,036	1.5
4	15	5.9	5.84	25 (25 elasticity)	0/0/0/0/0/0/19/46/123	8316	1.5
6	15	7.1	7.06	28 (27 elasticity)	0/0/0/0/0/0/29/71/179	11,676	1.5

TOL denotes the prescribed tolerance for the condition number, max cond the maximal condition number in the Newton iterations, max it the maximal number of preconditioned conjugate gradient iterations, and constraints/timestep the amount of constraints in each timestep. The tolerances TOL were chosen from considering the condition numbers of corresponding linear elastic problems. The number in brackets in the "max it" column refers to the iteration counts of these corresponding elasticity problems. We can also use the condition number of the first few time steps, where the material still behaves elastically, as a reference. It can be seen that the results are not very sensitive to small changes in the tolerance

References

1. C. Carstensen, R. Klose, Elastoviscoplastic finite element analysis in 100 lines of Matlab. J. Numer. Math. **10**(3), 157–192 (2002)
2. C. Farhat, M. Lesoinne, P. LeTallec, K. Pierson, D. Rixen, FETI-DP: a dual-primal unified FETI method, I. A faster alternative to the two-level FETI method. Int. J. Numer. Methods Eng. **50**(7), 1523–1544 (2001)
3. S. Gippert, A. Klawonn, O. Rheinbach, Analysis of FETI-DP and BDDC for linear elasticity in 3D with almost incompressible components and varying coefficients inside subdomains. SIAM J. Numer. Anal. **50**(5), 2208–2236 (2012)
4. W. Han, B. Daya Reddy, in *Plasticity. Mathematical Theory and Numerical Analysis*, 2nd edn. Interdisciplinary Applied Mathematics, vol. 9 (Springer, New York, 2013).
5. C.T. Kelley, in *Iterative Methods for Linear and Nonlinear Equations*. Frontiers in Applied Mathematics, vol. 16 (Society for Industrial and Applied Mathematics (SIAM), Philadelphia, 1995)
6. A. Klawonn, O. Rheinbach, Deflation, projector preconditioning, and balancing in iterative substructuring methods: connections and new results. SIAM J. Sci. Comput. **34**(1), A459–A484 (2012)
7. A. Klawonn, O.B. Widlund, M. Dryja, Dual-primal FETI methods for three-dimensional elliptic problems with heterogeneous coefficients. SIAM J. Numer. Anal. **40**, 159–179 (2002) (electronic)
8. A. Klawonn, O. Rheinbach, O.B. Widlund, An analysis of a FETI-DP algorithm on irregular subdomains in the plane. SIAM J. Numer. Anal. **46**(5), 2484–2504 (2008)
9. J. Mandel, B. Sousedík, Adaptive selection of face coarse degrees of freedom in the BDDC and the FETI-DP iterative substructuring methods. Comput. Methods Appl. Mech. Eng. **196**(8), 1389–1399 (2007)
10. J.C. Simo, T.J.R. Hughes, in *Computational Inelasticity*. Interdisciplinary Applied Mathematics, vol. 7 (Springer, New York, 1998)
11. A. Toselli, O. Widlund, in *Domain Decomposition Methods—Algorithms and Theory*. Springer Series in Computational Mathematics, vol. 34 (Springer, Berlin, 2005)

Adaptive Coarse Spaces for BDDC with a Transformation of Basis

Axel Klawonn, Patrick Radtke, and Oliver Rheinbach

1 Introduction

We describe a BDDC algorithm, see e.g., [1], and an adaptive coarse space enforced by a transformation of basis for the iterative solution of scalar diffusion problems with a discontinuous diffusion coefficient. The coefficient varies over several orders of magnitude both inside of the subdomains and along the interface. A related algorithm for FETI-DP with a balancing preconditioner has been already described in [6, 7]. Other adaptive coarse space constructions for FETI, FETI-DP, and BDDC methods have been proposed in [8, 10]. We also present some preliminary numerical results for different scalings, including the recent deluxe scaling; cf., [2].

We consider the following model problem. Let $\Omega \subset \mathbb{R}^2$ be a bounded polyhedral domain. We subdivide $\partial\Omega$ into a subset of positive measure $\partial\Omega_D$ where Dirichlet boundary conditions are imposed and $\partial\Omega_N = \partial\Omega \setminus \partial\Omega_D$ where general Neumann boundary conditions are prescribed. Define the Sobolev space $H_0^1(\Omega, \partial\Omega_D) = \{v \in H^1(\Omega) : v = 0 \text{ on } \partial\Omega_D\}$ and consider the piecewise linear finite element approximation of the scalar diffusion problem: Find $u \in H_0^1(\Omega, \partial\Omega_D)$, such that

A. Klawonn (✉) • P. Radtke
Mathematisches Institut, Universität zu Köln, Weyertal 86-90, 50931 Köln, Germany
e-mail: axel.klawonn@uni-koeln.de; patrick.radtke@uni-koeln.de

O. Rheinbach
Fakultät für Mathematik und Informatus, Institut für Numerische Mathematik und Optimierung, Technical Universität Bergakademie Freiberg, Akademiestr. 6, 09596 Freiberg, Germany
e-mail: oliver.rheinbach@math.tu-freiberg.de

T. Dickopf et al. (eds.), *Domain Decomposition Methods in Science and Engineering XXII*, Lecture Notes in Computational Science and Engineering 104, DOI 10.1007/978-3-319-18827-0_29

$a(u,v) = f(v)$ holds for all $v \in H_0^1(\Omega, \partial\Omega_D)$. The bilinear form $a(u,v)$ and the functional $f(v)$ are defined by

$$a(u,v) = \int_\Omega \rho(x) \nabla u \nabla v \, dx \quad \text{and} \quad f(v) = \int_\Omega f v \, dx + \int_{\partial\Omega_N} g_N v \, ds,$$

where g_N is the Neumann boundary data on $\partial\Omega_N$. The model problem is discretized with linear finite elements. We assume $\rho(x)$ to be positive and piecewise constant on Ω and constant on single elements of the triangulation.

The remainder of the paper is organized as follows. We describe the transformation of basis which is performed in our BDDC algorithm to introduce additional coarse constraints in Sect. 2. The characterization how these constraints are chosen via the solution of local eigenvalue problems and an overview over our theoretical results is given in Sect. 3. For a more detailed analysis, see [7]. In Sect. 4 we consider some examples and present numerical results.

2 Transformation of Basis and Scaling in the BDDC Algorithm

As a domain decomposition method we use BDDC. Due to space limitation, for a description of the algorithm and the notation, we refer the reader to [5]. Given a set of primal vertex variables, in the next section, we describe a way to obtain adaptively additional primal variables in the form of weighted edge averages. To implement these edge averages, we transform our local stiffness matrices $K^{(i)}$ and right hand sides $f^{(i)}$ with a transformation matrix $T^{(i)}$. The resulting transformed stiffness matrices $\overline{K}^{(i)} = T^{(i)T} K^{(i)} T^{(i)}$ and right hand sides $\overline{f}^{(i)} = T^{(i)T} f^{(i)}$ then replace $K^{(i)}$ and $f^{(i)}$ in the BDDC algorithm; see, e.g., [5] for more details. We construct the transformation matrices $T^{(i)}$ edge by edge. Consider an edge E of Ω_i and the restriction of $T^{(i)}$ to this edge, denoted by T_E. Suppose we have selected a set of weighted edge averages with weights described by orthonormal column vectors $\{v_{E,1}^{(i)}, \ldots, v_{E,m}^{(i)}\}$. We augment this set to an orthonormal basis $\{v_{E,1}^{(i)}, \ldots, v_{E,m}^{(i)}, v_{E,m+1}^{(i)}, \ldots, v_{E,n_E}^{(i)}\}$ of $\mathbb{R}^{n_E}$, where n_E denotes the number of nodes of the edge E. The transformation matrix T_E is defined by $T_E = [v_{E,1}^{(i)}, \ldots, v_{E,m}^{(i)}, v_{E,m+1}^{(i)}, \ldots, v_{E,n_E}^{(i)}]$ and describes the change of basis from the new to the original nodal basis. The first m columns of T_E correspond to the new additional primal variables and the remaining columns correspond to the new dual unknowns. Denoting the edge unknowns in the new basis by $\hat{u}_E$ and the unknowns in the original basis by u_E, we have $u_E = T_E \hat{u}_E$. We denote by $T_E^{(i)}$ the transformation matrix which operates on all edges of $\partial\Omega_i$. The transformation matrix $T^{(i)}$ is then defined by $T^{(i)} = \operatorname{diag}(I_I, I_V, T_E^{(i)})$, where I_I and I_V denote the identity on inner variables and on vertex variables, respectively. The transformed stiffness matrices are of the form

$$T^{(i)T}K^{(i)}T^{(i)} = \begin{bmatrix} K_{II}^{(i)} & K_{IV}^{(i)} & K_{IE}^{(i)}T_E^{(i)} \\ K_{VI}^{(i)} & K_{VV}^{(i)} & K_{VE}^{(i)}T_E^{(i)} \\ T_E^{(i)T}K_{EI}^{(i)} & T_E^{(i)T}K_{EV}^{(i)} & T_E^{(i)T}K_{EE}^{(i)}T_E^{(i)} \end{bmatrix},$$

with right hand sides $T^{(i)T}f^{(i)} = [f_I^{(i)T} \; f_V^{(i)T} \; f_E^{(i)T}T_E^{(i)}]^T$. We can now perform our BDDC algorithm with the transformed problem; see, e.g., [5] for a detailed description. In our algorithm we will use two different scalings. Let φ_i be the nodal finite element function associated with the node x_i and define $\hat{\rho}_j(x_i) = \max_{T \in \mathrm{supp}(\varphi_i) \cap \Omega_j} \rho_{j|T}(x_i)$. Our scaling weights are now defined as $\delta_j^\dagger(x) = \hat{\rho}_j(x) / \sum_{k \in N_x} \hat{\rho}_k(x)$, where N_x is the set of indices of the subdomains that have the node x on their boundary. The scaling matrices $D^{(j)}$ are diagonal matrices in this case with the weights $\delta_j^\dagger(x)$ on the diagonal. This approach is usually referred to as ρ-scaling. We consider another scaling variant, also known as deluxe scaling, see e.g., [2]. In this case the restriction $D_{E_{ij}}^{(k)}$ of $D^{(k)}$ to an edge $\mathcal{E}_{ij}$ is defined by $D_{E_{ij}}^{(k)} = (S_{E_{ij}E_{ij}}^{(i)} + S_{E_{ij}E_{ij}}^{(j)})^{-1} S_{E_{ij}E_{ij}}^{(k)}$, $k = i, j$, where $S_{E_{ij}E_{ij}}^{(k)}$ is the restriction of $S^{(k)}$ to the edge $\mathcal{E}_{ij}$ after the transformation of basis.

3 Choice of Weighted Edge Averages

In the following we will consider two different eigenvalue problems to compute weighted edge averages for our algorithm; see also [7]. The first eigenvalue problem is a replacement for the weighted Poincaré inequalities in the case of non-quasimonotone coefficient functions; see [6, 7]. The second is related to an extension theorem; see [7]. For a common edge $\mathcal{E}_{ij}$ of the subdomains Ω_i and Ω_j we define $S_{\mathcal{E}_{ij},\rho}^{(l)}$, $l = i, j$, as the Schur complement which is obtained after eliminating all variables of $K^{(l)}$ except of the variables on the closure of $\mathcal{E}_{ij}$, denoted by $\overline{\mathcal{E}}_{ij}$. We define the mass matrix $(M_{\mathcal{E}_{ij},\rho}^{(l)})_{pq} := \int_{\mathcal{E}_{ij}} \rho_l \varphi_p \varphi_q ds$, $p, q = 1, \ldots, n_{\mathcal{E}_{ij}}$, where $n_{\mathcal{E}_{ij}}$ denotes the number of degrees of freedom on $\overline{\mathcal{E}}_{ij}$ and φ_p is the nodal finite element basis function associated with a node $x_p \in \overline{\mathcal{E}}_{ij}$. We also introduce the bilinear forms $s_{\mathcal{E}_{ij},\rho}^{(l)}(u, v) := u^T S_{\mathcal{E}_{ij},\rho}^{(l)} v$ and $m_{\mathcal{E}_{ij},\rho}^{(l)}(u, v) := u^T M_{\mathcal{E}_{ij},\rho}^{(l)} v$. If the coefficient $\rho(x)$ of the diffusion problem varies over several orders of magnitude inside of subdomains and over the interface of the decomposition and is non-quasimonotone the constant in the Poincaré inequality is polluted by the contrast of the coefficient. For a definition of quasimonotone coefficients and a detailed analysis of weighted Poincaré inequalities, see [9]. The Poincaré constant also appears in the bound of the condition number estimate of substructuring methods equipped with a classical coarse space, e.g., a coarse space consisting of vertices and standard edge averages only. To circumvent this problem we introduce a generalized eigenvalue problem to compute new weighted averages which will be used to enhance our coarse space. Note, that related eigenvalue problems are also used in [3, 4] in the context of overlapping Schwarz methods. However, our approach is more local. We denote the finite element trace space on $\mathcal{E}_{ij}$ by $W^h(\mathcal{E}_{ij})$.

Eigenvalue Problem 1 (EVP 1) *Find* $(u_k^{(i)}, \mu_k^{(i)}) \in W^h(\mathcal{E}_{ij}) \times \mathbb{R}$ *such that*

$$s_{\mathcal{E}_{ij},\rho}^{(i)}(u_k^{(i)}, v) = \mu_k^{(i)} m_{\mathcal{E}_{ij},\rho}^{(i)}(u_k^{(i)}, v) \qquad \forall v \in W^h(\mathcal{E}_{ij}). \tag{1}$$

For $L \in \{1, \ldots, n_{\mathcal{E}_{ij}}\}$, where $n_{\mathcal{E}_{ij}}$ is the number of degrees of freedom on $\overline{\mathcal{E}}_{ij}$, and for $l = i, j$ we introduce the projection

$$I_L^{\mathcal{E}_{ij},(l)} = \sum_{k=1}^{L} m_{\mathcal{E}_{ij},\rho}^{(l)}(u_k^{(l)}, v) u_k^{(l)}, \quad l = i, j,$$

with the eigenvectors $u_k^{(l)}$ of Eigenvalue Problem 1. The next lemma provides a generalized Poincaré inequality and is needed to estimate weighted L^2-norms of projected finite element functions on edges; for a proof, see [7].

Lemma 1 *For* $v \in W^h(\mathcal{E}_{ij})$ *and* $w := \left(v - I_L^{\mathcal{E}_{ij},(l)} v\right) \in W^h(\mathcal{E}_{ij})$, *we have*

$$||v - I_L^{\mathcal{E}_{ij},(l)} v||_{L^2_{\rho_l}(\mathcal{E}_{ij})}^2 = m_{\mathcal{E}_{ij},\rho}^{(l)}(w, w) \leq \frac{1}{\mu_{L+1}^{(l)}} s_{\mathcal{E}_{ij},\rho}^{(l)}(v, v) \tag{2}$$

$$\text{and} \qquad s_{\mathcal{E}_{ij},\rho}^{(l)}(w, w) \leq s_{\mathcal{E}_{ij},\rho}^{(l)}(v, v). \tag{3}$$

In our BDDC coarse space we will enforce the equality of the projected functions $I_L^{\mathcal{E}_{ij},(i)} v^{(i)} = I_L^{\mathcal{E}_{ij},(i)} v^{(j)}$ and $I_L^{\mathcal{E}_{ij},(j)} v^{(i)} = I_L^{\mathcal{E}_{ij},(j)} v^{(j)}$ on the interface. We cannot directly enforce this equality, but instead we guarantee that $m_{\mathcal{E}_{ij},\rho}^{(l)}(u_k^{(l)}, v_{\mathcal{E}_{ij}}^{(i)}) = m_{\mathcal{E}_{ij},\rho}^{(l)}(u_k^{(l)}, v_{\mathcal{E}_{ij}}^{(j)})$, for $k = 1, .., L$, by a transformation of basis. To do so, we first build $M_{\mathcal{E}_{ij},\rho}^{(l)} u_k^{(l)}$ and discard the entries related to primal vertices. Then, this vector defines those columns of the local transformation matrices $T_E^{(i)}$ and $T_E^{(j)}$ which are related to the corresponding primal variable in the new basis. We choose all eigenvectors of Eigenvalue Problem 1 whose corresponding eigenvalues satisfy $\mu \leq \tau_\mu$ with a chosen tolerance τ_μ.

To guarantee that certain extensions can be bounded with constants independent of coefficient jumps, we introduce a second eigenvalue problem.

Eigenvalue Problem 2 (EVP 2)

$$s^{(j)}_{\mathcal{E}_{ij},\rho_j}(v, w_\kappa) = \nu^{(i)}_\kappa s^{(i)}_{\mathcal{E}_{ij},\rho_i}(v, w_\kappa), \quad \kappa = 1, \ldots, n_{\mathcal{E}_{ij}}. \tag{4}$$

Remark 1 If $\ker(s^{(j)}_{\mathcal{E}_{ij},\rho_j}) = \ker(s^{(i)}_{\mathcal{E}_{ij},\rho_i})$, instead of solving Eigenvalue Problem 2 on $\mathrm{range}(s^{(j)}_{\mathcal{E}_{ij},\rho_j})$, we solve

$$\overline{\Pi} S^{(j)}_{\mathcal{E}_{ij},\rho_j} \overline{\Pi}\, \overline{w} = \nu \left(\overline{\Pi} S^{(i)}_{\mathcal{E}_{ij},\rho_i} \overline{\Pi} + \sigma \left(I - \overline{\Pi} \right) \right) \overline{w},$$

where σ is any positive constant and $\overline{\Pi}$ is an orthogonal projection onto $\mathrm{range}(s^{(i)}_{\mathcal{E}_{ij},\rho_i})$. In our computations we have chosen σ as the maximum diagonal entry of $\overline{\Pi} S^{(i)}_{\mathcal{E}_{ij},\rho_i} \overline{\Pi}$. The right-hand side of this problem is positive definite; see also [8].

We introduce a second projection operator

$$\Pi^{(l)}_K v := \sum_{k=1}^{K} s^{(l)}_{\mathcal{E}_{ij},\rho}(w^{(l)}_k, v) w^{(l)}_k, \quad l = i, j,$$

with $K \in \{1, \ldots, n_{\mathcal{E}_{ij}}\}$ and obtain the following lemma; see [7] for a proof.

Lemma 2 *We have* $\forall\, w^{(j)} \in W^h(\mathcal{E}_{ij})$

$$s^{(i)}_{\mathcal{E}_{ij},\rho_i}\left(w^{(j)} - \Pi^{(i)}_K w^{(j)}, w^{(j)} - \Pi^{(i)}_K w^{(j)} \right) \leq \frac{1}{\nu^{(i)}_{K+1}} s^{(j)}_{\mathcal{E}_{ij},\rho_j}\left(w^{(j)}, w^{(j)} \right).$$

To take advantage of Lemma 2 we need to introduce a second set of primal constraints of the form $\Pi^{(i)}_K w^{(i)} = \Pi^{(i)}_K w^{(j)}$ and $\Pi^{(j)}_K w^{(i)} = \Pi^{(j)}_K w^{(j)}$. For both generalized eigenvalue problems 1 and 2 we introduce tolerances to decide which eigenvectors are chosen to enhance our coarse space. Additionally to the eigenvectors of Eigenvalue Problem 1 we choose all eigenvectors of Eigenvalue Problem 2 whose corresponding eigenvalues satisfy $\nu \leq \tau_\nu$. with a chosen tolerance τ_ν.

Definition 1 By an η-patch $\omega \subset \Omega$ we denote an open set which can be represented as a union of shape regular finite elements and which has $\mathrm{diam}(\omega) \in \mathcal{O}(\eta)$ and a measure of $\mathcal{O}(\eta^2)$. Let $\mathcal{E}_{ij} \subset \partial\Omega_i$ be an edge. Then, a slab $\tilde{\Omega}_{i\eta}$ is a subset of Ω_i of width η with $\mathcal{E}_{ij} \subset \partial\tilde{\Omega}_{i\eta}$ which can be represented as the union of η-patches ω_{ik}, $k = 1, \ldots, n$, such that $\mathcal{E}^{(k)}_{ij} := (\partial\omega_{ik} \cap \mathcal{E}_{ij})^\circ \neq \emptyset$, $k = 1, \ldots, n$.

For each edge $\mathcal{E}_{ij}$ let $\tilde{\Omega}_{i\eta} \subset \Omega_i$ be a slab of width η, such that $\mathcal{E}_{ij} \subset \partial\tilde{\Omega}_{i\eta}$. Let $\omega_{ik} \subset \tilde{\Omega}_{i\eta}$, $k = 1, \ldots, n$, be a set of η-patches such that $\tilde{\Omega}_{i\eta} = \cup^n_{k=1}\omega_{ik}$, and the coefficient function $\rho_{i|\omega_{ik}} = \rho_{ik}$ is constant on each ω_{ik}. Let $\omega_{ik} \cap \omega_{il} = \emptyset$, $k \neq l$. We obtain the following condition number estimate which is proven in [7].

Theorem 1 *The condition number for our BDDC algorithm satisfies*

$$\kappa(M_{BDDC}^{-1}S) \leq C\left(1+\log\left(\frac{\eta}{h}\right)\right)^2 \frac{1}{\nu_{K+1}}\left(1+\frac{1}{\eta\mu_{L+1}}\right).$$

Here, $C > 0$ is a constant independent of H, h, and η and

$$\frac{1}{\mu_{L+1}} = \max_{k=1,\ldots,N}\left\{\frac{1}{\mu_{L_k+1}^{(k)}}\right\}, \quad \frac{1}{\nu_{K+1}} = \max\left\{1, \max_{k=1,\ldots,N}\frac{1}{\nu_{K+1}^{(k)}}\right\}.$$

4 Numerical Results

We now present a few numerical examples that support our theory. We choose $\Omega = [0,1]^2$ with Dirichlet boundary conditions on $\partial\Omega$ and a constant right hand side $f = 0.1$. The coefficient distributions are depicted in Fig. 1. Algorithm A corresponds to a FETI-DP method using only vertex constraints. In Table 1 we vary the number of elements for each subdomain. In Table 2 we vary the coefficient in the channels. In both cases the coefficient distribution is symmetric with respect to the interface, and thus the extension from EVP 2 is not needed. Indeed, the results in Tables 1 and 2 support that EVP 1 is sufficient, here. In Table 3 we vary the number of subdomains. In Table 4 we apply the adaptive method using EVP 1 for the coefficient distribution in Fig. 1 (middle) using standard ρ-scaling and deluxe scaling. The coefficient distribution is mildly unsymmetric and a good condition number is obtained using only EVP 1. This is different for Fig. 1 (right); see Table 5. Here, EVP 2 seems to be necessary. It interesting to note that, in Table 5, using deluxe scaling a relatively low condition number can be obtained using Algorithm A. This is not the case in Table 4.

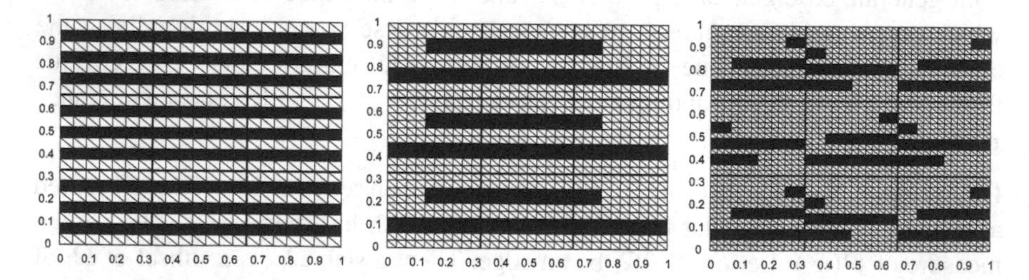

Fig. 1 Coefficient distribution for 3×3 domain decomposition: three channels (*left*), two shorter and displaced channels (*middle*), three shorter and displaced channels (*right*). *Black* corresponds to a high coefficient $\rho = 1e + 06$, *white* corresponds to $\rho = 1$

Table 1 Three channels for each subdomain; see Fig. 1 (left)

	Algorithm A ($\tau_\mu = -\infty$, $\tau_\nu = -\infty$)			Adaptive method EVP 1 ($\tau_\mu = 1$)			Adaptive method EVP 1+2 ($\tau_\mu = 1$, $\tau_\nu = 1e-01$)		
H/h	Cond	Its	# primal	Cond	Its	# primal	Cond	Its	# primal
14	1.227e05	13	4	1.0387	2	24	1.0387	2	24
28	1.545e05	17	4	1.1507	3	24	1.1507	3	24
42	1.730e05	16	4	1.2471	3	24	1.2462	4	28
56	1.861e05	16	4	1.3272	3	24	1.3272	3	24
70	1.962e05	16	4	1.3954	3	24	1.3954	5	28

We have $\rho_1 = 1e06$ in the channel, and $\rho_2 = 1$ elsewhere. The number of additional constraints is clearly determined by the structure of the heterogeneity and independent of the mesh size. $1/H = 3$

Table 2 Three channels for each subdomain; see Fig. 1 (left)

	Algorithm A ($\tau_\mu = -\infty$, $\tau_\nu = -\infty$)			Adaptive method EVP 1 ($\tau_\mu = 1$)			Adaptive method EVP 1+2 ($\tau_\mu = 1$, $\tau_\nu = 1e-01$)		
ρ_2/ρ_1	Cond	# its	# primal	Cond	# its	# primal	Cond	# its	# primal
1e00	3.207	5	4	1.6376	5	8	1.6376	5	8
1e01	5.581	7	4	1.5663	7	8	1.5663	7	8
1e02	1.998$e+01$	9	4	1.4599	7	12	1.4567	7	16
1e03	1.591$e+02$	10	4	1.1505	4	24	1.1505	4	32
1e04	1.550$e+03$	13	4	1.1507	3	24	1.1476	4	31
1e05	1.545$e+04$	15	4	1.1507	3	24	1.1507	3	28
1e06	1.545$e+05$	17	4	1.1507	3	24	1.1507	3	24

Adaptive method using Eigenvalue Problem 1+2. We have ρ_2 in the channels, and $\rho_1 = 1$ elsewhere. $H/h = 28$. The number of additional constraints is bounded for increasing contrast ρ_2/ρ_1. $1/H = 3$

Table 3 Three channels for each subdomain; see Fig. 1 (left)

	Algorithm A ($\tau_\mu = -\infty$, $\tau_\nu = -\infty$)			Adaptive method EVP 1 ($\tau_\mu = 1$)			Adaptive method EVP 1+2 ($\tau_\mu = 1$, $\tau_\nu = 1e-01$)		
$1/H$	Cond	# its	# primal	Cond	# its	# primal	Cond	# its	# primal
2	1	1	1	1.0000	1	1	1.0000	1	1
3	1.545$e+05$	17	4	1.1507	3	24	1.1507	3	24
4	2.734$e+05$	26	9	1.1507	3	51	1.1502	4	59
5	3.475$e+05$	65	16	1.1507	3	88	1.1507	3	90
6	4.078$e+05$	65	25	1.1507	3	135	1.1507	3	152

Increasing number of subdomains and channels. We have $\rho_2 = 1e06$ in the channels, and $\rho_1 = 1$ elsewhere. $H/h = 28$

Table 4 Adaptive method for the coefficient distribution in Fig. 1 (middle)

H/h	Algorithm A ($\tau_\mu = -\infty$, $\tau_\nu = -\infty$)					Adaptive method EVP 1 ($\tau_\mu = 1$)				
	ρ-scaling		Deluxe			ρ-scaling		Deluxe		
	Cond	Its	Cond	Its	# primal	Cond	Its	Cond	Its	# primal
10	6.201e4	25	6.200e4	20	4	1.1480	6	1.1421	5	24
20	7.684e4	25	7.683e4	20	4	1.1978	7	1.1948	6	24
30	8.544e4	25	8.544e4	23	4	1.2630	7	1.2618	6	24

$1/H = 3$. Deluxe scaling and standard ρ-scaling is used

Table 5 Adaptive method for the heterogenous problem from the image in Fig. 1 (right) with a coefficient of 10^6 (black) and 1 (white) respectively

	τ_μ	τ_ν	H/h	Multiplicity-scaling		Deluxe-scaling		
				Cond	# its	Cond	# its	# primal
Algorithm A	$-\infty$	$-\infty$	42	2.492*e*5	161	24.4261	17	4
EVP 1	1	$-\infty$	42	2.496*e*5	128	9.760*e*4	40	24
EVP 1+2	1	1/10	42	1.5184	10	1.4306	9	126

$1/H = 3$. Either multiplicity or deluxe scaling are used

References

1. C.R. Dohrmann, A preconditioner for substructuring based on constrained energy minimization. SIAM J. Sci. Comput. **25**(1), 246–258 (2003)
2. C.R. Dohrmann, O.B. Widlund, Some recent tools and a BDDC algorithm for 3D problems in H(curl), in *Proceedings of the 20th International Conference on Domain Decomposition Methods*. Springer Lecture Notes in Computational Science and Engineering, vol. 95 (Springer, Heidelberg, 2013)
3. V. Dolean, F. Nataf, R. Scheichl, N. Spillane, Analysis of a two-level Schwarz method with coarse spaces based on local Dirichlet-to-Neumann maps. Comput. Methods Appl. Math. **12**(4), 391–414 (2012) [ISSN 1609-4840]
4. J. Galvis, Y. Efendiev, Domain decomposition preconditioners for multiscale flows in high contrast media: reduced dimension coarse spaces. Multiscale Model. Simul. **8**(5), 1621–1644 (2010) [ISSN 1540-3459]
5. A. Klawonn, L.F. Pavarino, O. Rheinbach, Spectral element FETI-DP and BDDC preconditioners with multi-element subdomains. Comput. Meth. Appl. Mech. Eng. **198**(3), 511–523 (2008)
6. A. Klawonn, M. Lanser, P. Radtke, O. Rheinbach, Nonlinear domain decomposition and an adaptive coarse space, in *Proceedings of the 21st International Conference on Domain Decomposition Methods*. Springer Lecture Notes in Computational Science and Engineering, Rennes, 25–29 June 2012 (2013a)
7. A. Klawonn, P. Radtke, O. Rheinbach, FETI-DP methods with an adaptive coarse space. SIAM J. Numer. Anal. **53**(1), 297–320 (2015) [ISSN 0036-1429]. doi:10.1137/130939675. http://dx.doi.org/10.1137/130939675
8. J. Mandel, B. Sousedík, Adaptive selection of face coarse degrees of freedom in the BDDC and the FETI-DP iterative substructuring methods. Comput. Methods Appl. Mech. Eng. **196**, 1389–1399 (2007)

9. C. Pechstein, R. Scheichl, Weighted Poincaré inequalities. IMA J. Numer. Anal. **33**(2), 652–686 (2013) [ISSN 0272-4979]
10. N. Spillane, D.J. Rixen. Automatic spectral coarse spaces for robust finite element tearing and interconnecting and balanced domain decomposition algorithms. Int. J. Numer. Meth. Eng. **95**, 953–990 (2013) [ISSN 0029-5981]

A Massive Parallel Fast Marching Method

Petr Kotas, Roberto Croce, Valentina Poletti, Vit Vondrak, and Rolf Krause

1 Introduction

In this paper we present a novel technique based on domain decomposition which enables us to perform the fast marching method (FMM) [4] on massive parallel high performance computers (HPC) for given triangulated geometries. The FMM is a widely used numerical method and one of the fastest serial state-of-the-art techniques for computing the solution to the Eikonal equation.

For clarification we define an open set $\Omega = \Omega_I \cup \Omega_E \cup \Gamma \subset \mathbb{R}^2 (\text{or } \mathbb{R}^3)$ where Ω_I is the interior and Ω_E the exterior of the domain enclosed by Γ and the bounding box itself, as shown in Fig. 1. Then the resulting problem for the Eikonal equation reads as the following boundary value formulation

$$\begin{aligned} |\nabla T(x)| F(x) &= 1 \text{ for } x \in \Omega, \\ T|_\Gamma &= 0, \\ T(x) > 0 \text{ for } x \in \Omega_E &\text{ and } T(x) < 0 \text{ for } x \in \Omega_I. \end{aligned} \tag{1}$$

with $F(x)$ as speed function. The solution to this problem with $F(x) = 1$ leads to the well-known signed distance function T with respect to Γ.

P. Kotas (✉) • V. Vondrak
Department of Applied Mathematics, VSB-Technical University of Ostrava, Ostrava-Poruba, 70833, Czech Republic
e-mail: petr.kotas@vsb.cz; vit.vondrak@vsb.cz

R. Croce • V. Poletti • R. Krause
Institute of Computational Science, University of Lugano, CH-6904 Lugano, Switzerland
e-mail: roberto.croce@usi.ch; rolf.krause@usi.ch; valentina.poletti@usi.ch

T. Dickopf et al. (eds.), *Domain Decomposition Methods in Science and Engineering XXII*, Lecture Notes in Computational Science and Engineering 104, DOI 10.1007/978-3-319-18827-0_30

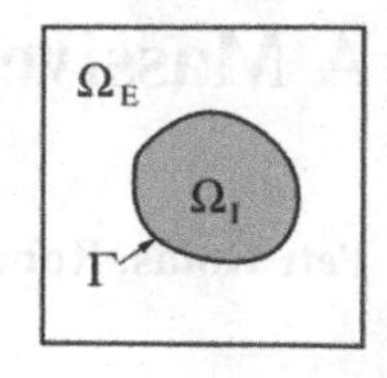

Fig. 1 Set-up of the Eikonal equation problem

Signed distance functions are indispensable in varied fields such as seismic imaging, approximation of geodesic distances and computational fluid dynamics [5]. Hence, finding a fast solution method for the Eikonal equation is of high interest. Several attempts lead to different approaches for a fast and reliable solver. Among various techniques are iterative schemes, expanding box schemes, expanding wavefront schemes, and sweeping schemes [2].

Though the FMM is a very efficient algorithm of complexity $O(N \log N)$, its major drawback is that it cannot be easily parallelized due to its inherently serial nature. Attempts to parallelize it either modify its underlying scheme, losing some of its agility, or have limited scalability. Nevertheless various authors tried to achieve a faster and more efficient scheme.

In [7] a modification of the FMM algorithm is introduced to make it parallelizable. The implementation relies heavily on memory shareability, and the maximal number of processes is limited by the size of the updating stencil. Other authors have relied on different schemes all together, such as level set methods or variations thereof [1, 2, 6]. The speed-up gained by the scalability of these methods comes at the loss of serial algorithmic efficiency as the complexity of the underlying algorithms is higher.

In this paper we will present a parallel algorithm for the computation of the FMM on distributed memory machines via MPI. Another strength of our parallel algorithm is that inter-processor communication does not exist during the parallel FMM computation, each core is basically computing independently the level-set function on its subdomain. This is possible, because each subdomain computes accurate boundary values for its local dataset before starting the parallel FMM.

The remainder of this paper is organized as follows: the sequential fast marching method is shortly explained in Sect. 2 and our extensions towards a Massive Parallel Fast Marching Method (MPFMM) are presented in Sect. 3. Finally, we present several numerical results in Sect. 4 which investigate the performance of the new MPFMM-algorithm. We conclude with some remarks in Sect. 5.

Algorithm 1 Sequential Fast Marching Method (SFMM)

1: Compute the distance $T(x)$ to all node values that are directly adjacent to the interface and tag them as *accepted*. Tag all nodes adjacent to these *accepted* nodes as *narrow* nodes and all others as *away* nodes.
2: Compute $T(x)$ of all *narrow* nodes via Eq. (1), treating $T(x)$ in any adjacent *narrow* or *away* node as ∞. Set the loop index $n = 1$.
3: **repeat**
4: Mark as *accepted* the *narrow* node i, j, k with the smallest $T(x)$ value, denoted by $T(x)^n = T(x)_{i,j,k}$.
5: Mark all *away* nodes adjacent to $T(x)_{i,j,k}$ as *narrow*.
6: Recompute the $T(x)$ values of all *narrow* nodes adjacent to $T(x)_{i,j,k}$ by Eq. (1), treating $T(x)$ in any adjacent *narrow* or *away* node as ∞.
7: Set $n = n + 1$.
8: **until** All nodes are tagged as *accepted*

2 Sequential Fast Marching Method (SFMM)

The Fast Marching Method [4, 5] is designed to efficiently solve the Eikonal equation (1). To do so, the FMM uses the first order Godunov scheme to approximate the gradient term $|\nabla T(x)|$, thus the Eikonal equation is given as

$$F(x)\begin{bmatrix} \max(D^{-x}_{ijk}T(x), -D^{+x}_{ijk}T(x), 0)^2 + \\ \max(D^{-y}_{ijk}T(x), -D^{+y}_{ijk}T(x), 0)^2 + \\ \max(D^{-z}_{ijk}T(x), -D^{+z}_{ijk}T(x), 0)^2 \end{bmatrix}^{\frac{1}{2}} = 1 \qquad (2)$$

where D^{-}_{ijk} is the first order backward and D^{+}_{ijk} the first order forward finite difference operator. Equation (2) utilizes the upwind technique for approximating the Eikonal equation (1). This works, because the front Γ propagates forward and visits each cell only once.

In the core of the FMM there are three lists preserving the state of each cell in the computation domain. Nodes marked as *accepted* are nodes for which the singed distance function has already been computed. Nodes within the vicinity of known nodes are marked *narrow band* and are updated according to equation (2). Finally, nodes with unknown distance are marked as *away*. The complete and easy to follow description of FMM algorithm is given by Sethian in his book [5].

3 Massive Parallel Fast Marching Method (MPFMM)

A few existing strategies for parallelizing the FMM exist, with varying level of success. The simplest approach is to split the interface in two disjoint regions and to run the FMM on both of them at once. This method however lacks larger parallelism and it cannot achieve ideal load-balancing. To overcome this problem another natural approach of decomposing the computational region into a group of sub regions was studied in [6]. This implementation utilizes the strategy of overlapping domain decomposition. To exchange the boundary information an iterative update

Algorithm 2 Parallel Fast Marching Method (MPFMM)

1: Divide the given gridded domain into N sub-domains $D_0, D_1, \dots D_N$
2: On each domain D_i chose initial points on the local narrow band around the geometry and on the domain boundary.
3: Load balance initial data points among domains.
4: Compute the initial data in parallel on each domain using closest point projection to triangulated geometry.
5: Compute the SDF in parallel on each domain using the SFMM algorithm

strategy is used. The overall algorithm is designed in an iterative fashion, since after the boundary update, the FMM needs to be re-run on each sub-domain. This process repeats until convergence is achieved. Finally, methods decomposing the computational domain in such a way that each sub-domain contains part of the initial interface are shown for instance in [3]. With this strategy global minima need to be exchanged in each FMM run, therefore the algorithm is not entirely parallel in nature. Furthermore, none of the existing methods is able to provide reasonable scalability and performance needed by large datasets.

In our approach we use domain decomposition with a combination of exact boundary conditions on each sub-domain. Our decomposition scheme does not require the initial interface to be present in each sub-domain, however the exact distance between the initial interface and the sub-domain boundaries is necessary. This property allows us to loosen the strict limit on scalability that the above mentioned methods possess and allow even very large computational domains to be processed.

The features of our algorithm can be summarized as:

- Easy to implement since it is the SFMM with according boundary conditions on the subdomains and narrow band.
- The algorithm works for massive parallel computations.
- Excellent FMM-speedup on fine grids since no communication is needed.
- Parallelization improves accuracy, for as the number of processes increases, the number of points computed directly with closest point projection increases.
- The parallel algorithm works also for second order schemes.

The parallelization of the entire algorithm basically consists of the parallelization of its two main subroutines, i.e.

1. Narrow band initialization.
2. Ghostcell boundary data computation.

4 Numerical Experiments

In the following section we designed a series of test cases to exploit the numerical features of our algorithm. We show numerically the performance and scalability of our new massive parallel fast marching algorithm, as well as to check the accuracy of the computed signed distance function. In particular we will show that it maintains first order error, as can be easily deduced from the type of evolution scheme we use.

4.1 MPFMM Performance for "Analytical Circle/Sphere"

At first, we investigate the parallel error propagation of our MPFMM algorithm. We make use of a 2D circle with its center $x_c = (0, 0, 0)$, radius $R = 3.0$, and set in a computational domain with the sizc $[-10, 10] \times [-10, 10]$. The signed distance function for this geometry can be computed analytically via the following equation

$$T(\mathbf{x}) = R_0 - \|\mathbf{x} - \mathbf{x_c}\|, x \in R^N. \tag{3}$$

This simplifies both the initialization procedure and the computation of the error. We used two different grid resolutions: 41×41 and 81×81 grid cells on a uniform grid with grid cell sizes $dx = 0.48780$ and $dx = 0.24691$. The computation is performed on 16 cores. Figure 2 shows the error distribution through the global domain subdivided into 16 subdomains.

As expected, the MPFMM algorithm aggregates the error in diagonal direction. This is because the fast marching method computes the discrete gradient in the horizontal and vertical coordinate directions. The maximum global error on each grid is 0.21 and 0.14. This is less than the grid sizes $dx = 0.48780$ and $dx = 0.24691$. Thus this experiment shows that the MPFMM algorithm maintains first order accuracy on all local subdomains.

In order to further exploit the nice initialization properties of the analytical sphere, we set up the 3D problem described in [1]. In this problem, the sphere is located inside a unit cube, with $\mathbf{x_c} = (\frac{1}{2}, \frac{1}{2}, \frac{1}{2})$ and radius $R_0 = 0.25$. The signed distance function in this problem, is defined similarly to Eq. (3), thus we provide the initial data for the MPFMM algorithm using this equation. Again, we make use of two grid sizes: $192 \times 192 \times 192$ and $384 \times 384 \times 384$. We run our algorithm using up to 8192 parallel cores. Figure 3, shows the MPFMM algorithm's super linear

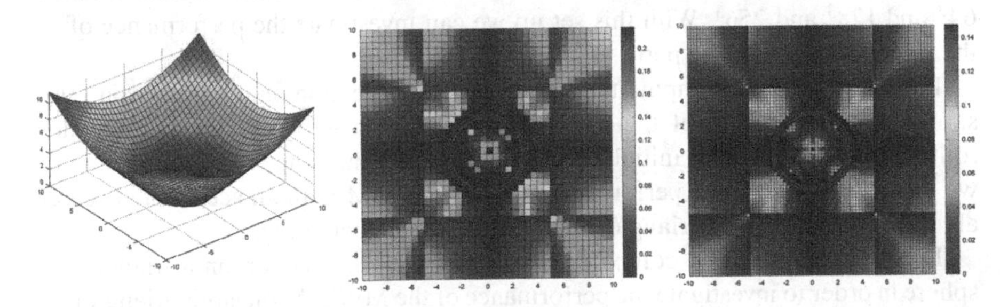

Fig. 2 On the *left*: parallel signed distance computation on 16 cores for a *circle*. In the *middle* and *right*: error-evolution on a 41×41 and a 81×81 grid

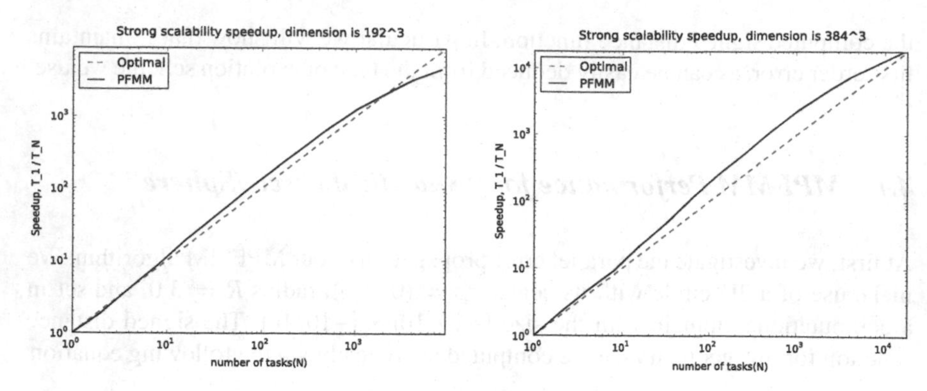

Fig. 3 Speedup for "analytical sphere" for up to 8192 cores (*left* 192^3, *right* 384^3)

scalability over all processor ranges. This is due to the logarithmic complexity of the sequential Fast Marching, which therefore scales logarithmically. It is worth noting that the only limiting factor in the number of subdomains on which the MPFMM algorithm can be parallelized on is the number of global grid cells. This does not present a problem particularly when dealing with larger domains. Thus it is a highly scalable algorithm for computing the signed distance function on large domains.

4.2 MPFMM for Triangulated Surfaces

Here we show some numerical tests targeting the overall performance of the MPFMM algorithm. Thus we run our MPFMM algorithm together with the data initialization routine. We run our algorithm on two different benchmark geometries, each of which is composed of a different number of triangles:

- Tetrahedron (Fig. 4) consisting of 4 triangles,
- Sphere (Fig. 4) comprising 840 triangles.

We run our algorithm on the benchmark geometries using three different meshes: 64^3 and 128^3 and 256^3. With this set up we can investigate the performance of all the important algorithmic parts.

In the first test we compute the signed distance function on a tetrahedron. This shows the performance of the MPFMM algorithm paired with the initialization routine for a very simple initialization. Due to the simple nature of the geometry, we can easily deduce the performance of the MPFMM with an accelerated search algorithm for the closest triangle. Such results are shown in Fig. 5.

In the second test, we compute the signed distance function on a triangulated sphere in order to investigate the performance of the MPFMM on larger triangulated meshes. Results depicted in Fig. 6 show that the MPFMM maintains good scalability and performance in this case as well.

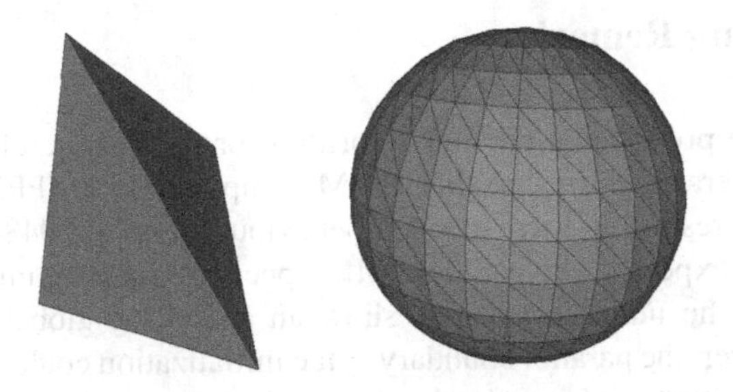

Fig. 4 A tetrahedron consisting of four *triangles* (*left*) and a triangulated sphere consisting of 840 *triangles* (*right*) are used for our speedup investigations

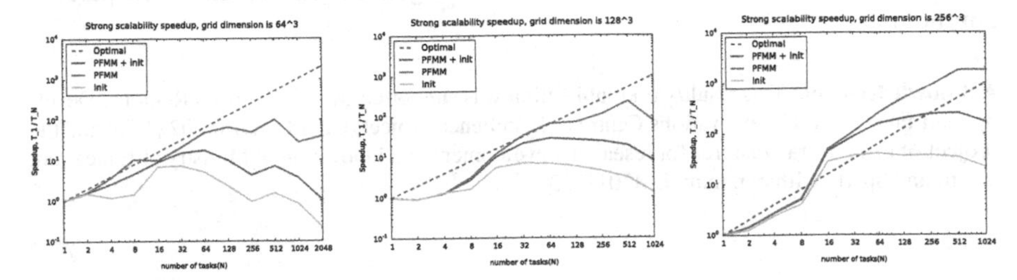

Fig. 5 Speedup for up to 2048 cores performed for the triangulated tetrahedron with three grid-resolutions: 64^3 (*left*) and 128^3 (*middle*) and 256^3 (*right*) gridcells

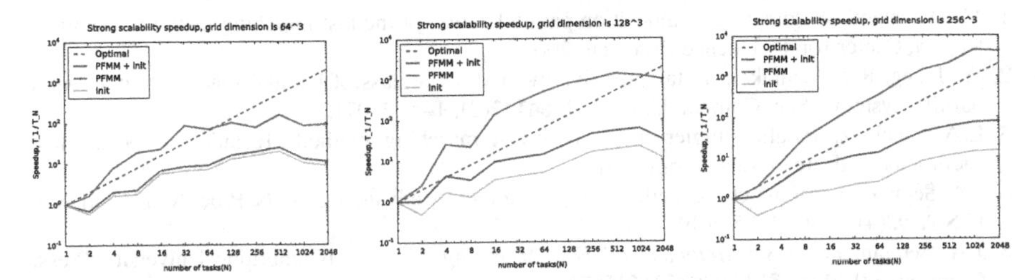

Fig. 6 Speedup for up to 2048 cores performed for the triangulated sphere with three grid-resolutions: 64^3 (*left*) and 128^3 (*middle*) and 256^3 (*right*) gridcells

Both tests show similar scaling properties and performance, suggesting that the latter are maintained through larger meshes. However, the algorithm performs more poorly for the smaller mesh of size 64^3. This is due to the fact that the ratio of set-up time to PFMM-computation time is higher, as the number of degrees of freedom is too small to obtain reasonably efficient parallel computation. These benchmark tests therefore show that scaling is limited only on the lower size of the mesh.

5 Concluding Remarks

In this paper we presented a parallel algorithm for the fast marching method. We investigated several massive parallel FMM-computations (MPFMM) for simple geometries with respect to their speedup behaviour on up to 2048- and 8192 cores respectively. As expected, the parallel FMM-speedup scales optimally for fine grid resolutions and the numerical results show an according global signed distance function. However, the parallel boundary value initialization could still be improved by storing the geometry information in a tree and use a triangle search with a special partition on the tree, instead of distributing the entire geometry to each process. Furthermore, we showed that the order of convergence is conserved for the parallel computations.

Acknowledgement This result/work/publication was supported by the European Regional Development Fund in the IT4Innovations Centre of Excellence project (CZ.1.05/1.1.00/02.0070) and the project of major infrastructures for research, development and innovation of Ministry of Education, Youth and Sports with reg. num. LM2011033.

References

1. M. Herrmann, A domain decomposition parallelization of the fast marching method. Technical Report, Center for Turbulence Research, 2003
2. W. Jeong, R.T. Whitaker, A fast iterative method for a class of hamilton-jacobi equations on parallel systems. Sch. Comput. Univ. Utah **84112**(2), 1–4 (2007)
3. L.A.Z. Núnez, Parallel implementation of fast marching method. Technical Report, Massachusetts Institute of Technology, 2011
4. J.A. Sethian, A fast marching method for monotonically advancing fronts. Proc. Natl. Acad. Sci. U.S.A. **93**(4), 1591–1595 (1996)
5. J.A. Sethian, *Level Set Methods and Fast Marching Methods* (Cambridge University Press, Cambridge, 1999). ISBN 9780521645577
6. M.C. Tugurlan, *Fast marching methods - Parallel implementation and analysis*. Dissetation, Louisiana State University, 2008
7. O. Weber, Y.S. Devir, A.M. Bronstein, M.M. Bronstein, R. Kimmel, Parallel algorithms for approximation of distance maps on parametric surfaces. ACM Trans. Graph. **27**(4), 104 (2008)

Discontinuous Galerkin Isogeometric Analysis of Elliptic PDEs on Surfaces

Ulrich Langer and Stephen E. Moore

1 Introduction

The Isogeometric Analysis (IGA), that was introduced by Hughes et al. [9] and has since been developed intensively, see also monograph [4], is a very suitable framework for representing and discretizing Partial Differential Equations (PDEs) on surfaces. We refer the reader to the survey paper by Dziuk and Elliot [7] where different finite element approaches to the numerical solution of PDEs on surfaces are discussed. Very recently, Dedner et al. [6] have used and analyzed the Discontinuous Galerkin (DG) finite element method for solving elliptic problems on surfaces. The IGA of second-order PDEs on surfaces has been introduced and numerically studied by Dede and Quarteroni [5] for the single-patch case. Brunero [3] presented some discretization error analysis of the DG-IGA applied to plane (2d) diffusion problems that carries over to plane linear elasticity problems which have recently been studied numerically in [1]. Evans and Hughes [8] used the DG technology in order to handle no-slip boundary conditions and multi-patch geometries for IGA of Darcy-Stokes-Brinkman equations. The efficient generation of the IGA equations, their fast solution, and the implementation of adaptive IGA schemes are currently

U. Langer (✉)
Institute for Computational Mathematics, Johannes Kepler University, Altenbergerstr. 69, A-4040 Linz, Austria
e-mail: ulanger@numa.uni-linz.ac.at

S.E. Moore
Johann Radon Institute for Computational and Applied Mathematics, Austrian Academy of Sciences, Altenbergerstr. 69, A-4040 Linz, Austria
e-mail: stephen.moore@ricam.oeaw.ac.at

T. Dickopf et al. (eds.), *Domain Decomposition Methods in Science and Engineering XXII*, Lecture Notes in Computational Science and Engineering 104, DOI 10.1007/978-3-319-18827-0_31

hot research topics. The use of DG technologies will certainly facilitate the handling of the multi-patch case.

In this paper, we use the DG method to handle the IGA of diffusion problems on closed or open, multi-patch NURBS surfaces. The DG technology easily allows us to handle non-homogeneous Dirichlet boundary conditions and multi-patch NURBS spaces which can be discontinuous across the patch boundaries. We also derive discretization error estimates in the DG- and L_2-norms. Finally, we present some numerical results confirming our theoretical estimates.

2 Surface Diffusion Model Problem

Let us assume that the physical (computational) domain Ω, where we are going to solve our diffusion problem, is a sufficiently smooth, two-dimensional generic (Riemannian) manifold (surface) defined in the physical space $\mathbb{R}^3$ by means of a smooth multi-patch NURBS mapping that is defined as follows. Let $\mathcal{T}_H = \{\Omega^{(i)}\}_{i=1}^N$ be a partition of our physical computational domain Ω into non-overlapping patches (sub-domains) $\Omega^{(i)}$ such that $\overline{\Omega} = \bigcup_{i=1}^N \overline{\Omega}^{(i)}$ and $\Omega^{(i)} \cap \Omega^{(j)} = \emptyset$ for $i \neq j$, and let each patch $\Omega^{(i)}$ be the image of the parameter domain $\hat{\Omega} = (0,1)^2 \subset \mathbb{R}^2$ by some NURBS mapping $G^{(i)} : \hat{\Omega} \to \Omega^{(i)} \subset \mathbb{R}^3, \boldsymbol{\xi} = (\boldsymbol{\xi}_1, \boldsymbol{\xi}_2) \mapsto \mathbf{x} = (\mathbf{x}_1, \mathbf{x}_2, \mathbf{x}_3) = G^{(i)}(\boldsymbol{\xi})$, which can be represented in the form

$$G^{(i)}(\xi_1, \xi_2) = \sum_{k_1=1}^{n_1} \sum_{k_2=1}^{n_2} \mathbf{P}^{(i)}_{(k_1,k_2)} \hat{R}^{(i)}_{(k_1,k_2)}(\xi_1, \xi_2) \tag{1}$$

where $\{\hat{R}^{(i)}_{(k_1,k_2)}\}$ are the bivariate NURBS basis functions, and $\{\mathbf{P}^{(i)}_{(k_1,k_2)}\}$ are the control points, see [4] for a detailed description.

Let us now consider a diffusion problem on the surface Ω, the weak formulation of which can be written as follows: find $u \in V_g$ such that

$$a(u, v) = \langle F, v \rangle \quad \forall v \in V_0, \tag{2}$$

with the bilinear and linear forms are given by the relations

$$a(u, v) = \int_\Omega \alpha\, \nabla_\Omega u \cdot \nabla_\Omega v \, d\Omega \quad \text{and} \quad \langle F, v \rangle = \int_\Omega f v \, d\Omega + \int_{\Gamma_N} g_N v \, d\Gamma,$$

respectively, where ∇_Ω denotes the so-called tangential or surface gradient, see e.g. Definition 2.3 in [7] for its precise description. The hyperplane V_g and the test space V_0 are given by $V_g = \{v \in V = H^1(\Omega) : v = g_D \text{ on } \Gamma_D\}$ and $V_0 = \{v \in V : v = 0 \text{ on } \Gamma_D\}$ for the case of an open surface Ω with the boundary $\Gamma = \overline{\Gamma}_D \cup \overline{\Gamma}_N$ such that $\text{meas}_1(\Gamma_D) > 0$, whereas $V_g = V_0 = \{v \in V : \int_\Omega v \, d\Omega = 0\}$ in the case of a pure Neumann problem ($\Gamma_N = \Gamma$) as well as in the case of closed surfaces unless there is a reaction term. In case of closed surfaces there is of course no integral over Γ_N in the linear functional on the right-hand side of (2). In the remainder of the paper, we will mainly discuss the case of mixed boundary value problems on an open surface under appropriate assumptions (e.g., $\text{meas}_1(\Gamma_D) > 0$, α—uniformly positive and bounded, $f \in L_2(\Omega)$, $g_D \in H^{\frac{1}{2}}(\Gamma_D)$ and $g_N \in L_2(\Gamma_N)$) ensuring existence and uniqueness of the solution of (2). For simplicity, we assume that the diffusion coefficient α is patch-wise constant, i.e. $\alpha = \alpha_i$ on $\Omega^{(i)}$ for $i = 1, 2, \ldots, N$. The other cases including the reaction-diffusion case can be treated in the same way and yield the same results like presented below.

3 DG-IGA Schemes and Their Properties

The DG-IGA variational identity

$$a_{DG}(u, v) = \langle F_{DG}, v \rangle \quad \forall v \in \mathcal{V} = H^{1+s}(\mathcal{T}_H), \tag{3}$$

which corresponds to (2), can be derived in the same way as their FE counterpart, where $H^{1+s}(\mathcal{T}_H) = \{v \in L_2(\Omega) : v|_{\Omega^{(i)}} \in H^{1+s}(\Omega^{(i)}),\ \forall\, i = 1, \ldots, N\}$ with some $s > 1/2$. The DG bilinear and linear forms in the *Symmetric Interior Penalty Galerkin* (SIPG) version, that is considered throughout this paper for definiteness, are defined by the relationships

$$\begin{aligned} a_{DG}(u, v) = & \sum_{i=1}^{N} \int_{\Omega^{(i)}} \alpha_i \nabla_\Omega u \cdot \nabla_\Omega v \, d\Omega \\ & - \sum_{\gamma \in \mathcal{E}_I \cup \mathcal{E}_D} \int_\gamma (\{\alpha \nabla_\Omega u \cdot \mathbf{n}\}[v] + \{\alpha \nabla_\Omega v \cdot \mathbf{n}\}[u]) \, d\Gamma \\ & + \sum_{\gamma \in \mathcal{E}_I \cup \mathcal{E}_D} \frac{\delta}{h_\gamma} \int_\gamma \alpha_\gamma [u][v] \, d\Gamma \end{aligned} \tag{4}$$

and

$$\begin{aligned} \langle F_{DG}, v \rangle = & \int_\Omega f v \, d\Omega + \sum_{\gamma \in \mathcal{E}_N} \int_\gamma g_N v \, d\Gamma \\ & + \sum_{\gamma \in \mathcal{E}_D} \int_\gamma \alpha_\gamma \left(-\nabla_\Omega v \cdot \mathbf{n} + \frac{\delta}{h_\gamma} v \right) g_D \, d\Gamma, \end{aligned} \tag{5}$$

respectively, where the usual DG notations for the averages $\{v\} = 1/2(v_i + v_j)$, jumps $[v] = v_i - v_j$ and $\alpha_\gamma = (\alpha_i + \alpha_j)/2$ on $\mathcal{E}_I$, with the corresponding modifications $\{v\} := v_i =: [v]$ and $\alpha_\gamma = \alpha_i$ on $\mathcal{E}_D$, are used, where i and j correspond to the indices of the patches to which the edge γ belongs, see, e.g., [12]. The sets $\mathcal{E}_I$, $\mathcal{E}_D$ and $\mathcal{E}_N$ denote the sets of edges γ of the patches belonging to $\Gamma_I = \cup \partial\Omega^{(i)} \setminus \{\Gamma_D \cup \Gamma_N\}$, Γ_D and Γ_N, respectively, whereas h_γ is the mesh-size on γ. The penalty parameter δ must be chosen such that the ellipticity of the DG bilinear on the DG space $\mathcal{V}_h$ can be ensured. The relationship between our model problem (2) and the DG variational identity (3) is given by the consistency theorem that can easily be verified.

Theorem 1 *If the solution u of the variational problem (2) belongs to $V_g \cap H^{1+s}(\mathcal{T}_H)$ with some $s > 1/2$, then u satisfies the DG variational identity (3). Conversely, if $u \in H^{1+s}(\mathcal{T}_H)$ satisfies (3), then u is the solution of our original variational problem (2).*

Now we consider the finite-dimensional Multi-Patch NURBS subspace

$$\mathcal{V}_h = \{v \in L_2(\Omega) : v|_{\Omega^{(i)}} \in V_h^i(\Omega^{(i)}), \ i = 1, \ldots, N\}$$

of our DG space $\mathcal{V}$, where $V_h^i(\Omega^{(i)}) = \text{span}\{R_{\mathbf{k}}^{(i)}\}$ denotes the space of NURBS functions on each single-patch $\Omega^{(i)}$, $i = 1, \ldots, N$, and the NURBS basis functions $R_{\mathbf{k}}^{(i)} = \hat{R}_{\mathbf{k}}^{(i)} \circ G^{(i)^{-1}}$ are given by the push-forward of the NURBS functions $\hat{R}_{\mathbf{k}}^{(i)}$ to their corresponding physical sub-domains $\Omega^{(i)}$ on the surface Ω. Finally, the DG scheme for our model problem (2) reads as follows: find $u_h \in \mathcal{V}_h$ such that

$$a_{DG}(u_h, v_h) = \langle F_{DG}, v_h \rangle, \quad \forall v_h \in \mathcal{V}_h. \tag{6}$$

For simplicity of our analysis, we assume matching meshes, see, e.g., [10]. Using special trace and inverse inequalities in the NURBS spaces $\mathcal{V}_h$ and Young's inequality, for sufficiently large DG penalty parameter δ, we can easily establish $\mathcal{V}_h$ coercivity and boundedness of the DG bilinear form with respect to the DG energy norm

$$\|v\|_{DG}^2 = \sum_{i=1}^{N} \alpha_i \|\nabla_\Omega v_i\|_{L_2(\Omega^{(i)})}^2 + \sum_{\gamma \in \mathcal{E}_I \cup \mathcal{E}_D} \alpha_\gamma \frac{\delta}{h_\gamma} \|[v]\|_{L_2(\gamma)}^2, \tag{7}$$

yielding existence and uniqueness of the DG solution $u_h \in \mathcal{V}_h$ of (6) that can be determined by the solution of a linear system of algebraic equations.

4 Discretization Error Estimates

Theorem 2 *Let $u \in V_g \cap H^{1+s}(\mathcal{T}_H)$ with some $s > 1/2$ be the solution of (2), $u_h \in \mathcal{V}_h$ be the solution of (6), and the penalty parameter δ be chosen large enough. Then there exists a positive constant c that is independent of u, the discretization* *parameters and the jumps in the diffusion coefficients such that the DG-norm error estimate*

$$\|u - u_h\|_{DG}^2 \le c \sum_{i=1}^{N} \alpha_i h_i^{2t} \|u\|_{H^{1+t}(\Omega^{(i)})}^2, \tag{8}$$

holds with $t := \min\{s, p\}$*, where the discretization parameter* h_i *characterizes the mesh-size in the patch* $\Omega^{(i)}$*, and* p *always denotes the underlying polynomial degree of the NURBS.*

Proof Let us give a sketch of the proof. By the triangle inequality, we have

$$\|u - u_h\|_{DG} \le \|u - \Pi_h u\|_{DG} + \|\Pi_h u - u_h\|_{DG} \tag{9}$$

with some quasi-interpolation operator $\Pi_h : \mathcal{V} \mapsto \mathcal{V}_h$ such that the first term can be estimated with optimal order, i.e. by the term on the right-hand side of (8) with some other constant c. This is possible due to the approximation results known for NURBS, see, e.g., [2, 4]. Now it remains to estimate the second term in the same way. Using the Galerkin orthogonality $a_{DG}(u - u_h, v_h) = 0$ for all $v_h \in \mathcal{V}_h$, the $\mathcal{V}_h$ coercitivity of the bilinear form $a_{DG}(\cdot,\cdot)$, the scaled trace inequality

$$\|v\|_{L_2(e)} \le C h_E^{-1/2} \left(\|v\|_{L_2(E)} + h_E^{1/2+\epsilon} |v|_{H^{1/2+\epsilon}(E)} \right), \tag{10}$$

that holds for all $v \in H^{1/2+\epsilon}(E)$, for all IGA mesh elements E, for all edges $e \subset \partial E$, and for $\epsilon > 0$, where h_E denotes the mesh-size of E or the length of e, Young's inequality, and again the approximation properties of the quasi-interpolation operator Π_h, we can estimate the second term by the same term $c \left(\sum_{i=1}^{N} \alpha_i h_i^{2t} \|u\|_{H^{1+t}(\Omega^{(i)})}^2 \right)^{1/2}$ with some (other) constant c. This completes the proof of the theorem, cf. [12] for the finite element case. □

Using duality arguments, we can also derive L_2-norm error estimates that depend on the elliptic regularity. Under the assumption of full elliptic regularity, we get $\|u - u_h\|_{L_2(\Omega)} \le c\, h^{p+1} \|u\|_{H^{p+1}(\Omega)}$ that is nicely confirmed by our numerical experiments presented in the next section for $p = 1, 2, 3, 4$.

5 Numerical Results

The DG IGA method presented in this paper as well as its continuous Galerkin counterpart have been implemented in the object oriented C++ IGA library "Geometry + Simulation Modules" (G+SMO).[1] We present some first numerical results for testing the numerical behavior of the discretization error with respect to the mesh

[1]G+SMO: www.gs.jku.at.

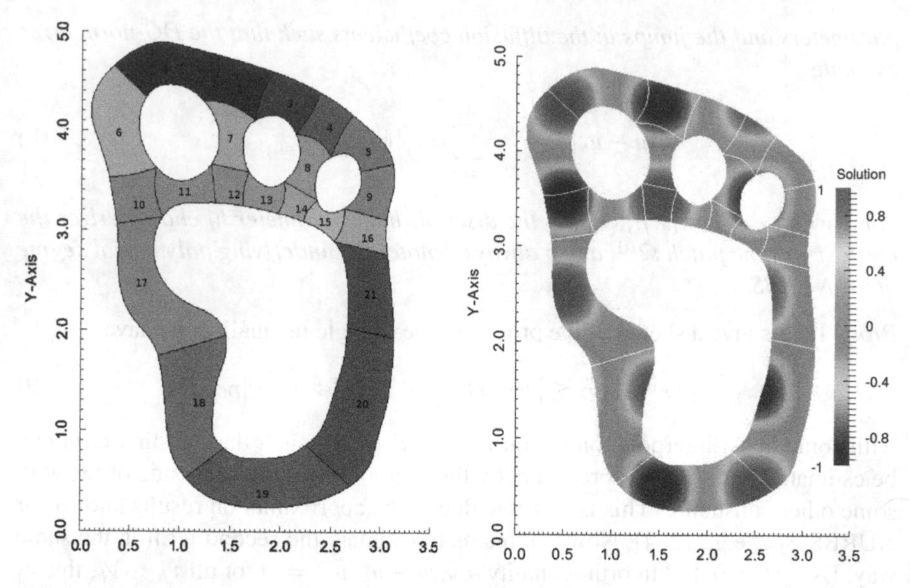

Fig. 1 Yeti foot: geometry (*left*) and DG-IGA solution (*right*)

parameter h and the polynomial degree p. Concerning the choice of the penalty parameter, we used $\delta = 2(p+2)(p+1)$.

As a first example, we consider a non-homogeneous Dirichlet problem for the Poisson equation in the 2d computational domain $\Omega \subset \mathbb{R}^2$ called Yeti's footprint, see also [10], where the right-hand side f and the Dirichlet data g_D are chosen such that $u(x_1, x_2) = \sin(\pi x_1)\sin(\pi x_2)$ is the solution of the boundary value problem. The computational domain (left) and the solution (right) can be seen in Fig. 1. The Yeti footprint consists of 21 patches with varying open knot vectors Ξ describing the NURBS discretization in a short and precise way, see, e.g., [4] for a detailed definition. The open knot vectors for building the patches 1 to 16 and 21 are given by $\Xi = (0,0,0,0.5,1,1,1)$ in both directions, whereas the knot vectors for the patches 17 to 20 are given by $\Xi_1 = (0,0,0,0.5,1,1,1)$ and $\Xi_2 = (0,0,0,0.25,0.5,0.75,1,1,1)$. In Fig. 2, the errors in the L_2-norm and in the DG energy norm (7) are plotted against the degree of freedom (DOFs) with polynomial degrees from 1 to 4. It can be observed that we have convergence rates of $\mathcal{O}(h^{p+1})$ and $\mathcal{O}(h^p)$ respectively. This corresponds to our theory in Sect. 4.

In the second example, we apply the DG-IGA to the same Laplace-Beltrami problem on an open surface as described in [5], Section 5.1, where Ω is a quarter cylinder represented by four patches in our computations, see Fig. 3 (left). The open knot vectors $\Xi_1 := (0,0,0,1,1,1)$ and $\Xi_2 := (0,0,1,1)$ are used to build the patches. The L_2-norm errors plotted on the right side of Fig. 3 exhibit the same numerical behavior as in the plane case of the Yeti foot. The same is true for the DG-norm.

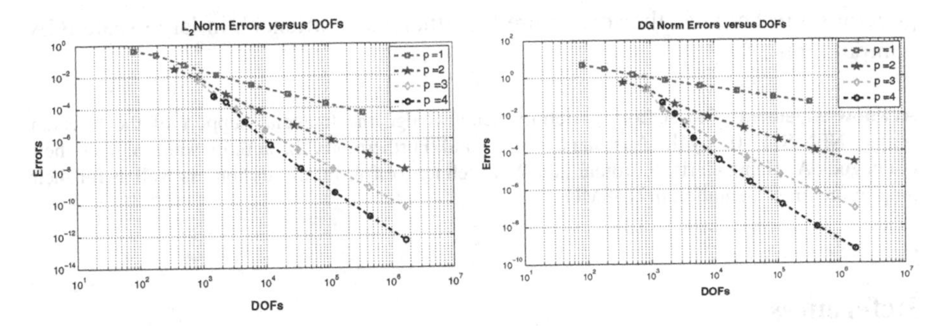

Fig. 2 Yeti foot: L_2- and DG-norm errors with polynomial degree p

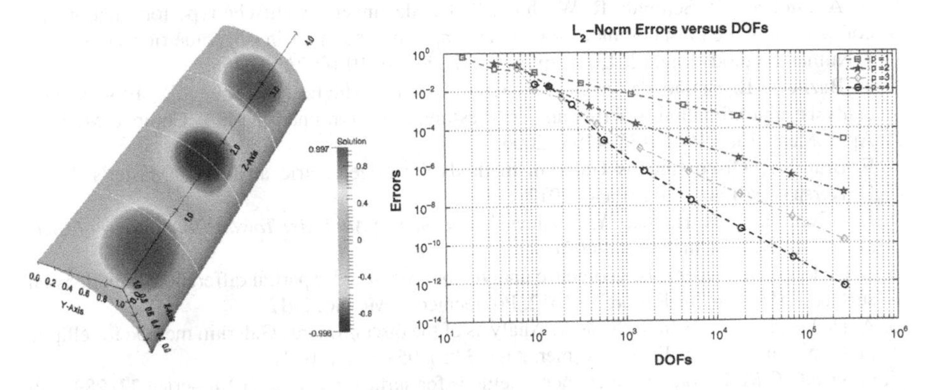

Fig. 3 Quarter cylinder: geometry with the solution (*left*) and L_2 norm errors (*right*)

6 Conclusions

We have developed and analyzed a new method for the numerical approximation of diffusion problems on open and closed surfaces by combining the discontinuous Galerkin technique with isogeometric analysis. We refer to our approach as the Discontinuous Galerkin Isogeometric Analysis (DG-IGA). In our DG approach we allow discontinuities only across the boundaries of the patches, into which the computational domain is decomposed, and enforce the interface conditions in the DG framework. For simplicity of presentation, we assume that the meshes are matching across the patches, and the solution u is at least patch-wise in H^{1+s}, i.e. $u \in H^{1+s}(\mathcal{T}_H)$, with some $s > 1/2$. The cases of non-matching meshes and low-regularity solution, that are technically more involved and that were investigated, e.g., by Di Pietro and Ern [11], will be considered in a forthcoming paper. The parallel solution of the DG-IGA equations can efficiently be performed by Domain Decomposition (DD) solvers like the IETI technique proposed by Kleiss et al. [10], see also [1] for other DD solvers. The construction and analysis of efficient solution strategies is currently a hot research topic since, beside efficient generation techniques, the solvers are the efficiency bottleneck in large-scale IGA computations.

Acknowledgement The authors gratefully acknowledge the financial support of the research project NFN S117-03 by the Austrian Science Fund. Furthermore, the authors want to thank their colleagues A. Mantzaflaris, S. Tomar and W. Zulehner for fruitful discussions as well as for their help in the implementation in G+SMO.

References

1. A. Apostolatos, R. Schmidt, R. Wüchner, K.-U. Bletzinger, A Nitsche-type formulation and comparison of the most common domain decomposition methods in isogeometric analysis. Int. J. Numer. Methods Eng. (2013). http://www.dx.doi.org/10.1002/nme.4568
2. Y. Bazilevs, L. Beirão da Veiga, J.A. Cottrell, T.J.R. Hughes, G. Sangalli, Isogeometric analysis: approximation, stability and error estimates for h-refined meshes. Comput. Methods Appl. Mech. Eng. **194**, 4135–4195 (2006)
3. F. Brunero, Discontinuous Galerkin methods for isogeometric analysis. Master's thesis, Università degli Studi di Milano, 2012
4. J.A. Cottrell, T.J.R. Hughes, Y. Bazilevs, *Isogeometric Analysis: Toward Integration of CAD and FEA* (Wiley, Chichester, 2009)
5. L. Dede, A. Quarteroni, Isogeometric analyis for second order partial differential equations on surfaces. MATHICSE Report 36.2012, Politecnico di Milano, 2012
6. A. Dedner, P. Madhavan, B. Stinner, Analysis of the discontinuous Galerkin method for elliptic problems on surfaces. IMA J. Numer. Anal **33**(3), 952–973 (2013)
7. G. Dziuk, C.M. Elliot, Finite element methods for surface pdes. Acta Numerica **22**, 289–396 (2013)
8. J.A. Evans, T.J.R. Hughes, Isogeometric divergence-conforming B-splines for the Darcy-Stokes-Brinkman equations. Math. Models Meth. Appl. Sci. **23**(4), 671–741 (2013)
9. T.J.R. Hughes, J.A. Cottrell, Y. Bazilevs, Isogeometric analysis: CAD, finite elements, NURBS, exact geometry and mesh refinement. Comput. Methods Appl. Mech. Eng. **194**, 4135–4195 (2005)
10. S.K. Kleiss, C. Pechstein, B. Jüttler, S. Tomar, IETI – isogeometric tearing and interconnecting. Comput. Methods Appl. Mech. Eng. **247–248**, 201–215 (2012)
11. D.A. Di Pietro, A. Ern, *Mathematical Aspects of Discontinous Galerkin Methods*. Mathématiques et Applications, vol. 69 (Springer, Heidelberg, Dordrecht, London, New York, 2012)
12. B. Rivière, *Discontinuous Galerkin Methods for Solving Elliptic and Parabolic Equations: Theory and Implementation* (SIAM, Philadelphia, 2008)

A FETI-DP Algorithm for Saddle Point Problems in Three Dimensions

Xuemin Tu and Jing Li

1 Introduction

In [2, 6, 7], a new class of FETI-DP type domain decomposition algorithms was introduced and analyzed by the authors for solving incompressible Stokes equations in two dimensions. Both discontinuous and continuous pressures can be used in the mixed finite element discretization. In both cases, the indefinite system of linear equations can be reduced to a symmetric positive semi-definite system. Therefore, the preconditioned conjugate gradient method can be applied.

Both lumped and Dirichlet preconditioners have been studied in [2, 6, 7]. For the lumped preconditioner, it has been proved in [2] that the coarse level space can be chosen as simple as for solving scalar elliptic problems corresponding to each velocity component to achieve a scalable convergence rate. However, for the Dirichlet preconditioner, most existing FETI-DP and BDDC type algorithms [1, 3, 4] for Stokes problems use subdomain Stokes extensions in the preconditioners and the coarse level velocity space has to contain sufficient components to enforce divergence free subdomain boundary velocity conditions. Due to this divergence free requirement, the coarse space becomes very complicated, especially for three-dimensional problems as discussed in [3]. For the Dirichlet preconditioner

X. Tu (✉)
Department of Mathematics, University of Kansas, 1460 Jayhawk Blvd, Lawrence, Kansas 66045-7594, USA
e-mail: xtu@math.ku.edu

J. Li
Department of Mathematical Sciences, Kent State University, Kent, Ohio 44242, USA
e-mail: li@math.kent.edu

T. Dickopf et al. (eds.), *Domain Decomposition Methods in Science and Engineering XXII*, Lecture Notes in Computational Science and Engineering 104, DOI 10.1007/978-3-319-18827-0_32

introduced in [6, 7], an application of subdomain harmonic extension instead of Stoke extension in the preconditioner makes it possible to remove the divergence free constraints for the coarse level velocity space. Unfortunately, the analysis provided for the algorithms in [6, 7] still requires the divergence free constraints.

In this paper, we provide a new analysis for the algorithms in [6, 7], which can not only analyze both lumped and Dirichlet preconditioners in a same framework, but also remove the divergence free constraints for the Dirichlet preconditioner. We then extended this class of algorithms [2, 6, 7] to three dimensional problems; see [8] for more details.

2 Discretization, Domain Decomposition, and a Reduced Interface System

Let Ω be a bounded, three-dimensional polyhedral domain. We consider solving the following saddle point problem: find $\mathbf{u}^* \in \left(H_0^1(\Omega)\right)^3 = \{\mathbf{v} \in (H^1(\Omega))^3 \mid \mathbf{v} = \mathbf{0} \text{ on } \partial\Omega\}$ and $p^* \in L^2(\Omega)$, such that

$$\begin{cases} a(\mathbf{u}^*, \mathbf{v}) + b(\mathbf{v}, p^*) = (\mathbf{f}, \mathbf{v}), \ \forall \mathbf{v} \in \left(H_0^1(\Omega)\right)^3, \\ b(\mathbf{u}^*, q) \qquad\qquad\quad = 0, \qquad \forall q \in L^2(\Omega), \end{cases} \tag{1}$$

where

$$a(\mathbf{u}^*, \mathbf{v}) = \int_\Omega \nabla \mathbf{u}^* \cdot \nabla \mathbf{v}, \quad b(\mathbf{u}^*, q) = -\int_\Omega (\nabla \cdot \mathbf{u}^*) q, \quad (\mathbf{f}, \mathbf{v}) = \int_\Omega \mathbf{f} \cdot \mathbf{v}.$$

The solution of (1) is not unique and the pressure p^* is determined up to an additive constant.

The domain Ω is partitioned into shape-regular rectangular elements of characteristic size h, and the Q_2-Q_1 Taylor-Hood mixed finite element is used to solve (1). The pressure finite element space, $Q \subset L^2(\Omega)$, is taken as the space of continuous piecewise trilinear functions while the velocity finite element space, $\mathbf{W} \in \left(H_0^1(\Omega)\right)^3$, is formed by the continuous piecewise triquadratic functions.

The finite element solution $(\mathbf{u}, p) \in \mathbf{W} \bigoplus Q$ of (1) satisfies

$$\begin{bmatrix} A & B^T \\ B & 0 \end{bmatrix} \begin{bmatrix} \mathbf{u} \\ p \end{bmatrix} = \begin{bmatrix} \mathbf{f} \\ 0 \end{bmatrix}, \tag{2}$$

where A, B, and $\mathbf{f}$ represent, respectively, the restrictions of $a(\cdot,\cdot)$, $b(\cdot,\cdot)$ and $(\mathbf{f},\cdot)$ to the finite-dimensional spaces $\mathbf{W}$ and Q. The solution of (2) always exists and is uniquely determined when the pressure is required to have a zero average.

The Q_2-Q_1 Taylor-Hood mixed finite element space $\mathbf{W} \times Q$ is inf-sup stable in the sense that there exists a positive constant β, independent of h, such that, in matrix/vector form,

$$\sup_{\mathbf{w} \in \mathbf{W}} \frac{\langle q, B\mathbf{w} \rangle^2}{\langle \mathbf{w}, A\mathbf{w} \rangle} \geq \beta^2 \langle q, Zq \rangle, \qquad \forall q \in Q/Ker(B^T). \tag{3}$$

Here the matrix Z represents the mass matrix defined on the pressure finite element space Q, i.e., for any $q \in Q$, $\|q\|_{L^2}^2 = \langle q, Zq \rangle$. It is easy to see, cf. [5, Lemma B.31], that Z is spectrally equivalent to $h^3 I$ for three-dimensional problems, where I represents the identity matrix of the same dimension.

The domain Ω is decomposed into N non-overlapping polyhedral subdomains Ω_i, $i = 1, 2, \ldots, N$. Each subdomain is the union of a bounded number of elements, with the diameter of the subdomain in the order of H. The nodes on the interface Γ of neighboring subdomains match across the subdomain boundaries and Γ is composed of subdomain faces, which are regarded as open subsets of Γ shared by two subdomains, subdomain edges, which are regarded as open subsets of Γ shared by more than two subdomains, and of the subdomain vertices, which are end points of edges.

The velocity and pressure finite element spaces $\mathbf{W}$ and Q are decomposed into

$$\mathbf{W} = \mathbf{W}_I \bigoplus \mathbf{W}_\Gamma, \quad Q = Q_I \bigoplus Q_\Gamma,$$

where $\mathbf{W}_I$ and Q_I are direct sums of independent subdomain interior velocity spaces $\mathbf{W}_I^{(i)}$, and interior pressure spaces $Q_I^{(i)}$, respectively. $\mathbf{W}_\Gamma$ and Q_Γ are subdomain interface velocity and pressure spaces, respectively. All functions in $\mathbf{W}_\Gamma$ and Q_Γ are continuous across Γ; their degrees of freedom are shared by neighboring subdomains. A partially sub-assembled subdomain interface velocity space $\widetilde{\mathbf{W}}_\Gamma$ is defined as

$$\widetilde{\mathbf{W}}_\Gamma = \mathbf{W}_\Delta \bigoplus \mathbf{W}_\Pi = \left(\bigoplus_{i=1}^{N} \mathbf{W}_\Delta^{(i)} \right) \bigoplus \mathbf{W}_\Pi.$$

$\mathbf{W}_\Pi$ is the continuous, coarse level, primal velocity space which is typically spanned by subdomain vertex nodal basis functions, and/or by interface edge/face-cutoff functions with constant nodal values on each edge/face, or with values of positive weights on these edges/faces. The primal, coarse level velocity degrees of freedom are shared by neighboring subdomains. The complimentary space $\mathbf{W}_\Delta$ is the direct sum of independent subdomain dual interface velocity spaces $\mathbf{W}_\Delta^{(i)}$, which correspond to the remaining subdomain interface velocity degrees of freedom and are spanned by basis functions which vanish at the primal degrees of freedom. Thus, an element in $\widetilde{\mathbf{W}}_\Gamma$ typically has a continuous primal velocity component and a discontinuous dual velocity component.

We construct a matrix B_Δ from $\{0,1,-1\}$ to enforce the continuity for dual velocity components. For any $\mathbf{w}_\Delta$ in $\mathbf{W}_\Delta$, each row of $B_\Delta \mathbf{w}_\Delta = 0$ implies that the two independent degrees of freedom from the neighboring subdomains be the same. The range of B_Δ applied on $\mathbf{W}_\Delta$ is a vector space of the Lagrange multipliers, denoted by Λ. For each node x on the subdomain boundary Γ, we define a positive scaling factor $\delta^\dagger(x) = 1/\mathcal{N}_x$, where $\mathcal{N}_x$ represents the number of subdomains sharing x. Multiplying the entries on each row of B_Δ by the corresponding scaling factor $\delta^\dagger(x)$ gives us $B_{\Delta,D}$.

The original system (2) is equivalent to: find $(\mathbf{u}_I, p_I, \mathbf{u}_\Delta, \mathbf{u}_\Pi, p_\Gamma, \lambda) \in \mathbf{W}_I \bigoplus Q_I \bigoplus \mathbf{W}_\Delta \bigoplus \mathbf{W}_\Pi \bigoplus Q_\Gamma \bigoplus \Lambda$, such that

$$\begin{bmatrix} A_{II} & B_{II}^T & A_{I\Delta} & A_{I\Pi} & B_{\Gamma I}^T & 0 \\ B_{II} & 0 & B_{I\Delta} & B_{I\Pi} & 0 & 0 \\ A_{\Delta I} & B_{I\Delta}^T & A_{\Delta\Delta} & A_{\Delta\Pi} & B_{\Gamma\Delta}^T & B_\Delta^T \\ A_{\Pi I} & B_{I\Pi}^T & A_{\Pi\Delta} & A_{\Pi\Pi} & B_{\Gamma\Pi}^T & 0 \\ B_{\Gamma I} & 0 & B_{\Gamma\Delta} & B_{\Gamma\Pi} & 0 & 0 \\ 0 & 0 & B_\Delta & 0 & 0 & 0 \end{bmatrix} \begin{bmatrix} \mathbf{u}_I \\ p_I \\ \mathbf{u}_\Delta \\ \mathbf{u}_\Pi \\ p_\Gamma \\ \lambda \end{bmatrix} = \begin{bmatrix} \mathbf{f}_I \\ 0 \\ \mathbf{f}_\Delta \\ \mathbf{f}_\Pi \\ 0 \\ 0 \end{bmatrix}, \tag{4}$$

where the sub-blocks in the coefficient matrix represent the restrictions of A and B of (2) to appropriate subspaces. The leading three-by-three block can be ordered to become block diagonal with each diagonal block representing one independent subdomain problem.

Lemma 1 ([8, Lemma 4]) *The basis vector in the null space of (4), corresponding to the one-dimensional null space of the original incompressible Stokes system (2), is*

$$\left(\mathbf{0},\ 1_{p_I},\ \mathbf{0},\ \mathbf{0},\ 1_{p_\Gamma},\ -B_{\Delta,D}[B_{I\Delta}^T\ B_{\Gamma\Delta}^T] \begin{bmatrix} 1_{p_I} \\ 1_{p_\Gamma} \end{bmatrix} \right). \tag{5}$$

Here $1_{p_I} \in Q_I$ and $1_{p_\Gamma} \in Q_\Gamma$ represent vectors with each entry equal to 1.

System (4) can be reduced to a Schur complement problem for the variables (p_Γ, λ)

$$G \begin{bmatrix} p_\Gamma \\ \lambda \end{bmatrix} = g, \tag{6}$$

where

$$G = B_C \tilde{A}^{-1} B_C^T, \qquad g = B_C \tilde{A}^{-1} \begin{bmatrix} \mathbf{f}_I \\ 0 \\ \mathbf{f}_\Delta \\ \mathbf{f}_\Pi \end{bmatrix}, \tag{7}$$

with

$$\tilde{A} = \begin{bmatrix} A_{II} & B_{II}^T & A_{I\Delta} & A_{I\Pi} \\ B_{II} & 0 & B_{I\Delta} & B_{I\Pi} \\ A_{\Delta I} & B_{I\Delta}^T & A_{\Delta\Delta} & A_{\Delta\Pi} \\ A_{\Pi I} & B_{I\Pi}^T & A_{\Pi\Delta} & A_{\Pi\Pi} \end{bmatrix} \quad \text{and} \quad B_C = \begin{bmatrix} B_{\Gamma I} & 0 & B_{\Gamma\Delta} & B_{\Gamma\Pi} \\ 0 & 0 & B_{\Delta} & 0 \end{bmatrix}. \tag{8}$$

G is symmetric positive semi-definite. The null space of G can be derived from Lemma 1, and its basis has the form

$$\left(1_{p_\Gamma}, -B_{\Delta,D}[B_{I\Delta}^T \; B_{\Gamma\Delta}^T]\begin{bmatrix} 1_{p_I} \\ 1_{p_\Gamma} \end{bmatrix}\right).$$

Let $X = Q_\Gamma \bigoplus \Lambda$. The range of G, denoted by R_G, is the subspace of X, which is orthogonal to the null space of G and has the form

$$R_G = \left\{ \begin{bmatrix} g_{p_\Gamma} \\ g_\lambda \end{bmatrix} \in X \;\middle|\; g_{p_\Gamma}^T 1_{p_\Gamma} - g_\lambda^T \left(B_{\Delta,D}[B_{I\Delta}^T \; B_{\Gamma\Delta}^T] \begin{bmatrix} 1_{p_I} \\ 1_{p_\Gamma} \end{bmatrix}\right) = 0 \right\}. \tag{9}$$

The restriction of G to its range R_G is positive definite. The conjugate gradient method will be used to solve (6), with preconditioners given in the next section.

We denote

$$A_{rr} = \begin{bmatrix} A_{II} & B_{II}^T & A_{I\Delta} \\ B_{II} & 0 & B_{I\Delta} \\ A_{\Delta I} & B_{I\Delta}^T & A_{\Delta\Delta} \end{bmatrix}, \; A_{\Pi r} = A_{r\Pi}^T = \begin{bmatrix} A_{\Pi I} & B_{I\Pi}^T & A_{\Pi\Delta} \end{bmatrix}, \; f_r = \begin{bmatrix} \mathbf{f}_I \\ 0 \\ \mathbf{f}_\Delta \end{bmatrix},$$

and define the Schur complement $S_\Pi = A_{\Pi\Pi} - A_{\Pi r}A_{rr}^{-1}A_{r\Pi}$, which is symmetric positive definite and defines the coarse level problem of this algorithm.

The main operation in the implementation of multiplying G by a vector is the product of $\tilde{A}^{-1}$ with a vector consisting of f_r and $\mathbf{f}_\Pi$. This product can be represented by

$$\begin{bmatrix} A_{rr}^{-1} f_r \\ \mathbf{0} \end{bmatrix} + \begin{bmatrix} -A_{rr}^{-1}A_{r\Pi} \\ I_\Pi \end{bmatrix} S_\Pi^{-1} \left(\mathbf{f}_\Pi - A_{\Pi r}A_{rr}^{-1} f_r\right),$$

which requires solving the coarse level problem once and independent subdomain Stokes problems with Neumann type boundary conditions twice.

3 Preconditioners and Condition Number Bounds

We define $\tilde{V} = \mathbf{W}_I \bigoplus Q_I \bigoplus \mathbf{W}_\Delta \bigoplus \mathbf{W}_\Pi$, and its subspace

$$\tilde{V}_0 = \left\{ w = (\mathbf{w}_I,\, p_I,\, \mathbf{w}_\Delta,\, \mathbf{w}_\Pi) \in \tilde{V} : B_{II}\mathbf{w}_I + B_{I\Delta}\mathbf{w}_\Delta + B_{I\Pi}\mathbf{w}_\Pi = 0 \right\}.$$

For any $v \in \tilde{V}_0$, the value $\langle v, v \rangle_{\tilde{A}} = v^T \tilde{A} v$ is independent of its pressure component p_I. $\langle \cdot, \cdot \rangle_{\tilde{A}}$ defines a semi-inner product on $\tilde{V}_0$; $\langle v, v \rangle_{\tilde{A}} = 0$ if and only if the velocity component of v is zero while its pressure component can be arbitrary. We denote the restriction operator from $\tilde{V}$ onto $\mathbf{W}_\Delta$ by $\tilde{R}_\Delta$ such that for any $v = (\mathbf{w}_I,\, p_I,\, \mathbf{w}_\Delta,\, \mathbf{w}_\Pi) \in \tilde{V}$, $\tilde{R}_\Delta v = \mathbf{w}_\Delta$.

Let H_Δ represent the direct sum of discrete subdomain harmonic extension operators. Let $M_{L,\lambda}^{-1} = B_{\Delta,D} \tilde{R}_\Delta \tilde{A} \tilde{R}_\Delta^T B_{\Delta,D}^T$ and $M_{D,\lambda}^{-1} = B_{\Delta,D} H_\Delta B_{\Delta,D}^T$. The lumped and Dirichlet preconditioners M_L^{-1} and M_D^{-1} for solving (6) are given by

$$M_L^{-1} = \begin{bmatrix} \frac{\alpha}{h^3} I_{p_\Gamma} & \\ & M_{L,\lambda}^{-1} \end{bmatrix} \quad \text{and} \quad M_D^{-1} = \begin{bmatrix} \frac{\alpha}{h^3} I_{p_\Gamma} & \\ & M_{D,\lambda}^{-1} \end{bmatrix}.$$

Here I_{p_Γ} is the identity matrix of the same length as p_Γ. α is a given constant, whose value is typically taken as 1. We introduce α in the preconditioner just for the convenience in the numerical experiments to demonstrate the convergence rates of the proposed algorithm.

For both lumped and Dirichlet preconditioners, the coarse space includes only subdomain corner and edge-average variables for each velocity component, just as for solving scalar elliptic problems. Such coarse space is sufficient for this algorithm to achieve scalable convergence rates as given in the following theorem for both type preconditioners, denoted here by M^{-1}.

Lemma 2 ([8, Lemma 10]) *There exists a constant C, such that for all $v \in \tilde{V}_0$,*

$$\left\langle M^{-1} B_C v, B_C v \right\rangle \leq C\,(\alpha + \Phi(H/h)) \left\langle \tilde{A} v, v \right\rangle.$$

Here, for the lumped preconditioner, $\Phi(H,h) = C(H/h)(1 + \log(H/h))$, and for the Dirichlet preconditioner, $\Phi(H,h) = C(1 + \log(H/h))^2$.

Lemma 3 ([8, Lemma 11]) *There exists a constant C, such that for any nonzero $y = (g_{p_\Gamma}, g_\lambda) \in R_G$, there exits $v \in \tilde{V}_0$, which satisfies $B_C v = y$, $\langle v, v \rangle_{\tilde{A}} \neq 0$, and*

$$\left\langle \tilde{A} v, v \right\rangle \leq C \max\left\{1, \tfrac{1}{\alpha}\right\} \left(1 + \tfrac{1}{\beta^2}\right) \left\langle M^{-1} y, y \right\rangle.$$

Theorem 1 ([8, Theorem 1]) *There exist positive constants c and C, such that for all x in the range of $M^{-1}G$,*

$$\min\{1, \alpha\} \frac{c\beta^2}{(1+\beta^2)} \langle Mx, x \rangle \leq \langle Gx, x \rangle \leq C\,(\alpha + \Phi(H/h))\,\langle Mx, x \rangle.$$

4 Numerical Experiments

We solve the saddle point problem (1) on the cube $\Omega = [0, 1]^3$ with a zero Dirichlet boundary condition. The right-hand side $\mathbf{f}$ is chosen such that the exact solution is

$$\mathbf{u} = \begin{bmatrix} \sin^2(\pi x)\,(\sin(2\pi y)\sin(\pi z) - \sin(\pi y)\sin(2\pi z)) \\ \sin^2(\pi y)\,(\sin(2\pi z)\sin(\pi x) - \sin(\pi z)\sin(2\pi x)) \\ \sin^2(\pi z)\,(\sin(2\pi x)\sin(\pi y) - \sin(\pi x)\sin(2\pi y)) \end{bmatrix}, \quad p = xyz - \frac{1}{8}.$$

The Q_2-Q_1 Taylor-Hood mixed finite element is used and the preconditioned system is solved by a conjugate gradient (CG) method. The CG iteration is stopped when the L^2–norm of the residual is reduced by a factor of 10^{-6}. We use the tridiagonal Lanczos matrix generated in the iteration to estimate the extreme eigenvalues of $M^{-1}G$.

For both preconditioners, the coarse level velocity space is the same as for solving scalar elliptic problems in [5, Algorithm 6.25] corresponding to each velocity component, which is spanned by the subdomain vertex nodal basis functions and subdomain edge-cutoff functions.

We take $\alpha = 1$ in Table 1 and $\alpha = 1/2$ in Table 2, to demonstrate more clearly the upper eigenvalue bound in Theorem 1. Using the Dirichlet preconditioner can reduce $\Phi(H/h)$ compared with the lumped preconditioner. However, for a small value of H/h, $\alpha = 1$ will be dominant in the upper bound and the effect of $\Phi(H/h)$ on the convergence rate is not visible in Table 1. When α is reduced to $1/2$, $\Phi(H/h)$ becomes visible and the upper eigenvalue bounds in Table 2 exhibit the pattern of $\Phi(H/h)$ for both preconditioners. They are independent of the number of subdomains for fixed H/h; for fixed number of subdomains, they depend on H/h in the order of $(H/h)(1 + \log(H/h))$ for the lumped preconditioner, and $(1 + \log(H/h))^2$ for the Dirichlet preconditioner. The lower eigenvalue bounds in

Table 1 Performance of solving three-dimensional problem on $[0, 1]^3$, $\alpha = 1$

		Lumped			Dirichlet		
H/h	#sub	λ_{min}	λ_{max}	Iteration	λ_{min}	λ_{max}	Iteration
4	$3 \times 3 \times 3$	0.0776	9.13	56	0.0776	8.97	56
	$4 \times 4 \times 4$	0.0775	9.35	54	0.0774	9.19	55
	$6 \times 6 \times 6$	0.0773	9.41	58	0.0773	9.23	59
	$8 \times 8 \times 8$	0.0773	9.51	57	0.0772	9.34	61
#sub	H/h	λ_{min}	λ_{max}	Iteration	λ_{min}	λ_{max}	Iteration
$3 \times 3 \times 3$	3	0.0760	8.06	54	0.0760	7.96	54
	4	0.0776	9.13	56	0.0776	8.97	56
	6	0.0780	11.88	53	0.0780	9.35	55
	8	0.0780	16.64	57	0.0780	9.44	55

Table 2 Performance of solving three-dimensional problem on $[0, 1]^3$, $\alpha = 1/2$

		Lumped			Dirichlet		
H/h	#sub	λ_{min}	λ_{max}	Iteration	λ_{min}	λ_{max}	Iteration
4	$3\times3\times3$	0.0395	7.20	59	0.0395	4.89	54
	$4\times4\times4$	0.0394	8.15	66	0.0394	5.01	53
	$6\times6\times6$	0.0393	8.85	70	0.0393	5.03	55
	$8\times8\times8$	0.0393	9.09	72	0.0393	5.09	56
#sub	H/h	λ_{min}	λ_{max}	Iteration	λ_{min}	λ_{max}	Iteration
$3\times3\times3$	3	0.0387	5.15	55	0.0387	4.35	53
	4	0.0395	7.20	57	0.0395	4.89	54
	6	0.0397	11.70	63	0.0397	5.11	52
	8	0.0397	16.52	73	0.0397	5.17	52

Table 2 are half of those in Table 1 since α is reduced by half, and they are also independent of the mesh size, consistent with Theorem 1.

We also comment that the inf-sup stability constant β of the mixed finite element space determines the lower eigenvalue bound in Theorem 1, which is quite small as shown in Tables 1 and 2 for this example. Some mixed finite element spaces with discontinuous pressures have better inf-sup stability and as a result give better lower eigenvalue bounds in Theorem 1.

Acknowledgement This work was supported in part by National Science Foundation Contracts No. DMS-1115759 and DMS-1419069.

References

1. J. Li, A Dual-Primal FETI method for incompressible Stokes equations. Numer. Math. **102**, 257–275 (2005)
2. J. Li, X. Tu, A nonoverlapping domain decomposition method for incompressible Stokes equations with continuous pressures. SIAM J. Numer. Anal. **51**(2), 1235–1253 (2013)
3. J. Li, O.B. Widlund, BDDC algorithms for incompressible Stokes equations. SIAM J. Numer. Anal. **44**(6), 2432–2455 (2006)
4. L.F. Pavarino, O.B. Widlund, Balancing Neumann-Neumann methods for incompressible Stokes equations. Comm. Pure Appl. Math. **55**(3), 302–335 (2002)
5. A. Toselli, O.B. Widlund, in Domain Decomposition Methods - Algorithms and Theory, *Springer Series in Computational Mathematics*, vol. 34 (Springer, Berlin-Heidelberg-New York, 2005)
6. X. Tu, J. Li, A FETI-DP algorithm for incompressible stokes equations with continuous pressures, in *Proceedings of the 21th Domain International Conference on Domain Decomposition Methods*, 2012
7. X. Tu, J. Li, A unified dual-primal finite element tearing and interconnecting approach for incompressible Stokes equations. Int. J. Numer. Methods Eng. **94**(2), 128–149 (2013)
8. X. Tu, J. Li, A FETI-DP type domain decomposition algorithm for three dimensional incompressible Stokes equations. SIAM J. Numer. Anal. **53**(2), 720–742 (2015)

Error of an eFDDM: What Do Matched Asymptotic Expansions Teach Us?

Jérôme Michaud and Pierre-Henri Cocquet

1 Introduction

In this paper, we are interested in heterogeneous decomposition methods. For complex problems, it may be useful to rely on approximations on subdomains and obtain an approximate global solution through appropriate coupling conditions on the interface. For an overview of such techniques, see [5] and references therein. In particular, we want to look at methods that neglect diffusion in a subdomain of non-zero measure. Gander and Martin [6] have compared the existing coupling methods with respect to their order in the small parameter in the different subdomains. An example of such a method is the χ-method, see [1, 2]. We want to extend these results to the Fuzzy Domain Decomposition Methods developed by Gander and Michaud [7]. This method is interesting as it provides an adaptive coupling method that allows for a tracking of domain of validity of different approximations. In [7], the authors show an approximation error analysis for a very simple problem that does not seem to generalize to higher dimensions. We develop a more general analysis based on matched asymptotic expansions [3] that show the convergence of an explicit FDDM (eFDDM) [7] method. For the comparison with the result of Gander and Martin [6], we note that our results compare with their $a < 0$ case. They show that the coupling is usually of order $\mathcal{O}(\nu)$, unless a factorization of the operator is done, in which case, the result can be improved to get an order of $\mathcal{O}(\nu^m)$. We show that an eFDDM is of order $\mathcal{O}(\nu)$ and have numerical evidence that (in 1D at least) this method is of order $\mathcal{O}(\nu^{3/2})$ in the subdomain where diffusion is taken into account.

J. Michaud (✉) • P.-H. Cocquet
Université de Genève, 2-4 rue du Lièvre, CP 64, CH-1211 Genève 4, Switzerland
e-mail: Jerome.Michaud@unige.ch; Pierre-Henri.Cocquet@univ-reunion.fr

T. Dickopf et al. (eds.), *Domain Decomposition Methods in Science and Engineering XXII*, Lecture Notes in Computational Science and Engineering 104, DOI 10.1007/978-3-319-18827-0_33

Basic facts about eFDDMs: Following [7], we recall that an eFDDM is a numerical method based on a FDD $\Omega = \Omega_1 + \cdots + \Omega_n$, where Ω_i are fuzzy sets of membership functions h_i and $\sum_{i=1}^n h_i = 1$. In this paper, we will work with a FDD of two subdomains Ω_1 and Ω_2 of membership function $h_1 = h$ and $h_2 = 1 - h$ respectively.

We approximate the linear problem with zero Dirichlet boundary condition

$$\mathcal{L}(u) = f \quad \text{on } \Omega, \quad u|_{\partial\Omega} = 0, \tag{1}$$

using two approximations $\mathcal{L}_i$, $i = 1, 2$, valid in a fuzzy sense in Ω_i.

We have the *global approximation*

$$h\mathcal{L}_1(u) + (1-h)\mathcal{L}_2(u) = f, \quad \text{on } \Omega, \quad u|_{\partial\Omega} = 0, \tag{2}$$

equivalent to the *eFDD approximation*

$$\begin{cases} \tilde{\mathcal{L}}_1(u_1) = hf + \mathcal{L}_{12}(u_2) & \text{on Supp}(\Omega_1), \quad u_i|_{\partial\Omega} = 0, \\ \tilde{\mathcal{L}}_2(u_2) = (1-h)f + \mathcal{L}_{21}(u_1) & \text{on Supp}(\Omega_2), \end{cases} \tag{3}$$

with $u_i = h_i u$ and $\tilde{\mathcal{L}}_i$ and $\mathcal{L}_{ij}$ are linear operators coming from the application of the product rule to exchange h with the operators $\mathcal{L}_i$, see [7] for details.

2 Model Problem

We are interested in the reaction diffusion model problem

$$\begin{cases} \mathcal{L}_{h\nu}(u_{h\nu}) := -h\nu\Delta u_{h\nu} - a \cdot \nabla u_{h\nu} + cu_{h\nu} = f, & \text{in } \Omega \\ \qquad\qquad\qquad\qquad\qquad\qquad\qquad\quad u_{h\nu} = 0, & \text{on } \partial\Omega \end{cases} \tag{4}$$

where $\nu > 0$, $a > 0$ and $c(x) + \text{div } a(x)/2 - \nu\Delta h/2 \geq \alpha > 0$ a.e. in a smooth domain Ω, $0 \leq h \leq 1$ is a smooth function with $\nabla(h^{1/2}) \in L^2(\Omega)$.[1]

We want to study the approximation error of an eFDDM for an approximation of $\mathcal{L}_{1\nu}(u_{1\nu}) = \mathcal{L}_\nu(u_\nu) = f$ by the global approximation $h\mathcal{L}_\nu(u) + (1-h)\mathcal{L}_{0\nu}(u) = \mathcal{L}_{h\nu}(u_{h\nu}) = f$, which can be written in the eFDDM as in (3).

We multiply (4) by $v \in H_0^h = \{u \in L^2(\Omega), h^{1/2}\nabla u \in L^2(\Omega), (h^{1/2}u)|_{\partial\Omega} = 0\}$ (this is a Hilbert space for the inner product $(u, v)_{L^2} + (h^{1/2}\nabla u, h^{1/2}\nabla v)_{L^2}$) and

[1]This is only a technicality to guaranty the wellposedness of the trace $h^{1/2}u$ on $\partial\Omega$. Typical smooth "plateau" functions satisfy this condition.

integrate by parts to obtain the following variational formulation

$$\begin{cases} \text{Find } u_{h\nu} := \tilde{u} \in H_0^h \text{ such that for every } v \in H_0^h, \\ a_{h\nu}(\tilde{u}, v) := \nu \int_\Omega h\nabla\tilde{u}\cdot\nabla v dx - \int_\Omega[((a-\nu\nabla h)\cdot\nabla\tilde{u})v + c\tilde{u}v]dx = \int_\Omega f v dx. \end{cases} \quad (5)$$

In order to see that problem (5) is well-posed, we need the following lemma.

Lemma 1 *If* $c(x) + \operatorname{div} a(x)/2 - \nu\Delta h/2 \geq \alpha > 0$ *a.e., where* α *is independent of* ν, *we have:*

$$a_{h\nu}(u,u) \geq \nu\|h^{\frac{1}{2}}\nabla u\|^2_{L^2(\Omega)} + \alpha\|u\|^2_{L^2(\Omega)}, \quad (6)$$

$$\|u_{h\nu}\|_{L^2(\Omega)} \leq \frac{1}{\alpha}\|f\|_{L^2(\Omega)}. \quad (7)$$

Proof In order to obtain a lower bound of the bilinear form we use

$$\begin{aligned} a_{h\nu}(u,u) &= \nu\|h^{\frac{1}{2}}\nabla u\|^2_{L^2(\Omega)} + \int_\Omega (c + \frac{1}{2}\operatorname{div} a)|u|^2 dx - \frac{\nu}{2}\int_\Omega \Delta h|u|^2 dx \\ &\geq \nu\|h^{\frac{1}{2}}\nabla u\|^2_{L^2(\Omega)} + \alpha\|u\|^2_{L^2(\Omega)}. \end{aligned} \quad (8)$$

The first equality follows from the definition of the bilinear form using an integration by parts and the divergence theorem to rewrite $\int_\Omega u(a\cdot\nabla u)dx = -\frac{1}{2}\int_\Omega(\operatorname{div} a)|u|^2 dx$.

The a priori estimate (7) follows from the fact that $a_{h\nu}(u_{h\nu}, u_{h\nu}) \leq \|u_{h\nu}\|_{L^2(\Omega)}\|f\|_{L^2(\Omega)}$ and using (6). □

Remark 1 We want the constant $\alpha > 0$ to be independent on ν. In general, this induces a restriction on h since $\nu\Delta h/2$ needs to be small. For example this is achieved if h is independent of ν.

We assume that (4) has a solution in H_0^h at least, then the a priori estimate (7) ensures the uniqueness and the stability of the solution whenever the assumptions of Lemma 1 holds.

3 Matched Asymptotic Expansion

From now on, we restrict ourselves to a 1D problem with constant coefficient on $\Omega = (0,1)$. We want to use matched asymptotic expansions to study the approximation error of the eFDDM. Therefore we compute a matched asymptotic expansions solution of (4) assuming that the membership function $h = 1$ at least in the boundary layer of size of order ν forming near 0 [3].

To obtain a matched asymptotic expansions solution, we use:

1. For the external field we assume that $u(x) \approx \sum_{k\geq 0}\nu^k\varphi_k(x),\ x \in (0,1]$.
2. For the internal field, we zoom in the boundary layer by rescaling x. This is done by setting $X = x/\nu$ and assuming that $u(\nu X) = \Phi(X) \approx \sum_{k\geq 0}\Phi_k(X)\nu^k$.

The zeroth-order approximation, to which we will restrict our analysis, is obtained by solving the following system [3]

$$\begin{cases} -a\varphi_0' + c\varphi_0 = f, \quad \varphi_0(1) = 0, \\ -\Phi_0'' - a\Phi_0' = f(0), \quad \Phi_0(0) = 0, \\ \lim_{X\to\infty} \Phi_0(X) = \varphi_0(0). \end{cases} \tag{9}$$

If $f(0) = 0$, the solution of this system is given by

$$\Phi_0(X) = \frac{1 - e^{-aX}}{a} \int_0^1 f(y) e^{-\frac{cy}{a}} dy, \quad \varphi_0(x) = \frac{1}{a} e^{\frac{cx}{a}} \int_x^1 f(y)\, e^{-\frac{cy}{a}} dy; \tag{10}$$

otherwise the matching fails and the system does not have any solution.

We obtain a globally valid approximation by merging the two solutions using a partition of unity $\{\chi, 1-\chi\}$

$$\tilde{u}_{\nu,\chi}(x) := \chi(x)\Phi_0\left(\frac{x}{\nu}\right) + (1 - \chi(x))\varphi_0(x); \tag{11}$$

$$\chi(x) := \begin{cases} 1, & \text{if } x < d_1\nu^s \\ \chi^* \in [0,1], & \text{if } d_1\nu^s \le x \le d_2\nu^s, \\ 0, & \text{otherwise} \end{cases} \qquad 0 < s < 1, \tag{12}$$

is smooth. Note that if we scale the χ function $\chi(x\nu^s)$, then χ and its derivatives become independent of ν.

Lemma 2 *For every function χ defined as in* (12), *we have*

$$\|\chi^{(n)}\|_{L^\infty(\Omega)} = \mathcal{O}(\nu^{-ns}). \tag{13}$$

Proof This result is a direct consequence of the independence of $\chi(x\nu^s)$ on ν. We change the variable in the function χ and every derivative leads to an additional factor of ν^{-s}, hence the result. □

4 Approximation Error Estimates

We use a membership function similar to χ to simplify the computations

$$h(x) := \begin{cases} 1, & \text{if } x < c_1\nu^t \\ h^*(x) \in [0,1], & \text{if } c_1\nu^t \le x \le c_2\nu^t \\ 0, & \text{otherwise,} \end{cases} \tag{14}$$

and have the following result:

Theorem 1 *Let $u_{h\nu}$ be the weak solution of* (4) *with constant $a \neq 0$ and c, h defined in* (14) *with $0 \leq t < 1$ such that $c - \nu\Delta h/2 \geq \alpha > 0$ a.e. and $\tilde{u}_{\nu,\chi}$ the globally valid approximation of the corresponding first term in the matched asymptotic expansions. Assume also that $f(0) = 0$ and $f \in W^{1,\infty}(\Omega)$. For $\Omega = (0, 1)$ and $s = 2/3 + t/3$ in* (12), *we have*

$$\|u_{h\nu} - \tilde{u}_{\nu,\chi}\|_{L^2(\Omega)} = \mathcal{O}(\nu^{1+t/2}). \tag{15}$$

Proof We look at the equation for the error and use the fact that the internal and external fields satisfy (9) and $\mathcal{L}_{h\nu}(\tilde{u}_{\nu,\chi}) = (-h\Delta - a \cdot \nabla + c)(\chi\Phi_0 + (1-\chi)\varphi_0)$. The triangle inequality implies

$$\begin{aligned}
&\|\mathcal{L}_{h\nu}(u_{h\nu} - \tilde{u}_{\nu,\chi})\|_{L^2(\Omega)} = \|f - \mathcal{L}_{h\nu}(\tilde{u}_{\nu,\chi})\|_{L^2(\Omega)} \\
&\leq \|\chi f\|_{L^2(\Omega)} + \left\|c\chi\Phi_0(\frac{\cdot}{\nu})\right\|_{L^2(\Omega)} + \left\|(\Phi_0(\frac{\cdot}{\nu}) - \varphi_0(\cdot))(h\nu\chi'' + a\chi')\right\|_{L^2(\Omega)} \\
&\quad + \left\|2h\nu\chi'(\Phi_0'(\frac{\cdot}{\nu}) - \varphi_0'(\cdot))\right\|_{L^2(\Omega)} + \left\|\nu(1-h)\chi(\Phi_0''(\frac{\cdot}{\nu}) - \varphi_0''(\cdot))\right\|_{L^2(\Omega)} \\
&\quad + \|\nu(h-\chi)\varphi_0''\|_{L^2(\Omega)} \\
&\leq \|f\|_{L^2(0,d_2\nu^s)} + c\left\|\Phi_0(\frac{\cdot}{\nu})\right\|_{L^2(0,d_2\nu^s)} \\
&\quad + \left\|(\Phi_0(\frac{\cdot}{\nu}) - \varphi_0(\cdot))\right\|_{L^2(d_1\nu^s,d_2\nu^s)} \|(\nu\chi'' + a\chi')\|_{L^\infty(d_1\nu^s,d_2\nu^s)} \\
&\quad + 2\nu\,\|\chi'\|_{L^\infty(d_1\nu^s,d_2\nu^s)} \left\|(\Phi_0'(\frac{\cdot}{\nu}) - \varphi_0'(\cdot))\right\|_{L^2(d_1\nu^s,d_2\nu^s)} \\
&\quad + \nu\left\|(\Phi_0''(\frac{\cdot}{\nu}) - \varphi_0''(\cdot))\right\|_{L^2(0,d_2\nu^s)} + \nu\,\|\varphi_0''\|_{L^2(d_1\nu^s,c_2\nu^t)}\,.
\end{aligned} \tag{16}$$

The second inequality follows from the definition of χ using the support of its derivatives. In order to finish the proof, we need a technical lemma.

Lemma 3 *Let $s < 1$, $\Omega^s = (\kappa_1\nu^s, \kappa_2\nu^s)$ and $f(0) = 0$. For $n = 0, 1, 2$ we have the following estimates*

$$\left\|\frac{d^n}{dx^n}\left(\Phi_0(\frac{\cdot}{\nu}) - \varphi_0(\cdot)\right)\right\|_{L^2(\Omega^s)} = \mathcal{O}(\nu^{\frac{5}{2}s - ns}). \tag{17}$$

Proof We start by computing the derivatives of $\Phi_0(\frac{x}{\nu}) - \varphi_0(x)$:

$$\Phi_0(\frac{x}{\nu}) - \varphi_0(x) = \frac{1}{a}\left[\int_0^x f(y)e^{-\frac{cy}{a}}dy - e^{-\frac{ax}{\nu}}\int_0^1 f(y)e^{-\frac{cy}{a}}dy\right],$$

$$\frac{d}{dx}\left(\Phi_0\left(\frac{x}{\nu}\right)-\varphi_0(x)\right)=\frac{1}{a}f(x)e^{-\frac{cx}{a}}+\frac{1}{\nu}e^{-\frac{ax}{\nu}}\int_0^1 f(y)e^{-\frac{cy}{a}}dy,$$

$$\frac{d^2}{dx^2}\left(\Phi_0\left(\frac{x}{\nu}\right)-\varphi_0(x)\right)=\left(\frac{f'(x)}{a}-\frac{cf(x)}{a^2}\right)e^{-\frac{cx}{a}}-\frac{a}{\nu^2}e^{-\frac{ax}{\nu}}\int_0^1 f(y)e^{-\frac{cy}{a}}dy.$$

In order to estimate the L^2-norm of these expressions, we use the fact that $\|\int_0^x f(y)dy\|_{L^2(\Omega^s)} \leq \sqrt{3}\nu^{s/2}(\kappa_2^3-\kappa_1^3)^{1/2}\|f\|_{L^\infty(\Omega^s)}/3$, $\|f\|_{L^2(\Omega^s)} \leq \nu^{s/2}(\kappa_2-\kappa_1)^{1/2}\|f\|_{L^\infty(\Omega^s)}$ and the fact that $e^{-\frac{cx}{a}}<1$, for all $x\in(0,1)$. Furthermore, as $f(0)=0$, we have $\|f\|_{L^\infty(\Omega^s)} \leq \nu^s\kappa_2\|f'\|_{L^\infty(0,\kappa_2)}$, hence we have

$$\begin{aligned}\left\|\Phi_0\left(\frac{\cdot}{\nu}\right)-\varphi_0(\cdot)\right\|_{L^2(\Omega^s)} &\leq \frac{\sqrt{3}\|f\|_{L^\infty(\Omega^s)}}{3a}\nu^{\frac{3s}{2}}(\kappa_2^3-\kappa_1^3)^{\frac{1}{2}}+\mathcal{O}(\nu^{\frac{1}{2}}e^{\frac{-a\kappa_1}{\nu^{1-s}}})\\ &\leq C_1\nu^s\|f'\|_{L^\infty(\Omega^s)}\nu^{\frac{3s}{2}}+\mathcal{O}(\nu^{\frac{1}{2}}e^{\frac{-a\kappa_1}{\nu^{1-s}}})\\ &= \mathcal{O}(\nu^{5s/2}+\nu^{\frac{1}{2}}e^{\frac{-a\kappa_1}{\nu^{1-s}}}),\end{aligned}$$

$$\begin{aligned}\left\|\frac{d}{dx}\left(\Phi_0\left(\frac{\cdot}{\nu}\right)-\varphi_0(\cdot)\right)\right\|_{L^2(\Omega^s)} &\leq \frac{\sqrt{\kappa_2-\kappa_1}}{a}\nu^{s/2}\|f\|_{L^\infty(\Omega^s)}+\mathcal{O}(\nu^{-\frac{1}{2}}e^{\frac{-a\kappa_1}{\nu^{1-s}}})\\ &\leq C_2\nu^s\|f'\|_{L^\infty(\Omega^s)}\nu^{\frac{s}{2}}+\mathcal{O}(\nu^{-\frac{1}{2}}e^{\frac{-a\kappa_1}{\nu^{1-s}}})\\ &= \mathcal{O}(\nu^{3s/2}+\nu^{-\frac{1}{2}}e^{\frac{-a\kappa_1}{\nu^{1-s}}}),\end{aligned}$$

$$\begin{aligned}\left\|\frac{d^2}{dx^2}\left(\Phi_0\left(\frac{\cdot}{\nu}\right)-\varphi_0(\cdot)\right)\right\|_{L^2(\Omega^s)} &\leq \sqrt{\kappa_2-\kappa_1}\nu^{s/2}\left\|\left(\frac{f'}{a}-\frac{cf}{a^2}\right)\right\|_{L^\infty(\Omega^s)}+\mathcal{O}(\nu^{-\frac{3}{2}}e^{\frac{-a\kappa_1}{\nu^{1-s}}})\\ &= \mathcal{O}(\nu^{s/2}+\nu^{-\frac{3}{2}}e^{\frac{-a\kappa_1}{\nu^{1-s}}}).\end{aligned}$$

We obtain the desired result noting that if $s<1$ then the exponential terms are negligible and can be neglected in the $\mathcal{O}$. □

We can now finish the proof of Theorem 1. Using Eqs. (13) and (17) and estimates previously used for the norms of f and φ_0''. Equation (16) becomes

$$\begin{aligned}\|f-\mathcal{L}_{h\nu}(\tilde{u}_{\nu,\chi})\|_{L^2(\Omega)} &= \mathcal{O}(\nu^{3s/2})+\mathcal{O}(\nu^{3s/2})+\mathcal{O}(\nu^{5s/2})\left(\mathcal{O}(\nu^{1-2s})+\mathcal{O}(\nu^{-s})\right)\\ &\quad+\nu\mathcal{O}(\nu^{-s})\mathcal{O}(\nu^{3s/2})+\nu\mathcal{O}(\nu^{s/2})+\nu\mathcal{O}(\nu^{t/2})\\ &= \mathcal{O}(\nu^{3s/2}+\nu^{1+s/2}+\nu^{1+t/2})\end{aligned}$$

We know that $t<s$ by hypothesis so that the second term is subdominant, choosing s such that $3s/2=1+t/2$ gives the condition on s in Theorem 1. We conclude the proof using the a priori estimate (7). □

Corollary 1 *The approximation error done by the use of an eFDDM as described in Sect. 2 is of order 1 in* ν*, that is*

$$\|u_\nu - u_{h\nu}\|_{L^2(\Omega)} = \mathcal{O}(\nu). \tag{18}$$

Proof This follows from Theorem 1 by the triangle inequality, noting that $h = 1$ implies $t = 0$. □

The approximation error obtained here is global. We now show a numerical example that illustrates the local convergence of the approximation error of the method.

Numerical experiment: We show here that an eFDDM for the problem $\mathcal{L}_\nu(u) = f$ on $\Omega = (0, 1)$ is of order $\mathcal{O}(\nu)$ as predicted by Corollary 1 and that it is numerically of order $\mathcal{O}(\nu^{3/2})$ in the subdomain where diffusion is taken into account. For this, we solve the corresponding eFDD approximation (3) with $\mathcal{L}_1 := \mathcal{L}_\nu$ and $\mathcal{L}_2 := \mathcal{L}_{0\nu}$ and $a = c = 1$, see [7] for the definition of the operators $\tilde{\mathcal{L}}_k$ and $\mathcal{L}_{kl}$, $k, l = 1, 2$.

We define h as in (14) with h^* a cubic spline on $(c_1\nu^t, c_2\nu^t)$,

$$h^*(x) = \delta^{-3}(2x^3 - 3\nu^t(c_1 + c_2)x^2 + 6\nu^{2t}c_1c_2x - c_2^2\nu^{3t}(3c_1 - c_2)),$$

with $\delta := (c_2 - c_1)\nu^t$ and $0 < c_1\nu^t \leq c_2\nu^t \leq 1$.

In order to satisfy the hypothesis of Theorem 1, we need to have $\alpha > 0$. In our case, we have $\|h''\|_{L^\infty(\Omega)} = 6/\delta^2$ which implies the condition $\nu^{t-1/2} > (3/c)^{1/2}/(c_2 - c_1)$. Choosing $t = 1/2$, $c_1 = 6$ and $c_2 = 8$, we satisfy this condition and we expect an order of convergence of $\mathcal{O}(\nu^{5/4})$ in the diffusive domain. Intuitively we can understand this result by Theorem 1, as both u_ν and $u_{h\nu}$ have $h = 1$ in this domain. A triangle inequality then implies the result. This order of convergence is better than the order of most of the available methods [6], but not optimal. Using the same reasoning, we can hope for a $\mathcal{O}(\nu^{3/2-\varepsilon})$ for $t = 1 - \varepsilon$.

We now show a numerical example with $t = 0.99$ that realizes an order $\mathcal{O}(\nu^{3/2-\varepsilon})$. Even if we can not prove the corresponding hypothesis in this case, the numerical example behaves as expected.

We introduce a set of equidistant points $x_i = i \cdot \Delta x$, $i = 0, \ldots, n + 1$ and $\Delta x = 1/(n+1)$ and discretize the eFDDM with an upwind 3-point finite difference scheme. This gives us a system of $2n$ coupled equations. For each component u_j, $j = 1, 2$, we remove from the system all the irrelevant equations, those for which $h_j(x_i) = 0$; this corresponds to the restriction to $\mathrm{Supp}(\Omega_j)$. In order to obtain an approximation error curve, we let ν tends to 0 keeping $n\nu$ constant to insure the resolution of the boundary layer. This is just to test the behavior of the method. In Fig. 1 we display the L^2 relative error between the numerical approximations of u_ν and $u_{h\nu}$ computed with the eFDDM scheme for three choices of f.

We see that for the three choices of f the method behaves as predicted by Corollary 1, that is the error is of order $\mathcal{O}(\nu)$ in the advective subdomain. And we see numerically that the error curves are of order $\mathcal{O}(\nu^{3/2})$ in the diffusive subdomain, as expected.

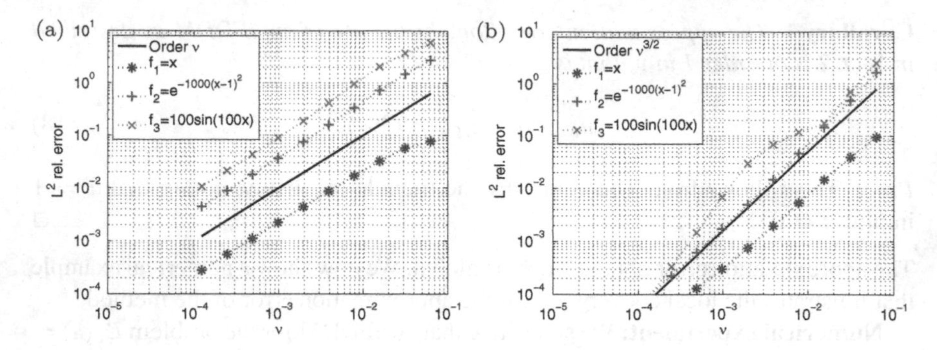

Fig. 1 Approximation errors where we refined the grid keeping $n\nu$ constant. (**a**) Results for the pure advective subdomain. Approximation error of order 1. (**b**) Results for the diffusive subdomain. Approximation error of order $3/2$

5 Conclusion

In this paper we have shown that matched asymptotic expansions are useful for the analytical study of approximation error of an eFDDM. We have proved that the error is of order ν by taking advantage of the similarities between the two approaches. The first is based on a decomposition of the operator whereas the second is based on a decomposition of the solution.

Our results compare those for $a < 0$ in Gander and Martin [6] with Dirichlet boundary conditions. We have proven that an eFDDM is not worse than the other coupling methods and our numerical example shows that we are in fact better inside the diffusive subdomain. The justification of the order $\mathcal{O}(\nu^{3/2})$ in the diffusive subdomain is only heuristic as we have not been able to prove it yet. We will address this problem and get local estimates in future work. The only other known method that achieves an order better than $\mathcal{O}(\nu)$ is the one based on the factorization of the operator, which does not generalize to higher dimensions. Our method generalize to higher dimensions and we are working on extension of this work to 2D, 3D and time-dependent problems. We also want to generalize the method to more complicated problem such as the kinetic equations. This has been done for example in the work of Degond et al. [4], but without any approximation error analysis.

References

1. Y. Achdou, O. Pironneau, The χ-method for the Navier-Stokes equations. IMA J. Numer. Anal. **13**(4), 537–558 (1993)
2. F. Brezzi, C. Canuto, A. Russo, A self-adaptive formulation for the Euler/Navier-Stokes coupling. Comput. Methods Appl. Mech. Eng. **73**(3), 317–330 (1989)
3. J. Cousteix, J. Mauss, *Asymptotic Analysis and Boundary Layers* (Springer, Berlin, 2007)

4. P. Degond, G. Dimarco, L. Mieussens, A multiscale kinetic-fluid solver with dynamic localization of kinetic effects. J. Comput. Phys. **229**, 4907–4933 (2010)
5. M. Discacciati, P. Gervasio, A. Quarteroni, Heterogeneous mathematical models in fluid dynamics and associated solution algorithms, in *Multiscale and Adaptivity: Modeling, Numerics and Applications*, ed. by G. Naldi, G. Russo. Lecture Notes in Mathematics, vol. 2040 (Springer, Berlin, 2012), pp. 57–123
6. M.J. Gander, V. Martin, An asymptotic approach to compare coupling mechanisms for different partial differential equations, in *Domain Decomposition Methods in Science and Engineering XX*, ed. by R. Bank, M. Holst, O.B. Widlund, J. Xu. Lecture Notes in Computational Science and Engineering (Springer, Berlin, 2012)
7. M.J. Gander, J. Michaud, Fuzzy domain decomposition: a new perspective on heterogeneous DD methods, in *Domain Decomposition Methods in Science and Engineering XXI*, ed. by J. Erhel, M.J. Gander, L. Halpern, G. Pichot, T. Sassi, O.B. Widlund. Lecture Notes in Computational Science and Engineering (Springer, Berlin, 2013)

A Comparison of Additive Schwarz Preconditioners for Parallel Adaptive Finite Elements

Sébastien Loisel and Hieu Nguyen

1 Introduction

We consider a second order elliptic boundary value problem in the variational form: find $u^* \in H_0^1(\Omega)$, for a given polygonal (polyhedral) domain $\Omega \subset \mathbb{R}^d$, $d = 2, 3$ and a source term $f \in L^2(\Omega)$, such that

$$\underbrace{\int_\Omega \nabla u^*(x) \cdot \nabla v(x)\, dx}_{\equiv a(u^*,v)} = \underbrace{\int_\Omega f(x) v(x)\, dx}_{\equiv (f,v)}, \quad \text{for all } v \in H_0^1(\Omega). \tag{1}$$

The Bank–Holst parallel adaptive meshing paradigm [1–3] is utilised to solve (1) in a combination of domain decomposition and adaptivity. It can be summarised as follows:

Step I—Mesh Partition: Starting with a coarse mesh $\mathcal{T}_H$, the domain is partitioned into non-overlapping subdomains: $\Omega = \cup_{i=1}^p \Omega_i$.

Step II—Adaptive Meshing: Each processor i is provided with $\mathcal{T}_H$ and instructed to sequentially solve the *entire* problem, with the stipulation that its adaptive enrichment should be limited largely to Ω_i. At the end of this step, the local mesh $\mathcal{T}_i$ on processor i are regularised such that the global fine mesh described in Step III is conforming.

Step III—Global Solve: A final finite element solution is computed on the mesh $\mathcal{T}_h = \cup_{i=1}^p \mathcal{T}_i|_{\Omega_i}$, which is the union of the refined submeshes.

S. Loisel • H. Nguyen (✉)
Department of Mathematics, Heriot-Watt University, Riccarton, Edinburgh, EH14 4additive Schwarz, UK
e-mail: S.Loisel@hw.ac.uk; H.Nguyen@hw.ac.uk; hnguyen@cimne.upc.edu

T. Dickopf et al. (eds.), *Domain Decomposition Methods in Science and Engineering XXII*, Lecture Notes in Computational Science and Engineering 104, DOI 10.1007/978-3-319-18827-0_34

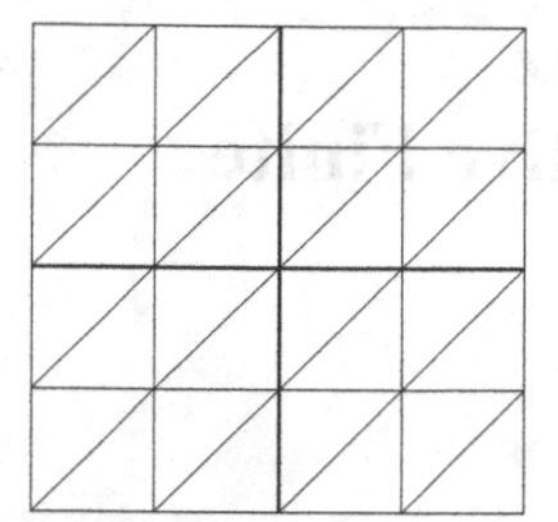 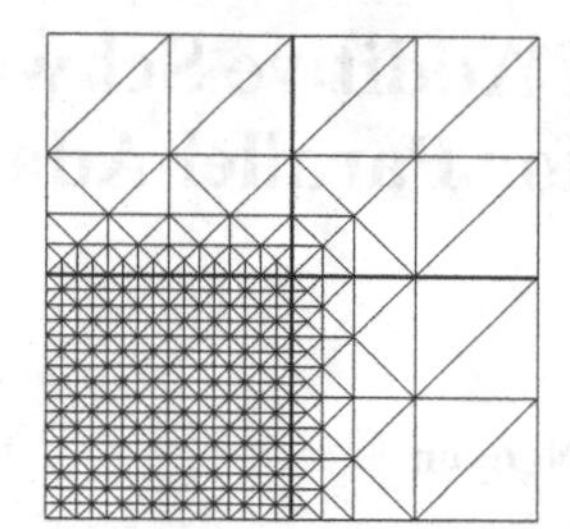 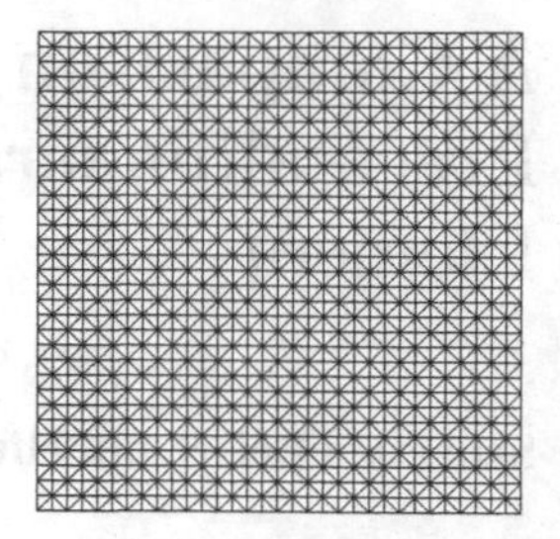

Fig. 1 A coarse mesh with partition (*left*), a local mesh on a processor (*middle*) and the global mesh (*right*)

An example of meshes in different steps of the Bank–Holst paradigm is illustrated in Fig. 1.

Discretizing (1) using linear finite elements on the global mesh $\mathcal{T}_h$, we arrive at the following system of linear equations:

$$Au = f, \quad A \in \mathbb{R}^{n \times n}, \ u, f \in \mathbb{R}^n. \tag{2}$$

The purpose of this paper is to formulate and compare three additive Schwarz preconditioners that can be used to accelerate Krylov methods in solving (2). The improved convergence analysis will be reported somewhere else. The considered preconditioners are: the two-level additive Schwarz preconditioner with small overlap [5, 8], two-level additive Schwarz preconditioner with weakly overlapping [4] and optimal one-level additive Schwarz preconditioner based on full-domain decomposition [6].

2 Preconditioners Formulation

As all of the considered preconditioners are additive Schwarz preconditioners, they can be formulated and analyzed using the abstract theory of Schwarz methods (cf. [8]) which is summarized as follows.

Assume the global finite element space V_h associated with $\mathcal{T}_h$ admits the decomposition

$$V_h = \sum_{i=i_0}^{p} V_i, \tag{3}$$

where V_i are subspaces of V_h and $i_0 = 0$ or 1. The subspace V_0 is usually related to a coarse problem, built on a coarse mesh ($\mathcal{T}_H$ in the Bank–Holst paradigm). The subspaces V_i, on the other hand, are often related to a partition in subdomains and are associated with local submeshes. But, this is not the case for the third considered preconditioner, which is proposed in [6].

Now let $\{\psi_1^{(i)}(x), \ldots, \psi_{n_i}^{(i)}(x)\}$ be a basis of V_i and let $x_1, \ldots, x_n$ be the nodal points of the global mesh $\mathcal{T}_h$. We define

$$R_i = \begin{bmatrix} \psi_1^{(i)}(x_1) & \cdots & \psi_1^{(i)}(x_n) \\ \vdots & \cdots & \vdots \\ \psi_{n_i}^{(i)}(x_1) & \cdots & \psi_{n_i}^{(i)}(x_n) \end{bmatrix}. \tag{4}$$

It can be noted that R_i is the matrix representation of the restriction operator from V_i to V. Using this operator, the local stiffness matrix associated with subspace V_i is defined by

$$A_i = R_i A R_i^T. \tag{5}$$

Then the additive Schwarz preconditioner associated with the decomposition (3) is

$$P = \sum_{i=i_0}^{p} R_i^T A_i^{-1} R_i \tag{6}$$

The preconditioner P is said to be two-level when i_0 is 0 (coarse level: V_0, fine level: $\{V_i\}_{i=1}^p$) or one-level when i_0 is 1.

Next we will formulate three different additive Schwarz preconditioners for the Bank–Holst paradigm using different decomposition (3) with different choices of V_i. For clarity, we will use different variations of the notations V_i, R_i and A_i to denote the subspace, its corresponding restriction matrix and local stiffness matrix.

Two-level additive Schwarz preconditioner with small overlap: This is the standard and most popular version of additive Schwarz preconditioner. It is introduced in a general context without adaptivity. However, it can be used for the Bank–Holst paradigm and we present it here for comparison. For this preconditioner, each subdomain Ω_i is extended to a larger region $\hat{\Omega}_i$ by adding a small number of layers of elements **in the global (fine) mesh** $\mathcal{T}_h$ (see Fig. 2, left). The subspaces $\hat{V}_i$ are then defined as

$$\hat{V}_i = \{v(x) \in H_0^1(\hat{\Omega}_i) | \quad v(x)|_T \in \mathbb{P}_1(T), \ \forall T \in \mathcal{T}_h\}. \tag{7}$$

The two-level additive Schwarz preconditioner with small overlap is simply

$$P_{SO} = R_0^T A_0^{-1} R_0 + \sum_{i=1}^{p} \hat{R}_i^T \hat{A}_i^{-1} \hat{R}_i. \tag{8}$$

The condition number of the preconditioned system associated with P_{SO} is bounded from above by $C(1 + (H/\delta))$, where C is a constant independent of the mesh sizes,

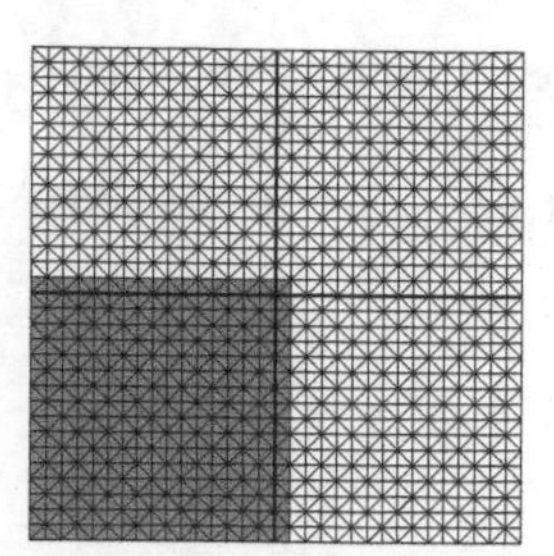 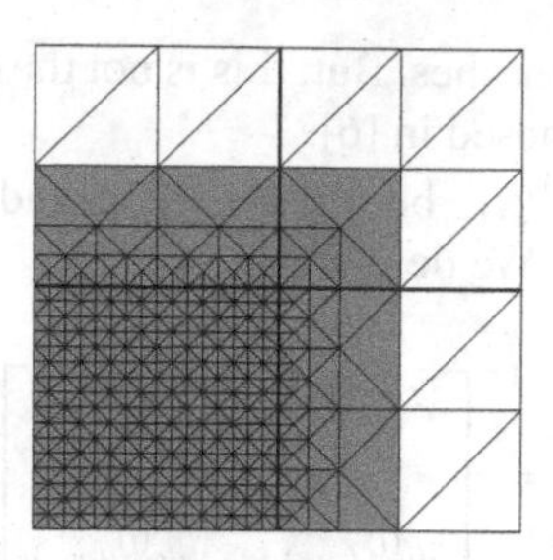 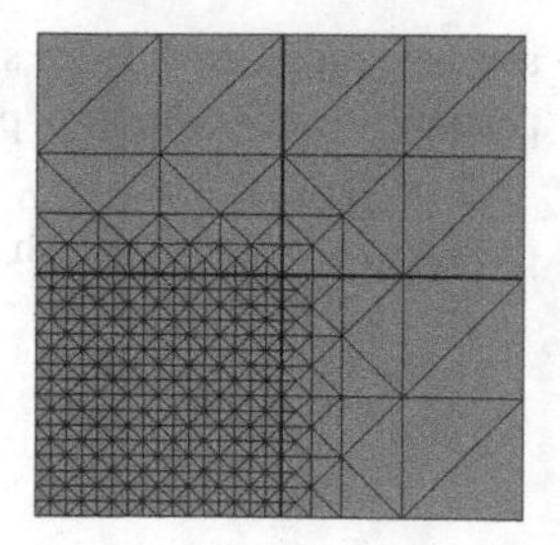

Fig. 2 Extension regions (*shaded areas*) and their associated meshes in cases of small overlap (*left*), weak overlap (*center*), full domain overlap (*right*)

H is the coarse mesh size and δ is the width of the overlap (cf. [5, 8] and references therein). If δ is of size $O(h)$, the usual case in practice, the condition number of the preconditioned system in the Bank–Holst paradigm will increase linearly as the level of refinement increases. In case the overlap is "generous", δ is of size $O(H)$, the condition number is bounded by a constant, i.e. $O(1)$, independent of the mesh sizes H, h and the number of subdomains p. But, there is an important practical concern that the cost of using generous overlap is too expensive as the number of vertices in the overlapping region would be $O(h^{-2})$ in 2D and $O(h^{-3})$ in 3D.

Weakly overlapping two-level additive Schwarz preconditioner: The formulation of this preconditioner is very much similar to that of P_{SO}. The only difference is that each subdomain Ω_i is extended to a larger region $\tilde{\Omega}_i$ by adding layers of elements **in the adaptive mesh $\mathcal{T}_i$ so that the overlap is of size** $O(H)$ (see Fig. 2, center). Then the subspace $\tilde{V}_i$ is defined by

$$\tilde{V}_i = \{v(x) \in H_0^1(\tilde{\Omega}_i) | \quad v(x)|_T \in \mathbb{P}_1(T), \ \forall T \in \mathcal{T}_i\}, \tag{9}$$

and the weakly overlapping two-level additive Schwarz preconditioner is defined by

$$P_{WO} = R_0^T A_0^{-1} R_0 + \sum_{i=1}^{p} \tilde{R}_i^T \tilde{A}_i^{-1} \tilde{R}_i. \tag{10}$$

By using adaptive mesh $\mathcal{T}_i$ instead of $\mathcal{T}_h$, the number of vertices in the overlapping region is reduced to $O(h^{-1})$ in 2D and $O(h^{-2})$ in 3D. In addition, the condition number of the preconditioned system associated with P_{WO} can be bounded independently of the mesh sizes H, h and the number of subdomains p, i.e. is $O(1)$ (see [4]).

Optimal one-level additive Schwarz preconditioner: In order to take full advantage the Bank–Holst paradigm, [6] formulate an additive Schwarz preconditioner that utilises the subspaces associated with the local adaptive meshes in the paradigm. These are meshes of the whole domain Ω residing locally on each processor. They form a "full domain overlap" partition of the domain (see Fig. 2, right). In this case, the local subspaces are:

$$V_i = \{v(x) \in H_0^1(\Omega)| \quad v(x)|_T \in \mathbb{P}_1(T),\ \forall T \in \mathcal{T}_i\}. \tag{11}$$

And the optimal one-level additive Schwarz preconditioner is

$$P_{O_1} = \sum_{i=1}^{p} R_i^T A_i^{-1} R_i. \tag{12}$$

Here, we should emphasize that explicit coarse component (in two-level formulation) is not needed in this case because the coarse space V_0 is contained in each and every subspace V_i.

It is shown in [6] that the condition number of the preconditioned system associated with P_{O_1} can also be bounded independently of the mesh sizes H, h and the number of subdomains p, i.e. is O(1).

3 Remarks on the Implementation

In order to compute the restriction matrices as defined in (4), one usually uses the nodal basis functions corresponding to the submeshes/meshes associated with V_i for $\{\psi_1^{(i)}(x), \ldots, \psi_{n_i}^{(i)}(x)\}$. In cases of P_{SO}, the nodal points in the submesh associated with $\hat{V}_i$ form a subset of the fine nodal points $\{x_1, \ldots, x_n\}$. Consequently, $\hat{R}_i$, $i > 0$, are rectangular matrices of zeros and ones, which extracts the nodal points that lie in the extension region $\hat{\Omega}_i$. In case of P_{WO} and P_{O_1}, the nodal points associated with $\tilde{V}_i$ and V_i that lie outside Ω_i does not belong to the fine mesh $\mathcal{T}_h$. Therefore, the corresponding rows of $\tilde{R}_i$ and R_i can have values in $(0, 1)$. For simplex elements, one can compute these rows using the fact that $\psi_j^{(i)}(x_k)$ equals either zero or the barycentric coordinate of x_k with respect to the coarse element containing x_k and having $x_j^{(i)}$ as one of its vertices. Here $x_j^{(i)}$ is the nodal point in $\mathcal{T}_i$ associated with $\psi_j^{(i)}$. The same technique can be used to compute the restriction matrix R_0.

For P_{O_1}, if minimal refinement is allowed outside the local subdomain in each local adaptive mesh, the rows of R_i associated with nodal points far away from Ω_i are the same with the corresponding rows of R_0. Computing these rows requires only the knowledge of the coarse mesh $\mathcal{T}_H$ and the local submesh of $\mathcal{T}_h$ which is available locally on each processor. Therefore, each processor can compute parts of R_0 locally and exchange the information with others to construct the full R_i.

In case of P_{SO} and P_{WO}, the only way of obtaining the local stiffness matrices A_i is via (5), which has the computational cost of $O(N_i^2)$. Here N_i is the number of degrees of freedom in Ω_i. If the global matrix A is assembled but distributed, there will also be communication cost that can be expensive. For P_{O1}, one is able to assemble A_i with the computational cost of roughly $O(N_i)$. The assembling requires no communication as $\mathcal{T}_i$ are available locally and are meshes of the whole domain Ω. In addition, the communication cost can be reduced further as A is not needed to be assembled.

4 Numerical Experiments

In this section, we present numerical experiments for the following problem

$$\begin{aligned} -\Delta u &= 1 \quad \text{in } \Omega, \\ u &= 0 \quad \text{on } \partial\Omega, \end{aligned} \tag{13}$$

where Ω is a L-shaped domain (the unit square missing the lower right quarter). The solution of this problem is shown in Fig. 3 (left).

We start with an unstructured triangular (coarse) mesh of 436 vertices and 1026 elements. Then, we partition it into p subdomains, $p = 16, 32, 64, 128$. Each processor gets exactly the same copy of this mesh. The coarse mesh with a partition of 16 subdomains are shown in Fig. 3. In Step 2, local adaptive meshes are obtained by refining elements inside and surrounding local subdomains. In this experiment, we limit outside refinement by refining only ones which share at least one point with the local subdomain. Hanging nodes are allowed even though they are not considered as real nodal points. When an element is refined, it is split into four similar elements having half of its size. We use l levels of refinement for each local mesh, $l = 4, 5, 6$. The preconditioners P_{SO}, P_{WO} and P_{O_1} are implemented with the first two having the overlap of size h (one layer of fine elements) and H (equivalent to one layer of coarse elements) respectively.

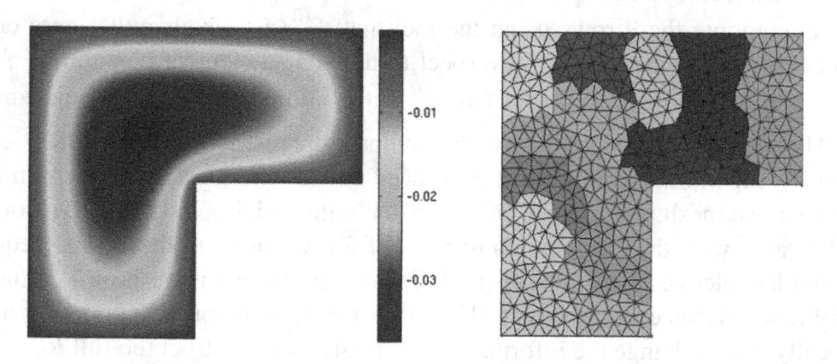

Fig. 3 Solution (*left*) and a coarse mesh with a partition of 16 subdomains (*right*)

Since all of the three preconditioners are symmetric positive definite, they are suitable to use with the CG method. However, it is well-known [7] that the convergence of CG in finite precision departs significantly from the theoretical convergence of CG in exact arithmetic. Therefore, we also use the GMRES method, which is slightly more numerically robust, in our experiments.

Table 1 reports the number of CG and GMRES iterations to bring the relative residual below 10^{-6}. The number of degrees of freedom and the average of elapsed time required to apply the preconditioners on a vector are also provided for comparison.

It can be seen that P_{SO} requires the most iterations for both CG and GMRES to converge. The iteration counts are clearly increasing as h becomes smaller (higher level of refinement). For GMRES, P_{O_1} is the best performer. It requires just half the number of iterations needed in case of P_{WO}. The numbers of GMRES iterations for these two preconditioners appear to be bounded by a constant, as predicted by theory. For CG, the number of iterations increases when l increases in case of P_{WO}, and when p increases in case of P_{O_1}. Between the two preconditioners, P_{O_1} has more wining cases.

In term of elapsed time, P_{O_1} and P_{WO} are roughly the same. Even though they are more expensive to apply, they are more efficient than P_{SO} because they require fewer number of iterations.

In the second experiment, we study whether it is beneficial to refine local meshes in the region outside local subdomains. Now instead of using minimal outside refinement, we perform at least one level of refinement for elements that do not belong to the local subdomain. It should be noted that the global mesh $\mathcal{T}_h$ and the global stiffness matrix A are the same with those in the previous experiments. The restriction matrices and local stiffness matrix, however, are changed.

We do not see any improvement in term of iterations count for P_{WO}. Perhaps, this is due to the fact that a coarse space is already incorporated in this preconditioner. We do see clear improvement for P_{O_1} with significant reduction in iteration counts and slight increase of time. However, care must be taken when using generous refinement outside subdomains as this would require more memory and time to calculate restriction matrices (Table 2).

Table 1 Number of CG and GMRES iterations to bring the relative residual below 10^{-6} and average of elapsed time (in seconds) to apply the preconditioners on a vector—minimal outside refinement

	l=4, $N = 101{,}761$				l=5, $N = 405{,}761$				l =6, $N = 1{,}620{,}481$			
p=	16	32	64	128	16	32	64	128	16	32	64	128
$P = P_{SO}$												
CG no. it.	23	24	24	26	31	32	33	36	43	44	45	49
GMRES no. it.	13	14	15	15	16	17	18	18	20	20	21	22
time p. mult.	1.3	1.1	0.4	0.5	6.1	5.9	6.0	5.3	27.3	25.9	25.8	26.9
$P = P_{WO}$												
CG no. it.	18	18	18	20	19	20	21	22	25	26	26	28
GMRES no. it.	12	12	12	13	12	12	12	13	13	14	15	15
time. p. mult.	1.5	1.6	1.6	1.7	7.4	7.3	7.3	7.4	32.3	31.0	31.3	31.8
$P = P_{O_1}$												
CG no. it.	15	17	20	23	15	17	21	23	15	17	21	23
GMRES no. it.	6	7	8	8	6	7	8	8	6	7	8	8
time. p. mult.	1.7	1.9	2.3	2.7	7.2	7.5	8.3	9.4	32.7	32.3	33.2	34.4

Table 2 Number of CG and GMRES iterations to bring the relative residual below 10^{-6} and average of elapsed time (in seconds) to apply the preconditioners on a vector—extra outside refinement

	$l=4, N=101{,}761$				$l=5, N=405{,}761$				$l=6, N=1{,}620{,}481$			
p=	16	32	64	128	16	32	64	128	16	32	64	128
$P = P_{WO}$												
CG no. it.	18	18	18	20	19	20	21	22	25	26	26	28
GMRES no. it.	12	12	12	13	12	12	12	13	13	14	15	15
time. p. mult.	1.7	1.7	1.8	1.8	7.7	7.6	7.6	7.4	35.6	32.6	33.0	32.5
$P = P_{O_1}$												
CG no. it.	13	14	15	18	13	14	15	18	13	15	16	19
GMRES no. it.	5	5	5	5	5	5	5	5	5	5	5	5
time. p. mult.	2.3	2.6	3.8	5.5	8.2	9.1	10.2	13.0	34.2	34.6	38.4	42.6

Acknowledgements This work was supported by the Numerical Algorithms and Intelligent Software Centre funded by the UK EPSRC grant EP/G036136 and the Scottish Funding Council.

References

1. R.E. Bank, Some variants of the Bank-Holst parallel adaptive meshing paradigm. Comput. Vis. Sci. **9**(3), 133–144 (2006)
2. R.E. Bank, M. Holst, A new paradigm for parallel adaptive meshing algorithms. SIAM J. Sci. Comput. **22**(4), 1411–1443 (2000) (electronic)
3. R.E. Bank, M. Holst, A new paradigm for parallel adaptive meshing algorithms. SIAM Rev. **45**(2), 291–323 (2003) (electronic). Reprinted from SIAM J. Sci. Comput. **22**(4), 1411–1443 (2000) [MR1797889]
4. R.E. Bank, P.K. Jimack, S.A. Nadeem, S.V. Nepomnyaschikh, A weakly overlapping domain decomposition preconditioner for the finite element solution of elliptic partial differential equations. SIAM J. Sci. Comput. **23**(6), 1817–1841 (2002) (electronic)
5. M. Dryja, O.B. Widlund, Domain decomposition algorithms with small overlap. SIAM J. Sci. Comput. **15**(3), 604–620 (1994). Iterative methods in numerical linear algebra (Copper Mountain Resort, CO, 1992)
6. S. Loisel, H. Nguyen, An optimal schwarz preconditioner for a class of parallel adaptive finite elements (submitted)
7. G. Meurant, *The Lanczos and Conjugate Gradient Algorithms*. Software, Environments, and Tools, vol. 19 (Society for Industrial and Applied Mathematics, Philadelphia, PA, 2006). From theory to finite precision computations
8. A. Toselli, O. Widlund, *Domain Decomposition Methods—Algorithms and Theory*. Springer Series in Computational Mathematics, vol. 34 (Springer, Berlin, 2005)

A BDDC Preconditioner for Problems Posed in H (div) with Deluxe Scaling

Duk-Soon Oh

1 Introduction

Let Ω be a bounded polyhedral domain in $\mathbb{R}^3$. We will work with the Hilbert space $H(\text{div}; \Omega)$, the subspace of vector valued functions $\boldsymbol{u} \in (L^2(\Omega))^3$ with $\text{div}\, \boldsymbol{u} \in L^2(\Omega)$. The space $H_0(\text{div}; \Omega)$ is the subspace of $H(\text{div}; \Omega)$ with a vanishing normal component on the boundary $\partial\Omega$.

We will consider the following problem: Find $\boldsymbol{u} \in H_0(\text{div}; \Omega)$, such that

$$a(\boldsymbol{u}, \boldsymbol{v}) := \int_{\Omega} (\alpha \,\text{div}\, \boldsymbol{u} \,\text{div}\, \boldsymbol{v} + \beta\, \boldsymbol{u} \cdot \boldsymbol{v}) dx = \int_{\Omega} \boldsymbol{f} \cdot \boldsymbol{v}\, dx, \quad \boldsymbol{v} \in H_0(\text{div}; \Omega). \quad (1)$$

We will assume that the coefficient $\alpha \in L^{\infty}(\Omega)$ is nonnegative, that $\beta \in L^{\infty}(\Omega)$ is strictly positive, and that the right hand side $\boldsymbol{f} \in (L^2(\Omega))^3$.

The model problem (1) is equivalent to the variational forms of mixed or first order system least-squares formulations as in [3]. There are also other applications of $H(\text{div})$, e.g., in the sequential regularization method for the Navier-Stokes equations; see [12].

The main purpose of this paper is to construct a BDDC preconditioner for vector field problems discretized with Raviart-Thomas finite elements. Iterative substructuring methods for such problems were first considered in [25]. Other iterative substructuring methods for these types of problems have been developed

D.-S. Oh (✉)
Department of Mathematics, Rutgers University, Piscataway, NJ 08854, USA
e-mail: duksoon@math.rutgers.edu

T. Dickopf et al. (eds.), *Domain Decomposition Methods in Science and Engineering XXII*, Lecture Notes in Computational Science and Engineering 104, DOI 10.1007/978-3-319-18827-0_35

in [19]. Overlapping Schwarz methods have also been introduced; see [1, 14–16]. Other methods such as multigrid methods have been applied successfully in [2, 8, 10]. We also remark that domain decomposition methods for $H(\mathbf{curl})$ problems were introduced in [5, 7, 9, 20–22]. BDDC methods for other problems related to $H(\mathrm{div})$ can be found in [18, 23, 24].

In the construction of a BDDC preconditioners, a set of primal constraints and a weighted averaging technique have to be chosen and these choices will very directly affect the performance. Effective primal constraints are very simple for the Raviart-Thomas elements; we choose the average value of the normal component over the subdomain faces as primal variables. However, the choice of averaging is much more intricate. We will use a new type of weighted averaging technique introduced in [6] for three dimensional $H(\mathbf{curl})$ problems.

2 Preliminary

We first introduce a triangulation $\mathcal{T}_h$ of Ω of hexahedral elements. We will consider the lowest order Raviart-Thomas elements on mesh $\mathcal{T}_h$. We then decompose the domain Ω into N nonoverlapping subdomains Ω_i. We also define the global interface Γ and the local interface Γ_i by

$$\Gamma := \left(\bigcup_{i=1}^{N} \partial\Omega_i\right) \backslash \partial\Omega, \qquad \Gamma_i := \Gamma \cap \partial\Omega_i,$$

respectively.

Let $W^{(i)}$ be the space of the finite elements on Ω_i with a zero normal component on $\partial\Omega \cap \partial\Omega_i$. We decompose $W^{(i)}$ into two subspaces, $W_\Gamma^{(i)}$ and $W_I^{(i)}$. Here, $W_\Gamma^{(i)}$ is the interface space which consists of degrees of freedom corresponding to Γ_i and $W_I^{(i)}$ is the space of discrete unknowns of the interior of Ω_i. The space $W_\Gamma^{(i)}$ can be decomposed into a primal space $W_\Pi^{(i)}$ and a dual space $W_\Delta^{(i)}$. In general, the functions in $W_\Gamma := \prod_{i=1}^{N} W_\Gamma^{(i)}$ have discontinuous normal components across the interface while those of the finite element solutions are continuous. We denote the continuous subspace by $\hat{W}_\Gamma$ ($\subset W_\Gamma$). We next define operators $R_\Gamma^{(i)} : \hat{W}_\Gamma \rightarrow W_\Gamma^{(i)}$ which extract the degrees of freedom associated with Γ_i. Similarly, we define a space $\tilde{W}_\Gamma$, for which all the primal constraints are enforced. We next define local operators $\overline{R}_\Gamma^{(i)} : \tilde{W}_\Gamma \rightarrow W_\Gamma^{(i)}$ which extract the degrees of freedom corresponding to Γ_i. We also define the global operator $\tilde{R}_\Gamma : \hat{W}_\Gamma \rightarrow \tilde{W}_\Gamma$. Finally, we introduce the scaled operator $\tilde{R}_{D,\Gamma} : \hat{W}_\Gamma \rightarrow \tilde{W}_\Gamma$ obtained by pre-multiplying the entries of $\tilde{R}_\Gamma$ associated with $W_\Delta^{(i)}$ by a scaling matrix $D^{(i)}$. The discrete form of our problem is written in terms of local stiffness matrices as

$$\begin{bmatrix} A_{II} & A_{I\Gamma} \\ A_{\Gamma I} & A_{\Gamma\Gamma} \end{bmatrix} \begin{bmatrix} \boldsymbol{u}_I \\ \boldsymbol{u}_\Gamma \end{bmatrix} = \sum_{i=1}^{N} \begin{bmatrix} A_{II}^{(i)} & A_{I\Gamma}^{(i)} \\ A_{\Gamma I}^{(i)} & A_{\Gamma\Gamma}^{(i)} \end{bmatrix} \begin{bmatrix} \boldsymbol{u}_I^{(i)} \\ \boldsymbol{u}_\Gamma^{(i)} \end{bmatrix} = \sum_{i=1}^{N} \begin{bmatrix} \boldsymbol{f}_I^{(i)} \\ \boldsymbol{f}_\Gamma^{(i)} \end{bmatrix}. \tag{2}$$

Before we introduce the BDDC algorithm, we eliminate all interior unknowns locally. After this step, we obtain these local Schur complements:

$$S_\Gamma^{(i)} := A_{\Gamma\Gamma}^{(i)} - A_{\Gamma I}^{(i)} A_{II}^{(i)-1} A_{I\Gamma}^{(i)}.$$

By using the local Schur complements, we can build a reduced interface problem. The global problem is given by

$$\hat{S}_\Gamma \boldsymbol{u}_\Gamma = \boldsymbol{g}_\Gamma, \tag{3}$$

where

$$\hat{S}_\Gamma := \sum_{i=1}^{N} R_\Gamma^{(i)T} S_\Gamma^{(i)} R_\Gamma^{(i)} \quad \text{and} \quad \boldsymbol{g}_\Gamma := \sum_{i=1}^{N} R_\Gamma^{(i)T} \left(\boldsymbol{f}_\Gamma - A_{\Gamma I}^{(i)} A_{II}^{(i)-1} \boldsymbol{f}_I^{(i)} \right).$$

Moreover, we have the partially assembled Schur complement $\tilde{S}_\Gamma$:

$$\tilde{S}_\Gamma = \sum_{i=1}^{N} \overline{R}_\Gamma^{(i)^T} S_\Gamma^{(i)} \overline{R}_\Gamma^{(i)}. \tag{4}$$

3 BDDC

We consider a BDDC preconditioner to solve the interface problem (3). We can find background information and a description of the algorithm in [4, 11]. The BDDC preconditioner has the following form:

$$M^{-1} = \tilde{R}_{D,\Gamma}^T \tilde{S}_\Gamma^{-1} \tilde{R}_{D,\Gamma}. \tag{5}$$

It is convenient to make a change of variables by introducing a basis for the primal degrees of freedom and a complementary basis for the dual subspace $W_\Delta^{(i)}$. Here we can follow the recipes of [11, Sect. 3.3] closely. For our problem, the only primal variables will be the averages of the normal component over the subdomain faces.

In order to specify the algorithm completely, we need to define the weighted averaging operator $D^{(i)}$. Conventional weighted averaging techniques, known as stiffness and ρ scalings, are described in [4, 13]. However, these methods are designed for constant coefficients or for one variable coefficient. For more than one variable coefficient, we need a different approach and we will use the new weighted averaging technique introduced in [6] for $H(\mathbf{curl})$ problems.

Let F_{ij} be the common face of two adjacent subdomains Ω_i and Ω_j. Moreover, let $R_{F_{ij}}^{(i)}$ be the restriction operator which extracts the degrees of freedom on F_{ij} from those on Γ_i. Then, the two Schur complements associated with F_{ij} are given by $S_{F_{ij}}^{(i)} = R_{F_{ij}}^{(i)} S_{\Gamma}^{(i)} R_{F_{ij}}^{(i)T}$ and $S_{F_{ij}}^{(j)} = R_{F_{ij}}^{(j)} S_{\Gamma}^{(j)} R_{F_{ij}}^{(j)T}$. We will use the scaling matrices $D_j^{(i)} := \left(S_{F_{ij}}^{(i)} + S_{F_{ij}}^{(j)}\right)^{-1} S_{F_{ij}}^{(i)}$. We note that we can apply the operator $\left(S_{F_{ij}}^{(i)} + S_{F_{ij}}^{(j)}\right)^{-1}$ by solving a Dirichlet problem on $\Omega_i \cup F_{ij} \cup \Omega_j$ with zero Dirichlet boundary conditions. The scaling operator $D^{(i)}$ is then given by a block diagonal matrix with the diagonal components $D_{j_1}^{(i)}, D_{j_2}^{(i)}, \cdots, D_{j_k}^{(i)}$, where $j_1, j_2, \ldots, j_k \in \mathcal{N}_i$ and $\mathcal{N}_i$ is the set of indices of the Ω_j's ($i \neq j$) which share a subdomain face with Ω_i.

The condition number of $M^{-1}\hat{S}_{\Gamma}$ is bounded by $C(1 + \log H/h)^2$, where the constant C does not depend on the size of subdomain and mesh size as well as the coefficients and their jumps between subdomains. Due to space restriction, a detailed analysis will not be reported here. Further details are provided in [17].

4 Numerical Results

We have applied the BDDC algorithm to our model problem (1). For algorithmic details, we follow [11]. We set $\Omega = (0, 1)^3$ and decompose the unit cube into N^3 identical cubic subdomains. Each subdomain has a side length $H = 1/N$. Moreover, we assume that the coefficients α and β have jumps across the interface between the subdomains with a checkerboard pattern in which (α, β) for a subdomain is either (α_b, β_b) or (α_w, β_w). We discretize the model problem (1) by using the lowest order hexahedral Raviart-Thomas finite elements and use the preconditioned conjugate gradient method to solve the discretized problem. The iteration is stopped when the l^2-norm of the residual has been reduced by a factor of 10^{-6}.

We first fix the value of β and vary α. Second, we fix the value of α and vary β. Tables 1 and 2 show the first two sets of results. We next use a different distribution, instead of the checkerboard distribution. We first generate $2N^3$ random numbers $\{r_{\alpha_i}\}_{i=1,\ldots,N^2}$ and $\{r_{\beta_i}\}_{i=1,\ldots,N^2}$ in $[-3, 3]$ with a uniform distribution. We then assign $10^{r_{\alpha_i}}$ and $10^{r_{\beta_i}}$ for α_i and β_i, respectively. The third set of results can be found in Table 3. We see that the condition number is insensitive to the jumps of coefficients.

We next report on numerical experiments for the case where coefficients have jumps inside the subdomains. For each subdomain Ω_i, we let $\Omega_i^o = \{(x, y, z) \mid 1/4 \leq x^o, y^o, z^o \leq 1/2, \text{where} x^o = x/H - \lfloor x/H \rfloor, y^o = y/H - \lfloor y/H \rfloor, \text{and} z^o = z/H - \lfloor z/H \rfloor.\}$. Here, $\lfloor x \rfloor = \max\{m \in \mathbb{Z} \mid m \leq x\}$, where $\mathbb{Z}$ is the set of integers. We use the α_i and β_i specified in Tables 1 and 2 as coefficients for $\Omega_i \backslash \Omega_i^o$. For Ω_i^o, we assign $100\alpha_i$ and $100\beta_i$ and with α_i and β_i in a checkerboard pattern. Tables 4

Table 1 Condition numbers and iteration counts (in parentheses)

(α_b, β_b)	(α_w, β_w)	$H/h = 2$	$H/h = 4$	$H/h = 8$	$H/h = 16$
$(10^{-2}, 1)$	$(1, 1)$	1.64 (7)	2.32 (9)	3.26 (11)	4.37 (13)
$(10^{-1}, 1)$	$(1, 1)$	1.80 (7)	2.64 (9)	3.70 (12)	4.94 (13)
$(1, 1)$	$(1, 1)$	1.83 (7)	2.69 (10)	3.75 (11)	5.01 (14)
$(10^{1}, 1)$	$(1, 1)$	1.83 (7)	2.69 (10)	3.76 (11)	5.02 (14)
$(10^{2}, 1)$	$(1, 1)$	1.83 (7)	2.69 (10)	3.76 (11)	5.02 (14)

Checkerboard constant β pattern and $N = 4$

Table 2 Condition numbers and iteration counts (in parentheses)

(α_b, β_b)	(α_w, β_w)	$H/h = 2$	$H/h = 4$	$H/h = 8$	$H/h = 16$
$(1, 10^{-2})$	$(1, 1)$	1.03 (3)	1.06 (4)	1.09 (4)	1.12 (4)
$(1, 10^{-1})$	$(1, 1)$	1.28 (5)	1.53 (6)	1.89 (8)	2.31 (9)
$(1, 10^{1})$	$(1, 1)$	1.27 (5)	1.51 (6)	1.85 (7)	2.27 (9)
$(1, 10^{2})$	$(1, 1)$	1.02 (3)	1.05 (4)	1.08 (4)	1.12 (4)

Checkerboard constant α pattern and $N = 4$

Table 3 Condition numbers and iteration counts (in parentheses)

	$H/h = 2$	$H/h = 4$	$H/h = 8$	$H/h = 16$
Set 1	1.80 (8)	2.69 (11)	3.76 (13)	5.01 (16)
Set 2	1.65 (8)	2.37 (9)	3.39 (11)	4.61 (14)
Set 3	1.78 (8)	2.50 (10)	3.49 (12)	4.82 (14)
Set 4	1.67 (8)	2.50 (10)	3.50 (12)	4.68 (14)
Set 5	1.74 (8)	2.49 (10)	3.45 (13)	4.54 (15)

Random coefficients and $N = 4$

Table 4 Condition numbers and iteration counts (in parentheses)

	$H/h = 4$	$H/h = 8$	$H/h = 16$
$\alpha_b = 10^{-2}$	2.32 (9)	3.34 (11)	4.41 (13)
$\alpha_b = 10^{-1}$	2.64 (9)	3.83 (12)	5.05 (14)
$\alpha_b = 10^{0}$	2.69 (10)	3.90 (12)	5.16 (14)
$\alpha_b = 10^{1}$	2.69 (10)	3.91 (12)	5.17 (14)
$\alpha_b = 10^{2}$	2.69 (10)	3.91 (12)	5.17 (14)

Specified values as indicated in Table 1 with jumps inside subdomains and $N = 4$

and 5 show the results. We see that our method works well even though we have discontinuities inside the subdomains.

Finally, for a comparison, we report on some numerical experiments using conventional techniques. We have performed three different types of experiments with the same set of coefficient distributions. The first set of experiments, named "deluxe", is based on the deluxe scaling techniques. In the second, "diag", we use the conventional methods described in [4, 13]. In this case, the scaling is based on the diagonal entries of each subdomain matrix. We use the cardinality in the last set, "card". For Raviart-Thomas elements, only two subdomains share a subdomain face

Table 5 Condition numbers and iteration counts (in parentheses)

	$H/h = 4$	$H/h = 8$	$H/h = 16$
$\beta_b = 10^{-2}$	1.05 (4)	1.09 (4)	1.13 (4)
$\beta_b = 10^{-1}$	1.51 (6)	1.90 (8)	2.34 (9)
$\beta_b = 10^{1}$	1.53 (6)	1.95 (8)	2.39 (9)
$\beta_b = 10^{2}$	1.06 (4)	1.09 (4)	1.13 (4)

Specified values as indicated in Table 2 with jumps inside subdomains and $N = 4$

Table 6 Condition numbers and iteration counts (in parentheses)

(α_b, β_b)	(α_w, β_w)	Deluxe	Diag	Card
$(10^{-3}, 10^{3})$	$(1, 1)$	1.05e0 (3)	9.03e2 (47)	2.66e2 (43)
$(10^{-2}, 10^{2})$	$(1, 1)$	1.17e0 (4)	1.88e2 (36)	5.13e1 (31)
$(10^{-1}, 10^{1})$	$(1, 1)$	1.82e0 (7)	7.22e1 (43)	2.19e1 (30)
$(10^{1}, 10^{-1})$	$(1, 1)$	1.89e0 (8)	8.63e1 (48)	2.61e1 (32)
$(10^{2}, 10^{-2})$	$(1, 1)$	1.09e0 (4)	1.01e3 (74)	2.58e2 (66)
$(10^{3}, 10^{-3})$	$(1, 1)$	1.01e0 (3)	1.48e4 (130)	3.71e3 (120)

Checkerboard pattern, $N = 4$, and $H/h = 8$

in common. Hence, we use $1/2$ as scaling factors. As we see in Table 6, our weighted averaging technique works well while the others are sensitive to the discontinuities across the interface.

We remark that the deluxe scaling technique requires additional computational costs for solving local subproblems on each subdomain face. Experimentally, conventional methods are approximately 5–6 times faster than deluxe scaling in each iteration. However, deluxe scaling requires much less iteration counts especially for the case where we have large jumps between subdomains. Hence, we can expect a better performance. We note that a more computationally efficient version of deluxe scaling is introduced in [7].

Acknowledgement This work was completed while the author was working at Louisiana State University. This material is based upon work supported by the HPC@LSU computing resources and the Louisiana Optical Network Institute (LONI).

References

1. D.N. Arnold, R.S. Falk, R. Winther, Preconditioning in H(div) and applications. Math. Comput. **66**(219), 957–984 (1997)
2. D.N. Arnold, R.S. Falk, R. Winther, Multigrid in H(div) and H(curl). Numer. Math. **85**(2), 197–217 (2000)
3. Z. Cai, R.D. Lazarov, T.A. Manteuffel, S.F. McCormick, First-order system least squares for second-order partial differential equations: part I. SIAM J. Numer. Anal. **31**(6), 1785–1799 (1994)
4. C.R. Dohrmann, A preconditioner for substructuring based on constrained energy minimization. SIAM J. Sci. Comput. **25**(1), 246–258 (2003)

5. C.R. Dohrmann, O.B. Widlund, An iterative substructuring algorithm for two-dimensional problems in H(curl). SIAM J. Numer. Anal. **50**(3), 1004–1028 (2012)
6. C.R. Dohrmann, O.B. Widlund, Some recent tools and a BDDC algorithm for 3D problems in H(curl), in *Domain Decomposition Methods in Science and Engineering XX*. Lecture Notes in Computational Science and Engineering, vol. 91 (Springer, Berlin, 2013), pp. 15–25
7. C.R. Dohrmann, O.B. Widlund, A BDDC algorithm with deluxe scaling for three-dimensional H(curl) problems. Technical Report, TR2014-964, Courant Institue of Mathematical Sciences, Department of Computer Science (2014)
8. R. Hiptmair, Multigrid method for **H**(div) in three dimensions. Electron. Trans. Numer. Anal. **6**, 133–152 (1997) [Special issue on multilevel methods (Copper Mountain, CO, 1997)]
9. R. Hiptmair, A. Toselli, Overlapping and multilevel Schwarz methods for vector valued elliptic problems in three dimensions, in *Parallel Solution of Partial Differential Equations (Minneapolis, MN, 1997)*. The IMA Volumes in Mathematics and Its Applications, vol. 120 (Springer, New York, 2000), pp. 181–208
10. T.V. Kolev, P.S. Vassilevski, Parallel auxiliary space AMG solver for H(div) problems. SIAM J. Sci. Comput. **34**(6), A3079–A3098 (2012)
11. J. Li, O.B. Widlund, FETI-DP, BDDC, and block Cholesky methods. Int. J. Numer. Methods Eng. **66**(2), 250–271 (2006)
12. P. Lin, A sequential regularization method for time-dependent incompressible Navier-Stokes equations. SIAM J. Numer. Anal. **34**(3), 1051–1071 (1997)
13. J. Mandel, C.R. Dohrmann, R. Tezaur, An algebraic theory for primal and dual substructuring methods by constraints. Appl. Numer. Math. **54**(2), 167–193 (2005)
14. D.-S. Oh, Domain decomposition methods for Raviart-Thomas vector fields. Ph.D. thesis, Courant Institue of Mathematical Sciences, 2011. TR2011-942. http://cs.nyu.edu/web/Research/TechReports/TR2011-942/TR2011-942.pdf
15. D.-S. Oh, An alternative coarse space method for overlapping Schwarz preconditioners for Raviart-Thomas vector fields, in *Domain Decomposition Methods in Science and Engineering XX*. Lecture Notes in Computational Science and Engineering, vol. 91 (Springer, Berlin, 2013a), pp. 361–367
16. D.-S. Oh, An overlapping schwarz algorithm for Raviart–Thomas vector fields with discontinuous coefficients. SIAM J. Numer. Anal. **51**(1), 297–321 (2013b)
17. D.-S. Oh, O.B. Widlund, C.R. Dohrmann, A BDDC algorithm for Raviart-Thomas vector fields. Technical Report, TR2013-951, Courant Institue of Mathematical Sciences, Department of Computer Science (2013)
18. B. Sousedík, Nested BDDC for a saddle-point problem. Numer. Math. **125**(4), 761–783 (2013)
19. A. Toselli, Neumann-Neumann methods for vector field problems. Electron. Trans. Numer. Anal. **11**, 1–24 (2000a)
20. A. Toselli, Overlapping Schwarz methods for Maxwell's equations in three dimensions. Numer. Math. **86**(4), 733–752 (2000b)
21. A. Toselli, Dual-primal FETI algorithms for edge finite-element approximations in 3D. IMA J. Numer. Anal. **26**(1), 96–130 (2006)
22. A. Toselli, A. Klawonn, A FETI domain decomposition method for edge element approximations in two dimensions with discontinuous coefficients. SIAM J. Numer. Anal. **39**(3), 932–956 (2001)
23. X. Tu, A BDDC algorithm for a mixed formulation of flow in porous media. Electron. Trans. Numer. Anal. **20**, 164–179 (electronic) (2005)
24. X. Tu, A BDDC algorithm for flow in porous media with a hybrid finite element discretization. Electron. Trans. Numer. Anal. **26**, 146–160 (2007)
25. B.I. Wohlmuth, A. Toselli, O.B. Widlund, An iterative substructuring method for Raviart-Thomas vector fields in three dimensions. SIAM J. Numer. Anal. **37**(5), 1657–1676 (2000)

Pipeline Schwarz Waveform Relaxation

Benjamin Ong, Scott High, and Felix Kwok

1 Introduction

Schwarz Waveform Relaxation (SWR) introduced in [2] has been analyzed for a wide range of time-dependent problems, including the parabolic heat equation [6], wave equation and advection-diffusion equations [7, 8], Maxwell's equations [4], and the porous medium equation [9]. In contrast to classical Schwarz iterations, where the time-dependent PDE is discretized in time and domain-decomposition is applied to the sequence of steady-state problems, SWR solves *time-dependent* sub-problems; this relaxes synchronization of the sub-problems and provides a means to couple disparate solvers applied to individual sub-problems, as shown in [10] for example. SWR has also been shown in [1, 8] to have superlinear convergence for small time windows. This paper outlines a framework that reformulates SWR so that successive waveform iterates can be computed in a pipeline fashion, allowing for increased concurrency and hence, increased scalability for SWR-type algorithms. In Sect. 2, we review the SWR algorithm before introducing and comparing several

B. Ong (✉)
Michigan Technological University, Houghton, MI 49945, USA
e-mail: ongbw@mtu.edu

S. High
University of Illinois at Urbana-Champaign, IL 61801, USA
e-mail: highscot@gmail.com

F. Kwok
Hong Kong Baptist University, Kowloon Tong, Hong Kong
e-mail: felix_kwok@hkbu.edu.hk

T. Dickopf et al. (eds.), *Domain Decomposition Methods in Science and Engineering XXII*, Lecture Notes in Computational Science and Engineering 104, DOI 10.1007/978-3-319-18827-0_36

Pipeline Schwarz Waveform Relaxation algorithms (PSWR) in Sect. 3. Numerical scaling results for the linear heat equation are presented in Sect. 4.

2 Schwarz Waveform Relaxation

Denote the PDE of interest as

$$\begin{aligned} u_t = \mathcal{L}(t, u), \quad (x, t) &\in \Omega \times [0, T] \\ u(x, 0) = f(x), \quad x &\in \Omega \\ u(z, t) = g(z, t), \quad z &\in \partial\Omega. \end{aligned} \tag{1}$$

Consider a partitioning of the domain, $\Omega = \cup_i \Omega_i$. The domains in the partition may be overlapping or non-overlapping. Let u_i denote the solution on sub-domain Ω_i. Then, Eq. (1) can be decomposed into a coupled system of equations,

$$\begin{aligned} (u_i)_t = \mathcal{L}(t, u_i), \quad (x, t) &\in \Omega_i \times [0, T] \\ u_i(x, 0) = f(x), \quad x &\in \Omega_i \\ u_i(z, t) = g(z, t), \quad z &\in \partial\Omega_i \cap \partial\Omega, \\ \mathcal{T}_{ij}(u_i(z, t)) = \mathcal{T}_{ij}(u_j(z, t)), \quad z &\in \partial\Omega_i \cap \partial\Omega_j, \end{aligned} \tag{2}$$

where T are transmission operators appropriate to the Eq. (1). SWR decouples the system of PDEs in Eq. (2). Let $u_i^{[k]}$ denote the kth waveform iterate on sub-domain Ω_i. After specifying an initial estimate for the sub-domain solution on the interfaces, $u_i^{[0]}(z, t), z \in \partial\Omega_i \backslash \partial\Omega$, the SWR algorithm iteratively solves PDEs (3) for waveform iterates $k = 1, 2, \ldots$ until convergence,

$$\begin{aligned} (u_i^{[k]})_t = \mathcal{L}(t, u_i^{[k]}), \quad (x, t) &\in \Omega_i \times [0, T] \\ u_i^{[k]}(x, 0) = f(x), \quad x &\in \Omega_i \\ u_i^{[k]}(z, t) = g(z, t), \quad z &\in \partial\Omega_i \cap \partial\Omega, \\ \mathcal{T}_{ij}(u_i^{[k]}(z, t)) = \mathcal{T}_{ij}(u_j^{[k-1]}(z, t)), \quad z &\in \partial\Omega_i \cap \partial\Omega_j. \end{aligned} \tag{3}$$

A pseudo-code for the algorithm is presented on the next page. Observe that SWR allows for each sub-domain to independently compute time-dependent solutions on their respective sub-domains (lines 9–11) During each waveform iteration, transmission data on each sub-domain is aggregated for the entire computational time interval before boundary data is exchanged between neighboring sub-domains (lines 12–14).

Schwarz Waveform Relaxation Algorithm

1. `MPI Initialization`
2. `parallel for` $i = 1 \ldots N$ *(Sub-domain)*
3. `for` $t = \Delta t \ldots T$
4. `Guess` $u_i^{[0]}(z,t), \quad z \in \partial\Omega_i \cap \partial\Omega_j$
5. `end`
6. `end`
7. `for` $k = 1 \ldots K$ *(Waveform iteration)*
8. `parallel for` $i = 1 \ldots N$ *(Sub-domain)*
9. `for` $t = \Delta t \ldots T$
10. `Solve for` $u_i^{[k]}(x,t)$
11. `end`
12. `for` $t = \Delta t \ldots T$
13. `Exchange transmission data` $\mathcal{T}(u_i^{[k]}(z,t))$
14. `end`
15. `Check convergence`
16. `end`
17. `end`

3 Pipeline Schwarz Waveform Relaxation

Using a similar approach described in [3, 12], the relaxation framework can be rewritten so that after initial start-up costs, multiple waveform iterations can be computed in a pipeline-parallel fashion. A graphical example of the PSWR algorithm for two subdomains is shown in Fig. 1. To simplify the presentation, we

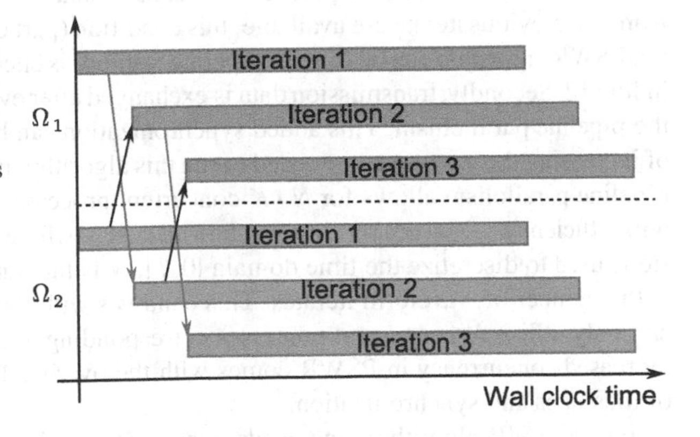

Fig. 1 The proposed PSWR algorithm allows for multiple Schwarz waveform iterations to be simultaneously computed. After an initial start-up cost, multiple iterates are computed in a pipeline fashion

present the algorithm for the simplified case where the *same* time discretization is used for all sub-problems (PSWR Algorithm 1).

Pipeline Schwarz Waveform Relaxation Algorithm 1

```
1. MPI Initialization
2. parallel for  i = 1...N (Sub-domain)
3.      for t = Δt...T
4.           Guess u_i^[0](z, t),   z ∈ ∂Ω_i ∩ ∂Ω_j
5.      end
6.      Set t^[0] = T
7. end
8. parallel for  k = 1...K (Waveform iteration)
9.      parallel for  i = 1...N (Sub-domain)
10.          set t^[k] = Δt
11.          While t^[k] ≤ T
12.               If  t^[k] < t^[k-1]
13.                    Solve for  u_i^[k](x, t^[k])
14.                    Exchange transmission data T(u_i^[k](z, t^[k]))
15.                    t^[k] ← t^[k] + Δt
16.               end
17.          end
18.          Check convergence
19.      end
20. end
```

Several observations should be made about the proposed PSWR algorithm. First, a Schwarz iteration can only proceed if boundary data (i.e. transmission conditions) from the previous iterate are available; this condition (part of the start-up cost before the PSWR algorithm can be run in a pipeline fashion) is checked by the `if` statement in line 12. Secondly, transmission data is exchanged after every time step to facilitate the pipeline parallellism. This added synchronization can be relaxed at the expense of increasing the start-up cost needed to run this algorithm in a pipeline fashion. This pipeline parallelism allows for $N \cdot K$ concurrent processes in the PSWR algorithm with efficiency $\frac{N_t}{K+N_t}$ (accounting for start-up costs), where N_t is the number of time steps used to discretize the time domain $[0, T]$, N is the number of sub-domains, K is the number of waveform iterates. This contrasts with the SWR algorithm, which can only utilize N concurrent processes corresponding to the N sub-domains. This increased concurrency in PSWR comes with the overhead of an increased number of messages and synchronization.

For the SWR algorithm, one needs to send $O(K-1)$ message of size $O(N_t)$. If $N \cdot K$ processors are used in a pipeline parallel fashion as described in Pipeline

Schwarz Waveform Relaxation Algorithm 1, $O((K-1)\cdot N_t)$ messages of size $O(1)$ are needed. More generally, if $N\cdot p$ processors are used in the PSWR algorithm, where $p<K$ is a multiple of K, then $O((p-1)/p\cdot K\cdot N_t)$ messages of size $O(1)$, and $O(K/p-1)$ messages of size $O(N_t)$, are needed. We note that the PSWR algorithm can also be implemented using a framework that naturally reduces the number of messages in a system. Assuming a heterogeneous computing platform (where each socket has multiple cores), one can use the MPI-3 framework [11] or the OpenMP protocol in the outer "parallel for" statement in line 8, to aggregate transmission data from line 14 naturally before exchanging transmission data with neighboring nodes. Alternatively, because nodes working on waveform iterate k only need to communicate with waveform iterates $k-1$, the PSWR algorithm allows for a natural grouping of nodes so that one can (in principle) use multiple overlapping communicators to leverage data/network-topology and software defined networking advances (see [5]) to add scalability.

Generalizations to allow for disparate time discretizations in each sub-problem are possible. We list the algorithm (Algorithm 2) without implementation. Unlike PSWR Algorithm 1, it is not possible to keep the "pipe" full, i.e. domain i might necessarily need to wait for its neighboring domains to provide boundary data. Additionally, solving for $u_i^{[k]}(x,t_i^{[k]})$ in line 14 requires an interpolation algorithm to obtain the correct transmission condition to be used in the solution of (3). Lastly, an implementation decision has to be made on how to collect and store the data from neighboring domains before the interpolation is used to obtain the transmission conditions for an update in line 14.

4 Numerical Experiments

We present results from scaling studies, which vary the number of computational cores used to compute the PSWR algorithm while keeping total discretized problem size constant. The diffusion equation $u_t = k(u_{xx}+u_{yy})$ is solved in $\mathbb{R}^2$ using a centered five point finite-difference approximation in space, and a backward Euler time integrator. The convergence of the waveform iterates is shown in Fig. 2. In our first scaling study, 400×400 grid points are decomposed into 4×4 non-overlapping domains for 400 total time steps. Optimized robin transmission conditions of the form

$$\mathcal{T}_{ij}[\cdot] = \left(\frac{d}{d\hat{n}}+p\right)[\cdot], \quad \mathcal{T}_{ji}[\cdot] = \left(\frac{d}{d\hat{n}}-p\right)[\cdot],$$

are used, where $\frac{d}{d\hat{n}}$ is the derivative in the normal direction, and $p=1$. (A recursive formula is used to compute the transmission condition in lieu of discretizing the derivative in the normal direction.) In each experiment a total of 16 full waveform iterations are completed. Timing results are obtained using the Stampede, a supercomputer at the Texas Advanced Computing Center. Good parallel efficiency

Pipeline Schwarz Waveform Relaxation Algorithm 2

1. `MPI Initialization`
2. `parallel for` $i = \ldots 1..N$ *(Sub-domain)*
3. `for` $t_i = \Delta t_i \ldots T$
4. `Guess` $u_i^{[0]}(z,t), \quad z \in \partial\Omega_i \cap \partial\Omega_j$
5. `end`
6. `Set` $t_i^{[0]} = T$
7. `end`
8. `parallel for` $k = 1 \ldots K$ *(Waveform iteration)*
9. `parallel for` $i = 1 \ldots N$ *(Sub-domain)*
10. `initialize` $\Delta t_i^{[k]}$
11. `set` $t_i^{[k]} = \Delta t_i^{[k]}$
12. `While` $t_i^{[k]} \leq T$
13. `If` $t_i^{[k]} < t_j^{[k-1]}$ for all neighbors j
14. `Solve for` $u_i^{[k]}(x, t_i^{[k]})$
15. `Send transmission data` $\mathcal{T}(u_i^{[k]}(z, t_i^{[k]}))$ to neighbor nodes
16. $t_i^{[k]} \leftarrow t_i^{[k]} + \Delta t_i^{[k]}$
17. `end`
18. `end`
19. `Check convergence`
20. `end`
21. `end`

and speedup is observed in spite of the increase in the number of messages required by the PSWR algorithm. Note that the 4×4×1 case is identically the SWR algorithm.

In our second scaling study, 1600×1600 grid points are decomposed into 16×16 non-overlapping domains for 400 total time steps. Again, a centered five point finite difference stencil, a backward Euler time integrator, and optimized transmission conditions are used. Good parallel efficiency and speedup is observed even with the increased synchronization/number of messages in the system.

$N_x \times N_y \times N_k$	# Cores	Walltime (s)	Speedup	Efficiency
$4 \times 4 \times 1$	16	293.02	1.00 ×	1.00
$4 \times 4 \times 2$	32	149.92	1.95 ×	0.98
$4 \times 4 \times 4$	64	75.48	3.89 ×	0.97
$4 \times 4 \times 8$	128	38.71	7.57 ×	0.95
$4 \times 4 \times 16$	256	23.90	12.26 ×	0.77

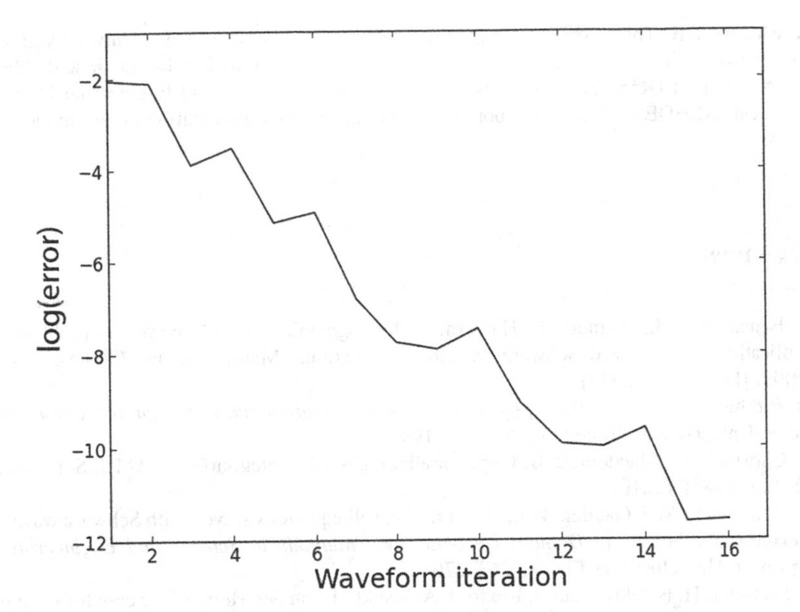

Fig. 2 The error of the waveform iterates at time T is computed relative to monodomain solution for a 4×4 decomposition of the problem using optimized transmission conditions. The convergence behavior of the PSWR algorithm is identical to the convergence behavior of the SWR algorithm

In the above computations, a linear solve on a sub-domain takes $O(10^{-2})$ s. This relatively small problems size was chosen (100 × 100 on each sub-domain) so that communications would play a substantial role in the timing studies. The presented efficiencies can be improved by partitioning the problem to be more computationally expensive (i.e. more time is spent in the linear solve).

$N_x \times N_y \times N_k$	# Cores	Walltime (s)	Speedup	Efficiency
16 × 16 × 1	256	295.86	1.00 ×	1.00
16 × 16 × 2	512	155.98	1.90 ×	0.95
16 × 16 × 4	1024	77.10	3.84 ×	0.96
16 × 16 × 8	2048	43.20	6.85 ×	0.86
16 × 16 × 16	4096	26.65	11.10 ×	0.69

5 Conclusions

In this paper, we have reformulated classical Schwarz waveform relaxation to allow for pipeline-parallel computation of the waveform iterates, after an initial startup cost. Theoretical estimates for the parallel speedup and communication overhead are presented, along with scaling studies to show the effectiveness of the pipeline Schwarz waveform relaxation algorithm.

Acknowledgement This work was supported in part by Michigan State University through computational resources provided by the Institute for Cyber-Enabled Research and AFOSR Grant FA9550-12-1-0455. This work also used the Extreme Science and Engineering Discovery Environment (XSEDE), which is supported by National Science Foundation grant number OCI-1053575.

References

1. D. Bennequin, M.J. Gander, L. Halpern, A homographic best approximation problem with application to optimized Schwarz waveform relaxation. Math. Comput. **78**(265), 185–223 (2009) [ISSN 0025-5718]
2. M. Bjørhus, *On domain decomposition, subdomain iteration and waveform relaxation*. Ph.D. thesis, University of Trondheim, Norway, 1995
3. A. Christlieb, C. Macdonald, B. Ong, Parallel high-order integrators. SIAM J. Sci. Comput. **32**(2), 818–835 (2010)
4. Y. Courvoisier, M.J. Gander, Time domain Maxwell equations solved with Schwarz waveform relaxation methods, in *Domain Decomposition Methods in Science and Engineering XX* (Springer, Heidelberg, 2013), pp. 263–270
5. N. Feamster, H. Balakrishnan, J. Rexford, A. Shaikh, J. van der Merwe, The case for separating routing from routers, in *Proceedings of the ACM SIGCOMM Workshop on Future Directions in Network Architecture (FDNA '04)* (ACM, New York, 2004), pp. 5–12. [ISBN 1-58113-942-X]
6. M.J. Gander, A.M. Stuart, Space-time continuous analysis of waveform relaxation for the heat equation. SIAM J. Sci. Comput. **19**(6), 2014–2031 (1998) [ISSN 1064-8275]
7. M.J. Gander, L. Halpern, F. Nataf et al., Optimal convergence for overlapping and non-overlapping Schwarz waveform relaxation, in *The Eleventh International Conference on Domain Decomposition Methods*, Citeseer, 1999, ed. by C.H. Lai, P. Bjørstad, M. Cross, O. Widlund (1999), pp. 27–36
8. E. Giladi, H.B. Keller, Space-time domain decomposition for parabolic problems. Numer. Math. **93**(2), 279–313 (2002) [ISSN 0029-599X]
9. C. Japhet, P. Omnes, Optimized Schwarz waveform relaxation for porous media applications, in *Domain Decomposition Methods in Science and Engineering XX* (Springer, Heidelberg, 2013), pp. 585–592
10. F. Lemarié, M. Patrick, L. Debreu, E. Blayo, Sensitivity of an Ocean-Atmosphere Coupled Model to the Coupling Method: Study of Tropical Cyclone Erica (2014). http://hal.inria.fr/hal-00872496
11. V. Tipparaju, W. Gropp, H. Ritzdorf, R. Thakur, J.L. Traff, Investigating high performance RMA interfaces for the MPI-3 standard, in *International Conference on Parallel Processing, 2009*, September 2009, pp. 293–300
12. S.G. Vandewalle, E.F. Van de Velde, Space-time concurrent multigrid waveform relaxation. Ann. Numer. Math. **1**(1–4), 347–360 (1994) [ISSN 1021-2655]

Parareal for Diffusion Problems with Space- and Time-Dependent Coefficients

Daniel Ruprecht, Robert Speck, and Rolf Krause

1 Introduction

The very rapidly increasing number of cores in state-of-the-art supercomputers fuels both the need for and the interest in novel numerical algorithms inherently designed to feature concurrency. In addition to the mature field of space-parallel approaches (e.g. domain decomposition techniques), time-parallel methods that allow concurrency along the temporal dimension are now an increasingly active field of research, although first ideas, like in [12], go back several decades. A prominent and widely studied algorithm in this area is Parareal, introduced in [10], which has the advantage that one can couple and reuse classical time-stepping schemes in an iterative fashion to parallelize in time. However, there also exist a number of other approaches, e.g. the "parallel implicit time algorithm" (PITA) from [5], the "parallel full approximation scheme in space and time" (PFASST) from [4] or "revisionist integral deferred corrections" (RIDC) from [3] to name a few. Parareal in particular and temporal parallelism in general has been considered early as an addition to spatial parallelism in order to extend strong scaling limits, see [11]. Efficacy of this approach in large-scale parallel simulations on hundreds of thousands of cores has been demonstrated for the PFASST algorithm in [14].

D. Ruprecht (✉) • R. Krause
Institute of Computational Science, Università della Svizzera italiana, Lugano, Switzerland
e-mail: daniel.ruprecht@usi.ch; rolf.krause@usi.ch

R. Speck
Institute of Computational Science, Università della Svizzera italiana, Lugano, Switzerland

Jülich Supercomputing Centre, Forschungszentrum Jülich GmbH, Jülich, Germany
e-mail: r.speck@fz-juelich.de

T. Dickopf et al. (eds.), *Domain Decomposition Methods in Science and Engineering XXII*, Lecture Notes in Computational Science and Engineering 104, DOI 10.1007/978-3-319-18827-0_37

For Parareal, multiple works exist that demonstrate its efficiency for diffusion problems: Gander and Vandewalle [9] prove super-linear convergence of Parareal for the standard 1D heat equation. A more general theorem showing super-linear convergence for nonlinear ODEs is proven by Gander and Hairer [7], while [2] presents a convergence theorem for linear parabolic PDEs with constant coefficients. The present paper investigates the effect of space- and time-dependent coefficients in the two-dimensional heat equation on the convergence of Parareal. This is done by means of numerical examples, including one that shows how convergence of Parareal can be estimated by the maximum singular value of a Parareal iteration matrix.

2 Parareal

To match the numerical examples in Sect. 3, the presentation of Parareal given here starts with an initial value problem

$$My_t(t) = f(y(t), t),\ y(0) = b \in \mathbb{R}^d,\ t \in [0, T], \tag{1}$$

with a mass matrix M and right-hand side f arising from a finite element discretization of a partial differential equation. Let $(t_n)_{n=0}^N$ with $t_0 = 0$ and $t_N = T$ be a decomposition of $[0, T]$ into N so-called time-slices $[t_n, t_{n+1}]$ which, for the sake of simplicity, are assumed to be of equal length here. Furthermore, let y_n be an approximation to the solution at t_n, that is $y_n \approx y(t_n)$.

Denote by $\mathcal{F}$ a "fine", computationally expensive and accurate integration method with a time step δt (e.g. a higher-order Runge-Kutta method) and by $\mathcal{G}$ a "coarse", computationally cheap and probably inaccurate method with a time step $\Delta t \gg \delta t$ (e.g. implicit Euler). Assume here that the constant length of the time-slices is a multiple of both δt and Δt, so that the fine as well as the coarse method can integrate over one time-slice using a fixed integer number of time-steps. Denote the result of integrating over the slice $[t_n, t_{n+1}]$, starting from an initial value y at t_n, using the fine or coarse method as $\mathcal{F}(y, t_{n+1}, t_n)$ and $\mathcal{G}(y, t_{n+1}, t_n)$ respectively. Serial integration using the fine method would then correspond to computing

$$y_{n+1} = \mathcal{F}(y_n, t_{n+1}, t_n),\ n = 0, \ldots, N-1, \tag{2}$$

step-by-step with $y_0 := b$. Instead, Parareal computes the iteration given by

$$y_{n+1}^{k+1} = \mathcal{G}(y_n^{k+1}, t_{n+1}, t_n) + \mathcal{F}(y_n^k, t_{n+1}, t_n) - \mathcal{G}(y_n^k, t_{n+1}, t_n) \tag{3}$$

where the evaluation of the fine method over the N time-slices can be distributed over N processors (see [10] for details). The iteration converges to the serial fine solution as $k \to N$. Speedup can be achieved if $\mathcal{G}$ is sufficiently cheap compared to $\mathcal{F}$ and if the iteration converges in $K \ll N$ iterations. Therefore, rapid convergence is critical for Parareal to be efficient. In the examples below, the defect

$$d^k := \max_{i=0,\ldots,N} \left\| y_i - y_i^k \right\|_\infty \tag{4}$$

between the solution provided by the Parareal iteration (3) after k iterations and the serial fine solution (2) is used to measure convergence.

3 Heat Equation with Non-constant Coefficients

The test problem used here to study the convergence of Parareal for non-constant coefficients is the two-dimensional heat equation

$$u_t(x,y,t) = \nu(t)\nabla \cdot (a(x,y)\nabla u(x,y,t)) \tag{5}$$

on a square $\Omega = [0,1]^2$. The initial values are given by

$$u_0(x,y) = \exp\left[-\left((x-0.5)^2 + (y-0.5)^2\right)/\sigma^2\right],\ \sigma = 0.35, \tag{6}$$

and the problem is run until $T = 4.0$. The interval $[0,T]$ is divided into $N = 40$ time-slices and an implicit Euler method with $\Delta t = 1/100$ is used for $\mathcal{G}$ and a third order RadauIIA(3) method with $\delta t = 1/200$ for $\mathcal{F}$. The spatial domain Ω is divided into three "strips"

$$\Omega_1 = [0, x_0) \times [0,1], \tag{7}$$

$$\Omega_2 = [x_0, x_0 + w) \times [0,1], \tag{8}$$

$$\Omega_3 = [x_0 + w, 1] \times [0,1], \tag{9}$$

and a different constant value for a is prescribed on every strip, that is

$$a(x,y) = \begin{cases} a_1 : (x,y) \in \Omega_1 \\ a_2 : (x,y) \in \Omega_2 \\ a_3 : (x,y) \in \Omega_3. \end{cases} \tag{10}$$

Furthermore, the effect of varying the width w of the middle strip Ω_2 is investigated. Conforming triangle meshes aligned with the strips Ω_i are generated for values of $w \in \{0.2, 0.1, 0.05, 0.02\}$. Then, for every value of w, a number of uniform refinement steps is performed in order to produce meshes of comparable mesh width. After refinement, the minimum element sizes for the different values of w range from $h_{\min} = 0.01$ to $h_{\min} = 0.005$ and the maximum element sizes from $h_{\max} = 0.02$ to $h_{\max} = 0.035$, so that the resolutions are comparable. All experiments reported below use linear finite elements, but preliminary tests not

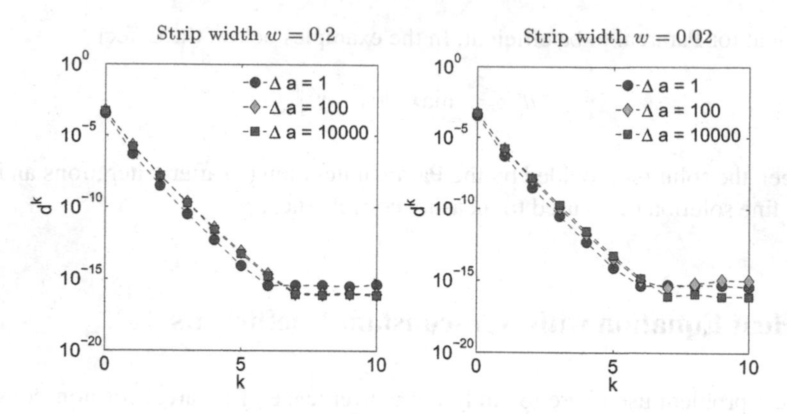

Fig. 1 Defect d^k between Parareal and the serial fine solution versus the iteration number k depending on the magnitude of the jump in the diffusion coefficient from Ω_1, Ω_3 to Ω_2

documented here suggest that the results are not significantly affected by the use of higher-order FEM. Homogeneous Dirichlet boundary conditions are employed. Simulations are run with $a_1 = a_3 = 0.01$ fixed and $a_2 \in \{0.01, 1.0, 100\}$, resulting in ratios $\Delta a = a_2/a_3 = a_2/a_1 \in \{1, 100, 10000\}$.

3.1 Space-Dependent Coefficients

First, set $\nu \equiv 1$ in order to study only the effect of spatially varying coefficients. Figure 1 shows the resulting convergence of Parareal for the different values of Δa and $w = 0.2$ (left) and $w = 0.02$ (right). Convergence in the cases with jumping coefficients is slightly slower, but the effect is very small. Also, the reduction in convergence speed seems to be rather independent of the magnitude of the jump in the diffusion coefficient: In both plots, the lines for $\Delta a = 100$ and $\Delta a = 10{,}000$ are more or less indistinguishable.

Convergence of Parareal is utterly oblivious to the width w of the middle strip Ω_2: When plotting the defects for fixed Δa and different values of w, the resulting data points all essentially coincide so that the corresponding plots are rather uninteresting and are therefore omitted.

3.2 Space- and Time-Dependent Coefficients

To investigate the effect of a time-dependent diffusion coefficient on the convergence of Parareal, fix the strip width to $w = 0.2$ and the coefficient jump to $\Delta a = 100$. Furthermore, use the following three different profiles for the time-

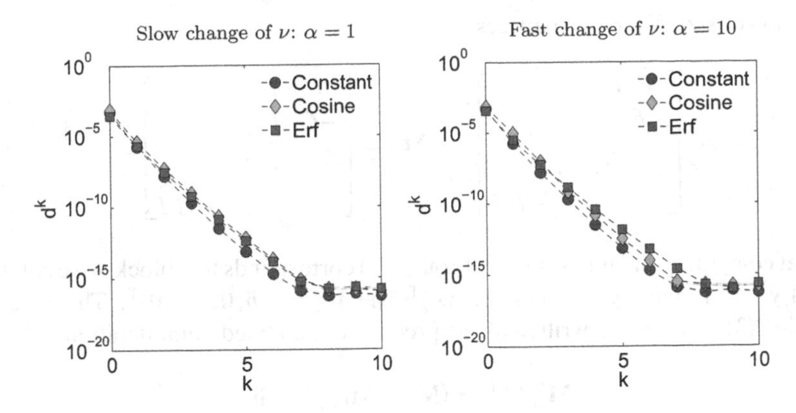

Fig. 2 Defect d^k of Parareal versus the iteration number k for different time-dependent ν-profiles with $\Delta a = 100$, $\alpha = 1$ (*left*) and $\alpha = 10$ (*right*)

dependent diffusion coefficient ν:

$$\nu(t) = 1 \qquad \text{("constant")}, \tag{11}$$

$$\nu(t) = \frac{1}{2}\left(1 + \cos\left(\alpha\frac{\pi}{2}t\right)\right) \qquad \text{("cosine")}, \tag{12}$$

$$\nu(t) = \frac{1}{2}(1 + \operatorname{erf}(\alpha(t-2))) \qquad \text{("erf")}. \tag{13}$$

Initial value and boundary conditions are set as described above. Two sets of simulations are performed, one with $\alpha = 1$ corresponding to a very slowly changing ν and one with $\alpha = 10$ corresponding to a more rapid change. The resulting convergence of Parareal is shown in Fig. 2. In both cases, the slow as well as the fast varying one, Parareal's convergence is only marginally affected by the space- and time-dependent diffusion coefficients. The resulting defects are slightly larger than for the reference case and the difference is a little more pronounced for $\alpha = 10$, but the overall effect is not drastic: In the fast varying case with the error function profile (13), Parareal requires only a single additional iteration compared to the constant reference in order to reach the same defect level.

3.3 *Error Bound from Singular Values*

Parareal can be considered as a fixed point iteration, see e.g. [1] or [6] for more detailed explanations. For $\nu \equiv 1$ and the linear problem considered here, the action of the propagators $\mathcal{F}$ and $\mathcal{G}$ can be expressed as multiplication by matrices G or F. Then, running the fine or coarse method over all N time-slices can be expressed as inversion of size $Nd \times Nd$ matrices

$$\mathbf{M}_f = \begin{bmatrix} I & \dots & & \\ -F & I & & \\ & \ddots & \ddots & \\ & & -F & I \end{bmatrix}, \quad \mathbf{M}_g = \begin{bmatrix} I & \dots & & \\ -G & I & & \\ & \ddots & \ddots & \\ & & -G & I \end{bmatrix}, \tag{14}$$

so that computing the fine solution through (2) corresponds to a block-wise solution of $\mathbf{M}_f\mathbf{y} = \mathbf{b}$ with $\mathbf{y} = (y_0, \dots, y_N)^{\mathrm{T}}$ and $\mathbf{b} = (b, 0, \dots, 0)^{\mathrm{T}}$. The Parareal iteration (3) can then be written as the preconditioned fixed point iteration

$$\mathbf{M}_g \mathbf{y}^{k+1} = (\mathbf{M}_g - \mathbf{M}_f)\,\mathbf{y}^k + \mathbf{b}, \tag{15}$$

where inverting $\mathbf{M}_g$ corresponds to running the coarse method. A straightforward computation shows that the iteration matrix $\mathbf{I} - \mathbf{M}_g^{-1}\mathbf{M}_f$ is nilpotent and thus that its spectral radius is zero, corresponding to the well-known fact that Parareal always converges to the fine solution after N iterations, see e.g. [9] (although Parareal won't provide any speedup in this case). A bound for the convergence rate can be obtained by computing the maximum singular value instead. In order to keep the size of the iteration matrix manageable, the example studied here uses only the $w = 0.2$ geometry with a coarser grid with $h_{\min} = 0.04$, $h_{\max} = 0.068$ and only $N = 20$ time-slices. The maximum singular values of the iteration matrix are computed with Matlab's SVDS function and are $\sigma_{\max} \approx 0.162$ for $\Delta a = 1$ and $\sigma_{\max} \approx 0.163$ for $\Delta a = 10{,}000$: The minimal difference gives an additional indication that the coefficient jump should not influence Parareal's convergence. Figure 3 shows the convergence rates of Parareal for this example for $\Delta a = 1$, i.e. with a constant coefficient (left) and with $\Delta a = 10{,}000$ (right), as well as the estimate $d^0 \times (\sigma_{\max})^k$ resulting from the maximum singular value.

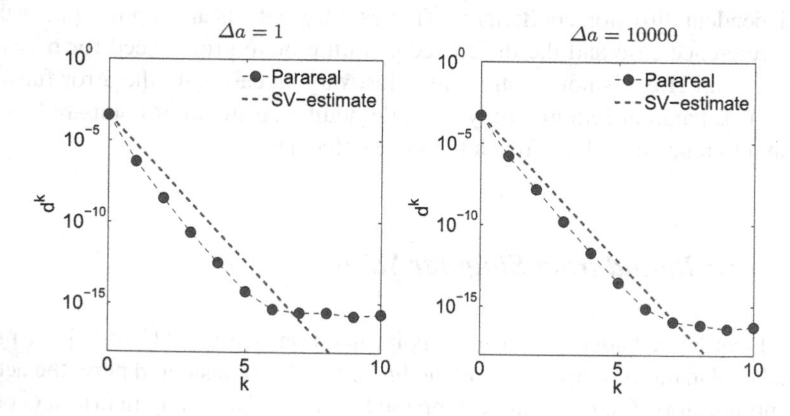

Fig. 3 Convergence of Parareal and error estimate from the largest singular value of the Parareal iteration matrix (*dashed line*)

In both cases, actual convergence is a little better than expected but $\sigma_{\max}$ gives a reasonable estimate. Again, the jumping coefficients affect Parareal's convergence only marginally. Note that interpreting variants like the "Krylov-subspace-enhanced Parareal", introduced in [8] and studied further in [13], as a non-stationary fixed point iteration could be an interesting approach for a mathematical analysis.

4 Conclusions

The paper presents a numerical study of the convergence behavior of the time-parallel Parareal method for the heat equation with space- and time-dependent coefficients. It demonstrates that the good convergence of Parareal for diffusive problems is only marginally affected by both jumps in the diffusion coefficients and a diffusion coefficient that changes in time. For linear problems, Parareal can be interpreted as a preconditioned fixed point iteration and, at least for small enough problems, the iteration matrix and its maximum singular value can be computed numerically. An example is shown that demonstrates that the largest singular value gives a reasonable estimate for the convergence of Parareal. Extending the analysis presented here to more complicated cases e.g. in three dimensions with complicated geometries, with coefficient jumps not aligned with the mesh or cases that also include advection would be an interesting direction of future research.

Acknowledgements This work was supported by the Swiss National Science Foundation (SNSF) under the lead agency agreement through the project "ExaSolvers" within the Priority Programme 1648 "Software for Exascale Computing" (SPPEXA) of the Deutsche Forschungsgemeinschaft (DFG). The authors thankfully acknowledge support from Achim Schädle, who provided parts of the used code.

References

1. P. Amodio, L. Brugnano, Parallel solution in time of odes: some achievements and perspectives. Appl. Numer. Math. **59**(3–4), 424–435 (2009)
2. G. Bal, On the convergence and the stability of the parareal algorithm to solve partial differential equations, in *Domain Decomposition Methods in Science and Engineering*, ed. by R. Kornhuber et al. Lecture Notes in Computational Science and Engineering, vol. 40 (Springer, Berlin, 2005), pp. 426–432
3. A.J. Christlieb, C.B. Macdonald, B.W. Ong, Parallel high-order integrators. SIAM J. Sci. Comput. **32**(2), 818–835 (2010)
4. M. Emmett, M.L. Minion, Toward an efficient parallel in time method for partial differential equations. Commun. Appl. Math. Comput. Sci. **7**, 105–132 (2012)
5. C. Farhat, M. Chandesris, Time-decomposed parallel time-integrators: theory and feasibility studies for fluid, structure, and fluid-structure applications. Int. J. Numer. Methods Eng. **58**(9), 1397–1434 (2003)

6. S. Friedhoff, R.D. Falgout, T.V. Kolev, S. MacLachlan, J.B. Schroder, A multigrid-in-time algorithm for solving evolution equations in parallel, in *Sixteenth Copper Mountain Conference on Multigrid Methods*, Copper Mountain, 17–22 March 2013
7. M. Gander, E. Hairer, Nonlinear convergence analysis for the parareal algorithm, in *Domain Decomposition Methods in Science and Engineering*, ed. by U. Langer, O. Widlund, D. Keyes. Lecture Notes in Computational Science and Engineering, vol. 60 (Springer, Berlin/Heidelberg, 2008), pp. 45–56
8. M. Gander, M. Petcu, Analysis of a Krylov subspace enhanced parareal algorithm for linear problems. ESAIM Proc. **25**, 114–129 (2008)
9. M.J. Gander, S. Vandewalle, Analysis of the parareal time-parallel time-integration method. SIAM J. Sci. Comput. **29**(2), 556–578 (2007)
10. J.-L. Lions, Y. Maday, G. Turinici, A "parareal" in time discretization of PDE's. C. R. Acad. Sci. Ser. I Math. **332**, 661–668 (2001)
11. Y. Maday, G. Turinici, The parareal in time iterative solver: a further direction to parallel implementation, in *Domain Decomposition Methods in Science and Engineering*, ed. by R. Kornhuber et al. Lecture Notes in Computational Science and Engineering, vol. 40 (Springer, Berlin, 2005), pp. 441–448
12. J. Nievergelt, Parallel methods for integrating ordinary differential equations. Commun. ACM **7**(12), 731–733 (1964)
13. D. Ruprecht, R. Krause, Explicit parallel-in-time integration of a linear acoustic-advection system. Comput. Fluids **59**, 72–83 (2012)
14. R. Speck, D. Ruprecht, R. Krause, M. Emmett, M. Minion, M. Winkel, P. Gibbon, A massively space-time parallel N-body solver, in *Proceedings of the International Conference on High Performance Computing, Networking, Storage and Analysis*, SC '12 (IEEE Computer Society, Los Alamitos, 2012), pp. 92:1–92:11

A Discontinuous Coarse Space (DCS) Algorithm for Cell Centered Finite Volume Based Domain Decomposition Methods: The DCS-RJMin Algorithm

Kévin Santugini

1 Introduction

Due to the ever increasing parallelism in modern computers, and the ever increasing affordability of massively parallel calculators, it is of utmost importance to develop algorithms that are not only parallel but scalable. In this paper, we are interested in Domain Decomposition Methods (DDMs), which is one way to parallelize the numerical resolution of Partial Differential Equations (PDE).

In Domain Decomposition Methods, the whole domain is subdivided in several subdomains and a computation unit is assigned to each subdomain. In this paper, we only consider non-overlapping domain decompositions. The numerical solution is then computed in parallel inside each subdomain with artificial boundary conditions. Then, subdomains exchange information between each other. This process is reapplied until convergence. In practice, such a scheme, called iterative DDM, should be accelerated using Krylov methods. However, for the purpose of analyzing an algorithm, it can be interesting to work directly with the iterative algorithm itself as Krylov acceleration is so efficient it can hide small design problems in the algorithm.

In one-level DDMs, only neighboring subdomains exchange information. Most classical DDM are one-level. While one-level DDMs can be very efficient and can converge in few iterations, they are not scalable: convergence can never occur before information has propagated between the two furthest apart subdomains, i.e., a one level DDM must iterate at least as many times as the diameter of the connectivity

K. Santugini (✉)
INP Bordeaux, IMB, CNRS UMR 5251, MC2, INRIA Bordeaux -Sud-Ouest, Bordeaux, France
e-mail: Kevin.Santugini@math.u-bordeaux1.fr

T. Dickopf et al. (eds.), *Domain Decomposition Methods in Science and Engineering XXII*, Lecture Notes in Computational Science and Engineering 104, DOI 10.1007/978-3-319-18827-0_38

graph of the domain decomposition. Typically, if N is the number of subdomains, this means at least $O(N)$ iterations for one-dimensional problems, $O(\sqrt{N})$ for two-dimensional ones and $O(\sqrt[3]{N})$ for three-dimensional ones. For DDMs to be scalable, some kind of global information exchange is needed. The traditional approach to achieve such global information exchange is adding a coarse space to a pre-existing one-level DDM.

To the author's knowledge, the first use of coarse spaces in Domain Decomposition Methods can be found in [19]. Because coarse spaces enable global information exchange, scalability becomes possible. Well known methods with coarse spaces are the two-level Additive Schwarz method [4], the FETI method [16], and the balancing Neumann-Neumann method [5, 15, 17]. Coarse spaces are also an active area of research, see for example [3, 9, 18, 21] for high contrast problems. It is not trivial to add an effective coarse space to one-level DDMs that produce discontinuous iterates such as Optimized Schwarz Methods, see [7, 8], and [6, Chap. 5].

In [11], the authors introduced the idea of using discontinuous coarse spaces. Since many DDM algorithms produce discontinuous iterates, the use of discontinuous coarse corrections is needed to correct the discontinuities between subdomains, where the iterates of the one-level OSM are discontinuous. In [11], one possible algorithm, the DCS-DMNV (Discontinuous Coarse Space Dirichlet Minimizer Neumann Variational), was described at the continuous level and at the discrete level for Finite Element Methods on a non-overlapping Domain Decomposition. In [20], a similar method, the DCS-DGLC algorithm was proposed. Both the DCS-DMNV and the DCS-DGLC are well suited to finite element discretizations. Also, a similar approach was proposed in [10] for Restricted Additive Schwarz (RAS), an overlapping DDM.

The proof of convergence for Schwarz found in [2, 14] can be extended to the Discrete Optimized Schwarz algorithm with cell centered finite volume methods, see [1, 12, 13]. It would be interesting to have a discontinuous coarse space algorithm that is suited to cell centered finite volumes. Unfortunately, neither the DCS-DMNV algorithm nor the DCS-DGLC algorithm are practical for cell centered-finite volume methods: the stiffness matrix necessary to compute the coarse correction is not as sparse as one would intuitively believe. In this paper, our main goal is to describe one family of algorithms making use of discontinuous coarse spaces suitable for cell centered finite volumes discretizations.

In Sect. 2, we briefly recall the motivations behind the use of discontinuous coarse spaces. In Sect. 3, we present the DCS-RJMin algorithm. In Sect. 4, we prove that under some conditions on the algorithm parameter, the L^2-norm of the difference between two consecutive iterates goes to zero. Finally, we present numerical results in Sect. 5.

2 Optimized Schwarz and Discontinuous Coarse Spaces

Let us consider a polygonal domain Ω in $\mathbb{R}^2$. As a simple test case, we wish to solve

$$\begin{aligned} \eta u - \Delta u &= f \text{ in } \Omega, \\ u &= 0 \text{ on } \partial\Omega. \end{aligned}$$

Without a coarse space, the Optimized Schwarz Method is defined as

Algorithm 1 (One-Level OSM)

1. *Set u_i^0 to either the null function or to a first approximation.*
2. *Until convergence*

   a. *Set u_i^{n+1} as the unique solution to*

$$\begin{aligned} \eta u_i^{n+1} - \Delta u_i^{n+1} &= f \text{ in } \Omega_i, \\ \frac{\partial u_i^{n+1}}{\partial \boldsymbol{n}_i} + p u_i^{n+1} &= \frac{\partial u_j^n}{\partial \boldsymbol{n}_i} + p u_j^n \text{ on } \partial\Omega_i \cap \partial\Omega_j, \\ u^{n+1} &= 0 \text{ on } \partial\Omega_i \cap \partial\Omega. \end{aligned}$$

The main shortcoming of the one-level Optimized Schwarz method is the absence of direct communication between distant subdomains. To get a scalable algorithm, one can use a coarse space. A general version of a coarse space method for the OSM is

Algorithm 2 (Generic Two-Level OSM)

1. *Set u_i^0 to either the null function or to the coarse solution.*
2. *Until convergence*

   a. *Set u_i^{n+1} as the unique solution to*

$$\begin{aligned} \eta u_i^{n+1/2} - \Delta u_i^{n+1} &= f \text{ in } \Omega_i, \\ \frac{\partial u_i^{n+1/2}}{\partial \boldsymbol{n}_i} + p u_i^{n+1/2} &= \frac{\partial u_j^n}{\partial \boldsymbol{n}_i} + p u_j^n \text{ on } \partial\Omega_i \cap \partial\Omega_j, \\ u^{n+1/2} &= 0 \text{ on } \partial\Omega_i \cap \partial\Omega. \end{aligned}$$

   b. *Compute in some way a coarse corrector U^{n+1} belonging to the coarse space X, then set*

$$u^{n+1} = u^{n+1/2} + U^{n+1}.$$

More important than the algorithm used to compute the coarse correction U^{n+1} is the choice of an adequate coarse space itself. The ideas presented in [11] still apply. In particular, the coarse space should contain discontinuous functions and the discontinuities of the coarse corrector should be located at the interfaces between subdomains. For these reasons, we suppose the whole domain Ω is meshed by either a coarse triangular mesh or a cartesian mesh $\mathscr{T}_H$ and we use each coarse cell of $\mathscr{T}_H$ as a subdomain Ω_i of Ω. The optimal theoretical coarse space $\mathscr{A}$ is the set of all functions that are solutions to the homogenous equation inside each subdomain: for linear problems, the errors made by any iterate are guaranteed to belong to that space. With an adequate algorithm to compute U^{n+1}, the coarse space $\mathscr{A}$ gives a convergence in a single coarse iteration. Unfortunately this complete coarse space is only practical for one dimensional problems as it is of infinite dimension in higher dimensions. One should therefore choose a finite dimensional subspace X_d of $\mathscr{A}$.

The choice of the coarse space X_d is primordial. It should have a dimension that is a small multiple of the number of subdomains. To choose X_d, one only needs to choose boundary conditions on every subdomain, then fill the interior of each subdomain by solving the homogenous equation in each subdomain. In this paper, we have not tried to optimize X_d and for the sake of simplicity have chosen X_d as the set of all functions in $\mathscr{A}$ with linear Dirichlet boundary conditions on each interface between any two adjacent subdomains.

3 The DCS-RJMin Algorithm

We now describe the DCS-Robin Jump Minimizer algorithm:

Algorithm 3 (DCS-RJMin)

1. *Set $p > 0$ and $q > 0$ and X_d a finite dimensional subspace of $\mathscr{A}$.*
2. *Set u^0 to either 0 or to the coarse space solution.*
3. *Until Convergence*

   a. *Set $u^{n+\frac{1}{2}}$ as the unique solution to*

$$\eta u^{n+\frac{1}{2}} - \Delta u^{n+\frac{1}{2}} = f \text{ in } \Omega_i,$$

$$\frac{\partial u_i^{n+\frac{1}{2}}}{\partial \boldsymbol{\nu}_{ij}} + p u_i^{n+\frac{1}{2}} = \frac{\partial u_j^n}{\partial \boldsymbol{\nu}_{ij}} + p u_j^n \text{ on } \partial\Omega_i \cap \partial\Omega_j,$$

$$u_i = 0 \text{ on } \partial\Omega_i \cap \partial\Omega_j.$$

b. *Set U^{n+1} in X_d as the unique coarse function that minimizes*

$$\sum_{i=1}^{N}\sum_{j\in\mathcal{N}(i)}\left\|\frac{\partial(u_i^{n+\frac{1}{2}}+U_i^{n+1})}{\partial\boldsymbol{\nu}_i}+q(u_i^{n+\frac{1}{2}}+U_i^{n+1})\right.$$
$$\left.-\frac{\partial(u_j^{n+\frac{1}{2}}+U_j^{n+1})}{\partial\boldsymbol{\nu}_i}-q(u_j^{n+\frac{1}{2}}+U_j^{n+1})\right\|^2_{L^2(\partial\Omega_i\cap\partial\Omega_j)},$$

where $\boldsymbol{\nu}_i$ is the outward normal to subdomain Ω_i and $\mathcal{N}(i)$ is the set of all j such that Ω_j and Ω_i are adjacent.

c. *Set $u^{n+1} := u^{n+1/2} + U^{n+1}$.*

4 Partial Convergence Results for DCS-RJMin

We do not have a complete convergence theorem for the DCS-RJMin algorithm. However, we can prove the following results concerning the iterates of the DCS-RJMin algorithm when $p = q$:

Proposition 1 *If $q = p$, then the iterates produced by the DCS-RJMin Algorithm 3 satisfy $\lim_{n\to+\infty}\|u_i^{n+1/2} - u_i^n\|_{L^2} = 0$.*

Proof Let u be the mono-domain solution and set $e_i^n = u_i^n - u_i$. Then, following Lions energy estimates [14], we compute

$$\eta\int_{\Omega_i}|e_i^{n+1/2}-e_i^n|^2\mathrm{d}\boldsymbol{x}+\int_{\Omega_i}|\nabla(e_i^{n+1/2}-e_i^n)|^2\mathrm{d}\boldsymbol{x}$$
$$=\int_{\partial\Omega_i}\frac{\partial(e_i^{n+1/2}-e_i^n)}{\partial\boldsymbol{\nu}}\cdot(e_i^{n+1/2}-e_i^n)$$
$$=\frac{1}{4p}\left(\int_{\partial\Omega_i}\left|\frac{\partial(e_i^{n+1/2}-e_i^n)}{\partial\boldsymbol{\nu}}+p(e_i^{n+1/2}-e_i^n)\right|^2\right.$$
$$\left.-\left|\frac{\partial(e_i^{n+1/2}-e_i^n)}{\partial\boldsymbol{\nu}}-p(e_i^{n+1/2}-e_i^n)\right|^2\right)$$
$$=\frac{1}{4p}\left(\int_{\partial\Omega_i}\left|\frac{\partial(e_i^{n+1/2}-e_i^n)}{\partial\boldsymbol{\nu}}+p(e_i^{n+1/2}-e_i^n)\right|^2\right.$$
$$\left.-\int_{\partial\Omega_i}\left|\frac{\partial(e_i^{n+1/2}-e_i^n)}{\partial\boldsymbol{\nu}}-p(e_i^{n+1/2}-e_i^n)\right|^2\right)$$

$$= \frac{1}{4p}\Bigg(\sum_j \int_{\partial\Omega_i\cap\partial\Omega_j} \left|\left(\frac{\partial e_j^n}{\partial \boldsymbol{\nu}_i} + pe_j^n\right) - \left(\frac{\partial e_i^{n)}}{\partial \boldsymbol{\nu}_i} + pe_i^n\right)\right|^2$$

$$- \sum_j \int_{\partial\Omega_i\cap\partial\Omega_j} \left|\left(\frac{\partial e_i^{n+1/2}}{\partial \boldsymbol{\nu}_i} - pe_i^{n+1/2}\right) - \left(\frac{\partial e_j^{n+1/2)}}{\partial \boldsymbol{\nu}_i} - pe_j^{n+1/2}\right)\right|^2\Bigg).$$

We sum the above equality over all subdomains Ω_i and get

$$\eta \sum_i \int_{\Omega_i} |e_i^{n+1/2} - e_i^n|^2 \mathrm{d}\boldsymbol{x} + \int_{\Omega_i} |\nabla(e_i^{n+1/2} - e_i^n)|^2 \mathrm{d}\boldsymbol{x} =$$

$$= \sum_{(i,j)} \frac{1}{4p}\left(\int_{\Gamma_{ij}} \left|\left[\frac{\partial e^n}{\partial \boldsymbol{\nu}_i} + pe^n\right]\right|^2 - \int_{\Gamma_{ij}} \left|\left[\frac{\partial e^{n+1/2}}{\partial \boldsymbol{\nu}_i} + pe^{n+1/2}\right]\right|^2\right),$$

where $[\cdot]$ represents a jump across the interface. Since the coarse step of the DCS-RJMin algorithm minimizes the Robin Jumps, we have

$$\eta \sum_i \int_{\Omega_i} |e_i^{n+1/2} - e_i^n|^2 \mathrm{d}\boldsymbol{x} + \int_{\Omega_i} |\nabla(e_i^{n+1/2} - e_i^n)|^2 \mathrm{d}\boldsymbol{x} \leq$$

$$\leq \sum_{(i,j)} \frac{1}{4p}\left(\int_{\Gamma_{ij}} \left|\left[\frac{\partial e^n}{\partial \boldsymbol{\nu}_i} + pe^n\right]\right|^2 - \int_{\Gamma_{ij}} \left|\left[\frac{\partial e^{n+1}}{\partial \boldsymbol{\nu}_i} + pe^{n+1}\right]\right|^2\right).$$

Summing over $n \geq 0$ yields the stated result.

Remark 1 For $q \neq p$, convergence can be proven in the two subdomain case if each subdomain is obtained by reflection of the other with respect to the common interface.

5 Numerical Results

We have implemented the DCS-RJMin algorithm in C++ for cell-centered finite volumes on a cartesian grid. We chose $\Omega =]0,4[\times]0,4[$, $\eta = 0$ and iterated directly on the errors by choosing $f = 0$. We initialized the Robin transmission conditions at the interfaces between subdomains at random and performed multiple runs of the DCS-RJMin algorithm for various values of p, q and of the number of subdomains. We had p vary from 1.0 to 20.0 with 0.5 increments and q took the values $q_m \times 10^{q_e}$ with q_m in $\{1.0, 2.0, 4.0, 8.0\}$ and q_e in $\{0, 1\}$. We consider 2×2, 4×4, 6×6 and 8×8 subdomains. There are always 20×20 cells per subdomains. In Fig. 1, we plot

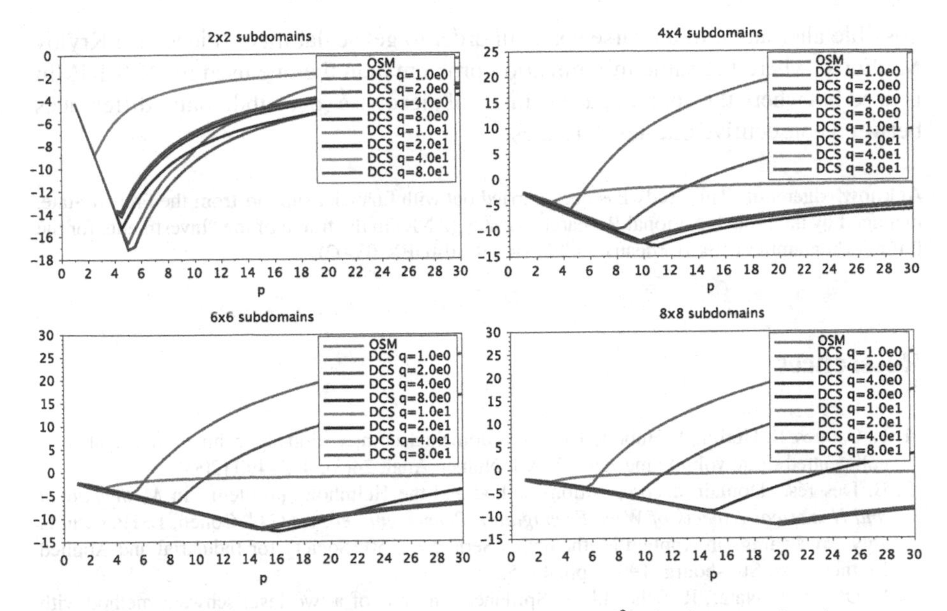

Fig. 1 Convergence for OSM and DCS-RJmin with $\Omega = [0, 4]^2, f(x, y) = 0$ and random guess. Plotting $\log(\|e_{50}\|_\infty / \|e_0\|_\infty)$

$\log(\|e_{50}\|_\infty / \|e_0\|_\infty)$ as a function of p for various values of q. First, we notice that for each value of q, the convergence deteriorates above a certain p_q. In fact, for low values of q and high values of p, the iterates diverge. For two different values of q, the curves are very close when p is smaller than both p_q. We also notice that even though we could only prove Proposition 1 for the case $p = q$, we observe numerical convergence even when $p \neq q$. In fact $p = q$ is not the numerical optimum. This is to be expected intuitively: for a theoretical proof of convergence, we want the algorithm to keep lowering some functional. The existence of such a functional is likely only if all the substeps of the algorithm are optimized for the same kind of errors. If $p = q$, both the coarse step or the local step will either remove low frequency errors (small p and q) or high frequency ones (high p and q). An efficient numerical algorithm should have substeps optimized for completely different kind of errors.

6 Conclusion

In this paper, we have introduced a new discontinuous coarse space algorithm, the DCS-RJMin, which is suitable for cell-centered finite volume discretizations. The coarse space greatly improves numerical convergence. It would be of great interest to study which is the optimal low-dimensional subspace of all piecewise discontinuous piecewise harmonic functions. Future work also includes the development of a possible alternative to a coarse space in order to get scalability: "Piecewise Krylov Methods" where the same minimization problem than the one used in DCS-RJMin is used but where the coarse space is made of piecewise (per subdomain) differences between consecutive one-level iterates.

Acknowledgement This study has been carried out with financial support from the French State, managed by the French National Research Agency (ANR) in the frame of the "Investments for the future" Programme IdEx Bordeaux—CPU (ANR-10-IDEX-03-02).

References

1. R. Cautres, R. Herbin, F. Hubert, The Lions domain decomposition algorithm on non-matching cell centred finite volume meshes. IMA J. Numer. Anal. **24**(3), 465–490 (2004)
2. B. Després, Domain decomposition method and the Helmholtz problem, in *Mathematical and Numerical Aspects of Wave Propagation Phenomena*, ed. by G.C. Cohen, L. Halpern, P. Joly. Proceedings in Applied Mathematics Series, vol. 50 (Society for Industrial and Applied Mathematics, Strasbourg, 1991), pp. 44–52
3. V. Dolean, F. Nataf, R. Scheichl, N. Spillane, Analysis of a two-level schwarz method with coarse spaces based on local dirichlet to neumann maps. Comput. Methods Appl. Math. **12**(4), 391–414 (2012)
4. M. Dryja, O.B. Widlund, An additive variant of the Schwarz alternating method for the case of many subregions. Technical Report 339, also Ultracomputer Note 131, Department of Computer Science, Courant Institute (1987)
5. M. Dryja, O.B. Widlund, Schwarz methods of Neumann-Neumann type for three-dimensional elliptic finite element problems. Commun. Pure Appl. Math. **48**(2), 121–155 (1995)
6. O. Dubois, Optimized Schwarz methods for the advection-diffusion equation and for problems with discontinuous coefficients. Ph.D. thesis, McGill University, 2007
7. O. Dubois, M.J. Gander, Convergence behavior of a two-level optimized Schwarz preconditioner, in *Domain Decomposition Methods in Science and Engineering, XVIII*, ed. by M. Bercovier, M.J. Gander, R. Kornhuber, O. Widlund. Lecture Notes in Computational Science and Engineering (Springer, Berlin, 2009)
8. O. Dubois, M.J. Gander, S. Loisel, A. St-Cyr, D. Szyld, The optimized Schwarz method with a coarse grid correction. SIAM J. Sci. Comput. **34**(1), A421–A458 (2012)
9. J. Galvis, Y. Efendiev, Domain decomposition preconditioners for multiscale flows in high contrast media: reduced dimension coarse spaces. Multiscale Model. Simul. **8**(5), 1621–1644 (2010)
10. M.J. Gander, L. Halpern, K. Santugini Repiquet, A new coarse grid correction for RAS/AS, in *Domain Decomposition Methods in Science and Engineering, XXI*, ed. by J. Erhel, M.J. Gander, L. Halpern, G. Pichot, T. Sassi, O. Widlund. Lecture Notes in Computational Science and Engineering (Springer, Switzerland, 2014), pp. 275–283
11. M.J. Gander, L. Halpern, K. Santugini Repiquet, Discontinuous coarse spaces for DD-methods with discontinuous iterates, in *Domain Decomposition Methods in Science and Engineering XXI*, ed. by J. Erhel, M.J. Gander, L. Halpern, G. Pichot, T. Sassi, O. Widlund. Lecture Notes in Computational Science and Engineering (Springer, Switzerland, 2014), pp. 607–615
12. M.J. Gander, F. Kwok, K. Santugini, Optimized Schwarz at cross points: finite volume case (2015, in preparation)
13. L. Halpern, F. Hubert, A finite volume Ventcell-Schwarz algorithm for advection-diffusion equations. SIAM J. Numer. Anal. **52**(3), 1269–1291 (2014)

14. P.-L. Lions, On the Schwarz alternating method. III: a variant for nonoverlapping subdomains, in *Third International Symposium on Domain Decomposition Methods for Partial Differential Equations*, ed. by T.F. Chan, R. Glowinski, J. Périaux, O. Widlund (SIAM, Philadelphia, 1990), pp. 202–223
15. J. Mandel, Balancing domain decomposition. Commun. Numer. Methods Eng. **9**(3), 233–241 (1993)
16. J. Mandel, M. Brezina, Balancing domain decomposition for problems with large jumps in coefficients. Math. Comput. **65**, 1387–1401 (1996)
17. J. Mandel, R. Tezaur, Convergence of a substructuring method with Lagrange multipliers. Numer. Math. **73**, 473–487 (1996)
18. F. Nataf, H. Xiang, V. Dolean, N. Spillane, A coarse sparse construction based on local Dirichlet-to-Neumann maps. SIAM J. Sci. Comput. **33**(4), 1623–1642 (2011)
19. R.A. Nicolaides, Deflation conjugate gradients with application to boundary value problems. SIAM J. Numer. Anal. **24**(2), 355–365 (1987)
20. K. Santugini, A discontinuous galerkin like coarse space correction for domain decomposition methods with continuous local spaces: the DCS-DGLC algorithm. ESAIM Proc. Surv. **45**, 275–284 (2014)
21. N. Spillane, V. Dolean, P. Hauret, F. Nataf, C. Pechstein, R. Scheichl, Abstract robust coarse spaces for systems of PDEs via generalized eigenproblems in the overlaps. Numer. Math. **126**(4), 741–770 (2013)

Inexact Spectral Deferred Corrections

Robert Speck, Daniel Ruprecht, Michael Minion, Matthew Emmett, and Rolf Krause

1 Introduction

Implicit integration methods based on collocation are attractive for a number of reasons, e.g. their ideal (for Gauss-Legendre nodes) or near ideal (Gauss-Radau or Gauss-Lobatto nodes) order and stability properties. However, straightforward application of a collocation formula with M nodes to an initial value problem with dimension d requires the solution of one large $Md \times Md$ system of nonlinear equations.

Spectral deferred correction (SDC) methods, introduced by Dutt et al. [4], are an attractive approach for iteratively computing the solution to the collocation problem using a low-order method (like implicit or IMEX Euler) as a building block. Instead of solving one huge system of size $Md \times Md$, SDC iteratively solves M smaller $d \times d$ systems to approximate the solution of the full system (see also the discussion

R. Speck (✉)
Jülich Supercomputing Centre, Forschungszentrum Jülich GmbH, Germany
e-mail: r.speck@fz-juelich.de

R. Speck • D. Ruprecht • R. Krause
Institute of Computational Science, Università della Svizzera italiana, Lugano, Switzerland
e-mail: daniel.ruprecht@usi.ch; rolf.krause@usi.ch

M. Minion
Institute for Computational and Mathematical Engineering, Stanford University, Stanford, CA 94305, USA
e-mail: mlminion@stanford.edu

M. Emmett
Center for Computational Sciences and Engineering, Lawrence Berkeley National Laboratory, CA 94720, USA
e-mail: mwemmett@lbl.gov

T. Dickopf et al. (eds.), *Domain Decomposition Methods in Science and Engineering XXII*, Lecture Notes in Computational Science and Engineering 104, DOI 10.1007/978-3-319-18827-0_39

in [7]). It has been shown e.g. by Xia et al. [17] that each iteration/sweep of SDC raises the order by one, so that SDC with k iterations and a first-order base method is of order k, up to the order of the underlying collocation formula. Therefore, to achieve formal order p, SDC requires $p/2$ nodes and p iterations and thus $p^2/2$ solves of a $d \times d$ system (for Gauss-Legendre nodes).

Considering the number of solves required to achieve a certain order, one might conclude that, notwithstanding the results presented here, SDC is less efficient than e.g. diagonally implicit Runge Kutta (DIRK) methods, see e.g. [1], which only require $p - 1$ solves. However, the flexibility of the choice of the base propagator in SDC and the very favorable stability properties make it an attractive method nevertheless. In particular, semi-implicit methods of high order can easily be constructed with SDC which make it competitive for complex applications, see [2, 10]. Further extensions to SDC allow it to integrate processes with different time scales, see [3, 8], efficiently; and the iterative nature of SDC also allows it to be extended to a multigrid-like multi-level algorithm, where work is shifted to coarser, computationally cheaper levels, see [15].

In the present paper, we introduce another strategy to improve the efficiency of SDC, which is similar to ideas from [12] where a single V-cycle of a multigrid method is used as a preconditioner. We show here that the iterative nature of SDC allows us to use incomplete solves of the linear systems arising in each sweep. In the resulting *inexact spectral deferred corrections* (ISDC), the linear problem in each Euler step is solved only approximately using a small number (two in the examples presented here) of multigrid V-cycles. It is numerically shown that this strategy results in only a small increase of the number of required sweeps while reducing the cost for each sweep. We demonstrate that ISDC can provide a significant reduction of the overall number of multigrid V-cycles required to complete an SDC time step.

2 Semi-Implicit Spectral Deferred Corrections

We consider an initial value problem in Picard form

$$u(t) = u_0 + \int_{T_0}^{t} f(u(s))\, \mathrm{d}s \tag{1}$$

where $t \in [T_0, T]$ and $u, f(u) \in \mathbb{R}^N$. Subdividing a time interval $[T_n, T_{n+1}]$ into M intermediate substeps $T_n = t_0 \leq t_1 < \ldots < t_M \leq T_{n+1}$, the integrals from t_m to t_{m+1} can be approximated by

$$I_m^{m+1} = \int_{t_m}^{t_{m+1}} f(u(s))\, \mathrm{d}s \approx \Delta t \sum_{j=0}^{M} s_{m,j} f(u_j) = S_m^{m+1} F(u) \tag{2}$$

where $u_m \approx u(t_m), m = 0, 1, \ldots, M, F(u) = (f(u_1), \ldots, f(u_M))^T, \Delta t = T_{n+1} - T_n$, and $s_{m,j}$ are quadrature weights. The nodes t_m correspond to quadrature nodes of a spectral collocation rule like Gauss-Legendre or Gauss-Lobatto quadrature rule. The basic implicit SDC update formula at node $m + 1$ in iteration $k + 1$ can be written as

$$u_{m+1}^{k+1} = u_m^{k+1} + \Delta t_m \left[f(u_{m+1}^{k+1}) - f(u_{m+1}^k) \right] + S_m^{m+1} F(u^k), \tag{3}$$

where $\Delta t_m = t_{m+1} - t_m$, for $m = 0, \ldots, M - 1$. Alternatively, if f can be split into a stiff part f^I and a non-stiff part f^E, a semi-implicit update is easily constructed for SDC using

$$\begin{aligned} u_{m+1}^{k+1} = u_m^{k+1} &+ \Delta t_m \left[f^I(u_{m+1}^{k+1}) - f^I(u_{m+1}^k) \right] \\ &+ \Delta t_m \left[f^E(u_m^{k+1}) - f^E(u_m^k) \right] + S_m^{m+1} F(u^k). \end{aligned} \tag{4}$$

Here, only the f^I-part is treated implicitly, while f^E is explicit. We refer to [10] for the details on semi-implicit spectral deferred corrections.

3 Inexact Spectral Deferred Corrections

In the following, we consider the linearly implicit case $f^I(u) = Au$, where A is a discretization of the Laplacian operator. Here, spatial multigrid is a natural choice for solving the implicit part in (4). As in [15], we use a high-order compact finite difference stencil to discretize the Laplacian (see e.g. [16]). This results in a weighting matrix W for the right-hand side of the implicit system and, with the notation $\tilde{f}^I(u) = Wf^I(u)$, the semi-implicit SDC update (4) becomes

$$\begin{aligned} (W - \Delta t_m A) u_{m+1}^{k+1} = W u_m^{k+1} &+ \Delta t_m W \left[f^E(u_m^{k+1}) - f^E(u_m^k) \right] \\ &- \Delta t_m \tilde{f}^I(u_{m+1}^k) + S_m^{m+1} \tilde{F}(u^k), \end{aligned} \tag{5}$$

where $\tilde{f} = Wf^E + \tilde{f}^I$ and $\tilde{F}(u^k) = (\tilde{f}(u_1^k), \ldots, \tilde{f}(u_M^k))^T$. Thus, instead of inverting the operator $I - \Delta t_m A$ in (4), the right-hand side of (4) is modified by W and the operator $W - \Delta t_m A$ needs to be inverted. We note that for calculating the residual during the SDC iteration, the weighting matrix needs to be inverted once per node, which can be done using multigrid as well.

For classical SDC, each computation of u_{m+1}^{k+1} includes a full inversion of $W - \Delta t_m A$ using e.g. a multigrid solver in space. For K iterations and M nodes, the multigrid solver is executed $K(M - 1)$ times, each time until a predefined tolerance is reached. In order to reduce the overall number of required multigrid V-cycles, ISDC replaces this full solve with a small fixed number L of V-cycles, leading to an accumulated number of $\tilde{K}(M - 1)L$ V-cycles in total. Naturally, the number of iterations in ISDC will be larger than the number of SDC, that is $K \leq \tilde{K}$. However, if $\tilde{K}$ is small enough so that $\tilde{K}(M-1)L$ is below the total number of multigrid V-cycles required for $K(M-1)$ full multigrid solves, inexact SDC will be more efficient than classical SDC.

Convergence is monitored using the maximum norm of the SDC residual, a discrete analogue of $u^k(t) - u_0 - \int_{T_0}^{t} f(u^k(s))\, ds$, that measures how well our iterative solution satisfies the discrete collocation problem. See [15] for definition and details. In the tests below, sweeps are performed until the SDC residual is below a set threshold.

4 Numerical Tests

In order to illustrate the performance of ISDC, we consider two different numerical examples, the 2D diffusion equation and 2D viscous Burgers' equation. As described above, in both cases the diffusion term is discretized using a 4th-order compact stencil with weighting matrix and a spatial mesh with 64 points. For Burgers' equation, the advection term is discretized using a fifth order WENO scheme.

4.1 Setups

The first test problem is the 2D heat equation on the unit square, namely

$$u_t(\mathbf{x}, t) = \nu \Delta u(\mathbf{x}, t), \quad \mathbf{x} \in \Omega = (0, 1)^2 \tag{6}$$

$$u(\mathbf{x}, 0) = \sin(\pi x) \sin(\pi y) \tag{7}$$

$$u(\mathbf{x}, t) = 0 \text{ on } \partial\Omega \tag{8}$$

with $\mathbf{x} = (x, y)$. The exact solution is $u(x, t) = \exp(-2\pi^2 \nu t) \sin(\pi x) \sin(\pi y)$. An implicit Euler is used here as base method in SDC.

The second test problem is the nonlinear viscous Burgers' equation

$$u_t(\mathbf{x}, t) + u(\mathbf{x}, t)u_x(\mathbf{x}, t) + u(\mathbf{x}, t)u_y(\mathbf{x}, t) = \nu \Delta u(\mathbf{x}, t), \mathbf{x} \in (-1, 1)^2, \tag{9}$$

$$u(x, 0) = \exp\left(-\frac{x^2}{\sigma^2}\right), \quad \sigma = 0.1 \tag{10}$$

with periodic boundary conditions. Here, an IMEX Euler is used as base method, i.e. the Laplacian is treated implicitly, while the advection term is integrated explicitly.

Table 1 Accumulated multigrid V-cycles for (a) the heat equation and (b) the viscous Burgers' equation with different values for the diffusion coefficients ν and the number of quadrature nodes M. Cycles are accumulated over all sweeps required to reduce the SDC or ISDC residual below 5×10^{-8}. The number of deferred correction sweeps is shown in parentheses. Saving indicates the amount of V-cycles saved by ISDC in percent of the cycles required by SDC

(a) Heat equation					(b) Viscous Burgers' equation				
ν	M	SDC	ISDC	Savings (%)	ν	M	SDC	ISDC	Savings (%)
1	3	16(4)	12(4)	25	10^{-1}	3	21(8)	21(8)	0
	5	23(3)	20(3)	13		5	26(6)	26(6)	0
	7	32(3)	28(3)	13		7	33(5)	33(5)	0
10	3	36(5)	20(5)	44	1.0	3	97(17)	66(17)	32
	5	61(5)	40(5)	34		5	140(17)	117(17)	16
	7	79(4)	47(4)	41		7	160(15)	143(15)	11
100	3	106(13)	52(13)	51	10	3	207(25)	100(25)	52
	5	150(10)	104(13)	31		5	523(38)	298(38)	43
	7	187(9)	167(14)	11		7	902(50)	578(50)	36

In both examples the diffusion parameter ν controls the stiffness of the term f^I: for a given spatial resolution, the shifted Laplacian $W - \nu \Delta t A$, and therefore the performance of the multigrid solver, depends critically on ν. We choose three different values of ν for each example to measure the impact of stiffness on the performance of ISDC: $\nu = 1, 10, 100$ for the heat equation and $\nu = 0.1, 1, 10$ for Burgers' equation. For ISDC, each implicit solve is approximated using $L = 2$ V-cycles. A single time-step of length $\Delta t = 0.001$ is analyzed for a spatial discretization with $\Delta x = \Delta y = 1/64$ in both cases, leading to CFL numbers for the diffusive term of approximately 4.1, 41 and 410 for the heat equation and 0.41, 4.1 and 41 for Burgers' equation.

4.2 Results

Table 1 shows the total number of multigrid V-cycles for the heat equation (left) and for Burgers' equation (right) for three different numbers of collocation nodes M and different values of ν. The number of SDC or ISDC sweeps performed is shown in parentheses. In each case, sweeps are performed until the SDC or ISDC residual is below 5×10^{-8}. To simplify the analysis in the presence of the weighting matrix, the V-cycles required to invert the weighting matrix are not counted here. In the last row, the amount of V-cycles saved by ISDC is given in percent of the required SDC cycles.

In most cases, ISDC provides a substantial reduction of the total number of required multigrid V-cycles and requires only slightly more sweeps to converge than SDC. The most savings can be obtained if the number of multigrid V-cycles in SDC is high but ISDC does not lead to a significant increase in sweeps, which is the case for mildly stiff problems (e.g. $\nu = 10$ for heat equation and $\nu = 1$ and $\nu = 10$ for Burgers) or stiff problems with small values for M. For stiff problems with large M (e.g. heat equation with $\nu = 100$ and $M = 7$), however, ISDC leads to a more significant increase in required sweeps, therefore only resulting in small savings. For the non-stiff cases, particularly for Burger's equation, ISDC does not provide much benefit, but also does no harm: the multigrid solves in SDC take only very few V-cycles to converge, so that SDC and ISDC are almost identical (for Burgers with $\nu = 0.1$, SDC and ISDC are actually identical). In a sense, for simple problems where the stopping criterion of the multigrid solver is reached after one or two V-cycle anyhow, SDC automatically reduces to ISDC.

In summary, the tests presented here suggest that replacing full multigrid solves by a small number of V-cycles in SDC only leads to a small increase in the total number of SDC sweeps required for convergence but can significantly reduce the computational cost of each sweep. The savings in the overall number of multigrid V-cycles of ISDC directly translates into faster run times of ISDC runs compared to classical SDC. Preliminary numerical tests not document here suggest that, as long as the approximate solution of the linear system is sufficiently accurate, the order of ISDC still increases by each iteration, as shown for SDC in [17]. A detailed study confirming this, including a possible extension of the proof, is left for future work.

4.3 *Interpretation*

The good performance of ISDC in the examples presented above is mainly due to the choice of the starting values for the multigrid solver. When performing the implicit Euler step to compute u_{m+1}^{k+1}, the value u_{m+1}^{k} from the previous SDC sweep gives a very good starting value, particularly in later sweeps. Therefore, even two multigrid V-cycles are sufficient to approximate the real solution of the linear system of equations reasonably well. This effect can be observed by monitoring the number of V-cycles in classical SDC. During the first sweep, many more V-cycles are typically required for multigrid to converge than in later sweeps where the initial guess becomes very accurate as the SDC iterations converge. In fact, during the last sweeps of SDC, a single V-cycle is often sufficient for solving the implicit system. Hence, the additional sweeps required by ISDC are mainly due to the less accurate approximations during the first sweeps. As soon as the initial guess u_{m+1}^{k} for u_{m+1}^{k+1} is good enough, ISDC basically proceeds like SDC. A computational experiment that confirms this is as follows: if, when solving for u_{m+1}^{k+1}, we replace the initial guess with the zero vector, or even u_{m}^{k+1}, then ISDC fails to converge altogether. On the other hand, SDC still convergences in this scenario, but the number of required multigrid V-cycles increases dramatically.

It is important to contrast this behavior to non-iterative schemes like diagonally implicit Runge-Kutta, where usually only the value from the previous time step or stage is available to be used as starting value. Our experience with SDC methods suggests that more multigrid V-cycles would be required to solve each stage in a DIRK scheme than in later SDC iterations. Hence, simply counting the number of implicit function evaluations required could be a misleading way to compare the cost of SDC and DIRK schemes.

5 Conclusion and Outlook

The paper presents a variant of spectral deferred corrections called *inexact spectral deferred corrections* that can significantly reduce the computational cost of SDC. In ISDC, full spatial solves within SDC sweeps with an implicit or semi-implicit Euler are replaced by only a few V-cycles of a multigrid. In the two investigated examples, ISDC saves up to 52 % of the total multigrid V-cycles required by SDC with full linear solves in each step, while only minimally increasing the number of sweeps required to reduce the SDC residual below some set tolerance. The main reason for the good performance of ISDC is that the iterative nature of SDC provides very accurate initial guesses for the multigrid solver. Besides providing significant speedup, ISDC essentially removes the need to define a tolerance or maximum number of iterations for the spatial solver.

A natural extension of the work presented in this paper is the application of ISDC sweeps in MLSDC, the multi-level version of SDC. MLSDC performs SDC sweeps in a multigrid-like way on multiple levels. The levels are connected through an FAS correction term in forming the coarsened spatial representation of the problem on upper levels of the hierarchy. Using ISDC corresponds to the "reduced implicit solve" strategy mentioned in [15] and incorporating it into MLSDC could further improve its performance. Finally, the "parallel full approximation scheme in space and time" (PFASST, see [5, 6, 11] for details) performs SDC sweeps on multiple levels combined with a forward transfer of updated initial values in a manner similar to Parareal (see [9]). Instead of performing a full time integration as done in Parareal, PFASST interweaves SDC sweeps with Parareal iterations so that on each time level, only a single SDC sweep is performed (i.e. an inexact time integrator is applied), leading to a time-parallel method with good parallel efficiency (see e.g. [13, 14]). Integrating ISDC into PFASST could further improve its parallel efficiency.

Acknowledgement This work was supported by the Swiss National Science Foundation (SNSF) under the lead agency agreement through the project "ExaSolvers" within the Priority Programme 1648 "Software for Exascale Computing" (SPPEXA) of the Deutsche Forschungsgemeinschaft (DFG). Matthew Emmett and Michael Minion were supported by the Applied Mathematics Program of the DOE Office of Advanced Scientific Computing Research under the U.S. Department of Energy under contract DE-AC02-05CH11231. Michael Minion was also supported by the U.S. National Science Foundation grant DMS-1217080. The authors acknowledge support from Matthias Bolten, who provided the employed multigrid solver.

References

1. R. Alexander, Diagonally implicit Runge-Kutta methods for stiff O.D.E.'s. SIAM J. Numer. Anal. **14**(6), 1006–1021 (1977)
2. A. Bourlioux, A. Layton, M. Minion, High-order multi-implicit spectral deferred correction methods for problems of reactive flow. J. Comput. Phys. **189**(2), 651–675 (2003)
3. E. Bouzarth, M. Minion, A multirate time integrator for regularized stokeslets. J. Comput. Phys. **229**(11), 4208–4224 (2010)
4. A. Dutt, L. Greengard, V. Rokhlin, Spectral deferred correction methods for ordinary differential equations. BIT Numer. Math. **40**(2), 241–266 (2000)
5. M. Emmett, M. Minion, Toward an efficient parallel in time method for partial differential equations. Commun. Appl. Math. Comput. Sci. **7**, 105–132 (2012)
6. M. Emmett, M.L. Minion, *Efficient Implementation of a Multi-Level Parallel in Time Algorithm*. Domain Decomposition Methods in Science and Engineering XXI, Lecture Notes in Computational Science and Engineering, vol. 98 (Springer, Switzerland, 2014), pp. 359–366
7. J. Huang, J. Jia, M. Minion, Accelerating the convergence of spectral deferred correction methods. J. Comput. Phys. **214**(2), 633–656 (2006)
8. A. Layton, M. Minion, Conservative multi-implicit spectral deferred correction methods for reacting gas dynamics. J. Comput. Phys. **194**(2), 697–715 (2004)
9. J.-L. Lions, Y. Maday, G. Turinici, A "parareal" in time discretization of PDE's. C. R. l'Académie Sci. Math. **332**, 661–668 (2001)
10. M. Minion, Semi-implicit spectral deferred correction methods for ordinary differential equations. Commun. Math. Sci. **1**(3), 471–500 (2003)
11. M. Minion, A hybrid parareal spectral deferred corrections method. Commun. Appl. Math. Comput. Sci. **5**(2), 265–301 (2010)
12. C. Oosterlee, T. Washio, On the use of multigrid as a preconditioner, in *Proceedings of 9th International Conference on Domain Decomposition Methods*, pp. 441–448 (1996)
13. D. Ruprecht, R. Speck, M. Emmett, M. Bolten, R. Krause, Poster: Extreme-scale space-time parallelism, in *Proceedings of the 2013 Supercomputing Companion, SC '13 Companion, 2013*
14. R. Speck, D. Ruprecht, M. Emmett, M. Bolten, R. Krause. A space-time parallel solver for the three-dimensional heat equation. in *Parallel Computing: Accelerating Computational Science and Engineering (CSE)*, Advances in Parallel Computing, vol. 25 (IOS Press, 2014), pp. 263–272. doi:10.3233/978-1-61499-381-0-263
15. R. Speck, D. Ruprecht, M. Emmett, M. Minion, M. Bolten, R. Krause, A multi-level spectral deferred correction method. BIT Numer. Math. (2014)
16. W. Spotz, G. Carey, A high-order compact formulation for the 3D Poisson equation. Numer. Meth. PDEs **12**(2), 235–243 (1996)
17. Y. Xia, Y. Xu, C.-W. Shu, Efficient time discretization for local discontinuous Galerkin methods. Disc. Cont. Dyn. Syst. **8**(3), 677–693 (2007)

Schwarz Preconditioner for the Stochastic Finite Element Method

Waad Subber and Sébastien Loisel

1 Introduction

For large-scale problems, domain decomposition techniques are a natural way to split the problem into smaller subproblems that can be solved in parallel on multiprocessors computers. To this end, stochastic versions of FETI-DP and BDDC domain decomposition techniques for uncertainty quantification of large-scale problems have been recently proposed in [2, 6, 7]. In this paper, we formulate two-level Schwarz domain decomposition technique for the solution of the large-scale linear system arising from the SSFEM discretization. In the stochastic Schwarz preconditioner, we partition the spatial domain and preserve all the couplings along the stochastic directions. Consequently, stochastic Dirichlet problems are defined and solved on each subdomain concurrently. The solution of these local problems are used to define the first level of the preconditioner. A coarse grid correction is added to the one-level preconditioner to provide a global mechanism to propagate information over the subdomains. This global exchange of information across the spacial and stochastic directions leads to a scalable preconditioner. It turns out that the one-level stochastic Schwarz preconditioner based on the mean properties can be viewed as a parallel generalization of the block-diagonal mean based preconditioner [3], whereby the associated deterministic problems are solved in parallel using the deterministic Schwarz preconditioner. For the numerical illustrations, a two dimensional stochastic elliptic PDE with spatially varying random coefficients is considered. The numerical scalability of the algorithm is investigated with respect

W. Subber (✉) • S. Loisel
School of Mathematical and Computer Sciences, Heriot-Watt University, Edinburgh EH14 4AS, UK
e-mail: w.subber@hw.ac.uk; s.loisel@hw.ac.uk

T. Dickopf et al. (eds.), *Domain Decomposition Methods in Science and Engineering XXII*, Lecture Notes in Computational Science and Engineering 104, DOI 10.1007/978-3-319-18827-0_40

to the geometric parameters and the strength of the input uncertainty, dimension and order of the stochastic expansion.

2 Mathematical Formulations

We consider the case of finite dimensional noise in a suitable probability space $(\Theta, \Sigma, \mathcal{P})$ [1]. That is we assume that there exist a finite set of independent and identically distributed random variables $\boldsymbol{\xi}(\theta) = \{\xi_1(\theta), \xi_2(\theta), \cdots, \xi_M(\theta)\}$ with joint probability density function $p(\boldsymbol{\xi}) = p_1(\xi_1)p_2(\xi_2)\cdots p_M(\xi_M)$ which can be used to parametrize the input uncertainty. Consider the following stochastic boundary value problem: Find a random function $u(\mathbf{x}, \xi(\theta)) : \Omega \times \Gamma \rightarrow \mathbb{R}$ such that:

$$\begin{aligned} -\nabla \cdot (\kappa(\mathbf{x}, \xi(\theta)) \nabla u(\mathbf{x}, \xi(\theta))) &= f(\mathbf{x}), \text{ in } \quad \Omega \times \Gamma, \\ u(\mathbf{x}, \xi(\theta)) &= 0, \qquad \text{on } \partial\Omega \times \Gamma, \end{aligned} \tag{1}$$

where $(\Omega \subset \mathbb{R}^d, d = 1, 2, 3)$ denotes a bounded domain with Lipschitz boundary $\partial\Omega$ and $\Gamma = \Gamma_1 \times \Gamma_2 \cdots \times \Gamma_M \subset \mathbb{R}^M$ is the support of the joint probability density function $p(\boldsymbol{\xi})$ of the random vector $\boldsymbol{\xi}(\theta)$. Here we assume that the input uncertainty $\kappa(\mathbf{x}, \boldsymbol{\xi}(\theta)) : \Omega \times \Gamma \rightarrow \mathbb{R}$ is a $\mathcal{P}$-almost surely bounded and strictly positive random field, that is

$$0 < \kappa_{min} \leq \kappa(\mathbf{x}, \boldsymbol{\xi}(\theta)) \leq \kappa_{max} < +\infty, \quad \text{a.e. in} \quad \Omega \times \Gamma. \tag{2}$$

The weak form of the stochastic boundary value problem (1), can be stated as: Find $u(\mathbf{x}, \boldsymbol{\xi}) \in V$ such that for all $v \in V$

$$\int_\Gamma \left(\int_\Omega \kappa(\mathbf{x}, \boldsymbol{\xi}) \nabla u(\mathbf{x}, \boldsymbol{\xi}) \nabla v(\mathbf{x}, \boldsymbol{\xi}) \mathrm{d}\mathbf{x} \right) p(\boldsymbol{\xi}) \mathrm{d}\boldsymbol{\xi} = \int_\Gamma \left(\int_\Omega f(\mathbf{x}) v(\mathbf{x}, \boldsymbol{\xi}) \mathrm{d}\mathbf{x} \right) p(\boldsymbol{\xi}) \mathrm{d}\boldsymbol{\xi}$$

where the tensor product function space $V = H_0^1(\Omega) \otimes L^2(\Gamma)$ is defined as

$$V = \{v(\mathbf{x}, \boldsymbol{\xi}(\theta)) : \Omega \times \Gamma \rightarrow \mathbb{R} \mid \|v\|_V^2 < \infty\} \subset H_0^1(\Omega) \otimes L^2(\Gamma), \tag{3}$$

here $H_0^1(\Omega)$ and $L^2(\Gamma)$ represent the deterministic Hilbert space and the space of second-order random variables, respectively. The energy norm $\| \cdot \|_V^2$ is given by

$$\|v(\mathbf{x}, \boldsymbol{\xi}(\theta))\|_V^2 = \int_\Gamma \left(\int_\Omega \kappa(\mathbf{x}, \boldsymbol{\xi}) |\nabla v(\mathbf{x}, \boldsymbol{\xi})|^2 \mathrm{d}\mathbf{x} \right) p(\boldsymbol{\xi}) \mathrm{d}\boldsymbol{\xi}. \tag{4}$$

3 Stochastic Process Representation

Let $\kappa_0(\mathbf{x})$ and $C_{\kappa\kappa}(\mathbf{x}_1, \mathbf{x}_2)$ denote the mean and covariance function of the input uncertainty, then the Karhunen-Loéve expansion (KLE) can be used to represent $\kappa(\mathbf{x}, \boldsymbol{\xi})$ as

$$\kappa(\mathbf{x}, \theta) = \sum_{i=0}^{M} \kappa_i(\mathbf{x})\xi_i(\theta), \tag{5}$$

where $\xi_0(\theta) = 1$ and $\kappa_i(\mathbf{x}) = \sigma\sqrt{\lambda_i}\phi_i(\mathbf{x})$; $i \geq 1$, here σ denotes the standard deviation of the input process and λ_i and $\phi_i(\mathbf{x})$ are the eigenpairs of the covariance kernel and can be obtained from the solution of the following integral equation

$$\int_{\Omega} C_{\kappa\kappa}(\mathbf{x}_1, \mathbf{x}_2)\phi_i(\mathbf{x}_1)\mathrm{d}\mathbf{x}_1 = \lambda_i\phi_i(\mathbf{x}_2). \tag{6}$$

The solution process (with a priori unknown mean and covariance function) can be approximated using the PC expansion as

$$u(\mathbf{x}, \theta) = \sum_{j=0}^{N} u_j(\mathbf{x})\Psi_j(\boldsymbol{\xi}(\theta)), \tag{7}$$

where $N+1$ denote the total number of terms in PCE and $u_j(\mathbf{x})$ are the deterministic PC coefficients to be determined and $\Psi_j(\boldsymbol{\xi})$ are a set of multivariate orthogonal random polynomials with the following properties

$$\langle\Psi_0\rangle = \int_{\Gamma} \Psi_0(\boldsymbol{\xi})p(\boldsymbol{\xi})\mathrm{d}\boldsymbol{\xi} = 1, \quad \langle\Psi_j\rangle = 0, j > 0, \quad \text{and} \quad \langle\Psi_j\Psi_k\rangle = \delta_{jk}\langle\Psi_j^2\rangle.$$

4 The Stochastic Finite Element Discretization

Let $\mathcal{T}_h$ denote the triangulation of the physical domain Ω with a maximum element size h, and let the associated finite element space $\mathcal{X}_h \subset H_0^1(\Omega)$ be spanned by the traditional nodal basis functions $\{\phi_l(\mathbf{x})\}_{l=1}^{L}$. Further, for the stochastic discretization, let $\mathcal{Y}_p \subset L_2(\Gamma)$ be a finite dimensional space spanned by the PC basis functions $\{\Psi_j(\boldsymbol{\xi})\}_{j=0}^{N}$ in the random variables $\boldsymbol{\xi}$. Thus, the approximate SSFEM solution u_{hp} in the discrete tensor product space $\mathcal{X}_h \otimes \mathcal{Y}_p \subset H_0^1(\Omega) \otimes L_2(\Gamma)$ can be expressed as

$$u_{hp}(\mathbf{x}, \boldsymbol{\xi}) = \sum_{j=0}^{N}\sum_{l=1}^{L} u_{jl}\phi_l(\mathbf{x})\Psi_j(\boldsymbol{\xi}). \tag{8}$$

Using (5) and (8), we can translate the stochastic weak form defined in (3) into the following coupled set of deterministic linear system

$$\sum_{j=0}^{N}\sum_{i=0}^{M}\sum_{l=1}^{L} u_{jl}\left(\int_{\Gamma} \xi_i \Psi_j(\boldsymbol{\xi})\Psi_k(\boldsymbol{\xi})p(\boldsymbol{\xi})\mathrm{d}\boldsymbol{\xi}\right)\left(\int_{\Omega} \kappa_i(\mathbf{x})\nabla\phi_l(\mathbf{x})\cdot\nabla\phi_m(\mathbf{x})\mathrm{d}\mathbf{x}\right) =$$
$$\int_{\Gamma}\left(\int_{\Omega} f(\mathbf{x})\phi_m(\mathbf{x})\mathrm{d}\mathbf{x}\right)\Psi_k(\boldsymbol{\xi})p(\boldsymbol{\xi})\mathrm{d}\boldsymbol{\xi}, \quad m=1,\cdots,L, \quad k=0,\cdots,N \tag{9}$$

The linear system arising from (9) can be expressed as follows

$$\sum_{i=0}^{M} \mathbf{A}^{(i)}\mathbf{U}\mathbf{C}^{(i)} = \mathbf{F}, \tag{10}$$

where we define

$$\mathbf{A}_{lm}^{(i)} = \int_{\Omega} \kappa_i \nabla\phi_l \cdot \nabla\phi_m \mathrm{d}\mathbf{x}, \quad \mathbf{C}_{jk}^{(i)} = \int_{\Gamma} \xi_i \Psi_j(\boldsymbol{\xi})\Psi_k(\boldsymbol{\xi})p(\boldsymbol{\xi})\mathrm{d}\boldsymbol{\xi}. \tag{11}$$

$$F_{mk} = \int_{\Gamma}\left(\int_{\Omega} f(\mathbf{x})\phi_m(\mathbf{x})\mathrm{d}\mathbf{x}\right)\Psi_k(\boldsymbol{\xi})p(\boldsymbol{\xi})\mathrm{d}\boldsymbol{\xi}. \tag{12}$$

Equation (10) can be vectorized by taking the vec$(\cdot)$ operator for the both sides leading to the following concise form

$$\mathcal{A}\mathcal{U} = \mathcal{F}, \tag{13}$$

where

$$\mathcal{A} = \sum_{i=0}^{M} \mathbf{C}^{(i)} \otimes \mathbf{A}^{(i)}, \quad \mathcal{U} = \mathrm{vec}(\mathbf{U}) \quad \text{and} \quad \mathcal{F} = \mathrm{vec}(\mathbf{F}). \tag{14}$$

5 Schwarz Preconditioner for Stochastic PDEs

In the Schwarz preconditioner for the stochastic problem, the physical domain Ω is partitioned into a number of overlapping subdomain $\{\Omega_s, 1 \leq s \leq S\}$ by splitting the vertices of the computational mesh. For each subdomain $\Omega_s \subset \Omega$, let $\mathbf{R}_s$ be a restriction matrix of size $n_s \times n$ (where n_s and n are the size of the subdomain and global unknowns) to extract the local nodal values from the global unknowns vector as

$$\mathbf{U}_s = \mathbf{R}_s\mathbf{U}, \tag{15}$$

applying the vec(·) operator to (15), leads to

$$\text{vec}(\mathbf{U}_s) = (\mathbf{I} \otimes \mathbf{R}_s)\text{vec}(\mathbf{U}), \tag{16}$$

here $\mathbf{I}$ is $(N+1) \times (N+1)$ identity matrix. Let $\mathcal{U}_s = \text{vec}(\mathbf{U}_s)$ and $\mathcal{R}_s = (\mathbf{I} \otimes \mathbf{R}_s)$ denote the stochastic subdomain nodal values and the stochastic restriction matrix, then (16) becomes

$$\mathcal{U}_s = \mathcal{R}_s \mathcal{U}. \tag{17}$$

Consequently, the stochastic stiffness matrix for subdomain Ω_s can be defined as a block extracted from the global stiffness matrix $\mathcal{A}$ as

$$\mathcal{A}_s = \mathcal{R}_s \mathcal{A} \mathcal{R}_s^T, \tag{18}$$

$$= (\mathbf{I} \otimes \mathbf{R}_s) \left(\sum_{i=0}^{M} \mathbf{C}^{(i)} \otimes \mathbf{A}^{(i)} \right) (\mathbf{I} \otimes \mathbf{R}_s^T), \tag{19}$$

$$= \sum_{i=0}^{M} \mathbf{C}^{(i)} \otimes \mathbf{A}_s^{(i)}. \tag{20}$$

Next, we define the one-level stochastic Schwarz preconditioner as a direct sum of the solution of the local stochastic Dirichlet problems as:

$$\mathcal{M}^{-1} = \sum_{s=1}^{S} \mathcal{R}_s^T \mathcal{A}_s^{-1} \mathcal{R}_s, \tag{21}$$

which can be expressed as follows

$$\mathcal{M}^{-1} = \sum_{s=1}^{S} (\mathbf{I} \otimes \mathbf{R}_s^T) \left(\sum_{i=0}^{M} \mathbf{C}^{(i)} \otimes \mathbf{A}_s^{(i)} \right)^{-1} (\mathbf{I} \otimes \mathbf{R}_s). \tag{22}$$

Remark 1 The stochastic Schwarz preconditioner has the same structure as the stochastic Neumann-Neumann preconditioner in [5].

Remark 2 The stochastic Schwarz preconditioner based on the mean properties can be obtained from (22) by setting $i = 0$ which gives

$$\mathcal{M}_0^{-1} = [\mathbf{C}^{(0)}]^{-1} \otimes \sum_{s=1}^{S} \mathbf{R}_s^T [\mathbf{A}_s^{(0)}]^{-1} \mathbf{R}_s, \tag{23}$$

where $\mathbf{A}_s^{(0)} = \mathbf{R}_s \mathbf{A}^{(0)} \mathbf{R}_s^T$ and $\mathbf{C}^{(0)} = \delta_{ij} \langle \Psi_i^2 \rangle$.

Remark 3 For one subdomain $S = 1$ and normalized PC basis functions, $\mathbf{C}^{(0)} = \mathbf{I}$, the mean-based Schwarz preconditioner defined in (23) becomes

$$\mathcal{M}_0^{-1} = \mathbf{I} \otimes [\mathbf{A}^{(0)}]^{-1}. \tag{24}$$

Remark 4 The one-level stochastic Schwarz preconditioner based on the mean properties is a generalization of the block-diagonal mean based preconditioner [3] whereby the associated deterministic problem is solved in parallel using the deterministic Schwarz preconditioner.

6 Coarse Grid Correction

Domain decomposition preconditioners can achieve a scalable performance provided that they are equipped with a coarse grid correction for global communication. To define a coarse problem for the stochastic Schwarz preconditioner, let $\mathbf{R}_0^T \in \mathbb{R}^{n_i \times n_0}$ be an interpolation matrix defined as

$$\mathbf{R}_0^T = \begin{bmatrix} \psi_1(\mathbf{x}_1) & \psi_2(\mathbf{x}_1) & \cdots & \psi_{n_0}(\mathbf{x}_1) \\ \psi_1(\mathbf{x}_2) & \psi_2(\mathbf{x}_2) & \cdots & \psi_{n_0}(\mathbf{x}_2) \\ \vdots & \vdots & \cdots & \vdots \\ \psi_1(\mathbf{x}_{n_i}) & \psi_2(\mathbf{x}_{n_i}) & \cdots & \psi_{n_0}(\mathbf{x}_{n_i}) \end{bmatrix} \tag{25}$$

where $\{\psi_i(\mathbf{x})\}_{i=1}^{n_0}$ is a set of linear basis functions, here n_0 denotes the dimension of the coarse space and $(\mathbf{x}_1, \mathbf{x}_2, \cdots, \mathbf{x}_{n_i})$ are the coordinates of the nodal points of the fine mesh. The corresponding *stochastic coarse space interpolation operator* can be defined as

$$\mathcal{R}_0 = \mathbf{I} \otimes \mathbf{R}_0, \tag{26}$$

and thus the coarse grid correction for the stochastic problem can be obtained as

$$\mathcal{A}_0 = \mathcal{R}_0^T \mathcal{A} \mathcal{R}_0. \tag{27}$$

According, the two-level stochastic Schwarz preconditioner can be defined by adding the coarse grid correction to the one-level preconditioner in (21) leading to

$$\mathcal{M}^{-1} = \mathcal{R}_0^T \mathcal{A}_0^{-1} \mathcal{R}_0 + \sum_{s=1}^{S} \mathcal{R}_s^T \mathcal{A}_s^{-1} \mathcal{R}_s. \tag{28}$$

Theorem 1 *There exists positive constants C and d that are independent of the geometric parameters (i.e. mesh size h, subdomain size H and the overlap distance δ) and the stochastic parameters (i.e. strength of randomness σ, dimension M and order p of the stochastic expansion), such that*

$$cond(\mathcal{M}^{-1}\mathcal{A}) \leq C(d+1)^2 \left(\frac{\kappa_{max}}{\kappa_{min}}\right)^2 \frac{H}{\delta}. \tag{29}$$

Proof See [4]. □

7 Numerical Results

In this section, we illustrate the performance of the two-level stochastic Schwarz preconditioner defined in (28). In particular, we consider the following elliptic SPDE

$$\begin{aligned} -\nabla \cdot (\kappa(\mathbf{x},\theta)\nabla u(\mathbf{x},\theta)) &= f(\mathbf{x}), \text{ in } \quad \Omega \times \Theta, \\ u(\mathbf{x},\theta) &= 0, \quad \text{ on } \partial\Omega \times \Theta, \end{aligned} \tag{30}$$

where $f(\mathbf{x})$ denotes the source term taken as unity. The diffusivity coefficient $\kappa(\mathbf{x},\theta)$ is modelled as a uniform random field with an invariant mean and the following exponential covariance function

$$C_{\kappa\kappa}(\mathbf{x},\mathbf{y}) = \sigma^2 \exp\left(\frac{-|x_1 - y_1|}{b_1} + \frac{-|x_2 - y_2|}{b_2}\right), \tag{31}$$

Fig. 1a, b show the mean and standard deviation of the solution process. In Fig. 2a, b, we show the condition number growth of the Schwarz preconditioner for fixed

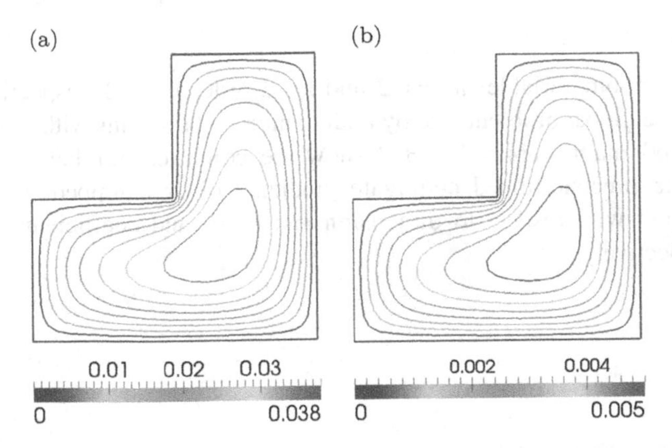

Fig. 1 The mean and standard deviation of the solution process. (**a**) Mean; (**b**) st. deviation

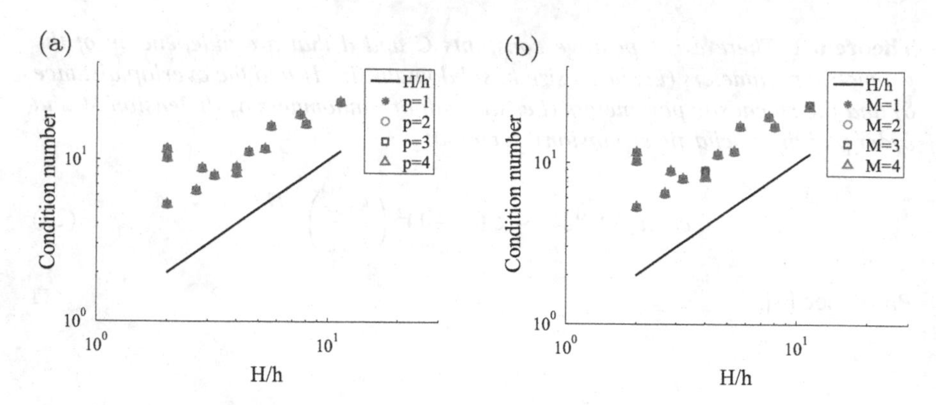

Fig. 2 Condition number growth with respect to fixed problem size per subdomain. **(a)** $M = 2$; **(b)** $p = 2$

Table 1 Condition number and iterations count with respect to M and p

M	p	*Cond*	*Iter*
1	1	10.1642	17
	2	10.1706	19
	3	10.1725	19
	4	10.1733	19
2	1	10.1781	19
	2	10.1834	19
	3	10.1861	19
	4	10.1876	19
3	1	10.1785	19
	2	10.1842	19
	3	10.1873	19
	4	10.1892	19
4	1	10.1816	19
	2	10.1887	20
	3	10.1926	20
	4	10.1951	20

number of random variables $M = 2$ and fixed order $p = 2$, respectively, while increasing the global problem size by adding more subdomains with fixed problem size per subdomain. Tables 1 and 2 show the condition number and iterations count of the preconditioned conjugate gradient solver equipped with Schwarz preconditioner with respect to dimension and order and coefficient of variation (CoV), respectively.

Table 2 Condition number and iterations count with respect to the CoV

$\frac{\sigma}{\mu}$	p	*Cond*	*Iter*
0.2	1	10.1760	19
	2	10.1812	19
	3	10.1841	19
	4	10.1860	19
0.3	1	10.1816	19
	2	10.1887	20
	3	10.1926	20
	4	10.1951	20
0.4	1	10.1871	19
	2	10.1959	20
	3	10.2006	20
	4	10.2035	20
0.5	1	10.1925	20
	2	10.2030	20
	3	10.2085	20
	4	10.2122	20

8 Conclusion

A two-level Schwarz domain decomposition preconditioner is proposed for the iterative solution of the large-scale linear system arising from the stochastic finite element discretization. The proposed preconditioner demonstrates a scalable performance with respect to the mesh parameters, strength of randomness, dimension and order of the stochastic expansion.

References

1. I. Babuska, P. Chatzipantelidis, On solving elliptic stochastic partial differential equations. CMAME **191**, 4093–4122 (2002)
2. D. Ghosh, Ph. Avery, Ch. Farhat, A FETI-preconditioned conjugate gradient method for large-scale stochastic finite element problems. IJNME **80**, 914–931 (2009)
3. C. Powell, H. Elman, Block-diagonal preconditioning for spectral stochastic finite-element systems. IMA-JNA **29**, 350–375 (2009)
4. W. Subber, S. Loisel, Schwarz preconditioners for stochastic elliptic PDEs. CMAME **272**, 34–57 (2014)
5. W. Subber, A. Sarkar, Domain decomposition of stochastic PDEs: a novel preconditioner and its parallel performance, in *HPCS*, vol. 5976 (Springer, New York, 2010), pp. 251–268
6. W. Subber, A. Sarkar, Dual-primal domain decomposition method for uncertainty quantification. CMAME **266**, 112–124 (2013)
7. W. Subber, A. Sarkar, A domain decomposition method of stochastic PDEs: an iterative solution techniques using a two-level scalable preconditioner. JCP Part A **257**, 298–317 (2014)

Domain Decomposition for a Hybrid Fully 3D Fluid Dynamics and Geophysical Fluid Dynamics Modeling System: A Numerical Experiment on Transient Sill Flow

H.S. Tang, K. Qu, X.G. Wu, and Z.K. Zhang

1 Introduction

Now it has become necessary to simulate multiphysics coastal ocean flow phenomena at distinct scales, especially those at small scales, in many emerging problems such as hydrodynamic impact on coastal bridges during Hurricane Katrina in 2005 and oil spill at the Deepwater Horizon in the Gulf of Mexico in 2010 (e.g., [2, 4]). Efforts using numerical simulations to predict coastal ocean flows have been greatly successful but strictly speaking, until now, are limited to large spatial scales in range O(10)–O(10,000) km and individual phenomena such as circulation currents and surface waves.

A natural and actually the most effective and feasible approach to simulate multiphysics coastal ocean flows at an affordable computational expense will be an integration of a fully three dimensional fluid dynamics (F3DFD) model, which is commonly referred as to a computational fluid dynamics (CFD) model in the coastal ocean community, and a geophysical fluid dynamics (GFD) model into a single modeling system using a domain decomposition method (DDM). In this

H.S. Tang (✉) • K. Qu
Department of Civil Engineering, City College, The City University of New York, New York, NY 10031, USA
e-mail: htang@ccny.cuny.edu

X.G. Wu
Department of Civil Engineering, City College, The City University of New York, New York, NY 10031, USA

Zhejiang Institute of Hydraulics and Estuary, Hangzhou, Zhejiang 310020, China

Z.K. Zhang
National Key Laboratory of Science and Technology on Aerodynamic Design and Research, Northwestern Polytechnical University, Xian, Shaanxi 710072, China

T. Dickopf et al. (eds.), *Domain Decomposition Methods in Science and Engineering XXII*, Lecture Notes in Computational Science and Engineering 104, DOI 10.1007/978-3-319-18827-0_41

approach, a flow field is divided into many subdomains dominated with different physical phenomena, and each subdomain will be assigned either with a F3DFD or a GFD model, whichever appropriate. With this idea, the authors proposed to couple the Solver for Incompressible Flow on Overset Meshes (SIFOM) developed by ourselves, which is a F3DFD model, and the unstructured grid Finite Volume Coastal Ocean Model (FVCOM), which is a GFD model, and demonstrated its capabilities and performance in capturing multiple physical phenomena [6, 9]. The SIFOM–FVCOM system is the first-of-its-kind system for coastal ocean flows, and it is able to simulate many flows that are beyond the reach of any other existing models.

This paper makes a numerical experiment to evaluate the performance of the SIFOM–FVCOM system, illustrating its promise in simulation of complicated flows as well as difficulties to be overcome. A comprehensive study on the SIFOM–FVCOM system has been presented in [8].

2 Governing Equations and Discretization

The governing equations of the F3DFD model are the continuity and the Reynolds-averaged Navier–Stokes equations that read as

$$\nabla \cdot \mathbf{u} = 0, \tag{1}$$

$$\mathbf{u}_t + \nabla \cdot \mathbf{u}\mathbf{u} = -\frac{1}{\rho}\nabla p + \nabla \cdot ((\nu + \nu_t)\nabla \mathbf{u}), \tag{2}$$

where $\mathbf{u}$ is the velocity, with component u, v, and w in x, y, and z direction respectively. Here x and y are in the horizontal direction, respectively, and z is in the vertical direction. p is the pressure, ρ the density, ν the viscosity, and ν_t the turbulence viscosity. Different turbulence closures are available to evaluate the turbulence viscosity, such as the mixing length model, $k - \epsilon$ model, and detached eddy simulation.

SIFOM has been developed by the first author and co-workers to solve the above governing equations (e.g., [3, 7]). In SIFOM, the governing equations are discretized using a second-order accurate, implicit, finite difference method in curvilinear coordinates, and they are solved using a dual time-stepping artificial compressibility method. The time derivative is approximated using a three-point backward difference, the convective terms are discretized using the QUICK scheme, and the other terms are treated using central difference. A DDM approach in conjunction with Chimera grids is implemented to deal with complex geometry. An effective mass conservation algorithm, which is a mass-flux based interpolation (MFBI), is proposed to achieve seamless transition of solutions between subdomains. For details about the technical aspects of the model, the reader is referred to [3, 5, 7].

FVCOM is a popular GFD model, and it has an external and an internal mode. The governing equations for the external mode in the hydrostatic version of the model are the two dimensional continuity and momentum equations [1]:

$$\eta_t + \nabla_H \cdot (\mathbf{V}D) = 0, \tag{3}$$

$$(\mathbf{V}D)_t + \nabla_H \cdot (\mathbf{VV}D) = -gD\nabla_H \eta + \frac{\boldsymbol{\tau}_\mathbf{s} - \boldsymbol{\tau}_\mathbf{b}}{\rho} + \mathbf{G}. \tag{4}$$

For the internal mode in that version, the governing equations are the three dimensional continuity and momentum equations:

$$\eta_t + \nabla_H \cdot (\mathbf{v}D) + \omega_\sigma = 0, \tag{5}$$

$$\begin{aligned}(\mathbf{v}D)_t + \nabla_H \cdot (\mathbf{vv}D) + (\mathbf{v}\omega)_\sigma &= -gD\nabla_H \eta \\ &\quad + \nabla_H \cdot (\alpha \mathbf{e}) + \frac{1}{D}(\beta \mathbf{v}_\sigma)_\sigma + \mathbf{H}.\end{aligned} \tag{6}$$

In the external mode, $\mathbf{V}$ is the depth-averaged horizontal velocity, D is the water depth, $\boldsymbol{\tau}_\mathbf{s}$ and $\boldsymbol{\tau}_\mathbf{b}$ are the shear stress on water surface and seabed, respectively, g is the gravity, and $\mathbf{G}$ includes the rest terms such as the Coriolis force. ∇_H is the gradient operator in the horizontal plane. In the internal mode, σ is the vertical coordinate, η is water surface elevation, $\mathbf{v}$ is the horizontal velocity, $\mathbf{e}$ is the strain rate, ω is the vertical velocity in σ-coordinate, and $\mathbf{H}$ represents the other terms. Subscript σ stands for the derivative in σ-direction. α and β are diffusion coefficients, which are evaluated by the Mellor and Yamada level-2.5 turbulent closure [1].

3 Methodology

A schematic representation of the hybrid SIFOM–FVCOM system is depicted in Fig. 1a, in which SIFOM is employed within a subdomain that covers the local flow around a seamount and FVCOM is used for the large-scale background flow. An overlapping zone is arranged between the regions of SIFOM and FVCOM, and it is assumed that the overlapping zone is located at a place where the hydrostatic assumption holds. Since variable $\mathbf{u}$ in the governing equations of SIFOM and variables $\mathbf{v}$ and ω in the internal mode of FVCOM are essentially the same, or, they are the three components of velocity, the two models will exchange solutions for them (Fig. 1b). In addition, pressure p at the boundary of SIFOM can be determined by the value of η obtained with FVCOM using the hydrostatic assumption, which states that pressure is proportional to water depth. Chimera grids, or overset grids, will be used to couple SIFOM and FVCOM, and grid connectivity and solutions exchange at interfaces of them will be implemented using interpolation. The

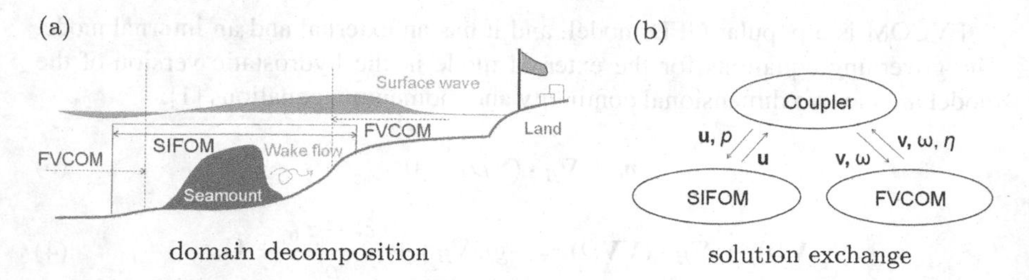

Fig. 1 A schematic representation of the hybrid SIFOM–FVCOM system. (**a**) Domain decomposition. (**b**) Solution exchange

Schwarz alternative iteration method is employed for the iteration between the solutions of the two models, and, in advancing solution of the hybrid system at time step n to that at $n + 1$, it reads as

Step I. Assign solution at time step n to all grid nodes/elements.
Step II. Exchange solution at model interfaces by interpolation.
Step III. Solve SIFOM and FVCOM.
Step IV. If SIFOM and FVCOM solutions converge, go to Step V. If not, return to Step II.
Step V. Assign the convergent solution as the solution at time step $n + 1$.

SIFOM and FVCOM will exchange solution at each time step. It is noted that the two models may use different time steps. In addition, the external and internal mode in FVCOM also permit different time steps. For details of the modeling system, the reader is referred to [6, 8, 9].

4 Numerical Experiment

Sill is a typical form of topography at bottom of oceans, and flows over it are rich in physical phenomena. Previous investigation indicates that it is necessary to include non-hydrostatic effects to adequately reproduce mixing and other processes involved in the flows (e.g., [10]). The hybrid SIFOM–FVCOM system is applied to simulate a transient flow over sill with configuration

$$\begin{cases} -1500 < x < 2000, \\ y = \pm 200(1 - 0.8e^{-4\times 10^{-6}x^2}), \quad x < 0; \ y = \pm 40, \ x > 0, \\ z = -150 + \dfrac{140}{1 + (x/500)^4}, \ x < 0; \ z = -120 + \dfrac{110}{1 + (x/500)^4}, \ x > 0, \end{cases} \tag{7}$$

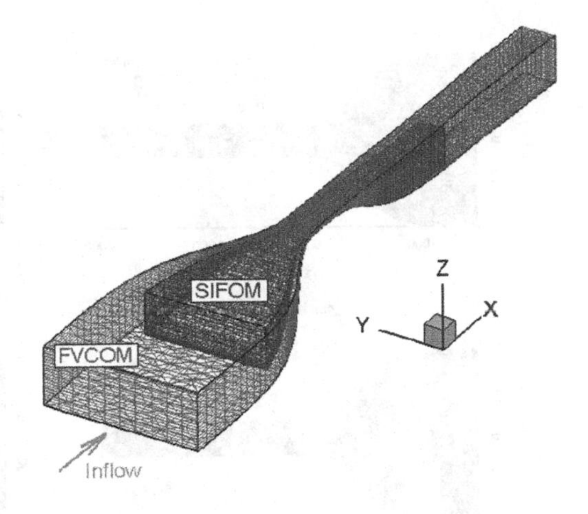

Fig. 2 Meshes for the sill flow

and initial and boundary condition

$$\begin{cases} u, v, w = 0,\ t = 0, \\ \eta u = 0.9175(1 - e^{-0.01t}),\ x = -1500;\ \eta = 0,\ x = 2000. \end{cases} \tag{8}$$

In above expressions, length is in m, time in s, and velocity in m/s.

The grid of FVCOM covers the whole flow field, and that of SIFOM is located over the sill (Fig. 2). The grid of FVCOM has 10,400 triangle elements in the horizontal plane and 41 σ-layers in the vertical direction, and the number of grid nodes of SIFOM is $161 \times 33 \times 49$ in the longitudinal, lateral, and vertical direction, respectively. The time step of the SIFOM–FVCOM system is 0.5.

As indicated in the initial condition in Eq. (8), the water body is initially stationary. Because of the imposition of an inflow at the entrance, a current occurs and moves to the right. Instantaneous solutions for the current at different moments are shown in Fig. 3. Figure 3a–f illustrate the current roughly when its front approaches, arrives at, and passes the top of the sill. Figure 3g and h present solutions for the flow after the front exits the computational domain, and it seems that it is at its equilibrium state at this moment. The solutions presented in the figures are reasonable in large scales in aspect of streamlines and velocity distribution.

Nevertheless, a detailed examination of the solutions in Fig. 3 finds problems with them. Figure 4a, e indicate that, near the front of the current, where water bodies in motion and at rest are adjacent to each other, the solutions of the SIFOM model cannot react simultaneously to those obtained with the FVCOM model, and there is a delay in them. In addition, there is a pronounced difference, in both magnitude and direction, between the velocity solutions provided by the two models in their overlapping regions (Fig. 4a, c, e). In Fig. 4c, near the top of the sill, FVCOM provides a forward flow in the upper layer, while SIFOM produces a reverse flow in the lower layer. It is expected that there is no physical mechanism to generate such reverse flow, and apparently it is an artifact. It is not clear what causes the delay and difference in solutions and the reverse flow. All of these indicate that the hybrid

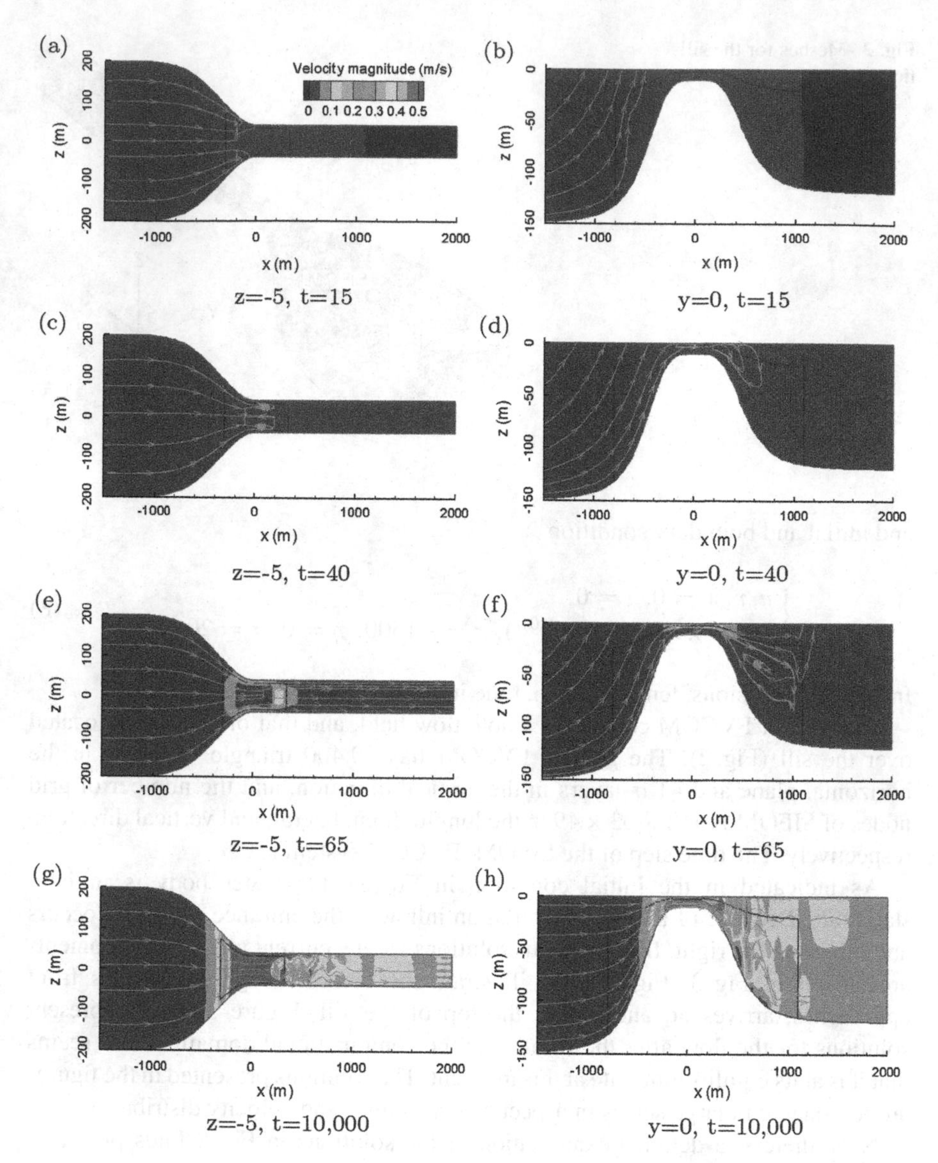

Fig. 3 Simulation of the sill flow at different moments. The *solid lines* around the sill indicate the boundary of SIFOM

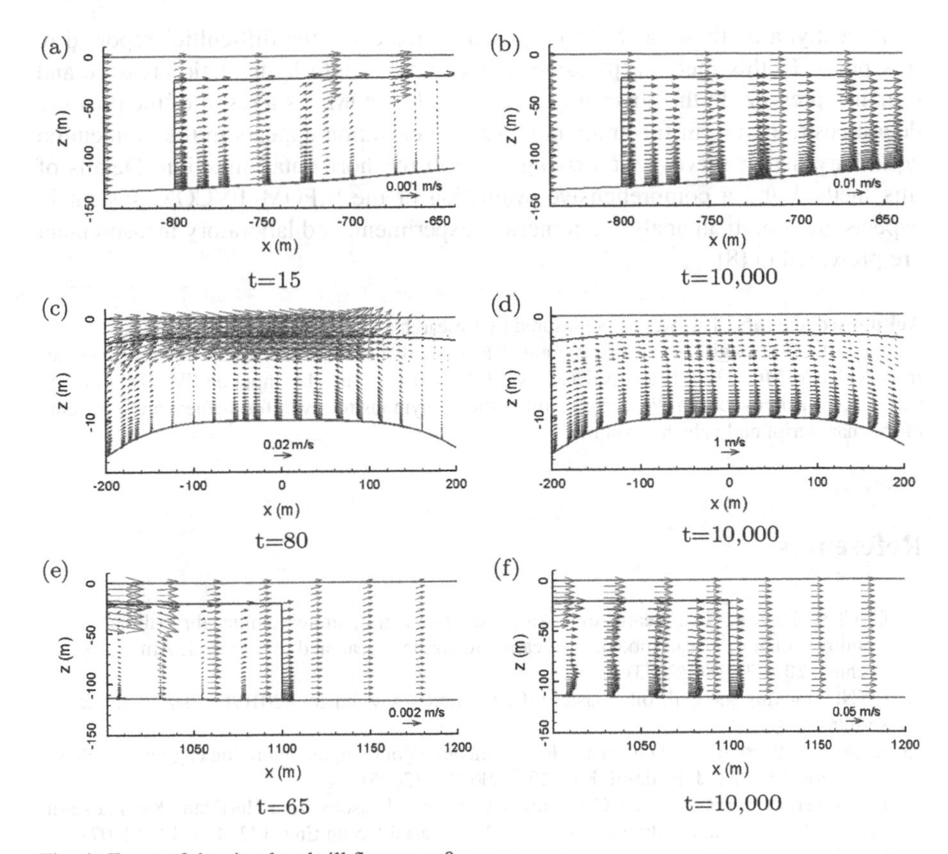

Fig. 4 Zoom of the simulated sill flow at y=0

system in its current form has difficulties to correctly resolve the flow at the current front, which is complicated and involves strong unsteadiness and non-hydrostatic effect. After the front passes, the solution becomes normal and above mentioned problems disappear, see Fig. 4b, d, f.

5 Concluding Remarks

Hybrid of F3DFD and GFD models based on domain decomposition is a feasible approach to simulate multiscale and multiphysics coastal processes. However, this approach is challenging in view it involves coupling of different governing equations, distinct numerical methods, and dissimilar computational meshes. This paper indicates that, while its performance is promising, such approach faces difficulties to correctly resolve a current front associated with strong transient effect and it needs discretion in implementation.

Recently, a method has been proposed to overcome the difficulties reported in this paper. In this method, pressure is decomposed into hydrostatic pressure and dynamic pressure in the governing equations of SIFOM. As a result of the pressure decomposition, a term for gradient of surface elevation appears in the momentum equations, and it serves as a driving force in the horizontal direction. Details of this method and a comprehensive evaluation of the SIFOM–FVCOM system in aspects of theoretical analysis, numerical experiment, and laboratory measurement are presented in [8].

Acknowledgements This work is sponsored by Research and Innovative Technology Administration of USDOT through the UTRC program (RFCUNY 49111-15-23). Partial support also comes from NSF (CMMI-1334551) and NJDOT (NJDOT 2010-15). Valuable input on FVCOM from Dr. C.S. Chen is acknowledged. We are grateful to the anonymous reviewer for his/her careful reading of the manuscript and valuable comments.

References

1. C. Chen, H. Liu, R.C. Beardsley, An unstructured grid, finite-volume, three-dimensional, primitive equation ocean model: application to coastal ocean and estuaries. J. Atmos. Ocean. Technol. **20**, 159–186 (2003)
2. CNN, Tracking the Gulf oil disaster (2010), http://www.cnn.com/2010/US/04/29/interactive.spill.tracker/
3. L. Ge, F. Sotiropoulos, 3D unsteady RANS modeling of complex hydraulic engineering flows. I: numerical model. J. Hydraul. Eng. **131**, 800–808 (2005)
4. I.N. Robertson, H.R. Riggs, S.C.S. Yim, Y.L. Young, Lessons from Hurricane Katrina storm surge on bridges and buildings. J. Waterw. Port Coastal Ocean Eng. **133**, 463–483 (2007)
5. H.S. Tang, Study on a grid interface algorithm for solutions of incompressible Navier-Stokes equations. Comput. Fluids, **35**, 1372–1383 (2006)
6. H.S. Tang, X.G. Wu, CFD and GFD hybrid approach for simulation of multi-scale coastal ocean flow, *iEMS's 2010 Int. Congress on Environmental Modelling and software, Fifth Biennial Meeting*, Ottawa, 5–8 July 2010. In Modelling for Environmental's Sake, D.A. Swayne et al. (Eds.)
7. H.S. Tang, C. Jones, F. Sotiropoulos, An overset grid method for 3D unsteady incompressible flows. J. Comput. Phys. **191**, 567–600 (2003)
8. H.S. Tang, K. Qu, X.G. Wu, An overset grid method for integration of fully 3D fluid dynamics and geophysics fluid dynamics models to simulate multiphysics coastal ocean flows. J. Comput. Phys. **273**, 548–571 (2014)
9. X.G. Wu, H.S. Tang, Coupling of CFD model and FVCOM to predict small-scale coastal flows. J. Hydrodyn. **22**, 284–289 (2010)
10. J.X. Xing, A.M. Davies, On the importance of non-hydrostatic processes in determining tidally induced mixing in sill regions. Continental Shelf Res. **27**, 2162–2185 (2007)

A Domain Decomposition Based Jacobi-Davidson Algorithm for Quantum Dot Simulation

Tao Zhao, Feng-Nan Hwang, and Xiao-Chuan Cai

1 Introduction

Quantum dot (QD) is a semiconducting nanostructure where electrons are confined in all three spatial dimensions [8], as shown in Fig. 1. The quantum states of the pyramidal quantum dot with a single electron can be described by the time-independent 3D Schrödinger equation

$$-\nabla \cdot \left(\frac{\hbar^2}{2m(\mathbf{r},\lambda)} \nabla u \right) + V(\mathbf{r})u = \lambda u, \tag{1}$$

defined on a cuboid Ω, where λ is called an energy state or eigenvalue, and u is the corresponding wave function or eigenvector. In (1), $\hbar$ is the reduced Plank constant, $\mathbf{r}$ is the space variable, $m(\mathbf{r},\lambda)$ is the effective electron mass, and $V(\mathbf{r})$ is the confinement potential.

The Ben Daniel-Duke interface condition

$$\left(\frac{1}{m(\mathbf{r},\lambda)} \frac{\partial u}{\partial \mathbf{n}} \right)\Bigg|_{\partial D_-} = \left(\frac{1}{m(\mathbf{r},\lambda)} \frac{\partial u}{\partial \mathbf{n}} \right)\Bigg|_{\partial D_+}$$

is imposed on the interface, where D denotes the domain of the pyramid dot and $\mathbf{n}$ is the unit outward normal of ∂D. We impose the homogeneous Dirichlet boundary

T. Zhao (✉) • X.-C. Cai
Department of Computer Science, University of Colorado Boulder, CO 80309, USA
e-mail: tao.zhao@colorado.edu; cai@cs.colorado.edu

F.-N. Hwang
Department of Mathematics, National Central University, Jhongli 320, Taiwan
e-mail: hwangf@math.ncu.edu.tw

T. Dickopf et al. (eds.), *Domain Decomposition Methods in Science and Engineering XXII*, Lecture Notes in Computational Science and Engineering 104, DOI 10.1007/978-3-319-18827-0_42

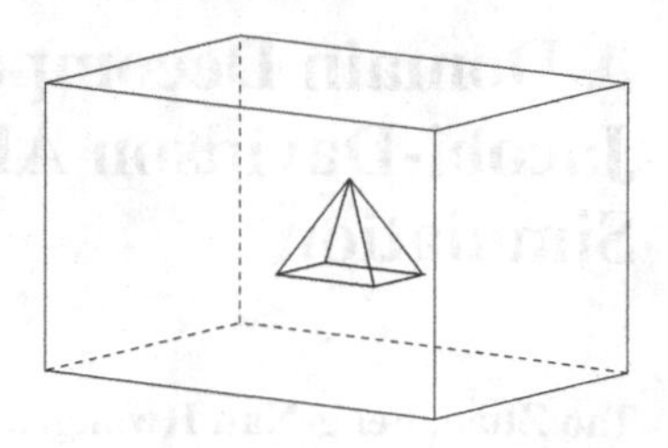

Fig. 1 Structure of a pyramidal quantum dot embedded in a cuboid. The size of the cuboid is 24.8×24.8×18.6 nm; the width of the pyramid base is 12.4 nm and the height of the pyramid is 6.2 nm

condition $u = 0$ on the boundary of the cuboid. For details, see [5, 8] and references therein.

A cell-centered finite volume method on an uniform mesh in Cartesian coordinates is applied to discretize the Schrödinger equation with non-parabolic effective mass model [5]. Then we obtain the polynomial eigenvalue problem

$$(\lambda^5 A_5 + \lambda^4 A_4 + \lambda^3 A_3 + \lambda^2 A_2 + \lambda A_1 + A_0)x = 0, \tag{2}$$

where $\lambda \in \mathbb{C}$, $x \in \mathbb{C}^N$, $A_i \in \mathbb{R}^{N\times N}$, and N is the total number of unknowns. The matrices A_0 and A_1 are diagonal, and all other matrices are nonsymmetric.

The rest of the paper is organized as follows. In Sect. 2, we first recall the convergence of the single-vector version of Jacobi-Davidson (JD) based on the residual of the approximate eigenpair for solving the general polynomial eigenvalue problem of degree m. Then we propose a three-grid parallel domain decomposition based JD algorithm for computing the relevant quantum dot eigenvalues and the corresponding eigenvectors. Numerical results are reported in Sect. 3. Some finial remarks are given in Sect. 4.

2 Jacobi-Davidson Algorithm and Domain Decomposition Based Preconditioners

For given $A_i \in \mathbb{C}^{N\times N}$, $i = 0, 1, \cdots, m$, we define $\mathcal{A}_\phi = \sum_{i=0}^{m} \phi^i A_i$ as a matrix polynomial of $\phi \in \mathbb{C}$. If there exist $\lambda \in \mathbb{C}$ and $x \in \mathbb{C}^N$ such that $\mathcal{A}_\lambda x = 0$, then λ is called an eigenvalue of $\mathcal{A}_\phi$ and x is the eigenvector of $\mathcal{A}_\phi$ associated with the eigenvalue λ. There are several versions of JD for solving eigenvalue problems; see [4] and references therein. A relatively simple version referred to as JD1 in this paper is summarized in Algorithm 1 below.

Algorithm 1 JD1 for polynomial eigenvalue problems

Input: A_i for $i = 0, \cdots, m$, and the maximum number of iterations k.

1: Choose an initial eigenvector u_0 with $\|u_0\|_2 = 1$.
  For $n = 0, \cdots, k$
2: Solve $u_n^* \mathcal{A}_{\phi_n} u_n = 0$ for a new eigenvalue approximation ϕ_n.
3: Compute the residual $r_n = \mathcal{A}_{\phi_n} u_n$.

4: If the stopping criteria is satisfied, then stop.
5: Compute $p_n = \mathcal{A}'_{\phi_n} u_n = \left(\sum_{i=1}^{m} i\phi_n^{i-1} A_i\right) u_n$.
6: Solve $\left\| \left(I - (p_n u_n^*)/(u_n^* p_n)\right) \mathcal{A}_{\phi_n} z_n + r_n \right\|_2 \leq \varepsilon_n \|r_n\|_2,\ z_n \perp u_n$.
7: Compute a new eigenvector approximation $u_{n+1} = (u_n + z_n)/\|u_n + z_n\|_2$.
End for

Theorem 1 *Let $P_n = p_n u_n^*/(u_n^* p_n)$ and (λ, x) be an eigenpair of $\mathcal{A}_\phi$. There exists $\mathbb{D}(\lambda, x, d) = \{\phi \in \mathbb{C},\ u \in \mathbb{C}^N : \mathcal{A}_\phi$ is nonsingular and $\|\mathcal{A}_\phi u\|_2 < d\}$. If the initial eigenpair (ϕ_0, u_0) and any eigenpair (ϕ_n, u_n) generated by JD1 are all in $\mathbb{D}(\lambda, x, d)$, then the residuals satisfy*

$$\|(I - P_n) r_{n+1}\|_2 \leq \varepsilon_n \|r_n\|_2 + \xi_n \|r_n\|_2^2 + \mathcal{O}(\|r_n\|_2^3) \tag{3}$$

if $\mathcal{A}_{\phi_n}$ is non-Hermitian, and

$$\|(I - P_n) r_{n+1}\|_2 \leq \varepsilon_n \|r_n\|_2 + \zeta_n \|r_n\|_2^3 + \mathcal{O}(\|r_n\|_2^4) \tag{4}$$

if $\mathcal{A}_{\phi_n}$ is Hermitian. Here, ξ_n and ζ_n depend on ε_n, ϕ_n and A_i's.

Because of the page limit, the proof of the theorem is not shown. Theorem 1 implies that if r_{n+1} is orthogonal to u_n, then $I - P_n$ can be removed from the left-hand sides without changing the right-hands sides of (3) and (4). The authors of [4] suggest that a subspace $\mathcal{V}_{n+2}$ is built by all the correction vectors $u_n' s$ and the initial vector u_0. Then a new approximate eigenvector u_{n+1} is extracted from the subspace with the Galerkin condition $r_{n+1} \perp \mathcal{V}_{n+2}$. We will refer the resulting method as the JD algorithm described in Algorithm 2 below. In the JD algorithm, r_{n+1} is orthogonal to any u_i for $i \leq n + 1$, which leads to $(I - P_n) r_{n+1} = r_{n+1}$. Thus one can reasonably expect that the JD algorithm has a quadratic or cubic convergence if ε_n is sufficiently small relative to the residual.

Algorithm 2 JD for polynomial eigenvalue problems

Input: A_i for $i = 0, \cdots, m$, and the maximum number of iterations k.
1: Let $V = [v]$, where v is an initial eigenvector such that $\|v\|_2 = 1$.
For $n = 0, \cdots, k$
2: Compute $W_i = A_i V$ and $M_i = V^H W_i$ for $i = 0, \cdots, m$.
3: Solve the projected polynomial eigenvalue problem $\left(\sum_{i=0}^{m} \phi^i M_i\right) s = 0$, then obtain the desired eigenpair (ϕ, s) such that $\|s\|_2 = 1$.
4: Compute the Ritz vector $u = Vs$, and the residual vector $r = \mathcal{A}_\phi u$.
5: If the stopping criteria is satisfied, then stop.
6: Compute $p = \mathcal{A}'_\phi u = \left(\sum_{i=1}^{m} i\phi^{i-1} A_i\right) u$.
7: Solve $\left\| (I - (pu^*)/(u^*p)) \mathcal{A}_\phi (I - uu^*) t + r \right\|_2 \leq \varepsilon_n \|r\|_2,\ t \perp u$.
8: Orthogonalize t against V, set $v = v/\|t\|_2$, then expand $V \leftarrow [V, v]$.
End for

Remark 1 If the initial guess is good enough, and if the tolerance of the correction equation satisfies $\varepsilon_n \leq \mathcal{O}(\|r_n\|_2)$, then the JD algorithm may converge quadratically, but we are unable to theoretically prove this.

Remark 2 If ε_n is chosen as a constant independent of n, then the convergence can only be linear in theory, however, in practice, if the constant tolerance is reasonably small, quadratic or near quadratic convergence has been observed in our numerical experiments.

In the entire JD approach, the linear correction equation is the most expensive part of the calculation since it is in the inner most loop. In earlier work, people often restrict the number of iterations to be carried out for the correction equation to be a small number (5 or 10) without considering how large the residual is when the iteration is stopped. This does cut down the computational cost per iteration, but as a result, the outer JD iteration may not have a quadratic convergence. In this paper, we make sure the correction equation is solved to a certain accuracy. A two-level preconditioner with a sufficiently fine coarse grid is used to control the number of iterations and scalability of the correction equation solver.

In JD, the preconditioner is of the form $\tilde{M} = (I - (pu^*)/(u^*p))M(I - uu^*)$, where M is an approximation of $\mathcal{A}_\phi$. We assume that the Krylov subspace method starts with an initial vector $t = 0$, and is preconditioned by $\tilde{M}$ from the right with a fixed ϕ. In the Krylov solver, we have to compute $x = \tilde{M}^{-1}y$ at each iteration. To avoid forming $\tilde{M}^{-1}$ explicitly, we solve a linear system $\tilde{M}x = y$ for x. Assume that x is orthogonal to u. It is straightforward to show that x takes the following form $x = M^{-1}y - (u^*M^{-1}y)/(u^*M^{-1}p)M^{-1}p$. Thus, for solving each correction equation, we need to compute $s = M^{-1}y$ at each iteration of Krylov subspace method, while compute $M^{-1}p$ only once.

In our method, we let M^{-1} be a two-level multiplicative type Schwarz preconditioner [1, 3]. Its multiplication with a vector y requires two steps:

$$s \leftarrow I_c^h M_c^{-1} R_h^c y,$$
$$s \leftarrow s + M_f^{-1}(y - \mathcal{A}_\phi s).$$

Here, M_c is a preconditioner defined on the coarse mesh Ω_c. To obtain M_c, we discretize the Schrödinger equation (1) on Ω_c by the finite volume method mentioned in Sect. 1 and then obtain a coarse mesh polynomial eigenvalue problem $\left(\sum_{i=0}^m \lambda^i B_i\right) x = 0$. For this particular quantum dot simulation that we are interested, m is equal to 5. The matrix B_i $(i = 0, \cdots, m)$ is much smaller than A_i in (2), but has the same nonzero structure pattern as A_i. Then we define $M_c = \sum_{i=0}^m \phi^i B_i$, where ϕ is the Ritz value computed on the fine mesh Ω_f.

In the first step of the two-level Schwarz preconditioner, I_c^h is an interpolation from Ω_c to Ω_f, and R_h^c is a restriction from Ω_f to Ω_c. As before, computing $w = M_c^{-1}(R_h^c y)$ is equivalent to solving a linear system $M_c w = R_h^c y$. In practice, we solve it approximately using a Krylov subspace method preconditioned by a one-level RAS preconditioner defined on the coarse mesh Ω_c using the same number of processors as on the fine mesh. In the second step of the two-level preconditioner, M_f is the RAS preconditioner defined on the fine mesh Ω_f.

To build the RAS preconditioners, we partition the cuboid into non-overlapping subdomains ω_i, $i = 1, \cdots, np$, then generate the overlapping subdomain ω_i^δ by including the δ layers of mesh cells in the neighboring subdomains of ω_i, i.e., $\omega_i \subset \omega_i^\delta$. Here, np is the number of processors that is the same as the number of subdomains, and δ is the size of overlap. Let R_i^0 and R_i^δ be restriction operators to non-overlapping and overlapping subdomains, respectively. With R_i^δ, we define the matrix $\mathcal{J}_i = R_i^\delta \mathcal{A}_\phi (R_i^\delta)^T$. Then the one-level RAS preconditioner reads as $M_{RAS}^{-1} = \sum_{i=1}^{np} (R_i^0)^T \mathcal{J}_i^{-1} R_i^\delta$. In practice, $\mathcal{J}_i^{-1}$ is not formed explicitly, instead it is approximated by ILU factorization.

In theory, a good initial guess implies good convergence of JD, but in practice, it is a nontrivial issue to find the right initial guess, especially when both the accuracy and the computational cost need to be balanced since our goal is to achieve near linear speedup measured by the total compute time. For convenience (less coding, less memory required, and computationally cheaper), the coarse mesh for finding the initials is usually chosen to be the coarse mesh of the two-level Schwarz method. However, as is shown in the next section, the coarse mesh of the two-level Schwarz preconditioner in this paper is not suitable to generate the initial guess since it is still very large. Note that only several eigenpairs around the ground state are of interests in this simulation. As a result, we have to generate another much coarser mesh Ω_o for computing the initials.

On Ω_o, we discretize the Schrödinger equation using the finite volume method mentioned in Sect. 1. Next, the resulting polynomial eigenvalue problem is solved using, for instance, the QZ method with linearization. Once we obtain the desired eigenpair (ϕ_o, v_o), ϕ_o is used as the initial eigenvalue and v_o is interpolated to the fine mesh Ω_f to generate the initial eigenvector on the fine mesh $v_h \leftarrow I_o^h v_o$, where I_o^h is an interpolation operator from Ω_o to Ω_f. Due to the small size, (ϕ_o, v_o) is computed redundantly on all processors.

With the coarse and fine grids, we describe the three-grid parallel domain decomposition based JD algorithm in Algorithm 3.

Algorithm 3 Three-grid parallel domain decomposition based JD algorithm for polynomial eigenvalue problems

Input: Coefficient matrices on Ω_o, Ω_c and Ω_f for $i = 0, \cdots, m$.

1: On Ω_c, solve the polynomial eigenvalue problem roughly and obtain the desired eigenpair (ϕ_o, v_o).
2: Obtain the initial eigenvector on the fine mesh $v_h \leftarrow I_o^h v_o$.
3: Solve the polynomial eigenvalue problem on Ω_f by Algorithm 2. At each iteration, the correction equation is solved to a modest accuracy by Krylov subspace method with either one-level or two-level preconditioner.

3 Numerical Results

We use Algorithm 3 to compute 6 smallest positive eigenvalues and the corresponding eigenvectors of the pyramidal quantum dot problem as shown in Fig. 1. The physical parameters in the non-parabolic effective mass model are the same as described in [5]. The software is implemented using PETSc [2], SLEPc [6] and PJDPack [5].

The fine mesh Ω_f is 600×600×450 with 161,101,649 unknowns. The coarse mesh Ω_o to generate the initial guess is 12×12×9 with 968 unknowns. Due to the small size of Ω_o, the Schrödinger equation discretized on Ω_o is solved redundantly on all processors using JD with the one-vector as the initial guess. The JD iteration is stopped when either the absolute or the relative residual norm is below 10^{-8}. The eigenvectors on Ω_o are interpolated to the fine mesh by trilinear interpolation.

On Ω_f, we stop the JD iteration when either the absolute or the relative residual norm is below 10^{-10}. The correction equation is solved by the flexible GMRES (FGMRES) without restarting [7] preconditioned by either one-level or two-level preconditioners. The stopping criteria of FGMRES on Ω_f is 10^{-4}. For the two-level Schwarz preconditioner, we solve the linear system on Ω_c by FGMRES with the RAS preconditioner. The stopping criteria of FGMRES on Ω_c is 10^{-1}. For the RAS preconditioners on Ω_c and Ω_f, ILU(0) is applied to solve the linear system on each subdomain; the size of overlap is 1.

Tables 1, 2, and 3 show the numerical performance of JD with the one-level and two-level preconditioners in terms of the number of JD iterations, the average

Table 1 The ground state $e_0 = 0.4162094856604$ and Ω_c is $56 \times 56 \times 42$

	One-level			Two-level		
np	JD	FGMRES	Time	JD	FGMRES	Time
5120	4	185.25	42.48	4	38.75	7.48
7168	4	185.25	29.20	4	39.50	6.36
9216	4	186.00	23.23	4	38.75	5.34
10,240	4	186.00	22.46	4	39.75	4.84

Table 2 The first excited state $e_1 = 5.990754117523$ and Ω_c is $80 \times 80 \times 60$

	One-level			Two-level		
np	JD	FGMRES	Time	JD	FGMRES	Time
5120	4	251.75	71.10	4	34.75	9.47
7168	4	255.75	47.96	4	34.25	6.20
9216	4	254.50	39.29	4	34.50	5.64
10,240	4	256.50	37.37	4	34.00	5.29

Table 3 The second excited state $e_2 = 0.5990754117522$ and Ω_c is $80 \times 80 \times 60$

	One-level			Two-level		
np	JD	FGMRES	Time	JD	FGMRES	Time
5120	4	258.25	71.90	4	34.25	9.99
7168	4	243.25	45.42	4	34.25	6.71
9216	4	258.00	40.99	4	35.50	5.92
10,240	4	250.50	35.42	4	34.50	5.44

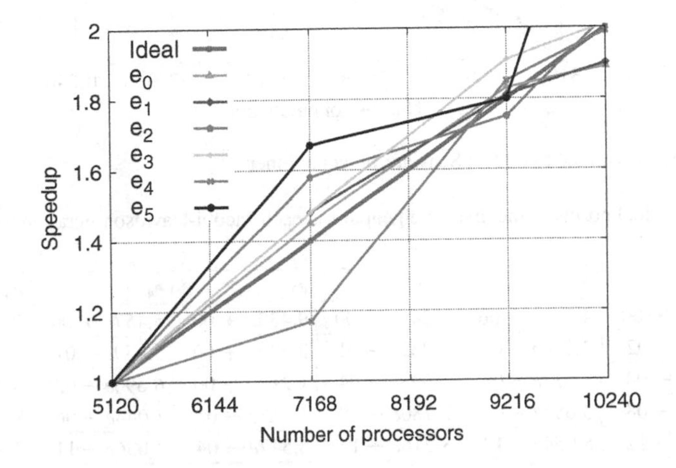

Fig. 2 Speedup with the one-level preconditioner

number of FGMRES for solving the correction equations and the compute time. Since the imaginary parts of the computed eigenvalues are less than 10^{-13}, we report the real parts only. Consider the compute time and the average number of FGMRES iterations, the two-level preconditioner is much better than the one-level preconditioner.

Figures 2 and 3 plot the speedup curves of JD with one-level and two-level preconditioners. Obviously, JD with both preconditioners are scalable, and the two-level approach is faster in terms of the total compute time (Table. 4).

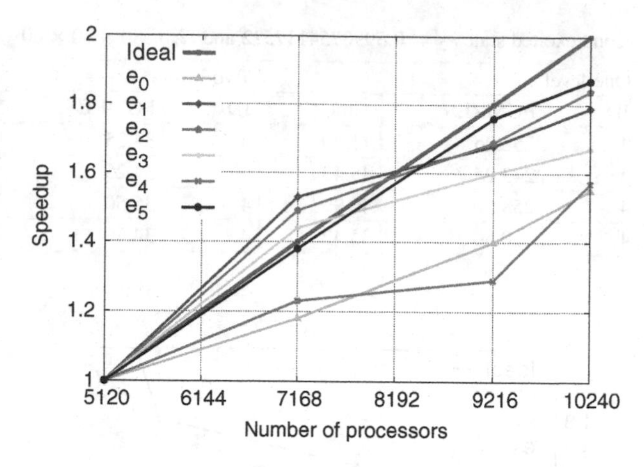

Fig. 3 Speedup with the two-level Schwarz preconditioner

Table 4 Residual norms of the first six eigenpairs at each Jacobi-Davidson iteration using 10,240 processors

it	e_0	e_1	e_2	e_3	e_4	e_5
0	$2.590e+00$	$4.249e+00$	$4.249e+00$	$9.430e+00$	$5.457e+00$	$7.339e+00$
1	$7.614e-02$	$2.236e-01$	$2.233e-01$	$3.926e+00$	$5.959e-01$	$1.465e+00$
2	$1.911e-04$	$3.433e-04$	$3.505e-04$	$1.211e+00$	$6.696e-03$	$8.375e-02$
3	$1.640e-08$	$3.657e-08$	$3.780e-08$	$1.418e-01$	$1.044e-06$	$8.642e-05$
4	$1.780e-12$	$8.054e-12$	$8.216e-12$	$3.547e-04$	$1.036e-11$	$3.270e-08$
5				$6.083e-08$		$7.659e-12$
6				$8.533e-12$		

The correction equations are preconditioned by the two-level Schwarz preconditioner. "it" is the index for the JD iteration

4 Conclusions

A parallel domain decomposition based Jacobi-Davidson algorithm with three meshes was introduced and studied for the pyramidal quantum dot simulation. The proposed method requires three meshes; one fine mesh that determines the accuracy of the solution and two coarse meshes to accelerate the convergence of the inner and outer iterations. Numerical results confirmed that our method converges quadratically with the proposed strategy for computing the initial guess, and also is scalable for problems with over 160 millions unknowns on a parallel computer with over 10,000 processors.

References

1. P.E. Bjørstad, B.F. Smith, W.D. Gropp, *Domain Decomposition: Parallel Multilevel Methods for Elliptic Partial Differential Equations*. (Cambridge University Press, New York, 1996)
2. K. Buschelman, V. Eijkhout, W.D. Gropp, D. Kaushik, M.G. Knepley, L.C. McInnes, B.F. Smith, S. Balay, J. Brown, H. Zhang, Petsc users manual (Argonne National Laboratory, 2013)
3. X.-C. Cai, M. Sarkis, A restricted additive Schwarz preconditioner for general sparse linear systems. SIAM J. Sci. Comput. **21**, 239–247 (1999)
4. D.R. Fokkema, G.L.G. Sleijpen, A.G.L. Booten, H. van der Vorst, Jacobi-davidson type methods for generalized eigenproblems and polynomial eigenproblems. BIT **36**, 595–633 (1996)
5. T.-M. Huang, F.-N. Hwang, Z.-H. Wei, W. Wang, A parallel additive schwarz preconditioned jacobi-davidson algorithm for polynomial eigenvalue problems in quantum dot simulation. J. Comput. Phys. **229**, 2923–2947 (2010)
6. J.E. Roman, V. Hernandez, V. Vidal, SLEPc: A scalable and flexible toolkit for the solution of eigenvalue problems. ACM Trans. Math. Softw. **31**, 351–362 (2005)
7. Y. Saad, *Iterative Methods for Sparse Linear Systems*, 2nd edn. (SIAM, Philadelphia, 2003)
8. N. Vukmirović, L.-W. Wang, Quantum dots: theory. Technical Report, Lawrence Berkeley National Laboratory, 2009

Part III
Contributed Talks (CT) and Posters

Globally Convergent Multigrid Method for Variational Inequalities with a Nonlinear Term

Lori Badea

1 Introduction

In [1], one- and two-level Schwarz methods have been proposed for variational inequalities with contraction operators. This type of inequalities generalizes the problems modeled by quasi-linear or semilinear inequalities. It is proved there that the convergence rates of the two-level methods are almost independent of the mesh and overlapping parameters. However, the original convex set, which is defined on the fine grid, is used to find the corrections on the coarse grid, too. This leads to a suboptimal computing complexity. A remedy can be found in adopting minimization techniques from the construction of multigrid methods for the constrained minimization of functionals. In this case, to avoid visiting the fine grid, some level convex sets for the corrections on the coarse levels have been proposed in [4, 7–10] and the review article [6] for complementarity problems, and in [2] for two two-obstacle problems. In this paper, we introduce and investigate the convergence of a new multigrid algorithm for the inequalities with contraction operators, and we have adopted the construction of the level convex sets which has been introduced in [2]. In this way, the introduced multigrid method has an optimal computing complexity of the iterations. Also, the convergence theorems for the methods introduced in [1] contain a convergence condition depending on the total number of the subdomains in the decompositions of the domain. The convergence condition of a direct extension of these methods to more than two-levels will introduce an upper bound for the number of mesh levels which can be used in the method. In comparison with these methods, the convergence condition of

L. Badea (✉)
Institute of Mathematics of the Romanian Academy, Research Unit 6, P.O. Box 1-764,
RO-014700 Bucharest, Romania
e-mail: lori.badea@imar.ro

T. Dickopf et al. (eds.), *Domain Decomposition Methods in Science and Engineering XXII*, Lecture Notes in Computational Science and Engineering 104, DOI 10.1007/978-3-319-18827-0_43

the algorithm introduced in this paper is less restrictive and depends neither on the number of the subdomains in the decompositions of the domain nor on the number of levels. Moreover, this convergence condition is very similar with the condition of existence and uniqueness of the solution of the problem.

The paper is organized as follows. In Sect. 2, the method is introduced as a subspace correction algorithm in a general reflexive Banach space. Under the same assumptions in [2] concerning the level convex sets where we are looking for the corrections, we prove that the algorithm is globally convergent and estimate the global convergence rate, provided that the convergence condition is satisfied. In Sect. 3, we show that the algorithm can be viewed as multilevel or multigrid methods if we associate finite element spaces to the level meshes and to the domain decompositions at each level. In [2], it has been proved that the assumptions made in the previous section hold for problems having the convex set of two-obstacle type. For this type of problems, we write the convergence rate of the proposed multigrid method in function of the number of level meshes.

2 Abstract Convergence Results

We consider a reflexive Banach space V and let $K \subset V$ be a nonempty closed convex set. Let $F : V \to \mathbf{R}$ be a Gâteaux differentiable functional and we assume that there exist two constants $\alpha,\ \beta > 0$ for which

$$\alpha||v-u||^2 \le \langle F'(v) - F'(u), v-u\rangle \text{ and } ||F'(v) - F'(u)||_{V'} \le \beta||v-u||, \quad (1)$$

for any $u, v \in V$. Above, we have denoted by F' the Gâteaux derivative of F, and V' is the dual space of V. Following the way in [5], we can prove that

$$\langle F'(u), v-u\rangle + \frac{\alpha}{2}||v-u||^2 \le F(v) - F(u) \le \langle F'(u), v-u\rangle + \frac{\beta}{2}||v-u||^2, \quad (2)$$

for any $u, v \in V$. We point out that since F is Gâteaux differentiable and satisfies (1), then F is a convex functional (see Proposition 5.5 in [3], p. 25). Also, let $T : V \to V'$ be an operator with the property that there exists a constant $\gamma > 0$ such that

$$||T(v) - T(u)||_{V'} \le \gamma||v-u|| \text{ for any } v, u \in V. \quad (3)$$

Now, we consider the quasi-variational inequality

$$u \in K : \langle F'(u), v-u\rangle + \langle T(u), v-u\rangle \ge 0 \text{ for any } v \in K. \quad (4)$$

Using (2), we get

$$\frac{\alpha}{2}||v-u||^2 \leq F(v) - F(u) + \langle T(u), v-u\rangle \text{ for any } v \in K. \tag{5}$$

Problem (4) has a solution and it is unique (see [1], for instance) if

$$\gamma/\alpha < 1. \tag{6}$$

Now, let us assume that we have J closed subspaces of V, $V_1, \ldots, V_J$, and let V_{ji}, $i = 1, \ldots I_j$ be some closed subspaces of V_j, $j = J, \ldots, 1$. The subspaces V_j, $j = J, \ldots, 1$, will be associated with the grid levels, and, for each level $j = J, \ldots, 1$, V_{ji}, $i = 1, \ldots I_j$, will be associated with a domain decomposition. Let us write $I = \max_{j=J,\ldots,1} I_j$.

To introduce the algorithm, we make an assumption on choice of the convex sets $\mathcal{K}_j$, $j = 1, \ldots, J$, where we look for the level corrections. The chosen level convex sets depend on the current approximation in the algorithms.

Assumption 1 *For a given $w \in K$, we recursively introduce the convex sets $\mathcal{K}_j$, $j = J, J-1, \ldots, 1$, as*

- *at level J: we assume that $0 \in \mathcal{K}_J$, $\mathcal{K}_J \subset \{v_J \in V_J : w + v_J \in K\}$ and consider a $w_J \in \mathcal{K}_J$,*
- *at a level $J-1 \geq j \geq 1$: we assume that $0 \in \mathcal{K}_j$ and $\mathcal{K}_j \subset \{v_j \in V_j : w + w_J + \ldots + w_{j+1} + v_j \in K\}$, and consider a $w_j \in \mathcal{K}_j$.*

We now introduce the algorithm, which is of multiplicative type, and where the argument of T is kept unchanged for several iterations.

Algorithm 1 *We start the algorithm with an arbitrary $u^0 \in K$. Assuming that at iteration $n \geq 0$ we have $u^n \in K$, we write $\tilde{u}^n = u^n$ and carry out the following two steps:*

1. *We perform $\kappa \geq 1$ multiplicative iterations, keeping the argument of T equal with u^n. We start with $\tilde{u}^n$ and having $\tilde{u}^{n+k-1}$ at iteration $1 \leq k \leq \kappa$, we successively calculate level corrections and compute $\tilde{u}^{n+k}$:*
   - *at the level J, we construct the convex set $\mathcal{K}_J$ as in Assumption 1, with $w = \tilde{u}^{n+k-1}$. Then, we first write $w_J^k = 0$, and, for $i = 1, \ldots, I_J$, we successively calculate $w_{Ji}^{k+1} \in V_{Ji}$, $w_J^{k+\frac{i-1}{I_J}} + w_{Ji}^{k+1} \in \mathcal{K}_J$, the solution of the inequality*

$$\langle F'(\tilde{u}^{n+k-1} + w_J^{k+\frac{i-1}{I_J}} + w_{Ji}^{k+1}), v_{Ji} - w_{Ji}^{k+1}\rangle$$
$$+ \langle T(u^n), v_{Ji} - w_{Ji}^{k+1}\rangle \geq 0,$$

for any $v_{Ji} \in V_{Ji}$, $w_J^{k+\frac{i-1}{I_J}} + v_{Ji} \in \mathcal{K}_J$, and write $w_J^{k+\frac{i}{I_J}} = w_J^{k+\frac{i-1}{I_J}} + w_{Ji}^{k+1}$,

– *at a level $J-1 \geq j \geq 1$, we construct the convex set $\mathcal{K}_j$ as in Assumption 1 with $w = \tilde{u}^{n+k-1}$ and $w_J = w_J^{k+1}, \dots, w_{j+1} = w_{j+1}^{k+1}$. Then, we write $w_j^{k+1} = 0$, and for $i = 1, \dots, I_j$, we successively calculate $w_{ji}^{k+1} \in V_{ji}$, $w_j^{k+\frac{i-1}{I_j}} + w_{ji}^{k+1} \in \mathcal{K}_j$, the solution of the inequality*

$$\langle F'(\tilde{u}^{n+k-1} + \sum_{l=j+1}^{J} w_l^{k+1} + w_j^{k+\frac{i-1}{I_j}} + w_{ji}^{k+1}), v_{ji} - w_{ji}^{k+1}\rangle$$
$$+ \langle T(u^n), v_{ji} - w_{ji}^{k+1}\rangle \geq 0,$$

for any $v_{ji} \in V_{ji}$, $w_j^{k+\frac{i-1}{I_j}} + v_{ji} \in \mathcal{K}_j$, and write $w_j^{k+\frac{i}{I_j}} = w_J^{k+\frac{i-1}{I_j}} + w_{ji}^{k+1}$,

– *we write $\tilde{u}^{n+k} = \tilde{u}^{n+k-1} + \sum_{j=1}^{J} w_j^{k+1}$.*

2. *We write $u^{n+1} = \tilde{u}^{n+k}$.*

In order to prove the convergence of the above algorithm, we shall make two new assumptions. In the case of the multigrid decompositions, the constants of some inequalities can be taken independent of the number J of levels, the classical Cauchy–Schwarz inequality can be strengthened, for instance. In this sense we make the following assumption.

Assumption 2

1. *There exist some constants $0 < \beta_{jk} \leq 1$, $\beta_{jk} = \beta_{kj}$, $j, k = J, \dots, 1$, such that $\langle F'(v + v_{ji}) - F'(v), v_{kl}\rangle \leq \beta\beta_{jk}||v_{ji}||||v_{kl}||$, for any $v \in V$, $v_{ji} \in V_{ji}$, $v_{kl} \in V_{kl}$, $i = 1, \dots, I_j$ and $l = 1, \dots, I_k$.*
2. *There exists a constant C_1 such that $||\sum_{j=1}^{J}\sum_{i=1}^{I_j} w_{ji}|| \leq C_1(\sum_{j=1}^{J}\sum_{i=1}^{I_j}||w_{ji}||^2)^{\frac{1}{2}}$, for any $w_{ji} \in V_{ji}$, $j = J, \dots, 1$, $i = 1, \dots, I_j$.*

Evidently, for the moment, we can consider $C_1 = (IJ)^{\frac{1}{2}}$ and $\beta_{jk} = 1$, $j, k = J, \dots, 1$. The second new assumption refers to additional properties asked to the convex sets $\mathcal{K}_j$, $j = 1, \dots, J$, introduced in Assumption 1.

Assumption 3 *There exists a constant $C_2 > 0$ such that for any $w \in K$, $w_{ji} \in V_{ji}$, $w_{j1} + \ldots + w_{ji} \in \mathcal{K}_j$, $j = J, \dots, 1$, $i = 1, \dots, I_j$, and $u \in K$, there exist $u_{ji} \in V_{ji}$, $j = J, \dots, 1$, $i = 1, \dots, I_j$, which satisfy*

$$u_{j1} \in \mathcal{K}_j \text{ and } w_{j1} + \ldots + w_{ji-1} + u_{ji} \in \mathcal{K}_j, i = 2, \dots, I_j,\ j = J, \dots, 1,$$

$$u - w = \sum_{j=1}^{J}\sum_{i=1}^{I_j} u_{ji}, \text{ and}$$

$$\sum_{j=1}^{J}\sum_{i=1}^{I_j}||u_{ji}||^2 \leq C_2^2\left(||u-w||^2 + \sum_{j=1}^{J}\sum_{i=1}^{I_j}||w_{ji}||^2\right).$$

The convex sets $\mathcal{K}_j$, $j = J, \dots, 1$, are constructed as in Assumption 1 with the above w and $w_j = \sum_{i=1}^{I_j} w_{ji}$, $j = J, \dots, 1$.

The global convergence of Algorithm 1 is proved by

Theorem 1 *Let V be a reflexive Banach space, V_j, $j = 1, \dots, J$, closed subspaces of V, and V_{ji}, $i = 1, \dots, I_j$, some closed subspaces of V_j, $j = 1, \dots, J$. Let K be a non empty closed convex subset of V, and we suppose that Assumptions 1–3 hold. Also, we assume that F is a Gâteaux differentiable functional which satisfies (1) and the operator T satisfies (3). On these conditions, if*

$$\gamma/\alpha < 1/2 \tag{7}$$

and κ satisfies

$$\left(\frac{\tilde{C}}{\tilde{C}+1}\right)^{\kappa} < \frac{1-2\frac{\gamma}{\alpha}}{1+3\frac{\gamma}{\alpha}+4\frac{\gamma^2}{\alpha^2}+\frac{\gamma^3}{\alpha^3}}, \tag{8}$$

where constant $\tilde{C}$ is given by

$$\tilde{C} = \frac{1}{C_2\varepsilon}\left[1 + C_2 + C_1C_2 + \frac{C_2}{\varepsilon}\right],\ \varepsilon = \frac{\alpha}{2\beta I(\max_{k=1,\cdots,J}\sum_{j=1}^{J}\beta_{kj})C_2}, \tag{9}$$

then Algorithm 1 is convergent and we have the following error estimations:

$$\begin{aligned} &F(u^{n+1}) + \langle T(u), u^{n+1}\rangle - F(u) - \langle T(u), u\rangle \\ &\le [2\tfrac{\gamma}{\alpha} + (\tfrac{\tilde{C}}{\tilde{C}+1})^{\kappa}(1 + 3\tfrac{\gamma}{\alpha} + 4\tfrac{\gamma^2}{\alpha^2} + \tfrac{\gamma^3}{\alpha^3})]^n \\ &\cdot[F(u^0) + \langle T(u), u^0\rangle - F(u) - \langle T(u), u\rangle], \end{aligned} \tag{10}$$

$$\begin{aligned} &\|u^n - u\|^2 \le \tfrac{2}{\alpha}[2\tfrac{\gamma}{\alpha} + (\tfrac{\tilde{C}}{\tilde{C}+1})^{\kappa}(1 + 3\tfrac{\gamma}{\alpha} + 4\tfrac{\gamma^2}{\alpha^2} + \tfrac{\gamma^3}{\alpha^3})]^n \\ &\cdot[F(u^0) + \langle T(u), u^0\rangle - F(u) - \langle T(u), u\rangle]. \end{aligned} \tag{11}$$

Proof First, we see that in view of (5), (11) can be obtained from (10). Now, for a fixed $n \ge 0$, let us consider the problem

$$\tilde{u} \in K : \langle F'(\tilde{u}), v - \tilde{u}\rangle + \langle T(\tilde{u}^n), v - \tilde{u}\rangle \ge 0, \text{ for any } v \in K, \tag{12}$$

where $\tilde{u}^n = u^n \in K$ is the approximation obtained from Algorithm 1 after n iterations. By applying Theorem 2.2 in [2] to variational inequality (12) we get that after κ iterations the following error estimation holds

$$\begin{aligned} &F(\tilde{u}^{n+\kappa}) + \langle T(\tilde{u}^n), \tilde{u}^{n+\kappa}\rangle - F(\tilde{u}) - \langle T(\tilde{u}^n), \tilde{u}\rangle \\ &\le (\tfrac{\tilde{C}}{\tilde{C}+1})^{\kappa}[F(\tilde{u}^n) + \langle T(\tilde{u}^n), \tilde{u}^n\rangle - F(\tilde{u}) - \langle T(\tilde{u}^n), \tilde{u}\rangle] \end{aligned}$$

or

$$
\begin{aligned}
&F(u^{n+1}) + \langle T(u^n), u^{n+1}\rangle - F(\tilde{u}) - \langle T(u^n), \tilde{u}\rangle \\
&\le (\tfrac{\tilde{C}}{\tilde{C}+1})^{\kappa}[F(u^n) + \langle T(u^n), u^n\rangle - F(\tilde{u}) - \langle T(u^n), \tilde{u}\rangle],
\end{aligned}
\tag{13}
$$

where $\tilde{C}$ is given in (9). From (2), (12) and (3), we have

$$
\begin{aligned}
&F(\tilde{u}) + \langle T(u), \tilde{u}\rangle - F(u) - \langle T(u), u\rangle + \tfrac{\alpha}{2}||\tilde{u} - u||^2 \\
&\le \langle F'(\tilde{u}), \tilde{u} - u\rangle + \langle T(u^n), \tilde{u} - u\rangle + \langle T(u) - T(u^n), \tilde{u} - u\rangle \\
&\le \langle T(u) - T(u^n), \tilde{u} - u\rangle \le \gamma||u - u^n||\,||u - \tilde{u}|| \le \tfrac{\gamma}{2}||u - u^n||^2 + \tfrac{\gamma}{2}||u - \tilde{u}||^2.
\end{aligned}
$$

From (4) and using again (2), we get

$$
\begin{aligned}
&\tfrac{\alpha}{2}||u - u^n||^2 \le \langle F'(u), u - u^n\rangle + F(u^n) - F(u) \\
&\le F(u^n) + \langle T(u), u^n\rangle - F(u) - \langle T(u), u\rangle.
\end{aligned}
\tag{14}
$$

From the last two equations, in view of (7), we get

$$
\begin{aligned}
&F(\tilde{u}) + \langle T(u), \tilde{u}\rangle - F(u) - \langle T(u), u\rangle \\
&\le \tfrac{\gamma}{\alpha}[F(u^n) + \langle T(u), u^n\rangle - F(u) - \langle T(u), u\rangle].
\end{aligned}
\tag{15}
$$

Now, we have

$$
\begin{aligned}
&F(u^{n+1}) + \langle T(u), u^{n+1}\rangle - F(u) - \langle T(u), u\rangle \\
&= F(u^{n+1}) + \langle T(u^n), u^{n+1}\rangle - F(\tilde{u}) - \langle T(u^n), \tilde{u}\rangle \\
&+F(\tilde{u}) + \langle T(u), \tilde{u}\rangle - F(u) - \langle T(u), u\rangle \\
&+\langle T(u) - T(u^n), u^{n+1} - \tilde{u}\rangle.
\end{aligned}
\tag{16}
$$

But, in view of (13), we get

$$
\begin{aligned}
&F(u^{n+1}) + \langle T(u^n), u^{n+1}\rangle - F(\tilde{u}) - \langle T(u^n), \tilde{u}\rangle \\
&\le (\tfrac{\tilde{C}}{\tilde{C}+1})^{\kappa}[F(u^n) + \langle T(u^n), u^n\rangle - F(\tilde{u}) - \langle T(u^n), \tilde{u}\rangle] \\
&= (\tfrac{\tilde{C}}{\tilde{C}+1})^{\kappa}[F(u^n) + \langle T(u), u^n\rangle - F(u) - \langle T(u), u\rangle \\
&+F(u) + \langle T(u), u\rangle - F(\tilde{u}) - \langle T(u), \tilde{u}\rangle] \\
&+(\tfrac{\tilde{C}}{\tilde{C}+1})^{\kappa}\langle T(u^n) - T(u), u^n - \tilde{u}\rangle.
\end{aligned}
\tag{17}
$$

It follows from (16), (17), (15) and (3) that

$$
\begin{aligned}
&F(u^{n+1}) + \langle T(u), u^{n+1}\rangle - F(u) - \langle T(u), u\rangle \\
&\le (\frac{\tilde{C}}{\tilde{C}+1})^{\kappa}[F(u^n) + \langle T(u), u^n\rangle - F(u) - \langle T(u), u\rangle]
\end{aligned}
$$

$$+[1-(\frac{\tilde{C}}{\tilde{C}+1})^{\kappa}][F(\tilde{u})+\langle T(u),\tilde{u}\rangle-F(u)-\langle T(u),u\rangle]$$
$$+(\frac{\tilde{C}}{\tilde{C}+1})^{\kappa}\langle T(u^n)-T(u),u^n-\tilde{u}\rangle+\langle T(u)-T(u^n),u^{n+1}-\tilde{u}\rangle$$
$$\leq[(\frac{\tilde{C}}{\tilde{C}+1})^{\kappa}-\frac{\gamma}{\alpha}(\frac{\tilde{C}}{\tilde{C}+1})^{\kappa}+\frac{\gamma}{\alpha}][F(u^n)+\langle T(u),u^n\rangle-F(u)-\langle T(u),u\rangle]$$
$$+\gamma(\frac{\tilde{C}}{\tilde{C}+1})^{\kappa}||u^n-u||||u^n-\tilde{u}||+\gamma||u^n-u||||u^{n+1}-\tilde{u}||.$$

Also, we have

$$\begin{aligned}
&(\tfrac{\tilde{C}}{\tilde{C}+1})^{\kappa}||u^n-u||||u^n-\tilde{u}||+||u^n-u||||u^{n+1}-\tilde{u}||\\
&\leq(\tfrac{\tilde{C}}{\tilde{C}+1})^{\kappa}(||u^n-u||^2+||u^n-u|||u-\tilde{u}||)+||u^n-u||||u^{n+1}-\tilde{u}||\\
&\leq\tfrac{1}{2}[3(\tfrac{\tilde{C}}{\tilde{C}+1})^{\kappa}+1]||u^n-u||^2+\tfrac{1}{2}(\tfrac{\tilde{C}}{\tilde{C}+1})^{\kappa}||u-\tilde{u}||^2+\tfrac{1}{2}||u^{n+1}-\tilde{u}||^2.
\end{aligned}$$

Therefore, from the last two equation, we have

$$\begin{aligned}
&F(u^{n+1})+\langle T(u),u^{n+1}\rangle-F(u)-\langle T(u),u\rangle\\
&\leq[(\tfrac{\tilde{C}}{\tilde{C}+1})^{\kappa}-\tfrac{\gamma}{\alpha}(\tfrac{\tilde{C}}{\tilde{C}+1})^{\kappa}+\tfrac{\gamma}{\alpha}][F(u^n)+\langle T(u),u^n\rangle-F(u)-\langle T(u),u\rangle]\\
&+\tfrac{\gamma}{2}[3(\tfrac{\tilde{C}}{\tilde{C}+1})^{\kappa}+1]||u^n-u||^2+\tfrac{\gamma}{2}(\tfrac{\tilde{C}}{\tilde{C}+1})^{\kappa}||u-\tilde{u}||^2+\tfrac{\gamma}{2}||u^{n+1}-\tilde{u}||^2.
\end{aligned}\tag{18}$$

From (2), (4) and (15) we have

$$\begin{aligned}
&\tfrac{\alpha}{2}||\tilde{u}-u||^2\leq\langle F'(u),u-\tilde{u}\rangle+F(\tilde{u})-F(u)\leq F(\tilde{u})+\langle T(u),\tilde{u}\rangle\\
&-F(u)-\langle T(u),u\rangle\leq\tfrac{\gamma}{\alpha}[F(u^n)+\langle T(u),u^n\rangle-F(u)-\langle T(u),u\rangle].
\end{aligned}\tag{19}$$

In view of (2), (12), (17) and (3), we get

$$\begin{aligned}
&\tfrac{\alpha}{2}||u^{n+1}-\tilde{u}||^2\leq\langle F'(\tilde{u}),\tilde{u}-u^{n+1}\rangle+F(u^{n+1})-F(\tilde{u})\\
&\leq(\tfrac{\tilde{C}}{\tilde{C}+1})^{\kappa}[F(u^n)+\langle T(u),u^n\rangle-F(u)-\langle T(u),u\rangle\\
&+F(u)+\langle T(u),u\rangle-F(\tilde{u})-\langle T(u),\tilde{u}\rangle]+\gamma(\tfrac{\tilde{C}}{\tilde{C}+1})^{\kappa}||u^n-u||||u^n-\tilde{u}||.
\end{aligned}$$

As previously, using (14) and (19), we get

$$\begin{aligned}
&||u^n-u||||u^n-\tilde{u}||\leq\tfrac{3}{2}||u^n-u||^2+\tfrac{1}{2}||u-\tilde{u}||^2\\
&\leq[\tfrac{3}{\alpha}+\tfrac{\gamma}{\alpha^2}][F(u^n)+\langle T(u),u^n\rangle-F(u)-\langle T(u),u\rangle].
\end{aligned}$$

From the last two equations, since $F(u) - F(\tilde{u}) + \langle T(u), u - \tilde{u}\rangle \leq 0$, we have

$$\begin{aligned}\tfrac{\alpha}{2}||u^{n+1} - \tilde{u}||^2 \leq (\tfrac{\tilde{C}}{\tilde{C}+1})^{\kappa}[1 + 3\tfrac{\gamma}{\alpha} + \tfrac{\gamma^2}{\alpha^2}] \\ \cdot[F(u^n) + \langle T(u), u^n\rangle - F(u) - \langle T(u), u\rangle].\end{aligned} \tag{20}$$

Finally, from (18), (14), (19) and (20), we get

$$\begin{aligned}&F(u^{n+1}) + \langle T(u), u^{n+1}\rangle - F(u) - \langle T(u), u\rangle \\ &\leq [2\tfrac{\gamma}{\alpha} + (\tfrac{\tilde{C}}{\tilde{C}+1})^{\kappa}(1 + 3\tfrac{\gamma}{\alpha} + 4\tfrac{\gamma^2}{\alpha^2} + \tfrac{\gamma^3}{\alpha^3})][F(u^n) + \langle T(u, u^n\rangle - F(u) - \langle T(u, u)\rangle].\end{aligned}$$

Remark 1 Theorem 1 shows that if the convergence condition (7) is satisfied and the number κ of the intermediate iterations is sufficiently large then Algorithm 1 converges and error estimation (11) holds.

3 Multilevel and Multigrid Methods

We consider a family of regular meshes $\mathcal{T}_{h_j}$ of mesh sizes $h_j, j = 1, \ldots, J$ over the domain $\Omega \subset \mathbf{R}^d$ and assume that $\mathcal{T}_{h_{j+1}}$ is a refinement of $\mathcal{T}_{h_j}, j = 1, \ldots, J-1$. Also, at each level $j = 1, \ldots, J$, we consider an overlapping decomposition $\{\Omega_j^i\}_{1\leq i\leq I_j}$ of Ω, and assume that the mesh partition $\mathcal{T}_{h_j}$ supplies a mesh partition for each Ω_j^i, $1 \leq i \leq I_j$.

At each level $j = 1, \ldots, J$, we introduce the linear finite element spaces V_{h_j} whose elements vanish on $\partial\Omega$. Also, for $i = 1, \ldots, I_j$, we consider the subspaces $V_{h_j}^i$ of V_{h_j} whose elements vanish on $\Omega\backslash\Omega_j^i$. With these spaces, Algorithm 1 becomes a multilevel method. In [2], for a problem of two-obstacle type, $K = [\varphi, \psi]$, level convex sets $\mathcal{K}_j = [\varphi_j, \psi_j]$, $j = 1, \ldots, J$, satisfying Assumption 1 have been constructed. Also, it has been proved there that Assumption 3 holds with the constant

$$C_2 = CI^2(J-1)^{\frac{1}{2}}[\sum_{j=2}^{J} C_d(h_{j-1}, h_J)^2]^{\frac{1}{2}},$$

where

$$C_d(H, h) := 1 \text{ if } d = 1,\ (\ln\frac{H}{h} + 1)^{\frac{1}{2}} \text{ if } d = 2 \text{ and } (\frac{H}{h})^{\frac{d-2}{2}} \text{ if } 2 < d,$$

d being the Euclidean dimension of the space where the domain Ω lies and C is a constant independent of J and I_j, $i = 1, \cdots, J$. Consequently, Theorem 1 shows that the multilevel method corresponding to Algorithm 1 is convergent and we can explicitly write its convergence rate.

If the level decompositions of the domain are given by the supports of the nodal basis functions of the spaces V_{h_j}, $j = J, \dots, 1$, Algorithm 1 becomes a multigrid method. In this case, it is proved in [2] that we can take $C_1 = C$ and $\max_{k=1,\dots,J} \sum_{j=1}^{J} \beta_{kj} = C$, where $C \geq 1$ is a constant independent of the number of meshes. By expressing the constant C_2 only in function of J, the following result is a direct consequence of Theorem 1,

Corollary 1 *As a function of the number J of levels, the error estimate of the multigrid method obtained from Algorithm 1 can be written as*

$$\|u^n - u\|_1^2 \leq C\left[2\frac{\gamma}{\alpha} + \left(\frac{\tilde{C}(J)}{\tilde{C}(J)+1}\right)^{\kappa}\left(1 + 3\frac{\gamma}{\alpha} + 4\frac{\gamma^2}{\alpha^2} + \frac{\gamma^3}{\alpha^3}\right)\right]^n,$$

where $\|\cdot\|_1$ *is the norm of* $H^1(\Omega)$ *and* $\tilde{C}(J) = CJS_d(J)^2$, *in which* $S_d(J)$ *is* $\left[\sum_{j=2}^{J} C_d(h_{j-1}, h_J)^2\right]^{\frac{1}{2}}$ *expressed in function of J,*

$$S_d(J) := (J-1)^{\frac{1}{2}} \text{ if } d = 1,\ CJ \text{ if } d = 2 \text{ and } C^J \text{ if } d = 3,$$

constant C being independent of the number of levels J.

Acknowledgement The author acknowledges the support of this work by "Laboratoire Euroéen Associé CNRS Franco-Roumain de Matématiques et Moélisation" LEA Math-Mode.

References

1. L. Badea, Schwarz methods for inequalities with contraction operators. J. Comput. Appl. Math. **215**(1), 196–219 (2008)
2. L. Badea, Global convergence rate of a standard multigrid method for variational inequalities. IMA J. Numer. Anal. **34**(1), 197–216 (2014). doi:10.1093/imanum/drs054
3. I. Ekeland, R. Temam, *Analyse Convexe et Problèmes Variationnels* (Dunod, Paris, 1974)
4. E. Gelman, J. Mandel, On multilevel iterative method for optimization problems. Math. Program. **48**, 1–17 (1990)
5. R. Glowinski, J.L. Lions, R. Trémolières, *Analyse Numérique des Inéquations Variationnelles* (Dunod, Paris, 1976)
6. C. Graser, R. Kornhuber, Multigrid methods for obstacle problems. J. Comput. Math. **27**(1), 1–44 (2009)
7. R. Kornhuber, Monotone multigrid methods for elliptic variational inequalities I. Numer. Math. **69**, 167–184 (1994)
8. R. Kornhuber, Monotone multigrid methods for elliptic variational inequalities II. Numer. Math. **72**, 481–499 (1996)
9. J. Mandel, A multilevel iterative method for symmetric, positive definite linear complementarity problems. Appl. Math. Optim. **11**, 77–95 (1984a)
10. J. Mandel, Etude algébrique d'une méthode multigrille pour quelques problèmes de frontière libre. C.R. Acad. Sci., Ser. I **298**, 469–472 (1984b)

Partially Updated Restricted Additive Schwarz Preconditioner

Laurent Berenguer and Damien Tromeur-Dervout

1 Introduction

The solution of differential equations with implicit methods requires the solution of a nonlinear problem at each time step. We consider Newton-Krylov ([8], Chap. 3) methods to solve these nonlinear problems: the linearized system of each Newton iteration of each time step is solved by a Krylov method. Generally speaking, the most time-consuming part of the numerical simulation is the solution of the sequence of linear systems by the Krylov method. Then, providing a good preconditioner is a critical point: a balance must be found between the ability of the preconditioner to reduce the number of Krylov iterations, and its computational cost. The method that combines a Newton-Krylov method with a Schwarz domain decomposition preconditioner is called Newton-Krylov-Schwarz (NKS) [5]. In this paper, we deal with the Restricted Additive Schwarz (RAS) preconditioner [4]. We propose to freeze this preconditioner for a few time steps, and to partially update it. Here, the partial update of the preconditioner consists in recomputing some parts of the preconditioner associated to certain subdomains, keeping the other ones frozen. These partial updates improve the efficiency and the longevity of the frozen preconditioner. Furthermore, they can be computed asynchronously in order to improve the parallelism.

The remainder of this paper is organized as follows: Sect. 2 presents the partial update of the Restricted Additive Schwarz (RAS) preconditioner. In Sect. 3, we propose to compute this partial update asynchronously on additional devices in order to achieve a parallel algorithm. The third section is devoted to numerical

L. Berenguer (✉) • D. Tromeur-Dervout
Université de Lyon, Université Lyon 1, CNRS, Institut Camille Jordan (UMR 5208), 69622 Villeurbanne, France
e-mail: laurent.berenguer@univ-lyon1.fr; damien.tromeur-dervout@univ-lyon1.fr

T. Dickopf et al. (eds.), *Domain Decomposition Methods in Science and Engineering XXII*, Lecture Notes in Computational Science and Engineering 104, DOI 10.1007/978-3-319-18827-0_44

experiments on a reaction-diffusion problem. They show that the partially updated preconditioner is more robust than the frozen preconditioner, and that a superlinear speed-up can be achieved.

2 The Partial Update of the RAS Preconditioner

We consider ordinary differential equations of the form $\dot{x} = f(x,t)$ where $x \in \mathbb{R}^n$ is the solution and the function f from $\mathbb{R}^{n+1}$ to $\mathbb{R}^n$ is nonlinear. The problem $\dot{x} = f(x,t)$ is solved for a given initial condition $x(0) = x_0$ and suitable boundary conditions. If an implicit method is used for the time integration, then a nonlinear problem of the form $F(x^l, t^l) = 0$ must be solved in x^l at each time step t^l. This nonlinear problem is generally solved by Newton-like methods that require the solution of linear systems of the form

$$J(x_k^l, t^l)\delta x_k^l = -F(x_k^l, t^l) \tag{1}$$

where the subscript k stands for the number of the Newton iteration, $J(x_k^l, t^l) \in \mathbb{R}^{n\times n}$ is the Jacobian matrix of $F(\cdot, t^l)$ at the solution x_k^l. The Newton-Krylov method can be viewed as an inexact Newton method if Eq. (1) is solved by a Krylov method. A good preconditioning method is needed to accelerate the convergence of Krylov methods. The preconditioning matrix M_k^l should approximate $J(x_k^l, t^l)$ and its inverse must be computed easily. Preconditioners based on domain decomposition methods are often used because their application to vectors requires only the solution of subdomain problems. The domain of n unknowns is split in N overlapping subdomains. Each subdomain i has n_i unknowns if we include the overlap, and $\tilde{n}_i$ if we exclude the overlap (i.e. $n = \sum_i^N \tilde{n}_i$). Let $R_i \in \mathbb{R}^{n_i\times n}$ denote the operator that restricts a vector to the ith subdomain, including the overlap. We also denote by $\tilde{R}_i \in \mathbb{R}^{n_i\times n}$ the restriction operator to the ith subdomain that excludes the overlap by setting to zero the lines corresponding to the overlap. For simplicity of notation, we write J instead of $J(x,t)$ when no confusion can arise. Thus, the RAS preconditioner of the matrix J is given by Eq. (2).

$$M_{RAS}^{-1} = \sum_{i=1}^{N} \tilde{R}_i^T \left(R_i J R_i^T\right)^{-1} R_i \tag{2}$$

In Eq. (2), we assumed that the local Jacobian matrices $J_i = R_i J R_i^T$ are invertible. This is not necessary to compute explicitly the matrices J_i^{-1} because the Krylov method requires only the application of the preconditioner to vectors. This application is computed by the parallel solution of N local linear systems. In the following, we solve this local linear system using the LU factorizations of the matrices J_i.

Several methods have been proposed to optimize the solution of a sequence of slightly changing linear systems, and all of them consist in reusing some computations done at the previous linear systems. Then, several ways to reuse the Krylov subspace have been considered. An overview of these techniques is given in [9] but it is worth mentioning that the information provided by the Krylov subspace can be used to update a preconditioner. Hence, in [6] a preconditioner based on deflation is updated at each restart of GMRES. In order to save computational time, we consider the reuse of the same RAS preconditioner for a few successive linear systems: this frozen preconditioner is called Lagged RAS (LRAS) in the following. In this case, it may be relevant to update the preconditioner from one linear system to another, instead of recomputing it. Several ways to update a preconditioning matrix have already been considered. It has been proposed in [3] to update the preconditioner, adding a low rank matrix that corresponds to the quasi-Newton update. The update of a factorized preconditioner has also been considered: the update AINV preconditioner has been studied in [2] for a sequence of diagonally shifted matrices. In [12], an algebraic formula is derived to update the ILU preconditioner from the difference of two successive linear operators. This idea has been extended to Jacobian-free methods in [13]. The frozen preconditioner is expected to become less and less efficient from one linear system to another. Then, a recomputation of the preconditioner may be needed to prevent convergence failures of the Krylov method. It is a difficult task to decide when a frozen preconditioner needs recomputed, but two heuristic criteria are often used:

- The preconditioner can be recomputed if the previous linear system has needed more than K_{max} Krylov iterations.
- The preconditioner can also be updated every L linear systems.

In this paper, we choose the first approach that seems more flexible: it can allow to save numerous of unnecessary global updates. On the other hand, the number of needed Krylov iterations can vary during the simulation because it does not only depend on the age of the preconditioner. Then, to be optimal, K_{max} should be adapted during the simulation. However, this topic exceeds the scope of this paper.

When LRAS is used, the update of the preconditioner is global: all the local LU factorizations are computed simultaneously. We can extend this idea to a partial update: only some parts of the preconditioner are updated. Hence, the preconditioner can be written as in Eq. (3), where *AsRAS* stands for *Asynchronous Restricted Additive Schwarz*. The preconditioner is now compounded of local Jacobian matrices evaluated at different Newton iterations or time steps. It is worth pointing out that if $t_i = t$ and $k_i = k$ for $i = 1 \dots N$ then $M^{-1}_{AsRAS} = M^{-1}_{LRAS}$.

$$M^{-1}_{AsRAS} = \sum_{i=1}^{N} \tilde{R}_i^T J_i(x_{k_i}^{l_i}, t^{l_i})^{-1} R_i \tag{3}$$

In order to avoid idle time, asynchronous solvers have been studied for linear and nonlinear problems [7, 11]. The disadvantage of asynchronous solvers lies in the fact that one needs to make extra assumptions on the problem and its splitting to ensure the convergence. Here, the updates of the local parts of the preconditioner are asynchronous, but the communications between subdomains are synchronous. Then, the theoretical framework of Newton-Krylov solvers applies directly: the exact solution of preconditioned linear systems is the same regardless of the preconditioning matrix. That being said, Krylov methods approximate the solution to a given tolerance, the digits of the solution beyond this tolerance might differ from one preconditioner to another.

One can expect that the partial update allows to save Krylov iterations during the simulation. Numerical results will confirm this idea from a global point of view, but we do not assume that every single partial update improves the condition number of the linear systems.

The sequential implementation of the AsRAS preconditioner is straightforward. The implementation on a parallel computer is a much more difficult task because idle time may arise when only some processors compute the LU factorization. To circumvent this difficulty, in the next section we propose to dedicate additional processes to the LU factorizations.

3 Parallel Implementation of the Asynchronous RAS Preconditioner

The method presented in the previous section does not seem suitable for parallel computing because the load is not balanced: some of the processors will have to wait while the other ones compute the LU factorizations since the Krylov method entails synchronizations. In the following, we present an efficient algorithm where the LU factorizations are computed by processors that are not in charge of a subdomain.

The key point of the asynchronous partial update of the RAS preconditioner is to define two kinds of tasks that communicate: the first one is the solution of a subdomain problem (i.e. the classical Newton-Krylov method). The second one is the LU factorizations of local Jacobian matrices. Then, in order to solve the physical problem, one should assign most of the CPU cores to the first task.

Algorithm 1 describes how to implement the method in a client-server approach. The client processes are those assigned to subdomains, while the server processes are devoted to the computation of LU factorizations. The client processes must be able to continue the computation between the sending of the local Jacobian matrix and the reception of the factorized matrix. The reception of the factorized matrix is the partial update since the new LU factorization is received in the memory space of the previous one. In our MPI implementation, the communication pattern is the following: client processes check if the server is ready to receive the local Jacobian matrix before they send it. This checking is implemented using

Algorithm 1 Asynchronous update of the preconditioner

1: // *Client process*
2: **for** each time step **do**
3: // *Newton iterations:*
4: **repeat**
5: if a LU factorization is available, partially update M^{-1}_{AsRAS}
6: **if** global update **then**
7: $M^{-1}_{AsRAS} = \sum_{i=1}^{N} \tilde{R}_i^T J_i^{-1} R_i$
8: **end if**
9: Krylov method to solve $J\delta x = -F(x)$ preconditioned by M^{-1}_{AsRAS}
10: $x \leftarrow x + \delta x$
11: if needed, send J
12: **until** convergence
13: **end for**

1: // *Server process*
2: **repeat**
3: receive the matrix
4: compute the LU factorization
5: send the factorization
6: **until** the end of the integration

MPI_Test. Likewise, client processes check if the LU factorization is ready before they start the reception. The reception can be done between two Newton iterations or even between two Krylov iterations if the Krylov method allows variable preconditioners [10]. Finally, Algorithm 1 can be viewed as an improvement of LRAS: both algorithms are equivalent if no processes are assigned to the partial update of the preconditioner. As stated above, there are more client processes than server processes. As a consequence, only few partial updates can be computed simultaneously. Then, one needs to decide in which order the requests will be treated. In the remainder of this paper, we limit ourselves to a cyclic update: we first update the first subdomains, then the second ones and so on. This approach is not optimal, because it does not update more frequently the subdomains where highly nonlinear phenomena appear. Since we proposed an asynchronous implementation of AsRAS in Algorithm 1, we cannot assume that the partial updates will be received it time to prevent convergence failure of the Krylov method. That is the reason why, there is a global restart in Algorithm 1, step 7, as in classical LRAS implementations.

4 Numerical Tests

This section presents some tests that highlight the behavior of the methods. The computer cluster used for the numerical experiments is an SGI Altix XE 1300, with two six-cores Intel Xeon 5650 per node. The PETSc library [1] was used for the implementation. Generally speaking, one should avoid the exchange of factorized matrix through the network, then a MPI library that performs efficient intranode communications is required. The tasks must be distributed in such a way that client processes and their server share some memory. For example, one core per processor hosts the server process and the other cores host its client processes. We compare the behavior of AsRAS to LRAS for a reaction-diffusion problem given in Eq. (4). The domain is the unit square with periodic boundary conditions.

$$\begin{cases} \dot{u} - \alpha_1 \Delta u = A + u^2 v - (B+1)u \\ \dot{v} - \alpha_2 \Delta v = Bu - u^2 v \end{cases} \tag{4}$$

We consider the following parameters:

- The domain is discretized in 100×100 points, using a five-point stencil and decomposed in 16 subdomains.
- The problem is solved for $t \in [0, 10]$ using the backward Euler scheme with a variable time step. The solution at $t = 0$ is $u(x, y) = A + 10^{-2} \times r(x, y)$ and $v(x, y) = A/B + 10^{-2} \times r(x, y)$ where $r(x, y) \in [-0.5, 0.5]$ are random numbers.
- The coefficients of the reaction are $A = 3.5$, $B = 12$, $\alpha_1 = 1 \times h^2$ and $\alpha_2 = 2.6 \times h^2$, where h is the spatial length scale.
- The nonlinear solver is a Newton method with line search where Jacobian matrices are approximated by finite differences. The linear solver is a right-preconditioned BiCGSTAB [14].
- The RAS preconditioner, with an overlap of one, is recomputed if the number of Krylov iterations of the previous linear system has exceed K_{max}.
- LRAS has run on 16 cores, and AsRAS on 20 cores (16 cores associated to subdomains, and 4 cores dedicated to the partial update of the preconditioner).

Let us remark that, because of the asynchronous behaviour of our implementation of AsRAS, the number of Krylov iterations and the last digits of the solution may vary from one run to another. That is the reason why the execution time and the number of Krylov iterations given in Figs. 1 and 2 are averages over five runs. The cumulate number of Krylov iterations is given in Fig. 1. In both cases, the number of total Krylov iterations increases with respect to K_{max}, that is to say when the number of global updates decreases. However, the partial update limits this increase. The total execution times are plotted in Fig. 2. The minimum wall time is given by the best balance between the Krylov iterations and the computational cost of the preconditioner. The minimum wall time is 2.46 for LRAS and 2.15 for AsRAS. If we compare these minimum wall times, the speedup is 1.14 which is lower than the theoretical linear speedup 1.25. On the other hand, if we consider all the tests, the average speedup is about 1.5 because AsRAS is less sensitive to K_{max} than LRAS. In practice, one will solve the problem only once with a poor approximation of the best K_{max}. In that case a superlinear speedup can be obtained.

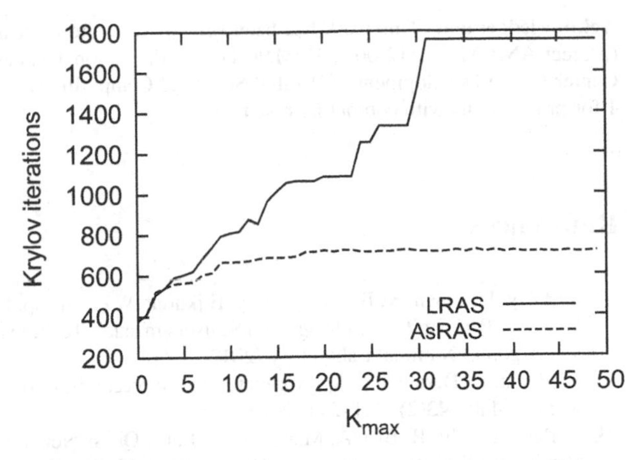

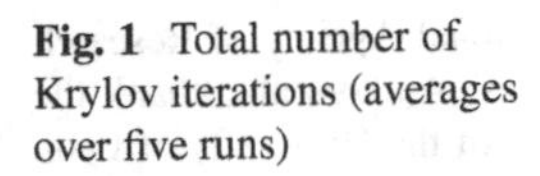
Fig. 1 Total number of Krylov iterations (averages over five runs)

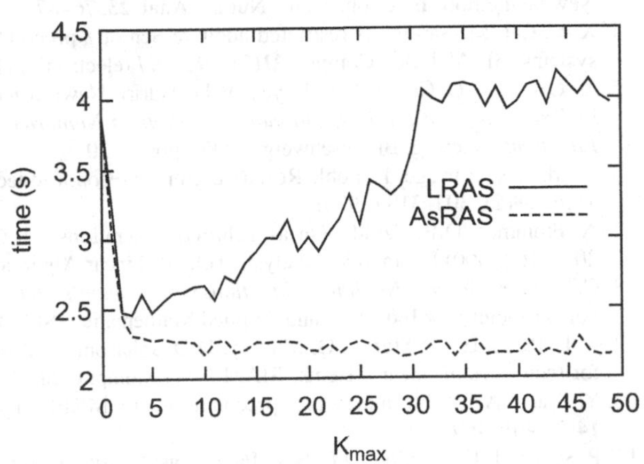

Fig. 2 Wall times in seconds (averages over five runs)

5 Conclusions

The utilization of domain decomposition preconditioners allows us to update only certain parts of the preconditioner, keeping the other ones constant. In the context of parallel computing, this partial update can be computed asynchronously, that is to say that the time-stepper computations are not stopped if the update is not available. Finally, numerical results showed that superlinear speedups can be obtained by adding processes dedicated to the LU factorizations. Furthermore, the preconditioner is continually updated which makes the results less sensitive to the frequency of global update. In this paper, all subdomain parts of the preconditioner are successively updated, but it would be relevant to update more often the LU factorizations associated to subdomains with high local nonlinearities. Then, the AsRAS preconditioner should benefit from a numerical criterion that helps to choose which subdomains need the more an update.

Acknowledgement This work has been supported by the French National Agency of Research (project ANR-MONU12-0012 H2MNO4), and the région Rhône-Alpes. Authors also thank the Center for the Development of Parallel Scientific Computing (CDCSP) of the University of Lyon 1 for providing us with computing resources.

References

1. S. Balay, J. Brown, K. Buschelman, V. Eijkhout, W.D. Gropp, D. Kaushik, M.G. Knepley, L.C. McInnes, B.F. Smith, H. Zhang, PETSc users manual. Technical Report ANL-95/11 - Revision 3.3, Argonne National Laboratory (2012)
2. M. Benzi, D. Bertaccini, Approximate inverse preconditioning for shifted linear systems. BIT Numer. Math. **43**(2), 231–244 (2003)
3. L. Bergamaschi, R. Bru, A. Martínez, M. Putti, Quasi-Newton preconditioners for the inexact Newton method. Electron. Trans. Numer. Anal. **23**, 76–87 (electronic) (2006)
4. X.-C. Cai, M. Sarkis, A restricted additive Schwarz preconditioner for general sparse linear systems. SIAM J. Sci. Comput. **21**(2), 792–797 (electronic) (1999)
5. X.-C. Cai, W.D. Gropp, D.E. Keyes, M.D. Tidriri, Newton-Krylov-Schwarz methods in CFD, in *Proceedings of the International Workshop on Numerical Methods for the Navier-Stokes Equations* (Vieweg, Braunschweig, 1995), pp. 17–30
6. J. Erhel, K. Burrage, B. Pohl, Restarted gmres preconditioned by deflation. J. Comput. Appl. Math. **69**(2), 303–318 (1996)
7. A. Frommer, D.B. Szyld, On asynchronous iterations. J. Comput. Appl. Math. **123**(1–2), 201–216 (2000) [Numerical Analysis, vol. III. Linear Algebra (2000)]
8. C.T. Kelley, *Solving Nonlinear Equations with Newton's Method*. Fundamentals of Algorithms, vol. 1 (Society for Industrial and Applied Mathematics (SIAM), Philadelphia, 2003)
9. M.L. Parks, E. De Sturler, G. Mackey, D.D. Johnson, S. Maiti, Recycling Krylov subspaces for sequences of linear systems. SIAM J. Sci. Comput. **28**(5), 1651–1674 (2006)
10. Y. Saad, A flexible inner-outer preconditioned GMRES algorithm. SIAM J. Sci. Comput. **14**(2), 461–469 (1993)
11. P. Spiteri, J.-C. Miellou, D.E. Baz, Parallel asynchronous Schwarz and multisplitting methods for a nonlinear diffusion problem. Numer. Algorithms **33**(1–4), 461–474 (2003). [International Conference on Numerical Algorithms, vol. I (Marrakesh, 2001)]
12. J.D. Tebbens, M. Tuma, Efficient preconditioning of sequences of nonsymmetric linear systems. SIAM J. Sci. Comput. **29**(5), 1918–1941 (2007)
13. J.D. Tebbens, M. Tuma, Preconditioner updates for solving sequences of linear systems in matrix-free environment. Numer. Linear Algebra Appl. **17**, 997–1019 (2010)
14. H.A. van der Vorst, Bi-CGSTAB: a fast and smoothly converging variant of Bi-CG for the solution of nonsymetric linear systems. SIAM J. Sci. Stat. Comput. **13**, 631–644 (1992)

Coupling Finite and Boundary Element Methods Using a Localized Adaptive Radiation Condition for the Helmholtz's Equation

Y. Boubendir, A. Bendali, and N. Zerbib

1 Introduction

In this paper, we are interested in impenetrable surfaces with relatively large size on which a heterogeneous object of relatively small size is posed. In this case, a straightforward FEM-BEM (finite and boundary element methods) coupling leads to a linear system of very large scale difficult to solve [7]. In this work, we propose an alternative method derived from a modification of the adaptive radiation condition approach [1, 11, 12]. This technique consists of enclosing the computational domain by an artificial truncating surface on which the adaptive radiation condition is posed. This condition is expressed using integral operators acting as a correction term of the absorbing boundary condition. However, enclosing completely the computational domain by an artificial surface in this range leads to problems with very large size, and results in very slow convergence of the iterative procedure. We propose to localize this surface only around the heterogenous region, which will generates a relatively small bounded domain dealt with by a FEM, and suitably coupled with a BEM expressing the solution on the impenetrable surface. The resulting

Y. Boubendir (✉)
Department of Mathematical Sciences and Center for Applied Mathematics and Statistics, NJIT, University Heights. 323 Dr. M. L. King Jr. Blvd, Newark, NJ 07102, USA
e-mail: boubendi@njit.edu

A. Bendali
University of Toulouse, INSA de Toulouse, Institut Mathématique de Toulouse, UMR CNRS 5219, 135 avenue de Rangueil F31077, Toulouse cedex 1, France
e-mail: abendali@insa-toulouse.fr

N. Zerbib
ESI Group, 20 rue du Fonds Pernant, 60471 Compiegne Cedex, France
e-mail: nicolas.zerbib@esi-group.com

T. Dickopf et al. (eds.), *Domain Decomposition Methods in Science and Engineering XXII*, Lecture Notes in Computational Science and Engineering 104, DOI 10.1007/978-3-319-18827-0_45

formulation, based on a particular overlapping domain decomposition method, is solved iteratively where FEM and BEM linear systems are solved separately. The wave problem considered in this paper is stated as follows

$$\begin{cases} \nabla \cdot (\chi \nabla u) + \chi \kappa^2 n^2 u = 0 \quad \text{in } \Omega, \\ \chi \partial_{\mathbf{n}} u = -f \text{ on } \Gamma, \\ \lim_{|x|\to\infty} |x|^{1/2} (\partial_{|x|} u - i\kappa u) = 0, \end{cases} \tag{1}$$

where Ω is the complement of the impenetrable obstacle. We indicate by Ω_1 a bounded domain filled by a possibly heterogeneous material and posed on a slot Γ_{slot} on which are applied the sources producing the radiated wave u. The interface Σ separates Ω_1 from the free propagation domain Ω_0, $\mathbf{n}$ denotes the normal to Γ or to Σ directed outwards respectively the impenetrable obstacle enclosed by Γ or the domain Ω_1 (see Fig. 1), χ and n indicate, respectively, the relative dielectric permittivity and the relative magnetic permeability, and κ is the wave number. Let us note finally that $\chi = n = 1$ in Ω_0. For the sake of presentation, we express problem (1) in the form of the following system

$$\begin{cases} \Delta u_0 + \kappa^2 u_0 = 0 \text{ in } \Omega_0, \\ \partial_{\mathbf{n}} u_0 = 0 \text{ on } \Gamma \cap \partial\Omega_0, \\ \lim_{|x|\to\infty} |x|^{1/2} (\partial_{|x|} u_0 - i\kappa u_0) = 0, \end{cases} \tag{2}$$

$$\begin{cases} \nabla \cdot (\chi \nabla u_1) + \chi \kappa^2 n^2 u_1 = 0 \text{ in } \Omega_1, \\ \chi \partial_{\mathbf{n}} u_1 = -f \text{ on } \Gamma \cap \partial\Omega_1. \end{cases} \tag{3}$$

These boundary-value problems are coupled on Σ through the transmission conditions

$$u_0 = u_1, \quad \partial_{\mathbf{n}} u_0 = \chi \partial_{\mathbf{n}} u_1. \tag{4}$$

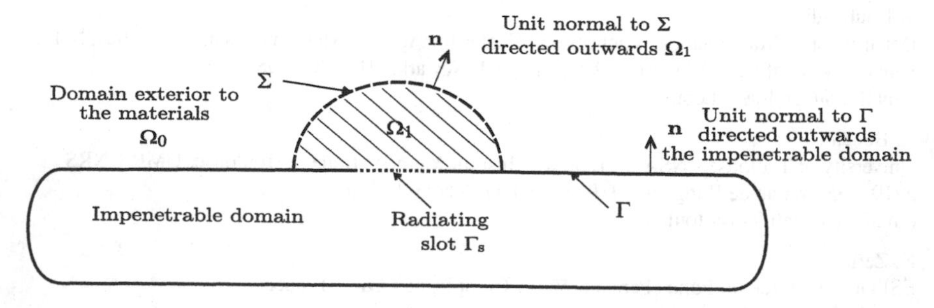

Fig. 1 Non-overlapping decomposition of the exterior domain Ω into Ω_0 and Ω_1

2 The Adaptive Localized Radiation Condition

To localize the truncating interface only around the penetrable material, Fig. 2, we introduce a fictitious boundary S which in turn produces the bounded domain Ω_S limited by S and the impenetrable zone. The goal is to derive a formulation of problem (1) as a coupled system composed of two equations with two unknowns u_0 and u_S where the function $u_S = u|_{\Omega_S}$ is approximated by a FEM, and u_0, already defined above, is computed using an integral equation on Γ_Σ (Fig. 3). The integral representation of the function u_0 is given in terms of a single- and a double-layer potential created by densities on Γ_Σ, and as a result can be seen as the restriction to Ω_0 of the solution of a transmission problem posed on all of the plane $\mathbb{R}^2$ (cf., e.g., [10, 13, 14]). In view of the equations that are set in Ω_S, we are in the case of a particular decomposition with an overlap of the computational domain (see similar ideas in [3, 4] for the usual adaptive radiation condition). However, it will be more convenient not to distinguish u_0 from u_S and to refer to them as the same function u in $H^1_{\text{loc}}(\overline{\Omega})$. Simply by restricting u to Ω_S, we get from (1) that u satisfies

$$\begin{cases} \nabla \cdot (\chi \nabla u) + \chi \kappa^2 n^2 u = 0, & \text{in } \Omega_S, \\ \chi \partial_{\mathbf{n}} u = -f & \text{on } \Gamma \cap \partial \Omega_S. \end{cases} \tag{5}$$

In Ω_0, we use the integral representations of the solutions to the Helmholtz equation satisfying the Sommerfeld radiation condition (cf., e.g., [5, 8, 9, 14])

$$u(x) = V^{*,\Sigma} p(x) - N^{*,\Gamma_\Sigma} u(x), \quad x \in \Omega_0, \tag{6}$$

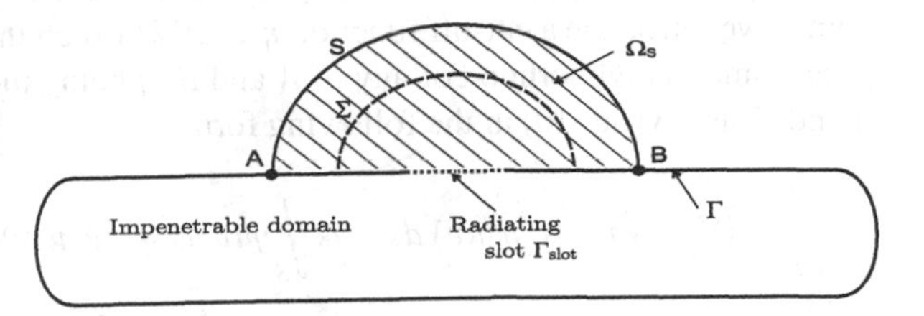

Fig. 2 The bounded domain Ω_S and the fictitious boundary S on which is posed the adaptive radiation condition

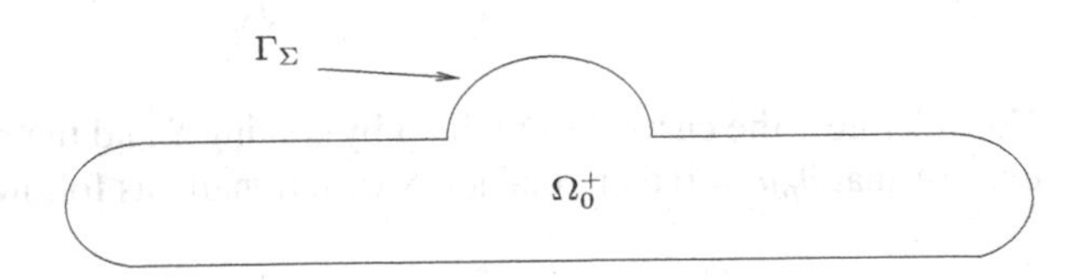

Fig. 3 Representation of Ω_0^+ and its boundary Γ_Σ

with

$$V^{*,\Sigma}p(x) = \int_{\Sigma} G(x,y)p(y)\, ds_y \tag{7}$$

$$p = -\chi \partial_{\mathbf{n}} u|_{\Sigma} \tag{8}$$

$$N^{*,\Gamma_\Sigma}u(x) = -\int_{\Gamma_\Sigma} \partial_{n_y} G(x,y)u(y)\, ds_y \tag{9}$$

where $G(x,y) = (i/4)H_0^{(1)}(\kappa|x-y|)$ for $x \neq y \in \mathbb{R}^2$.

The derivation of the FEM-BEM coupling procedure can be introduced starting from the following Green formula

$$\int_{\Omega_S} \chi \left(\nabla u \cdot \nabla v - \kappa^2 n^2 uv\right) dx = \langle \partial_{\mathbf{n}} u, v\rangle_{\tilde{H}^{-1/2}(S),H^{1/2}(S)} + \langle f, v\rangle_{\tilde{H}^{-1/2}(\Gamma_{\text{slot}}),H^{1/2}(\Gamma_{\text{slot}})} \tag{10}$$

where $\langle \cdot,\cdot\rangle_{\tilde{H}^{-1/2}(S),H^{1/2}(S)}$ denotes the duality pairing between $\tilde{H}^{-1/2}(S)$ and $H^{1/2}(S)$, and v is an arbitrary test function in $H^1_{\text{loc}}(\overline{\Omega})$. The space $\tilde{H}^{-1/2}(S)$ is defined similarly to $\tilde{H}^{-1/2}(\Gamma_{\text{slot}})$ (cf. [6, 13] for the definition of Sobolev spaces).

The localized adaptive radiation condition approach (LRC) uses an iterative method to solve problem (10) where the term $\partial_{\mathbf{n}} u$, at the right-hand side, is updated at each iteration. However, there is no guarantee that problem (10) can be safely solved. To avoid these kinds of difficulties, we introduce the stabilization term $-i\kappa \int_S uv\, ds$ in both sides of (10). On the other hand, S is an open curve having its end-points A and B on Γ (see Fig. 2). To prevent singular integrals near these points, we introduce a cut-off function $\eta \in \mathcal{D}(\mathbb{R}^2)$ such that $0 \leq \eta \leq 1$, $\eta = 1$ on S, except small neighborhood of any of A and B, η being moreover equal to 0 around A and B, and write (10) in the following form

$$\int_{\Omega_S} \chi \left(\nabla u \cdot \nabla v - \kappa^2 n^2 uv\right) dx - i\kappa \int_S \eta uv\, ds = \langle \partial_{\mathbf{n}} u, v\rangle_{\tilde{H}^{-1/2}(S),H^{1/2}(S)} - i\kappa \int_S \eta uv\, ds + \langle f, v\rangle_{\tilde{H}^{-1/2}(\Gamma_{\text{slot}}),H^{1/2}(\Gamma_{\text{slot}})} . \tag{11}$$

Consider now the curve Γ_S obtained by joining S and the part of Γ outside Ω_S and express that $\partial_{\mathbf{n}} u = 0$ there outside S variationally as follows

$$\langle \partial_{\mathbf{n}} u, v\rangle_{\tilde{H}^{-1/2}(S),H^{1/2}(S)} = \langle \partial_{\mathbf{n}} u, v\rangle_{H^{-1/2}(\Gamma_S),H^{1/2}(\Gamma_S)} , \tag{12}$$

for all test function v. We then get

$$\int_{\Omega_S} \chi\left(\nabla u \cdot \nabla v - \kappa^2 n^2 uv\right) dx - i\kappa \int_S \eta uv \, ds = \\ \langle \partial_{\mathbf{n}} u, v \rangle_{H^{-1/2}(\Gamma_S), H^{1/2}(\Gamma_S)} - i\kappa \int_S \eta uv \, ds + \langle f, v \rangle_{\tilde{H}^{-1/2}(\Gamma_{\mathrm{slot}}), H^{1/2}(\Gamma_{\mathrm{slot}})} \tag{13}$$

where the traces in the right-hand side are expressed from the integral representation (6) of u

$$\begin{cases} \eta u|_S = \eta V^{S,\Sigma} p - \eta N^{S,\Gamma_\Sigma} u \\ \partial_{\mathbf{n}} u|_{\Gamma_S} = \partial_{\mathbf{n}} V^{\Gamma_S,\Sigma} p - \partial_{\mathbf{n}} N^{\Gamma_S,\Gamma_\Sigma} u. \end{cases} \tag{14}$$

Clearly, since Σ and S share no common point and η is zero in the proximity of the end-points of S, if p and u are sufficiently smooth functions, say for example continuous, only the integral corresponding to $\partial_{\mathbf{n}} N^{*,\Gamma_\Sigma} u$ in (14) is an improper integral which can be expressed by means of a weakly singular kernel as follows

$$\begin{aligned} \left\langle \partial_{\mathbf{n}} N^{\Gamma_S,\Gamma_\Sigma} u, v \right\rangle_{H^{-1/2}(\Gamma_S), H^{1/2}(\Gamma_S)} &= \left\langle \partial_s v, V^{\Gamma_S,\Gamma_\Sigma} \partial_s u \right\rangle_{H^{-1/2}(\Gamma_S), H^{1/2}(\Gamma_S)} \\ &\quad - \kappa^2 \left\langle v\boldsymbol{\tau}, V^{\Gamma_S,\Gamma_\Sigma} (u\boldsymbol{\tau}) \right\rangle_{H^{-1/2}(\Gamma_S), H^{1/2}(\Gamma_S)} \end{aligned} \tag{15}$$

from a slight adaptation of the case where $\Gamma_S = \Gamma_\Sigma$ (cf., e.g., [10], p. 5). The superscripts in the integral operators indicate that they correspond to a potential created by a density on Γ_Σ and evaluated on Γ_S, and $\boldsymbol{\tau}$ is the unit tangent vector pointing in the growth direction of the arc length s.

In order to be able to use a nodal approximation of (13), we use a standard technique for gluing finite element approximations of different kinds or associated with non-conforming meshes generally called mortar FEM (cf., e.g., [2]). It is worth mentioning that here only standard meshes and finite element methods of the same kind are used. This way to proceed is just considered as a tool providing an approximation for the additional unknown p in the framework of a nodal finite element method. This technique consists in breaking the continuity across Σ that u is compelled to satisfy a priori and to express it as a constraint. The Lagrange multiplier corresponding to this constraint will be precisely the unknown p. It is hence more convenient to denote by separate symbols: u_0 for the restriction of u to $\Omega_0 \cap \Omega_S$ and Γ_Σ and u_1 for its restriction to Ω_1. More precisely, we will use the following functional framework

$$\begin{cases} X_0 = \{u_0 \text{ defined (a.e.) on } \Omega_0 \cap \Omega_S \text{ and } \Gamma_\Sigma; \\ \qquad\qquad \exists U \in H^1(\Omega_0),\, U|_{\Gamma_\Sigma} = u_0|_{\Gamma_\Sigma} \text{ and } U|_{\Omega_0 \cap \Omega_S} = u_0|_{\Omega_0 \cap \Omega_S}\} \\ X_1 = H^1(\Omega_1), \quad X = X_0 \times X_1, \end{cases} \tag{16}$$

relation (8) and (13) to write

$$\begin{aligned}&\int_{\Omega_S\cap\Omega_0}\left(\nabla u_0\cdot\nabla v_0-\kappa^2 u_0 v_0\right)dx - i\kappa\int_S \eta u_0 v\, ds\\ &\quad+\int_{\Omega_1}\chi\left(\nabla u_1\cdot\nabla v_1-\kappa^2 n^2 u_1 v_1\right)dx+\langle p, v_1-v_0\rangle_{\tilde{H}^{-1/2}(\Sigma),H^{1/2}(\Sigma)}=\\ &\quad\langle\partial_{\mathbf{n}}u, v_0\rangle_{H^{-1/2}(\Gamma_S),H^{1/2}(\Gamma_S)} - i\kappa\int_S \eta u v_0\, ds+\langle f, v_1\rangle_{\tilde{H}^{-1/2}(\Gamma_{\text{slot}}),H^{1/2}(\Gamma_{\text{slot}})}\end{aligned}$$

for all $(v_0, v_1) \in X$. Using then the integral representation of $\partial_{\mathbf{n}}u|_{\Gamma_S}$ and $u|_S$ given above in (14), we readily arrive to the formulation effectively used to solve problem (1) numerically

$$\begin{cases}(u,p)\in X\times M,\ \forall\,(v,q)\in X\times M\\ a(u,v)+d(u_0,v_0)+b(p,v)+r(p,v_0)=\langle f,v_1\rangle_{\tilde{H}^{-1/2}(\Gamma_{\text{slot}}),H^{1/2}(\Gamma_{\text{slot}})}\\ b(q,u)=0\end{cases}\tag{17}$$

with the following notation

$$\begin{cases}a_0(u_0,v_0)=\displaystyle\int_{\Omega_S\cap\Omega_0}\left(\nabla u_0\cdot\nabla v_0-\kappa^2 u_0 v_0\right)dx - i\kappa\int_S \eta u_0 v_0 ds,\\ a_1(u_1,v_1)=\displaystyle\int_{\Omega_1}\chi\left(\nabla u_1\cdot\nabla v_1-\kappa^2 n^2 u_1 v_1\right)dx,\\ a(u,v)=a_0(u_0,v_0)+a_1(u_1,v_1),\\ d(u_0,v_0)=\left\langle\partial_{\mathbf{n}}N^{\Gamma_S,\Gamma_\Sigma}u_0,v_0\right\rangle_{H^{-1/2}(\Gamma_S),H^{1/2}(\Gamma_S)}-i\kappa\displaystyle\int_S\eta v_0 N^{S,\Sigma}u_0 ds\\ r(p,v_0)=-\displaystyle\int_{\Gamma_S}v_0\partial_{\mathbf{n}}V^{\Gamma_S,\Sigma}p\, ds+i\kappa\int_S\eta v_0 V^{S,\Sigma}p\, ds,\\ b(p,v)=\langle p,v_1-v_0\rangle_{\tilde{H}^{-1/2}(\Sigma),H^{1/2}(\Sigma)},\end{cases}\tag{18}$$

and $M = \tilde{H}^{-1/2}(\Sigma)$. We refer to [6] for the analysis of the well-posedness and the stability of (17).

3 Numerical Results

To validate the LRC method, we will compare it with a direct FEM-BEM coupling and a domain decomposition one noted P-DDM (see [6] for more details about this method). The reference solution will be given by BE formulation (boundary elements) known to be the less dispersive. The geometry considered here (Fig. 4) depends on a parameter L used to set a large size for the impenetrable domain relatively to the zone meshed in triangles as shown in Fig. 4. By varying this parameter, we test each numerical technique in terms of accuracy, CPU time, and convergence for the iterative ones. The lengths are expressed in wavelength units.

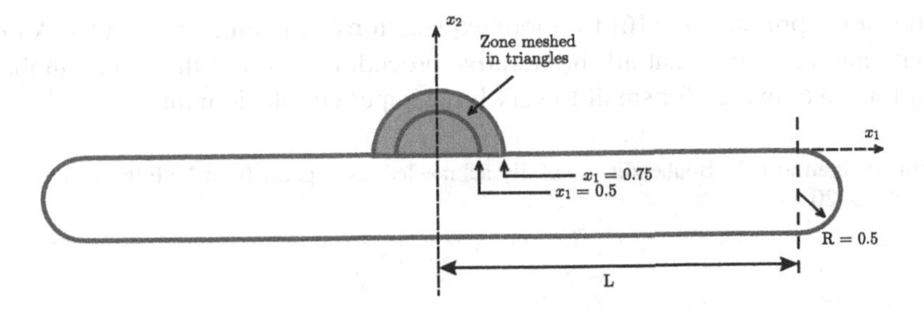

Fig. 4 Geometry of the test-case

Table 1 Comparison of the various formulations in terms of accuracy, CPU time, and number of iterations

	L	CPU	$\mathcal{E}$	Iter
BE	4	3	–	–
	50	94	–	–
FEBE	4	19	0.14	–
	50	169	0.14	–
LRC	4	38	0.17	11
	50	148	0.23	11
P-DDM	4	13	0.14	30
	50	127	0.14	30

To be able to compare the LRC formulation with the BE one, we suppose χ and n constant in Ω_1. More precisely, we choose $\chi = 1/4$ and $n = 2(1+i)$, which correspond to a magnetic material in electromagnetism. The sources are located on the segment $\{x_2 = 0, -0.25 < x_1 < 0.25\}$ and are given by the Gaussian function $f(x_1) = -\exp(-(10x_1)^2)$.

The mesh used is of 20 points by wavelength in the free propagation zone and 15 points by wavelength in the material for the FEM-BEM formulations. The BE formulation is meshed using 20 points by wavelength for both the free propagation zone and the material. All the iterative methods are solved using the GMRES algorithm (cf., e.g. [15]). The first test concerns the case of a moderately elongated impenetrable domain corresponding to $L = 4$ and the second, much more elongated, is obtained for $L = 50$. Table 1 summarizes the numerical for each method in terms of accuracy and CPU time. For the iterative methods, we also compute the iteration number, noted "Iter" in Table 1, obtained by reducing the residual by a factor 10^{-6}. To measure the accuracy, we use the quantity $\mathcal{E} = \max |s(\theta) - s_{\mathrm{BE}}(\theta)|$ where $s_{\mathrm{BE}}(\theta)$ is the far field computed by the BE approach.

The results reported in Table 1 confirm the robustness of the LRC formulation, it keeps the accuracy of the FEBE and P-DDM approaches. The CPU time used by the different methods also clearly shows the advantage of decoupling the solution of the sparse and the dense parts in the problem. Even if usual DDMs exhibit the same efficiency in terms of number of iterations and accuracy, their related iterative procedures may break down if the corresponding boundary-value problems set in the interior domain Ω_1 present a resonance at the considered frequency, contrary to the LRC approach, see [6] for more explanations and numerical results. Another remarkable feature is that all the iterative procedures require the same number of iterations to converge for small to very large impenetrable domains.

Acknowledgement Y. Boubendir gratefully acknowledges support from NSF through grant No. DMS-1319720.

References

1. S. Alfonzetti, G. Borzì, N. Salerno, Iteratively-improved Robin boundary conditions for the finite element solution of scattering problems in unbounded domains. Int. J. Numer. Methods Eng. **42**, 601–629 (1998)
2. F.B. Belgacem, The mortar finite element methodwith Lagrange multipliers. Numer. Math. **84**, 173–197 (1999)
3. F.B. Belgacem, M. Fournié, N. Gmati, F. Jelassi, On the Schwarz algorithms for the elliptic exterior boundary value problems. M2AN. Math. Model. Numer. Anal. **39**(4), 693–714 (2005)
4. F. Ben Belgacem, N. Gmati, F. Jelassi, Convergence bounds of GMRES with Schwarz' preconditioner for the scattering problem. Int. J. Numer. Methods Eng. **80**, 191–209 (2009)
5. A. Bendali, M. Fares, *Boundary Integral Equations Methods in Acoustics* (Saxe-Coburg Publications, Kippen, Stirlingshire, 2008)
6. A. Bendali, Y. Boubendir, N. Zerbib, Localized adaptive radiation condition for coupling boundary with finite element methods applied to wave propagation problems. IMA Numer. Anal. **34**(3), 1240–1265 (2014)
7. Y. Boubendir, A. Bendali, M. Fares, Coupling of a non-overlapping domain decomposition method for a nodal finite element method with a boundary element method. Int. J. Numer. Methods Eng **73**(11), 1624–1650 (2008)
8. D. Colton, R. Kress, *Integral Equation Methods in Scattering Theory* (Wiley, New York, 1983)
9. D. Colton, R. Kress, *Inverse Acoustic and Electromagnetic Scattering Theory*. Series in Applied Mathematics, vol. 93 (Springer, New York, 1992)
10. G.C. Hsiao, W.L. Wendland, *Boundary Iintegral Equations* (Springer, Berlin, 2008)
11. J. Jin, *The Finite Element Method in Electromagnetics*, 2nd edn. (Wiley, New York, 2002)
12. Y. Li, Z. Cendes, High-accuracy absorbing boundary conditions. IEEE Trans. Magn. **31**, 1524–1529 (1995)
13. W. McLean, *Strongly Elliptic Systems and Boundary Integral Equations* (Cambridge University Press, Cambridge/New York, 2000)
14. J.-C. Nédélec, *Acoustic and Electromagnetic Equations: Integral Representations for Harmonic Problems* (Springer, Berlin, 2001)
15. Y. Saad, *Iterative Methods for Sparse Linear Systems* (PWS Publishing Company, Boston, 1996)

Simulating Flows Passing a Wind Turbine with a Fully Implicit Domain Decomposition Method

Rongliang Chen, Zhengzheng Yan, Yubo Zhao, and Xiao-Chuan Cai

1 Introduction

Wind power is an increasingly popular renewable energy. In the design process of the wind turbine blade, the accurate aerodynamic simulation is important. In the past, most of the wind turbine simulations were carried out with some low fidelity methods, such as the blade element momentum method [9]. Recently, with the rapid development of the supercomputers, high fidelity simulations based on 3D unsteady Navier-Stokes (N-S) equations become more popular. For example, Sorensen et al. studied the 3D wind turbine rotor using the Reynolds-Averaged Navier-Stokes (RANS) framework where a finite volume method and a semi-implicit method are used for the spatial and temporal discretization, respectively [17]. Bazilevs et al. investigated the aerodynamic of the NREL 5 MW offshore baseline wind turbine rotor using large eddy simulation built with a deforming-spatial-domain/stabilized space-time formulation [3, 11] and later extended the simulation to the full wind turbine including both the rotor and the tower [10]. Li et al. conducted dynamic overset CFD simulations for the NREL phase VI wind turbine using RANS and detached eddy models [15].

In this paper, we study a scalable parallel method based on the 3D unsteady incompressible N-S equations and its application to a NREL S-series wind turbine with realistic geometry and Reynolds number. In this simulation, the main

R. Chen (✉) • Z. Yan • Y. Zhao
Shenzhen Institutes of Advanced Technology, Chinese Academy of Sciences, Shenzhen 518055, P.R. China
e-mail: rl.chen@siat.ac.cn; zz.yan@siat.ac.cn; yb.zhao@siat.ac.cn

X.-C. Cai
Department of Computer Science, University of Colorado Boulder, Boulder, CO 80309, USA
e-mail: cai@cs.colorado.edu

T. Dickopf et al. (eds.), *Domain Decomposition Methods in Science and Engineering XXII*, Lecture Notes in Computational Science and Engineering 104, DOI 10.1007/978-3-319-18827-0_46

challenges are: (1) the moving of the computation domain because of the rotation of the rotor; (2) the complex geometry; (3) the large computational meshes; and (4) the high nonlinearity resulting from high Reynolds number. To answer these challenges, an Arbitrary-Lagrange-Eulerian (ALE) method is used to handle the mesh movement, an unstructured tetrahedron mesh with a stabilized finite element method and a fully implicit backward difference scheme are employed to discretize the N-S equations [6, 18] and a parallel Newton-Krylov-Schwarz (NKS) method [5, 12] is used to solve the large sparse nonlinear system at each time step. In NKS, an inexact Newton method with analytic Jacobian is employed as the nonlinear solver, a Krylov subspace method is used as the linear Jacobian system solver in the Newton steps, and an overlapping domain decomposition method is used as a preconditioner to accelerate the convergence of the linear solver [4, 14]. For the rotor-only simulation, one can either fix the computation domain and apply a given velocity on the surface of the rotor, or let the domain move with the rotating rotor and apply a no-slip boundary condition on the rotor surface. We choose the latter one in this paper. We mainly focus on the solution method, including the robustness and parallel scalability.

The rest of the paper is organized as follows. In Sect. 2, we briefly introduce the governing equations and their discretization. In Sect. 3, the Newton-Krylov-Schwarz algorithm is discussed, and some numerical results are presented in Sect. 4. In Sect. 5, we draw some conclusions.

2 Governing Equations and a Fully Implicit Discretization

We model the flow around the wind turbine using the 3D unsteady incompressible N-S equations. Since the computational domain moves during the simulation, a moving mesh method is introduced to handle the change of the flow domain. In this paper, we use the ALE method. Let $\mathbf{Y}$ be the ALE coordinate, $\mathbf{X}$ the Eulerian coordinate. Then the N-S equations read as [7]:

$$\begin{aligned} \rho\left(\left.\frac{\partial \mathbf{u}}{\partial t}\right|_{\mathbf{Y}} + (\mathbf{u}-\omega)\cdot\nabla\mathbf{u}\right) + \nabla\cdot\sigma &= \mathbf{f} \ \text{ in } \Omega^t, \\ \nabla\cdot\mathbf{u} &= 0 \ \text{ in } \Omega^t, \\ \mathbf{u} &= \mathbf{g} \ \text{ on } \Gamma_{inlet} \\ \sigma\cdot\mathbf{n} &= \mathbf{0} \ \text{ on } \Gamma_{outlet}, \\ \mathbf{u} &= \mathbf{0} \ \text{ on } \Gamma_{wall}, \\ \mathbf{u} &= \mathbf{u_0} \ \text{ in } \Omega^t \text{ at } t=0, \end{aligned} \tag{1}$$

where Ω^t is the computational domain at time t and $\omega = \frac{\partial \mathbf{x}}{\partial t}$ is the velocity of the rotating flow domain which is equal to the rotor speed since we let the whole computation domain rotate with the rotor. $\sigma = -p\mathbf{I} + \mu(\nabla\mathbf{u} + (\nabla\mathbf{u})^T)$ is the Cauchy stress tensor. $\mathbf{u}$ and p are the velocity and pressure of the flow. ρ and μ are the density and viscosity of the fluid, respectively. $\mathbf{f}$ refers to the source term and $\mathbf{g}$ is a given function defined at the inlet boundary. $\mathbf{u_0}$ is a given initial condition which is zero in our test cases. Γ_{inlet}, Γ_{outlet} and Γ_{wall} refer to the inlet, outlet and wall boundaries, respectively.

A $P1 - P1$ finite element method is used to discretize (1) on an unstructured tetrahedral mesh $\mathcal{T}^h = \{K\}$. Since this finite element method is not stable for the N-S equations because it does not satisfy the Ladyzenskaja-Babuska-Brezzi (LBB) condition, additional stabilization terms are needed in the formulation as described in [2]. We denote the finite element spaces of the trial and weighting functions for the velocity and pressure as $\mathcal{U}^h$, $\mathcal{U}^{0,h}$, and $\mathcal{P}^h$, respectively. Then the semi-discrete stabilized finite element formulation of (1) is given as follows: Find $\mathbf{u}^h \in \mathcal{U}^h$ and $p^h \in \mathcal{P}^h$, such that for any $\Phi^h \in \mathcal{U}^{0,h}$ and $\psi^h \in \mathcal{P}^h$,

$$\mathbf{B}^h(\mathbf{u}^h, p^h; \Phi^h, \psi^h) - \mathbf{F}^h(\Phi^h, \psi^h) = \mathbf{0}, \tag{2}$$

where $\mathbf{u}^h$, p^h are the nodal values of the velocity and pressure functions, φ^h and each of the three components of Φ^h are the basis functions which are piecewise continuous linear functions, and

$$\begin{aligned}
\mathbf{B}^h(\mathbf{u}^h, p^h; \Phi^h, \psi^h) = \rho \int_{\Omega^t} \left.\frac{\partial \mathbf{u}^h}{\partial t}\right|_{\mathbf{Y}} \cdot \Phi^h d\Omega^t + \mu \int_{\Omega^t} \nabla \mathbf{u}^h : \nabla \Phi^h d\Omega^t \\
+\rho \int_{\Omega^t} ((\mathbf{u}^h - \omega) \cdot \nabla)\mathbf{u}^h \cdot \Phi^h d\Omega^t - \int_{\Omega^t} p^h \nabla \cdot \Phi^h d\Omega^t \\
+ \int_{\Omega^t} (\nabla \cdot \mathbf{u}^h)\varphi^h d\Omega^t + \sum_{K \in \mathcal{T}} (\nabla \cdot \mathbf{u}^h,\ \tau_c \nabla \cdot \Phi^h)_K \\
+ \sum_{K \in \mathcal{T}} \left(\left.\frac{\partial \mathbf{u}^h}{\partial t}\right|_{\mathbf{Y}} + ((\mathbf{u}^h - \omega) \cdot \nabla)\mathbf{u}^h + \nabla p^h,\ \tau_m(\mathbf{u}^h \cdot \nabla \Phi^h + \nabla \varphi^h) \right)_K,
\end{aligned}$$

$$\mathbf{F}^h(\Phi^h, \psi^h) = \int_{\Omega^t} \mathbf{f} \cdot \Phi^h d\Omega^t + \sum_{K \in \mathcal{T}} (\mathbf{f},\ \tau_m(\mathbf{u}^h \cdot \nabla \Phi^h + \nabla \varphi^h))_K.$$

Here the parameters τ_c and τ_m are defined as in [2].

For the temporal discretization, we use an implicit backward finite difference formula with a fixed time step size Δt. In the implicit method, one needs to solve a nonlinear system at each time step (the nth time step):

$$\mathbf{F}^n(\mathbf{U}^n) = \mathbf{0}, \tag{3}$$

to obtain the solution of the nth time step $\mathbf{U}^n$, which consists of the nodal values of the velocity and pressure.

3 Monolithic Newton-Krylov-Schwarz Algorithm

In most N-S solvers, such as the projection methods, the operator is split into the velocity component and pressure component, and the algorithm takes the form of a nonlinear Gauss-Seidel iteration with two large blocks. In the monolithic approach that we consider in this paper, the velocity and pressure variables associated with a grid point stay together throughout the computation. In this approach, the two critically important ingredients, namely the monolithic Schwarz preconditioner, and the robustness and scalability are realized with the point-block ILU based subdomain solver.

The nonlinear system (3) is solved by a Newton-Krylov-Schwarz method which reads as

(a) *Let* $\mathbf{U}^0$ *be the given initial condition and set* $n = 0$
(b) *For* $n = 1, 2, \cdots$ *, do*

- *Using an initial guess* $\mathbf{U}_0^n = \mathbf{U}^{n-1}$ *and set* $k = 0$
- *Move the computational domain* ($\Omega^{n-1} \rightarrow \Omega^n$) *and the mesh* $\mathcal{T}_h^n$ *(the coordinate of each mesh point at the current time step* $\mathbf{x}^n$ *is obtained by rotating the initial mesh* $\mathbf{x}^0$*):*

$$\mathbf{x}^n = \begin{bmatrix} cos(\omega n \Delta t) & -sin(\omega n \Delta t) & 0 \\ sin(\omega n \Delta t) & cos(\omega n \Delta t) & 0 \\ 0 & 0 & 1 \end{bmatrix} \mathbf{x}^0$$

- *For* $k = 1, 2, \cdots$ *, until converges, do*
  - *Find* d_k^n *such that*

$$\| \nabla \mathbf{F}^n(\mathbf{U}_{k-1}^n)(\mathbf{M}_k^n)^{-1}(\mathbf{M}_k^n \mathbf{d}_k^n) + \mathbf{F}^n(\mathbf{U}_{k-1}^n) \| \leq \eta \| \mathbf{F}^n(\mathbf{U}_{k-1}^n) \| \tag{4}$$

  - *Set* $\mathbf{U}_k^n = \mathbf{U}_{k-1}^n + \tau_k^n \mathbf{d}_k^n$
- *Set* $\mathbf{U}^n = \mathbf{U}_k^n$

Here $(\mathbf{M}_k^n)^{-1}$ is an additive Schwarz preconditioner to be defined shortly, ω is the angular speed of the rotor, and η is the relative tolerance for the linear solver [8]. Note that, in the wind turbine simulation, we simply let the whole computational domain rotate at the same angular speed as the rotor, so the current mesh can be obtained by rotating the initial mesh and the connectivity of the mesh does not change. For simplicity, we ignore the scripts n and k for the rest of the paper.

In NKS, the most difficult and time-consuming step is the solution of the large, sparse, and nonsymmetric Jacobian system (4) by a preconditioned GMRES method. In the Jacobian solver, the most important component is the preconditioner, without which GMRES doesn't converge or converges very slowly, and a good preconditioner accelerates the convergence significantly. In this paper, we use an overlapping restricted additive Schwarz preconditioner introduced in [4]:

$$\mathbf{M}_{RAS} = \sum_{l=1}^{n_p} (R_l^0)^T \mathbf{J}_l^{-1} R_l^\delta, \tag{5}$$

where $\mathbf{J}_l$ is the local Jacobian matrix defined on the overlapping subdomain, n_p is the number of subdomains, R_l^δ and R_l^0 are the restriction operators from the whole domain to the overlapping and non-overlapping subdomain, respectively. In practice, we only need the application of $\mathbf{J}_l^{-1}$ to a given vector, which can be obtained by solving a subdomain linear system. Since $\mathbf{J}_l^{-1}$ is used as a preconditioner here, the subdomain linear system can be solved exactly or approximately by using LU factorization or incomplete LU factorization (ILU) in the point-block format [16].

4 Numerical Experiments

In this section, we report some numerical experiments using the proposed algorithm. Our solver is implemented on top of the Portable Extensible Toolkit for Scientific computation (PETSc) [1]. The computations are carried out on the Dawning Nebulae supercomputer at the China National Supercomputer Center at Shenzhen. The geometry of the wind turbine is provided by GrabCAD[1] (we scale the size to that of a 5 MW wind turbine) and meshed by ANSYS; see Fig. 1 for details. The mesh partitions for the additive Schwarz preconditioner are obtained with

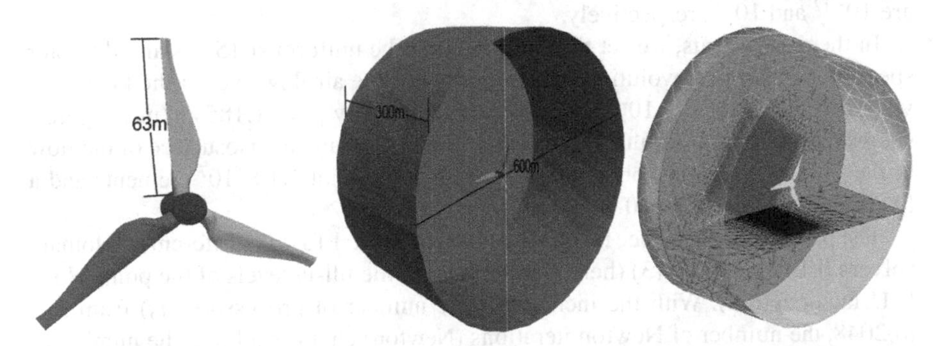

Fig. 1 A three-blade wind turbine with NREL S807 root region airfoil and NREL S806 tip region airfoil from GrabCAD (*left*), the computational domain (*mid*), and the computational mesh (*right*)

[1] www.grabcad.com.

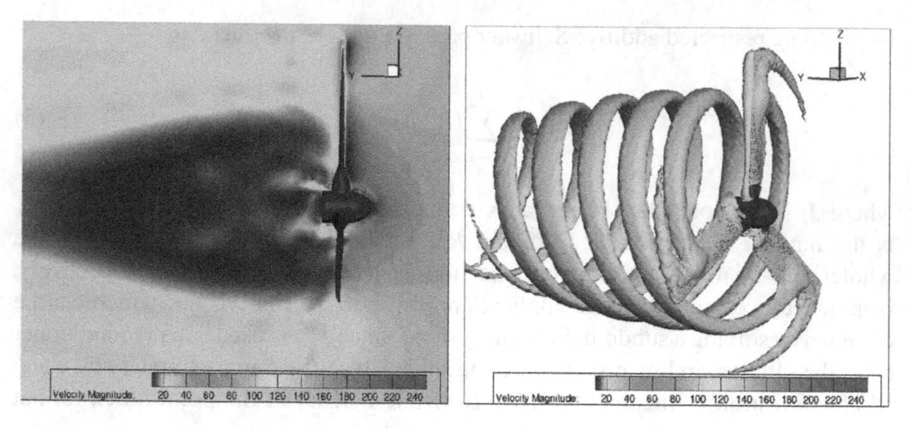

Fig. 2 The velocity distribution (*left*) and the isosurface (*right*) of the simulation

Table 1 Parallel performance of the algorithm

n_p	ILU(3)			ILU(2)		
	Newton	GMRES	Time(s)	Newton	GMRES	Time(s)
512	3.0	51.72	127.3	3.0	64.02	75.0
1024	3.0	52.77	77.7	3.0	66.40	45.8
1536	3.1	53.94	67.5	3.0	67.60	35.6
2048	3.0	57.42	53.0	3.0	67.75	29.3

Here the degrees of freedom (DOF) is about 8.4×10^6 and the overlapping size is 4. The "Time" refers to the average compute time in seconds at each time step

ParMETIS [13]. The relative stopping conditions for the nonlinear and linear solvers are 10^{-12} and 10^{-6}, respectively.

In the experiments, we set the wind speed to be uniform at 15 m/s and the rotor speed to be 22 rpm (revolutions per minute). For the air flow, we set the kinematic viscosity $\mu = 1.831 \times 10^{-5}$ kg/(ms) and the density $\rho = 1.185$ kg/m^3. Figure 2 shows the simulation results: the velocity distribution and the isosurface of the flow at $t = 10.0$ s, which are obtained on a mesh with about 1.1×10^7 elements and a fixed time step size $\Delta t = 0.01$ s.

The parallel performance results are given in Table 1 for two different subdomain solvers ILU(2) and ILU(3) (here 2 and 3 refer to the fill-in levels of the point-block ILU factorization). With the increase of the number of processors (n_p) from 512 to 2048, the number of Newton iterations (Newton) changes a little, the number of GMRES iterations (GMRES) increases reasonably, and the compute time (Time) decreases. These results show that the algorithm scales well when n_p is around 1024 or less and the efficiency reduces with the increase of n_p, which is reasonable because we use a one-level method. The result also suggests that for large number of processors, in order to obtain a good scalability, multilevel methods are necessary.

5 Concluding Remarks

A domain decomposition based fully implicit parallel algorithm for the numerical simulation of the flow around a wind turbine rotor was introduced and studied in this paper. The algorithm begins with a fully implicit discretization of the unsteady incompressible N-S equations on a moving unstructured mesh with a stabilized finite element method, then an inexact Newton method is employed to solve the large nonlinear system at each time step, and a preconditioned GMRES method is employed to solve the linear Jacobian system in each Newton step with a one-level restricted additive Schwarz preconditioner. We tested the algorithm for a flow around a 5 MW wind turbine with more than eight million degrees of freedom on a supercomputer with up to 2048 processors. The algorithm scales well when the number of processors is around 1024 or less. We plan to develop a multilevel version of the algorithm in order to obtain better scalability results when the number of processors is larger.

Acknowledgement The research was supported in part by the NSF of China under 11401564, the Chinese National 863 Plan Program under 2015AA01A302, the Knowledge Innovation Program of the Chinese Academy of Sciences under KJCX2-EW-L01 and the Shenzhen Peacock Plan (China) under KQCX20130628112914303.

References

1. S. Balay, S. Abhyankar, M.F. Adams, J. Brown, P. Brune, K. Buschelman, V. Eijkhout, W.D. Gropp, D. Kaushik, M.G. Knepley, L.C. McInnes, K. Rupp, B.F. Smith, H. Zhang, PETSc users manual. Technical Report, Argonne National Laboratory (2014)
2. Y. Bazilevs, V. Calo, T. Hughes, Y. Zhang, Isogeometric fluid-structure interaction: theory, algorithms, and computations. Comput. Mech. **43**, 3–37 (2008)
3. Y. Bazilevs, M. Hsu, I. Akkerman, S. Wright, K. Takizawa, B. Henicke, T. Spielman, T. Tezduyar, 3D simulation of wind turbine rotors at full scale. Part I: geometry modeling and aerodynamics. Int. J. Numer. Methods Fluids **65**, 207–235 (2011)
4. X.-C. Cai, M. Sarkis, A restricted additive Schwarz preconditioner for general sparse linear systems. SIAM J. Sci. Comput. **21**, 792–797 (1999)
5. X.-C. Cai, W. Gropp, D. Keyes, R. Melvin, D. Young, Parallel Newton-Krylov-Schwarz algorithms for the transonic full potential equation. SIAM J. Sci. Comput. **19**, 246–265 (1998)
6. R. Chen, Q. Wu, Z. Yan, Y. Zhao, X.-C. Cai, A parallel domain decomposition method for 3D unsteady incompressible flows at high Reynolds number. J. Sci. Comput. **58**, 275–289 (2014)
7. F. Duarte, R. Gormaz, S. Natesan, Arbitrary Lagrangian-Eulerian method for Navier-Stokes equations with moving boundaries. Comput. Methods Appl. Mech. Eng. **45**, 4819–4836 (2004)
8. S. Eisenstat, H. Walker, Choosing the forcing terms in an inexact Newton method. SIAM J. Sci. Comput. **17**, 16–32 (1996)
9. M. Hansen, *Aerodynamics of Wind Turbines*, 2nd edn. (Erthscan, London, 2008)
10. M. Hsu, Y. Bazilevs, Fluid-structure interaction modeling of wind turbines: simulating the full machine. Comput. Mech. **50**, 821–833 (2012)
11. M. Hsu, I. Akkerman, Y. Bazilevs, High-performance computing of wind turbine aerodynamics using isogeometric analysis. Comput. Fluids **49**, 93–100 (2011)

12. F.-N. Hwang, C.-Y. Wu, X.-C. Cai, Numerical simulation of three-dimensional blood flows using domain decomposition method on parallel computer. J. Chin. Soc. Mech. Eng. **31**, 199–208 (2010)
13. G. Karypis, METIS/ParMETIS webpage, University of Minnesota (2013), http://glaros.dtc.umn.edu/gkhome/views/metis
14. A. Klawonn, L. Pavarino, Overlapping Schwarz methods for mixed linear elasticity and Stokes problems. Comput. Methods Appl. Mech. Eng. **165**, 233–245 (1998)
15. Y. Li, K. Paik, T. Xing, P. Carrica, Dynamic overset CFD simulations of wind turbine aerodynamics. Renew. Energy **37**, 285–298 (2012)
16. Y. Saad, *Iterative Methods for Sparse Linear Systems* (PWS Publishing Company, Boston, 1996)
17. N. Sorensen, J. Michelsen, S. Schreck, Navier-Stokes predictions of the NREL phase VI rotor in the NASA Ames 80 ft × 120 ft wind tunnel. Wind Energy **5**, 151–169 (2002)
18. Y. Wu, X.-C. Cai, A fully implicit domain decomposition based ALE framework for three-dimensional fluid-structure interaction with application in blood flow computation. J. Comput. Phys. **258**, 524–537 (2014)

Overlapping Domain Decomposition Applied to the Navier–Stokes Equations

Oana Ciobanu, Laurence Halpern, Xavier Juvigny, and Juliette Ryan

1 Introduction

This article focuses on the research field of laminar flow of an ideal gas, on the resolution of aerodynamic multi-scale problems that are costly and difficult to solve in their original form. In order to solve these large data systems several techniques of parallel computing have been developed but some convergence problems may occur for large number of sub-domains.

Robust and fast methods are now available, which combine non-linear and linear solvers requiring less memory capacity. In the context of long term simulations, global implicit approaches have proven their superiority as they are able to simulate a quasi-steady-state behaviour without being restricted to short time steps to ensure convergence. Implementing these approaches on GPUs can certainly improve the efficiency versus a simple CPU implementation, as will be shown below, but by combining this implementation with domain decomposition another scale of efficiency could be achieved. In this paper, we propose an improved parallel time-space method for steady/unsteady problems modelled by Euler and Navier-Stokes equations for a direct numerical simulation.

Domain decomposition methods split large problems into smaller sub-problems that can be solved in parallel. Usually, only space domain decomposition method is used to provide high-performing algorithms in many fields of numerical applications. To achieve full performance on large clusters with up to 100,000 nodes (such

O. Ciobanu (✉) • X. Juvigny • J. Ryan
ONERA, BP72-29 avenue de la Divison Leclerc, 92322 Chatillon, France
e-mail: oana.ciobanu@onera.fr

L. Halpern
Université Paris 13, LAGA, 93430 Villetaneuse, France
e-mail: halpern@math.univ-paris13.fr

T. Dickopf et al. (eds.), *Domain Decomposition Methods in Science and Engineering XXII*, Lecture Notes in Computational Science and Engineering 104, DOI 10.1007/978-3-319-18827-0_47

as recently the IBM Sequoia, or GPUs) the time dimension has to be taken into account. An essential gain to be obtained from time-space domain decomposition is the ability to apply different time-space discretisation on sub-domains thus improving efficiency and convergence of implicit schemes.

In practice, we are often working with large computational domains where only a small part is highly interactive and a wide region of the domain is close to equilibrium state. What is usually done is that the sub-domains are balanced in space so that each processor finishes the simulation at the same moment and the computation is done, on each sub-domain with the same time step, the global one. The time step depends on the CFL condition, the value of the flow velocity and the space step. This means that the part of the simulation domain which is not dominated by strong non-linearities is solved with a much higher precision than is needed. Some sub-domains are over-solved. The sub-domain close to the equilibrium state converges in fewer iterations and it is less costly, but it has to wait for the high reactive sub-domain to end in order to continue the simulation. To avoid this loss of efficiency and optimize the computational cost, the time step should be computed locally and the distribution of flow in sub-domains should take into consideration several factors: closeness to equilibrium region, strong non-linearities region and time step influence.

Our work focuses on the improvement of the Schwarz waveform relaxation (SWR) Method introduced under this name by Gander [4] at the 10th Domain Decomposition Conference to solve parabolic equations. It was previously presented by Gander and Stuart [5] as a multi-splitting formulation on overlapping sub-domains [9] combined with a waveform relaxation algorithm [12] in space-time for the heat equation. The purpose is to solve the space-time partial differential equation in each sub-domain in parallel, and to transmit domain boundary information to the neighbours at the end of the time interval. Originally applied to linear PDEs, the SWR algorithm was extended and optimised to the non-linear reactive transport equations by Haeberlein [6] and Haeberlein and Halpern [7]. With the SWR method different time-space discretisation can be applied on sub-domains thus improving efficiency and convergence of the schemes.

2 Navier–Stokes Solvers

The Navier–Stokes equations are given by three conservation laws.

- Mass conservation: $$\frac{\partial \rho}{\partial t} + \nabla.(\rho \mathbf{u}) = 0$$
- Momentum conservation: $$\frac{\partial \rho \mathbf{u}}{\partial t} + \nabla.(\mathbf{u} \otimes (\rho \mathbf{u})) + \nabla.pI - \nabla.\tau = 0$$
- Energy conservation: $$\frac{\partial \rho E}{\partial t} + \nabla.(\mathbf{u}(\rho E + p)) - \nabla.(\tau \mathbf{u} - q) = 0$$

where $\rho, \mathbf{u}, E, \tau, q$ are, respectively density, velocity, energy, viscous tensor and heat flux. Three algorithms are presented. They are all based on the same time discretisation (second order implicit Backward Differentiation Formula), the non-linear problem is solved with the Newton method and linear problems are solved directly ($(L + D)D^{-1}(D + U)$ factorisation). The first method is a classical non-linear domain decomposition method [10, 11] which consists in semi-discretising uniformly in time the system, in applying a global Newton linearisation, then dividing the linear system in several local overlapping subsystems that we can solve in parallel. This algorithm is referred to as the Newton-Schwarz algorithm.

Newton-Schwarz Algorithm:
- Semi-discretisation in time
- Linearisation (Newton)
- Space Schwarz DDM
  –Solve the local linear system

In some cases, one Schwarz iteration is sufficient to achieve convergence of Newton to the solution of the problem. Space decomposition and linearisation are independent. The next idea is to first do the decomposition and then solve in each sub-domain the non-linear system. This algorithm is the same as the one introduced by Cai and Keyes [1], but using a different linear solver.

Schwarz-Newton Algorithm:
- Semi-discretisation in time
- Space Schwarz DDM
  –Solve the local non-linear system

To achieve full speed-up performance, a SWR method is used, as it allows local space and time stepping. The whole time interval of study is split into sub-intervals or time windows, then space is decomposed into sub-domains. For each time window the space-time Navier-Stokes equations are solved in each sub-domain in parallel. Boundary conditions are transmitted at the end of the time window.

SWR Algorithm:
- Schwarz DDM over time windows
- For each sub-domain:
  –Semi-discretisation in time
  –Solve the local non-linear system

SWR uses time windowing techniques that doesn't degrade the solution and exchanges less information between processors. After each iteration we proceed to the improvement of the interface condition in each sub-domain. This can lead to a completely different time step to satisfy either a stability criteria (for explicit schemes) or an accuracy bound, both based on the CFL number, thus the necessity to locally recompute the time step which is an improvement of the classical SWR algorithm. In this paper we propose, within the SWR iterative process, an adaptive time stepping technique to improve the scheme consistency, thus different time steps in each sub-domain and inside each time window. In the following we shall test the scalability of these three algorithms.

3 Numerical Results

Space discretisation is achieved with finite volumes on cartesian non-conforming grids. The Euler fluxes are computed using the MUSCL-Hancock (Monotone Upstream centred Scheme for Conservative Laws) second order scheme combined with the AUSM^{+}-UP (Advection Upstream Splitting Method) scheme. The advantage of the AUSM^{+}-UP developed by Liou [13] is that it was conceived to be uniformly valid for all speed regimes. The viscous fluxes are computed with a second order Finite Difference scheme. First, we solve the global domain for a simple configuration on CPU and we compare the results with those found using exactly the same second order algorithm, but on GPUs. Then, performances of the different parallel computing strategies (using OpenMP, MPI) are compared on the inviscid and viscous motion of a 2D isolated vortex in an uniform free-stream based on [15] and on the case of the mixing layer. The sub-domains overlap region has the stencil size. We use a second order projection method to exchange data in time and in space. All implicit algorithms are second order in time and space.

3.1 GPU Versus CPU for Euler Equations

First, these algorithms can be accelerated using GPUs. GPUs are used to solve a global problem or a local one using a massive parallel architecture. We start by solving the global problem on a GPU (NVidia Corporation GF110 [Geforce GTX 580] Compute Capability 2.0) with CUDA [3] launched from a CPU and compare its computational cost with one running on a CPU (7.8 GB, 2 Cores at 3.33 GHz) with OpenMP. The computational domain is a rectangular one with an imposed inflow velocity at each time step. On Table 1 is shown the ratio of the computation on a CPU with OpenMP with the computation on CPU-GPU. As can be seen there is a definite gain to be obtained on the CPU-GPU configuration with one domain, and the greater the number of points the better is the ratio. GPU code is portable on any NVidia GPUs using CUDA programming model, though, it should be noted that performances on GPUs vary a great deal depending on the GPU specifications.

Table 1 CPU-OPENMP time cost/CPU-GPU time cost

Grid size	Time step	2D fluxes	Update step	Boundary update	Total
130 × 130	43.08	1.63	8.62	0.31	3.72
260 × 260	109.26	1.71	15.90	1.58	4.65
525 × 525	164.83	2.81	40.37	1.38	6.88
1050 × 1050	392.72	2.58	321.21	2.39	7.80

3.2 2D Isentropic Vortex for Euler Equations

We present results on a convective vortex with $(u_\infty, v_\infty) = (1, 1)$ for a perfect gas: $\gamma = 1.4$, $\frac{p}{\rho^\gamma} = 1$. The computational domain is $[-5., 5.] \times [-5., 5.]$. The initial condition equals the mean flow field plus an isentropic vortex with no perturbation in entropy. We use periodic boundary conditions and Dirichlet transmission conditions. This test is interesting as the isentropic vortex is an exact solution of the Euler equations. At the end of each cycle that lasts 10 s the vortex equals the initial solution.

$$\begin{aligned}
\rho &= (T_\infty + \delta T)^{\frac{1}{\gamma-1}} = (1 - \tfrac{(\gamma-1)\beta^2}{8\gamma\pi} e^{1-(x^2+y^2)})^{\frac{1}{\gamma-1}} \\
\rho u &= \rho(u_\infty + \delta u) = \rho(1 - \tfrac{\beta}{2\pi} e^{\frac{1-(x^2+y^2)}{2}}) \\
\rho v &= \rho(v_\infty + \delta v) = \rho(1 + \tfrac{\beta}{2\pi} e^{\frac{1-(x^2+y^2)}{2}}) \\
p &= \rho^\gamma \\
e &= \tfrac{p}{\gamma-1} + \tfrac{1}{2}\rho(u^2 + v^2)
\end{aligned}$$

3.2.1 Accuracy Study

Let us begin with an accuracy study of the Euler equations computing L^2 and L^∞ slopes of errors in the case of non adaptive time steps. First, let us fix the number of sub-domains to 2×2 and a common time step. We increase the global number of space cells from 40×40 cells to 60×60 cells and 80×80 cells (the time step varies in the same ratio as the space step) We consider that we have converged when we reach an error less than a tolerance equal to $1.e-6$ for both Newton and Schwarz stopping criteria.

As can be seen in Fig. 1, all presented methods are second order in time and close to second order in space, depending on the Van Albada limiter chosen in MUSCL scheme. Velocity, pressure and energy errors behave similarly for all presented methods. The method denoted as Newton in Fig. 1 is the Newton-Schwarz method using only one Schwarz iteration, it only has order one accuracy showing that Schwarz is a good preconditioner for our scheme.

3.2.2 Computational Cost

To evaluate the cost (machine independent), a good indicator is the number of local linear solves, given by the product between the number of Newton iterations and the number of Schwarz iterations. This cost is a linear function of the number of cells. On Table 2 are shown the average number of local linear solves per time step for the Newton-Schwarz (NS) method, the Schwarz-Newton (SN) method and for the

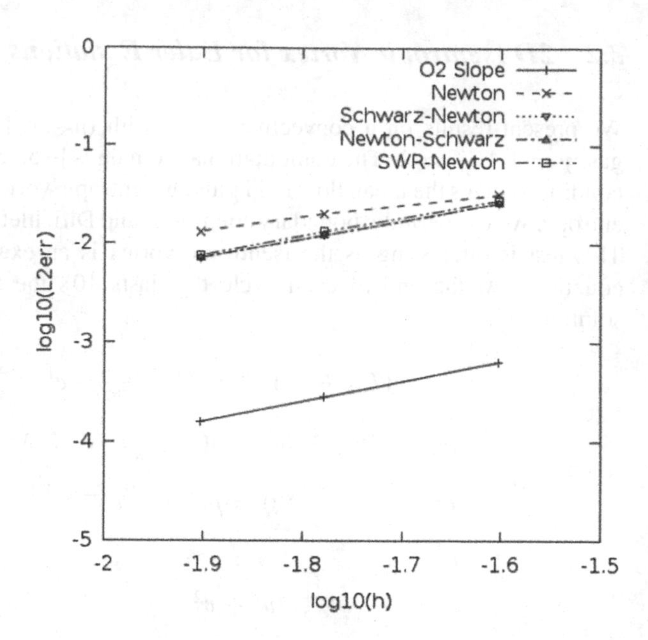

Fig. 1 L^2 error over density field

SWR-Newton (SWR) method for an increasing number of sub-domains with a fixed number of size cells in each sub-domain (weak scalability). The sub-domain size is fixed to 20×20 points and the CFL number has the value 0.5. For the SWR-Newton scheme, we choose δt the same time step on each sub-domain and $\Delta T = 5\delta t$ the time window. The Newton stopping tolerance is set to $1e-6$. The Schwarz convergence tolerance is varying as shown on Table 2. This table shows the good weak scalability of all considered methods. Moreover, it proves that a tolerance of $1e-2$ in the Schwarz stopping criteria decreases the number of linear solves without affecting the precision of the non-linear system. Thus, we can conclude that there is no need to achieve convergence in Schwarz. The SWR method is competitive with the Newton-Schwarz, but two times less efficient than the Schwarz-Newton scheme. On Table 2, in order to compute one time window the four processors communicate in average over all time windows 18.6 times (average number of Schwarz iterations per window) when a SWR-Newton scheme is chosen. In order to reach the same time window the Schwarz-Newton scheme communicates in average 35.45 times (Schwarz iterations $\times$ window size) and the Newton-Schwarz scheme communicates in average 250 times (Newton iterations $\times$ Schwarz iterations $\times$ window size). The SWR method is thus ideal for clusters with high latencies. Note: It should be mentioned that higher order coupling conditions like unsteady Robin type conditions can improve the efficiency of the algorithm and should positively influence the number of Schwarz iterates (cf. [7]).

The adaptive time step SWR method converges to the solution in exactly the same way as the fixed time step SWR method. The gain of the SWR method comes from the improved stability of the scheme since the time step is recomputed at each iteration thus less communication between the sub-domains as it appears that when

Table 2 Weak scaling of the schemes

		Schwarz tol. = $1.e-6$			Schwarz tol. = $1.e-3$			Schwarz tol. = $1.e-2$		
Scheme	Iterations\Sub-domains	4	9	16	4	9	16	4	9	16
NS	Newton iterations	8.48	8.11	6.94	8.48	8.11	6.93	8.48	8.11	9.29
NS	Schwarz iterations	5.89	6.28	6.41	5.26	5.70	5.78	4.71	5.00	5.66
NS	Linear solvers	50.00	51.02	44.51	44.64	46.33	40.15	39.89	40.63	52.62
SN	Newton iterations	3.07	2.72	2.28	5.00	4.18	3.28	5.93	4.90	5.31
SN	Schwarz iterations	7.09	7.49	7.15	3.55	4.0	4.0	2.46	3.0	3.0
SN	Linear solvers	21.79	20.40	16.31	17.78	16.75	13.15	14.63	14.72	15.93
SWR	Newton iterations	7.54	6.62	5.41	7.52	6.61	5.37	7.49	6.56	7.13
SWR	Schwarz iterations	18.6	14.28	19.25	7.08	7.98	7.44	4.36	4.77	5.66
SWR	Linear solvers	140.6	95.3	104.28	53.28	52.78	39.99	32.75	31.43	40.40

Computational costs

the coupling conditions are improved, larger times steps are usually needed. This also leads to less CPU memory when fewer coupling conditions need to be stored.

3.3 *Sound Generation in a 2D Low-Reynolds Mixing Layer*

The second case presented here if the case of a 2D low-Reynolds mixing layer where a high precision scheme is required. It is studied especially focusing on the acoustic waves emitted by the vortex pairings in a perturbed mixing layer. The flow configuration is the same as the one proposed by Colonius et al. [2] consisting in a slightly perturbed hyperbolic tangential shape velocity profile, $u = \bar{u} + 0.125\, tanh(2y)$, with $\bar{u} = (u_{\infty} + u_{-\infty})/2$ and $u_{\infty} = 0.5$, $u_{-\infty} = 0.25$, and $\rho_{\infty} = \rho_{-\infty} = 1$ and $p_{\infty} = p_{-\infty} = 1/\gamma$, respectively, with $\gamma = 1.4$. We fix the Reynolds number at 250 and add a sponge layer as shown in Fig. 2 to absorb the flow. This is a particularly sensitive case in acoustics and phenomena are quite different within each subdomain. The results presented on Table 3 are for simulations between $t = 200$ s and $t = 250$ s, interval inside which all sub-domains are interacting. The initial solution was computed with an explicit second order Runge-Kutta method. We have fixed the stopping criterion in the Newton algorithm to a tolerance of $1.e-4$ and the stopping criterion of the Schwarz decomposition to a tolerance of $1.e-2$ (cf. Sect. 3.2) which gives a good solution. The time window inside the SWR methods equal 5 times the smallest global time step and the global domain was divided in 22 sub-domains : 18 sub-domains of equal size 107×21 cells in the middle region and 4 sponge sub-domains with 107×41 in the sponge area. The number of linear solves is no longer a good measure since sub-domains with different size have been computed and we adapt the time step after each iteration for SWR and for all time steps in SWRA the adaptive SWR. On Table 3 we vary the

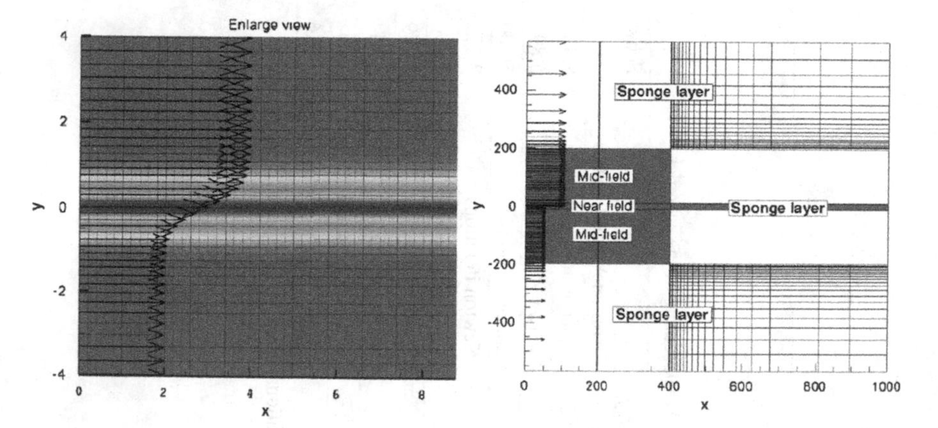

Fig. 2 Mixing layer acoustic pressure field. Initial condition (*left*) and computational domain with sponge layer (*right*)

Table 3 Global computational costs for $\Delta T = 5\delta t$, Schwarz tol. = $1.e-2$ and Newton tol. = $1.e-4$

Scheme\CFL	0.5	1	2	5	10
NS	333.56	167.65	116.45	24.20	18.79
SN	129.97	76.07	89.41	26.76	10.45
SWR	189.13	189.65	121.77	21.90	5.87
SWRA	189.82	191.08	121.54	21.86	5.12

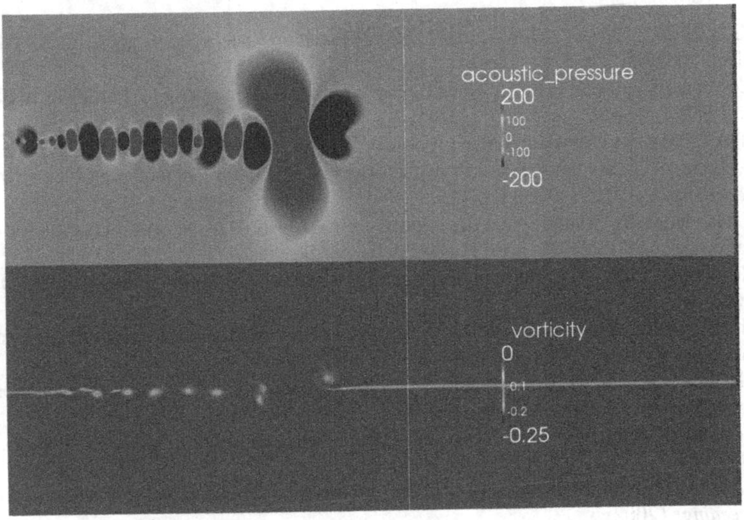

Fig. 3 Mixing layer acoustic pressure field (*top*) and vorticity (*bottom*)

time window length and show only the total computational time cost for all three methods. For low CFL (less than 2) SWR is less efficient than SN. For higher CFL, SWR becomes the most efficient, the SWR with adaptive step becoming the leader in terms of performance.

Results obtained with the time adaptive SWR scheme (see Fig. 3) compare well with those obtained with an explicit third order Runge Kutta Discontinuous Galerkin solver developed by Halpern et al. [8].

4 Conclusion and Remarks

A variation on the non-linear SWR algorithm has been developed using an adaptive time stepping approach to simulate 2D multi-scale Euler and Navier-Stokes problems. The above results show that the method has the ability to treat large data systems without loss of parallel efficiency. This SWR algorithm has similar computational efficiency as the original SWR and adds a new flexibility to the SWR method. There are at least three ways to improve the SWR technique. One is to optimize the time space interface condition, another is to implement the pipeline SWR iterations as presented by Ong et al. [14] and of course the use of GPUs that can considerably improve the efficiency.

References

1. X.-C. Cai, D.E. Keyes, Nonlinearly preconditioned inexact Newton algorithms. SIAM **24** (1),183–200 (2002)
2. T. Colonius, S.K. Lele, P. Moin, Sound generation in a mixing layer. J. Fluid Mech. **330**,375–409 (1997)
3. CUDA (2015), http://www.nvidia.com/object/cuda_home_new.html
4. M.J. Gander, Overlapping Schwarz waveform relaxation for parabolic problems, in *DD10 Proceedings*, vol. 218 (1998), pp. 425–431
5. M.J. Gander, A.M. Stuart, Space-time continuous analysis of waveform relaxation for the heat equations. SIAM **19**(6), 2014–2031 (1998)
6. F. Haeberlein, Time-space domain decomposition methods for reactive transport. Ph.D. thesis, University Paris 13, 2011
7. F. Haeberlein, L. Halpern, Optimized Schwarz waveform relaxation for nonlinear systems of parabolic type, in *DD21 Proceedings* (2012)
8. L. Halpern, J. Ryan, M. Borrel, Domain decomposition vs. overset Chimera grid approaches for coupling CFD and CAA, in *ICCFD7* (2012)
9. R. Jeltsch, B. Pohl, Waveform relaxation with overlapping splittings. SIAM **16**(1), 40–49 (1995)
10. D.E. Keyes, Domain decomposition in the mainstream of computational science, in *DD14 Proceedings* (2002)
11. D.A. Knoll, D.E. Keyes, Jacobian-free Newton–Krylov methods: a survey of approaches and applications. J. Comput. Phys. **193**(2), 357–397 (2004)
12. E. Lelarasmee, A.E. Ruehli, A.L. Sangiovanni-Vincentelli, The waveform relaxation method for time-domain analysis of large scale integrated circuits. IEEE **1**(3), 131–145 (1982)
13. M.-S. Liou, A sequel to AUSM, part II: AUSM$^+$-up for all speeds. J. Comput. Phys. **214**(1), 137–170 (2006)
14. B. Ong, S. High, F. Kwok, Pipeline Schwarz waveform relaxation, in *22nd DDM Conference* (2013, submitted)
15. H.C. Yee, N.D. Sandham, M.J. Djomehri, Low-dissipative high-order shock-capturing methods using characteristic-based filters. J. Comput. Phys. **150**(1), 199–238 (1999)

Schwarz Methods for Second Order Maxwell Equations in 3D with Coefficient Jumps

Victorita Dolean, Martin J. Gander, and Erwin Veneros

1 Introduction

Classical Schwarz methods need in general overlap to converge, but in the case of hyperbolic problems, they can also be convergent without overlap, see [5]. For the first order formulation of Maxwell equations, we have proved however in [11] that the classical Schwarz method without overlap does not converge in most cases in the presence of coefficient jumps aligned with interfaces.

Optimized Schwarz methods have been developed for Maxwell equations in first order form without conductivity in [8], and with conductivity in [9, 13]. These methods use modified transmission conditions, and often converge much faster than classical Schwarz methods. For DG discretizations of Maxwell equations, optimized Schwarz methods can be found in [6, 7, 10]. Optimized Schwarz methods were also developed for the second order formulation of Maxwell equations, see [1, 17, 18] for scattering problems with applications [13], see also the earlier work by [2–4, 15, 16].

While usually coefficient jumps hamper the convergence of domain decomposition methods, this is very different for optimized Schwarz methods. For diffusive problems, it was shown in [12] that jumps in the coefficients can actually lead to faster iterations, when they are taken into account correctly in the transmission conditions: optimized Schwarz methods benefit from jumps in the coefficients at interfaces. We had shown in [11] that this also holds for the special case of transverse magnetic modes (TMz) in the two dimensional first order Maxwell equations. We show in this short paper that these results for the TMz modes [and the corresponding ones for the transverse electric modes (TEz)] can be used to formulate optimized

V. Dolean • M.J. Gander • E. Veneros (✉)
Section de mathématiques, Université de Genève, 1211 Genève 4, Switzerland
e-mail: victorita.dolean@unige.ch; martin.gander@unige.ch; erwin.veneros@unige.ch

T. Dickopf et al. (eds.), *Domain Decomposition Methods in Science and Engineering XXII*, Lecture Notes in Computational Science and Engineering 104, DOI 10.1007/978-3-319-18827-0_48

Schwarz methods for the 3D second order Maxwell equations which then in some cases converge faster, the bigger the coefficient jumps are.

2 Classical Schwarz for Second Order Maxwell Equations

The time dependent Maxwell equations in their second order formulation are

$$\epsilon \partial_t^2 \mathscr{E} + \nabla \times (\mu^{-1} \nabla \times \mathscr{E}) = \partial_t \mathscr{J}, \tag{1}$$

where $\mathscr{E} = (\mathscr{E}_1, \mathscr{E}_2, \mathscr{E}_3)^T$ is the electric field, ϵ is the *electric permittivity*, μ is the *magnetic permeability*, and $\mathscr{J}$ is the applied current density. We assume that the applied current density is divergence free, $\operatorname{div} \mathscr{J} = 0$. There is a similar system also for the magnetic field $\mathscr{H} = (\mathscr{H}_1, \mathscr{H}_2, \mathscr{H}_3)^T$,

$$\mu \partial_t^2 \mathscr{H} + \nabla \times (\epsilon^{-1} \nabla \times \mathscr{H}) = \nabla \times \epsilon^{-1} \mathscr{J}, \tag{2}$$

but we will only consider the Eq. (1) for the electric field in this short paper.

The time dependent Maxwell equations (1) form a system of hyperbolic partial differential equations [8]. Imposing incoming characteristics is equivalent to imposing the impedance condition

$$\mathscr{B}_{\mathbf{n}_j}(\mathscr{E}^{m,n}) = \frac{1}{\mu_m}(\nabla \times \mathscr{E}^{m,n} \times \mathbf{n}_j) \times \mathbf{n}_j + \frac{i\omega}{Z_m}(\mathscr{E}^{m,n} \times \mathbf{n}_j) = \mathbf{s}, \tag{3}$$

where $Z_m = \sqrt{\frac{\mu_m}{\epsilon_m}}$. We are interested here in the time-harmonic Maxwell equations, which are obtained by supposing that $\mathscr{E}(x,t) = e^{i\omega t}\mathbf{E}(x)$ for a fixed frequency ω. After some simplifications, we obtain from Eq. (1) the time harmonic second order Maxwell equation

$$\epsilon \omega^2 \mathbf{E} - \nabla \times (\mu^{-1} \nabla \times \mathbf{E}) = -i\omega \mathbf{J}. \tag{4}$$

We are interested here in the heterogeneous case, where the domain Ω of interest consists of two non-overlapping subdomains Ω_1 and Ω_2 with interface Γ, and piecewise constant parameters ϵ_j and μ_j in Ω_j, $j = 1, 2$. We want to solve such problems using the Schwarz algorithm

$$\begin{cases} \epsilon_1 \omega^2 \mathbf{E}^{1,n} - \nabla \times (\mu_1^{-1} \nabla \times \mathbf{E}^{1,n}) = -i\omega \mathbf{J}, & \text{in } \Omega_1, \\ \mathscr{T}_{\mathbf{n}_1}(\mathbf{E}^{1,n}) = \mathscr{T}_{\mathbf{n}_1}(\mathbf{E}^{2,n-1}) & \text{on } \Gamma, \end{cases}$$
$$\begin{cases} \epsilon_2 \omega^2 \mathbf{E}^{2,n} - \nabla \times (\mu_2^{-1} \nabla \times \mathbf{E}^{2,n}) = -i\omega \mathbf{J}, & \text{in } \Omega_2, \\ \mathscr{T}_{\mathbf{n}_2}(\mathbf{E}^{2,n}) = \mathscr{T}_{\mathbf{n}_2}(\mathbf{E}^{1,n-1}) & \text{on } \Gamma, \end{cases} \tag{5}$$

with the transmission condition

$$\mathscr{T}_{\mathbf{n}_j}(\mathbf{E}^{m,n}) = (Id - A_j)(\frac{1}{\mu_m}\mathbf{n}_j \times \nabla \times \mathbf{E}^{m,n}) - \frac{i\omega_j}{\mu_j}(Id + A_j)(\mathbf{n}_j \times (\mathbf{E}^{m,n} \times \mathbf{n}_j)). \quad (6)$$

with $\omega_j = \omega\sqrt{\epsilon_j\mu_j}$, $j = 1, 2$. The classical Schwarz algorithm is obtained for the choice $A_j = 0$, for $j = 1, 2$. We see that the classical Schwarz algorithm is exchanging characteristic information at the interfaces between subdomains, i.e. $\mathscr{T}_{\mathbf{n}_j}(\mathbf{E}^{m,n}) = \mathscr{B}_{\mathbf{n}_j}(\mathbf{E}^{m,n})$ where $\mathscr{B}$ is defined in (3).

In [11], we studied the classical Schwarz algorithm for the first order Maxwell equations on the domain $\Omega = \mathbb{R}^3$, with subdomains $\Omega_1 = (-\infty, 0] \times \mathbb{R}^2$ and $\Omega_2 = [0, \infty) \times \mathbb{R}^2$ and interface $\Gamma = \{0\} \times \mathbb{R}^2$ and the Silver-Müller radiation condition. We showed that the convergence factor of the classical Schwarz algorithm in 3D is $\rho_{\text{cla}} = \max\{\rho_{E\text{cla}}, \rho_{M\text{cla}}\}$, where $\rho_{E\text{cla}}$ and $\rho_{M\text{cla}}$ are the convergence factors of the TEz and TMz cases in 2D. We then proved that if there are coefficient jumps along the interface Γ, i.e. $\mu_1 \neq \mu_2$ and/or $\epsilon_1 \neq \epsilon_2$, the classical Schwarz algorithm is divergent in 3D if $\mu_1\epsilon_2 \neq \mu_2\epsilon_1$. If $\mu_1\epsilon_2 = \mu_2\epsilon_1$, we obtained $\rho_{E\text{cla}} = \rho_{M\text{cla}}$, and $\rho_{cla} < 1$ for the propagative modes, $|\mathbf{k}| < \omega_j$, $j = 1, 2$, but $\rho_{cla}(|\mathbf{k}|) = 1$ for the evanescent modes, $|\mathbf{k}| > \omega_j$, $j = 1, 2$, so the algorithm is stagnating for all evanescent modes. It is thus never convergent in 3D. We then investigated in [11] the 2D case of TMz modes in more detail, and found that the classical Schwarz algorithm in the presence of coefficient jumps is convergent in certain situations, depending on the jumps in ϵ and μ.

These results also hold for the second order Maxwell equations when the Schwarz algorithm (5), (6) with classical transmission conditions is applied, and for the convergent cases from [11] in 2D, we have the following new contraction estimate:

Theorem 1 (Classical Schwarz in 2D) *If the classical Schwarz algorithm (5,6) in 2D converges, then we have the asymptotic convergence factor estimate*

$$\rho_{Mcla}(k, \omega_1, \omega_2, Z) = \rho_{Ecla}(k, \omega_1, \omega_2, Z) = 1 - O(h^2)$$

with $Z = \sqrt{\frac{\mu_1\epsilon_2}{\mu_2\epsilon_1}}$ *and h the uniform mesh size.*

Proof As in [11], we can write the convergence factors for the TMz case as

$$\rho_{M\text{cla}}(k, \omega_1, \omega_2, Z) = \left| \frac{\left(\sqrt{k^2 - \omega_1^2} - i\omega_1 Z\right)\left(\sqrt{k^2 - \omega_2^2} - i\omega_2/Z\right)}{\left(\sqrt{k^2 - \omega_1^2} + i\omega_1\right)\left(\sqrt{k^2 - \omega_2^2} + i\omega_2\right)} \right|^{\frac{1}{2}}, \quad (7)$$

and for evanescent modes ($k > \omega_1, \omega_2$), Eq. (7) is equal to

$$\rho_{M\mathrm{cla}}(k, \omega_1, \omega_2, Z) = 1 + \frac{(Z^2-1)\omega_1^2}{Z^2 k^2}\left(Z^2 - Y^2 - \frac{(Z^2-1)\omega_2^2}{k^2}\right), \tag{8}$$

with $Y = \frac{\omega_2}{\omega_1}$. From Eq. (8) we see that $\lim_{k\to\infty} \rho_{M\mathrm{cla}} = 1$. If the classical Schwarz algorithm is convergent then $\rho_{M\mathrm{cla}} < 1$, $\forall k$, the previous remark permits us to conclude that the maximum over all the frequencies must be at $k = k_{\max} = \frac{c_{\max}}{h}$, the largest frequency supported by the numerical grid, where h is the mesh size and $c_{\max}$ is a constant depending on the geometry. To conclude the proof, we just insert $k = c_{\max}/h$ into (8) and the result follows by expansion. The proof for the TEz case is similar.

3 Optimized Schwarz for Second Order Maxwell Equations

Since the classical Schwarz method is not an effective solver for Maxwell equations in the presence of coefficient jumps, we introduce now more effective transmission conditions which take the coefficient jumps into account. We consider algorithm (5), (6) with the particular choice

$$A_j := \gamma_{jM} S_{TM} + \gamma_{jE} S_{TE}, \quad S_{TM} = \nabla_\tau \nabla_\tau \cdot, \quad S_{TE} = \nabla_\tau \times \nabla_\tau \times,$$

where τ is the tangential direction to the interface. We note that $S_{TM} - S_{TE} = \Delta_\tau I$, where Δ_τ is the Laplace-Beltrami operator in the tangential plane (for example , $\Delta_\tau = \partial_{yy} + \partial_{zz}$ when $\mathbf{n} = (1,0,0)$). The constants γ_{1E}, γ_{2E} and γ_{1M}, γ_{2M} can be chosen in order to optimize the algorithm.

Performing a Fourier transform in the yz plane, we find after a lengthy calculation the iteration matrix of the optimized Schwarz algorithm to be

$$IT = \begin{pmatrix} C_E & 0 \\ 0 & C_M \end{pmatrix} \tag{9}$$

with the coefficients

$$C_E = \frac{((\lambda_1 - i\omega_1/Z) - \gamma_{2M}|k|^2(\lambda_1 + i\omega_1/Z))((\lambda_2 - i\omega_2 Z) - \gamma_{1M}|k|^2(\lambda_2 + i\omega_2 Z))}{(2\omega_1 - i(\lambda_1 - i\omega_1)(1-\gamma_{1M}|k|^2))(2\omega_2 - i(\lambda_2 - i\omega_2)(1-\gamma_{2M}|k|^2))},$$

$$C_M = \frac{((\lambda_1 - i\omega_1 Z) - \gamma_{2E}|k|^2(\lambda_1 + i\omega_1 Z))((\lambda_2 - i\omega_2/Z) - \gamma_{1E}|k|^2(\lambda_2 + i\omega_2/Z))}{((1-\gamma_{1E}|k|^2)(\lambda_1 - i\omega_1) + 2i\omega_1)((1-\gamma_{2E}|k|^2)(\lambda_2 - i\omega_2) + 2i\omega_2)}, \tag{10}$$

with $\lambda_j = \sqrt{|\mathbf{k}|^2 - \omega_j^2}$, $j = 1, 2$. If we choose for the parameters the values

$$\begin{aligned} \gamma_{1M} &= \frac{\lambda_2 - i\omega_2 Z}{|k|^2(\lambda_2 + i\omega_2 Z)}, \; \gamma_{1E} = \frac{\lambda_2 - i\omega_2/Z}{|k|^2(\lambda_2 + i\omega_2/Z)}, \\ \gamma_{2E} &= \frac{\lambda_1 - i\omega_1 Z}{|k|^2(\lambda_1 + i\omega_1 Z)}, \; \gamma_{2M} = \frac{\lambda_1 - i\omega_1/Z}{|k|^2(\lambda_1 + i\omega_1/Z)}, \end{aligned} \tag{11}$$

then the iteration matrix IT in (9) vanishes and we have convergence in two iterations. The corresponding transmission conditions are called transparent conditions, and are optimal, since they lead to a direct solver. But the operators corresponding to the symbols in (11) are non local and thus costly to use. We therefore propose to replace λ_1 and λ_2 in (11) by zeroth order approximations s_{1E}, s_{1M}, s_{2E} and s_{2E}. The convergence factor of the method is then the maximum of the spectral radius of (9) over all Fourier frequencies. We obtain

$$\rho_{\text{opt}} = \max\{\rho_{E\text{opt}}, \rho_{M\text{opt}}\}, \tag{12}$$

with

$$\begin{aligned} \rho_{E\text{opt}}(|\mathbf{k}|, \omega, \epsilon_1, \epsilon_2, \mu_1, \mu_2, s_{1M}, s_{2M}) &= \left| \frac{(\lambda_2 - s_{2M})(\lambda_1 - s_{1M})}{(\lambda_2 + s_{1M}\epsilon_2/\epsilon_1)(\lambda_1 + s_{2M}\epsilon_1/\epsilon_2)} \right|^{1/2}, \\ \rho_{M\text{opt}}(|\mathbf{k}|, \omega, \epsilon_1, \epsilon_2, \mu_1, \mu_2, s_{1E}, s_{2E}) &= \left| \frac{(\lambda_1 - s_{1E})(\lambda_2 - s_{2E})}{(\lambda_2 + s_{1E}\mu_2/\mu_1)(\lambda_1 + s_{2E}\mu_1/\mu_2)} \right|^{1/2}. \end{aligned} \tag{13}$$

These factors can be optimized separately and they are once again the convergence factors of the TMz and TEz cases in 2D. In order to optimize we have to choose $s_{jE}, s_{jM}, j = 1, 2$ such that ρ_{opt} is as small as possible for all numerically relevant frequencies $k \in K := [k_{\min}, k_{\max}]$. Here $k_{\min}$ is the smallest frequency relevant to the subdomain, and $k_{\max} = \frac{c_{\max}}{h}$ is the largest frequency supported by the numerical grid, h being the mesh size, see for example [14]. We search for s_{jE} and s_{jM} of the form $s_{jE} = c_{jE}(1 + i)$, $s_{jM} = c_{jM}(1 + i)$ such that $s_{jE}, s_{jM}, j = 1, 2$ will be the solutions of the min-max problems

$$\min_{s_{1E}, s_{2E} \in \mathbb{C}} \max_{k \in K} \rho_{M\text{opt}}(|\mathbf{k}|, \omega, \epsilon_1, \epsilon_2, \mu_1, \mu_2, s_{1E}, s_{2E}), \tag{14}$$

$$\min_{s_{1M}, s_{2M} \in \mathbb{C}} \max_{k \in K} \rho_{E\text{opt}}(|\mathbf{k}|, \omega, \epsilon_1, \epsilon_2, \mu_1, \mu_2, s_{1M}, s_{2M}). \tag{15}$$

Since the optimization can be performed independently, we can use our results from [11] and obtain

Corollary 1 (2D Asymptotically Optimized Contraction Factor) *For TMz, the solution of (14) for $Y \neq 1$ gives the asymptotic convergence factor*

$$\rho^*_{Mopt} = \begin{cases} 1 - \mathcal{O}(h^{1/4}) & \text{if } Z = Y, \\ \sqrt{\frac{\mu_{min}}{\mu_{max}}} + \mathcal{O}(h) & \text{if } Z \leq Y < \sqrt{2}Z \text{ or } Y \leq Z < \sqrt{2}Y, \\ \sqrt[4]{\frac{1}{2}} + \mathcal{O}(h) & \text{if } Z < \sqrt{2}Y \text{ or } Y > \sqrt{2}Z. \end{cases} \tag{16}$$

If $Z \neq 1$ and $Y = 1$, we obtain after excluding the resonance frequency [8]

$$\rho^*_{Mopt} = \sqrt{\frac{\mu_{min}}{\mu_{max}}} + \mathcal{O}(h).$$

For the TEz case, the same conclusion holds if we replace Y by Y^{-1} and μ by ϵ.

The results in 3D follow now by a systematic consideration of both cases together:

Theorem 2 (3D Asymptotically Optimized Contraction Factor, Case A) *If $Z \neq Y, Y^{-1}$ and $Y \neq 1$, the optimized convergence factor ρ^*_{opt} in (12) has the asymptotic behavior:*

1. *If* $\min\left\{\max\{(ZY)^{-1}, ZY\}, \max\{Z/Y, Y/Z\}\right\} > \sqrt{2}$, *then*

$$\rho^*_{opt} = \sqrt[4]{1/2} + \mathcal{O}(h). \tag{17}$$

2. *If* $\min\left\{\max\{(ZY)^{-1}, ZY\}, \max\{Z/Y, Y/Z\}\right\} = \max\{Z/Y, Y/Z\} \leq \sqrt{2}$, *then*

$$\rho^*_{opt} = \sqrt{\frac{\mu_{min}}{\mu_{max}}} + \mathcal{O}(h). \tag{18}$$

3. *If* $\min\left\{\max\{(ZY)^{-1}, ZY\}, \max\{Z/Y, Y/Z\}\right\} = \max\{(YZ)^{-1}, YZ\} \leq \sqrt{2}$, *then*

$$\rho^*_{opt} = \sqrt{\frac{\epsilon_{min}}{\epsilon_{max}}} + \mathcal{O}(h). \tag{19}$$

Proof To prove 1. we use twice Corollary 1. If $\max\{Z/Y, Y/Z\} > \sqrt{2}$, we use the third result in (16) for the TMz case. Similarly if $\max\{ZY, (ZY)^{-1}\} > \sqrt{2}$ we use also the third result in (16) but for the TEz case. From Eq. (12) we know that ρ_{opt} is the maximum of $\rho_{E\text{opt}}$ and $\rho_{M\text{opt}}$, and if both of them have the asymptotic behaviour $\sqrt[4]{1/2} + \mathcal{O}(h)$, we get (17) as required.

For 2. we know that $\max\{Z/Y, Y/Z\} \leq \sqrt{2}$, which means that we can use the second result in (16), i.e. $\rho_{M\text{opt}} = \sqrt{\frac{\mu_{\min}}{\mu_{\max}}} + \mathcal{O}(h)$. We note that $Z/Y = \frac{\mu_1}{\mu_2}$ and $ZY = \frac{\epsilon_2}{\epsilon_1}$ which implies $1 \geq \sqrt{\frac{\mu_{\min}}{\mu_{\max}}} \geq \sqrt[4]{\frac{1}{2}}$. If $\max\{(ZY)^{-1}, ZY\} > \sqrt{2}$, by Corollary 1 we

have $\rho_{E\text{opt}} = \sqrt[4]{1/2} + \mathcal{O}(h)$, and we clearly see $\rho_{M\text{opt}} > \rho_{E\text{opt}}$. If $\max\{(ZY)^{-1}, ZY\} \leq \sqrt{2}$, we obtain by hypothesis the inequality $\max\{Z/Y, Y/Z\} \leq \max\{(ZY)^{-1}, ZY\} \leq \sqrt{2}$, and this implies $\frac{\mu_{\min}}{\mu_{\max}} \geq \frac{\epsilon_{\min}}{\epsilon_{\max}}$. Then we obtain $\rho_{M\text{opt}} \geq \rho_{E\text{opt}}$ and thus (18).
Finally, for 3., one can proceed as for 2 to obtain (19).

Theorem 3 (3D Asymptotically Optimized Contraction Factor, Case B) *If $Z = Y$ or $Z = Y^{-1}$, then the optimized convergence factor ρ^*_{opt} in (12) satisfies*

$$\rho^*_{opt} = 1 - \mathcal{O}(h^{1/4}). \tag{20}$$

Proof We use the first result in (16) of Corollary 1 and proceed as in Theorem 2.

Theorem 4 (3D Asymptotically Optimized Contraction Factor, Case C) *If $Y = 1$ and $Z \neq Y$, then the optimized convergence factor ρ^*_{opt} in (12) satisfies*

$$\rho^*_{opt} = \sqrt{\frac{\mu_{min}}{\mu_{max}}} + \mathcal{O}(\sqrt{h}). \tag{21}$$

Proof After excluding the resonance frequency, we apply the second part of Corollary 1. Note that in this case $\rho_{M\text{opt}} = \rho_{E\text{opt}}$.

Theorem 2 and 4 contain the important result that in the presence of jumps in the coefficients, the convergence of the optimized Schwarz method for Maxwell equations gets faster when the jump increases, the method benefits from the jumps! In the first part of Theorem 2, the convergence is independent of the jump in the coefficients, and in all these cases the nonoverlapping method converges independently of the mesh parameter, also unusual for optimized Schwarz methods without jumps in the coefficients. In the case of $Z = Y$ or $Z = Y^{-1}$ ($\mu_1 = \mu_2$ or $\epsilon_1 = \epsilon_2$) in Theorem 3 however, the convergence factor depends on h and deteriorates as h goes to zero, as in the case without jumps presented in [8].

We now illustrate graphically the improvement of the optimized Schwarz method over the classical one in 2D. We show in Fig. 1 in red the divergence regions and in blue the convergence regions for different values of Z and Y. In the left graphic the white part is still an open problem. In the right the light blue line have convergence dependant of the mesh size h, the light blue region have convergence dependent on the coefficients $\mu's$ and the dark blue region have convergence independent of the mesh size h and the coefficients $\mu's$, the red line is the zone of resonance corrected with Theorem 1. We clearly see that the optimization of the transmission conditions transforms an algorithm that fails for a large range of problems into one that works in all cases.

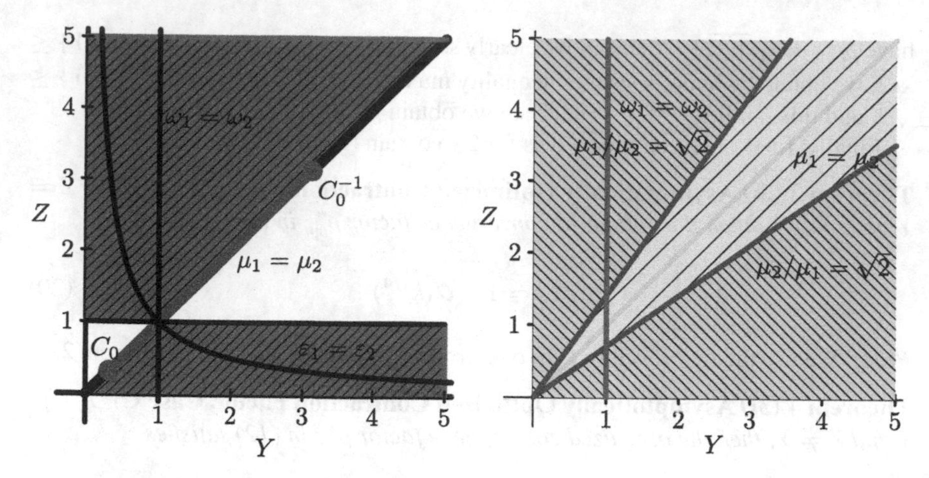

Fig. 1 Convergence regions in *blue* and divergence regions in *red* for Classical Schwarz (*left*, converges only for (Y, Z) on the line between C_0 and $C- =^{-1}$), and optimized Schwarz (*right*, converges everywhere, except on the line $Y = 1$)

4 Conclusions

Classical Schwarz methods applied to 3D Maxwell equations with jumps in the coefficients aligned with the interfaces do not converge, and this is also the case for the second order formulation of Maxwell equations. Using however optimized transmission conditions, we showed that one can obtain Schwarz methods for the 3D Maxwell equations that converge independently of the mesh parameter in some cases, and even become faster as the jumps get larger at the interfaces. These methods directly benefit from the jumps in the coefficients. We presented precise asymptotic convergence factor estimates for the many different cases of coefficient jumps, and are currently working on the numerical implementation of these methods in the full 3D setting.

References

1. A. Alonso-Rodriguez, L. Gerardo-Giorda, New nonoverlapping domain decomposition methods for the harmonic Maxwell system. SIAM J. Sci. Comput. **28**(1), 102–122 (2006)
2. P. Chevalier, F. Nataf, An OO2 (Optimized Order 2) method for the Helmholtz and Maxwell equations, in *10th International Conference on Domain Decomposition Methods in Science and in Engineering* (AMS, Providence, 1997), pp. 400–407
3. B. Després, Décomposition de domaine et problème de Helmholtz. C. R. Acad. Sci. Paris **1**(6), 313–316 (1990)
4. B. Després, P. Joly, J. Roberts, A domain decomposition method for the harmonic Maxwell equations, in *Iterative Methods in Linear Algebra* (North-Holland, Amsterdam, 1992), pp. 475–484

5. V. Dolean, M.J. Gander, Why classical Schwarz methods applied to hyperbolic systems can converge even without overlap, in *Domain Decomposition Methods in Science and Engineering XVII*. Lect. Notes Comput. Sci. Eng., vol. 60 (Springer, Heidelberg, 2007), pp. 467–475
6. V. Dolean, S. Lanteri, R. Perrussel, A domain decomposition method for solving the three-dimensional time-harmonic Maxwell equations discretized by discontinuous Galerkin methods. J. Comput. Phys. **227**(3), 2044–2072 (2008a)
7. V. Dolean, S. Lanteri, R. Perrussel, Optimized Schwarz algorithms for solving time-harmonic Maxwell's equations discretized by a discontinuous Galerkin method. IEEE. Trans. Magn. **44**(6), 954–957 (2008b)
8. V. Dolean, L. Gerardo-Giorda, M. Gander, Optimized Schwarz methods for Maxwell equations. SIAM J. Sci. Comput. **31**(3), 2193–2213 (2009)
9. V. Dolean, M. El Bouajaji, M.J. Gander, S. Lanteri, Optimized Schwarz methods for Maxwell's equations with non-zero electric conductivity, in *Domain Decomposition Methods in Science and Engineering XIX*. Lect. Notes Comput. Sci. Eng., vol. 78 (Springer, Heidelberg, 2011a), pp. 269–276
10. V. Dolean, M. El Bouajaji, M.J. Gander, S. Lanteri, R. Perrussel, Domain decomposition methods for electromagnetic wave propagation problems in heterogeneous media and complex domains, in *Domain Decomposition Methods in Science and Engineering XIX*, Lect. Notes Comput. Sci. Eng., vol. 78 (Springer, Heidelberg, 2011b), pp. 15–26
11. V. Dolean, M.J. Gander, E. Veneros, Optimized Schwarz methods for Maxwell equations with discontinuous coefficients, in *Domain Decomposition Methods in Science and Engineering XXI*, Lect. Notes Comput. Sci. Eng., (Springer, 2013), pp. 517–524
12. O. Dubois, Optimized Schwarz methods for the advection-diffusion equation and for problems with discontinuous coefficients. Ph.D. thesis, McGill University (2007)
13. M. El Bouajaji, V. Dolean, M.J. Gander, S. Lanteri, Optimized Schwarz methods for the time-harmonic Maxwell equations with damping. SIAM J. Sci. Comput. **34**(4), A2048–A2071 (2012)
14. M.J. Gander, Optimized Schwarz methods. SIAM J. Numer. Anal. **44**(2), 699–731 (2006)
15. M.J. Gander, F. Magoulès, F. Nataf, Optimized Schwarz methods without overlap for the Helmholtz equation. SIAM J. Sci. Comput. **24**(1), 38–60 (2002)
16. M.J. Gander, L. Halpern, F. Magoulès, An optimized Schwarz method with two-sided robin transmission conditions for the Helmholtz equation. Int. J. Numer. Methods Fluids **55**(2), 163–175 (2007)
17. Z. Peng, J.F. Lee, Non-conformal domain decomposition method with second-order transmission conditions for time-harmonic electromagnetics. J. Comput. Phys. **229**(16), 5615–5629 (2010)
18. Z. Peng, V. Rawat, J.F. Lee, One way domain decomposition method with second order transmission conditions for solving electromagnetic wave problems. J. Comput. Phys. **229**(4), 1181–1197 (2010)

DDFV Ventcell Schwarz Algorithms

Martin J. Gander, Laurence Halpern, Florence Hubert, and Stella Krell

1 Introduction

We are interested in this paper in anisotropic diffusion problems of the form

$$\mathcal{L}(u) := -\text{div}(A\nabla u) + \eta u = f \text{ in } \Omega, \quad u = 0 \quad \text{on } \partial\Omega, \tag{1}$$

$$\text{with } (x,y) \in \Omega \mapsto A(x,y) = \begin{pmatrix} A_{xx} & A_{xy} \\ A_{xy} & A_{yy} \end{pmatrix}. \tag{2}$$

Over the last 5 years, classical and optimized Schwarz methods have been developed for (1) discretized with Discrete Duality Finite Volume (DDFV) schemes. Like for Discontinuous Galerkin methods, it is not a priori clear how to appropriately discretize transmission conditions. Two versions have been proposed for Robin transmission conditions in [2, 4]. Only the second one leads to the expected

M.J. Gander
University of Geneva, 2-4 rue du Lièvre, CP 64, 1211 Genève, Switzerland
e-mail: martin.gander@unige.ch

L. Halpern
Université PARIS 13, LAGA, 93430 Villetaneuse, France
e-mail: halpern@math.univ-paris13.fr

F. Hubert
Aix-Marseille Université, CNRS, Centrale Marseille, I2M UMR 7373, 39 rue F. Joliot Curie, 13453 Marseille, Cedex 13, France
e-mail: florence.hubert@univ-amu.fr

S. Krell (✉)
Université de Nice, Parc Valrose, 28 avenue Valrose, 06108 Nice, Cedex 2, France
e-mail: krell@unice.fr

T. Dickopf et al. (eds.), *Domain Decomposition Methods in Science and Engineering XXII*, Lecture Notes in Computational Science and Engineering 104, DOI 10.1007/978-3-319-18827-0_49

rapid convergence rate of the optimized Schwarz algorithm, see [1] for parabolic problems.

The DDFV method needs a dual set of unknowns located on both vertices and "centers" of the initial mesh, which leads to two meshes, the primal and the dual one. This permits the reconstruction of two-dimensional discrete gradients located on a third partition of Ω, called the diamond mesh. A discrete divergence operator is also defined by duality. This method is particularly accurate in terms of gradient approximations, see the benchmark [6] for problem (1) with $\eta = 0$, and also an extensive bibliography.

A non-overlapping Schwarz method using Ventcell transmission conditions was first proposed in [7]. For the model problem (1), the algorithm with two non-overlapping subdomains, $\Omega = \Omega_1 \cup \Omega_2$, and iteration index $l = 0, 1, \ldots$ is

$$\mathcal{L}(u_j^{l+1}) = f \text{ in } \Omega_j, \quad u = 0 \quad \text{on } \partial\Omega_j \cap \partial\Omega, \tag{3}$$

$$(A\nabla u_j^{l+1}, \mathbf{n}_{ji}) + \Lambda u_j^{l+1} = -(A\nabla u_i^l, \mathbf{n}_{ij}) + \Lambda u_i^l \quad \text{on } \Gamma = \partial\Omega_i \cap \partial\Omega_j, \tag{4}$$

with $\Lambda u = pu - q\partial_y(A_{yy}\partial_y u)$ (assuming that $\Gamma = \{x = 0\}$) and $\mathbf{n}_{ji}$ is the unit normal directed from Ω_j to Ω_i. A FV4 finite volume discretization of this algorithm for an advection diffusion equation with isotropic diffusion is analyzed in [5]. We present here a DDFV discretization of (3) and (4), and prove convergence of the discretized algorithm.

2 DDFV Schemes

The Meshes We now describe the DDFV Schwarz algorithm for general subdomains and decompositions using the notation from [2], see Fig. 1. The primal mesh $\mathfrak{M}_j$ is a set of disjoint open polygonal control volumes $\mathrm{K} \subset \Omega_j$ such that $\cup\overline{\mathrm{K}} = \overline{\Omega_j}$. We denote by $\partial\mathfrak{M}_j$ the set of edges of the control volumes in $\mathfrak{M}_j$ included in $\partial\Omega_j$, and by $\partial\mathfrak{M}_{j,\Gamma}$ the set of edges of primal boundary cells related to the interface Γ. We use the same notations for the dual mesh, $\mathfrak{M}_j^*$, $\partial\mathfrak{M}_j^*$ and $\partial\mathfrak{M}_{j,\Gamma}^*$. We define the diamond cells $\mathrm{D}_{\sigma,\sigma^*}$ as the quadrangles whose diagonals are a primal edge $\sigma = \mathrm{K}|\mathrm{L} = (x_{\mathrm{K}^*}, x_{\mathrm{L}^*})$ and a corresponding dual edge $\sigma^* = \mathrm{K}^*|\mathrm{L}^* = (x_{\mathrm{K}}, x_{\mathrm{L}})$. The set of diamond cells is called the diamond mesh, denoted by $\mathfrak{D}_j$.

For any V in $\mathfrak{M}_j \cup \partial\mathfrak{M}_j$ or $\mathfrak{M}_j^* \cup \partial\mathfrak{M}_j^*$, we denote by m_V its Lebesgue measure, by $\mathcal{E}_V$ the set of its edges, and $\mathfrak{D}_V := \{\mathrm{D}_{\sigma,\sigma^*} \in \mathfrak{D}_j,\ \sigma \in \mathcal{E}_V\}$. For $\mathrm{D} = \mathrm{D}_{\sigma,\sigma^*}$ with vertices $(x_{\mathrm{K}}, x_{\mathrm{K}^*}, x_{\mathrm{L}}, x_{\mathrm{L}^*})$, we denote by x_{D} the center of D, that is the intersection of the primal edge σ and the dual edge σ^*, by m_{D} its measure, by m_σ the length of σ, by m_{σ^*} the length of σ^*, by $m_{\sigma_{\mathrm{K}^*}}$ the length of $\partial\mathrm{K}^* \cap \Gamma$, by $m_{\sigma_{\mathrm{L}}}$ the length of $\mathrm{D} \cap \Gamma$, and by $m_{\sigma_{\mathrm{K}}}$ the length of $[x_{\mathrm{K}}, x_{\mathrm{D}}]$. $\mathbf{n}_{\sigma\mathrm{K}}$ is the unit vector normal to σ oriented from x_{K} to x_{L}, and $\mathbf{n}_{\sigma^*\mathrm{K}^*}$ is the unit vector normal to σ^* oriented from x_{K^*} to x_{L^*}.

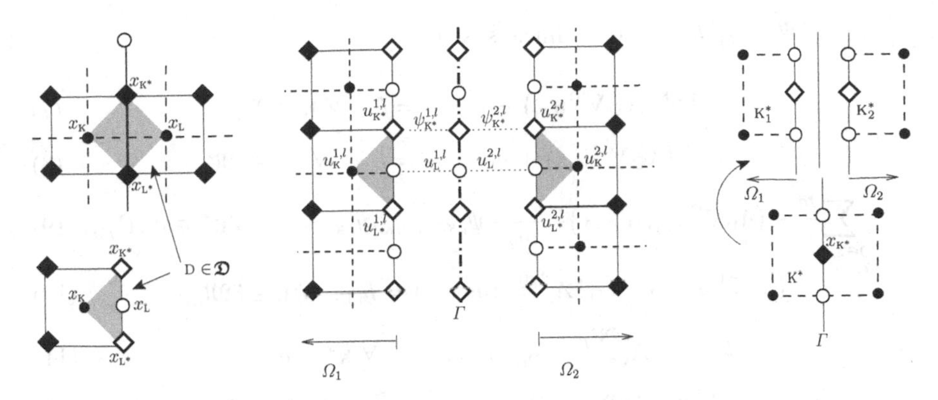

Fig. 1 *Diamond symbols* are vertices of primal cells, *circles* are vertices of dual cells. *Left*: zoom on diamond cells in *gray*. *Center*: zoom on the interface Γ, and new unknowns needed to describe the DDFV scheme as the limit of the Schwarz algorithm. *Right*: zoom on a dual cell K^* cut by Γ: $\mathrm{K}^* = \mathrm{K}_1^* \cup \mathrm{K}_2^*$ with $\mathrm{K}_i^* = \Omega_i \cap \mathrm{K}^*$

The Unknowns The DDFV method associates to all primal control volumes $\mathrm{K} \in \mathfrak{M}_j \cup \partial\mathfrak{M}_j$ an unknown value $u_{j,\mathrm{K}}$, and to all dual control volumes $\mathrm{K}^* \in \mathfrak{M}_j^* \cup \partial\mathfrak{M}_j^*$ an unknown value u_{j,K^*}. We denote the approximate solution on the mesh $\mathcal{T}_j$ by $u_{\mathcal{T}_j} = ((u_{j,\mathrm{K}})_{\mathrm{K}\in(\mathfrak{M}_j \cup \partial\mathfrak{M}_j)}, (u_{j,\mathrm{K}^*})_{\mathrm{K}^*\in(\mathfrak{M}_j^* \cup \partial\mathfrak{M}_j^*)}) \in \mathbb{R}^{\mathcal{T}_j}$. DDFV schemes are described by two operators: a discrete gradient $\nabla^{\mathfrak{D}}$ and a discrete divergence $\mathrm{div}^{\mathcal{T}}$, which are dual to each other, see [2]. We define the discrete gradient $\nabla^{\mathfrak{D}} : u_{\mathcal{T}} \in \mathbb{R}^{\mathcal{T}} \mapsto \left(\nabla^{\mathrm{D}} u_{\mathcal{T}}\right)_{\mathrm{D}\in\mathfrak{D}} \in (\mathbb{R}^2)^{\mathfrak{D}}$ by

$$\nabla^{\mathrm{D}} u_{\mathcal{T}} := \frac{1}{2m_{\mathrm{D}}} \left((u_{\mathrm{L}} - u_{\mathrm{K}}) m_\sigma \mathbf{n}_{\sigma\mathrm{K}} + (u_{\mathrm{L}^*} - u_{\mathrm{K}^*}) m_{\sigma^*} \mathbf{n}_{\sigma^*\mathrm{K}^*}\right), \quad \forall \mathrm{D} \in \mathfrak{D},$$

and the discrete divergence $\mathrm{div}^{\mathcal{T}} : \xi_{\mathfrak{D}} = (\xi_{\mathrm{D}})_{\mathrm{D}\in\mathfrak{D}} \mapsto \mathrm{div}^{\mathcal{T}} \xi_{\mathfrak{D}} \in \mathbb{R}^{\mathcal{T}}$ by

$$\mathrm{div}^{\mathrm{K}} \xi_{\mathfrak{D}} := \frac{1}{m_{\mathrm{K}}} \sum_{\mathrm{D}\in\mathfrak{D}_{\mathrm{K}}} m_\sigma (\xi_{\mathrm{D}}, \mathbf{n}_{\sigma\mathrm{K}}), \ \forall \mathrm{K} \in \mathfrak{M}, \text{ and } \mathrm{div}^{\mathrm{K}} \xi_{\mathfrak{D}} = 0, \forall \mathrm{K} \in \partial\mathfrak{M}, \quad (5)$$

$$\mathrm{div}^{\mathrm{K}^*} \xi_{\mathfrak{D}} := \frac{1}{m_{\mathrm{K}^*}} \sum_{\mathrm{D}\in\mathfrak{D}_{\mathrm{K}^*}} m_{\sigma^*} (\xi_{\mathrm{D}}, \mathbf{n}_{\sigma^*\mathrm{K}^*}), \ \forall \mathrm{K}^* \in \mathfrak{M}^* \cup \partial\mathfrak{M}^*. \quad (6)$$

We introduce additional flux unknowns ψ_{j,K^*} for $j = 1, 2$ on interface dual cells $\mathrm{K}^* \in \partial\mathfrak{M}_{j,\Gamma}^*$. Let N be the number of edges on Γ. We sort these edges $\sigma_1, \ldots, \sigma_N$ such that $\sigma_s \cap \sigma_{s+1} \neq \emptyset$, and $x_{\mathrm{K}_s^*}, x_{\mathrm{K}_{s+1}^*}$ are the vertices of σ_s, where $x_{\mathrm{K}_s^*} = \sigma_s \cap \sigma_{s-1}$. For $u_{\mathcal{T}_j} \in \mathbb{R}^{\mathcal{T}_j}$, $\Psi_{\mathcal{T}_j} \in \mathbb{R}^{\partial\mathfrak{M}_{j,\Gamma}^*}$, $f_{\mathcal{T}_j} \in \mathbb{R}^{\mathcal{T}_j}$ and $h_{\mathcal{T}_j} \in \mathbb{R}^{\partial\mathfrak{M}_{j,\Gamma} \cup \partial\mathfrak{M}_{j,\Gamma}^*}$, we denote by

$\mathcal{L}^{\mathcal{T}_j}_{\Omega_j,\Gamma}(u_{\mathcal{T}_j}, \Psi_{\mathcal{T}_j}, f_{\mathcal{T}_j}, h_{\mathcal{T}_j}) = 0$ the linear system

$$-\mathrm{div}^{\mathrm{K}}\left(A_{\mathfrak{D}}\nabla^{\mathfrak{D}}u_{\mathcal{T}_j}\right) + \eta_{\mathrm{K}}u_{j,\mathrm{K}} = f_{\mathrm{K}}, \quad \forall\, \mathrm{K} \in \mathfrak{M}_j, \tag{7}$$

$$-\mathrm{div}^{\mathrm{K}^*}\left(A_{\mathfrak{D}}\nabla^{\mathfrak{D}}u_{\mathcal{T}_j}\right) + \eta_{\mathrm{K}^*}u_{j,\mathrm{K}^*} = f_{\mathrm{K}^*}, \quad \forall\, \mathrm{K}^* \in \mathfrak{M}^*_j, \tag{8}$$

$$-\sum_{\mathrm{D}\in\mathfrak{D}_{\mathrm{K}^*}}\frac{m_{\sigma^*}}{m_{\mathrm{K}^*}}\left(A_{\mathrm{D}}\nabla^{\mathrm{D}}u_{\mathcal{T}_j}, \mathbf{n}_{\sigma^*\mathrm{K}^*}\right) - \frac{m_{\sigma_{\mathrm{K}^*}}}{m_{\mathrm{K}^*}}\psi_{j,\mathrm{K}^*} + \eta_{\mathrm{K}^*}u_{j,\mathrm{K}^*} = f_{\mathrm{K}^*}, \forall \mathrm{K}^* \in \partial\mathfrak{M}^*_{j,\Gamma}, \tag{9}$$

$$\left(A_{\mathrm{D}}\nabla^{\mathrm{D}}u_{\mathcal{T}_j}, \mathbf{n}_{\sigma\mathrm{L}_j}\right) + \Lambda^{\partial\mathfrak{M}_{j,\Gamma}}_{\mathrm{L}}(u_{\partial\mathfrak{M}_{j,\Gamma}}) = h_{j,\mathrm{L}}, \quad \forall\, \mathrm{L} \in \partial\mathfrak{M}_{j,\Gamma}, \tag{10}$$

$$\psi_{j,\mathrm{K}^*} + \Lambda^{\partial\mathfrak{M}^*_{j,\Gamma}}_{\mathrm{K}^*}(u_{\partial\mathfrak{M}^*_{j,\Gamma}}) = h_{j,\mathrm{K}^*}, \quad \forall\, \mathrm{K}^* \in \partial\mathfrak{M}^*_{j,\Gamma}, \tag{11}$$

$$u_{j,\mathrm{K}} = 0, \quad \forall\, \mathrm{K} \in \partial\mathfrak{M}_j \cap \partial\Omega, \qquad u_{j,\mathrm{K}^*} = 0, \quad \forall\, \mathrm{K}^* \in \partial\mathfrak{M}^*_j \cap \partial\Omega, \tag{12}$$

and for $s = 1, \cdots, N$

$$\Lambda^{\partial\mathfrak{M}_{j,\Gamma}}_{\mathrm{L}_s}(u_{\partial\mathfrak{M}_{j,\Gamma}}) = p u_{j,\mathrm{L}_s} - A_{yy}\frac{q}{m_{\sigma_s}}\left(\frac{u_{j,\mathrm{L}_{s+1}} - u_{j,\mathrm{L}_s}}{m_{\sigma_{\mathrm{K}^*_{s+1}}}} - \frac{u_{j,\mathrm{L}_s} - u_{j,\mathrm{L}_{s-1}}}{m_{\sigma_{\mathrm{K}^*_s}}}\right),$$

where $u_{j,\mathrm{L}_0} = u_{j,\mathrm{L}_{N+1}} = 0$, and for $s = 2, \cdots, N$

$$\Lambda^{\partial\mathfrak{M}^*_{j,\Gamma}}_{\mathrm{K}^*_s}(u_{\partial\mathfrak{M}^*_{j,\Gamma}}) = p u_{j,\mathrm{K}^*_s} - A_{yy}\frac{q}{m_{\sigma_{\mathrm{K}^*s}}}\left(\frac{u_{j,\mathrm{K}^*_{s+1}} - u_{j,\mathrm{K}^*_s}}{m_{\sigma_s}} - \frac{u_{j,\mathrm{K}^*_s} - u_{j,\mathrm{K}^*_{s-1}}}{m_{\sigma_{s-1}}}\right).$$

Note that $u_{j,\mathrm{K}^*_1} = u_{j,\mathrm{K}^*_{N+1}} = 0$ because of the homogeneous boundary condition on $\partial\Omega$. The unit normal $\mathbf{n}_{\sigma\mathrm{L}_j}$ is oriented from Ω_j to Ω_i.

Equations (7)–(9) correspond to approximations of the equation after integration on $\mathfrak{M}_j$, $\mathfrak{M}^*_j$ and $\partial\mathfrak{M}^*_j$; Eqs. (10) and (11) stem from the transmission condition on $\partial\mathfrak{M}_{j,\Gamma}$ and $\partial\mathfrak{M}^*_{j,\Gamma}$; Eq. (12) corresponds to the Dirichlet boundary condition on $\partial\Omega$.

The DDFV optimized Schwarz algorithm performs for an arbitrary initial guess $h^0_{\mathcal{T}_j} \in \mathbb{R}^{\partial\mathfrak{M}_{j,\Gamma}\cup\partial\mathfrak{M}^*_{j,\Gamma}}$, $j \in \{1,2\}$ and $l = 1, 2, \ldots$ the following steps:

- Compute for $j = 1, 2$ the solutions $(u^{l+1}_{\mathcal{T}_j}, \Psi^{l+1}_{\mathcal{T}_j}) \in \mathbb{R}^{\mathcal{T}_j} \times \mathbb{R}^{\partial\mathfrak{M}^*_{j,\Gamma}}$ of

$$\mathcal{L}^{\mathcal{T}_j}_{\Omega_j,\Gamma}(u^{l+1}_{\mathcal{T}_j}, \Psi^{l+1}_{\mathcal{T}_j}, f_{\mathcal{T}_j}, h^l_{\mathcal{T}_j}) = 0. \tag{13}$$

- Evaluate for $i, j \in \{1,2\}, j \neq i$ the new interface values $h^{l+1}_{\mathcal{T}_j}$ by

$$h^{l+1}_{j,\mathrm{L}} = -\left(A_{\mathrm{D}}\nabla^{\mathrm{D}}u^{l+1}_{\mathcal{T}_i}, \mathbf{n}_{\sigma\mathrm{L}_i}\right) + \Lambda^{\partial\mathfrak{M}_{j,\Gamma}}_{\mathrm{L}}(u^{l+1}_{\partial\mathfrak{M}_{i,\Gamma}}),\ \forall \mathrm{L} \in \partial\mathfrak{M}_{i,\Gamma}, \tag{14a}$$

$$h^{l+1}_{j,\mathrm{K}^*} = -\psi^{l+1}_{i,\mathrm{K}^*} + \Lambda^{\partial\mathfrak{M}^*_{j,\Gamma}}_{\mathrm{K}^*}(u^{l+1}_{\partial\mathfrak{M}^*_{i,\Gamma}}),\ \forall \mathrm{K}^* \in \partial\mathfrak{M}^*_{i,\Gamma}. \tag{14b}$$

Theorem 1 (Well-Posedness of Subdomain Problems) *For any* $f_{\mathcal{T}_j} \in \mathbb{R}^{\mathcal{T}_j}$ *and* $h_{\mathcal{T}_j} \in \mathbb{R}^{\partial\mathfrak{M}_{j,\Gamma}\cup\partial\mathfrak{M}^*_{j,\Gamma}}$, *there exists a unique solution* $(u_{\mathcal{T}_j}, \Psi_{\mathcal{T}_j}) \in \mathbb{R}^{\mathcal{T}_j} \times \mathbb{R}^{\partial\mathfrak{M}_{j,\Gamma}\cup\partial\mathfrak{M}^*_{j,\Gamma}}$ *of the linear system* $\mathcal{L}^{\mathcal{T}_j}_{\Omega_j,\Gamma}(u_{\mathcal{T}_j}, \Psi_{\mathcal{T}_j}, f_{\mathcal{T}_j}, h_{\mathcal{T}_j}) = 0$.

Proof By linearity, it is sufficient to prove that if $\mathcal{L}^{\mathcal{T}_j}_{\Omega_j,\Gamma}(u_{\mathcal{T}_j}, \Psi_{\mathcal{T}_j}, 0, 0) = 0$, then $u_{\mathcal{T}_j} = 0$ and $\Psi_{\mathcal{T}_j} = 0$. We multiply Eq. (7) by $m_K u_{j,K}$ and Eqs. (8) and (9) by $m_{K^*} u_{j,K^*}$ and sum the results over all control volumes in $\mathfrak{M}_j$ and $\mathfrak{M}^*j \cup \partial\mathfrak{M}^*_{j,\Gamma}$. Reordering the different contributions over all diamond cells, we obtain

$$2\sum_{D\in\mathfrak{D}} m_D(A_D\nabla^D u_{\mathcal{T}_j}, \nabla^D u_{\mathcal{T}_j}) + (\Lambda^{\partial\mathfrak{M}_\Gamma}(u_{\partial\mathfrak{M}_{j,\Gamma}}), u_{\partial\mathfrak{M}_{j,\Gamma}})$$
$$+ (\Lambda^{\partial\mathfrak{M}^*_\Gamma}(u_{\partial\mathfrak{M}^*_{j,\Gamma}}), u_{\partial\mathfrak{M}^*_{j,\Gamma}}) + \sum_{K\in\mathfrak{M}_j} m_K \eta_K u^2_{j,K} + \sum_{K^*\in\mathfrak{M}^*_j} m_{K^*}\eta_{K^*} u^2_{j,K^*} = 0.$$

The result thus follows by discrete Poincaré inequalities (see for example [2]) and the properties of $\Lambda^{\partial\mathfrak{M}_\Gamma}$ and $\Lambda^{\partial\mathfrak{M}^*_\Gamma}$. □

Theorem 2 (Convergence of the DDFV Schwarz Algorithm) *The solution of the Schwarz algorithm* (13) *and* (14) *converges as* l *goes to* ∞ *to the solution of the DDFV scheme on the entire domain* Ω.

Proof We follow the ideas of [5]: we first rewrite the DDFV scheme for the problem on Ω as the limit of the Schwarz algorithm. To this end, we introduce new unknowns near the boundary Γ, see Fig. 1:

- $\forall\, x_K \in \Omega_j$ and $x_{K^*} \in \Omega_j$, we set $u^\infty_{j,K} = u_K$ and $u^\infty_{j,K^*} = u_{K^*}$,
- $\forall\, x_K \in \partial\Omega$ and $x_{K^*} \in \partial\Omega$, we set $u^\infty_{j,K} = 0$ and $u^\infty_{j,K^*} = 0$,
- $\forall\, x_L \in \Gamma$, choose $u^\infty_{j,L}$ in such a way that $A_j\nabla^D u^\infty_{\mathcal{T}_j} \cdot \mathbf{n}_{\sigma K_j} = -A_i\nabla^D u^\infty_{\mathcal{T}_i} \cdot \mathbf{n}_{\sigma K_i}$:

$$u^\infty_{j,L} = u^\infty_{i,L} = \frac{m_{\sigma_{K_j}} m_{\sigma_{K_i}}}{\left(A_j m_{\sigma_{K_i}} + A_i m_{\sigma_{K_j}}\right)(\mathbf{n}_{\sigma K_j}, \mathbf{n}_{\sigma K_j})}\left[u_{K_j}\frac{(A_j\mathbf{n}_{\sigma K_j}, \mathbf{n}_{\sigma K_j})}{m_{\sigma_{K_j}}}\right.$$
$$\left. + u_{K_i}\frac{(A_i\mathbf{n}_{\sigma K_j}, \mathbf{n}_{\sigma K_j})}{m_{\sigma_{K_i}}} + \frac{u_{L^*} - u_{K^*}}{m_\sigma}\left(A_i - A_j\right)(\mathbf{n}_{\sigma^* K^*_j}, \mathbf{n}_{\sigma K_j})\right],$$

- $\forall\, x_{K^*} \in \Gamma$, $K^* = K^*_1 \cup K^*_2$ with $K^*_j \in \partial\mathfrak{M}^*_{j,\Gamma}$, choose $u^\infty_{j,K^*} = u^\infty_{i,K^*} = u_{K^*}$ and

$$\psi^\infty_{j,K^*} = -\psi^\infty_{i,K^*} = -\frac{1}{m_{\sigma_{K^*}}}\sum_{D\in\mathfrak{D}_{K^*_j}} m_{\sigma^*}\left(A_D\nabla^D u^\infty_{\mathcal{T}_j}, \mathbf{n}_{\sigma^* K^*_j}\right) + \frac{m_{K^*_j}}{m_{\sigma_{K^*}}}(\eta_{K^*}u_{K^*} - f_{K^*})$$
$$= \frac{1}{m_{\sigma_{K^*}}}\sum_{D\in\mathfrak{D}_{K^*_i}} m_{\sigma^*}\left(A_D\nabla^D u^\infty_{\mathcal{T}_i}, \mathbf{n}_{\sigma^* K^*_i}\right) - \frac{m_{K^*_i}}{m_{\sigma_{K^*}}}(\eta_{K^*}u_{K^*} - f_{K^*}).$$

By linearity, it suffices to prove convergence of the DDFV Schwarz algorithm (7) to 0. We have constructed $(u^{\infty}_{\mathcal{T}_j}, \psi^{\infty}_{\mathcal{T}_j})$ from the solution $u_{\mathcal{T}}$ of the DDFV scheme on Ω such that

$$\mathcal{L}^{\mathcal{T}_j}_{\Omega_j,\Gamma}(u^{\infty}_{\mathcal{T}_j}, \psi^{\infty}_{\mathcal{T}_j}, f_{\mathcal{T}_j}, h^{\infty}_{\mathcal{T}_j}) = 0.$$

Observe that the errors $e^{l+1}_{\mathcal{T}_j} = u^{l+1}_{\mathcal{T}_j} - u^{\infty}_{\mathcal{T}_j}$, $\Psi^{l+1}_{\mathcal{T}_j} = \psi^{l+1}_{\mathcal{T}_j} - \psi^{\infty}_{\mathcal{T}_j}$ satisfy

$$\mathcal{L}^{\mathcal{T}_j}_{\Omega_j,\Gamma}(e^{l+1}_{\mathcal{T}_j}, \Psi^{l+1}_{\mathcal{T}_j}, 0, H^{l}_{\mathcal{T}_j}) = 0,$$

with

$$\begin{aligned}
\forall\, \mathrm{K}^* \in \partial\mathfrak{M}^*_{i,\Gamma}, \quad H^{l}_{j,\mathrm{K}^*} &= -\Psi^{l}_{i,\mathrm{K}^*} + \Lambda^{\partial\mathfrak{M}^*_{i,\Gamma}}_{\mathrm{K}^*}(e^{l}_{\mathcal{T}_i}),\\
\forall\, \mathrm{Ł} \in \partial\mathfrak{M}_{i,\Gamma}, \quad H^{l}_{j,\mathrm{L}} &= -(A_{\mathrm{D}}\nabla^{\mathrm{D}} e^{l}_{\mathcal{T}_i}, \mathbf{n}_{\sigma\mathrm{L}_i}) + \Lambda^{\partial\mathfrak{M}_{i,\Gamma}}_{\mathrm{L}}(e^{l}_{\mathcal{T}_i}).
\end{aligned}$$

An a priori estimate using discrete duality leads to

$$\begin{aligned}
&2\sum_{\mathrm{D}\in\mathfrak{D}_j} m_{\mathrm{D}}(A_{\mathrm{D}}\nabla^{\mathrm{D}} e^{l+1}_{\mathcal{T}_j}, \nabla^{\mathrm{D}} e^{l+1}_{\mathcal{T}_j})\\
&\quad - \sum_{\mathrm{L}\in\partial\mathfrak{M}_{j,\Gamma}} m_{\sigma_{\mathrm{L}}}(A_{\mathrm{D}}\nabla^{\mathrm{D}} e^{l+1}_{\mathcal{T}_j}, \mathbf{n}_{\sigma\mathrm{L}_j}) e^{l+1}_{j,\mathrm{L}} - \sum_{\mathrm{K}^*\in\partial\mathfrak{M}^*_{j,\Gamma}} m_{\sigma_{\mathrm{K}^*}} \Psi^{l+1}_{j,\mathrm{K}^*} e^{l+1}_{j,\mathrm{K}^*}\\
&\qquad + \sum_{\mathrm{K}\in\mathfrak{M}_j} m_{\mathrm{K}}\eta_{\mathrm{K}}(e^{l+1}_{j,\mathrm{K}})^2 + \sum_{\mathrm{K}^*\in\mathfrak{M}^*_j\cup\partial\mathfrak{M}^*_{j,\Gamma}} m_{\mathrm{K}^*}\eta_{\mathrm{K}^*}(e^{l+1}_{j,\mathrm{K}^*})^2 = 0.
\end{aligned}$$

Using the scalar product defined by $(\Lambda^{\partial\mathfrak{M}_\Gamma})^{-1}$, we get

$$-\sum_{\mathrm{L}\in\partial\mathfrak{M}_{j,\Gamma}} m_{\sigma_{\mathrm{L}}}(A_{\mathrm{D}}\nabla^{\mathrm{D}} e^{l+1}_{\mathcal{T}_j}, \mathbf{n}_{\sigma\mathrm{L}_j}) e^{l+1}_{j,\mathrm{L}} = \Big((A_{\mathfrak{D}}\nabla^{\mathfrak{D}} e^{l+1}_{\mathcal{T}_j}, \mathbf{n}_j), \Lambda^{\partial\mathfrak{M}_\Gamma}(e^{l+1}_{\partial\mathfrak{M}_{j,\Gamma}})\Big)_{(\Lambda^{\partial\mathfrak{M}_\Gamma})^{-1}},$$

with $\mathbf{n}_j$ the unit outward normal of Ω_j. The formula $-4ab = (a-b)^2 - (a+b)^2$ now implies

$$\begin{aligned}
&-\sum_{\mathrm{L}\in\partial\mathfrak{M}_{j,\Gamma}} m_{\sigma_{\mathrm{L}}}(A_{\mathrm{D}}\nabla^{\mathrm{D}} e^{l+1}_{\mathcal{T}_j}, \mathbf{n}_{\sigma\mathrm{L}_j}) e^{l+1}_{j,\mathrm{L}}\\
&\qquad = \frac{1}{4}\left\| -(A_{\mathfrak{D}}\nabla^{\mathfrak{D}} e^{l+1}_{\mathcal{T}_j}, \mathbf{n}_j) + \Lambda^{\partial\mathfrak{M}_\Gamma}(e^{l+1}_{\partial\mathfrak{M}_{j,\Gamma}})\right\|^2_{(\Lambda^{\partial\mathfrak{M}_\Gamma})^{-1}}\\
&\qquad\quad - \frac{1}{4}\left\| (A_{\mathfrak{D}}\nabla^{\mathfrak{D}} e^{l+1}_{\mathcal{T}_j}, \mathbf{n}_j) + \Lambda^{\partial\mathfrak{M}_\Gamma}(e^{l+1}_{\partial\mathfrak{M}_{j,\Gamma}})\right\|^2_{(\Lambda^{\partial\mathfrak{M}_\Gamma})^{-1}}.
\end{aligned}$$

Using the Ventcell transmission condition, we now obtain

$$
\begin{aligned}
-\sum_{\mathrm{L}\in\partial\mathfrak{M}_{j,\Gamma}} & m_{\sigma_\mathrm{L}}(A_\mathrm{D}\nabla^\mathrm{D}e^{l+1}_{\mathcal{T}j}, \mathbf{n}_{\sigma\mathrm{L}_j})e^{l+1}_{j,\mathrm{L}} \\
&= \frac{1}{4}\left\| -(A_\mathfrak{D}\nabla^\mathfrak{D}e^{l+1}_{\mathcal{T}_j}, \mathbf{n}_j) + \Lambda^{\partial\mathfrak{M}_\Gamma}(e^{l+1}_{\partial\mathfrak{M}_{j,\Gamma}}) \right\|^2_{(\Lambda^{\partial\mathfrak{M}_\Gamma})^{-1}} \\
&\quad - \frac{1}{4}\left\| -(A_\mathfrak{D}\nabla^\mathfrak{D}e^{l}_{\mathcal{T}_i}, \mathbf{n}_i) + \Lambda^{\partial\mathfrak{M}_\Gamma}(e^{l}_{\partial\mathfrak{M}_{i,\Gamma}}) \right\|^2_{(\Lambda^{\partial\mathfrak{M}_\Gamma})^{-1}}.
\end{aligned}
$$

In a same way, we also obtain

$$
\begin{aligned}
-\sum_{\mathrm{K}^*\in\partial\mathfrak{M}^*_{j,\Gamma}} m_{\sigma_{\mathrm{K}^*}}\Psi^{l+1}_{j,\mathrm{K}^*}e^{l+1}_{j,\mathrm{K}^*} &= \frac{1}{4}\left\| -\Psi^{l+1}_{\mathcal{T}_j} + \Lambda^{\partial\mathfrak{M}^*_\Gamma}(e^{l+1}_{\partial\mathfrak{M}^*_{j,\Gamma}}) \right\|^2_{(\Lambda^{\partial\mathfrak{M}^*_\Gamma})^{-1}} \\
&\quad - \frac{1}{4}\left\| -\Psi^{l}_{\mathcal{T}_i} + \Lambda^{\partial\mathfrak{M}^*_\Gamma}(e^{l}_{\partial\mathfrak{M}^*_{i,\Gamma}}) \right\|^2_{(\Lambda^{\partial\mathfrak{M}^*_\Gamma})^{-1}}.
\end{aligned}
$$

Summing over l and j, the boundary terms cancel and we obtain the estimate

$$
\begin{aligned}
&2\sum_{l=0}^{l_{max}-1}\sum_{j=1,2}\sum_{\mathrm{D}\in\mathfrak{D}_j} m_\mathrm{D}(A_\mathrm{D}\nabla^\mathrm{D}e^{l+1}_{\mathcal{T}_j}, \nabla^\mathrm{D}e^{l+1}_{\mathcal{T}_j}) \\
&+ \sum_{n=0}^{l_{max}-1}\sum_{j=1,2}\sum_{\mathrm{K}\in\mathfrak{M}_j} m_\mathrm{K}\eta_\mathrm{K}(e^{l+1}_{j,\mathrm{K}})^2 + \sum_{n=0}^{l_{max}-1}\sum_{j=1,2}\sum_{\mathrm{K}^*\in\mathfrak{M}^*_j\cup\partial\mathfrak{M}^*_{j,\Gamma}} m_{\mathrm{K}^*}\eta_{\mathrm{K}^*}(e^{l+1}_{j,\mathrm{K}^*})^2 \\
&\le \sum_{j=1,2}\frac{1}{4}\left\| -(A_\mathfrak{D}\nabla^\mathfrak{D}e^{0}_{\mathcal{T}_j}, \mathbf{n}_j) + \Lambda^{\partial\mathfrak{M}_\Gamma}(e^{0}_{\partial\mathfrak{M}_{j,\Gamma}}) \right\|^2_{(\Lambda^{\partial\mathfrak{M}_\Gamma})^{-1}} \\
&+ \sum_{j=1,2}\frac{1}{4}\left\| -\Psi^{0}_{\mathcal{T}_j} + \Lambda^{\partial\mathfrak{M}^*_\Gamma}(e^{0}_{\partial\mathfrak{M}^*_{j,\Gamma}}) \right\|^2_{(\Lambda^{\partial\mathfrak{M}^*_\Gamma})^{-1}}.
\end{aligned}
$$

This shows that the total energy stays bounded as the iteration l goes to infinity, and hence the algorithm converges. □

3 Numerical Experiments

We use the domain $\Omega = (-1, 1) \times (0, 1)$ with the two subdomains $x > 0$ and $x < 0$. For the first experiment, we choose the data such that the exact solution is $u(x, y) = \cos(2.5\pi x)\cos(2.5\pi y)$, where we set $\eta := 1$ and

$$
A(x,y) := \begin{pmatrix} 1.5 & 0.5 \\ 0.5 & 1.5 \end{pmatrix} \text{ for } x < 0, \quad \text{and } A(x,y) := \begin{pmatrix} 1.5 & 0.5 \\ 0.5 & 1 \end{pmatrix} \text{ for } x > 0.
$$

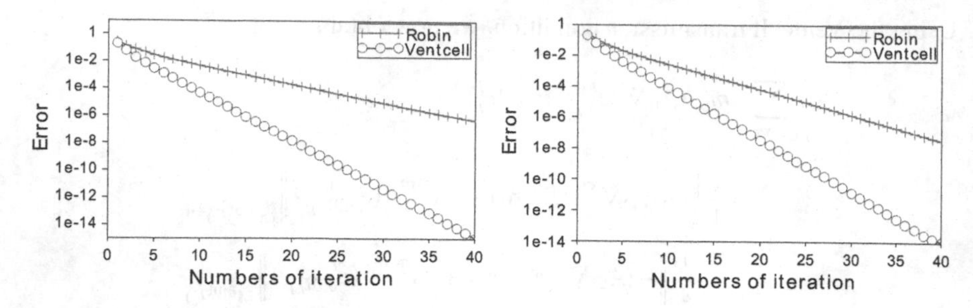

Fig. 2 Convergence history $\frac{\|u_n^{\mathcal{T}_i} - u^{\mathcal{T}_i}\|_2}{\|u^{\mathcal{T}_i}\|_2}$ for non-conforming square meshes (*left*) and a conforming triangle-square mesh configuration (*right*)

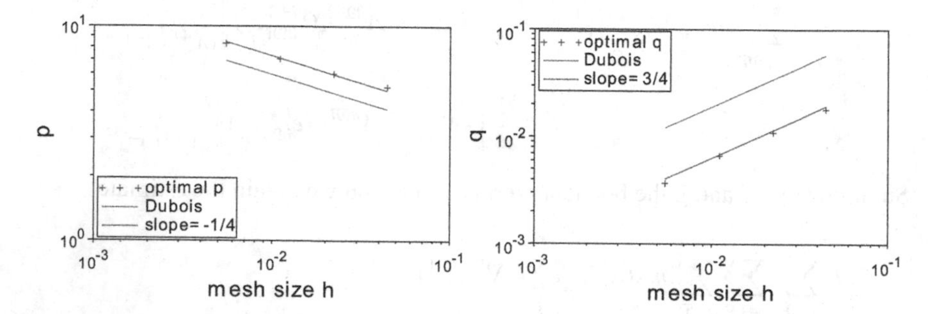

Fig. 3 Behavior of the numerically optimized parameter p on the *left*, q on the *right*

Starting with a random initial guess, Fig. 2 shows the convergence history of the algorithms using the Robin or Ventcell transmission conditions. For a fair comparison, the parameters p and q were numerically chosen to obtain the best convergence rate in each case. On the left, we used a non-conforming 32×32 square mesh on Ω_1 and a 48×48 square mesh on Ω_2 with $p = 11.2$ and $q = 0.007$ for the Ventcell transmission condition, and $p = 28$ and $q = 0$ for the Robin one. On the right, we used a conforming triangle-square mesh on Ω_1–Ω_2 with $p = 11.6$ and $q = 0.014$ for the Ventcell transmission condition, and $p = 23.5$ and $q = 0$ for the Robin one. We clearly see that the algorithm converges much faster with the Ventcell condition.

We next simulate the error equations, i.e. using homogeneous data, for a conforming square mesh ($2^i \times 2^i$ squares on Ω_j, $j = 1, 2$). We start again with a random initial guess. On the left in Fig. 3, we show the p that worked best as h is refined, and on the right the corresponding q. We also plot the asymptotic parameters from [3], which shows that the optimized parameters of the DDFV discretization behave asymptotically as expected.

In conclusion, we have shown how to discretize an optimized Schwarz algorithm with Ventcell transmission conditions using discrete duality finite volumes. Using energy estimates, we proved that the algorithm converges, and we showed in numerical experiments that the convergence is substantially faster than for Robin transmission conditions. We also showed that the optimized parameters behave asymptotically as expected from a continuous analysis. We are currently working on an asymptotic analysis for the optimized parameters and associated contraction factor of the algorithm.

References

1. P.M. Berthe, C. Japhet, P. Omnes, Space-time domain decomposition with finite volumes for porous media applications, in *Domain Decomposition Methods in Science and Engineering XXI*, vol. 98, ed. by J. Erhel, M. Gander, L. Halpern, G. Pichot, T. Sassi, O. Widlund (Springer, New York, 2014), pp. 479–486
2. F. Boyer, F. Hubert, S. Krell, Non-overlapping Schwarz algorithm for solving 2d m-DDFV schemes. IMA J. Numer. Anal. **30**, 1062–1100 (2009)
3. O. Dubois, Optimized Schwarz methods for the advection-diffusion equation and for problems with discontinuous coefficients. Ph.D. thesis, McGill University in Montréal, 2007
4. M.J. Gander, F. Hubert, S. Krell, Optimized Schwarz algorithms in the framework of DDFV schemes, in *Domain Decomposition Methods in Science and Engineering XXI*, vol. 98, ed. by J. Erhel, M. Gander, L. Halpern, G. Pichot, T. Sassi, O. Widlund (Springer, New York, 2014), pp. 391–398
5. L. Halpern, F. Hubert, A finite volume Ventcell-Schwarz algorithm for advection-diffusion equations. SIAM J. Numer. Anal. **52**(3) (2014)
6. R. Herbin, F. Hubert, Benchmark on discretization schemes for anisotropic diffusion problems on general grids, in *Proceedings of FVCA V*, ed. by R. Eymard, J.M. Hérard (ISTE Ltd and Wiley, 2008)
7. C. Japhet, Méthode de décomposition de domaine et conditions aux limites artificielles en mécanique des fluides: méthode Optimisée d'Ordre 2 (OO2). Ph.D. thesis, Université Paris 13, 1998

A Direct Solver for Time Parallelization

Martin J. Gander, Laurence Halpern, Juliet Ryan, and Thuy Thi Bich Tran

1 Introduction

Using the time direction in evolution problems for parallelization is an active field of research. Most of these methods are iterative, see for example the parareal algorithm analyzed in [3], a variant that became known under the name PFASST [10], and waveform relaxation methods based on domain decomposition [4, 5], see also [1] for a method called RIDC. Direct time parallel solvers are much more rare, see for example [2]. We present here a mathematical analysis of a different direct time parallel solver, proposed in [9]. We consider as our model partial differential equation (PDE) the heat equation on a rectangular domain Ω,

$$\frac{\partial u}{\partial t} - \Delta u = f \text{ in } \Omega \times (0, T), \quad u = g \text{ on } \partial\Omega, \quad \text{and } u(\cdot, 0) = u_0 \text{ in } \Omega. \tag{1}$$

Using a Backward Euler discretization on the time mesh $0 = t_0 < t_1 < t_2 < \cdots < t_N = T$, $k_n = t_n - t_{n-1}$, and a finite difference approximation Δ_h of Δ over a

M.J. Gander (✉)
University of Geneva, 2-4 rue du Lièvre, CP 64, 1211 Genève, Switzerland
e-mail: martin.gander@unige.ch

L. Halpern • T.T.B. Tran
Université Paris 13, LAGA, 93430 Villetaneuse, France
e-mail: halpern@math.univ-paris13.fr

J. Ryan
ONERA, BP72-29 avenue de la Divison Leclerc, 92322 Chatillon, France
e-mail: ryan@onera.fr

T. Dickopf et al. (eds.), *Domain Decomposition Methods in Science and Engineering XXII*, Lecture Notes in Computational Science and Engineering 104, DOI 10.1007/978-3-319-18827-0_50

rectangular grid of size $J = J_1 J_2$, we obtain the discrete problem

$$\frac{1}{k_n}(\mathbf{u}^n - \mathbf{u}^{n-1}) - \Delta_h \mathbf{u}^n = \mathbf{f}^n. \tag{2}$$

Let I_t be the $N \times N$ identity matrix associated with the time domain and I_x be the $J \times J$ identity matrix associated with the spatial domain. Setting $\mathbf{u} := (\mathbf{u}^1, \ldots, \mathbf{u}^N)$, $\mathbf{f} := (\mathbf{f}^1 + \frac{1}{k_1}\mathbf{u}_0, \mathbf{f}^2, \ldots, \mathbf{f}^N)$ and using the Kronecker symbol, (2) becomes

$$(B \otimes I_x - I_t \otimes \Delta_h)\mathbf{u} = \mathbf{f}, \qquad B := \begin{pmatrix} \frac{1}{k_1} & & & \\ -\frac{1}{k_2} & \frac{1}{k_2} & & 0 \\ 0 & \ddots & \ddots & \\ & & -\frac{1}{k_N} & \frac{1}{k_N} \end{pmatrix}. \tag{3}$$

If B is diagonalizable, $B = SDS^{-1}$, then (3) can be solved in 3 steps:

$$\begin{aligned} &(a) \quad (S \otimes I_x)\mathbf{g} = \mathbf{f}, \\ &(b) \ (\tfrac{1}{k_n} - \Delta_h)\mathbf{w}^n = \mathbf{g}^n, \quad 1 \le n \le N, \\ &(c) \ \ (S^{-1} \otimes I_x)\mathbf{u} = \mathbf{w}. \end{aligned} \tag{4}$$

The N equations in space in step (b) can now be solved in parallel. This interesting idea comes from [9], but its application requires some care: first, B is only diagonalizable if the time steps are all different, and this leads to a larger discretization error compared to using equidistant time steps, as we will see. Second, the condition number of S increases exponentially with N, which leads to inaccurate results in step (a) and (c) because of roundoff error. We accurately estimate these two errors, and then determine for a user tolerance the maximum N and optimal time step sequence of the form $k_{n+1} = qk_n$, which guarantees that errors stay below the user tolerance.

2 Error Estimate for Variable Time-Steps

We start by studying for $a > 0$ the ordinary differential equation (ODE)

$$\frac{du}{dt} + au = 0, \quad t \in (0, T), \quad u(0) = u_0 \quad \Longrightarrow \quad u(t) = u_0 a e^{-aT}. \tag{5}$$

For Backward Euler, $u_n = (1 + ak_n)^{-1}\, u_{n-1}$ with time steps from the division $\mathscr{T} = (k_1, \ldots, k_N)$ satisfying $T = \sum_1^N k_n$, we define the error propagator by

$$Err(\mathscr{T}; a, T, N) := \prod_{n=1}^{N}(1 + ak_n)^{-1} - e^{-aT}, \tag{6}$$

such that the error at time T equals $Err(\mathscr{T}; a, T, N)u^0$. We also define the equidistant division $\overline{\mathscr{T}} := (\bar{k}, \ldots, \bar{k})$, where $\bar{k} = T/N$.

Theorem 1 (Equidistant Partition Minimizes Error) *For any a, T and N, and any division $\mathscr{T}$, the error propagator is positive, and for the equidistant division $\overline{\mathscr{T}}$, the error is globally minimized.*

Proof Rewriting $Err(\mathscr{T}; a, T, N) = \prod_{n=1}^{N}(1 + ak_n)^{-1} - \prod_{n=1}^{N}(e^{ak_n})^{-1}$, we see that the error propagator is positive, since for all positive x, $e^x > 1 + x$. To minimize the error, we thus have to minimize $\Phi(\mathscr{T}) := \prod_{n=1}^{N}(1 + ak_n)^{-1}$ as a function of $\mathscr{T} \in \mathbb{R}^N$, with N inequality constraints $k_n \geq 0$, and one equality constraint $\sum_{n=1}^{N} k_n = T$. We compute the derivatives

$$\frac{\partial \Phi}{\partial k_i}(\mathscr{T}) = -\frac{a}{1 + ak_i}\Phi(\mathscr{T}), \quad \frac{\partial^2 \Phi}{\partial k_i k_j}(\mathscr{T}) = \frac{(1 + \delta_{ij})a^2}{(1 + ak_i)(1 + ak_j)}\Phi(\mathscr{T}),$$

and to show that Φ is convex, we evaluate for an arbitrary vector $\mathbf{x} = (x_1, \ldots, x_N)$

$$\sum_{i,j=1}^{N} \frac{\partial^2 \Phi}{\partial k_i k_j}(\mathscr{T})x_i x_j = \Big(\sum_{i \neq j} \frac{a^2}{(1 + ak_i)(1 + ak_j)} x_i x_j + \sum_{i} \frac{a^2}{(1 + ak_i)^2} x_i^2\Big)\Phi(\mathscr{T})$$
$$= \Big(\Big(\sum_{i=1}^{N} \frac{ax_i}{1 + ak_i}\Big)^2 + \sum_{i=1}^{N}\Big(\frac{ax_i}{1 + ak_i}\Big)^2\Big)\Phi(\mathscr{T}) > 0.$$

Therefore the Kuhn Tucker theorem applies, and the only minimum is given by the existence of a Lagrange multiplier p with $\Phi'(\mathscr{T}) + p\mathbf{1} = 0$, $\mathbf{1}$ the vector of all ones, whose only solution is $\mathscr{T} = \overline{\mathscr{T}}$, $p = a(1 + a\bar{k})^{-N-1}$.

We now consider a division $\mathscr{T}_q$ of geometric time steps $k_n := q^{n-1}k_1$ for $n = 1, \ldots, N$ as it was suggested in [8]. The constraint $\sum_{n=1}^{N} k_n = \sum_{n=1}^{N} q^{n-1}k_1 = T$ fixes k_1, and using this we get

$$k_n = \frac{q^n}{\sum_{j=1}^{N} q^j} T. \tag{7}$$

Since according to Theorem 1 the error is minimized for $q = 1$, one should not choose q very different from 1, and we now study the case $q = 1 + \varepsilon$ asymptotically.

Theorem 2 (Asymptotic Truncation Error Estimate) *Let $u_N(q) := \Phi(\mathscr{T}_q)u_0$ be the approximate solution obtained with the division $\mathscr{T}_q$ for $q = 1 + \varepsilon$. Then, for fixed a, T and N, the difference between the geometric mesh and fixed step mesh* *approximations satisfies for ε small*

$$u_N(q) - u_N(1) = \alpha(aT, N)u_0\varepsilon^2 + o(\varepsilon^2), \text{ with}$$
$$\alpha(x, N) = \frac{N(N^2-1)}{24}\left(\frac{x/N}{1+x/N}\right)^2 (1+x/N)^{-N}. \tag{8}$$

Proof Using a second order Taylor expansion, we obtain in the following two lemmas an expansion of $\Phi(\mathscr{T}_{1+\varepsilon})$ for small ε.

Lemma 1 *The time step k_n in (7) has for ε small the expansion $k_n = \bar{k}(1 + \alpha_n\varepsilon + \beta_n\varepsilon^2 + o(\varepsilon^2))$, with $\alpha_n = n - \frac{N+1}{2}$ and $\beta_n = n(n-N-2) + \frac{(N+1)(N+5)}{6}$. These coefficients satisfy the relations $\sum_n \alpha_n = \sum_n \beta_n = 0$, $\sum_n \alpha_n^2 = \frac{N(N-1)(N+1)}{12}$.*

Lemma 2 *For ε small, we have the expansion*

$$\prod_{n=1}^{N}(1 + ak_n) = (1 + a\bar{k})^N\Big(1 - \frac{b^2}{2}\sum_{n=1}^{N}\alpha_n^2\varepsilon^2 + o(\varepsilon^2)\Big), \text{with } b = \tfrac{a\bar{k}}{1+a\bar{k}}.$$

We can now apply Lemma 2 to obtain $\Phi(\mathscr{T}_{1+\varepsilon}) = \Phi(\mathscr{T}_1)(1 + \frac{b^2}{2}\sum_{n=1}^{N}\alpha_n^2\varepsilon^2 + o(\varepsilon^2))$, and replacing Φ in the definition of u_N concludes the proof.

3 Error Estimate for the Diagonalization of *B*

The matrix B is diagonalizable if and only if all time steps k_n are different. The eigenvalues are then $\frac{1}{k_n}$, and the eigenvectors form a basis of $\mathbb{R}^N$. We will see below that the matrix of eigenvectors is lower triangular. It can be chosen with unit diagonal, in which case it belongs to a special class of Toeplitz matrices:

Definition 1 A unipotent lower triangular Toeplitz matrix of size N is of the form

$$T(x_1, \ldots, x_{N-1}) = \begin{pmatrix} 1 & & & \\ x_1 & \ddots & & \\ \vdots & \ddots & \ddots & \\ x_{N-1} & \ldots & x_1 & 1 \end{pmatrix}. \tag{9}$$

Theorem 3 (Eigendecomposition of B**)** *If $k_n = q^{n-1}k_1$ as in (7), then B has the eigendecomposition $B = VDV^{-1}$, with $D := \mathrm{diag}(\frac{1}{k_n})$, and V and its inverse are* *unipotent lower triangular Toeplitz matrices given by*

$$V = T(p_1, \dots, p_{N-1}), \quad with \quad p_n := \frac{1}{\prod_{j=1}^{n}(1-q^j)}, \tag{10}$$

$$V^{-1} = T(q_1, \dots, q_{N-1}), \quad with \quad q_n := (-1)^n q^{\binom{n}{2}} p_n. \tag{11}$$

Proof Let $\mathbf{v}^{(n)}$ be the eigenvector with eigenvalue $\frac{1}{k_n}$. Since B is a lower bidiagonal matrix, a simple recursive argument shows that $v_j^{(n)} = 0$ for $j < n$. One may choose $v_n^{(n)} = 1$, which implies that for $j > n$ we have $v_j^{(n)} = (\prod_{i=1}^{n-j}(1 - \frac{k_{n+i}}{k_n}))^{-1}$, and the matrix $V = (\mathbf{v}^{(1)}, \dots, \mathbf{v}^{(N)})$ is lower triangular with unit diagonal. Furthermore, if $k_n = q^{n-1}k_1$, we obtain for $j = 1, 2, \dots, N-n$ that $v_{n+j}^{(n)} = (\prod_{i=1}^{j}(1-q^i))^{-1}$ which is independent of n and thus proves the Toeplitz structure in (10).

Consider now the inverse of V. First, it is easy to see that it is also unipotent Toeplitz. To establish (11) is equivalent to prove that

$$\text{for } 1 \le n \le N-1, \ \sum_{j=0}^{n} p_{n-j} q_j = 0, \text{ with the convention that } p_0 = q_0 = 1. \tag{12}$$

This result can be obtained using the $q-$analogue of the binomial formula, see [7]:

Theorem 4 (Simplified $q-$Binomial Theorem) *For any $q > 0$, $q \neq 1$, and for any $n \in \mathbb{N}$,*

$$\sum_{j=0}^{n}(-1)^j q^{\frac{j(j-1)}{2}} \frac{(1-q^{n-j+1})\cdots(1-q^n)}{(1-q)\cdots(1-q^j)} = 0. \tag{13}$$

Multiplying (13) by p_n then leads to (12).

In the steps (a) and (c) of the direct time parallel solver (4), the condition number of the eigenvector matrix S has a strong influence on the accuracy of the results. Normalizing the eigenvectors with respect to the ℓ^2 norm, $S := V\tilde{D}$, with $\tilde{D} = diag(\frac{1}{\sqrt{1+\sum_{i=1}^{N-n}|p_i|^2}})$, leads to an asymptotically better condition number:

Theorem 5 (Asymptotic Condition Number Estimate) *For $q = 1 + \varepsilon$, we have*

$$cond_\infty(V) \sim \left((N-1)!\varepsilon^{N-1}\right)^{-2}, \tag{14}$$

$$cond_\infty(S) \sim \frac{N}{\phi(N)}\varepsilon^{-(N-1)}, \quad \phi(N) = \begin{cases} \frac{N}{2}!(\frac{N}{2}-1)! & \textit{if } N \textit{ is even,} \\ (\frac{N-1}{2}!)^2 & \textit{if } N \textit{ is odd.} \end{cases} \tag{15}$$

Proof Note first that $|q_n| \sim |p_n| \sim (n!\,\varepsilon^n)^{-1}$. Therefore

$$\|V\|_\infty = 1 + |p_1| + |p_2| + \cdots + |p_{N-1}| \sim |p_{N-1}| \sim \left((N-1)!\varepsilon^{N-1}\right)^{-1}.$$

The same holds for V^{-1}, and gives the first result. We next define $\gamma_n := \sqrt{1+\sum_{j=1}^{N-n}|p_j|^2}$, $\tilde{d}_n := \frac{1}{\gamma_n}$, which implies $\tilde{D} = diag(\tilde{d}_n)$. Then $\gamma_n \sim |p_{N-n}|$, and we obtain

$$\|S\|_\infty = \sup_n \sum_{j=1}^{n} \frac{|p_{n-j}|}{\gamma_j} \sim \sup_n \sum_{j=1}^{n} \frac{|p_{n-j}|}{|p_{N-n}|} \sim \sup_n \sum_{j=1}^{n} \frac{(N-j)!}{(n-j)!}\varepsilon^{N-n} \sim N.$$

By definition $S^{-1} = \tilde{D}^{-1}V^{-1} = \tilde{D}^{-1}T(q_1,\cdots,q_{N-1})$, that is the line n of $T(q_1,\cdots,q_{N-1})$ is multiplied by γ_n. Therefore

$$\|S^{-1}\|_\infty = \sup_n \gamma_n \sum_{j=0}^{n-1}|q_j| \sim \sup_n \gamma_n|q_{n-1}| \sim \sup_n \gamma_n|p_{n-1}| \sim \sup_n |p_{N-n}||p_{n-1}|$$
$$\sim \varepsilon^{-(N-1)} \sup_n \frac{1}{(n-1)!(N-n)!} \sim \frac{1}{\phi(N)}\varepsilon^{-(N-1)}.$$

4 Relative Error Estimates for ODEs and PDEs

We first give an error estimate for the ODE (5):

Theorem 6 (Asymptotic Roundoff Error Estimate for ODEs) *Let* $\mathbf{u}$ *be the exact solution of* $B\mathbf{u} = \mathbf{f}$*, and* $\hat{\mathbf{u}}$ *be the computed solution from the direct time parallel solver (4) applied to (5), and* $\underline{u}$ *denote the machine precision. Then*

$$\frac{\|\mathbf{u}-\hat{\mathbf{u}}\|_\infty}{\|\mathbf{u}\|_\infty} \lesssim \underline{u}\,\frac{N^2(2N+1)(N+aT)}{\phi(N)}\,\varepsilon^{-(N-1)}. \tag{16}$$

Proof In the ODE case, $D = diag(\frac{1}{k_n}+a)$ and $\|D\|_\infty = 1/k_1 + a$. Using backward error analysis [6], the computed solution satisfies the perturbed systems

$$(S+\delta S_1)\hat{\mathbf{g}} = \mathbf{f}, \quad (D+\delta D)\hat{\mathbf{w}} = \hat{\mathbf{g}}, \quad (S^{-1}+\delta S_2)\hat{\mathbf{u}} = \hat{\mathbf{w}},$$

and since S and S^{-1} are triangular and D is diagonal we get (see [6])

$$\|\delta S_1\| \le N\underline{u}\|S\| + \mathcal{O}(\underline{u}^2), \quad \|\delta S_2\| \le N\underline{u}\|S^{-1}\| + \mathcal{O}(\underline{u}^2), \quad \|\delta D\| \le \underline{u}\|D\| + \mathcal{O}(\underline{u}^2).$$

Using Algorithm (4) to solve $B\mathbf{u} = \mathbf{f}$ by decomposition is equivalent to solving $(S+\delta S_1)(D+\delta D)(S^{-1}+\delta S_2)\,\hat{\mathbf{u}} = \mathbf{f}$, which is of the form

$$(B+\delta B)\hat{\mathbf{u}} = \mathbf{f}, \quad \|\delta B\| \le (2N+1)\underline{u}\|S\|\|S^{-1}\|\|D\| + \mathcal{O}(\underline{u}^2).$$

The relative error then satisfies (see [6])

$$\frac{\|\mathbf{u}-\hat{\mathbf{u}}\|}{\|\mathbf{u}\|} \le \mathrm{cond}(B)\frac{\|\delta B\|}{\|B\|} \le (2N+1)\underline{u}\,\|B^{-1}\|\,\|S\|\,\|S^{-1}\|\,\|D\|.$$

By a direct computation, we obtain for the inverse of B

$$B^{-1} = k_1 \begin{pmatrix} 1 & & & & \\ 1 & q & & & \\ 1 & \vdots & q^2 & & \\ \vdots & \vdots & \vdots & \ddots & \\ 1 & q & q^2 & \dots & q^{N-1} \end{pmatrix},$$

and hence $\|B^{-1}\|_\infty = k_1(1+q+\dots q^{N-1}) \sim Nk_1$. Since $k_1 \sim \overline{k} = T/N$, (16) is proved.

The error of the direct time parallel solver at time T can be estimated by

$$\frac{|e^{-aT}u_0 - \hat{u}_N|}{|u_0|} \le \frac{|e^{-aT}u_0 - u_N(1)|}{|u_0|} + \frac{|u_N(1) - u_N(q)|}{|u_0|} + \frac{|u_N(q) - \hat{u}_N|}{|u_0|}. \tag{17}$$

The first term on the right is the truncation error of the sequential method using equal time steps. The second term is due to the geometric mesh and was estimated asymptotically in Theorem 2 to be $\alpha\varepsilon^2$. The last term can be estimated by $\|\mathbf{u}(q) - \hat{\mathbf{u}}\|_\infty/|u_0|$ and thus Theorem 6, since $a > 0$ which implies $|u_0| = \|\mathbf{u}\|_\infty$. Because the second term is decreasing in ε and the last term is growing in ε, we equilibrate them asymptotically:

Theorem 7 (Optimized Geometric Time Mesh) *Suppose the time steps are geometric, $k_n = q^{n-1}k_1$, and $q = 1+\varepsilon$ with ε small. Let $\underline{u}$ be the machine precision. For $\varepsilon = \varepsilon_0(aT, N)$ with*

$$\varepsilon_0(aT, N) = \left(\underline{u}\frac{N^2(2N+1)(N+aT)}{\phi(N)\alpha(aT,N)}\right)^{\frac{1}{N+1}}, \tag{18}$$

where $\alpha(aT, N)$ is defined in (8) and $\phi(N)$ in (15), the error due to time parallelization is asymptotically comparable to the one produced by the geometric mesh.

Proof This is a direct consequence of Theorem 6.

We show in Fig. 1 on the left the optimized value $\varepsilon_0(aT, N)$ from Theorem 7. Choosing $\varepsilon = \varepsilon_0(aT, N)$, the ratio between the additional errors due to parallelization to the truncation error of the fixed time step method is shown in Fig. 1 on the right.

In order to obtain a PDE error estimate for the heat equation (1), one can argue as follows: expanding the solution in eigenfunctions of the Laplacian, we can apply our results for the ODE with $a = \lambda_\ell$, λ_ℓ for $\ell = 1, 2, \dots$ the eigenvalues of the negative Laplacian. One can show for all $N \ge 1$ and machine precision $\underline{u}$ small enough (single precision suffices in practice) that our error estimate (17) has its maximum for $aT < a_T^* = 2.5129$, and thus if $\lambda_{\min}$ and T are such that $\lambda_{\min}T > a_T^*$, one can

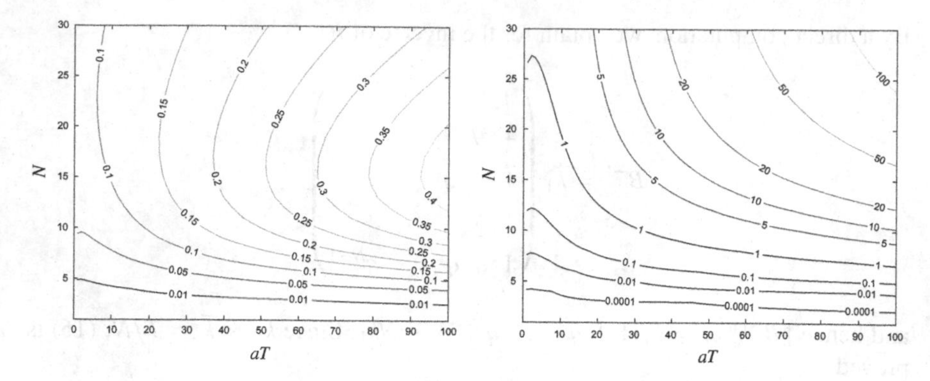

Fig. 1 Optimized choice $\varepsilon_0(aT, N)$ from Theorem 7 (*left*). Ratio of the additional errors due to parallelization to the truncation error of the fixed step method (17) with this choice of $\varepsilon_0(aT, N)$ (*right*)

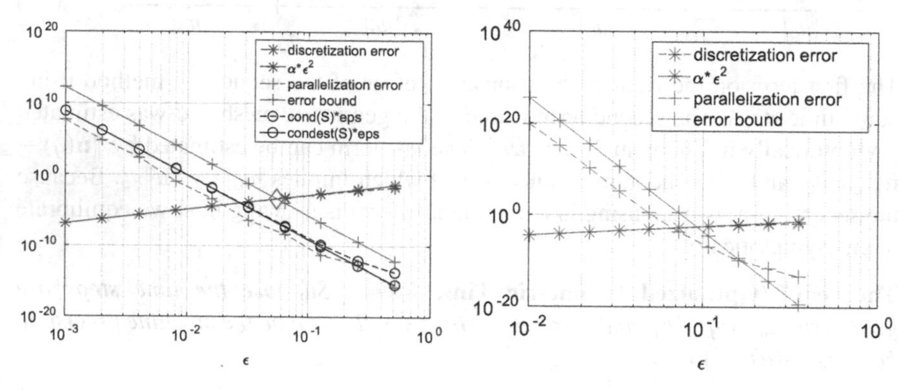

Fig. 2 Discretization and parallelization errors, and condition number of the eigenvector matrix, together with our theoretical bounds, *left* for the ODE, *right* for the PDE

read off the optimal choice ε_0 and resulting error estimate in Fig. 1 at $aT = \lambda_{\min} T$ for a given number of processors N. Similarly, if we have N processors and do not want to increase the error compared to a sequential computation by more than a given factor, we can read in Fig. 1 on the right the size of the time window T to use (knowing $a = \lambda_{\min}$), and the corresponding optimized ε_0 on the left.

5 Numerical Experiments

We first perform a numerical experiment for the scalar model problem (5) with $a = 1, T = 1$ and $N = 10$. We show in Fig. 2 on the left how the discretization error increases and the parallelization error decreases as a function of ε, together with our theoretical estimates, and also the condition number of the eigenvector matrix, and our theoretical bound. We can see that the theoretically predicted optimized ε marked by a rhombus is a good estimate, and on the safe side, of the optimal choice a bit to the left, where the dashed lines meet.

We next show an experiment for the heat equation in two dimensions on the unit square, with homogeneous Dirichlet boundary conditions and an initial condition $u_0(x, y) = \sin(\pi x)\sin(\pi y)$. We discretize with a standard five point finite difference method in space with mesh size $h_1 = h_2 = \frac{1}{10}$, and a Backward Euler discretization in time on the time interval $(0, \frac{1}{5})$ using $N = 30$ time steps. In Fig. 2 on the right we show again the measured discretization and parallelization errors compared to our theoretical bounds. As one can see from the graph, in this example, one could solve the problem using 30 processors, and would obtain an error which is within a factor two of the sequential computation.

References

1. A.J. Christlieb, C.B. Macdonald, B.W. Ong, Parallel high-order integrators. SIAM J. Sci. Comput. **32**(2), 818–835 (2010)
2. M.J. Gander, S. Güttel, Paraexp: a parallel integrator for linear initial-value problems. SIAM J. Sci. Comput. **35**(2), C123–C142 (2013)
3. M.J. Gander, E. Hairer, Nonlinear convergence analysis for the parareal algorithm, in *Domain Decomposition Methods in Science and Engineering XVII*, ed. by O.B. Widlund, D.E. Keyes, vol. 60 (Springer, Berlin, 2008), pp. 45–56
4. M.J. Gander, L. Halpern, Absorbing boundary conditions for the wave equation and parallel computing. Math. Comput. **74**, 153–176 (2005)
5. M.J. Gander, L. Halpern, Optimized Schwarz waveform relaxation methods for advection reaction diffusion problems. SIAM J. Numer. Anal. **45**(2), 666–697 (2007)
6. G.H. Golub, C.F. Van Loan, *Matrix Computations*, 4th edn. Johns Hopkins Studies in the Mathematical Sciences (Johns Hopkins University Press, Baltimore, 2013)
7. J. Haglund, *The q,t-Catalan Numbers and the Space of Diagonal Harmonics*. University Lecture Series, vol. 41 (American Mathematical Society, Providence, 2008)
8. Y. Maday, E.M. Rønquist, Fast tensor product solvers. Part II: spectral discretization in space and time. Technical report 7–9, Laboratoire Jacques-Louis Lions (2007)
9. Y. Maday, E.M. Rønquist, Parallelization in time through tensor-product space-time solvers. C. R. Math. Acad. Sci. Paris **346**(1–2), 113–118 (2008)
10. M.L. Minion, A hybrid parareal spectral deferred corrections method. Commun. Appl. Math. Comput. Sci. **5**(2), 265–301 (2010)

Dirichlet-Neumann and Neumann-Neumann Waveform Relaxation for the Wave Equation

Martin J. Gander, Felix Kwok, and Bankim C. Mandal

1 Introduction

We present two new types of Waveform Relaxation (WR) methods for hyperbolic problems based on the Dirichlet-Neumann and Neumann-Neumann algorithms, and present convergence results for these methods. The Dirichlet-Neumann algorithm for elliptic problems was first considered by Bjørstad and Widlund [2]; the Neumann-Neumann algorithm was introduced by Bourgat et al. [3]. The performance of these algorithms for elliptic problems is now well understood, see for example the book [13].

To solve time-dependent problems in parallel, one can either discretize in time to obtain a sequence of steady problems, and then apply domain decomposition algorithms to solve the steady problems at each time step in parallel, or one can first discretize in space and then apply WR to the large system of ordinary differential equations (ODEs) obtained from the spatial discretization. WR has its roots in the work of Picard and Lindelöf, who studied existence and uniqueness of solutions of ODEs in the late nineteenth century. Lelarasmee et al. [11] rediscovered WR as a parallel method for the solution of ODEs. The main computational advantage of WR is parallelization, and the possible use of different discretizations in different space-time subdomains.

Domain decomposition methods for elliptic PDEs can be extended to time-dependent problems by using the same decomposition in space. This leads to WR

M.J. Gander • B.C. Mandal (✉)
Department of Mathematics, University of Geneva, Geneva, Switzerland
e-mail: Martin.Gander@unige.ch; Bankim.Mandal@unige.ch

F. Kwok
Department of Mathematics, Hong Kong Baptist University, Kowloon, Hong Kong
e-mail: felix_kwok@hkbu.edu.hk

T. Dickopf et al. (eds.), *Domain Decomposition Methods in Science and Engineering XXII*, Lecture Notes in Computational Science and Engineering 104, DOI 10.1007/978-3-319-18827-0_51

type methods, see [1]. The systematic extension of the classical Schwarz method to time-dependent parabolic problems was started independently in [8, 9]. Like WR algorithms in general, the so-called Schwarz Waveform Relaxation algorithms (SWR) converge relatively slowly, except if the time window size is short. A remedy is to use optimized transmission conditions, which leads to much faster algorithms, see [4] for parabolic problems, and [5] for hyperbolic problems. More recently, we studied the WR extension of the Dirichlet-Neumann and Neumann-Neumann methods for parabolic problems [6, 10, 12]. We proved for the heat equation that on finite time intervals, the Dirichlet-Neumann Waveform Relaxation (DNWR) and the Neumann-Neumann Waveform Relaxation (NNWR) methods converge superlinearly for an optimal choice of the relaxation parameter. DNWR and NNWR also converge faster than classical and optimized SWR in this case.

In this paper, we define DNWR and NNWR for the second order wave equation

$$\begin{aligned} \partial_{tt}u - c^2\Delta u &= f(\boldsymbol{x},t), && \boldsymbol{x}\in\Omega, 0<t<T,\\ u(\boldsymbol{x},0) &= u_0(\boldsymbol{x}),\ u_t(\boldsymbol{x},0) = v_0(\boldsymbol{x}), && \boldsymbol{x}\in\Omega,\\ u(\boldsymbol{x},t) &= g(\boldsymbol{x},t), && \boldsymbol{x}\in\partial\Omega, 0<t<T, \end{aligned} \tag{1}$$

where $\Omega\subset\mathbb{R}^d$, $d = 1,2,3$, is a bounded domain with a smooth boundary, and c denotes the wave speed, and we analyze the convergence of both algorithms for the 1d wave equation.

2 Domain Decomposition and Algorithms

To explain the new algorithms, we assume for simplicity that the spatial domain Ω is partitioned into two non-overlapping subdomains Ω_1 and Ω_2. We denote by u_i the restriction of the solution u of (1) to Ω_i, $i = 1,2$, and by n_i the unit outward normal for Ω_i on the interface $\Gamma := \partial\Omega_1\cap\partial\Omega_2$.

The *Dirichlet-Neumann Waveform Relaxation algorithm (DNWR)* consists of the following steps: given an initial guess $h^0(x,t)$, $t\in(0,T)$ along the interface $\Gamma\times(0,T)$, compute for $k = 1,2,\ldots$ with $u_1^k = g$, on $\partial\Omega_1\setminus\Gamma$ and $u_2^k = g$, on $\partial\Omega_2\setminus\Gamma$ the approximations

$$\begin{aligned} \partial_{tt}u_1^k - c^2\Delta u_1^k &= f, &&\text{in } \Omega_1, & \partial_{tt}u_2^k - c^2\Delta u_2^k &= f, &&\text{in } \Omega_2,\\ u_1^k(\boldsymbol{x},0) &= u_0(\boldsymbol{x}), &&\text{in } \Omega_1, & u_2^k(\boldsymbol{x},0) &= u_0(\boldsymbol{x}), &&\text{in } \Omega_2,\\ \partial_t u_1^k(\boldsymbol{x},0) &= v_0(\boldsymbol{x}), &&\text{in } \Omega_1, & \partial_t u_2^k(\boldsymbol{x},0) &= v_0(\boldsymbol{x}), &&\text{in } \Omega_2,\\ u_1^k &= h^{k-1}, &&\text{on } \Gamma, & \partial_{n_2}u_2^k &= -\partial_{n_1}u_1^k, &&\text{on } \Gamma,\\ h^k(x,t) &= \theta u_2^k\big|_{\Gamma\times(0,T)} + (1-\theta)h^{k-1}(x,t), \end{aligned} \tag{2}$$

where $\theta\in(0,1]$ is a relaxation parameter.

The *Neumann-Neumann Waveform Relaxation algorithm (NNWR)* starts with an initial guess $w^0(x,t)$, $t \in (0,T)$ along the interface $\Gamma \times (0,T)$ and then computes for $\theta \in (0,1]$ simultaneously for $i = 1,2$ with $k = 1,2,\ldots$

$$\begin{array}{rlrl}
\partial_{tt}u_i^k - c^2\Delta u_i^k = f, & \text{in } \Omega_i, & \partial_{tt}\psi_i^k - c^2\Delta\psi_i^k = 0, & \text{in } \Omega_i,\\
u_i^k(\boldsymbol{x},0) = u_0(\boldsymbol{x}), & \text{in } \Omega_i, & \psi_i^k(\boldsymbol{x},0) = 0, & \text{in } \Omega_i,\\
\partial_t u_i^k(\boldsymbol{x},0) = v_0(\boldsymbol{x}), & \text{in } \Omega_i, & \partial_t\psi_i^k(\boldsymbol{x},0) = 0, & \text{in } \Omega_i,\\
u_i^k = g, & \text{on } \partial\Omega_i \setminus \Gamma, & \psi_i^k = 0, & \text{on } \partial\Omega_i \setminus \Gamma,\\
u_i^k = w^{k-1}, & \text{on } \Gamma, & \partial_{n_i}\psi_i^k = \partial_{n_1}u_1^k + \partial_{n_2}u_2^k, & \text{on } \Gamma,
\end{array}$$
$$w^k(x,t) = w^{k-1}(x,t) - \theta\left[\psi_1^k\left|_{\Gamma\times(0,T)} + \psi_2^k\right|_{\Gamma\times(0,T)}\right]. \tag{3}$$

3 Kernel Estimates and Convergence Analysis

We present the case $d = 1$, with $\Omega = (-a,b)$, $\Omega_1 = (-a,0)$ and $\Omega_2 = (0,b)$. By linearity, it suffices to study the error equations, $f(\boldsymbol{x},t) = 0$, $g(\boldsymbol{x},t) = 0$, $u_0(\boldsymbol{x}) = v_0(\boldsymbol{x}) = 0$ in (2) and (3), and to examine convergence to zero.

Our convergence analysis is based on Laplace transforms. The Laplace transform of a function $u(x,t)$ with respect to time t is defined by $\hat{u}(x,s) = \mathcal{L}\{u(x,t)\} := \int_0^\infty e^{-st}u(x,t)\,dt$, $s \in \mathbb{C}$. Applying a Laplace transform to the DNWR algorithm in (2) in 1d, we obtain for the transformed error equations

$$\begin{array}{rlrl}
(s^2 - c^2\partial_{xx})\hat{u}_1^k(x,s) = 0 & \text{in } (-a,0), & (s^2 - c^2\partial_{xx})\hat{u}_2^k(x,s) = 0 & \text{in } (0,b),\\
\hat{u}_1^k(-a,s) = 0, & & \partial_x\hat{u}_2^k(0,s) = \partial_x\hat{u}_1^k(0,s), &\\
\hat{u}_1^k(0,s) = \hat{h}^{k-1}(s), & & \hat{u}_2^k(b,s) = 0, &
\end{array}$$
$$\hat{h}^k(s) = \theta\hat{u}_2^k(0,s) + (1-\theta)\hat{h}^{k-1}(s). \tag{4}$$

Solving the two-point boundary value problems in (4), we get

$$\hat{u}_1^k = \frac{\hat{h}^{k-1}(s)}{\sinh(as/c)}\sinh\left((x+a)\tfrac{s}{c}\right),\ \hat{u}_2^k = \hat{h}^{k-1}(s)\frac{\coth(as/c)}{\cosh(bs/c)}\sinh\left((x-b)\tfrac{s}{c}\right),$$

and inserting them into the updating condition [last line in (4)], we get by induction

$$\hat{h}^k(s) = \left[1 - \theta - \theta\coth(as/c)\tanh(bs/c)\right]^k\hat{h}^0(s), \quad k = 1,2,\ldots \tag{5}$$

Similarly, the Laplace transform of the NNWR algorithm in (3) for the error equations yields for the subdomain solutions

$$\begin{aligned}
\hat{u}_1^k(x,s) &= \frac{\hat{w}^{k-1}(s)}{\sinh(as/c)}\sinh\left((x+a)\tfrac{s}{c}\right), & \hat{u}_2^k(x,s) &= -\frac{\hat{w}^{k-1}(s)}{\sinh(bs/c)}\sinh\left((x-b)\tfrac{s}{c}\right),\\
\hat{\psi}_1^k(x,s) &= \frac{\hat{w}^{k-1}(s)\Psi(s)}{\cosh(as/c)}\sinh\left((x+a)\tfrac{s}{c}\right), & \hat{\psi}_2^k(x,s) &= \frac{\hat{w}^{k-1}(s)\Psi(s)}{\cosh(bs/c)}\sinh\left((x-b)\tfrac{s}{c}\right),
\end{aligned}$$

where $\Psi(s) = [\coth(as/c) + \coth(bs/c)]$. Therefore, in Laplace space the updating condition in (3) becomes

$$\hat{w}^k(s) = \left[1 - \theta\left(2 + \frac{\coth(as/c)}{\coth(bs/c)} + \frac{\coth(bs/c)}{\coth(as/c)}\right)\right]^k \hat{w}^0(s), \quad k = 1, 2, \ldots \tag{6}$$

Theorem 1 (Convergence, Symmetric Decomposition) *For a symmetric decomposition, $a = b$, convergence is linear for DNWR (2) with $\theta \in (0, 1)$, $\theta \neq \frac{1}{2}$, and for NNWR (3) with $\theta \in (0, 1)$, $\theta \neq \frac{1}{4}$. If $\theta = \frac{1}{2}$ for DNWR, or $\theta = \frac{1}{4}$ for NNWR, convergence is achieved in two iterations.*

Proof For $a = b$, Eq. (5) reduces to $\hat{h}^k(s) = (1 - 2\theta)^k \hat{h}^0(s)$, which has the simple back transform $h^k(t) = (1 - 2\theta)^k h^0(t)$. Thus for the DNWR method, the convergence is linear for $0 < \theta < 1, \theta \neq \frac{1}{2}$. For $\theta = \frac{1}{2}$, we have $h^1(t) = 0$. Hence, one more iteration produces the desired solution on the whole domain.

For the NNWR algorithm, inserting $a = b$ into Eq. (6), we obtain similarly $w^k(t) = (1 - 4\theta)^k w^0(t)$, which leads to the second result. □

We next analyze the case of an asymmetric decomposition, $a \neq b$.

Lemma 1 *Let $a, b > 0$ and $s \in \mathbb{C}$, with $Re(s) > 0$. Then, we have the identity*

$$\begin{aligned} G_b^a(s) &:= \coth(as/c)\tanh(bs/c) - 1 \\ &= 2\sum_{m=1}^{\infty} e^{-2ams/c} - 2\sum_{n=1}^{\infty}(-1)^{n-1}e^{-2bns/c} - 4\sum_{n=1}^{\infty}\sum_{m=1}^{\infty}(-1)^{n-1}e^{-2(bn+am)s/c}. \end{aligned}$$

Proof Using that $|e^{-2bs/c}| < 1$ for $\mathrm{Re}(s) > 0$, we expand $\left(1 + e^{-2bs/c}\right)^{-1}$ into an infinite binomial series to obtain

$$\tanh\left(\frac{bs}{c}\right) = \frac{e^{\frac{bs}{c}} - e^{-\frac{bs}{c}}}{e^{\frac{bs}{c}} + e^{-\frac{bs}{c}}} = \left(1 - e^{-\frac{2bs}{c}}\right)\left(1 + e^{-\frac{2bs}{c}}\right)^{-1} = 1 - 2\sum_{n=1}^{\infty}(-1)^{n-1}e^{-\frac{2bns}{c}}.$$

Similarly, we get $\coth(as/c) = 1 + 2\sum_{m=1}^{\infty} e^{-\frac{2ams}{c}}$, and multiplying the two and subtracting 1, we obtain the expression for $G_b^a(s)$ in the Lemma. □

Using $G_b^a(s)$ from Lemma 1, we obtain for (5)

$$\hat{h}^k(s) = \left\{(1 - 2\theta) - \theta G_b^a(s)\right\}^k \hat{h}^0(s). \tag{7}$$

Now if $\theta = \frac{1}{2}$, we see that the linear factor in (7) vanishes, and convergence will be governed by convolutions of $G_b^a(s)$. We show next that this choice also gives finite step convergence, but the number of steps depends on the length of the time window T.

Theorem 2 (Convergence of DNWR, Asymmetric Decomposition) *Let $\theta = \frac{1}{2}$. Then the DNWR algorithm converges in at most $k+1$ iterations for two subdomains of lengths $a \neq b$, if the time window length T satisfies $T/k \leq 2\min\{a/c, b/c\}$, where c is the wave speed.*

Proof With $\theta = \frac{1}{2}$ we obtain from (7) for $k = 1, 2, \ldots$

$$\hat{h}^k(s) = \left(-\frac{1}{2}\right)^k \{G_b^a(s)\}^k \hat{h}^0(s) = \Bigg[-e^{-\frac{2as}{c}} + e^{-\frac{2bs}{c}} + \Bigg(\sum_{n>1}^{\infty}(-1)^{n-1}e^{-\frac{2bns}{c}}$$
$$-\sum_{m>1}^{\infty}e^{-\frac{2ams}{c}} + 2\sum_{m=1}^{\infty}\sum_{n=1}^{\infty}(-1)^{n-1}e^{-\frac{2(am+bn)s}{c}}\Bigg)\Bigg]^k \hat{h}^0(s) = (-1)^k e^{-\frac{2aks}{c}}\hat{h}^0(s)$$
$$+ e^{-\frac{2bks}{c}}\hat{h}^0(s) + \Bigg(\sum_{l>k}^{\infty}p_l^{(k)}e^{-\frac{2bls}{c}} + \sum_{l>k}^{\infty}q_l^{(k)}e^{-\frac{2als}{c}} + \sum_{m+n\geq k} r_{m,n}^{(k)}e^{-\frac{2(am+bn)s}{c}}\Bigg)\hat{h}^0(s), \tag{8}$$

$p_l^{(k)}, q_l^{(k)}, r_{m,n}^{(k)}$ being the corresponding coefficients. Using the inverse Laplace transform

$$\mathcal{L}^{-1}\{e^{-\alpha s}\hat{g}(s)\} = H(t-\alpha)g(t-\alpha), \tag{9}$$

$H(t)$ being the Heaviside step function, we obtain

$$h^k(t) = (-1)^k h^0(t-2ak/c)H(t-2ak/c) + h^0(t-2bk/c)H(t-2bk/c)$$
$$+\sum_{l>k}^{\infty}p_l^{(k)}h^0(t-2bl/c)H(t-2bl/c) + \sum_{l>k}^{\infty}q_l^{(k)}h^0(t-2al/c)H(t-2al/c)$$
$$+\sum_{m+n\geq k} r_{m,n}^{(k)}h^0(t-2(am+bn)/c)H(t-2(am+bn)/c).$$

Now if we choose our time window such that $T \leq 2k\min\left\{\frac{a}{c}, \frac{b}{c}\right\}$, then $h^k(t) = 0$, and therefore one more iteration produces the desired solution on the entire domain. □

Using $G_b^a(s)$ from Lemma 1, we can also rewrite (6) in the form

$$\hat{w}^k(s) = \left\{(1-4\theta) - \theta\left(G_b^a(s) + G_a^b(s)\right)\right\}^k \hat{w}^0(s), \quad k = 1, 2, \ldots, \tag{10}$$

and we see that for NNWR, the choice $\theta = \frac{1}{4}$ removes the linear factor.

Theorem 3 (Convergence of NNWR, Asymmetric Decomposition) *Let $\theta = \frac{1}{4}$. Then the NNWR algorithm converges in at most $k+1$ iterations for two subdomains* *of lengths $a \neq b$, if the time window length T satisfies $T/k \leq 4\min\{a/c, b/c\}$, c being again the wave speed.*

Proof With $\theta = \frac{1}{4}$ we obtain from (10) with a similar calculation as in (8)

$$\hat{w}^k(s) = \left(-\frac{1}{4}\right)^k \left[G_b^a(s) + G_a^b(s)\right]^k \hat{w}^0(s) = \left[-\sum_{m=1}^{\infty}\left(e^{-\frac{4ams}{c}} + e^{-\frac{4bms}{c}}\right)\right.$$

$$\left.+\sum_{m=1}^{\infty}\sum_{n=1}^{\infty}(-1)^{n-1}\left(e^{-\frac{2(am+bn)s}{c}} + e^{-\frac{2(an+bm)s}{c}}\right)\right]^k \hat{w}^0(s) = (-1)^k e^{-\frac{4aks}{c}}\hat{w}^0(s)$$

$$+\left[(-1)^k e^{-\frac{4bks}{c}} + \left(\sum_{l>k}^{\infty} d_l^{(k)} e^{-\frac{4als}{c}} + \sum_{l>k}^{\infty} z_l^{(k)} e^{-\frac{4bls}{c}} + \sum_{m+n\geq 2k} j_{m,n}^{(k)} e^{-\frac{2(am+bn)s}{c}}\right)\right]\hat{w}^0(s),$$

where $d_l^{(k)}, z_l^{(k)}, j_{m,n}^{(k)}$ are the corresponding coefficients. Now we use (9) to back transform and obtain

$$w^k(t) = (-1)^k w^0(t - 4ak/c)H(t - 4ak/c) + (-1)^k w^0(t - 4bk/c)H(t - 4bk/c)$$

$$+\sum_{l>k}^{\infty} d_l^{(k)} w^0(t - 4al/c)H(t - 4al/c) + \sum_{l>k}^{\infty} z_l^{(k)} w^0(t - 4bl/c)H(t - 4bl/c)$$

$$+\sum_{m+n\geq 2k} j_{m,n}^{(k)} w^0(t - 2(am + bn)/c)H(t - 2(am + bn)/c).$$

So for $T \leq 4k\min\left\{\frac{a}{c}, \frac{b}{c}\right\}$, we get $w^k(t) = 0$, and the conclusion follows. □

4 Numerical Experiments

We perform now numerical experiments to measure the actual convergence rate of the discretized DNWR and NNWR algorithms for the model problem

$$\begin{aligned} \partial_{tt}u - \partial_{xx}u &= 0, & x &\in (-3, 2), t > 0, \\ u(x, 0) &= 0,\ u_t(x, 0) = xe^{-x}, & -3 &< x < 2, \\ u(-3, t) &= -3e^3 t,\ u(2, t) = 2e^{-2}t, & t &> 0, \end{aligned} \tag{11}$$

with $\Omega_1 = (-3, 0)$ and $\Omega_2 = (0, 2)$, so that $a = 3$ and $b = 2$ in (4, 5, 6). We discretize the equation using the centered finite difference in both space and time (Leapfrog scheme) on a grid with $\Delta x = \Delta t = 2\times10^{-2}$. The error is calculated by $\|u - u^k\|_{L^\infty(0,T;L^2(\Omega))}$, where u is the discrete monodomain solution and u^k is the discrete solution in kth iteration.

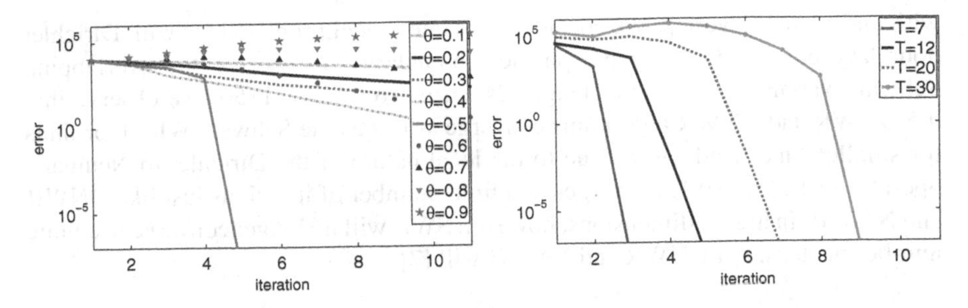

Fig. 1 Convergence of DNWR for various values of θ and $T = 16$ on the *left*, and for various lengths T of the time window and $\theta = \frac{1}{2}$ on the *right*

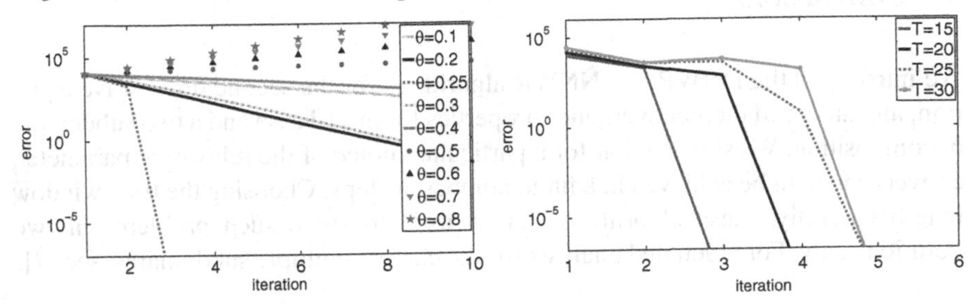

Fig. 2 Convergence of NNWR with various values of θ for $T = 16$ on the *left*, and for various lengths T of the time window and $\theta = \frac{1}{4}$ on the *right*

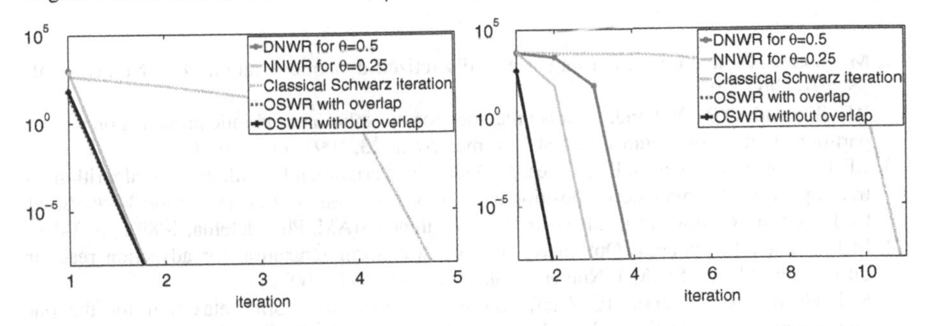

Fig. 3 Comparison of DNWR, NNWR, and SWR for $T = 4$ on the *left*, and $T = 10$ on the *right*

We test the DNWR algorithm by choosing $h^0(t) = t^2, t \in (0, T]$ as an initial guess. In Fig. 1, we show the convergence behavior for different values of the parameter θ for $T = 16$ on the left, and on the right for the best parameter $\theta = \frac{1}{2}$ for different time window length T. Note that for some values of θ (> 0.7) we get divergence. For the NNWR method, with the same initial guess, we show in Fig. 2 on the left the convergence curves for different values of θ for $T = 16$, and on the right the results for the best parameter $\theta = \frac{1}{4}$ for different time window lengths T.

We finally compare in Fig. 3 the performance of the DNWR and NNWR algorithms with the Schwarz Waveform Relaxation (SWR) algorithms from [5] with and without overlap. We consider the same model problem (11) with Dirichlet boundary conditions along the physical boundary. We use for the overlapping Schwarz variant an overlap of length $24\Delta x$, where $\Delta x = 1/50$. We observe that the DNWR and NNWR algorithms converge as fast as the Schwarz WR algorithms for smaller time windows T. Due to the local nature of the Dirichlet-to-Neumann operator in 1d [5], SWR converges in a finite number of iterations just like DNWR and NNWR. In higher dimensions, however, SWR will no longer converge in a finite number of steps, but DNWR and NNWR will [7].

5 Conclusions

We introduced the DNWR and NNWR algorithms for the second order wave equation, and analyzed their convergence properties for the 1d case and a two subdomain decomposition. We showed that for a particular choice of the relaxation parameter, convergence can be achieved in a finite number of steps. Choosing the time window length carefully, these algorithms can be used to solve such problems in two iterations only. For a detailed analysis for the case of multiple subdomains, see [7].

References

1. M. Bjørhus, A note on the convergence of discretized dynamic iteration. BIT Numer. Math. **35**, 291–296 (1995)
2. P.E. Bjørstad, O.B. Widlund, Iterative methods for the solution of elliptic problems on regions partitioned into substructures. SIAM J. Numer. Anal. **23**, 1097–1120 (1986)
3. J.F. Bourgat, R. Glowinski, P.L. Tallec, M. Vidrascu, Variational formulation and algorithm for trace operator in domain decomposition calculations, in *Domain Decomposition Methods*, ed. by T.F. Chan, R. Glowinski, J. Périaux, O.B. Widlund (SIAM, Philadelphia, 1989), pp. 3–16
4. M.J. Gander, L. Halpern, Optimized Schwarz waveform relaxation for advection reaction diffusion problems. SIAM J. Numer. Anal. **45**(2), 666–697 (2007)
5. M.J. Gander, L. Halpern, F. Nataf, Optimal Schwarz waveform relaxation for the one dimensional wave equation. SIAM J. Numer. Anal. **41**(5), 1643–1681 (2003)
6. M.J. Gander, F. Kwok, B.C. Mandal, Dirichlet-Neumann and Neumann-Neumann waveform relaxation algorithms for parabolic problems (2015) [arXiv:1311.2709]
7. B.C. Mandal, Convergence Analysis of Substructuring Waveform Relaxation Methods for Space-time Problems and Their Application to Optimal Control Problems, Ph.D. Thesis, University of Geneva, (2014)
8. M.J. Gander, A.M. Stuart, Space-time continuous analysis of waveform relaxation for the heat equation. SIAM J. Sci. Comput. **19**(6), 2014–2031 (1998)
9. E. Giladi, H.B. Keller, Space-time domain decomposition for parabolic problems, Numer. Math. **93**(2), pp. 279–313 (2002).
10. F. Kwok, Neumann-Neumann waveform relaxation for the time-dependent heat equation. *Domain Decomposition in Science and Engineering XXI* (Springer, Berlin, 2013)
11. E. Lelarasmee, A. Ruehli, A. Sangiovanni-Vincentelli, The waveform relaxation method for time-domain analysis of large scale integrated circuits. IEEE Trans. Comput. Aided Des. Integr. Circuits Syst. **1**(3), 131–145 (1982)

12. B.C. Mandal, A time-dependent Dirichlet-Neumann method for the heat equation. *Domain Decomposition in Science and Engineering XXI* (Springer, Berlin, 2013)
13. A. Toselli, O.B. Widlund, *Domain Decomposition Methods, Algorithms and Theory* (Springer, Berlin, 2005)

Binned Multilevel Monte Carlo for Bayesian Inverse Problems with Large Data

Robert N. Gantner, Claudia Schillings, and Christoph Schwab

1 Introduction

In recent years, various methods have been developed for solving parametric operator equations, mainly focusing on the *estimation of parameters given measurements* of the parametric solution, subject to a stochastic observation error model. A second objective is *prediction of a "most likely" response of the parametric system given noisy measurements*. The *Bayesian approach* to such inverse problems for partial differential equations (PDEs for short) has been the focus of numerous papers [7–10] and will be considered here. For multiple data points, averaging is often done with a standard Monte Carlo approach. We consider here the case where computational resources are limited and develop a multilevel Monte Carlo method (MLMC) achieving an error of the same order while requiring less work [1, 2, 5, 6].

2 Bayesian Inversion of Parametric Operator Equations

We assume an operator equation depending on a distributed, uncertain "parameter" u with values in a separable Banach space X. It takes the form of the operator equation

$$\text{Given } u \in \tilde{X} \subseteq X, \text{ find } q \in \mathscr{X} : \quad A(u;q) = F(u) \quad \text{in} \quad \mathscr{Y}' , \tag{1}$$

C. Schillings (✉)
Mathematics Institute, University of Warwick, Coventry CV4 7AL, England

Seminar for Applied Mathematics, ETH Zürich, CH-8092 Zürich, Switzerland
e-mail: c.schillings@warwick.ac.uk; claudia.schillings@sam.math.ethz.ch

R.N. Gantner • C. Schwab
Seminar for Applied Mathematics, ETH Zürich, CH-8092 Zürich, Switzerland
e-mail: schwab@math.ethz.ch

T. Dickopf et al. (eds.), *Domain Decomposition Methods in Science and Engineering XXII*, Lecture Notes in Computational Science and Engineering 104, DOI 10.1007/978-3-319-18827-0_52

where we denote by $\mathscr{X}$ and $\mathscr{Y}$ two reflexive Banach spaces over $\mathbb{R}$ with (topological) duals $\mathscr{X}'$ and $\mathscr{Y}'$, respectively and $A(u;\cdot) \in \mathscr{L}(\mathscr{X}, \mathscr{Y}')$. Assuming that the forcing function $F : \tilde{X} \mapsto \mathscr{Y}'$ is known, and the uncertain operator $A(u;\cdot) : \mathscr{X} \mapsto \mathscr{Y}'$ is locally boundedly invertible for uncertain input u in a sufficiently small neighborhood $\tilde{X}$, let the *uncertainty-to-observation map* $\mathscr{G} : \tilde{X} \mapsto \mathbb{R}^K$ have the structure

$$X \supseteq \tilde{X} \ni u \mapsto \mathscr{G}(u) := \mathscr{O}(G(u;F)) \in \mathbb{R}^K . \quad (2)$$

Here, $\tilde{X} \ni u \mapsto q(u) = G(u;F) \in \mathscr{X}$ denotes the (noise-free) response of the forward problem for a given instance of $u \in \tilde{X}$ and $\mathscr{O}$ a *bounded linear observation operator* $\mathscr{O} \in \mathscr{L}(\mathscr{X}, \mathbb{R}^K)$, $K < \infty$. The goal of computation is assumed to be the low-order statistics of a *quantity of interest* (QoI) given noisy observational data δ of the form

$$\delta = \mathscr{G}(u) + \eta , \quad (3)$$

where δ represents the observation of $\mathscr{G}(u)$ perturbed by the noise η, a random variable with given statistical properties. We restrict ourselves to the case where the measurement error is Gaussian and the covariance matrix symmetric positive definite, i.e. $\eta \sim \mathscr{N}(0, \Gamma)$ with $\Gamma \in \mathbb{R}^{K\times K}_{\text{spd}}$.

We work in the following under the assumption that the uncertainty u admits a parametric representation of the form

$$u = u(\mathbf{y}) := \langle u \rangle + \sum_{j\in\mathbb{J}} y_j \psi_j \in X$$

for some "nominal" value $\langle u \rangle \in X$ of the uncertain datum u, a countable sequence $(\psi_j)_{j\in\mathbb{J}}$ of X with $\mathbb{J} := \{1, \dots, J\}, J < \infty$ or $\mathbb{J} = \mathbb{N}$ and for some coefficient sequence $\mathbf{y} = (y_j)_{j\in\mathbb{J}}$ (after possibly rescaling the fluctuations) in the reference domain $U = [-1, 1]^{\mathbb{J}} = \bigotimes_{j\in\mathbb{J}}[-1, 1]$ with unconditional convergence. We assume $\mathbf{y}$ to be a random variable on the countable product probability space $(U, \mathscr{B}(U), \mu_0)$ with U as above and with $\mu_0(\mathrm{d}\mathbf{y}) = \prod_{j\in\mathbb{J}} \frac{1}{2}\lambda^1(\mathrm{d}y_j)$. This also makes δ a random variable; for a fixed value of $\mathbf{y}$, (3) gives an expression for $\delta(\mathbf{y})$, denoted by $\delta|\mathbf{y}$.

In general, our aim is to compute the "most likely" value of a QoI over all realizations of u, with the QoI defined as a function $\phi : U \to \mathscr{S}$ mapping from the parameter space U to some Banach space $\mathscr{S}$. Bayes' theorem characterizes this value as the mathematical expectation with respect to a probability measure μ_0 (the "Bayesian prior") on U which we choose as a countable product of uniform measures. In particular, we consider in what follows $\phi = G$, and $\mathscr{S} = \mathscr{X}$ the response of the system. To this end, we use Bayes' Theorem to obtain an expression for $\mathbf{y}|\delta$, as in [9, 10].

Theorem 1 (Bayes' Theorem) *Let* $\mathcal{G}\Big|_{u=\langle u\rangle+\sum_{j\in\mathbb{J}} y_j\psi_j} : U \to \mathbb{R}^K$ *be bounded and continuous. Then,* $\mu^\delta(d\mathbf{y})$*, the distribution of* $\mathbf{y}|\delta$*, is absolutely continuous with respect to* $\mu_0(d\mathbf{y})$*, and*

$$\frac{\mathrm{d}\mu^\delta(\mathbf{y})}{\mathrm{d}\mu_0(\mathbf{y})} = \frac{1}{Z_\delta}\exp\left(-\frac{1}{2}||\delta - \mathcal{G}(\mathbf{y})||_\Gamma^2\right)$$

with $Z_\delta := \int_U \exp(-\Phi(\mathbf{y};\delta))\ \mu_0(\mathrm{d}\mathbf{y}) > 0\,,\ \|\delta\|_\Gamma^2 = \delta^\top\Gamma^{-1}\delta.$

In the Bayesian setting, the distribution $\mathrm{d}\mu_0(\mathbf{y})$ is called the *prior distribution* and is assumed to be known and easily computable. Thus, we can write our desired expectation as an integral over the prior measure μ_0:

$$\mathbb{E}^{\mu^\delta}[\phi] = \int_U \phi(\mathbf{y})\,\mu^\delta(\mathrm{d}\mathbf{y}) = \frac{1}{Z_\delta}\int_U \phi(\mathbf{y})\exp\left(-\frac{1}{2}||\delta - \mathcal{G}(\mathbf{y})||_\Gamma^2\right)\mu_0(\mathrm{d}\mathbf{y}) =: \frac{Z_\delta'}{Z_\delta}\,. \tag{4}$$

This formulation of the expectation $\mathbb{E}^{\mu^\delta}[\cdot]$ is based on just one measurement δ. For a given model for the measurement errors η, we would like to additionally compute the expectation over all errors. Assuming that the perturbations η are normally distributed as above, this can be written as an expectation with respect to the measure $\gamma_\Gamma^K(\eta)$, the K-variate Gaussian measure with covariance $\Gamma > 0$. *Here, and throughout, we assume the observation noise* η *to be statistically independent from the uncertain parameter* u *in* (1). This yields the total expectation of the QoI ϕ in terms of Z_δ' and Z_δ as

$$\mathbb{E}^{\gamma_\Gamma^K}\left[\mathbb{E}^{\mu^\delta}[\phi]\right] = \int_{\mathbb{R}^K} \frac{Z_\delta'}{Z_\delta}\bigg|_{\delta=\mathcal{G}(\mathbf{y}_0)+\eta} \gamma_\Gamma^K(\mathrm{d}\eta)\,, \tag{5}$$

where $\mathcal{G}(\mathbf{y}_0)$ denotes the observation at the unknown, exact parameter $\mathbf{y}_0$.

In practice, we are given a set of measurements $\Delta := \{\delta_i,\ i = 1,\ldots,M\}$ with which this outer expectation should be approximated. The measurements can be taken at different positions, i.e. with respect to different observation maps $\mathcal{O}_i$ in (2). In the derivations below, we consider the notationally more convenient case where the measurements are all obtained using the same observation map. We do, however, impose the restriction that the measurements are homoscedastic, i.e. δ_i is Gaussian with the same covariance Γ for all $i = 1,\ldots,M$. In Sect. 4, we will approximate the outer expectation in (5) by a multilevel Monte Carlo averaging approach.

3 Approximation of Posterior Expectation

A first simplification of (5) is achieved by replacing the inner expectation over the posterior distribution μ^δ by an approximation $E_{\tau_L}^{\mu^\delta}[\phi]$ with tolerance parameter $\tau_L > 0$. We assume that the following bound holds for the considered QoI ϕ:

$$\left\| \mathbb{E}^{\mu^\delta}[\phi] - E_{\tau_L}^{\mu^\delta}[\phi] \right\|_{\mathscr{X}} \leq \tau_L \,. \tag{6}$$

Our method of choice is the adaptive Smolyak algorithm developed in [7], which adaptively constructs a sparse tensor quadrature rule that approximates Z_δ and Z'_δ. More precisely, the results in [7, 8] ensure existence of a monotone index set Λ with

$$\left\| \mathbb{E}^{\mu^\delta}[\phi] - E_{\tau_L}^{\mu^\delta}[\phi] \right\|_{\mathscr{X}} \leq C_\Gamma^{\mathrm{SM}} N_L^{-(\frac{1}{p}-1)} \,, \tag{7}$$

where N_L is the cardinality of the index set Λ assuming that the forward solution map $U \ni \mathbf{y} \mapsto q(\mathbf{y})$ is $(\mathbf{b}, p, \epsilon)$-analytic for some $0 < p < 1$ and $\epsilon > 0$, i.e. *Instance well-posedness of the forward problem:*

for each instance $\mathbf{y} \in U$, there exists a unique realization $u(\mathbf{y}) \in \tilde{X} \subseteq X$ of the uncertainty and a unique solution $q(\mathbf{y}) \in \mathscr{X}$ of the forward problem (1) satisfying $\|q(\mathbf{y})\|_{\mathscr{X}} \leq C_0$ for all $\mathbf{y} \in U$.

$(\boldsymbol{b}, p, \epsilon)$ Analyticity:

There exists a $0 \leq p \leq 1$ and a positive sequence $\mathbf{b} = (b_j)_{j \in \mathbb{J}} \in \ell^p(\mathbb{J})$ such that for every sequence $\boldsymbol{\rho} = (\rho_j)_{j \in \mathbb{J}}$ of poly-radii $\rho_j > 1$ with $\sum_{j \in \mathbb{J}} (\rho_j - 1) b_j \leq 1 - \epsilon$, the solution map $U \ni \mathbf{y} \mapsto q(\mathbf{y}) \in \mathscr{X}$ admits an analytic continuation to the open poly-ellipse $\mathscr{E}_\rho := \bigotimes_{j \in \mathbb{J}} \mathscr{E}_{\rho_j} \subset \mathbb{C}^{\mathbb{J}}$ and satisfies $\|q(z)\|_{\mathscr{X}} \leq C_\epsilon(\boldsymbol{\rho}),\ \forall z \in \mathscr{E}_{\boldsymbol{\rho}}$.

The concept of $(\mathbf{b}, p, \epsilon)$-analyticity allows to analyze the regions of analyticity $\mathscr{E}_\rho$ of the solution in each parameter and exploit the anisotropic smoothness of the problem reflected by the poly-radii $\boldsymbol{\rho}$. Sufficient conditions on the $(\mathbf{b}, p, \epsilon)$-analyticity of the forward problem (1) are given in [4, 8]. The results presented in [7, 8] suggest dimension robust convergence rates of the form (7) for adaptive Smolyak-based quadrature algorithms using a greedy-type approach to construct the monotone index set. The underlying quadrature points are symmetrized Léja sequences (see [3] and the references therein for more details), which allow us to relate the number of quadrature points to the prescribed tolerance τ_L as follows.

Proposition 1 *The work required for the evaluation of the adaptive Smolyak approximation up to the tolerance $\tau > 0$ based on symmetrized Léja quadrature is bounded by $C(\Gamma)\tau^{-\log_2 3 \cdot (\frac{1}{p}-1)^{-1}}$ with $C(\Gamma) > 0$ independent of τ.*

Proof For a multiindex ν in a monotone index set Λ_N with $\#\Lambda_N \leq N$, the bound $\#\{i \in \mathbb{J} : \nu \neq 0\} \leq \lfloor \log_2 N \rfloor$ holds as argued in the proof of Lemma 5.4 in [7]. A worst case bound can be derived by considering an isotropic refinement in the first $\lfloor \log_2 N \rfloor$ dimensions, i.e. it holds for the maximal number of quadrature points $M \leq 3^{\log_2 N} = N^{\log_2 3}$. Equating (7) to τ, solving for N and inserting into the above yields the claimed bound on the number of quadrature points.

Remark 1 Note that the result derived in Proposition 1 is based on a worst case bound on the number of quadrature points arising in the case of isotropic refinement.

4 Binned Multilevel Monte Carlo

In this section, we formulate our method for combining M realizations of δ, $\Delta = \{\delta_i : i = 1, \dots, M\}$ to compute an approximation to

$$\mathbb{E}^{\gamma_\Gamma^K}\left[\mathbb{E}^{\mu^\delta}[\phi]\right] = \int_{\mathbb{R}^K} \frac{1}{Z_\delta} \int_U \phi(\mathbf{y}) \exp\left(-\frac{1}{2}||\delta - \mathcal{G}(\mathbf{y})||_\Gamma^2\right)\bigg|_{\delta=\mathcal{G}(\mathbf{y}_0)+\eta} \mu_0(d\mathbf{y})\, \gamma_\Gamma^K(d\eta)\,. \tag{8}$$

Our approach is based on the multilevel Monte Carlo method originally applied by Heinrich [6] and Giles [5] and, in the current form, by Barth et al. [2].

Formulation of the Binned MLMC Algorithm We interpret the approximation $E_{\tau_L}^{\mu^\delta}[\phi]$ obtained by the method explained in Sect. 3 as corresponding to a discretization level L and write it as a telescopic sum over the levels $\ell = 0, \dots, L$. Using the convention $E_{\tau_{-1}}^{\mu^\delta} = 0$, we obtain the exact reformulation

$$E_{\tau_L}^{\mu^\delta}[\phi] = \sum_{\ell=0}^{L}\left(E_{\tau_\ell}^{\mu^\delta}[\phi] - E_{\tau_{\ell-1}}^{\mu^\delta}[\phi]\right)\,. \tag{9}$$

Inserting this back into (8) and applying the linearity of the expectation yields

$$\mathbb{E}^{\gamma_\Gamma^K}\left[\sum_{\ell=0}^{L}\left(E_{\tau_\ell}^{\mu^\delta}[\phi] - E_{\tau_{\ell-1}}^{\mu^\delta}[\phi]\right)\right] = \sum_{\ell=0}^{L} \mathbb{E}^{\gamma_\Gamma^K}\left[E_{\tau_\ell}^{\mu^\delta}[\phi] - E_{\tau_{\ell-1}}^{\mu^\delta}[\phi]\right]\,.$$

Replacing the expectations on each level by a sample mean over a level-dependent number of samples M_ℓ yields a full approximation to (8),

$$E_{\mathrm{ML},L}^{\gamma_\Gamma^K}[E_{\tau_L}^{\mu^\delta}[\phi]] := \sum_{\ell=0}^{L} E_{M_\ell}^{\gamma_\Gamma^K}\left[E_{\tau_\ell}^{\mu^\delta}[\phi] - E_{\tau_{\ell-1}}^{\mu^\delta}[\phi]\right]\,, \tag{10}$$

where we denote by $E_M[Y]$ the standard Monte Carlo estimator for realizations $\hat{Y}_i$ of a random variable $Y : \Omega \to S$, given by $E_M[Y] = \frac{1}{M}\sum_{i=1}^{M} \hat{Y}_i$.

A crucial aspect of this formulation is the choice of the number of samples per level $(M_\ell)_{\ell=0}^L$ and the tolerances per level $(\tau_\ell)_{\ell=0}^L$. Since the total number of samples is fixed, a natural approach is to make an ansatz for M_ℓ and then choose τ_ℓ optimally.

Number of Samples Per Level Thinking of the levels $\ell = 0, \dots, L$ as "bins" containing measurements over which we wish to average, we distribute the samples according to the ansatz $M_\ell = b^{L-\ell+1}$ with $b \in \mathbb{N}, b > 1$. The analysis presented can also be generalized to the case $b \in \mathbb{R}, b > 1$. The total number of samples is $\sum_{\ell=0}^L b^{L-\ell-1}$, which we assume to be the given number of measurements M.

Error Bounds For the computation of the error, we consider the Gaussian probability space $(\Omega, \mathscr{B}(\Omega), \gamma_\Gamma^K)$ and the random variable η. The approximation of the inner expectation is an $\mathscr{X}$-valued random variable whereas the full expectation is in $\mathscr{X}$. Clearly, $E_{\tau_\ell}^{\mu^\delta}[\phi] \in L^2(\Omega; \mathscr{X})$ and the error of (10) in the $L^2(\Omega; \mathscr{X})$ norm can be bounded by

$$\left\| \mathbb{E}^{\gamma_\Gamma^K}\left[\mathbb{E}^{\mu^\delta}[\phi]\right] - E_{\mathrm{ML},L}^{\gamma_\Gamma^K}[E_{\tau_L}^{\mu^\delta}[\phi]] \right\|_{L^2(\Omega;\mathscr{X})} \leq \left\| \mathbb{E}^{\gamma_\Gamma^K}\left[\mathbb{E}^{\mu^\delta}[\phi]\right] - \mathbb{E}^{\gamma_\Gamma^K}\left[E_{\tau_L}^{\mu^\delta}[\phi]\right] \right\|_{L^2(\Omega;\mathscr{X})}$$
$$+ \left\| \mathbb{E}^{\gamma_\Gamma^K}\left[E_{\tau_L}^{\mu^\delta}[\phi]\right] - \sum_{\ell=0}^{L} E_{M_\ell}^{\gamma_\Gamma^K}\left[E_{\tau_\ell}^{\mu^\delta}[\phi] - E_{\tau_{\ell-1}}^{\mu^\delta}[\phi]\right] \right\|_{L^2(\Omega;\mathscr{X})} . \tag{11}$$

Since the first term on the right in (11) already contains the expectation with respect to γ_Γ^K, we simply obtain the discretization error from (6), $\|\mathbb{E}^{\mu^\delta}[\phi] - E_{\tau_L}^{\mu^\delta}[\phi]\|_{\mathscr{X}} \leq \tau_L$. Inserting an expansion analogous to (9) into the second term of (11) yields

$$\left\| \sum_{\ell=0}^{L} \left(\mathbb{E}^{\gamma_\Gamma^K}\left[E_{\tau_\ell}^{\mu^\delta}[\phi] - E_{\tau_{\ell-1}}^{\mu^\delta}[\phi]\right] - E_{M_\ell}^{\gamma_\Gamma^K}\left[E_{\tau_\ell}^{\mu^\delta}[\phi] - E_{\tau_{\ell-1}}^{\mu^\delta}[\phi]\right] \right) \right\|_{L^2(\Omega;\mathscr{X})}$$
$$\leq \sum_{\ell=0}^{L} \left\| \mathbb{E}^{\gamma_\Gamma^K}\left[E_{\tau_\ell}^{\mu^\delta}[\phi] - E_{\tau_{\ell-1}}^{\mu^\delta}[\phi]\right] - E_{M_\ell}^{\gamma_\Gamma^K}\left[E_{\tau_\ell}^{\mu^\delta}[\phi] - E_{\tau_{\ell-1}}^{\mu^\delta}[\phi]\right] \right\|_{L^2(\Omega;\mathscr{X})} .$$

For each summand above, we use the standard Monte Carlo error bound that holds for any $M \in \mathbb{N}$, $Y \in L^2(\Omega; \mathscr{X})$, i.e. $\|\mathbb{E}[Y] - E_M^{\gamma_\Gamma^K}[Y]\|_{L^2(\Omega;\mathscr{X})} \leq \frac{1}{\sqrt{M}}\|Y\|_{L^2(\Omega;\mathscr{X})}$. Combining this with the given bound (7) as follows

$$\left\| E_{\tau_\ell}^{\mu^\delta}[\phi] - E_{\tau_{\ell-1}}^{\mu^\delta}[\phi] \right\|_{L^2(\Omega;\mathscr{X})} = \mathbb{E}^{\gamma_\Gamma^K}\left[\left\| E_{\tau_\ell}^{\mu^\delta}[\phi] - E_{\tau_{\ell-1}}^{\mu^\delta}[\phi] \right\|_{\mathscr{X}}^2 \right]^{\frac{1}{2}} \leq$$
$$\mathbb{E}^{\gamma_\Gamma^K}\left[\left(\left\| E_{\tau_\ell}^{\mu^\delta}[\phi] - \mathbb{E}^{\mu^\delta}[\phi] \right\|_{\mathscr{X}} + \left\| \mathbb{E}^{\mu^\delta}[\phi] - E_{\tau_{\ell-1}}^{\mu^\delta}[\phi] \right\|_{\mathscr{X}} \right)^2 \right]^{\frac{1}{2}} \leq C_\ell \tau_\ell + C_{\ell-1}\tau_{\ell-1} ,$$

and using $\tau_{-1} = 0$, we obtain a total sampling error bound of

$$\left\| \mathbb{E}^{\gamma_\Gamma^K}\left[E_{\tau_L}^{\mu^\delta}[\phi]\right] - E_{\mathrm{ML},L}^{\gamma_\Gamma^K}[E_{\tau_L}^{\mu^\delta}[\phi]]\right\|_{L^2(\Omega;\mathscr{X})} \leq \sum_{\ell=1}^{L} M_\ell^{-\frac{1}{2}}(C_\ell\tau_\ell + C_{\ell-1}\tau_{\ell-1}) + C_0 M_0^{-\frac{1}{2}}.$$

Combined with the discretization error, the total error is then bounded by

$$e_{\mathrm{tot}} \leq \tau_L + \sum_{\ell=1}^{L} M_\ell^{-\frac{1}{2}}(C_\ell\tau_\ell + C_{\ell-1}\tau_{\ell-1}) + C_0 M_0^{-\frac{1}{2}}. \tag{12}$$

Theorem 2 (Optimal Tolerances) *Given the sample distribution $M_\ell = b^{L-\ell+1}$, the optimal tolerances for the inner expectation that minimize the total work bound at given error are*

$$\tau_\ell = \frac{M_0^{-\frac{1}{2}}}{C(s,b,L)}\left(\frac{M_\ell}{D_\ell}\right)^{\frac{1}{s+1}}, \quad 0 \leq \ell \leq L,$$

for a constant $C(s,b,L)$ and $M_{-1} = D_{-1} = 0$, $D_0 = C_0 M_1^{-1/2}$, $D_L = C_L(1 + M_L^{-1/2})$ and for $0 < \ell < L$, $D_\ell = C_\ell(M_\ell^{-1/2} + M_{\ell+1}^{-1/2})$.

Proof The optimization problem we consider is the minimization of the total work subject to the constraint that the discretization and sampling errors are equilibrated,

$$\min \sum_{\ell=0}^{L} M_\ell w_\ell \quad \text{s.t.} \quad \tau_L + \sum_{\ell=1}^{L} \frac{C_\ell\tau_\ell + C_{\ell-1}\tau_{\ell-1}}{\sqrt{M_\ell}} = \frac{C_0}{\sqrt{M_0}},$$

where $w_\ell \sim \tau_\ell^{-s}$, $s > 0$ denotes the work on level ℓ (for the Smolyak approach mentioned above, we use $s = (\frac{1}{p} - 1)^{-1}$). Using Lagrange multipliers, one can impose the necessary condition that the partial derivatives of the Lagrange function $\mathscr{L}(\tau_0, \ldots, \tau_L, \lambda) = \sum_{\ell=1}^{L} M_\ell \frac{1}{\tau_\ell^s} + \lambda\left(\tau_L + \sum_{\ell=0}^{L} \frac{\tau_\ell + \tau_{\ell-1}}{\sqrt{M_\ell}} - \frac{C_0}{\sqrt{M_0}}\right)$ vanish in the optimum.

A straightforward calculation reveals that the total work when using M samples and the tolerances from above is bounded for $0 < s < 2$ and a constant $C(s,b)$ by $W_{\mathrm{tot}}^L \leq C(s,b)\, M^\gamma w_L$, $\gamma = \frac{2-s}{2(s+1)} \in (0,1)$. A slightly more involved computation yields an optimal error versus work relationship with exponent $-1/2$, independent of s, given by $e_{\mathrm{tot}} \sim C(s,b)\left(W_{\mathrm{tot}}^L\right)^{-1/2}$.

5 Numerical Experiment

As a model problem of the abstract, $(\mathbf{b}, p, \epsilon)$-analytic operator equation described in Sect. 2 we consider the diffusion equation $-\nabla \cdot (u \nabla p) = 100x$ in $D := [0,1]$, $p = 0$ on ∂D with stochastic diffusion coefficient u modeled as a random field described by $u = u(\mathbf{y}) := \langle u \rangle + \sum_{j=1}^{64} y_j \psi_j \in X$ with constant mean $\langle u \rangle = 1$, parameters $\mathbf{y} = (y_j)_{j=1}^{64} \in U := [-1,1]^{64}$ and basis functions $\psi_j = \frac{0.9}{j^3} \chi_{D_j}$, $D_j = [\frac{j-1}{64}, \frac{j}{64})$ describing the fluctuations and $X = \cup_{j=1}^{64} C^0(\overline{D_j})$. The problem is solved by a finite element approach with piecewise linear basis functions on a uniform mesh. The meshwidth is $h = 2^{-14}$ to avoid discretization error effects. Given a noisy measurement with $\eta \sim \mathcal{N}(0,1)$, our goal is to evaluate the conditional expectation $\mathbb{E}^{\mu^\delta}[\phi]$ of the QoI $\phi(u) = \mathcal{G}(u)$. The observation operator $\mathcal{O}$ consists of the system response at the point $x_1 = 0.5$ (note that here $\mathcal{S} = \mathbb{R}$).

For MLMC, the maximal level was chosen by numerically observing that (12) is convex in L and increasing the value of L until the error bound stops decreasing. For each L, b is computed such that $\sum_{\ell=0}^{L} M_\ell = M$ is satisfied. The reference solution is computed to high accuracy using 96-point Gauss-Hermite quadrature. The numerical results are summarized in Fig. 1.

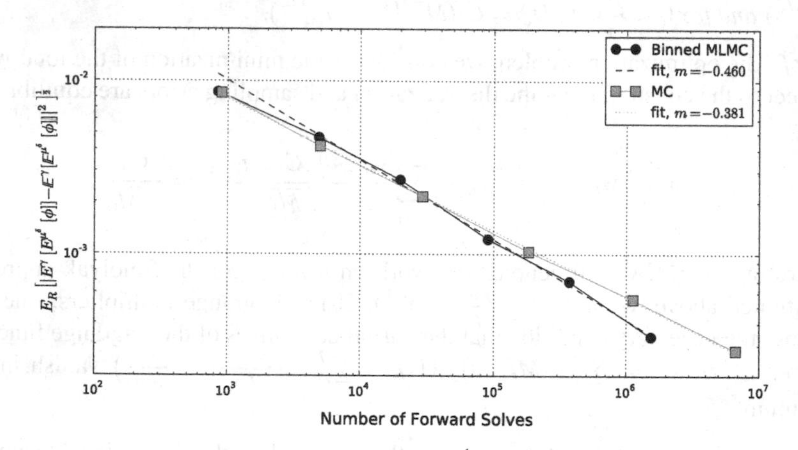

Fig. 1 Convergence of L^2 error approximation $E_R[|E_{\mathrm{ML},L}^{\gamma_1^1}[E_{\tau_\ell}^{\mu^\delta}[\phi]] - \mathbb{E}^{\gamma_1^1}[\mathbb{E}^{\mu^\delta}[\phi]]|^2]^{\frac{1}{2}}$ with $R = 200$ vs. the work, which is assumed proportional to the number of forward evaluations. The theoretically computed rates are $-1/3$ for Monte Carlo (MC) and $-1/2$ for multilevel Monte Carlo (MLMC). The number of measurements were $M = 16, 64, 256, 1024, 4096, 16{,}384$ and all tolerances were scaled with $C = 0.1$. For MLMC, the first point is not used in computing the slope, as $L = 0$ which corresponds to a MC simulation

6 Conclusion

Assuming a given set of measured responses of a forward problem, a multilevel Monte Carlo averaging method was derived by computing optimal values of forward map evaluation tolerances on each level. Numerical results based on Bayesian inversion of a parametric diffusion equation confirm the analytically derived optimal convergence rate of the error with respect to the work.

Acknowledgement This work is supported by the Swiss National Science Foundation (SNF) and the European Research Council (ERC) under FP7 Grant AdG247277.

References

1. A. Barth, A. Lang, C. Schwab, Multilevel Monte Carlo method for parabolic stochastic partial differential equations, in *BIT Numerical Mathematics* (Springer, New York, 2011a), pp. 3–27
2. A. Barth, C. Schwab, N. Zollinger, Multi-level Monte Carlo finite element method for elliptic PDEs with stochastic coefficients. Numer. Math. **119**(1), 123–161 (2011b)
3. A. Chkifa, On the Lebesgue constant of Leja sequences for the complex unit disk and of their real projection. J. Approx. Theory **166**, 176–200 (2013)
4. A. Cohen, A. Chkifa, C. Schwab, Breaking the curse of dimensionality in sparse polynomial approximation of parametric pdes. J. Math. Pures Appl. **103**(2), 400–428 (2014)
5. M.B. Giles, Multilevel Monte Carlo path simulation, Oper. Res. **56**(3), 607–617 (2008)
6. S. Heinrich, Multilevel Monte Carlo methods, in *Large-Scale Scientific Computing* (Springer, Berlin, 2001), pp. 58–67
7. C. Schillings, C. Schwab, Sparse, adaptive Smolyak quadratures for Bayesian inverse problems. Inverse Prob. **29**(6), 065011 (2013)
8. C. Schillings, C. Schwab, Sparsity in Bayesian inversion of parametric operator equations. Inverse Prob. **30**(6), 065007 (2014)
9. C. Schwab, A.M. Stuart, Sparse deterministic approximation of Bayesian inverse problems. Inverse Prob. **28**(4), 045003 (2012)
10. A.M. Stuart, Inverse problems: a Bayesian perspective. Acta Numer. **19**, 451–559 (2010)

Optimized Schwarz Method for the Fluid-Structure Interaction with Cylindrical Interfaces

Giacomo Gigante and Christian Vergara

1 Introduction

The Optimized Schwarz Method (OSM) is a domain decomposition method based on the introduction of generalized Robin interface conditions obtained by linearly combining the two physical interface conditions through the introduction of suitable symbols, and then on the optimization of such symbols within a proper subset, see [10, 13]. This method has been considered so far for many problems in the case of flat interfaces, see, e.g., [3, 5–7, 11, 16, 17]. Recently, OSM has been considered and analyzed for the case of cylindrical interfaces in [8, 9], and for the case of circular interfaces in [2]. In particular, in [8] we developed a general convergence analysis of the Schwarz method for elliptic problems and an optimization procedure within the constants, with application to the fluid-structure interaction (FSI) problem.

In this work, we provide a numerical study of the performance of the optimization procedure developed in [8] for the FSI problem when the solution is characterized by non-null angular frequencies, thus breaking the radial symmetry. The reported 3D numerical results for a cylindrical domain showed the effectiveness of the procedure proposed in [8] also in such a case.

G. Gigante
Dipartimento di Ingegneria Gestionale, dell'Informazione e della Produzione, Università di Bergamo, Viale Marconi 5, 24044 Dalmine (BG), Italy
e-mail: giacomo.gigante@unibg.it

C. Vergara (✉)
MOX, Dipartimento di Matematica, Politecnico di Milano, Piazza L. da Vinci 32, 20133 Milan, Italy
e-mail: christian.vergara@polimi.it

T. Dickopf et al. (eds.), *Domain Decomposition Methods in Science and Engineering XXII*, Lecture Notes in Computational Science and Engineering 104, DOI 10.1007/978-3-319-18827-0_53

The outline of this paper is as follows. In Sect. 2 we report the convergence result developed in [8] with application to the FSI problem, whereas in Sect. 3 we describe the optimization procedure. Finally, in Sect. 4 we show the numerical results.

2 Convergence Analysis

We consider the interaction between an incompressible, inviscid and linear fluid in the cylindrical domain $\Omega_f := \{(x, y, z) \in \mathbb{R}^3 : x^2 + y^2 < R^2\}$, for a given $R \in \mathbb{R}^+$, and a linear elastic structure described by the wave equation in the cylindrical crown $\Omega_s := \{(x, y, z) \in \mathbb{R}^3 : R^2 < x^2 + y^2 < (R + H)^2\}$, with H the structure thickness. The two subproblems interact at the common cylindrical interface $\Sigma_R = \{(x, y, z) \in \mathbb{R}^3 : x^2 + y^2 = R^2\}$. We also introduce the cylindrical variables r, φ defined by $x = r\cos\varphi$ and $y = r\sin\varphi$. After the time discretization obtained with a BDF1 scheme for both subproblems, the coupled problem at time $t^{n+1} := (n + 1)\Delta t$, Δt being the time discretization parameter, reads

$$\begin{cases} \rho_f \delta_t \boldsymbol{u} + \nabla p = \boldsymbol{0} & \text{in } \Omega_f, \\ \nabla \cdot \boldsymbol{u} = 0 & \text{in } \Omega_f, \\ \int_{-\infty}^{\infty}\int_0^{2\pi} |p\,(r\cos\varphi, r\sin\varphi, z)|\, d\varphi dz & \text{bounded as } r \to 0^+, \\ \boldsymbol{u}\cdot\boldsymbol{n} = \delta_t \boldsymbol{\eta}\cdot\boldsymbol{n} & \text{on } \Sigma_R, \\ -p\boldsymbol{n} = \lambda\, \nabla\boldsymbol{\eta}\,\boldsymbol{n} & \text{on } \Sigma_R, \\ \boldsymbol{\eta}\times\boldsymbol{n} = \boldsymbol{0} & \text{on } \Sigma_R, \\ \rho_s \delta_{tt}\boldsymbol{\eta} - \lambda\,\Delta\boldsymbol{\eta} = \boldsymbol{0} & \text{in } \Omega_s, \\ \gamma_{ST}\boldsymbol{\eta} + \lambda\nabla\boldsymbol{\eta}\,\boldsymbol{n} = P_{ext}\,\boldsymbol{n} & \text{on } \Sigma_{out}, \end{cases} \tag{1}$$

where ρ_f and ρ_s are the fluid and structure densities, λ the square of the wave propagation velocity, $\delta_t w := \frac{w-w^n}{\Delta t}$, $\delta_{tt} w := \frac{\delta_t w-\delta_t w^n}{\Delta t}$, $\Sigma_{out} = \{(x, y, z) \in \mathbb{R}^3 : x^2 + y^2 = (R + H)^2\}$, $\boldsymbol{n}$ is the unit vector orthogonal to the interfaces, and we have omitted the time index $n + 1$. Problem $(1)_{1-3}$ is the fluid problem, problem $(1)_{7-8}$ is the structure problem equipped with a Robin condition at the external surface to model the elastic surrounding tissue, P_{ext} is the external pressure, whereas $(1)_{4-6}$ are the coupling conditions at the FS interface, stating the continuity of the velocities and of the tractions along the normal direction. The fluid and the structure problems have to be completed with initial conditions and with the assumption of decay to zero for $|z| \to \infty$.

By combining linearly $(1)_4$ and $(1)_5$ through the introduction of the linear operators $\mathcal{S}_f$ and $\mathcal{S}_s$ acting in the tangential direction to Σ_R, we obtain the following generalized Robin interface conditions (see [8])

$$\begin{cases} \mathcal{S}_f\,\Delta t\,\delta_t u_r - p = \dfrac{\mathcal{S}_f}{\Delta t}\,\eta_r + \lambda\,\partial_r\eta_r, \\ \dfrac{\mathcal{S}_s}{\Delta t}\,\eta_r + \lambda\,\partial_r\eta_r = \mathcal{S}_s\,\Delta t\,\delta_t u_r - p, \end{cases}$$

where $u_r = \boldsymbol{u}\cdot\boldsymbol{n}$ and $\eta_r = \boldsymbol{\eta}\cdot\boldsymbol{n}$, and where we have set to zero the terms at previous time steps since we analyze the convergence to the zero solution. Then, at time t^{n+1}, the corresponding iterative Schwarz method reads:

Given $\boldsymbol{u}^0, p^0, \eta^0$, solve for $j \geq 0$ until convergence

1. Fluid problem

$$\begin{cases} \rho_f \delta_t \boldsymbol{u}^{j+1} + \nabla p^{j+1} = \boldsymbol{0} & \text{in } \Omega_f, \\ \nabla\cdot \boldsymbol{u}^{j+1} = 0 & \text{in } \Omega_f, \\ \int_{-\infty}^{\infty}\int_0^{2\pi} |p\,(r\cos\varphi, r\sin\varphi, z)|\,d\varphi dz & \text{bounded as } r\to 0^+, \\ \mathcal{S}_f\,\Delta t\,\delta_t u_r^{j+1} - p^{j+1} = \dfrac{\mathcal{S}_f}{\Delta t}\,\eta_r^j + \lambda\,\partial_r \eta_r^j & \text{on } \Sigma_R; \end{cases} \tag{2}$$

2. Structure problem

$$\begin{cases} \rho_s \delta_{tt}\boldsymbol{\eta}^{j+1} - \lambda\,\Delta\boldsymbol{\eta}^{j+1} = \boldsymbol{0} & \text{in } \Omega_s, \\ \gamma_{ST}\boldsymbol{\eta}^{j+1} + \lambda\nabla\boldsymbol{\eta}^{j+1}\,\boldsymbol{n} = \boldsymbol{0} & \text{on } \Sigma_{out}, \\ \dfrac{\mathcal{S}_s}{\Delta t}\,\eta_r^{j+1} + \lambda\,\partial_r\eta_r^{j+1} = \mathcal{S}_s\Delta t\,\delta_t u_r^{j+1} - p^{j+1} & \text{on } \Sigma_R, \\ \boldsymbol{\eta}^{j+1}\times\boldsymbol{n} = \boldsymbol{0} & \text{on } \Sigma_R, \end{cases} \tag{3}$$

where we have set to zero the forcing term P_{ext} since we analyze the convergence to the zero solution.

By introducing the Fourier transform with respect to z and φ, and the symbols σ_f and σ_s related to $\mathcal{S}_f$ and $\mathcal{S}_s$, we can write the previous iterations with respect to the variables (r, m, k), where m is the discrete frequency variable related to the angular variable φ and k is the continuous frequency variable related to z. Then, we have the following

Proposition 1 *Set*

$$A(m,k) = -\frac{\lambda\Delta t\beta\left(K_m'(\beta R) - \chi\, I_m'(\beta R)\right)}{K_m(\beta R) - \chi\, I_m(\beta R)}, \quad B(m,k) = -\frac{\rho_f\, I_m(kR)}{\Delta t\, k\, I_m'(kR)}, \tag{4}$$

where I_ν and K_ν are the modified Bessel functions, see [12], $\beta(k) = \sqrt{k^2 + \frac{\rho_s}{\lambda\Delta t^2}}$, and

$$\chi(m,k) := \frac{\gamma_{ST}K_m\left(\beta(R+H)\right) + \lambda\beta K_m'(\beta(R+H))}{\gamma_{ST}I_m(\beta(R+H)) + \lambda\beta I_m'(\beta(R+H))}. \tag{5}$$

Then, the reduction factor of iterations (2)–(3) *is given by*

$$\rho^j(m,k) \;=\; \rho(m,k) \;=\; \left|\frac{\sigma_f(m,k) - A(m,k)}{\sigma_s(m,k) - A(m,k)}\cdot\frac{\sigma_s(m,k) - B(m,k)}{\sigma_f(m,k) - B(m,k)}\right|. \tag{6}$$

Moreover, for any given frequency (m,k), *the convergence set is given by* $(\sigma_f,\sigma_s) \in \Theta_1(A,B) \cup \Theta_2(A,B)$, *where*

$$\Theta_1(A,B) = \left\{(\sigma_f,\sigma_s) : \sigma_s < \sigma_f \text{ and } \left(\sigma_f - \tfrac{A+B}{2}\right)\left(\sigma_s - \tfrac{A+B}{2}\right) < \left(\tfrac{B-A}{2}\right)^2\right\},$$
$$\Theta_2(A,B) = \left\{(\sigma_f,\sigma_s) : \sigma_s > \sigma_f \text{ and } \left(\sigma_f - \tfrac{A+B}{2}\right)\left(\sigma_s - \tfrac{A+B}{2}\right) > \left(\tfrac{B-A}{2}\right)^2\right\},$$

with A and B given by (4).

Proof See [8].

3 Optimization Procedure

From the previous results it follows that the reduction factor (6) is equal to zero for $\sigma_f^{opt} = A(m,k)$ and $\sigma_s^{opt} = B(m,k)$, with A and B given by (4). However, such choices are not implementable since they lead to non-local conditions. Then, a classical approach is to find the best values of the interface symbols within a suitable subset (Optimized Schwarz Method). In particular, in [8] it has been proposed to look for the best symbols belonging to a properly chosen curve parametrized with respect to a variable $q \in \mathbb{R}$, so that in fact we have a minimization problem over q. Let K be the set of the admissible frequencies. Then, we introduce the following quantities:

$$\overline{B} = \max_{(m,k)\in K} B(m,k), \; \overline{A} = \min_{(m,k)\in K} A(m,k), \; \overline{M} = \frac{1}{2}\left(\overline{A} + \overline{B}\right),$$

$$D(m,k) = \frac{1}{2}\left(A(m,k) - B(m,k)\right), \; M(m,k) = \frac{1}{2}\left(A(m,k) + B(m,k)\right),$$

$$Q(m,k) = \frac{\left|M(m,k) - \overline{M}\right|}{D(m,k)}, \; \overline{Q} = \sup_{(m,k)\in K} Q(m,k), \; N = \frac{\inf_{(m,k)\in K} D(m,k)}{\sup_{(m,k)\in K} D(m,k)}.$$

We have the following

Theorem 1 *Assume that* $A(m,k)$ *and* $B(m,k)$ *given by* (4) *are bounded on* K, *for all* $(m,k) \in K$. *Let*

$$\rho_0 = \max\left\{\left(\frac{1-\sqrt{N}}{1+\sqrt{N}}\right)^2; \left(\frac{1-\sqrt{1-\overline{Q}^2}}{\overline{Q}}\right)^2\right\}.$$

Then, for all $(m,k) \in K$, we have

$$\hat{\rho}(q,m,k) = \left| \frac{q - A(m,k)}{2\overline{M} - q - A(m,k)} \frac{2\overline{M} - q - B(m,k)}{q - B(m,k)} \right| \leq \rho_0, \tag{7}$$

if and only if $q \in [q_-, q_+]$ with

$$\begin{aligned} q_- &= \overline{M} \\ &+ \sup_{(m,k)\in K} \left\{ \tfrac{1+\rho_0}{1-\rho_0} D(m,k) - \sqrt{\left(\overline{M} - M(m,k)\right)^2 + \tfrac{4\rho_0}{(1-\rho_0)^2} (D(m,k))^2} \right\}, \\ q_+ &= \overline{M} \\ &+ \inf_{(m,k)\in K} \left\{ \tfrac{1+\rho_0}{1-\rho_0} D(m,k) + \sqrt{\left(\overline{M} - M(m,k)\right)^2 + \tfrac{4\rho_0}{(1-\rho_0)^2} (D(m,k))^2} \right\}. \end{aligned}$$

Proof In [8], a proof of a general result holding for any A and B satisfying $B < A$ for all $(m,k) \in K$, and $\overline{B} < \overline{A}$, has been provided. Here, we notice that these assumptions are automatically satisfied in our case. Indeed, first of all notice that $I, I', K > 0$ while $K' < 0$, so that $B < 0$ for any (m,k). As for A, we first observe that

$$\frac{K'_m(\beta R) - \chi I'_m(\beta R)}{K_m(\beta R) - \chi I_m(\beta R)} < 0$$

if and only if

$$\frac{K'_m(\beta R)}{I'_m(\beta R)} < \chi < \frac{K_m(\beta R)}{I_m(\beta R)},$$

which thanks to (5) becomes

$$\frac{K'_m(\beta R)}{I'_m(\beta R)} < \frac{\gamma_{ST} K_m(\beta(R+H)) + \lambda\beta K'_m(\beta(R+H))}{\gamma_{ST} I_m(\beta(R+H)) + \lambda\beta I'_m(\beta(R+H))} < \frac{K_m(\beta R)}{I_m(\beta R)}. \tag{8}$$

Since the denominators are positive, the first inequality becomes

$$\begin{aligned} &\gamma_{ST} K'_m(\beta R) I_m(\beta(R+H)) + \lambda\beta K'_m(\beta R) I'_m(\beta(R+H)) \\ &< \gamma_{ST} I'_m(\beta R) K_m(\beta(R+H)) + \lambda\beta I'_m(\beta R) K'_m(\beta(R+H)). \end{aligned}$$

Dividing by $I'_m(\beta R) K'_m(\beta R) < 0$, we obtain

$$\begin{aligned} &\gamma_{ST} \left(\frac{I_m(\beta(R+H))}{I'_m(\beta R)} - \frac{K_m(\beta(R+H))}{K'_m(\beta R)} \right) \\ &> -\lambda\beta \left(-\frac{K'_m(\beta(R+H))}{K'_m(\beta R)} + \frac{I'_m(\beta(R+H))}{I'_m(\beta R)} \right), \end{aligned}$$

and this inequality follows immediately, since

$$\frac{I_m(\beta(R+H))}{I'_m(\beta R)} > 0, \qquad \frac{K_m(\beta(R+H))}{K'_m(\beta R)} < 0,$$

and

$$\frac{I'_m(\beta(R+H))}{I'_m(\beta R)} > 1, \qquad \frac{K'_m(\beta(R+H))}{K'_m(\beta R)} < 1$$

(since I'_m is positive and increasing, while K'_m is negative and increasing). The second inequality in (8) is treated similarly. □

By comparing (7) with (6), we observe that moving along the line

$$\begin{cases} \sigma_f = q \\ \sigma_s = -q + 2\overline{M}, \end{cases}$$

$q \in [q_-, q_+]$, the reduction factor is guaranteed to be below a suitable value (ρ_0). This allows to choose properly the value of q in view of the numerical simulations.

4 Numerical Results

In [8] we studied the numerical performance of the proposed estimates for a real fluid-structure interaction problem, inspired by haemodynamics, where the solution was characterized by radial symmetry, thus to a null dominant angular frequency m. In particular, we showed that in this case, the estimates provided by Theorem 1 allowed to choose effective values of the interface parameters.

Here we want to study some cases where the solution features non null dominant angular frequencies. In particular, we considered the coupling between the incompressible Navier-Stokes equations written in the Arbitrary Lagrangian-Eulerian formulation, see [4], and the linear infinitesimal elasticity, see for example [15], and we used a Robin-Robin partitioned procedure for its numerical solution, see [1, 14]. In all the numerical experiments, we used the BDF scheme of order 1 for both the subproblems with a semi-implicit treatment of the fluid convective term. Moreover, we used the following data: $\rho_f = 1\,\text{g/cm}^3$, $\rho_s = 1.1\,\text{g/cm}^3$, $\gamma_{ST} = 3\times10^6\,\text{dyne/cm}^3$, fluid viscosity $\mu = 0.035\,\text{dyne/cm}^2$, Poisson ratio $\nu = 0.49$, Young modulus $E = 3\times10^6\,\text{dyne/cm}^2$. All these data are inspired from haemodynamic applications. We considered a cylinder with length $L = 5\,\text{cm}$, partitioned in an inner cylinder for the fluid problem with radius $R = 0.5\,\text{cm}$, 4680 tetrahedra and 1050 vertices, and an external cylindrical crown for the structure with thickness 0.1 cm and 1260 vertices. For the numerical discretization, we used $P1bubble - P1$ finite elements for the fluid subproblem and $P1$ finite elements for the structure subproblem, and a time discretization parameter $\Delta t = 0.001$ s. The space discretization parameter is $h = 0.25$ cm, so that the frequency k varies in the range $[0.6, 12.5]$. In all the numerical experiments we prescribed the pressure $P_{in} = 1000$ dyne/cm^2 at the inlet. All the numerical results have been obtained with the parallel Finite Element library LIFEV developed at MOX—Politecnico di Milano, INRIA—Paris, CMCS—EPF of Lausanne and Emory University—Atlanta.

We studied the following two cases, characterized by the following initial conditions for the velocity u_{0z} along the z direction:

1.

$$u_{0z}(x, y) = 10^3 \left(x^5 - 10x^3y^2 + 5xy^4\right) \text{ cm/s} = 10^3 r^5 \cos(5\varphi) \text{ cm/s}. \tag{9}$$

In this case the leading frequency is $m = 5$ and therefore we apply Theorem 1 with $m = 5$ and $0.6 \leq k \leq 12.5$, obtaining $\rho_0 = 0.05$, provided that $q \in [6684, 9586]$, with $\overline{M} = 3651$;

2.

$$u_{0z}(x, y) = \begin{cases} 10\left(x^2 + y^2\right) \dfrac{\sin\left(10.5 \arctan\left(\frac{y}{x}\right)\right)}{\sin\left(0.5 \arctan\left(\frac{y}{x}\right)\right)} \text{ cm/s} & x > 0, y \neq 0, \\ 210x^2 \text{ cm/s} & x > 0, y = 0, \\ 10\left(x^2 + y^2\right) \dfrac{\cos\left(10.5 \arctan\left(\frac{y}{x}\right)\right)}{\cos\left(0.5 \arctan\left(\frac{y}{x}\right)\right)} \text{ cm/s} & x < 0, \\ -10y^2 \text{ cm/s} & x = 0. \end{cases} \tag{10}$$

$$= 10r^2 \left(1 + 2 \sum_{m=1}^{10} \cos(m\varphi)\right) \text{ cm/s}.$$

This function is the Dirichlet kernel which is characterized by the fact that all the frequencies m between 0 and 10 are equally distributed. We then apply Theorem 1 with $0 \leq m \leq 10$ and $0.6 \leq k \leq 12.5$, obtaining $\rho_0 = 0.32$, provided that $q \in [1983, 7521]$, with $\overline{M} = 1323$.

Of course we prescribe a compatible initial condition for the displacement η along the z direction.

In both the cases the solution is supposed to feature, at least for the first time steps, the same leading frequencies as the initial condition, so that the application of the estimates provided by Theorem 1 are supposed to lead to excellent convergence properties. In particular, we run the numerical experiments for two time steps, that is we set $T = 0.002$ s as the final instant.

In Fig. 1 we depict the computed velocity along the z direction after the first time step.

We run the numerical simulations for a wide range of the parameter q. We found that the optimal value is $q = 9000$ for the first case and $q = 3000$ for the second case. In Table 1, we report the mean number of iterations for some of the couples of

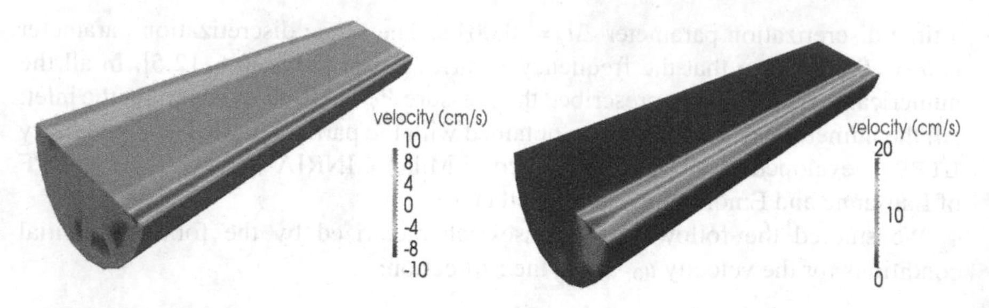

Fig. 1 Velocity along the z direction after the first time step. *Left*: case characterized by $m = 5$ as the leading frequency; *Right*: Dirichlet kernel

Table 1 Values of the interface parameters and mean number of iterations over the two time steps for the initial condition given by (9) (left) and by (10) (right)

σ_f/σ_s	u_{0z} given by (9)	σ_f/σ_s	u_{0z} given by (10)
6684/618	24.5	1000/-1646	X
8000/−698	12.0	**1983/663**	9.5
9000/−1698	8.5	**3000/−354**	4.5
9586/−2284	8.5	**5000/−2354**	8.0
12000/-4698	10.0	**7521/−4875**	10.5
15000/-7698	13.5	10000/-7354	13.0

In bold the couples of σ_f and σ_s within the optimal range estimated by Theorem 1. X means that no convergence has been achieved

the interface parameters used in the numerical simulations, some of them within the range estimated by Theorem 1 and some of them outside such a range.

These numerical results show that the optimal value of q falls in both the cases within the range estimated by Theorem 1 and that outside such a range the convergence properties deteriorate. In particular, we observe that the performance worsens faster going towards the left extreme of the optimal range. This could be explained by looking at Figure 4, left, in [8], where it could be observed that the level sets are denser for small values of σ_f and σ_s (that is of q), so that the performance is more sensitive to small perturbation of q when q is small.

In conclusion, our results showed the effectiveness of the estimates provided in [8] also when the dominant angular frequencies are different from zero.

References

1. S. Badia, F. Nobile, C. Vergara, Fluid-structure partitioned procedures based on Robin transmission conditions. J. Comput. Phys. **227**, 7027–7051 (2008)
2. H. Barucq, M.J. Gander, Y. Xu, On the Influence of Curvature on Transmission Conditions, in *Domain Decomposition Methods in Science and Engineering XXI*. Lecture Notes in Computational Science and Engineering (Springer, Berlin, 2013)

3. V. Dolean, M.J. Gander, L.G. Giorda, Optimized Schwarz Methods for Maxwell's equations. SIAM J. Sci. Comput. **31**(3), 2193–2213 (2009)
4. J. Donea, An arbitrary Lagrangian-Eulerian finite element method for transient dynamic fluid-structure interaction. Comput. Meth. Appl. Mech. Eng. **33**, 689–723 (1982)
5. M.J. Gander, Optimized Schwarz Methods. SIAM J. Numer. Anal. **44**(2), 699–731 (2006)
6. M.J. Gander, F. Magoulès, F. Nataf, Optimized Schwarz methods without overlap for the Helmholtz equation. SIAM J. Sci. Comput. **24**, 38–60 (2002)
7. L. Gerardo Giorda, F. Nobile, C. Vergara, Analysis and optimization of Robin-Robin partitioned procedures in fluid-structure interaction problems. SIAM J. Numer. Anal. **48**(6), 2091–2116 (2010)
8. G. Gigante, C. Vergara, Analysis and optimization of the generalized Schwarz method for elliptic problems with application to fluid-structure interaction. Numer. Math., doi:10.1007/s00211-014-0693-2
9. G. Gigante, M. Pozzoli, C. Vergara, Optimized Schwarz Methods for the diffusion-reaction problem with cylindrical interfaces. SIAM J. Numer. Anal. **51**(6), 3402–3420 (2013)
10. C. Japhet, Optimized Krylov-Ventcell method. Application to convection-diffusion problems, in *Proceedings of the Ninth International Conference on Domain Decomposition Methods*, ed. by P.E. Bjorstad, M.S. Espedal, D.E. Keyes (1998) pp. 382–389
11. C. Japhet, N. Nataf, F. Rogier, The optimized order 2 method. Application to convection-diffusion problems. Futur. Gener. Comput. Syst. **18**, 17–30 (2001)
12. N. Lebedev, *Special Functions and Their Applications* (Courier Dover Publications, New York, 1972)
13. P.L. Lions, On the Schwartz alternating method III, in *Proceedings of the Third International Symposium on Domain Decomposition Methods for PDE's*, ed. by T. Chan, R. Glowinki, J. Periaux, O.B. Widlund (SIAM, Philadelphia, 1990), pp. 202–223
14. F. Nobile, C. Vergara, Partitioned algorithms for fluid-structure interaction problems in haemodynamics. Milan J. Math. **80**(2), 443–467 (2012)
15. F. Nobile, M. Pozzoli, C. Vergara, Time accurate partitioned algorithms for the solution of fluid-structure interaction problems in haemodynamics. Comput. Fluids **86**, 470–482 (2013)
16. A. Qaddouria, L. Laayounib, S. Loiselc, J. Cotea, M.J. Gander, Optimized Schwarz methods with an overset grid for the shallow-water equations: preliminary results. Appl. Numer. Math. **58**, 459–471 (2008)
17. B. Stupfel, Improved transmission conditions for a one-dimensional domain decomposition method applied to the solution of the Helmhotz equation. J. Comput. Phys. **229**, 851–874 (2010)

Ventcell Conditions with Mixed Formulations for Flow in Porous Media

Thi Thao Phuong Hoang, Caroline Japhet, Michel Kern, and Jean Roberts

1 Introduction

The Optimized Schwarz method has been introduced and analyzed over the last decade, where the convergence speed of the Jacobi iteration is significantly enhanced by using general transmission conditions on the interfaces together with optimized parameters. In particular, Ventcell transmission conditions (see [3–6, 8–10]) have been studied for the primal formulation with different numerical schemes showing that the convergence of the Optimized Schwarz algorithm with Ventcell conditions is improved over that with Robin conditions. Ventcell conditions are second order differential conditions, see [12].

In this work, we study the Optimized Schwarz method with Ventcell conditions in the context of mixed formulations, which is not as straightforward as in the case of primal formulations and we have to introduce Lagrange multipliers on the interfaces to handle tangential derivatives involved in those conditions. Two dimensional numerical results for heterogeneous problems will be presented to compare the performance of the Ventcell transmission conditions with that of the Robin transmission conditions.

T.T.P. Hoang • M. Kern • J. Roberts
INRIA, Rocquencourt, France
e-mail: Phuong.Hoang_Thi_Thao@inria.fr; Michel.Kern@inria.fr; Jean.Roberts@inria.fr

C. Japhet (✉)
INRIA, Rocquencourt, France

Université Paris 13, LAGA, Villetaneuse, France
e-mail: japhet@math.univ-paris13.fr

T. Dickopf et al. (eds.), *Domain Decomposition Methods in Science and Engineering XXII*, Lecture Notes in Computational Science and Engineering 104, DOI 10.1007/978-3-319-18827-0_54

2 A Model Problem and Domain Decomposition with Ventcell Transmission Conditions

For an open, bounded domain $\Omega \subset \mathbb{R}^d$ $(d = 2, 3)$ with Lipschitz boundary $\partial\Omega$, consider single phase flow in a porous medium written in mixed form:

$$\begin{aligned} \operatorname{div} \boldsymbol{u} &= f \quad \text{in } \Omega, \\ \boldsymbol{K}^{-1}\boldsymbol{u} + \nabla p &= 0 \quad \text{in } \Omega, \\ p &= 0 \quad \text{on } \partial\Omega, \end{aligned} \tag{1}$$

where p is the pressure, $\boldsymbol{u}$ the Darcy velocity, f the source term and $\boldsymbol{K}$ a symmetric, positive definite, time independent, hydraulic conductivity (or permeability) tensor. For the sake of simplicity, we have imposed homogeneous Dirichlet condition on the boundary, other types of boundary conditions can be treated similarly. The well-posedness of problem (1) is well-known (see, e.g., [1, 2, 11]).

We consider a decomposition of Ω into two nonoverlapping subdomains, Ω_1 and Ω_2, separated by an interface Γ:

$$\Omega_1 \cap \Omega_2 = \emptyset; \quad \Gamma = \partial\Omega_1 \cap \partial\Omega_2 \cap \Omega, \quad \Omega = \Omega_1 \cup \Omega_2 \cup \Gamma.$$

Note that the same analysis can be extended to the case of many subdomains in bands. For $i = 1, 2$, let $\boldsymbol{n}_i$ denote the unit, outward pointing, normal on $\partial\Omega_i$, and for any scalar, vector or tensor valued function ψ defined on Ω, let ψ_i denote the restriction of ψ to Ω_i. In order to write the Ventcell transmission conditions, we use the notation ∇_τ and div_τ for the tangential gradient and divergence operators on Γ respectively. We denote by $\boldsymbol{K}_{i,\Gamma}$ the tangential component of the trace of $\boldsymbol{K}_i$, $i = 1, 2$, on Γ. A multidomain formulation equivalent to problem (1) is obtained by solving in each subdomain the following problem

$$\begin{aligned} \operatorname{div} \boldsymbol{u}_i &= f \quad \text{in } \Omega_i, \\ \boldsymbol{K}_i^{-1}\boldsymbol{u}_i + \nabla p_i &= 0 \quad \text{in } \Omega_i, \\ p_i &= 0 \quad \text{on } (\partial\Omega_i \cap \partial\Omega), \end{aligned}$$

for $i = 1, 2$, together with the transmission conditions

$$\left.\begin{aligned} p_1 &= p_2, \\ \boldsymbol{u}_1 \cdot \boldsymbol{n}_1 + \boldsymbol{u}_2 \cdot \boldsymbol{n}_2 &= 0, \end{aligned}\right\} \quad \text{on } \Gamma. \tag{2}$$

Under sufficient regularity, one may replace (2) by optimized Ventcell transmission conditions, which were introduced and analyzed for primal formulations in [8, 9]:

$$\begin{aligned} -\boldsymbol{u}_1 \cdot \boldsymbol{n}_1 + \alpha_{1,2}\, p_1 + \beta_{1,2} \operatorname{div}_\tau (-\boldsymbol{K}_{2,\Gamma} \nabla_\tau p_1) &= -\boldsymbol{u}_2 \cdot \boldsymbol{n}_1 + \alpha_{1,2}\, p_2 + \\ &\qquad \beta_{1,2} \operatorname{div}_\tau (-\boldsymbol{K}_{2,\Gamma} \nabla_\tau p_2) \quad \text{on } \Gamma, \\ -\boldsymbol{u}_2 \cdot \boldsymbol{n}_2 + \alpha_{2,1}\, p_2 + \beta_{2,1} \operatorname{div}_\tau (-\boldsymbol{K}_{1,\Gamma} \nabla_\tau p_2) &= -\boldsymbol{u}_1 \cdot \boldsymbol{n}_2 + \alpha_{2,1}\, p_1 + \\ &\qquad \beta_{2,1} \operatorname{div}_\tau (-\boldsymbol{K}_{1,\Gamma} \nabla_\tau p_1) \quad \text{on } \Gamma, \end{aligned} \tag{3}$$

where $\alpha_{i,j}$ and $\beta_{i,j}$, $i = 1, 2,\ j = (3 - i)$, are positive constants. The conditions (3) are derived in such a way that they are equivalent to the original ones given in (2) (cf. references above). These parameters are chosen to optimize the convergence factor, see [3, 4, 8, 9].

3 A Multidomain Formulation in Mixed Form

In this section, we study Ventcell transmission conditions with mixed formulations. In order to handle second order terms involved in (3), we introduce Lagrange multipliers on the interface Γ: $p_{i,\Gamma}, i = 1, 2$, representing the pressure trace p_i on Γ and a vector field $\boldsymbol{u}_{\Gamma,i} := -\boldsymbol{K}_{j,\Gamma} \nabla_\tau p_{i,\Gamma},\ i = 1, 2, j = (3 - i)$. We use the notation $\boldsymbol{u}_{\Gamma,i}$ instead of $\boldsymbol{u}_{i,\Gamma}$ to insist that $\boldsymbol{u}_{\Gamma,i}$ is **not** the tangential component of the trace of $\boldsymbol{u}_i$ on the interface. In fact $\boldsymbol{u}_{\Gamma,i}$ is used as an artificial tool for convergence purposes (it does not have a particular physical meaning). We rewrite (3) defined on Γ as follows, for $i = 1, 2, j = (3 - i)$:

$$\begin{aligned} -\boldsymbol{u}_i \cdot \boldsymbol{n}_i + \alpha_{i,j}\, p_{i,\Gamma} + \beta_{i,j} \operatorname{div}_\tau \boldsymbol{u}_{\Gamma,i} &= -\boldsymbol{u}_j \cdot \boldsymbol{n}_i + \alpha_{i,j}\, p_{j,\Gamma} \\ &\qquad + \beta_{i,j} \operatorname{div}_\tau \left(\boldsymbol{K}_{j,\Gamma} \boldsymbol{K}_{i,\Gamma}^{-1} \boldsymbol{u}_{\Gamma,j}\right), \\ \boldsymbol{K}_{j,\Gamma}^{-1}\, \boldsymbol{u}_{\Gamma,i} + \nabla_\tau p_{i,\Gamma} &= 0. \end{aligned} \tag{4}$$

The corresponding multidomain problem consists of solving in the subdomains the problem, for $i = 1, 2,\ j = (3 - i)$:

$$\begin{aligned} \operatorname{div} \boldsymbol{u}_i &= f && \text{in } \Omega_i, \\ \boldsymbol{K}_i^{-1} \boldsymbol{u}_i + \nabla p_i &= 0 && \text{in } \Omega_i, \\ p_i &= 0 && \text{on } (\partial\Omega_i \cap \partial\Omega), \\ -\boldsymbol{u}_i \cdot \boldsymbol{n}_i + \alpha_{i,j}\, p_{i,\Gamma} + \beta_{i,j} \operatorname{div}_\tau \boldsymbol{u}_{\Gamma,i} &= -\boldsymbol{u}_j \cdot \boldsymbol{n}_i + \alpha_{i,j}\, p_{j,\Gamma} && \\ &\quad + \beta_{i,j} \operatorname{div}_\tau (\boldsymbol{K}_{j,\Gamma} \boldsymbol{K}_{i,\Gamma}^{-1} \boldsymbol{u}_{\Gamma,j}) && \text{on } \Gamma, \\ \boldsymbol{K}_{j,\Gamma}^{-1}\, \boldsymbol{u}_{\Gamma,i} + \nabla_\tau p_{i,\Gamma} &= 0 && \text{on } \Gamma, \\ p_{i,\Gamma} &= 0 && \text{on } \partial\Gamma. \end{aligned} \tag{5}$$

This can be seen as a coupling problem between a d–dimensional PDE in the subdomain Ω_i and a $(d - 1)$-dimensional PDE on the interface Γ, and both PDEs are written in mixed form. Under a suitable regularity hypothesis the multidomain problem (5) is equivalent to the monodomain problem (1). Details of the proof can be found in [7, pp. 94–95].

3.1 Well-Posedness of the Ventcell Local Problem

For an open, bounded domain $\mathcal{O} \subset \mathbb{R}^d\,(d = 2, 3)$ with Lipschitz boundary $\partial\mathcal{O}$, consider the following elliptic problem written in mixed form with Ventcell boundary conditions

$$\begin{aligned} \operatorname{div} \boldsymbol{u}_{\mathcal{O}} &= f_{\mathcal{O}} && \text{in } \mathcal{O}, \\ \boldsymbol{K}^{-1}\boldsymbol{u}_{\mathcal{O}} + \nabla p_{\mathcal{O}} &= 0 && \text{in } \mathcal{O}, \\ -\boldsymbol{u}_{\mathcal{O}} \cdot \boldsymbol{n} + \alpha p_{\partial\mathcal{O}} + \beta \operatorname{div}_\tau \tilde{\boldsymbol{u}}_{\partial\mathcal{O}} &= f_{\partial\mathcal{O}} && \text{on } \partial\mathcal{O}, \\ \tilde{\boldsymbol{K}}_{\partial\mathcal{O}}^{-1} \tilde{\boldsymbol{u}}_{\partial\mathcal{O}} + \nabla_\tau p_{\partial\mathcal{O}} &= 0 && \text{on } \partial\mathcal{O}, \end{aligned} \tag{6}$$

where $\boldsymbol{n}$ is the unit, outward pointing, normal vector on $\partial\mathcal{O}$, $\boldsymbol{K}(\cdot) \in \mathbb{R}^{d^2}$ and $\tilde{\boldsymbol{K}}_{\partial\mathcal{O}}(\cdot) \in \mathbb{R}^{(d-1)^2}$ are given, and α and β are positive constants.

In order to write the weak formulation of problem (6), we define the following Hilbert spaces

$$\begin{aligned} M &= \left\{\mu = (\mu_{\mathcal{O}}, \mu_{\partial\mathcal{O}}) \in L^2(\mathcal{O}) \times L^2(\partial\mathcal{O})\right\}, \\ \Sigma &= \left\{\boldsymbol{v} = (\boldsymbol{v}_{\mathcal{O}}, \tilde{\boldsymbol{v}}_{\partial\mathcal{O}}) \in \boldsymbol{L}^2(\mathcal{O}) \times \boldsymbol{L}^2(\partial\mathcal{O}) : \operatorname{div} \boldsymbol{v}_{\mathcal{O}} \in L^2(\mathcal{O}) \text{ and} \right. \\ &\qquad \left. \beta \operatorname{div}_\tau \tilde{\boldsymbol{v}}_{\partial\mathcal{O}} - \boldsymbol{v}_{\mathcal{O}} \cdot \boldsymbol{n}_{|\partial\mathcal{O}} \in L^2(\partial\mathcal{O})\right\}, \end{aligned}$$

equipped with the norms

$$\begin{aligned} \|\mu\|_M^2 &= \|\mu_{\mathcal{O}}\|_{\mathcal{O}}^2 + \|\mu_{\partial\mathcal{O}}\|_{\partial\mathcal{O}}^2, \\ \|\boldsymbol{v}\|_\Sigma^2 &= \|\boldsymbol{v}_{\mathcal{O}}\|_{\mathcal{O}}^2 + \|\operatorname{div} \boldsymbol{v}_{\mathcal{O}}\|_{\mathcal{O}}^2 + \|\tilde{\boldsymbol{v}}_{\partial\mathcal{O}}\|_{\partial\mathcal{O}}^2 + \|\beta \operatorname{div} \tilde{\boldsymbol{v}}_{\partial\mathcal{O}} - \boldsymbol{v}_{\mathcal{O}} \cdot \boldsymbol{n}_{|\partial\mathcal{O}}\|_{\partial\mathcal{O}}^2, \end{aligned}$$

where $\|\cdot\|_{\mathcal{O}}$ and $\|\cdot\|_{\partial\mathcal{O}}$ are the $L^2(\mathcal{O})$ and $L^2(\partial\mathcal{O})$-norms, respectively. We denote by $(\cdot,\cdot)_{\mathcal{O}}$ and $(\cdot,\cdot)_{\partial\mathcal{O}}$ the inner products of $L^2(\mathcal{O})$ and $L^2(\partial\mathcal{O})$.

Next, define the following bilinear forms (recall that β is a positive constant) on $\Sigma \times \Sigma$, $\Sigma \times M$ and $M \times M$ respectively:

$$\begin{aligned} a(\boldsymbol{u}, \boldsymbol{v}) &= \left(\boldsymbol{K}^{-1}\boldsymbol{u}_{\mathcal{O}}, \boldsymbol{v}_{\mathcal{O}}\right)_{\mathcal{O}} + \left(\beta \tilde{\boldsymbol{K}}_{\partial\mathcal{O}}^{-1} \tilde{\boldsymbol{u}}_{\partial\mathcal{O}}, \tilde{\boldsymbol{v}}_{\partial\mathcal{O}}\right)_{\partial\mathcal{O}}, \\ b(\boldsymbol{u}, \mu) &= \left(\operatorname{div} \boldsymbol{u}_{\mathcal{O}}, \mu_{\mathcal{O}}\right)_{\mathcal{O}} + \left(\beta \operatorname{div}_\tau \tilde{\boldsymbol{u}}_{\partial\mathcal{O}} - \boldsymbol{u}_{\mathcal{O}} \cdot \boldsymbol{n}_{|\partial\mathcal{O}}, \mu_{\partial\mathcal{O}}\right)_{\partial\mathcal{O}}, \\ c(p, \mu) &= \left(\alpha p_{\partial\mathcal{O}}, \mu_{\partial\mathcal{O}}\right)_{\partial\mathcal{O}}, \end{aligned}$$

and the linear form defined on M by:

$$L_f(\mu) = (f_{\mathcal{O}}, \mu_{\mathcal{O}})_{\mathcal{O}} + (f_{\partial\mathcal{O}}, \mu_{\partial\mathcal{O}})_{\partial\mathcal{O}} \,.$$

With these spaces and forms, the weak form of (6) can be written as follows:

$$\begin{aligned} &\text{Find } (p, \boldsymbol{u}) \in M \times \Sigma \text{ such that} \\ &\quad a(\boldsymbol{u}, \boldsymbol{v}) - b(\boldsymbol{v}, p) = 0 \qquad \forall \boldsymbol{v} \in \Sigma, \\ &\quad -b(\boldsymbol{u}, \mu) - c(p, \mu) = -L_f(\mu) \ \forall \mu \in M. \end{aligned} \tag{7}$$

Theorem 1 *Assume that there exist positive constants K_- and K_+ such that $\varsigma^T \boldsymbol{K}^{-1}(\cdot)\varsigma \geq K_-|\varsigma|^2$, and $|\boldsymbol{K}(\cdot)\varsigma| \leq K_+|\varsigma|$ a.e. in $\mathcal{O}$ and $\forall \varsigma \in \mathbb{R}^d$; and that $\eta^T \tilde{\boldsymbol{K}}_{\partial\mathcal{O}}^{-1}(\cdot)\eta \geq K_-|\eta|^2$, and $|\tilde{\boldsymbol{K}}_{\partial\mathcal{O}}(\cdot)\eta| \leq K_+|\eta|$ a.e. in $\partial\mathcal{O}$ and $\forall \eta \in \mathbb{R}^{d-1}$. If $(f_{\mathcal{O}}, f_{\partial\mathcal{O}})$ is in M then there exists a unique solution $(p, \boldsymbol{u}) \in M \times \Sigma$ of problem (7).*

Proof The existence and uniqueness of the solution of (7) is a generalization of the classical case (see [2, pp. 47–50]; [11, pp. 572–573]). See [7, pp. 96–98] for details of the proof of Theorem 1.

3.2 An Interface Problem

In this subsection, we derive an interface problem associated with the multidomain problem (5). With this aim, we define the Ventcell-to-Ventcell operator $\mathcal{S}_i^{\text{VtV}}$ (note that we have assumed sufficient regularity of the solution as in Sect. 2), which depends on the parameters $\alpha_{i,j}$ and $\beta_{i,j}$, for $i = 1, 2$, and $j = (3 - i)$, as follows

$$\begin{aligned} \mathcal{S}_i^{\text{VtV}} &: L^2(\Gamma) \times L^2(\Omega_i) \to L^2(\Gamma) \\ (\vartheta, f) &\longmapsto \mathcal{S}_i^{\text{VtV}}(\vartheta, f) = -\boldsymbol{u}_i \cdot \boldsymbol{n}_{j|\Gamma} + \alpha_{j,i}\, p_{i,\Gamma} + \beta_{j,i} \operatorname{div}_\tau (\boldsymbol{K}_{i,\Gamma} \boldsymbol{K}_{j,\Gamma}^{-1} \boldsymbol{u}_{\Gamma,i}), \end{aligned}$$

where $(p_i, \boldsymbol{u}_i, p_{i,\Gamma}, \boldsymbol{u}_{\Gamma,i})$, $i = 1, 2$, is the solution of

$$\begin{aligned} \operatorname{div} \boldsymbol{u}_i &= f && \text{in } \Omega_i, \\ \boldsymbol{K}_i^{-1} \boldsymbol{u}_i + \nabla p_i &= 0 && \text{in } \Omega_i, \\ p_i &= 0 && \text{on } (\partial\Omega_i \cap \partial\Omega), \\ -\boldsymbol{u}_i \cdot \boldsymbol{n}_i + \alpha_{i,j}\, p_{i,\Gamma} + \beta_{i,j} \operatorname{div}_\tau \boldsymbol{u}_{\Gamma,i} &= \vartheta && \text{on } \Gamma, \\ \boldsymbol{K}_{j,\Gamma}^{-1}\, \boldsymbol{u}_{\Gamma,i} + \nabla_\tau p_{i,\Gamma} &= 0 && \text{on } \Gamma, \\ p_{i,\Gamma} &= 0 && \text{on } \partial\Gamma. \end{aligned} \tag{8}$$

The well-posedness of problem (8) is given by an extension of Theorem 1. The interface problem, corresponding to the Ventcell transmission conditions (4), is defined by

$$\begin{aligned} \vartheta_1 &= \mathcal{S}_2^{\mathrm{VtV}}(\vartheta_2, f) \\ \vartheta_2 &= \mathcal{S}_1^{\mathrm{VtV}}(\vartheta_1, f) \end{aligned} \quad \text{on } \Gamma, \tag{9}$$

or equivalently,

$$\mathcal{S}_{\mathbf{V}} \begin{pmatrix} \vartheta_1 \\ \vartheta_2 \end{pmatrix} = \chi_{\mathbf{V}}(f), \quad \text{on } \Gamma, \tag{10}$$

where

$$\begin{aligned} \mathcal{S}_{\mathbf{V}} : L^2(\Gamma) \times L^2(\Gamma) &\longrightarrow \quad L^2(\Gamma) \times L^2(\Gamma) \\ \begin{pmatrix} \vartheta_1 \\ \vartheta_2 \end{pmatrix} &\longmapsto \begin{pmatrix} \vartheta_1 - \mathcal{S}_2^{\mathrm{VtV}}(\vartheta_2, 0) \\ \vartheta_2 - \mathcal{S}_1^{\mathrm{VtV}}(\vartheta_1, 0) \end{pmatrix}, \end{aligned}$$

and

$$\begin{aligned} \chi_{\mathbf{V}} : L^2(\Gamma) &\longrightarrow L^2(\Gamma) \times L^2(\Gamma) \\ f &\longmapsto \begin{pmatrix} \mathcal{S}_2^{\mathrm{VtV}}(0, f) \\ \mathcal{S}_1^{\mathrm{VtV}}(0, f) \end{pmatrix}. \end{aligned}$$

One can solve problem (10) iteratively using Jacobi iteration or a Krylov method (e.g. GMRES, see for example [9]) the right hand side is computed (only once) by solving problem (8) in each subdomain with $\vartheta = 0$; then for a given pair of vectors $(\vartheta_1, \vartheta_2)$, the matrix vector product is obtained (at each iteration) by solving, for $i = 1, 2$, subdomain problem (8) in Ω_i with $\vartheta = \vartheta_i$ and with $f = 0$. If one uses Jacobi iteration for solving (10), the resulting algorithm is equivalent to the optimized Schwarz algorithm with Ventcell transmission conditions (see [4, 8]).

4 Numerical Results

We consider a domain $\Omega = (0, \pi)^2$ and its decomposition into two nonoverlapping subdomains $\Omega_1 = \left(0, \frac{\pi}{2}\right) \times (0, \pi)$ and $\Omega_2 = \left(\frac{\pi}{2}, \pi\right) \times (0, \pi)$. The permeability is $\boldsymbol{K} = \mathfrak{K}\boldsymbol{I}$ isotropic and constant on each subdomain, where $\boldsymbol{I}$ is the 2D identity matrix. We take $\mathfrak{K}_1 = 1/\mathcal{K}$ and $\mathfrak{K}_2 = 1$, where $\mathcal{K} = 1, 10$ or 100. The exact solution is $p(x, y) = \cos(\pi x)\sin(\pi y)$. For the spatial discretization, we use mixed finite elements (with interface Lagrange multipliers) with the lowest order Raviart-Thomas spaces on rectangles (see [2, 11]).

Table 1 Number of iterations required to reach an error reduction of 10^{-6} in p and in $\boldsymbol{u}$ (in square brackets) for different permeability ratios, and for different values of the discretization parameter h

	$\mathcal{K}=1$		$\mathcal{K}=10$		$\mathcal{K}=100$	
h	Jacobi	GMRES	Jacobi	GMRES	Jacobi	GMRES
$\pi/50$	15 [15]	10 [11]	11 [10]	9 [9]	7 [6]	7 [7]
$\pi/100$	17 [18]	11 [12]	11 [10]	9 [9]	7 [6]	7 [7]
$\pi/200$	21 [21]	13 [13]	11 [10]	9 [9]	7 [6]	7 [7]
$\pi/400$	25 [25]	14 [14]	11 [10]	10 [9]	7 [6]	8 [8]
$\pi/800$	29 [29]	15 [16]	13 [12]	10 [10]	7 [6]	8 [8]

Remark 1 In order to handle the discontinuous coefficients, we use the optimized, weighted Ventcell parameters defined by

$$\alpha_{1,2} = \mathfrak{K}_2\alpha, \qquad \alpha_{2,1} = \mathfrak{K}_1\alpha,$$
$$\beta_{1,2} = \mathfrak{K}_2\beta, \qquad \beta_{2,1} = \mathfrak{K}_1\beta.$$

The calculation of these parameters is done as in [3].

We first study the convergence behavior of the optimized Schwarz method with the optimized weighted Ventcell parameters. To that purpose, we consider the error equation, i.e. with $f = 0$ and homogeneous Dirichlet boundary conditions. We start with a random initial guess on the interface and compute the error in the $L^2(\Omega)$-norm of the pressure p and of the velocity $\boldsymbol{u}$. Table 1 gives the number of iterations needed to reach an error reduction of 10^{-6} first in p and then in $\boldsymbol{u}$ (in square brackets) as the mesh is refined. Both GMRES and Jacobi iterations are considered. For homogeneous case ($\mathcal{K} = 1$), GMRES significantly improves the convergence speed (by a factor of 2) and also the asymptotic results compared to the Jacobi iteration. These results are consistent with those obtained with primal formulations in [4] (where a finite difference scheme is used). As the ratio $\mathcal{K}$ increases, the number of iterations is smaller and GMRES does not greatly accelerate the convergence speed compared to Jacobi iterations. Also for large values of the contrast $\mathcal{K}$, the convergence rate of the algorithms with GMRES or Jacobi are almost independent of the mesh size (since the optimized parameters play the role of a preconditioner). This is also the case where a primal formulation and a finite volume method are used (cf. [3]).

Next we verify the performance of the optimized parameters, computed by numerically minimizing the continuous convergence factor. We take $h = \pi/100$, vary α and β, and compute the error in the velocity $\boldsymbol{u}$ after a fixed number of Jacobi iterations for different permeability ratios. The results are shown in Fig. 1 for $\mathcal{K} = 1$ (# iter $= 20$ iterations), $\mathcal{K} = 10$ (# iter $= 12$ iterations) and $\mathcal{K} = 100$ (# iter $= 8$ iterations) respectively. We see that, in all cases, the optimized weighted Ventcell parameters (the red star) are located close to those giving the smallest error after the same number of iterations.

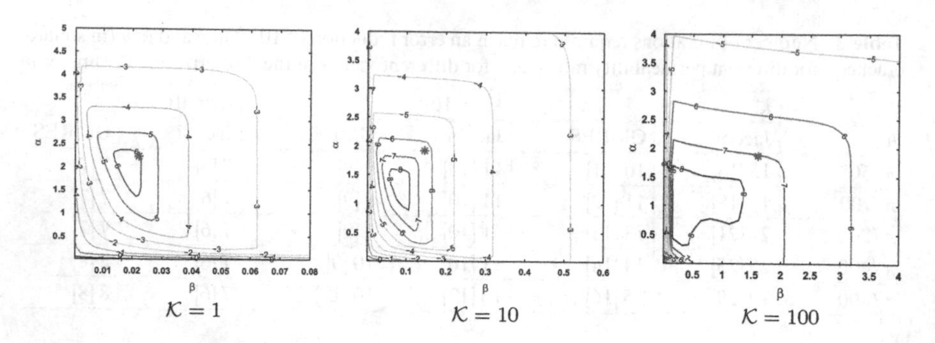

Fig. 1 Level curves for the error in the velocity (in logarithmic scale) after some fixed number of Jacobi iterations for various values of the parameters α and β and for different permeability ratios $\mathcal{K}$. The *red star* shows the optimized parameters (Color figure online)

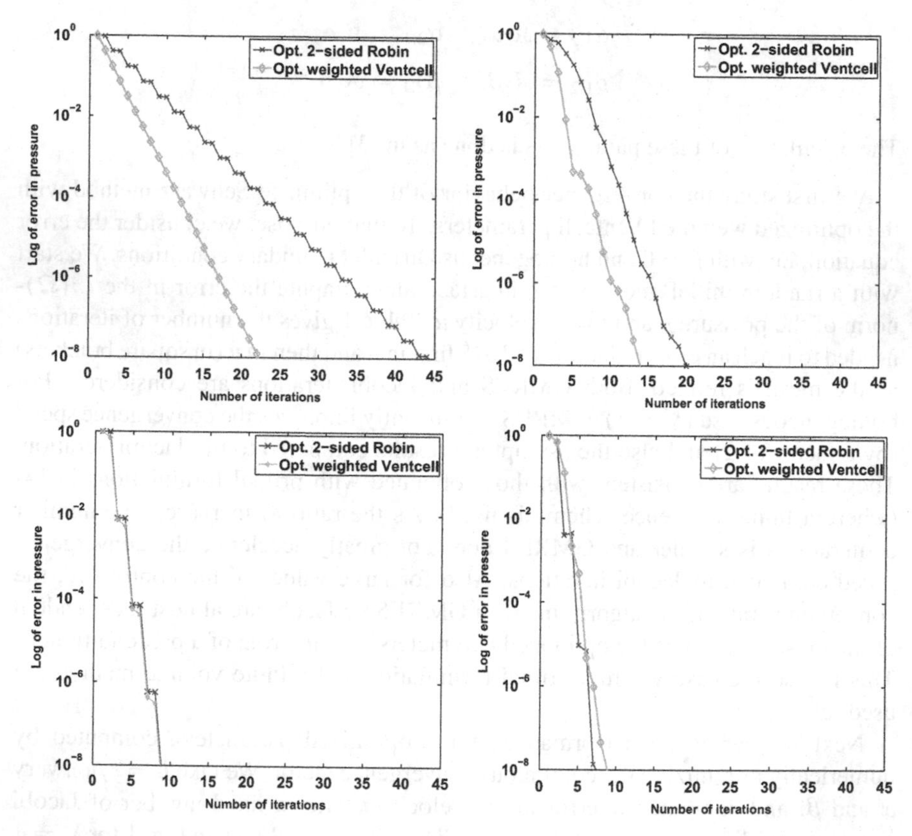

Fig. 2 L^2 error in the pressure p for $\mathcal{K} = 1$ (*top*) and $\mathcal{K} = 100$ (*bottom*): Jacobi (*left*) and GMRES (*right*)

Finally, we illustrate the improvement obtained using Ventcell transmission conditions over the Robin conditions (i.e. $\beta = 0$).

We consider the optimized 2-sided Robin parameters with $\alpha_{1,2} \neq \alpha_{2,1}$ and $\beta_{1,2} = \beta_{2,1} = 0$. Figure 2 shows the error in the pressure versus the number of iterations using Jacobi (on the left) and GMRES (on the right) for different diffusion ratios, $\mathcal{K} = 1$ and $\mathcal{K} = 100$ respectively.

We see that for the homogeneous case ($\mathcal{K} = 1$), with Jacobi iterations the optimized weighted Ventcell converges significantly faster than the optimized 2-sided Robin (by a factor of 2). As $\mathcal{K}$ increases, the optimized weighted Ventcell and the optimized 2-sided Robin become comparable. With GMRES, the difference in the convergence of the two types of optimized parameters becomes less significant, especially for high diffusion ratios. These results are for a symmetric two subdomain case with a conforming mesh, Ventcell transmission conditions may have a more important effect on the convergence (compared with Robin transmission conditions) when many subdomains and nonmatching grids are considered (cf. [10]).

Acknowledgement This work was supported by ANDRA, the French Agency for Nuclear Waste Management.

References

1. F. Brezzi, On the existence, uniqueness and approximation of saddle-point problems arising from Lagrangian multipliers. RAIRO Anal. Numér. **8**, 129–151 (1974)
2. F. Brezzi, M. Fortin, *Mixed and Hybrid Finite Elements Methods* (Springer, New York, 1991). ISBN 3540975829
3. O. Dubois, Optimized Schwarz methods for the advection-diffusion equation and for problems with discontinuous coefficients. Ph.D. thesis, McGill University, 2007
4. M.J. Gander, Optimized Schwarz methods. SIAM J. Numer. Anal. **44**(2), 699–731 (2006)
5. M.J. Gander, L. Halpern, F. Hubert, S. Krell, DDFV Ventcell Schwarz algorithms, in *Proceedings of the 22nd International Conference on Domain Decomposition Methods* to appear in these proceedings
6. L. Halpern, F. Hubert, Finite volume Ventcell-Schwarz algorithm for advection-diffusion equations. SIAM J. Numer. Anal. **52**, 1269–1291 (2014)
7. T.T.P. Hoang, Space-time domain decomposition methods for mixed formulations of flow and transport problems in porous media. Ph.D. thesis, University Paris 6, 2013
8. C. Japhet, Optimized Krylov-Ventcell method. Application to convection-diffusion problems, in *Domain Decomposition Methods in Science and Engineering IX*, ed. by U. Bjørstad, M. Espedal, D.E. Keyes (Wiley, New York, 1998), pp. 382–389
9. C. Japhet, F. Nataf, F. Rogier, The optimized order 2 method. Application to convection-diffusion problems. Futur. Gener. Comput. Syst. **18**(1), 17–30 (2001)
10. C. Japhet, Y. Maday, F. Nataf, A new interface cement equilibrated mortar method with Ventcel conditions, in *Domain Decomposition Methods in Science and Engineering XXI*, ed. by J. Erhel, M.J. Gander, L. Halpern, G. Pichot, T. Sassi, O. Widlund. Lecture Notes in Computational Science and Engineering, vol. 98 (Springer, Berlin, 2014), pp. 329–336

11. J.E. Roberts, J.-M. Thomas, Mixed and hybrid methods, in *Handbook of Numerical Analysis*, vol. 2 (North-Holland, Amsterdam, 1991), pp. 523–639
12. A.D. Ventcel', On boundary conditions for multi-dimensional diffusion processes. Theor. Probab. Appl. **4**, 164–177 (1959)

Mortar Methods with Optimized Transmission Conditions for Advection-Diffusion Problems

Caroline Japhet and Yvon Maday

1 Introduction

In many practical applications in fluid dynamics, a very large range of scales spanning many orders of magnitude are simultaneously present; one possibility to perform an economical and accurate approximation of the solution is to use different discretizations in different regions of the computational domain to match with the physical scales. The mortar element method introduced in [3] allows such a use of different discretizations in an optimal way in the sense that the error is bounded by the sum of the subregion-by-subregion approximation errors without constraint on the choice of the different discretizations. An extension to fluids is given in [1]. An alternative and simpler method, the New Interface Cement Equilibrated Mortar (NICEM) method proposed in [6] and analyzed in [8] for an elliptic problem, allows to optimally match Robin conditions on non-conforming grids. An extension to Ventcel conditions is given in [9]. The main feature of this approach is that, on each side of the interface, the jump of the Robin or Ventcel condition should be L^2-orthogonal to a well chosen finite element space on the interface (in that case there is no master and slave sides, which makes the method simpler). Thus, it allows

C. Japhet (✉)
Université Paris 13, LAGA, UMR 7539, F-93430 Villetaneuse, France

INRIA Paris-Rocquencourt, BP 105, 78153 Le Chesnay, France
e-mail: japhet@math.univ-paris13.fr

Y. Maday
Laboratoire Jacques-Louis Lions, Sorbonne Universités, UPMC Univ Paris 06 and CNRS, UMR 7598, F-75005 Paris, France

Division of Applied Maths, Institut Universitaire de France, Brown University, Providence, RI, USA
e-mail: maday@ann.jussieu.fr

T. Dickopf et al. (eds.), *Domain Decomposition Methods in Science and Engineering XXII*, Lecture Notes in Computational Science and Engineering 104, DOI 10.1007/978-3-319-18827-0_55

to combine different approximations in different subdomains in the framework of optimized Schwarz algorithms which are based on optimized Robin or Ventcel transmission conditions and lead to robust and fast algorithms (see [5, 7]).

In this paper we extend the NICEM method to advection-diffusion problems. For simplicity we consider the case of Robin conditions.

We first introduce the problem at the continuous level: find u such that

$$\eta u + \nabla \cdot (\boldsymbol{a}u) - \nabla \cdot (\nu \nabla u) = f \quad \text{in } \Omega \tag{1}$$

$$u = 0 \quad \text{on } \partial\Omega, \tag{2}$$

where Ω is a $\mathcal{C}^{1,1}$ (or convex polygon in 2D or polyhedron in 3D) domain of $\mathbb{R}^d$, $d = 2$ or 3, and f is given in $L^2(\Omega)$. We consider a decomposition of Ω into K non-overlapping subdomains: $\overline{\Omega} = \cup_{k=1}^{K} \overline{\Omega}^k$, where Ω^k, $1 \le k \le K$ are either $\mathcal{C}^{1,1}$ or polygons in 2D or polyhedrons in 3D. We suppose that this decomposition is geometrically conforming. Let $\boldsymbol{n}_k$ be the outward normal from Ω^k. Let $\Gamma^{k,\ell} := \partial\Omega^k \cap \partial\Omega^\ell$ denote the interface of two adjacent subdomains. An optimized Schwarz algorithm with Robin transmission conditions for problem (1)–(2) is

$$\begin{aligned} \eta u_k^{n+1} + \nabla \cdot (\boldsymbol{a}u_k^{n+1}) - \nabla \cdot (\nu \nabla u_k^{n+1}) &= \quad f \qquad \text{in } \Omega^k \\ u_k^{n+1} &= \quad 0 \qquad \text{on } \partial\Omega^k \cap \partial\Omega \\ \mathcal{B}_{k,\ell}(u_k^{n+1}) &= \mathcal{B}_{k,\ell}(u_\ell^n) \quad \text{on } \Gamma^{k,\ell} \end{aligned}$$

where $(\mathcal{B}_{k,\ell})_{1 \le k,\ell \le K, k \ne \ell}$ is the Robin transmission operator on the interface between subdomains Ω^k and Ω^ℓ: $\mathcal{B}_{k,\ell}\varphi = \nu \partial_{\boldsymbol{n}} \varphi - \frac{\boldsymbol{a} \cdot \boldsymbol{n}_k}{2} \varphi + \alpha \varphi$ with $\alpha > 0$ given. Following the ideas in [6, 8], we need to introduce a new independent entity representing the flux on the interface, in order to match the Robin conditions on non-conforming grids, and thus the method is of Petrov Galerkin type.

In Sect. 2 we introduce the method at the continuous level. Then in Sect. 3, we present the method in the non-conforming discrete case. The numerical analysis is given in Sect. 4. In Sect. 5 we give the discrete algorithm. In Sect. 6 we show simulations to illustrate the optimality of the method.

2 Definition of the Problem

The variational statement of problem (1)–(2) is: Find $u \in H_0^1(\Omega)$ such that

$$\int_\Omega \left(\nu \nabla u \cdot \nabla v + (\eta + \frac{1}{2} \nabla \cdot \boldsymbol{a}) uv + \frac{1}{2} ((\boldsymbol{a} \cdot \nabla u) v - (\boldsymbol{a} \cdot \nabla v) u) \right) dx$$

$$= \int_\Omega f v dx, \quad \forall v \in H_0^1(\Omega). \tag{3}$$

We suppose that $\nu \geq \nu_0 > 0$ *a.e.* in Ω and $\eta + \frac{1}{2}\nabla \cdot \boldsymbol{a} \geq \eta_0 > 0$ *a.e.* in Ω. Therefore problem (1)–(2) is coercive. We define the space $H^1_*(\Omega^k)$ by

$$H^1_*(\Omega^k) = \{\varphi \in H^1(\Omega^k),\ \varphi = 0 \text{ over } \partial\Omega \cap \partial\Omega^k\}.$$

In order to glue non-conforming grids with Robin conditions, denoting by $\underline{v}$ the K-tuple $(v_1, \ldots, v_K)$, we introduce the following constrained space,

$$\mathcal{V} = \{(\underline{v}, \underline{q}) \in \left(\prod_{k=1}^{K} H^1_*(\Omega^k)\right) \times \left(\prod_{k=1}^{K} H^{-1/2}(\partial\Omega^k)\right),$$
$$v_k = v_\ell \text{ and } q_k = -q_\ell \text{ over } \Gamma^{k,\ell},\ \forall k, \ell\}.$$

The following result is an extension of Lemma 1 in [8]: problem (3) is equivalent to the following one: Find $(\underline{u}, \underline{p}) \in \mathcal{V}$ such that

$$\sum_{k=1}^{K} \int_{\Omega^k} \left(\nu \nabla u_k \cdot \nabla v_k + (\eta + \frac{1}{2}\nabla \cdot \boldsymbol{a}) u_k v_k + \frac{1}{2}((\boldsymbol{a} \cdot \nabla u_k) v_k - (\boldsymbol{a} \cdot \nabla v_k) u_k)\right) dx$$
$$- \sum_{k=1}^{K} {}_{H^{-1/2}(\partial\Omega^k)} < p_k, v_k >_{H^{1/2}(\partial\Omega^k)} = \sum_{k=1}^{K} \int_{\Omega^k} f_k v_k dx, \quad \forall \underline{v} \in \prod_{k=1}^{K} H^1_*(\Omega^k).$$

Being equivalent to the original problem, with $p_k = \nu \frac{\partial u}{\partial \boldsymbol{n}_k} - \frac{\boldsymbol{a} \cdot \boldsymbol{n}_k}{2} u$ over $\partial\Omega^k$ (recall that f is assumed to be in $L^2(\Omega)$ so that $\frac{\partial u}{\partial \boldsymbol{n}_k}$ actually belongs to $H^{-1/2}(\partial\Omega^k)$), this problem is naturally well posed.

Let us describe the method in the non-conforming discrete case.

3 Non-conforming Discrete Formulation

We first introduce the discrete spaces. Each subdomain Ω^k is provided with its own mesh $\mathcal{T}_h^k$, such that $\overline{\Omega}^k = \cup_{T \in \mathcal{T}_h^k} T,\ 1 \leq k \leq K$. For $T \in \mathcal{T}_h^k$, let h_T be the diameter of T and h the discretization parameter: $h = \max_{1 \leq k \leq K} h_k$ with $h_k = \max_{T \in \mathcal{T}_h^k} h_T$. We suppose that $\mathcal{T}_h^k$ is uniformly regular and that the sets belonging to the meshes are of simplicial type (triangles or tetrahedra). Let $\mathcal{P}_M(T)$ denote the space of all polynomials defined over T of total degree less than or equal to M. The finite elements are of lagrangian type, of class $\mathcal{C}^0$. We define on Ω^k the spaces $Y_h^k = \{v_{h,k} \in \mathcal{C}^0(\overline{\Omega}^k),\ v_{h,k|T} \in \mathcal{P}_M(T),\ \forall T \in \mathcal{T}_h^k\}$ and $X_h^k = \{v_{h,k} \in Y_h^k,\ v_{h,k|\partial\Omega^k \cap \partial\Omega} = 0\}$. The space of traces over each $\Gamma^{k,\ell}$ of elements of Y_h^k is denoted by $\mathcal{Y}_h^{k,\ell}$. With each interface $\Gamma^{k,\ell}$, we associate a subspace $\tilde{W}_h^{k,\ell}$

of $\mathcal{Y}_h^{k,\ell}$ in the same spirit as in the mortar element method [3] in 2D or [2, 4] for a P_1-discretization in 3D.

More precisely, let $\mathcal{T}$ be the restriction to $\Gamma^{k,\ell}$ of the triangulation $\mathcal{T}_h^k$. In 2D, $\mathcal{T}$ is one-dimensional with vertices $x_0^{k,\ell}, x_1^{k,\ell}, \ldots, x_{n-1}^{k,\ell}, x_n^{k,\ell}$ and has two end points $x_0^{k,\ell}$ and $x_n^{k,\ell}$. Then $\tilde{W}_h^{k,\ell}$ is the subspace of $\mathcal{Y}_h^{k,\ell}$ of elements that are polynomials of degree $\leq M-1$ over both $[x_0^{k,\ell}, x_1^{k,\ell}]$ and $[x_{n-1}^{k,\ell}, x_n^{k,\ell}]$.

In 3D, we suppose that all the vertices of the boundary of $\Gamma^{k,\ell}$ are connected to zero, one, or two vertices in the interior of $\Gamma^{k,\ell}$. We denote by $\mathcal{V}$, $\mathcal{V}_0$, $\partial\mathcal{V}$ the sets of all the vertices of $\mathcal{T}$, the vertices in the interior of $\Gamma^{k,\ell}$, and the vertices on the boundary of $\Gamma^{k,\ell}$ respectively. Let $S(\mathcal{T})$ be the space of piecewise linear functions with respect to $\mathcal{T}$ which are continuous on $\Gamma^{k,\ell}$ and vanish on its boundary. Then $S(\mathcal{T}) =$ span $\{\Phi_a : a \in \mathcal{V}_0\}$ where Φ_a, $a \in \mathcal{V}$ are the finite element basis functions. For $a \in \mathcal{V}$, let $\sigma_a := \{T \in \mathcal{T} : a \in T\}$ denote the support of Φ_a, $\mathcal{N}_a := \{b \in \mathcal{V}_0 : b \in \sigma_a\}$, and $\mathcal{N} := \cup_{a\in\partial\mathcal{V}}\mathcal{N}_a$. Let $\mathcal{T}_c$ be the set of triangles $T \in \mathcal{T}$ which have all their vertices on the boundary of $\Gamma^{k,\ell}$. For $T \in \mathcal{T}_c$, we denote by c_T the only vertex of T that has no interior neighbor. Let $\mathcal{N}_c$ denote the vertices a_T of $\mathcal{N}$ which belong to a triangle adjacent to a triangle $T \in \mathcal{T}_c$. We introduce $\hat{\Phi}_a$ defined as follows: $\hat{\Phi}_a := \Phi_a$, $a \in \mathcal{V}_0 \setminus \mathcal{N}$, $\hat{\Phi}_a := \Phi_a + = \sum_{b\in\partial\mathcal{V}\cap\sigma_a} A_{b,a}\Phi_b$, $a \in \mathcal{N}\setminus\mathcal{N}_c$, and $\hat{\Phi}_a := \Phi_{a_T} + = \sum_{b\in\partial\mathcal{V}\cap\sigma_{a_T}} A_{b,a_T}\Phi_b + \Phi_{c_T}$, $a = a_T \in \mathcal{N}_c$. The weights are defined by : $A_{c,a} + A_{c,b} = 1$ and $|T_{2,b}|A_{c,a} = |T_{2,a}|A_{c,b}$, for all $c \in \partial\mathcal{V}$ connected to two interior nodes a and b, where $T_{2,a}$ (resp. $T_{2,b}$) denote the adjacent triangle to abc having a (resp. b) as a vertex and its two others vertices on $\partial\mathcal{V}$. For all $c \in \partial\mathcal{V}$ connected to only one interior node a, the weights are $A_{c,a} = 1$ (see [4]). The space $\tilde{W}_h^{k,\ell}$ is then defined by $\tilde{W}_h^{k,\ell} :=$ span $\{\hat{\Phi}_a, a \in \mathcal{V}_0\}$. Then $\tilde{W}_h^k := \prod_{\ell,\, \Gamma^{k,\ell}\neq\emptyset} \tilde{W}_h^{k,\ell}$.

We now define the discrete constrained space as follows:

$$\mathcal{V}_h = \{(\underline{u}_h, \underline{p}_h) \in \left(\prod_{k=1}^K X_h^k\right) \times \left(\prod_{k=1}^K \tilde{W}_h^k\right),$$

$$\int_{\Gamma^{k,\ell}} ((,p_{h,k} + \alpha u_{h,k}) - (-p_{h,\ell} + \alpha u_{h,\ell}))\psi_{h,k,\ell} = 0,\ \forall \psi_{h,k,\ell} \in \tilde{W}_h^{k,\ell},\ \ \forall k, \ell\}.$$

The discrete problem is the following one: Find $(\underline{u}_h, \underline{p}_h) \in \mathcal{V}_h$ such that

$$\begin{aligned}
&\sum_{k=1}^K \int_{\Omega^k} \left(\nu \nabla u_{h,k} \cdot \nabla v_{h,k} + (\eta + \frac{1}{2}\nabla \cdot \boldsymbol{a})u_{h,k}v_{h,k}\right)\right) dx \\
&\quad + \sum_{k=1}^K \int_{\Omega^k} \left(\frac{1}{2}((\boldsymbol{a}\cdot\nabla u_{h,k})v_{h,k} - (\boldsymbol{a}\cdot\nabla v_{h,k})u_{h,k})\right) dx \\
&\quad - \sum_{k=1}^K \int_{\partial\Omega^k} p_{h,k}v_{h,k}ds = \sum_{k=1}^K \int_{\Omega^k} f_k v_{h,k}dx, \quad \forall \underline{v}_h = (v_{h,1}, \ldots v_{h,K}) \in \prod_{k=1}^K X_h^k.
\end{aligned} \tag{4}$$

4 Best Approximation Error

In this part we give best approximation results of $(\underline{u}, \underline{p})$ by elements in $\mathcal{V}_h$ (see [8]). We define for any $\underline{p}$ in $\prod_{k=1}^K L^2(\partial\Omega_k)$ the norm $\|\underline{p}\|_{-\frac{1}{2},*} = (\sum_{k=1}^K \sum_{\substack{\ell=1\\ \ell\neq k}}^K \|p_k\|^2_{H_*^{-\frac{1}{2}}(\Gamma^{k,\ell})})^{\frac{1}{2}}$, where $\|.\|_{H_*^{-\frac{1}{2}}(\Gamma^{k,\ell})}$ stands for the dual norm of $H_{00}^{\frac{1}{2}}(\Gamma^{k,\ell})$ (recall that $H_{00}^{\frac{1}{2}}(\Gamma^{k,\ell})$ is the interpolated space of index $\frac{1}{2}$ between $H_0^1(\Gamma^{k,\ell})$ and $L^2(\Gamma^{k,\ell})$, see [10]).

Theorem 1 *Let us assume that $\alpha h \leq c$, for some small enough constant c. Then, the discrete problem (4) has a unique solution $(\underline{u}_h, \underline{p}_h) \in \mathcal{V}_h$.*

Assume that the solution u of (1)–(2) is in $H^2(\Omega) \cap H_0^1(\Omega)$, and $u_k = u_{|\Omega^k} \in H^{2+m}(\Omega^k)$, with $M-1 \geq m \geq 0$. Let $p_{k,\ell} = \nu\frac{\partial u}{\partial \boldsymbol{n}_k} - \frac{\boldsymbol{a}\cdot\boldsymbol{n}_k}{2}u$ on $\Gamma^{k,\ell}$. Then, there exists a constant c independent of h and α such that

$$\|\underline{u}_h - \underline{u}\|_* + \|\underline{p}_h - \underline{p}\|_{-\frac{1}{2},*} \leq c(\alpha h^{2+m} + h^{1+m}) \sum_{k=1}^K \|\underline{u}\|_{H^{2+m}(\Omega^k)} + c(\frac{h^m}{\alpha} + h^{1+m}) \sum_{k=1}^K \sum_{\ell} \|p_{k,\ell}\|_{H^{\frac{1}{2}+m}(\Gamma^{k,\ell})}. \tag{5}$$

Moreover, if $p_{k,\ell} = \nu\frac{\partial u}{\partial \boldsymbol{n}_k} - \frac{\boldsymbol{a}\cdot\boldsymbol{n}_k}{2}u$ is in $H^{\frac{3}{2}+m}(\Gamma_{k,\ell})$, with $M-1 \geq m \geq 0$, then there exists c independent of h and α such that

$$\|\underline{u}_h - \underline{u}\|_* + \|\underline{p}_h - \underline{p}\|_{-\frac{1}{2},*} \leq c(\alpha h^{2+m} + h^{1+m}) \sum_{k=1}^K \|\underline{u}\|_{H^{2+m}(\Omega^k)} + c(\frac{h^{1+m}}{\alpha} + h^{2+m})|\log h| \sum_{k=1}^K \sum_{\ell} \|p_{k,\ell}\|_{H^{\frac{3}{2}+m}(\Gamma^{k,\ell})} \tag{6}$$

5 Discrete Iterative Algorithm

The discrete algorithm to solve problem (4) is defined as follows: let $(u_{h,k}^n, p_{h,k}^n) \in X_h^k \times \tilde{W}_h^k$ be a discrete approximation of (u,p) in Ω^k at step n. Then, $(u_{h,k}^{n+1}, p_{h,k}^{n+1})$ is the solution in $X_h^k \times \tilde{W}_h^k$ of

$$\int_{\Omega^k} \left(\nu\nabla u_{h,k}^{n+1}\nabla v_{h,k} + (\eta + \frac{1}{2}\nabla\cdot\boldsymbol{a})u_{h,k}^{n+1}v_{h,k} + \frac{1}{2}((\boldsymbol{a}\cdot\nabla u_{h,k}^{n+1})v_{h,k} - (\boldsymbol{a}\cdot\nabla v_{h,k})u_{h,k}^{n+1})\right)dx - \int_{\partial\Omega^k} p_{h,k}^{n+1}v_{h,k}ds = \int_{\Omega^k} f_k v_{h,k}dx, \ \forall v_{h,k} \in X_h^k, \tag{7}$$

$$\int_{\Gamma^{k,\ell}} (p_{h,k}^{n+1} + \alpha u_{h,k}^{n+1})\psi_{h,k,\ell} = \int_{\Gamma^{k,\ell}} (-p_{h,\ell}^n + \alpha u_{h,\ell}^n)\psi_{h,k,\ell}, \quad \forall \psi_{h,k,\ell} \in \tilde{W}_h^{k,\ell}. \tag{8}$$

Using Lemma 2 in [8], we can prove the convergence of the iterative scheme

Theorem 2 *Under the hypothesis of Theorem 1, the algorithm (7)–(8) is well posed and converges in the sense that*

$$\lim_{n\longrightarrow\infty}(\|u^n_{h,k}-u_{h,k}\|_{H^1(\Omega^k)}+\sum_{\ell\neq k}\|p^n_{h,k,\ell}-p_{h,k,\ell}\|_{H_*^{-\frac{1}{2}}(\Gamma^{k,\ell})})=0,\ 1\leq k\leq K.$$

6 Numerical Results

We consider a P_1 finite element approximation. We study the numerical error analysis for problem (4). We consider the initial problem with exact solution $u(x, y) = x^3y^2 + sin(xy)$, $\eta = 1$ and $\nu = 0.01$. The domain is the unit square $\Omega = (0, 1) \times (0, 1)$. We decompose Ω into two non-overlapping subdomains with meshes generated in an independent manner as shown on Fig. 1.

The subdomain problems are solved using a direct solver. To observe the numerical error estimates for the discrete problem (4), one need to compute the converged solution of the algorithm (7)–(8) regardless of the algorithm used to compute it. Thus it is the solution at convergence of the algorithm (7)–(8) with a stopping criterion on the residual (i.e. the jumps of Robin conditions) that must be extremely small, e.g. smaller than 10^{-14}. The Robin parameter α is obtained by minimizing the convergence factor (see [5, 7]). For cases 1 and 2 below, this criterion is reached with an average of 40 and 45 iterations respectively (note that with Ventcell conditions, the number of iterations is almost independent of h (see [9]) and is 20 (case 1) and 26 (case 2). The error curves with Ventcell conditions are almost the same as the one on Figs. 2 and 3). Note that the regularity of the normal derivative of u along the interfaces enters most of the times in the frame of the error estimate (6) that allows a larger range of choice for α, compatible with the above chosen optimized choice.

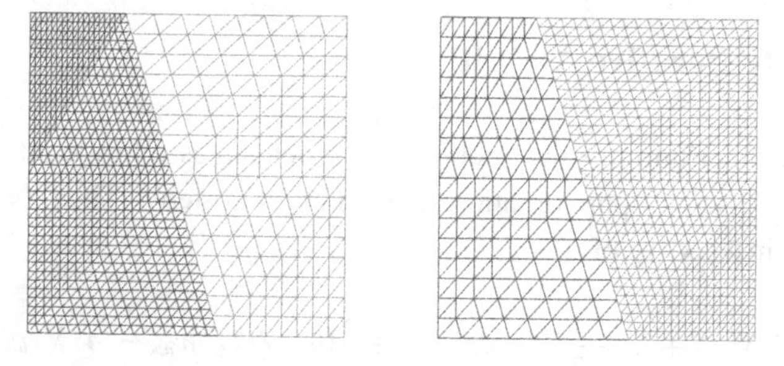

Fig. 1 Non-conforming meshes: mesh 2 (*on the left*), and mesh 3 (*on the right*)

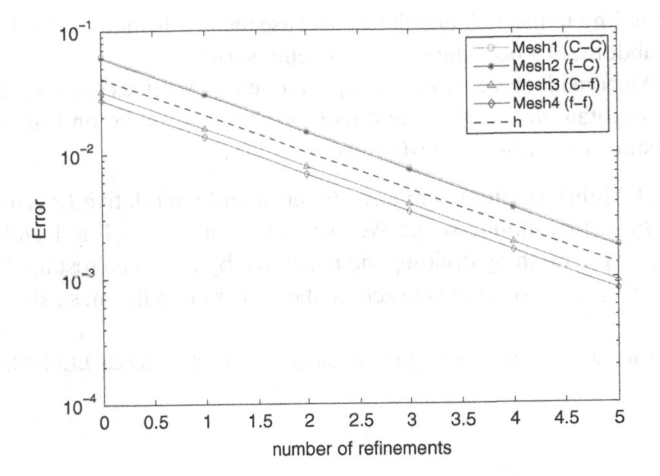

Fig. 2 Case 1. Relative H^1 error versus refinements for meshes 1–4

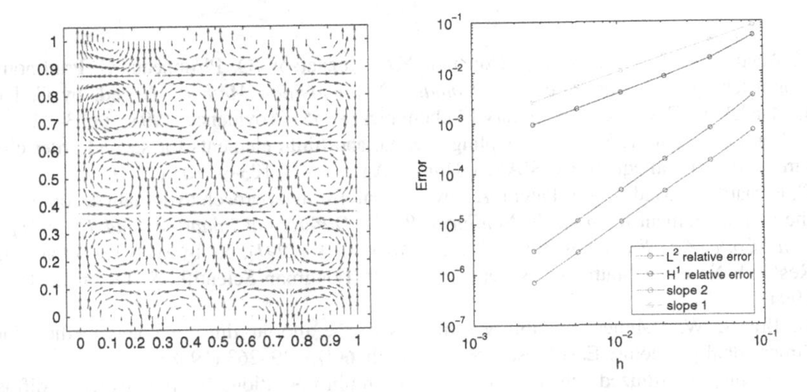

Fig. 3 Case 2. *Left*: velocity field. *Right*: Relative H^1 and L^2 errors versus h for mesh 3

Case 1 In this test case we have considered a rotating velocity defined by: $\boldsymbol{a} = (-\sin(\pi(y-\frac{1}{2}))\cos(\pi(x-\frac{1}{2})), \cos(\pi(y-\frac{1}{2}))\sin(\pi(x-\frac{1}{2})))$, and four initial meshes: meshes 1–4 where meshes 2 and 3 are the nonconforming meshes shown on Fig. 1, and mesh 1 (resp. mesh 4) is a conforming mesh obtained as the union of the coarse (resp. fine) sub-meshes of meshes 2 and 3.Figure 2 shows the relative H^1 error versus the number of refinement for these four meshes, and the mesh size h versus the number of refinement, in logarithmic scale. At each refinement, the mesh size is divided by two. The results of Fig. 2 show that the relative H^1 error tends to zero at the same rate than the mesh size, and this fits with the theoretical error estimates of Theorem 1. On the other hand, we observe that the two curves corresponding to the non-conforming meshes (meshes 2 and 3) are between the curves of the conforming meshes (meshes 1 and 4). The relative H^1 error for mesh 3 is smaller than the one corresponding to mesh 2, and this is because mesh 3 is more refined than mesh 2 in the subdomain where the solution steeply varies.

Case 2 We consider a velocity built up from sets of vortices such that their closest neighbors rotate in the opposite directions [11], as shown on Fig. 3 (left): $\boldsymbol{a} = 0.32\pi\,(\sin(4\pi x)\sin(4\pi y), \cos(4\pi y)\cos(4\pi x))$.

On Fig. 3 (right) we plot the relative H^1 error and the relative L^2 error versus the mesh size h, in logarithmic scale. We start from mesh 3 of Fig. 1 and then refine successively each mesh by dividing the mesh size by two. The results show that the relative H^1 (resp. L^2) error tends to zero at the same rate as the mesh size h (resp. h^2).

Acknowledgement This work was partially supported by the French LEFE/MANU-CoCOA project.

References

1. Y. Achdou, G. Abdoulaev, J.C. Hontand, Y.A. Kuznetsov, O. Pironneau, C. Prud'homme, Nonmatching grids for fluids, in *Domain Decomposition Methods 10*, ed. by C. Farhat J. Mandel, X.-C. Cai. Contemporary Mathematics (AMS, Providence, 1998), pp. 3–22
2. F. Ben Belgacem, Y. Maday, Coupling spectral and finite elements for second order elliptic three-dimensional equations. SIAM J. Numer. Anal. **36**(4), 1234–1263 (1999)
3. C. Bernardi, Y. Maday, A.T. Patera, A new nonconforming approach to domain decomposition: the mortar element method, in *Nonlinear Partial Differential Equations and Their Applications. Collège de France Seminar, Vol. XI (Paris, 1989–1991)*, ed. by H. Brezis et al. Pitman Research Notes in Mathematics Series, vol. 299 (Longman Scientific & Technical, Harlow, 1994)
4. D. Braess, W. Dahmen, Stability estimates of the mortar finite element method for 3-dimensional problems. East West J. Numer. Math. **6**(4), 249–263 (1998)
5. O. Dubois, Optimized schwarz methods with robin conditions for the advection-diffusion equation, in *Domain Decomposition Methods in Science and Engineering XVI*, ed. by O. Widlund, D.E. Keyes. Lecture Notes in Computational Science and Engineering, vol. 55 (Springer, Berlin, 2007), pp. 181–188
6. M.J. Gander, C. Japhet, Y. Maday, F. Nataf, A new cement to glue nonconforming grids with Robin interface conditions: the finite element case, in *Domain Decomposition Methods in Science and Engineering*, ed. by R. Kornhuber, R.H.W. Hoppe, J. Périaux, O. Pironneau, O.B. Widlund, J. Xu. Lecture Notes in Computational Science and Engineering, vol. 40 (Springer, Berlin, 2005), pp. 259–266
7. C. Japhet, Optimized Krylov-Ventcell method. Application to convection-diffusion problems, in *Domain Decomposition Methods 9, Domain Decomposition Methods in Sciences and Engineering*, ed. by U. Bjørstad, M. Espedal, D.E. Keyes (Wiley, New York, 1998), pp. 382–389
8. C. Japhet, Y. Maday, F. Nataf, A new interface cement equilibrated Mortar (NICEM) method with Robin interface conditions: the P_1 finite element case. Math. Models Methods Appl. Sci. **23**(12), 2253–2292 (2013)

9. C. Japhet, Y. Maday, F. Nataf, A new interface cement equilibrated mortar method with Ventcel conditions, in *Domain Decomposition Methods in Science and Engineering XXI*, ed. by J. Erhel, M.J. Gander, L. Halpern, G. Pichot, T. Sassi, O. Widlund. Lecture Notes in Computational Science and Engineering, vol. 98 (Springer, Berlin, 2014), pp. 329–336
10. J.-L. Lions, E. Magenes, *Problèmes aux Limites Non-homogènes*, vol. 1 (Dunod, Paris, 1968)
11. P.K. Smolarkiewicz, The multi-dimensional crowley advection scheme. Mon. Weather Rev. **110**, 1968–1983 (1982)

Augmented Lagrangian Domain Decomposition Method for Bonded Structures

J. Koko and T. Sassi

1 Introduction

Domain decomposition methods are subject to a greater interest, due to obvious implication for parallel computing. Non-overlapping methods are particularly well suited for coupled problems through an interface as bonded structures (e.g. [4]) air/water flows (e.g. [2]), two-body contact problems (e.g. [6, 9]), etc. For these coupled problems, the domain decomposition methods applied in a natural way, since the sub-domains are already defined.

Two types of domain decomposition methods exist for bonded structures: Lagrangian (dual) methods [1] and least-square methods [4, 7]. In Lagrangian methods, the objective functional is the energy functional and the constraint is the solution jump across the interface. In least-square methods, the original problem is reformulated as a constrained minimization problem for which the objective functional controls the solution jump across the interface. The constraints are the partial differential equations stated in each sub-domain with suitable boundary conditions. In a comparative study, Koko [8] shows that the least-square methods solve twice as many linear systems than the dual methods. But both methods fail if one of the subdomains allows rigid-body motions.

J. Koko (✉)
LIMOS, Université Blaise Pascal – CNRS UMR 6158, 63000 Clermont-Ferrand, France
e-mail: koko@isima.fr

T. Sassi
LMNO, Université de Caen – CNRS UMR, 14032 Caen, France
e-mail: sassi@univ-caen.fr

T. Dickopf et al. (eds.), *Domain Decomposition Methods in Science and Engineering XXII*, Lecture Notes in Computational Science and Engineering 104, DOI 10.1007/978-3-319-18827-0_56

The paper is organized as follows. In Sect. 2 we present the simplified model of bonded structures. The Uzawa block relaxation domain decomposition algorithm is described in Sect. 3. Some numerical experiments are carried out in Sect. 4.

2 Model Problem

We adopt the model problem described in [8]. To simplify, we present a model problem with two subdomains. A generalization to more than two subdomains is straightforward.

Consider a system of two isotropic elastic bodies each of which occupies, in the reference configuration, a bounded domain Ω_i in $\mathbb{R}^2$ $(i = 1, 2)$ (Fig. 1). Both elastic bodies are bonded along their common boundary S, assumed to be a nonempty surface of positive measure. Hooke's law is assumed for each elastic body, i.e.

$$\sigma^i_{\alpha\beta}(u_i) = 2\mu_i \varepsilon_{\alpha\beta}(u_i) + \lambda_i \mathrm{tr}(\varepsilon(u_i))\mathbb{I}_2, \quad \alpha, \beta = 1, 2,$$

where $\varepsilon(u_i) = (\nabla u_i + \nabla u_i^T)/2$, $\lambda_i \geq 0$ and $\mu_i > 0$ denote Lamé constants. Let u_i be the displacement field of the body Ω_i. We set $u = (u_1, u_2)$ the displacement field of the bonded structure and $[u] = (u_1 - u_2)_{|S}$ the relative tangential displacement along S. The simplified model of bonded structures we study in this paper can be formulated as follows

$$-\operatorname{div}\sigma^i(u_i) = f_i \text{ in } \Omega_i, \tag{1}$$

$$u_i = 0 \text{ on } \Gamma_i = \partial\Omega_i \setminus S, \tag{2}$$

$$\sigma^i(u_i) \cdot n_i = (-1)^i K[u] \text{ on } S, \tag{3}$$

where n_i is the unit outward normal to Ω_i, and K is the second order bonding tensor assumed to be symmetric and coercive with bounded coefficients. Equation (3) is the transmission condition for u_1 and u_2. The domain decomposition algorithms are (generally) parallel iterative procedures on (1)–(2) that tend to satisfy the transmission condition (3).

Let us introduce the subspaces $V_i = \{v \in H^1(\Omega_i);\ v = 0 \text{ on } \Gamma_i\}$, $V = V_1 \times V_2$ and the notations, for $u_i, v_i \in V_i$

$$a_i(u_i, v_i) = \int_{\Omega_i} \sigma^i(u_i)\varepsilon(v_i)\,\mathrm{d}x, \tag{4}$$

Fig. 1 Bonded structure : Ω_1 and Ω_2 the sub-domains (adherents), S the interface (*thin adhesive layer*)

Ω_2

S

Ω_1

$$(u_i, v_i)_{\Omega_i} = \int_{\Omega_i} u_i v_i \, \mathrm{d}x \quad \text{and} \quad (u_i, v_i)_{\Gamma_i} = \int_{\Gamma_i} u_i v_i \, \mathrm{d}\Gamma_i. \tag{5}$$

With the above notations, the total potential energy of the simplified model of a bonded structure we study is

$$F(v) = J(v) + \frac{1}{2}(K[v], [v])_S \quad \forall (v_1, v_2) \in V_1 \times V_2 \tag{6}$$

where

$$J(v) = \frac{1}{2}\sum_{i=1}^{2} a_i(v_i, v_i) - \sum_{i=1}^{2}(f_i, v_i)_{\Omega_i}.$$

The bonded structure problem can now be formulated as the following minimization problem.

Find $u \in V$ such that

$$F(u) \leq F(v), \quad \forall v \in V. \tag{7}$$

The functional J is convex and coercive (see, e.g. [3]) on V. Since K is symmetric and coercive, it follows that F is convex and coercive on V. Consequently, the minimization problem (7) has a unique solution.

In the method proposed by Bresch and Koko [1], the objective functional is the energy functional and the constraint is the solution jump across the interface. With the use of the Lagrangian functional, the resulting domain decomposition algorithm is of Uzawa type, precisely its conjugate gradient version. In the method proposed in [4, 7], the original problem is reformulated as a constrained minimization problem for which the objective functional controls the solution jump across the interface. The constraints are the partial differential equations stated in each sub-domain with suitable boundary conditions. Both methods fail if one of the subdomain allows rigid-body motions.

3 Augmented Lagrangian Domain Decomposition

Let us introduce the auxiliary interface unknowns $q_i = v_{i|S}$ so that the energy functional (6) becomes

$$F(v, q) = J(v) + \frac{1}{2}(K[q], [q])_S \quad \forall (v, q) \in V \times H,$$

where $H = L^2(S)^2$. We then replace the unconstrained minimization problem (7) by the following (equivalent) constrained minimization problem

Find $(u,p) \in V \times H$ such that

$$F(u,p) \leq F(v,q) \quad \forall (v,q) \in V \times H, \tag{8}$$

$$u_i = p_i \quad \text{on } S, \quad i = 1,2. \tag{9}$$

With (8)–(9), we associate the augmented Lagrangian functional

$$\mathcal{L}_r(v,q;\mu) = F(v,q) + \sum_{i=1}^{2} \left((\mu_i, v_i - q_i)_S + \frac{r}{2} \parallel v_i - q_i \parallel^2_{L^2(S)} \right); \tag{10}$$

where $r > 0$ is the penalty parameter. The saddle-point problem for the augmented Lagrangian functional is

Find $(u,p,\lambda) \in V \times H \times H$ such that:

$$\mathcal{L}_r(u,p,\mu) \leq \mathcal{L}_r(u,p,\lambda) \leq \mathcal{L}_r(v,q,\lambda) \quad \forall (v,q,\mu) \in V \times H \times H \tag{11}$$

The functional $\mathcal{L}_r$ is Gâteaux-differentiable on $V \times H \times H$, then the solution of (11) is characterized by the saddle-point (Euler-Lagrange) equations of the primal and dual problems as follows

Find $(u,p,\lambda) \in V \times H \times H$ such that

$$\frac{\partial \mathcal{L}_r}{\partial u}(u,p,\lambda) \cdot v = 0, \quad \forall v \in V, \tag{12}$$

$$\frac{\partial \mathcal{L}_r}{\partial p}(u,p,\lambda) \cdot q = 0, \quad \forall q \in H \tag{13}$$

$$\frac{\partial \mathcal{L}_r}{\partial \lambda}(u,p,\lambda) \cdot \mu = 0, \quad \forall \mu \in H. \tag{14}$$

Subdomain problems in u are uncoupled if the multipliers λ and the coordination variable p are known. We can use this property through a Uzawa algorithm associated with a block relaxation method.

Uzawa block/relaxation methods have been used in nonlinear mechanics for operator-splitting methods (see e.g. [5]). The idea is to minimize successively in u and p, in block Gauss-Seidel fashion. Applying a Uzawa block relaxation method to (12)–(14) we obtain the following algorithm, assuming p^0 and λ^0

$$u^{k+1} = \arg\min_{v} \mathcal{L}_r(v, p^k, \lambda^k), \tag{15}$$

$$p^{k+1} = \arg\min_{q} \mathcal{L}_r(u^{k+1}, q, \lambda^k), \tag{16}$$

$$\lambda^{k+1} = \lambda^k + r(u^{k+1} - p^{k+1}). \tag{17}$$

The minimization subproblem (15) is equivalent to the uncoupled subdomain problems

$$a_i(u_i^{k+1}, v_i) + r(u_i^{k+1}, v_i)_S = f_i(v_i) + (rp_i^k - \lambda_i^k, v), \quad \forall v_i \in V_i, \quad i = 1,2 \tag{18}$$

while (16) leads to the point-wise interface subproblem

$$(K + r\mathbb{I})p_1^{k+1} - Kp_2^{k+1} = \lambda_1^k + ru_1^{k+1}, \tag{19}$$

$$-Kp_1^{k+1} + (K + r\mathbb{I})p_2^{k+1} = \lambda_2^k + ru_2^{k+1}. \tag{20}$$

Gathering the results above, we obtain the Uzawa block relaxation method presented in Algorithm 1. We iterate until the relative error on (u^k, p^k) becomes sufficiently small.

Remark 1 The problem (18) always has a unique solution even without the Dirichlet condition (2). This property is useful for solving problems allowing rigid body motions.

Remark 2 Algorithm 1 is equivalent to the operator-splitting standard algorithm ALG 2 described in, e.g. [5, Chap. 3], applied to the minimization problem (8)–(9). Since F is convex and coercive and the constraints (9) are linear, the convergence of Algorithm 1 is guaranteed by, e.g. [5, theorem 3.4].

The discrete version of Algorithm 1 is straightforward using the finite element method (or the finite difference scheme). The only condition is the meshes compatibility on S. Assuming that Ω_{ih} is a triangulation of Ω_i, the meshes are compatible on S in the sense that $\bar{\Omega}_{1h} \cap S = \bar{\Omega}_{2h} \cap S$.

The uncoupled elasticity subproblems (18) lead to linear systems with symmetric positive definite matrices. Since these matrices do not change during the iterative

---

Algorithm 1 Uzawa block relaxation algorithm for a bonded structure

---

Initialization. p^0, λ^0 and $r > 0$ are given
Iteration $k \geq 0$. Compute successively u^{k+1}, p^{k+1} and λ^{k+1} as follows

1. Compute $u_i^{k+1} \in V_i$ such that

$$a_i(u_i^{k+1}, v_i) + r(u_i^{k+1}, v_i)_S = f_i(v_i) + (rp_i^k - \lambda_i^k, v_i)_S, \quad \forall v_i \in V_i, \quad i = 1,2.$$

2. Compute $(p_1^{k+1}, p_2^{k+1}) \in H$ such that

$$(K + r\mathbb{I})p_1^{k+1} - Kp_2^{k+1} = \lambda_1^k + ru_1^{k+1}$$
$$-Kp_1^{k+1} + (K + r\mathbb{I})p_2^{k+1} = \lambda_2^k + ru_2^{k+1}$$

3. Update Lagrange multipliers: $\lambda_i^{k+1} = \lambda_i^k + r(u_{i|S}^{k+1} - p_i^{k+1})$, $i = 1,2$.

---

process, a Cholesky factorization can be performed once and for all in the initialization step. Then forward/backward substitutions are performed in the rest of the iterative process. If a preconditioned iterative solver is used for solving (18), an incomplete factorization is performed once and for all in the initialization step.

The (linear) interface subproblem (19)–(20) is solved point-wise. At each point we have to invert a small size matrix (4×4 in 2D or 6×6 in 3D). We can therefore use a semi-analytical solution for (19)–(20). Indeed, direct Gaussian elimination yields to

$$\begin{aligned} p_1^{k+1} &= (K_r^2 - \mathbb{I})^{-1} K^{-1} \left(K_r(\lambda_1^k + r u_1^{k+1}) + \lambda_2^k + r u_2^{k+1} \right) \\ p_2^{k+1} &= K_r p_1^{k+1} - K^{-1}(\lambda_1^k + r u_1^{k+1}), \end{aligned}$$

where we have set $K_r = \mathbb{I} + rK^{-1}$. The size of K is 2×2 in 2D and 3×3 in 3D and in many applications, K is a diagonal matrix.

4 Numerical Experiments

Algorithm 1 was implemented in MATLAB 7 on a Linux workstation with 2.67 GHz clock frequency and 12 GB RAM. The test problem used is designed to illustrate the numerical behavior of the algorithm more than to model actual bonded structures. Setting $z = (u, p)$, the stopping criterion is

$$\| z^k - z^{k-1} \|_{L^2} < 10^{-6} \| z^k \|_{L^2} . \tag{21}$$

We are interested in the bonded structure of Fig. 2, made from three isotropic (linear) elastic bodies. The subdomains are $\Omega_1 = (0,\ 20) \times (5,\ 10) \cup (0,\ 20) \times (-10,\ 5)$ and $\Omega_2 = (0,\ 60) \times (-5,\ 5)$. The interface is therefore $S = (0,\ 20) \times \{5\} \cup (0,\ 20) \times \{-5\}$. The material constants of the adherents are $E_1 = 5 \times 10^4$ MPA, $\nu_1 = 0.3$, $E_2 = 2.5 \times 10^4$ MPA and $\nu_2 = 0.3$.

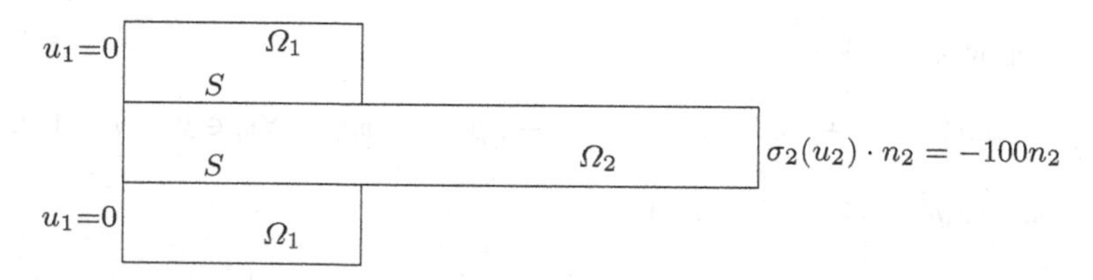

Fig. 2 Geometry of the bonded structure, $n_2 = (1, 0)^T$

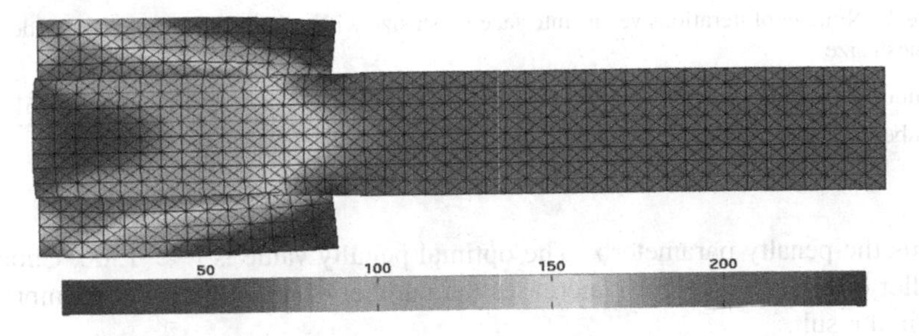

Fig. 3 Deformed configuration and Von Mises effective stress (magnification factor 20)

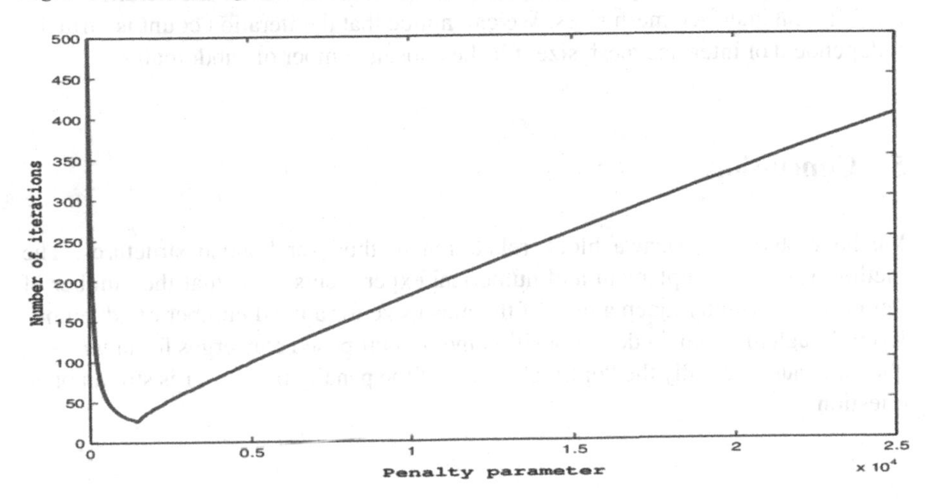

Fig. 4 Number of iterations versus penalty parameter

The material constants of the adhesive layer are $E^* = 1800\,\text{MPA}$, $\nu^* = 0.35$, $\tilde{\nu} = 2(1-\nu^*)/(1-2\nu^*)$

$$K = \frac{E^*}{2(1+\nu^*)}\text{diag}\,(\tilde{\nu},\,1)$$

Remark 3 Since the subproblem over Ω_2 allows rigid body motions, pure Lagrangian [1, 8] and least square methods [4, 7, 8] are not applicable.

The bonded structure is first modeled by a uniform mesh consisting of 2×121 nodes and 2×224 triangles (for Ω_1) and 723 nodes and 1344 triangles for Ω_2, with 2×15 nodes on S. We use piecewise linear finite element spaces. Applying Algorithm 1 with the penalty parameter $r = 1500$, (21) is satisfied after 27 iterations. Figure 3 shows the Von Mises effective stress distribution inside the tree-body system.

Augmented Lagrangian type algorithms are very sensitive to the choice of the penalty (or augmentation) parameter r. Figure 4 shows the number of iterations

Table 1 Number of iterations versus interface mesh size with $r = 1500$, chosen independently of the mesh size

Number of interface nodes	2×11	2×21	2×41	2×81	2×161
Number of iterations	27	27	27	27	27

versus the penalty parameter r. The optimal penalty value is $r \approx 1300$. Choosing smaller or larger values for r increases the number of iterations without improving the final result.

To study the scalability of our algorithm, we report in Table 1 the iteration count for different interface mesh sizes. We can notice that the iteration count is virtually independent of interface mesh size, for the chosen number of subdomains.

5 Conclusion

We have studied a Uzawa block relaxation method for bonded structures. The method is easy to implement and numerical experiments show that the number of iterations is virtually independent of the mesh size for a fixed number of adherents. Even though the domain decomposition method proposed converges for any $r > 0$, choosing automatically the "optimal" value of the penalty parameter is still an open question.

References

1. D. Bresch, J. Koko, An optimization-based domain decomposition method for nolinear wall laws in coupled systems. Math. Models Methods Appl. Sci. **14**, 1085–1101 (2004)
2. D. Bresch, J. Koko, Operator-splitting and Lagrange multiplier domain decomposition methods for numerical simulation of two coupled Navier-Stokes fluids. Int. J. Appl. Math. Comput. Sci. **16**, 101–113 (2006)
3. P.-G. Ciarlet, *Mathematical Elasticity I: Three-Dimensional Elasticity* (North-Holland, Amsterdam, 1988)
4. G. Geymonat, F. Krasucki, D. Marini, M. Vidrascu, A domain decomposition method for bonded structures. Math. Models Methods Appl. Sci. **8**, 1387–1402 (1998)
5. R. Glowinski, P. Le Tallec, *Augmented Lagrangian and Operator-splitting Methods in Nonlinear Mechanics* (SIAM, Philadelphia, 1989)
6. J. Haslinger, R. Kucera, J. Riton, T. Sassi, A domain decomposition method for two-body contact problems with Tresca friction. Adv. Comput. Math. **40**, 65–90 (2014)
7. J. Koko, An optimization based domain decomposition method for a bonded structure. Math. Models Methods Appl. Sci. **12**, 857–870 (2002)
8. J. Koko, Convergence analysis of optimization-based domain decomposition methods for a bonded structure. Appl. Numer. Math. **58**, 69–87 (2008)
9. J. Koko, Uzawa block relaxation domain decomposition method for the two-body contact problem with Tresca friction. Comput. Methods. Appl. Mech. Eng. **198**, 420–431 (2008)

Hierarchical Preconditioners for High-Order FEM

Sabine Le Borne

The finite element discretization of partial differential equations (PDEs) requires the selection of suitable finite element spaces. While high-order finite elements often lead to solutions of higher accuracy, their associated discrete linear systems of equations are often more difficult to solve (and to set up) compared to those of lower order elements.

We will present and compare preconditioners for these types of linear systems of equations. More specifically, we will use hierarchical ($\mathcal{H}$-) matrices to build block $\mathcal{H}$-LU preconditioners. $\mathcal{H}$-matrices provide a powerful technique to compute and store approximations to dense matrices in a data-sparse format. We distinguish between blackbox $\mathcal{H}$-LU preconditioners which factor the entire stiffness matrix and hybrid methods in which only certain subblocks of the matrix are factored after some problem-specific information has been exploited. We conclude with numerical results.

1 Introduction

This contribution is concerned with preconditioning the linear systems of equations arising in high-order finite element discretizations of PDEs [6, 11]. More specifically, we will introduce and analyse hybrid blackbox hierarchical matrix techniques in which

S. Le Borne (✉)
Hamburg University of Technology, Am Schwarzenberg-Campus 3, 21073 Hamburg, Germany
e-mail: leborne@tuhh.de

T. Dickopf et al. (eds.), *Domain Decomposition Methods in Science and Engineering XXII*, Lecture Notes in Computational Science and Engineering 104, DOI 10.1007/978-3-319-18827-0_57

- only the stiffness matrix is given (which excludes lower order preconditioning);
- the preconditioner might use some knowledge on expected matrix properties (e.g. sparsity structure) from the underlying problem.

The construction of efficient solution methods for these types of systems has been a very active recent research field. Many solution approaches are based on multigrid [9] and/or domain decomposition approaches [10], or try to construct sparse preconditioners for the dense matrices [1]. Here, we pursue a different approach in which we propose to use hierarchical matrix techniques to construct efficient preconditioners for these systems.

This paper is organized as follows: In Sect. 2, we will introduce a model problem and a particular type of high-order finite element discretization. While in this paper we develop preconditioners for this particular setting, the intention is in future work to extend these preconditioners to a wider range of PDEs and discretization schemes. In Sect. 3, we introduce two preconditioning approaches. The first one is blackbox, i.e., it only requires the (sparse) matrix as input and does not make any assumptions on the origin of the matrix. The second approach also requires just the matrix as input, but in addition it "knows" that it originates from some high-order finite element discretization and incorporates this knowledge into the construction of the preconditioner. In Sect. 4, we conclude with numerical results illustrating the performance of the proposed preconditioning approaches.

2 High-Order Q_p Finite Elements

The three-dimensional convection-diffusion equation

$$-\epsilon\Delta u + \mathbf{b}\cdot\nabla u = f \qquad \text{in } \Omega = (0,1)^3, \tag{1}$$

$$u = x^2 + y^2 + z^2 \qquad \text{on } \partial\Omega \tag{2}$$

serves as our test problem. In particular, we consider a small viscosity $\epsilon = 10^{-3}$ and a circular convection direction $\mathbf{b}(\mathbf{x},\mathbf{y},\mathbf{z}) = (0.5 - y,\ x - 0.5,\ 0.0)^T$, resulting in a convection-dominant problem.

We discretize this test problem using a finite element discretization with quadrilateral Q_p finite elements. To this end, we use a "triangulation" of a (regularly refined) quadrilateral grid (cubes), and then define the finite element space $V = Q_p$ of continuous, piecewise polynomial elements of (at most) order p in each coordinate direction. As a basis for this finite element space, we use Lagrange (tensor) basis functions $\{\phi_1, \cdots, \phi_n\}$ satisfying $\phi_i(x_j) = \delta_{ij}$. The finite element discretization results in a linear system of equations $Ax = b$ whose solution yields the finite element approximation $u_h = \sum x_i\phi_i \in V$.

The following Fig. 1 illustrates the sparsity structures of matrices obtained for linear ($h = \frac{1}{16}$), quadratic ($h = \frac{1}{8}$), fourth ($h = \frac{1}{4}$) and eighth ($h = \frac{1}{2}$) order basis functions, all having size 3375×3375.

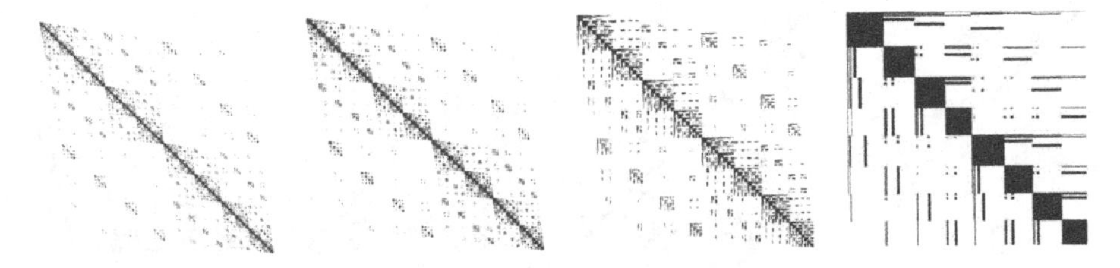

Fig. 1 Sparsity structures using linear, quadratic, fourth and eighth order polynomials, resulting in (on average) 24, 56, 186 and 850 nonzero entries per row, resp.

The advantage of high-order elements lies in their p'th order discretization error $\|u - u_h\|_{H^1} \sim h^p$ (where $\|\cdot\|_{H^1}$ denotes the usual Sobolev H^1-norm), while their disadvantages include the linear systems to be solved to be less sparse and in general worse conditioned compared to those obtained using lower order elements.

In the following section, we introduce several block preconditioners and analyse their performance as we increase the order of discretization.

3 Block-$\mathcal{H}$-LU Factorizations

The following block preconditioners are constructed through a combination of a row and column permutation of the (entire) matrix followed by an approximate LU factorization, either of the entire matrix or only of the blocks on the diagonal to be used in combination with a block Gauß-Seidel preconditioner.

3.1 Nested Dissection Based Block Structures

While high-order finite elements lead to less sparse matrices, they are usually still "sparse enough" for the well-known nested dissection ordering: Based on the matrix graph, the degrees of freedom are divided into three subsets D_1, D_2 and S. The two subsets D_1, D_2 are disconnected in the sense that $i \in D_1$ and $j \in D_2$ implies $a_{ij} = 0$ for the respective matrix entry. The subset S, the so-called "separator" which is preferably small (in the case of our three-dimensional model problem it is of order $\mathcal{O}(n^{2/3})$) consists of degrees of freedom with connections to both subsets D_1 and D_2. We recursively apply this ordering strategy to the two subsets D_1, D_2. Figure 2 shows the sparsity structures for the matrices of Fig. 1 after a nested dissection reordering.

The advantage of such a reordering prior to an (approximate) LU factorization lies in the fact that an (exact) LU factorization preserves the two off-diagonal zero blocks in the 1×2 and 2×1 block positions of the 3×3 block matrix. Therefore, the LU factorizations of the first two (approximately equal sized) diagonal blocks can be computed in parallel followed by the factorization of the last, smaller

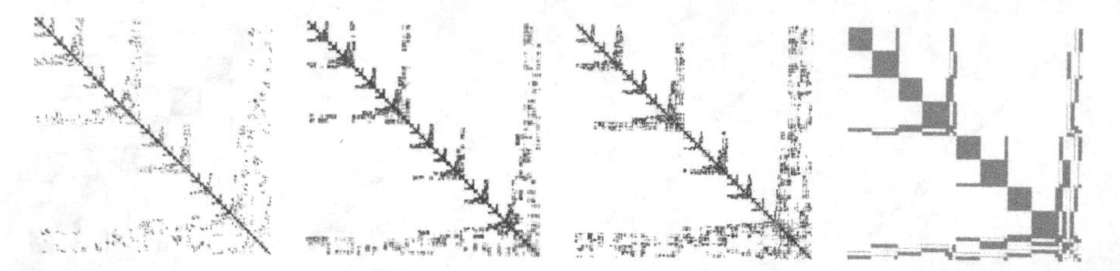

Fig. 2 Sparsity structures using linear, quadratic, fourth and eighth order polynomials after a nested dissection ordering

block corresponding to the separator. We make the following two observations with respect to the polynomial order p of the underlying finite element space:

- The relative size of the separator is independent of the polynomial order p of the finite element space (when using a Lagrange basis). This is important since it is typically the factorization of this separator block that dominates the work complexity of the factorization of the entire matrix.
- For larger p, the nested dissection "finds" the dense subblocks that correspond to so-called "bubble" functions—these are basis functions with support in a single element that can be eliminated without additional fill-in into the stiffness matrix.

The next step is the computation of an approximate LU-factorization in the hierarchical ($\mathcal{H}$-) matrix format. Hierarchical matrices have been introduced more than a decade ago [8]. They are based on a blockwise low rank approximation of off-diagonal matrix blocks. The efficient storage (of the rectangular matrix factors) of these low rank blocks reduces the storage complexity as well as the computational complexity for most matrix operations to almost optimal order (up to powers of logarithmic factors). For general details, we refer to the comprehensive lecture notes [5], and for nested dissection based $\mathcal{H}$-LU factorization that is used in the numerical results in Sect. 4 to [7]. Without going into the details of $\mathcal{H}$-matrices, here we only exploit the fact that the accuracy of $\mathcal{H}$-LU factorizations can be controlled adaptively through a parameter $\delta_{\mathcal{H}}$. As $\delta_{\mathcal{H}} \to 0$, the $\mathcal{H}$-LU factors converge toward the exact LU factors, although at the expense of increasing storage and computational costs.

3.2 *Degree of Freedom Type Based Block Structures*

We now propose to begin with an ordering of the degrees of freedom according to their association with vertices (V), edges (E), faces (F) or (the interior of) cells (C). Since we use a structured grid, this ordering can be obtained directly from the sparsity pattern without access to the grid geometry (through the number of nonzero entries per matrix row). Figure 3 shows the resulting reordered sparsity structures for the matrices of Fig. 1. The matrices still include the (Dirichlet) boundary vertices

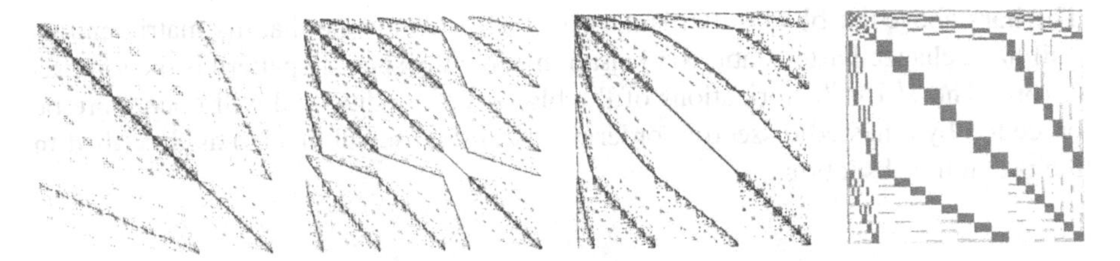

Fig. 3 Sparsity structures for linear, quadratic, fourth and eighth order polynomials after a reordering based on the types of degree of freedom (vertex, edge, face, cell or boundary)

(resulting in the last block on the diagonal of equal size for all four matrices, here about one third of the total number of matrix rows due to the small problem size) which will be eliminated before the iterative solution for the true degrees of freedom.

Besides this last block of boundary vertices, the matrices for quadratic, fourth and eighth degree polynomials have an additional common structure of four blocks on the diagonal according to their vertex, edge, face and (interior) cell degrees of freedom. The block corresponding to the (interior) cell degrees of freedom is the easiest one to recognize since it is of block diagonal structure, and its relative size, as well as the size of the dense blocks on the diagonal, increases with the increase in polynomial order. In theory, this block can also be eliminated without creating any additional fill-in in the remaining matrix. This process is called "static condensation" which is known to be sometimes ill-conditioned. For smaller polynomial order p, this can also be done in practice, but for high order p, the computational costs of such an elimination increase substantially and may dominate the overall computational costs for the solution of the linear system.

The resulting block structure can be used for Jacobi or Gauß-Seidel block preconditioners. These preconditioners require (approximate) solvers for the matrix blocks on the diagonal for which we will again use $\mathcal{H}$-LU factorizations. In particular, we will compute $\mathcal{H}$-LU factorizations for the two diagonal blocks in the matrices (3) and (4).

No static condensation:

$$\begin{pmatrix} A_{VV} & & & \\ & A_{EE} & A_{EF} & A_{EC} \\ & A_{FE} & A_{FF} & A_{FC} \\ & A_{CE} & A_{CF} & A_{CC} \end{pmatrix}, \quad \begin{pmatrix} A_{VV} & A_{VE} & & \\ A_{EV} & A_{EE} & & \\ & & A_{FF} & A_{FC} \\ & & A_{CF} & A_{CC} \end{pmatrix}, \tag{3}$$

After static condensation of interior cell (C) degrees of freedom:

$$\begin{pmatrix} \overline{A_{VV}} & & \\ & \overline{A_{EE}} & \overline{A_{EF}} \\ & \overline{A_{FE}} & \overline{A_{FF}} \end{pmatrix}, \quad \begin{pmatrix} \overline{A_{VV}} & \overline{A_{VE}} & \\ \overline{A_{EV}} & \overline{A_{EE}} & \\ & & \overline{A_{FF}} \end{pmatrix}. \tag{4}$$

The bars above the blocks after static condensation indicate that the matrix entries will have changed after static condensation while the sparsity pattern is the same as before. The $\mathcal{H}$-LU factorizations of the blocks on the diagonal will once more be preceded by a nested dissection ordering (within these subblocks) as described in the previous subsection.

4 Numerical Results and Conclusions

In this section, we will provide a small selection of numerical results for the various preconditioners that have been introduced in Sects. 3.1 and 3.2. All tests were performed on a DELL Latitude E6530 laptop (2.60 GHz, 16 GB). The discrete linear systems have been produced by the software package deal.II [2, 3], and we use the $\mathcal{H}$-arithmetic of the HLib package [4].

As an iterative solver, we use a preconditioned BiCGStab method, and we stop the iteration once the residual is reduced by a factor of 10^{-6}, i.e., when $\|r_k\|_2 \leq 10^{-6}\|r_0\|_2 = 10^{-6}\|b\|_2$ since we use the starting vector $x_0 = 0$. In Table 1, we show numerical results for linear finite elements: As we increase the level of grid refinement, the number of degrees of freedom (dofs) increases by a factor of approximately 8. A comparable factor of increase can be observed in the storage and set-up times of the $\mathcal{H}$-LU factorizations which have been performed for a relative accuracy of $\delta_{\mathcal{H}} \in \{0.1, 0.01\}$. The convergence rates of the preconditioned BiCGStab method deteriorate only modestly as the problem size increases, i.e., we obtain a robust solution method (the BiCGStab time for the problem size on level 7 for the higher $\mathcal{H}$-accuracy $\delta_{\mathcal{H}} = 0.01$ is slowed down by swapping and hence put into parentheses).

Table 2 shows respective results for finite element spaces using polynomial degrees $p = 4$, 6 and 9. We had to use higher accuracies for the $\mathcal{H}$-LU factorizations (now $\delta_{\mathcal{H}} \in \{10^{-1}, 10^{-2}\}$ for $p = 4, 6$ and even $\delta_{\mathcal{H}} \in \{10^{-4}, 10^{-5}\}$ for $p = 9$) in order to obtain convergence in the BiCGStab iteration, which is no surprise in view of the worse conditioning of the matrices for high-order elements. Once more, for an "accurate enough" $\mathcal{H}$-LU factorization as a preconditioner, we obtain an almost optimal iterative method.

Table 1 $\mathcal{H}$-LU: time (s) and storage (MB), linear elements

Dofs	3375	$29 \cdot 10^3$	$250 \cdot 10^3$	$2 \cdot 10^6$	3375	$29 \cdot 10^3$	$250 \cdot 10^3$	$2 \cdot 10^6$
Level	4	5	6	7	4	5	6	7
	$\mathcal{H}$-accuracy $\delta_{\mathcal{H}} = 0.1$				$\mathcal{H}$-accuracy $\delta_{\mathcal{H}} = 0.01$			
Storage (MB)	7.1	100	969	8611	10	151	1556	13,749
Set-up time (s)	0.3	6.6	91	796	0.7	17	271	2494
BiCGStab steps	5	7	10	23	2	3	4	4
Time (s)	0.0	0.2	3.2	67.5	0.0	0.1	1.7	(30.0)

Table 2 $\mathcal{H}$-LU: time (s) and storage (MB), polynomial degrees 4, 6, and 9

Level	3	4	2	3		1	2
$\mathcal{H}$-acc	Degree 4		Degree 6		$\mathcal{H}$-acc	Degree 9	
	Storage (MB)						
10^{-1}	117	1201	32	376	10^{-4}	33	485
10^{-2}	181	1977	61	733	10^{-5}	36	543
Stiffness matrix	82	654	82	654		90	716
$\mathcal{H}$-acc	Set-up time (s)						
10^{-1}	10.3	130	1.6	30.8	10^{-4}	3.5	110
10^{-2}	31	438	5.7	123	10^{-5}	3.7	142
$\mathcal{H}$-acc	BiCGStab (steps/time)						
10^{-1}	22/1.0	20/9.3	div	div	10^{-4}	4/0.1	8/1.7
10^{-2}	3/0.2	4/2.7	8/0.2	10/2.9	10^{-5}	2/0.04	3/0.7

Table 3 $\mathcal{H}$-LU: time (s) and storage (MB), polynomial degree 4, $\approx$ 250,000 dofs

Prec	$\begin{pmatrix} \text{vv} & \text{ve} & \text{vf} & \text{vc} \\ \text{ev} & \text{ee} & \text{ef} & \text{ec} \\ \text{fv} & \text{fe} & \text{ff} & \text{fc} \\ \text{cv} & \text{ce} & \text{cf} & \text{cc} \end{pmatrix}$	$\begin{pmatrix} \text{vv} & & & \\ & \text{ee} & \text{ef} & \text{ec} \\ & \text{fe} & \text{ff} & \text{fc} \\ & \text{ce} & \text{cf} & \text{cc} \end{pmatrix}$	$\begin{pmatrix} \text{vv} & \text{ve} & & \\ \text{ev} & \text{ee} & & \\ & & \text{ff} & \text{fc} \\ & & \text{cf} & \text{cc} \end{pmatrix}$	$\begin{pmatrix} \text{vv} & & & \\ & \text{ee} & & \\ & & \text{ff} & \text{fc} \\ & & \text{cf} & \text{cc} \end{pmatrix}$
$\mathcal{H}$-acc	Storage (MB)			
10^{-1}	1201	1184	1008	1001
10^{-2}	1977	1902	1516	1486
$\mathcal{H}$-acc	Set-up time (s)			
10^{-1}	129	123	72	70
10^{-2}	438	393	215	207
$\mathcal{H}$-acc	BiCGStab (steps/time)			
10^{-1}	20/9.3	26/12.0	56/22.9	59/24.1
10^{-2}	4/2.7	8/5.2	47/25.6	49/26.2

Table 3 shows a comparison of the preconditioners introduced in Sect. 3.1 versus those of Sect. 3.2, here shown for elements of polynomial degree 4 on a level-4 grid, resulting in a total of about 250,000 dofs. As we "drop" certain off-diagonal blocks from the matrix before its factorization, the required storage and set-up times are reduced. The preconditioned BiCGStab iteration now requires additional steps and hence more time, but the savings in set-up time appear to be more significant than the disadvantage in iteration time (as long as we still have a convergent method).

Finally, in Table 4 we show results where the matrix size has been reduced by eliminating the "bubble" dofs through static condensation before the system is solved iteratively. For fourth order polynomials on a level-4 grid (about 250,000 dofs before static condensation), the "naive" static condensation took 120 s, which is significantly more than the savings obtained through now solving a smaller system so that it is not recommended unless a more efficient implementation of static condensation can be found.

Table 4 BiCGStab: steps/time for method "static condensation" (top) and method "no static condensation" (bottom), "br" denotes a BiCGStab breakdown

Prec	$\begin{pmatrix} vv & ve & vf \\ ev & ee & ef \\ fv & fe & ff \end{pmatrix}$	$\begin{pmatrix} vv & & \\ & ee & ef \\ & fe & ff \end{pmatrix}$	$\begin{pmatrix} vv & ve & \\ ev & ee & \\ & & ff \end{pmatrix}$	$\begin{pmatrix} vv & & \\ & ee & \\ & & ff \end{pmatrix}$
10^{-1}	17/4.8	19/5.2	49/br	53/12.6
10^{-2}	4/1.9	8/3.5	42/15.6	44/15.7
Prec	$\begin{pmatrix} vv & ve & vf & vc \\ ev & ee & ef & ec \\ fv & fe & ff & fc \\ cv & ce & cf & cc \end{pmatrix}$	$\begin{pmatrix} vv & & & \\ & ee & ef & ec \\ & fe & ff & fc \\ & ce & cf & cc \end{pmatrix}$	$\begin{pmatrix} vv & ve & & \\ ev & ee & & \\ & & ff & fc \\ & & cf & cc \end{pmatrix}$	$\begin{pmatrix} vv & & & \\ & ee & & \\ & & ff & fc \\ & & cf & cc \end{pmatrix}$
10^{-1}	20/9.3	26/12.0	56/22.9	59/24.1
10^{-2}	4/2.7	8/5.2	47/25.6	49/26.2

References

1. T.M. Austin, M. Brezina, B. Jamroz, C. Jhurani, T.A. Manteuffel, J. Ruge, Semi-automatic sparse preconditioners for high-order finite element methods on non-uniform meshes. J. Comput. Phys. **231**(14), 4694–4708 (2012)
2. W. Bangerth, R. Hartmann, G. Kanschat, deal.II – a general purpose object oriented finite element library. ACM Trans. Math. Softw. **33**(4), 24/1–24/27 (2007)
3. W. Bangerth, T. Heister, L. Heltai, G. Kanschat, M. Kronbichler, M. Maier, B. Turcksin, T.D. Young, The deal.ii library, version 8.1. arXiv:http://arxiv.org/abs/1312.2266v4 (2013, preprint)
4. S. Börm, L. Grasedyck. HLIB version 1.3. Available at www.hlib.org
5. S. Börm, L. Grasedyck, W. Hackbusch, *Hierarchical Matrices, 2006*. Lecture Notes No. 21 (Max-Planck-Institute for Mathematics in the Sciences, Leipzig, 2006)
6. M.O. Deville, P.F. Fischer, E.H. Mund, *High-Order Methods for Incompressible Fluid flow*. Cambridge Monographs on Applied and Computational Mathematics, vol. 9 (Cambridge University Press, Cambridge, 2002)
7. L. Grasedyck, R. Kriemann, S. Le Borne, Domain decomposition based $\mathcal{H}$-LU preconditioning. Numer. Math. **112**, 565–600 (2009)
8. W. Hackbusch, A sparse matrix arithmetic based on $\mathcal{H}$-matrices. Part I: introduction to $\mathcal{H}$-matrices. Computing **62**, 89–108 (1999)
9. J.J. Heys, T.A. Manteuffel, S.F. McCormick, L.N. Olson, Algebraic multigrid for higher-order finite elements. J. Comput. Phys. **204**(2), 520–532 (2005)
10. J. Lottes, P. Fischer, Hybrid multigrid/Schwarz algorithms for the spectral element method. J. Sci. Comput. **24**, 45–78 (2005)
11. P. Šolín, K. Segeth, I. Doležel. *Higher-Order Finite Element Methods*. Studies in Advanced Mathematics (Chapman & Hall/CRC, Boca Raton, 2004)

A Domain Decomposition Method Based on Augmented Lagrangian with an Optimized Penalty Parameter

Chang-Ock Lee and Eun-Hee Park

1 A Non-overlapping DDM with a Penalty Parameter

A non-overlapping domain decomposition method based on augmented Lagrangian with a penalty term was introduced in the previous works by the authors [6, 7], which is a variant of the FETI-DP method. In this paper we present a further study focusing on the case of small penalty parameters in terms of condition number estimate and practical efficiency. The full analysis of the proposed method can be found in [8].

Throughout the paper, we denote by $\lambda_{\min}^{A}$ and $\lambda_{\max}^{A}$ the minimum eigenvalue and the maximum eigenvalue of a matrix A, respectively. To avoid the proliferation of constants, we will use $A \lesssim B$ and $A \gtrsim B$ to represent the statements that $A \le (\text{constant})B$ and $A \ge (\text{constant})B$, respectively, where the positive constant is independent of the mesh size, the subdomain size, and the number of subdomains. The statement $A \approx B$ is equivalent to $A \lesssim B$ and $A \gtrsim B$.

We first review the non-overlapping domain decomposition method with a penalty term in the previous works. Then, we state how we can enhance this method in terms of a better choice of a penalty parameter.

C.-O. Lee
KAIST, Daejeon 305-701, South Korea
e-mail: colee@kaist.edu

E.-H. Park (✉)
Kangwon National University, Gangwon-do 245-711, South Korea
e-mail: eh.park@kangwon.ac.kr

T. Dickopf et al. (eds.), *Domain Decomposition Methods in Science and Engineering XXII*, Lecture Notes in Computational Science and Engineering 104, DOI 10.1007/978-3-319-18827-0_58

We consider the following Poisson model problem with the homogeneous Dirichlet boundary condition

$$\begin{aligned} -\Delta u &= f \quad \text{in } \Omega, \\ u &= 0 \quad \text{on } \partial\Omega, \end{aligned} \tag{1}$$

where Ω is a bounded polygonal domain in $\mathbb{R}^2$ and f is a given function in $L_2(\Omega)$. Let $\mathcal{T}_h$ denote a quasi-uniform triangulation on Ω and $\hat{X}_h$ the space of the conforming $\mathbb{P}_1$ elements associated with $\mathcal{T}_h$. We are concerned with a discretized variational problem of (1) as follows: find $u_h \in \hat{X}_h$ such that

$$a(u_h, v_h) = (f, v_h) \quad \forall v_h \in \hat{X}_h, \tag{2}$$

where

$$a(u_h, v_h) = \int_\Omega \nabla u_h \cdot \nabla v_h \, dx, \quad (f, v_h) = \int_\Omega f v_h \, dx.$$

We start with recalling an iterative solver of (2) in [6, 7], which is a non-overlapping domain decomposition algorithm based on an augmented Lagrangian. We decompose Ω into non-overlapping subdomains $\{\Omega_j\}_{j=1}^J$ as open sets, where the boundary $\partial\Omega_j$ is aligned with $\mathcal{T}_h$ and the diameter of Ω_j is H_j. On each subdomain, the triangulation $\mathcal{T}_j$ is the triangulation of Ω_j inherited from $\mathcal{T}_h$ and matching grids are taken on the boundaries of neighboring subdomains across the interface Γ. Here the interface Γ is the union of the common interfaces among all subdomains, i.e., $\Gamma = \bigcup_{j<k} \Gamma_{jk}$, where Γ_{jk} denotes the common interface of two adjacent subdomains Ω_j and Ω_k.

Based on the non-overlapping subdomain decomposition, a partitioned problem is obtained as follows:

$$\min_{v \in \prod_{j=1}^J X_h^j} \left(\frac{1}{2} \sum_{j=1}^J \int_{\Omega_j} |\nabla v|^2 \, dx - (f, v) \right) \tag{3a}$$

$$\text{subject to } v^j = v^k \text{ on } \Gamma_{jk} \text{ for } j < k, \tag{3b}$$

where X_h^j is the restriction of $\hat{X}_h$ on a subdomain Ω_j. To make a localized minimization problem recover the original solution of (2), the continuity constraint (3b) needs to be satisfied on the interface Γ in an appropriate manner (e.g. [2–5]).

The FETI-DP method, one of the most advanced non-overlapping domain decomposition algorithms, imposes the continuity differently at vertices and the remaining interface nodes except vertices in terms of the choice of finite elements. The continuity at vertices is enforced strongly in a manner that subdomains sharing a vertex have the common value at the vertex while the continuity on the interface except vertices is enforced weakly by introducing Lagrange multipliers. Hence the FETI-DP method starts with the saddle-point problem

$$\mathcal{L}(u_h, \lambda_h) = \max_{\mu_h \in \mathbb{R}^M} \min_{v_h \in X_h^c} \mathcal{L}(v_h, \mu_h), \tag{4}$$

where a Lagrangian functional $\mathcal{L}$ is defined on $X_h^c \times \mathbb{R}^M$ as

$$\mathcal{L}(v, \mu) = \frac{1}{2} \sum_{j=1}^{J} \int_{\Omega_j} |\nabla v|^2 \, dx - (f, v) + \langle Bv, \mu \rangle.$$

Here, X_h^c denotes the subspace of $\prod_{j=1}^{J} X_h^j$ obtained by enforcing the vertex continuity, B is a signed Boolean matrix which plays a role in making values defined individually on the interface pointwise-matched, M represents the number of constraints used for imposing the pointwise matching on the interface and $\langle \cdot, \cdot \rangle$ is the Euclidean inner product in $\mathbb{R}^M$.

In [9], for the FETI-DP method accompanied by the Dirichlet preconditioner it is well-known that the condition number of the resulting dual problem from (4) grows asymptotically as $O(1 + \ln(H/h))^2$, where H is the subdomain size and h is the mesh size. It shows that the convergence slows down only due to the increase of H/h, where $(H/h)^2$ are the local problem size. Due to such a scalable property of the FETI-DP method, there seems to be nothing to improve as a parallel algorithm only if parallel machines with infinitely many CPUs or cores are available. But, keeping in mind that most of ordinary users have limited computing resources, the condition number growth with respect to the increase of H/h is unsatisfactory. In this view, Lee and Park [6, 7] proposed a dual iterative substructuring method with a penalty term which plays a key role in enhancing the convergence to the extent of the constant condition number bound independent of both H and h. A penalty term $\eta \mathcal{J}$ is considered, which consists of a positive penalty parameter η and a measure of the jump on the interface. The addition of a penalty term $\eta \mathcal{J}$ to the Lagrangian $\mathcal{L}$ yields a saddle-point problem for an augmented Lagrangian functional $\mathcal{L}_\eta$ such as

$$\mathcal{L}_\eta(u_h, \lambda_h) = \max_{\mu_h \in \mathbb{R}^M} \min_{v_h \in X_h^c} \mathcal{L}_\eta(v_h, \mu_h), \tag{5}$$

where

$$\mathcal{L}_\eta(v, \mu) = \mathcal{L}(v, \mu) + \frac{1}{2} \eta \mathcal{J}(v, v).$$

Here the penalty term $\mathcal{J}$ is a bilinear form on $X_h^c \times X_h^c$ defined as

$$\mathcal{J}(u, v) = \frac{1}{h} \sum_{j<k} \int_{\Gamma_{jk}} (u^j - u^k)(v^j - v^k) \, ds,$$

where $h = \max_{j=1,\cdots,J} h_j$ with the mesh size h_j of $\mathcal{T}_j$.

The problem (5) is expressed in the algebraic form

$$\begin{bmatrix} A_{\Pi\Pi} & A_{\Pi\Delta} & 0 \\ A_{\Pi\Delta}^T & A_{\Delta\Delta} & B_\Delta^T \\ 0 & B_\Delta & 0 \end{bmatrix} \begin{bmatrix} \boldsymbol{u}_\Pi \\ \boldsymbol{u}_\Delta \\ \lambda \end{bmatrix} = \begin{bmatrix} f_\Pi \\ f_\Delta \\ 0 \end{bmatrix},$$

where λ indicates the Lagrange multipliers introduced for imposing the continuity constraint across the interface, Π the degrees of freedom associated with both the interior nodes and the subdomain corners, and Δ the remaining part of the degrees of freedom on the interface. The matrix J results from the penalty term $\mathcal{J}$, which is written as

$$J = B_\Delta^T D_M B_\Delta, \tag{6}$$

where D_M is the block diagonal matrix with a diagonal block $\frac{1}{h} M_e$. Here M_e is the 1-D mass matrix on each edge. Eliminating $\boldsymbol{u}_\Pi$ and $\boldsymbol{u}_\Delta$ successively, we have a dual system

$$F_\eta \lambda = d_\eta \tag{7}$$

where

$$F_\eta = B_\Delta S_\eta^{-1} B_\Delta^T, \quad d_\eta = B_\Delta S_\eta^{-1}(f_\Delta - A_{\Pi\Delta}^T A_{\Pi\Pi}^{-1} f_\Pi)$$

with

$$S_\eta = S + \eta J = (A_{\Delta\Delta} - A_{\Pi\Delta}^T A_{\Pi\Pi}^{-1} A_{\Pi\Delta}) + \eta J. \tag{8}$$

For the proposed dual iterative substructuring method which results in the dual problem (7), we are concerned with two key properties: one is the convergence of the primal solution u_h of the saddle-point problem (5) from which (7) is originated, to the exact weak solution of (1) and the other is the condition number of F_η which determines the convergence rate of dual iterations on (7). In this context, we now discuss the choice of a penalty parameter in the proposed dual iterative substructuring method.

Let us first look over what effect the choice of the penalty parameter has on the convergence of the finite element solution to the weak solution of (1). In finite element formulations based on penalty methods for (3) (cf. [1, 2]), the choice of a sufficiently large penalty parameter is required for the stability of a concerning finite element formulation, which is necessary for the convergence of the finite element solution to the exact weak solution of (1). On the other hand, the penalty parameter η plays a different role in the saddle-point formulation (5) based on an augmented Lagrangian functional because Lagrange multipliers as well as a penalty term are introduced to enforce the continuity across the interface. More precisely, such a role difference was confirmed in [6] by the fact that the primal solution u_h of the saddle-point problem (5) is exactly equal to the finite element solution of (2) regardless of the choice of η. Hence there is no need to consider a right choice of η in the aspect of the convergence of a finite element solution to the solution of (1).

Let us next discuss the choice of the penalty parameter in terms of the condition number of F_η. The convergence study for dual iterations in [6, 7] shows that the dual system (7) has a constant condition number bound independent of H and h where a sufficiently large penalty parameter is taken. On the contrary, we have observed through numerical results that there might be an estimated parameter $\eta^* < 10$ with which the proposed dual iterative algorithm is almost optimal in terms of its condition number. Based on such observation, we shall focus on the case of small penalty parameters throughout the following sections.

2 Condition Number Estimate

In this section, we find the relationship between the standard FETI-DP operator and the proposed dual operator in algebraic form. Based on the relationship, we carry out convergence analysis in terms of the condition number of the dual system F_η without size limitation of the penalty parameter. As results, it is confirmed why a fast convergence of the dual iteration is attained even if a small η is taken.

We first have the following condition number estimate of the concerned dual system based on a key relationship between two matrices F_η and F, where F is the standard FETI-DP operator as $F = B_\Delta S^{-1} B_\Delta^T$.

Theorem 1 *For any $\eta > 0$, the condition number $\kappa(F)$ is estimated as*

$$\kappa(F_\eta) \le \frac{C_{F,D_M}}{\eta + C_{F,D_M}} \kappa(F) + \frac{\eta}{\eta + C_{F,D_M}} \kappa(D_M), \tag{9}$$

where $C_{F,D_M} = (\lambda_{\max}^F \lambda_{\min}^{D_M})^{-1}$.

Remark 1 Theorem 1 shows the change of $\kappa(F_\eta)$ with respect to a choice of η as well as the connection of $\kappa(F_\eta)$ with $\kappa(F)$. In particular, $\kappa(F_\eta)$ becomes close to $\kappa(F)$ as η decreases to zero. In addition, it follows from (9) that

$$\kappa(F_\eta) \le \kappa(D_M) + \frac{C_{F,D_M}(\kappa(F) - \kappa(D_M))}{\eta + C_{F,D_M}}, \tag{10}$$

which implies that the result shown in Fig. 1 in [6] is in agreement with (10) when $\kappa(F) > \kappa(D_M)$.

Then the extreme eigenvalues of matrices F and D_M can be estimated as

$$\lambda_{\min}^{D_M} \gtrsim 1, \quad \lambda_{\max}^{D_M} \lesssim 1$$

$$\lambda_{\min}^{F} \gtrsim 1, \quad \lambda_{\max}^{F} \lesssim \max_{j=1,\cdots,J} \left(\frac{H_j}{h_j} \left(1 + \ln \frac{H_j}{h_j} \right) \right),$$

which imply that

$$\kappa(D_M) \lesssim 1$$

$$\kappa(F) \lesssim \max_{j=1,\cdots,J} \left(\frac{H_j}{h_j} \left(1 + \ln \frac{H_j}{h_j} \right) \right).$$

Hence it is noted that either $\kappa(F) \leq \kappa(D_M)$ or $\kappa(F) > \kappa(D_M)$ holds according to the size of H/h. First, in the case of small H/h such that

$$\kappa(F) \leq \kappa(D_M),$$

it follows from Theorem 1 that, for any $\eta > 0$,

$$\kappa(F_\eta) \leq \kappa(D_M) \lesssim 1. \tag{11}$$

Next, in the following theorem we will see the case of large H/h such that

$$\kappa(F) > \kappa(D_M).$$

Using the estimated extreme eigenvalues of D_M and F, we can characterize bounds of the condition number of the concerned dual system as follows.

Theorem 2 *For any H/h such that*

$$\kappa(F) > \kappa(D_M),$$

there is a positive constant C_{opt} independent of H and h such that

$$\kappa(F_\eta) < \kappa(D_M) + C_{opt} \quad \textit{for any } \eta \geq C_{opt},$$

where $C_{opt} \approx 1$.

Remark 2 The convergence studies in [6, 7] for a dual iterative substructuring method with a penalty term were limited to the case when a sufficiently large penalty parameter η is used. The estimate (11) and Theorem 2 show why a faster convergence of the dual iteration in the proposed method is attained in comparison with the FETI-DP method even if a relatively small η is taken while Theorem 1 for a large η is identical to the previous results in [6, 7].

Remark 3 Due to length limitation, this paper is focused on the convergence analysis for the case of small penalty parameters in 2-D. More works for 3-D extension and computational issues such as the preconditioning of the subdomain problems can be found [8].

3 Numerical Results

In this section, computational results are presented, which are in agreement with the theoretical bound estimated in Sect. 2. We consider the model problem (1) with the exact solution

$$u(x,y) = \begin{cases} y(1-y)\sin(\pi x) & \text{in 2-D} \\ \sin(\pi x)\sin(\pi y)z(1-z) & \text{in 3-D} \end{cases}$$

for $\Omega = (0,1)^d, d = 2,3$. We use the conjugate gradient method with a constant initial guess ($\lambda_0 \equiv 1$). The stop criterion is the relative reduction of the initial residual by a chosen TOL

$$\frac{\|r_k\|_2}{\|r_0\|_2} \leq \text{TOL},$$

where r_k is the dual residual error on the kth CG iteration and TOL $= 10^{-8}$. Here, discretization parameters h, H, and J are used, which stand for the mesh size, the subdomain size, and the number of subdomains, respectively. Through numerical tests, Ω in 2-D is decomposed into J square subdomains with $J = 1/H \times 1/H$. Each subdomain is partitioned into $2 \times H/h \times H/h$ uniform triangular elements. In 3-D, Ω is decomposed into J cubic subdomains with $J = 1/H \times 1/H \times 1/H$ while each subdomain is partitioned into $H/h \times H/h \times H/h$ uniform cubic elements.

In Table 1 for the two-dimensional problem, the condition numbers of the dual system are presented in the cases with η in $[0, 10]$. In addition, for comparison with the case with a large η, the result for $\eta = 10^6$ is presented. For each $\eta > 0$, the condition number $\kappa(F_\eta)$ is bounded by a constant even if H/h increases. In Table 1, any penalty parameter chosen in $(1/2, 10)$ improves the condition number regardless of the increase of H/h. In addition, the condition numbers for the case with $\eta \in (1/2, 10)$ are less than that for the case with a large η. According to the condition number and the iteration count, $\eta = 2$ is regarded as an optimal one. Table 2 for 3-D shows similar results in 2-D; $\eta = 1$ seems to be optimal as H/h increases.

Table 1 Condition number of F_η for a small η where $J = 4 \times 4$ in 2-D

	$\frac{H}{h} = 4$		$\frac{H}{h} = 8$		$\frac{H}{h} = 16$		$\frac{H}{h} = 32$	
η	$\kappa(F_\eta)$	Iter. #	$\kappa(F_\eta)$	Iter. #	$\kappa(F_\eta)$	Iter. #	$\kappa(F_\eta)$	Iter. #
0	7.2033	14	2.2901e+1	23	5.9558e+1	33	1.4707e+2	48
0.2	3.7811	12	5.6829	15	6.4744	18	6.7436	19
0.4	2.6637	10	3.3617	13	3.5166	13	3.6410	14
0.6	2.0733	9	2.3969	10	2.5127	11	2.5753	12
0.8	1.6990	8	1.9367	9	1.9974	10	2.0247	10
1	1.5030	7	1.6468	8	1.6801	9	1.6957	9
2	1.1304	5	1.1067	5	1.1053	5	1.1050	5
4	1.3353	6	1.4469	7	1.4625	8	1.4477	8
6	1.5050	7	1.7008	9	1.7470	9	1.7378	9
8	1.6130	7	1.8691	9	1.9404	10	1.9387	10
10	1.6875	7	1.9945	10	2.0799	11	2.0868	11
10^6	2.0938	3	2.7170	7	2.9243	13	2.9771	14

Table 2 Condition number of F_η for a small η where $J = 4 \times 4 \times 4$ in 3-D

	$\frac{H}{h} = 4$		$\frac{H}{h} = 8$		$\frac{H}{h} = 16$		$\frac{H}{h} = 32$	
η	$\kappa(F_\eta)$	Iter. #	$\kappa(F_\eta)$	Iter. #	$\kappa(F_\eta)$	Iter. #	$\kappa(F_\eta)$	Iter. #
0	8.1805e+1	73	3.0183e+2	107	1.1892e+3	153	4.6946e+3	218
0.2	6.9551	22	6.8882	22	6.7708	21	6.6486	21
0.4	4.4201	18	4.6965	18	4.8197	18	4.8325	17
0.6	3.8658	16	4.3214	16	4.4810	17	4.4959	16
0.8	3.5613	15	4.0772	16	4.2515	16	4.2834	16
1	3.3611	15	3.9076	15	4.0901	15	4.1292	15
2	3.1992	14	4.0118	16	4.3020	16	4.3345	16
4	3.6343	15	4.8935	17	5.3381	18	5.4422	19
6	3.8905	15	5.4275	17	5.9726	19	6.1152	20
8	4.0564	15	5.7842	18	6.4011	20	6.5659	21
10	4.1740	15	6.0390	19	6.7099	21	6.8890	21
10^6	4.8585	7	7.5658	14	8.5609	16	8.8699	18

Acknowledgement The work of the first author was supported by NRF-2011-0015399. The second author was supported in part by Korea Research Council of Fundamental Science and Technology (KRCF) research fellowship for young scientists.

References

1. I. Babuška, The finite element method with penalty. Math. Comput. **27**, 221–228 (1973)
2. E. Burman, P. Zunino, A domain decomposition method based on weighted interior penalties for advection-diffusion-reaction problems. SIAM J. Numer. Anal. **44**, 1612–1638 (2006)
3. C. Farhat, F.-X. Roux, A method of finite element tearing and interconnecting and its parallel solution algorithm. Int. J. Numer. Methods Eng. **32**, 1205–1227 (1991)
4. C. Farhat, M. Lesoinne, K. Pierson, A scalable dual-primal domain decomposition method. Numer. Linear Algebra Appl. **7**, 687–714 (2000)
5. R. Glowinski, P. Le Tallec, Augmented Lagrangian interpretation of the nonoverlapping Schwarz alternating method, in *Third International Symposium on Domain Decomposition Methods for Partial Differential Equations (Houston, TX, 1989)* (SIAM, Philadelphia, 1990), pp. 224–231
6. C.-O. Lee, E.-H. Park, A dual iterative substructuring method with a penalty term. Numer. Math. **112**, 89–113 (2009)
7. C.-O. Lee, E.-H. Park, A dual iterative substructuring method with a penalty term in three dimensions. Comput. Math. Appl. **64**, 2787–2805 (2012)
8. C.-O. Lee, E.-H. Park, A dual iterative substructuring method with a small penalty parameter. Submitted (2013)
9. J. Mandel, R. Tezaur, On the convergence of a dual-primal substructuring method. Numer. Math. **88**, 543–558 (2001)

Dual Schur Method in Time for Nonlinear ODE

P. Linel and D. Tromeur-Dervout

1 Introduction

We developed parallel time domain decomposition methods to solve systems of linear ordinary differential equations (ODEs) based on the Aitken-Schwarz [5] or primal Schur complement domain decomposition methods [4]. The methods require the transformation of the initial value problem in time defined on $]0, T]$ into a time boundary values problem. Let $f(t, y(t))$ be a function belonging to $\mathscr{C}^1(\mathbb{R}^+, \mathbb{R}^d)$ and consider the Cauchy problem for the first order ODE:

$$\left\{\dot{y} = f(t, y(t)),\ t \in]0, T],\ y(0) = \alpha \in \mathbb{R}^d. \right. \tag{1}$$

The time interval $[0, T]$ is split into p time slices $S^{(i)} = [T_{i-1}^+, T_i^-]$, with $T_0^+ = 0$ and $T_p^- = T^-$. The difficulty is to match the solutions $y_i(t)$ defined on $S^{(i)}$ at the boundaries T_{i-1}^+ and T_i^-. Most of time domain decomposition methods are shooting methods [1] where the jumps $y_i(T_i^-) - y_{i+1}(T_i^+)$ are corrected by a sequential process which is propagated in the forward direction (i.e. the correction on the time slice $S^{(i-1)}$ is needed to compute the correction on time slice $S^{(i)}$). Our approach consists in breaking the sequentiality of the update of each time slice initial value. To this end, we transform the initial value problem (IVP) into a boundary values

P. Linel
Department of Biostatistics and Computational Biology, University of Rochester, Rochester, NY, USA
e-mail: Patrice_Linel@URMC.Rochester.edu

D. Tromeur-Dervout (✉)
University of Lyon, University Lyon 1, CNRS, UMR5208 Institut Camille Jordan, Lyon, France
e-mail: Damien.tromeur-dervout@univ-lyon1.fr

T. Dickopf et al. (eds.), *Domain Decomposition Methods in Science and Engineering XXII*, Lecture Notes in Computational Science and Engineering 104, DOI 10.1007/978-3-319-18827-0_59

problem (BVP) leading to a second order ODE:

$$\begin{cases} \ddot{y}(t) = g(t, y(t), \dot{y}(t)) \stackrel{def}{=} \dfrac{\partial f}{\partial t}(t, y(t)) + \dot{y}(t)\dfrac{\partial f}{\partial y}(t, y(t)),\ t \in]0, T[, & \text{(2a)} \\ y(0) = \alpha, & \text{(2b)} \\ \dot{y}(T) = \beta \stackrel{def}{=} f(T, y(T)) & \text{(2c)} \end{cases}$$

Then classical domain decomposition methods apply such as the multiplicative Schwarz method with no overlapping time slices and Dirichlet-Neumann transmission conditions (T.C.) for linear system of ODE (or PDE [6]). As proved in [5] the convergence/divergence of the error at the boundaries of this Schwarz time DDM can be accelerated by the Aitken technique to the right solution when $f(t, y(t))$ is linear. Nevertheless, the difficulty in solving Eq. (2) is that β is not given by the original IVP. In [7] when $f(t, y(t))$ is nonlinear with respect of $y(t)$ and scalar, we proposed to replace the end boundary condition (2c) by imposing, if $f(T, y(T)) \neq 0$, the invariant flux condition for $t = T$:

$$(f(T, y(T))^{-1}\dot{y}(T) = 1. \tag{3a}$$

We also showed that the right T.C. between time slices must involve the nonlinear flux condition $(f(T_i^-, y(T_i^-))^{-1}\dot{y}_i(T_i^-) = (f(T_{i+1}^+, y(T_{i+1}^+))^{-1}\dot{y}_{i+1}(T_{i+1}^+)$. In this case, we showed that the behavior of the Schwarz method with an appropriate nonlinear change of variable Θ is linear. Then, it is possible to apply the Aitken acceleration by using Θ if it is known. To overcome the lack of knowledge of Θ, we propose in this paper to replace the Schwarz method by a Schur complement method.

In Sect. 2, we recall some results on the existence and uniqueness of the proposed BVP. Section 3 gives the dual Schur complement method intimely related to the Newton step solving. The choice of T.C. to define the time slice function is discussed there. Some numerical results are given in Sect. 4 before the conclusion.

2 Existence and Uniqueness of the BVP Solution

The problem (2) with $d = 1$ is a particular case of the more general problem:

$$\begin{cases} \ddot{y} = g(t, y, \dot{y}), a \leq t \leq b, & \text{(4a)} \\ a_0\, y(a) - a_1\, \dot{y}(a) = \alpha, |a_0| + |a_1| \neq 0, & \text{(4b)} \\ b_0\, y(b) + b_1\, \dot{y}(b) = \beta, |b_0| + |b_1| \neq 0. & \text{(4c)} \end{cases}$$

Keller [3] has established the existence and uniqueness of a solution to problem (4) under the hypotheses of monotonicity and upper bound on the partial derivatives of g in the theorem that follows:

Theorem 1 (H.B. Keller) *Let $g(t, y, \dot{y})$ have continuous derivatives which satisfy:*

$$\frac{\partial g(t, y(t), \dot{y}(t))}{\partial y} > 0, |\frac{\partial g(t, y(t), \dot{y}(t))}{\partial \dot{y}}| \leq M, \tag{5}$$

for some $M \geq 0$, $a \leq t \leq b$ and all continuously differentiable functions $y(t)$. Let the constants a_i, b_i satisfy:

$$a_i \geq 0, b_i \geq 0, i = 0, 1; a_0 + b_0 > 0. \tag{6}$$

then a unique solution of (4a)–(4c) *exists for each* (α, β).

3 Dual Schur Complement Time DDM

3.1 BVP Discretizing and It Solution

Problem (2a), (2b), (3a), is discretized using a Störmer-Verlet implicit scheme [2] with $N_g + 1$ regular time steps with $\Delta t = T/N_g$ over the time interval $[0, T]$. Solving it requires to find the zero of the function $F(u) : \mathbb{R}^{N_g} \rightarrow \mathbb{R}^{N_g}$ with $u_j \simeq u(t_j)$, $t_j = (j-1)\Delta t$ and defined as:

$$F(u) = \begin{pmatrix} u_0 - \alpha \\ u_{j+1} - 2u_j + u_{j-1} - \Delta t^2 g(t_j, u_j), j = 1, \ldots, N_g - 1 \\ f^{-1}(t_{N_g}, u_{N_g}) B(u_{N_g}) - 1 \end{pmatrix} \tag{7}$$

where $g(t, u) \stackrel{def}{=} \frac{\partial f}{\partial t}(t, u) + f(t, u)\frac{\partial f}{\partial u}(t, u)$, and $B(u_{N_g})$ corresponds to the discretizing of $\dot{u}(T)$ as:

$$B(u_{N_g}) = \frac{3u_{N_g} - 4u_{N_g-1} + u_{N_g-2}}{2\Delta t} \simeq \dot{u}(T) + O(\Delta t^2) \tag{8a}$$

$$B(u_{N_g}) = \frac{11u_{N_g} - 18u_{N_g-1} + 9u_{N_g-2} - 2u_{N_g-3}}{6\Delta t} \simeq \dot{u}(T) + O(\Delta t^3) \tag{8b}$$

We applied the Newton method to find the zero of function $F(u)$. Starting from an initial guess, it writes if $||F(u^m)|| > \epsilon$ for the $(m + 1)$-th iteration:

$$h^m = -(\nabla_u F(u^m))^{-1} F(u^m), \ u^{m+1} = u^m + h^m. \tag{9}$$

Let us notice that the Newton method is sensitive to the initial solution. One can consider to search the initial solution by performing a few Newton iterations on different coarse levels of time grid discretizing. The approximate solution obtained on a previous coarse grid gives the initial guess solution for the next time grid after interpolating. There is no Courant-Friedrich-Lax stability condition because we use an implicit Störmer-Verlet scheme.

3.2 Dual Schur Complement in Time Formulation

For the time domain decomposition, we split the time interval $[0, T]$ in p slices $S^{(i)}, i = 1, \ldots, p$ and we denote by $u^{(i)}$ the solution on the i-th time slice $S^{(i)}$. For the sake of simplicity and without loss of generality we set all the time slices to have the same size and use $N + 1$ regular time steps on each such that $S^{(i)} = [t_0^{(i)}, t_N^{(i)}] \stackrel{def}{=} [(i-1)N\Delta t, iN\Delta t]$ (then the total number of time steps on $[0, T]$ is $N_g + 1 = p \times N + 1$). Here, the main idea consists in finding the zero of the local function F_i defined on the time slice $S^{(i)}$ under the continuity constraint of the solution at the time slices boundaries. Two strategies can be applied to define the transmission conditions (T.C.) of the local function $F_i(u^{(i)})$:

1. The first strategy S1 considers the original function F and split its components at the time slices boundaries in two parts. Each one corresponds to the contribution of the solution components belonging to the time slice under consideration (10b) at $j = 0$ for $S^{(i)}, i = 2, \ldots, p$ and (12b) at $j = N$ for $S^{(i)}, i = 1, \ldots, p-1$.
2. The second strategy S2 considers the T.C. corresponding to the nonlinear flux (10c) at $j = 0$ for $S^{(i)}, i = 2, \ldots, p$ and (12c) at $j = N$ for $S^{(i)}, i = 1, \ldots, p-1$.

$$\begin{cases} (F_1(u))_0 = u_0 - \alpha & \text{(10a)} \\ S1 : (F_i(u))_0 = u_1 - u_0 - \frac{1}{2}\Delta t^2 g(t_0^{(i)}, u_0), i = 2, \ldots, p & \text{(10b)} \\ S2 : (F_i(u))_0 = f^{-1}(t_0^{(i)}, u_0)B(u_0), i = 2, \ldots, p & \text{(10c)} \end{cases}$$

$$\begin{cases} (F_i(u))_j = u_{j+1} - 2u_j + u_{j-1} - \Delta t^2 g(t_j^{(i)}, u_j), & \text{(11a)} \\ j = 1, \ldots, N-1, \ i = 1, \ldots, p \end{cases}$$

$$\begin{cases} (F_p(u))_N = f^{-1}(t_N^{(p)}, u)B(u_N) - 1 & \text{(12a)} \\ S1 : (F_i(u))_N = -u_N + u_{N-1} - \frac{1}{2}\Delta t^2 g(t_N^{(i)}, u_N), \ i = 2, \ldots, p-1 & \text{(12b)} \\ S2 : (F_i(u))_N = f^{-1}(t_N^{(i)}, u_N)B(u_N), \ i = 2, \ldots, p-1 & \text{(12c)} \end{cases}$$

Then, we use the Newton method on each time slices $S^{(i)}$ and introduce the Lagrange multipliers $\lambda_i, i = 1, \ldots, p-1$ to ensure the continuity of the solution between the time slices (adding this Lagrange multiplier to (10b) [respectively (10c)] and subtracting it to (12b) [respectively (12c)]). It writes:

$$h^{(i),m} = u^{(i),m+1} - u^{(i),m} = -(\nabla F_i(u^{(i),m}))^{-1}(F(u^{(i),m}) + \underbrace{(\lambda_{i-1}, 0, \ldots, 0, -\lambda_i)^t}_{\in \mathbb{R}^{N+1}}) \tag{13}$$

with the constraints

$$u_0^{(i),m} + h_0^{(i),m} = u_N^{(i-1),m} + h_N^{(i-1),m}, i = 2, \ldots, p \tag{14}$$

Let us give the computing details. Introducing the Jacobian matrix $J^{(i)}$ corresponding to $\nabla F_i(u^{(i),m})$, the index I for the unknowns $[1, \ldots, N-1]$ and E for the unknowns $0, N$, the linearized system of the Newton step writes after a permutation of unknowns:

$$\begin{pmatrix} J_{II}^{(i)} & J_{I\Gamma}^{(i)} \\ J_{\Gamma I}^{(i)} & J_{\Gamma\Gamma}^{(i)} \end{pmatrix} \begin{pmatrix} h_I^{(i)} \\ h_E^{(i)} \end{pmatrix} = \begin{pmatrix} b_I^{(i)} \\ b_E^{(i)} \end{pmatrix} + \begin{pmatrix} 0 \\ \Lambda_i \end{pmatrix} \tag{15}$$

where

$$(\Lambda_i, b_I^{(i)}, b_E^{(i)}) = \begin{cases} (-\lambda_1, -(F_1(u^{(1),m})_{0,\ldots,N-1}, -(F_1(u^{(1),m})_N) & i = 1 \\ (\begin{pmatrix} \lambda_{i-1} \\ -\lambda_i \end{pmatrix}, -(F_i(u^{(i),m})_{1,\ldots,N-1}, -(F_i(u^{(i),m})_{[0,N]}) & i \neq \{1, p\} \\ \lambda_{p-1}, -(F_p(u^{(p),m})_{1,\ldots,N}, -(F_p(u^{(p),m})_0) & i = p \end{cases}$$

if $h_E^{(i)}$ is known then the first line of system (15) gives:

$$h_I^{(i)} = (J_{II}^{(i)})^{-1}(b_E^{(i)} - J_{I\Gamma}^{(i)} h_E^{(i)}) \tag{16}$$

Reporting $h_I^{(i)}$ in the second line of system (15), we obtain:

$$S_\Gamma^{(i)} h_E^{(i)} \overset{def}{=} (J_{\Gamma\Gamma}^{(i)} - J_{\Gamma I}^{(i)} (J_{II}^{(i)})^{-1} J_{I\Gamma}^{(i)}) h_E^{(i)} = (b_E^{(i)} - (J_{II}^{(i)})^{-1} b_I^{(i)}) + \Lambda_i \tag{17}$$

If Λ_i is known then $h_E^{(i)}$ can be computed. To compute Λ_i, we impose the continuity of the solution among the time slices:

$$\begin{pmatrix} u_0^{(i)} + h_0^{(i)} \\ u_N^{(i)} + h_N^{(i)} \end{pmatrix} = \begin{pmatrix} u_N^{(i-1)} + h_N^{(i-1)} \\ u_0^{(i+1)} + h_0^{(i+1)} \end{pmatrix} \tag{18}$$

$$\begin{pmatrix} h_0^{(i)} \\ h_N^{(i)} \end{pmatrix} = \begin{pmatrix} \bar{S}_{\Gamma,00}^{(i)} & \bar{S}_{\Gamma,0N}^{(i)} \\ \bar{S}_{\Gamma,N0}^{(i)} & \bar{S}_{\Gamma,NN}^{(i)} \end{pmatrix} \begin{pmatrix} g_0^{(i)} + \lambda_{i-1} \\ g_N^{(i)} - \lambda_i \end{pmatrix} \stackrel{def}{=} \bar{S}_\Gamma^{(i)} \begin{pmatrix} g_0^{(i)} + \lambda_{i-1} \\ g_N^{(i)} - \lambda_i \end{pmatrix} \tag{19}$$

where

$$\bar{S}_{\Gamma,N}^{(1)} \stackrel{def}{=} (S_\Gamma^{(1)})^{-1}, \begin{pmatrix} \bar{S}_{\Gamma,00}^{(i)} & \bar{S}_{\Gamma,0N}^{(i)} \\ \bar{S}_{\Gamma,N0}^{(i)} & \bar{S}_{\Gamma,NN}^{(i)} \end{pmatrix} \stackrel{def}{=} (S_\Gamma^{(i)})^{-1}, i = 2, \ldots, p-1, \bar{S}_{\Gamma,0}^{(p)} \stackrel{def}{=} (S_\Gamma^{(p)})^{-1}.$$

We obtain the Lagrange multipliers tridiagonal system (20) of the form $M(\lambda_1, \ldots, \lambda_{p-1})^t = (b_\Gamma^{(1)}, \ldots, b_\Gamma^{(p-1)})^t$ that links all the time slices and allows the instantaneous propagation of the information between all the time slices:

$$\begin{cases} -(\bar{S}_{\Gamma,N}^{(1)} + \bar{S}_{\Gamma,00}^{(2)})\lambda_1 + \bar{S}_{\Gamma,0N}^{(2)}\lambda_2 = b_\Gamma^{(1)} \\ \bar{S}_{\Gamma,N0}^{(i-1)}\lambda_{i-2} - (\bar{S}_{\Gamma,NN}^{(i-1)} + \bar{S}_{\Gamma,00}^{(i)})\lambda_{i-1} + \bar{S}_{\Gamma,0N}^{(i)}\lambda_i = b_\Gamma^{(i-1)}, i = 3, \ldots, p-1 \\ \bar{S}_{\Gamma,N0}^{(p-1)}\lambda_{p-2} - (\bar{S}_{\Gamma,NN}^{(p-1)} + \bar{S}_{\Gamma,0}^{(p)})\lambda_{p-1} = b_\Gamma^{(p-1)} \end{cases} \tag{20}$$

with

$$\begin{aligned} b_\Gamma^{(1)} &= u_0^{(2)} - u_N^{(1)} - S_{\Gamma,N}^{(1)} g_N^{(1)} + S_{\Gamma,00}^{(2)} g_0^{(2)} + S_{\Gamma,0N}^{(2)} g_N^{(2)} \\ b_\Gamma^{(i-1)} &= u_0^{(i)} - u_N^{(i-1)} - \bar{S}_{\Gamma,N0}^{(i-1)} g_0^{(i-1)} - \bar{S}_{\Gamma,NN}^{(i-1)} g_N^{(i-1)} + \bar{S}_{\Gamma,00}^{(i)} g_0^{(i)} \\ &\quad + \bar{S}_{\Gamma,0N}^{(i)} g_N^{(i)}, i = 3, \ldots, p-1, \\ b_\Gamma^{(p-1)} &= u_0^{(p)} - u_N^{(p-1)} - \bar{S}_{\Gamma,N0}^{(p-1)} g_0^{(p-1)} - \bar{S}_{\Gamma,NN}^{(p-1)} g_N^{(p-1)} + \bar{S}_{\Gamma,0}^{(p)} g_0^{(p)} \end{aligned}$$

4 Numerical Results of the Schur Time DDM

We tested our Schur time DDM on the IVP (1) with $f(t, y) = 1 + y^3(t)$ leading to $g(t, y, \dot{y}) = \dot{y}(t)(3y^2(t))$. The number of time steps is $N_g = 2000$ over $[0, 1]$ and $\alpha = 1$.The monotonicity hypothesis of Theorem 1 is satisfied because $y(t)$ is an increasing function on $[0, 1]$ and $\alpha > 0$. The upper bound hypothesis is satisfied on interval $[0, b]$ for b taken sufficiently small, because $f(t, y)$ is continuous in y. The initial guess is computed using two Newton iterations on each of the two coarse grids of 20 and 200 time steps respectively. The initial $||F||_2$ is then around 10^{-2}. Let us notice that Newton's method on the coarsest time mesh does not converge to the solution of the problem. Table 1 shows that both strategies for T.C. (10b) (12b) or (10c) (12c) work well until the number ten of time slices. The first strategy seems to be more robust until $p = 100$ time slices. For $p = 50$ and $p = 100$ time slices the method does not reach the convergence criterion and oscillates with $||F||_2$ around 10^{-5}. These oscillations are mainly due to the local Schur complement

Table 1 Number of Newton iterations #it, with respect to the number of time slices p, required to reach $log_{10}(||f||_2) < -7$ and $log_{10}(||f||_2) < -6$ and with the two discretizing of $B(u)$

$N_g = 2000$	T.C. : (10b) (12b), **B(u)** : (8b)							
p	1	2	4	8	10	25	50	100
#it	5	5	5	5	5(6)	5 (6)	5	5
$log_{10}(\|\|F\|\|_2)$	−13.02	−7.56	−7.55	−7.55	−7.49	−7.59	−7.52	−7.55
$log_{10}(\|\|h\|\|_2)$	−5.42	−6.33	−6.31	−6.28	−5.41	−5.94	−6.19	−6.18
$log_{10}(min(\kappa_2(\bar{S}_\Gamma^{(i)})))$	–	–	0.96	1.15	1.24	1.75	4.00	2.78
$log_{10}(max(\kappa_2(\bar{S}_\Gamma^{(i)})))$	–	–	1.40	2.35	2.65	3.85	5.65	5.65
$log_{10}(min(\kappa_2(M)))$	–	0	1.31	2.75	3.14	4.58	3.99	6.50
$log_{10}(max(\kappa_2(M))$	–	0	1.64	3.05	3.45	4.89	4.00	6.82
$N_g = 2000$	T.C. : (10c) (12c), B(u) : (8a)							
p	1	2	4	8	10	25	50	100
#it	5	5	5	5	5	5	9	−(8)
$log_{10}(\|\|F\|\|_2)$	−12.66	−10.62	−10.62	−9.73	−10.00	−8.37	−7.12	−6.04
$log_{10}(\|\|h\|\|_2)$	−7.29	−7.30	−7.11	−7.05	−6.97	−6.58	6.06	−5.47
$log_{10}(min(\kappa_2(\bar{S}_\Gamma^{(i)})))$	–	–	3.05	3.25	3.30	3.62	5.97	4.52
$log_{10}(max(\kappa_2(\bar{S}_\Gamma^{(i)})))$	–	–	5.70	6.23	7.31	8.29	10.68	10.68
$log_{10}(min(\kappa_2(M)))$	–	0	1.97	2.82	3.12	4.31	5.95	6.03
$log_{10}(max(\kappa_2(M))$	–	0	3.38	3.87	4.73	5.78	7.19	8.45
$N_g = 2000$	T.C. : (10c) (12c), **B(u)** : (8b)							
p	1	2	4	8	10	25	50	100
#it	5	6	6	6	6	15	–	–
$log_{10}(\|\|F\|\|_2)$	−13.02	−10.96	−9.67	−8.31	−8.60	−7.48	−6.76	−5.59
$log_{10}(\|\|h\|\|_2)$	−5.42	−8.11	−7.83	−7.52	−7.87	−5.97	−4.11	−3.56
$log_{10}(min(\kappa_2(\bar{S}_\Gamma^{(i)})))$	–	–	2.75	3.12	3.21	3.62	5.55	4.54
$log_{10}(max(\kappa_2(\bar{S}_\Gamma^{(i)})))$	–	–	6.35	6.52	7.08	9.05	11.08	11.08
$log_{10}(min(\kappa_2(M)))$	–	0	1.73	2.47	3.24	4.45	5.59	6.23
$log_{10}(max(\kappa_2(M))$	–	0	3.34	3.39	4.23	6.09	7.85	8.85

$log_{10}(min/max(\kappa_2(\bar{S}_\Gamma^{(i)})))$ [respectively $log_{10}(min/max(\kappa_2(M))$] refers to the minimum or maximum value of the condition number of the local Schur complement for the time slices 2 to $p-1$ (respectively of the Lagrange multipliers system) over the Newton iterations

of the time slices 2 to $p-1$ where its condition number maximum value, over all the Newton iterations, reaches around 10^{11} for some time slices. Even with this local bad condition numbers, the condition number for the Lagrange multipliers system is around 10^9 (symbol – in row #it means no convergence and (8) means the iteration number among 21 iterations where the minimum values of $||F||_2$ and $||h||_2$ have been reached).

Nevertheless the right T.C. are (10c) (12c) as shown in [7] and illustrated by the following results for $f(t, y) = \sqrt{y(t) + 2}$ on [0, 3] with $\alpha = 0.5$. The initial guess is computed with 2 (respectively 1) Newton iterations on the coarsest (respectively intermediate) time grid leading to $||F||_2 \simeq 10^{-4}$. Table 2 shows that T.C. (10b) (12b) do not lead to convergence, excepted for $p = 2$ where the interface system is reduced

Table 2 Number of Newton iterations #it, with respect to the number of time slices p, required to reach $log_{10}(||F||_2) < -6$ and $log_{10}(||h||_2) < -5$ and with the discretizing of $B(u)$ in $O(\Delta t^2)$

$N_g = 2000$	T.C. : (10b) (12b), B(u) : (8a)							
p	1	2	4	8	10	25	50	100
#it	3	3	–	–	–	–	–	–
$log_{10}(\|\|F\|\|_2)$	−13.02	−10.51	–	–	–	–	–	–
$log_{10}(\|\|h\|\|_2)$	−7.43	−7.85	–	–	–	–	–	–
$N_g = 2000$	T.C. : (10c) (12c), B(u) : (8a)							
p	1	2	4	8	10	25	50	100
#it	3	3	3	3	3	3	3	10
$log_{10}(\|\|F\|\|_2)$	−12.84	−10.58	−9.89	−8.96	−8.26	−6.00	−6.06	−6.06
$log_{10}(\|\|h\|\|_2)$	−7.43	−8.46	−7.05	−7.22	−6.10	−5.37	−5.27	−5.60
$log_{10}(min(\kappa_2(\bar{S}_\Gamma^{(i)})))$	–	–	4.98	5.25	5.25	5.28	6.33	5.90
$log_{10}(max(\kappa_2(\bar{S}_\Gamma^{(i)})))$	–	–	5.88	6.58	7.55	8.70	8.66	8.66
$log_{10}(min(\kappa_2(M)))$	–	0	1.67	2.25	2.55	4.49	6.36	4.65
$log_{10}(max(\kappa_2(M))$	–	0	2.27	2.84	3.84	5.19	7.96	5.91

$log_{10}(min/max(\kappa_2(\bar{S}_\Gamma^{(i)})))$ (respectively $log_{10}(min/max(\kappa_2(M)))$ refers to the minimum or maximum of the condition number of the local Schur complement of the time slices 2 to $p-1$ (respectively of the Lagrange multipliers system) over the Newton iterations

to one point. This lack of convergence is due to local Jacobian matrices that become singular because $g(t, y)$ is constant. However, T.C. (10c) (12c) lead to convergence in the same number of Newton iterations as for one time domain except for $p = 100$, where the condition number of local Schur complements increases.

5 Conclusions

We have extended the time domain decomposition that transforms the IVP into a BVP in order to introduce a Dual Schur complement inside the Newton method. This allows the Newton iterative solution to satisfy the continuity constraints at the time slices boundaries. Nevertheless, in this nonlinear framework the right transmission conditions for defining the local functions on time slices are those involving the flux even if the number of time slices that can be used reaches a limit due to the bad condition number of the local Schur complements.

Acknowledgements This work was supported by the French National Agency of Research through the project ANR MONU-12-0012 H2MNO4. This work was granted access to the HPC resources of CINES under the allocation 2014-c2014066099 made by GENCI (Grand Equipement National de Calcul Intensif) and used the HPC resources of Center for the Development of Parallel Scientific Computing (CDCSP) of University Lyon 1.

References

1. A. Bellen, M. Zennaro, Parallel algorithms for initial value problems for difference and differential equations. J. Comput. Appl. Math. **25**(3), 341–350 (1989)
2. E. Hairer, C. Lubich, G. Wanner, *Geometric Numerical Integration: Structure-Preserving Algorithms for Ordinary Differential Equations*. Springer Series in Computational Mathematics, 2nd edn, vol. 31 (Springer, Berlin, 2006)
3. H.B. Keller, Existence theory for two point boundary value problems. Bull. Am. Math. Soc. **72**, 728–731 (1966)
4. P. Linel, D. Tromeur-Dervout, Aitken-Schwarz and Schur complement methods for time domain decomposition, in *Parallel Computing: From Multicores and GPU's to Petascale*. Advances in Parallel Computing, vol. 19 (IOS Press, Amsterdam, 2010), pp. 75–82
5. P. Linel, D. Tromeur-Dervout, Une méthode de décomposition en temps avec des schémas d'intégration réversible pour la résolution de systèmes d'équations différentielles ordinaires. C. R. Math. Acad. Sci. Paris **349**(15–16), 911–914 (2011)
6. P. Linel, D. Tromeur-Dervout, Analysis of the time-Schwarz DDM on the heat PDE. Comput. Fluids **80**(0), 94–101 (2013)
7. P. Linel, D. Tromeur-Dervout, Nonlinear transmission conditions for time domain decomposition method, in *Domain Decomposition Methods in Science and Engineering XXI*, ed. by J. Erhel, M.J. Gander, L. Halpern, G. Pichot, T. Sassi, O. Widlund. Lecture Notes in Computational Science and Engineering, vol. 98 (Springer, New York, 2014), pp. 807–814

Additive Average Schwarz Method for a Crouzeix-Raviart Finite Volume Element Discretization of Elliptic Problems

Atle Loneland, Leszek Marcinkowski, and Talal Rahman

1 Introduction

In this paper we introduce an additive Schwarz method for a Crouzeix-Raviart Finite Volume Element (CRFVE) discretization of a second order elliptic problem with discontinuous coefficients, where the discontinuities are inside subdomains and across and along subdomain boundaries. For recent work addressing domain decomposition methods for such problems, cf. [7, 14] and references therein. Depending on the distribution of the coefficient in the model problem, the parameters describing the GMRES convergence rate of the proposed method depend linearly or quadratically on the mesh parameters H/h.

The CRFVE method was first introduced by Chatzipantelidis [4] and investigated further in [10].

Additive Schwarz Methods (ASM) for solving elliptic problems discretized by the finite element method have been studied thoroughly, cf. [13, 15], but ASMs for conforming FVE discretization have only been consider in [5, 16]. For the CR finite

A. Loneland (✉)
Department of Informatics, University of Bergen, 5020 Bergen, Norway
e-mail: Atle.Loneland@ii.uib.no

L. Marcinkowski
Faculty of Mathematics, University of Warsaw, Banacha 2, 02-097 Warszawa, Poland
e-mail: Leszek.Marcinkowski@mimuw.edu.pl

T. Rahman
Faculty of Engineering, Bergen University College, Nygrdsgaten 112, 5020 Bergen, Norway
e-mail: Talal.Rahman@hib.no

T. Dickopf et al. (eds.), *Domain Decomposition Methods in Science and Engineering XXII*, Lecture Notes in Computational Science and Engineering 104, DOI 10.1007/978-3-319-18827-0_60

element discretization, there exists several results for second order elliptic problems; cf. [1, 8, 9, 12]. In the CRFVE case, ASMs have not been studied.

2 Discrete Problem

We consider the following elliptic boundary value problem

$$\begin{aligned} -\nabla \cdot (\alpha(x)\nabla u) &= f \qquad \text{in } \Omega, \\ u &= 0 \qquad \text{on } \partial\Omega, \end{aligned} \tag{1}$$

where Ω is a bounded convex domain in $\mathbb{R}^2$ and $f \in L^2(\Omega)$. The coefficient $\alpha(x) > a_0 > 0$ has the property $\alpha \in W^{1,\infty}(D_j)$ with respect to a nonoverlapping partitioning of Ω into open, connected Lipschitz polytopes $\mathcal{D} := \{D_j : j = 1, \ldots, n\}$ such that $\bar{\Omega} = \bigcup_{j=1}^n \bar{D}_j$. We assume that the restriction of the coefficient α to D_j has the property $|\alpha|_{1,\infty,D_j} \leq C$ for all $j = 1, \ldots, n$, i.e., we assume that locally the coefficient is smooth and not too much varying. For simplicity of presentation we require that $\alpha \geq 1$. This last property can always be achieved by scaling (1).

3 The CRFVE Method

In this section we present the Crouzeix-Raviart finite element (CRFE) and finite volume (CRFVE) discretizations of a model second order elliptic problem with discontinuous coefficients inside and across prescribed substructures boundaries.

We assume that there exists another nonoverlapping partitioning of Ω into open, connected Lipschitz polytopes Ω_i such that $\overline{\Omega} = \bigcup_{i=1}^N \overline{\Omega}_i$. We also assume that these subdomains form a coarse triangulation of the domain which is shape regular as in [2]. We define the sets of Crouzeix-Raviart (CR) nodal points Ω_h^{CR}, $\partial\Omega_h^{CR}$, Ω_{ih}^{CR} and $\partial\Omega_{ih}^{CR}$ as the midpoints of edges of elements in T_h corresponding to Ω, $\partial\Omega$, Ω_i and $\partial\Omega_i$, respectively.

Now we introduce a quasi-uniform triangulation $\mathcal{T}_h$ of Ω consisting of closed triangle elements such that $\bar{\Omega} = \bigcup_{K\in\mathcal{T}_h} K$. Let h_K be the diameter of K and define $h = \max_{K\in\mathcal{T}_h} h_K$ as the largest diameter of the triangles $K \in \mathcal{T}_h$. We assume that the triangulation is defined in such way that ∂K's are aligned with ∂D_j's. This implies that the coefficient $\alpha(x)$ has the property that $\alpha \in W^{1,\infty}(K)$ for all $K \in \mathcal{T}_h$.

Using this triangulation $\mathcal{T}_h$ we may now introduce a dual mesh $\mathcal{T}_h^*$ with elements called control volumes. Let z_K be an interior point of $K \in \mathcal{T}_h$, we connect it with straight lines to the vertices of K such that K is partitioned into three subtriangles, K_e for each edge $e \in \partial K \cap \Omega$ interior to Ω. Denote this new finer triangulation of Ω by $\tilde{\mathcal{T}}_h$. With each edge e we associate a corresponding control volume b_e consisting

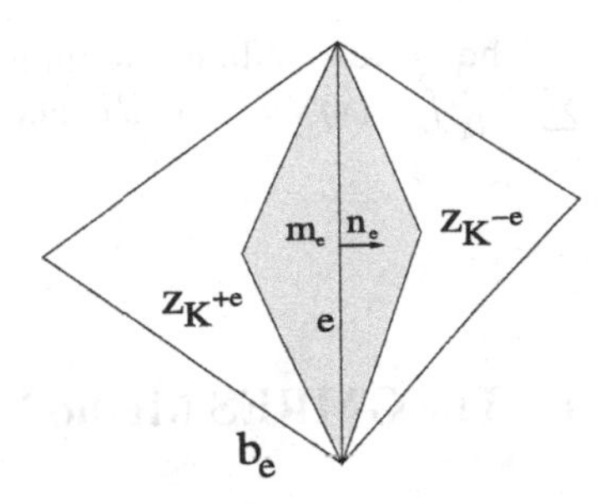

Fig. 1 The control volume b_e for an edge e which is the common edge to the triangles K^{+e} and K^{-e}. Here m_e is the midpoint of e, n_e normal unit vector to e, $z_{K^{+e}}$ and $z_{K^{-e}}$ are the interior points of the triangles K^{+e} and K^{-e} which share the edge e

of the two subtriangles of $\tilde{T}_h$ which have e as an common edge, cf Fig. 1. Define $T_h^* = \{b_e : e \in \mathrm{E}_h^{\mathrm{in}}\}$ to be the set of all such control volumes, where E_h^{in} is the set of all interior edges of the elements in T_h.

Let V_h be the nonconforming CR finite element space defined on the triangulation T_h,

$$V_h = V_h(\Omega) := \{u \in L^2(\Omega) : v_{|K} \in P_1, \quad K \in T_h \quad v(m) = 0 \quad m \in \partial\Omega_h^{CR}\},$$

and let V_h^* be its dual control volume space

$$V_h^* = V_h^*(\Omega) := \{u \in L^2(\Omega) : v_{|b_e} \in P_0, \quad b_e \in T_h^* \quad v(m) = 0 \quad m \in \partial\Omega_h^{CR}\}.$$

Obviously, $V_h^* = \mathrm{span}\{\chi_e(x) : e \in \mathrm{E}_h^{\mathrm{in}}\}$, where $\{\chi_e\}$ are the characteristic functions of the control volumes $\{b_e\}$. Now, let $I_h^* : V_h \to V_h^*$ be the standard interpolation operator, i.e.

$$I_h^* u = \sum_{e \in \mathrm{E}_h^{in}} u(m_e)\chi_e.$$

We may then define the CRFVE approximation u_h of (1) as the solution to the following problem: Find $u_h \in V_h$ such that

$$a_h(u_h, I_h^* v) = \left(f, I_h^* v\right), \qquad v \in V_h \tag{2}$$

where the bilinear form is defined as

$$a_h(u, v) = -\sum_{e \in \mathrm{E}_h^{in}} v(m_e) \int_{\partial b_e} \alpha(x) \nabla u \cdot \boldsymbol{n} \, ds \qquad u \in V_h, v \in V_h^*.$$

where $\boldsymbol{n}$ is the outward unit normal vector of the control volume b_e.

The corresponding CR finite element bilinear form is defined as: $a(u, v) = \sum_{K\in T_h} \int_K \alpha(x)\nabla u \cdot \nabla v \; dx$, and we define the energy norm induced by $a(\cdot,\cdot)$ as

$$\|\cdot\|_a = \sqrt{a(\cdot,\cdot)}. \tag{3}$$

4 The GMRES Method

The linear system of equations which arises from problem (2) is in general nonsymmetric. We may solve such a system using a preconditioned GMRES method; cf. [6, 11]. This method has proven to be quite powerful for a large class of nonsymmetric problems. The theory originally developed for $L^2(\Omega)$ in [6] can easily be extended to an arbitrary Hilbert space; see [3].

In this paper, we use GMRES to solve the linear system of equations

$$Tu_h = g, \tag{4}$$

where T is a nonsymmetric, nonsingular operator, $g \in V_h$ is the right hand side and $u_h \in V_h$ is the solution vector. The formulation of T will be given in the next section.

The main idea of the GMRES method is to solve a least square problem in each iteration, i.e. at step m we approximate the exact solution $u_h = T^{-1}g$ by a vector $u_m \in \mathcal{K}_m$ which minimizes the a-norm (energy norm) of the residual, cf. (3), where $\mathcal{K}_m$ is the m-th Krylov subspace defined as $\mathcal{K}_m = \text{span}\left\{r_0, Tr_0, \cdots T^{m-1}r_0\right\}$ and $r_0 = g - Tu_0$. In other words, z_m solves

$$\min_{z\in\mathcal{K}_m} \|g - T(u_0 + z)\|_a.$$

Thus, the m-th iterate is $u_m = u_0 + z_m$.

The convergence rate of the GMRES method is usually expressed in terms of the following two parameters

$$c_p = \inf_{u\neq 0} \frac{a(Tu, u)}{\|u\|_a^2} \text{ and } C_p = \sup_{u\neq 0} \frac{\|Tu\|_a}{\|u\|_a}.$$

The decrease of the norm of the residual in a single step is described in the next theorem.

Theorem 1 (Eisenstat-Elman, Schultz) *If $c_p > 0$, then the GMRES method converges and after m steps, the norm of the residual is bounded by*

$$\|r_m\|_a \leq \left(1 - \frac{c_p^2}{C_p^2}\right)^{m/2} \|r_0\|_a, \tag{5}$$

where $r_m = g - Tu_m$.

5 An Additive Average Method

In this section we introduce the additive Schwarz method for the discrete problem (2) and provide bounds on the convergence rate, both for the cases of symmetric and nonsymmetric preconditioners.

5.1 *Decomposition of $V_h(\Omega)$*

We decompose the original space into

$$V_h(\Omega) = V_0(\Omega) + V_1(\Omega) + \cdots + V_N(\Omega), \tag{6}$$

where for $i = 1, \ldots, N$ we have defined $V_i(\Omega)$ as the restriction of $V_h(\Omega)$ to Ω_i with functions vanishing on $\partial\Omega_{ih}^{CR}$ and as well as on the other subdomains. The coarse space $V_0(\Omega)$ is defined as the range of the interpolation operator I_A. For $u \in V_h(\Omega)$, we let $I_A u \in V_h(\Omega)$ be defined as

$$I_A u := \begin{cases} u(x), & x \in \partial\Omega_{ih}^{CR} \\ \bar{u}_i, & x \in \Omega_{ih}^{CR} \end{cases} \qquad i = 1, \ldots, N, \tag{7}$$

where

$$\bar{u}_i := \frac{1}{n_i} \sum_{x \in \partial\Omega_{ih}^{\mathrm{CR}}} u(x). \tag{8}$$

Here n_i is the number of nodal points of $\partial\Omega_{ih}^{CR}$.

We also assume that $\mathcal{T}_h(\Omega_i)$ inherits the shape regular and quasi-uniform triangulation for each Ω_i with mesh parameters h_i and $H_i = diam(\Omega_i)$. The layer along $\partial\Omega_i$ consisting of unions of triangles $K \in \mathcal{T}(\Omega_i)$ which touch $\partial\Omega_i$ is denoted as Ω_i^δ. Corresponding to each layer we define the maximum and minimum values of the coefficient α as

$$\overline{\alpha}_i := \sup_{x \in \bar{\Omega}_i^\delta} \alpha(x) \qquad \text{and} \qquad \underline{\alpha}_i := \inf_{x \in \bar{\Omega}_i^\delta} \alpha(x),$$

respectively.

For $i = 0, \ldots, N$ we define the two types of projection like operators $T_i^{(k)} \colon V_h \to V_i(\Omega)$, $k = 1, 2$ as

$$a(T_i^{(1)} u, v) = a_h(u, I_h^* v) \qquad \forall v \in V_i(\Omega), \tag{9}$$

for the symmetric preconditioner, and

$$a_h(T_i^{(2)}u, v) = a_h(u, I_h^* v) \qquad \forall v \in V_i(\Omega), \tag{10}$$

for the non-symmetric preconditioner. Each of these problems have a unique solution. We now introduce

$$T_A^{(k)} := T_0^{(k)} + T_1^{(k)} + \cdots + T_N^{(k)}, \qquad k = 1, 2, \tag{11}$$

which allow us to replace the original problem, respectively for $k = 1$ and $k = 2$, by the equation

$$T_A^{(k)} u = g^{(k)}, \tag{12}$$

where $g^{(k)} = \sum_{i=0}^{N} g_i$ and $g_i^{(k)} = T_i^{(k)} u$. Note that $g_i^{(k)}$ may be computed without knowing the solution u of (2).

Theorem 2 *There exists $h_0 > 0$ such that for all $h < h_0$, $k = 1, 2$, and $u \in V_h$*

$$\|T^{(k)}u\|_a \le C\|u\|_a,$$
$$a(T^{(k)}u, u) \ge c \max_i \frac{\overline{\alpha}_i}{\underline{\alpha}_i} \left(\frac{H_i}{h_i}\right)^{-2} a(u, u),$$

where C, c are positive constants independent of α, $\frac{\overline{\alpha}_i}{\underline{\alpha}_i}$, h_i and H_i for $i = 1, \ldots, N$.

For certain distributions of α we may improve the estimate.

Proposition 1 *There exists $h_0 > 0$ such that for all $h < h_0$, $u \in V_h$ and $\underline{\alpha}_i \le \alpha(x)$ in $\Omega_i \setminus \Omega_i^\delta$*

$$\|T^{(k)}u\|_a \le C\|u\|_a,$$
$$a(T^{(k)}u, u) \ge c \max_i \frac{\overline{\alpha}_i}{\underline{\alpha}_i} \left(\frac{H_i}{h_i}\right)^{-1} a(u, u) \qquad \forall u \in V_h,$$

where C, c are positive constants independent of α, $\frac{\overline{\alpha}_i}{\underline{\alpha}_i}$, h_i and H_i for $i = 1, \ldots, N$.

6 Numerical Results

In this section we present some preliminary numerical results for the proposed method with the symmetric preconditioner, i.e. for $k = 1$ in (12). All experiments are done for problem (1) on a unit square domain $\Omega = (0, 1)^2$. The coefficient α is

Table 1 Number of iterations until convergence for the solution of (1) for different values of α_1 in the distributions of the coefficient α given in Fig. 2a, b

α_1	1e0	1e1	1e2	1e3	1e4	1e5	1e6
Problem 1	40	40	40	40	40	40	40
Problem 2	40	66	108	177	233	276	316

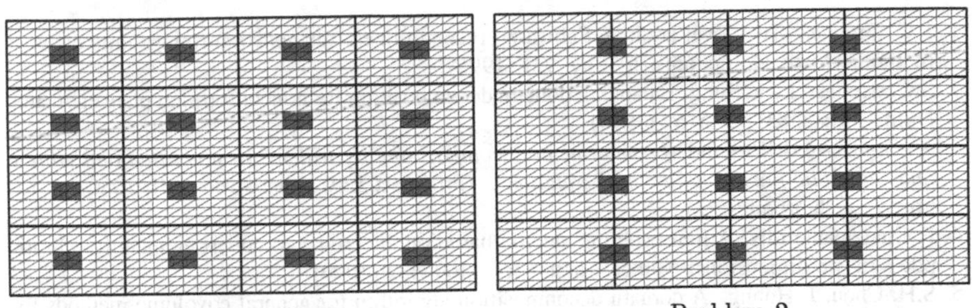

Fig. 2 Distributions of α corresponding to Problem 1 and Problem 2

equal to $2 + \sin(100\pi x)\sin(100\pi y)$ except for the areas marked with red where α equals $\alpha_1(2 + \sin(100\pi x)\sin(100\pi y))$. The right hand side is chosen as $f = 1$.

The numerical solution is found by using the Generalized minimal residual method (GMRES). We run the method until the l_2 norm of the residual is reduced by a factor 10^6, i.e., as soon as $\|r_i\|_2/\|r_0\|_2 \leq 10^{-6}$. For each of the problems under consideration the number of iterations until convergence for different values of α_1 are shown in Table 1.

The numerical results from our two examples shows that the performance of the method agrees with the theory. If the inclusions are in the interior of the subdomains the method is completely insensitive to any discontinuities in the coefficient, while if the inclusions are on the subdomain layer the method depends strongly on the jumps in the coefficient.

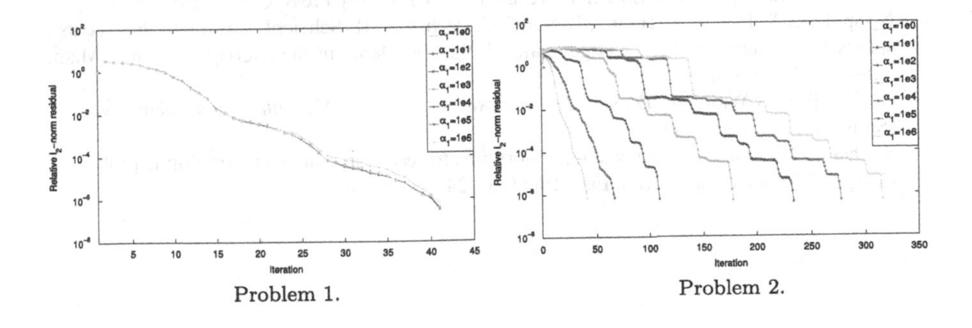

Acknowledgement This work was partially supported by Polish Scientific Grant 2011/01/B/ST1/01179 (Leszek Marcinkowski).

References

1. S.C. Brenner, Two-level additive Schwarz preconditioners for nonconforming finite element methods. Math. Comp. **65**(215), 897–921 (1996)
2. S.C. Brenner, L.-Y. Sung, Balancing domain decomposition for nonconforming plate elements. Numer. Math. **83**(1), 25–52 (1999)
3. X.-C. Cai, O.B Widlund, *Some domain decomposition algorithms for nonselfadjoint elliptic and parabolic partial differential equations*, (ProQuest LLC, Ann Arbor, MI, 1989), p. 82. http://search.proquest.com
4. P. Chatzipantelidis, A finite volume method based on the crouzeix–raviart element for elliptic pde's in two dimensions. Numer. Math. **82**(3), 409–432 (1999)
5. S.H. Chou, J. Huang, A domain decomposition algorithm for general covolume methods for elliptic problems. J. Numer. Math. jnma **11**(3), 179–194 (2003)
6. S.C Eisenstat, H.C Elman, M.H Schultz, Variational iterative methods for nonsymmetric systems of linear equations. SIAM J. Numer. Anal. **20**(2), 345–357 (1983)
7. J. Galvis, Y. Efendiev, Domain decomposition preconditioners for multiscale flows in high contrast media: reduced dimension coarse spaces. Multiscale Model. Simul. **8**(5), 1621–1644 (2010).
8. L. Marcinkowski, The mortar element method with locally nonconforming elements. BIT Numer. Math. **39**(4), 716–739 (1999)
9. T. Rahman, X. Xu, R. Hoppe, Additive schwarz methods for the crouzeix-raviart mortar finite element for elliptic problems with discontinuous coefficients. Numer. Math. **101**(3), 551–572 (2005)
10. H. Rui, C. Bi, Convergence analysis of an upwind finite volume element method with crouzeix-raviart element for non-selfadjoint and indefinite problems. Front. Math. China **3**(4), 563–579 (2008)
11. Y. Saad, M.H Schultz, Gmres: a generalized minimal residual algorithm for solving nonsymmetric linear systems. SIAM J. Sci. Stat. Comput. **7**(3), 856–869 (1986)
12. M. Sarkis, Nonstandard coarse spaces and schwarz methods for elliptic problems with discontinuous coefficients using non-conforming elements. Numer. Math. **77**(3), 383–406 (1997)
13. B. Smith, P. Bjorstad, W. Gropp, *Domain Decomposition: Parallel Multilevel Methods for Elliptic Partial Differential Equations* (Cambridge University Press, Cambridge, 1996)
14. N. Spillane, V. Dolean, P. Hauret, F. Nataf, C. Pechstein, R. Scheichl, Abstract robust coarse spaces for systems of PDEs via generalized eigenproblems in the overlaps. Numer. Math. **126**(4), 741–770 (2014)
15. A. Toselli, O.B Widlund, *Domain Decomposition Methods: Algorithms and Theory*, vol. 34 (Springer, New York, 2005)
16. S. Zhang, On domain decomposition algorithms for covolume methods for elliptic problems. Comput. Methods Appl. Mech. Eng. **196**(1–3), 24–32 (2006)

Schwarz Methods for a Crouzeix-Raviart Finite Volume Discretization of Elliptic Problems

Leszek Marcinkowski, Atle Loneland, and Talal Rahman

1 Introduction

In this paper, we present two variants of the Additive Schwarz Method (ASM) for a Crouzeix-Raviart finite volume (CRFV) discretization of the second order elliptic problem with discontinuous coefficients, where the discontinuities are only across subdomain boundaries. The resulting system, which is nonsymmetric, is solved using the preconditioned GMRES iteration, where in one variant of the ASM the preconditioner is symmetric while in the other variant it is nonsymmetric. The proposed methods are almost optimal, in the sense that the convergence of the GMRES iteration, in the both cases, depend only poly-logarithmically on the mesh parameters.

In the CRFV method, the equations are discretized on a mesh which is dual to a primal mesh where the nonconforming Crouzeix-Raviart finite element space is defined, it is the space in which we seek for an approximation of the solution, cf. [4].

There are many results concerning Additive Schwarz Methods (ASM) for solving symmetric systems, those arising from the finite element discretization of second order elliptic problems, cf. e.g. [12], but only a few papers that consider the FV discretization using the standard finite element space, cf. [5, 13]. There is also a

L. Marcinkowski (✉)
Faculty of Mathematics, University of Warsaw, Banacha 2, 02-097 Warszawa, Poland
e-mail: Leszek.Marcinkowski@mimuw.edu.pl

A. Loneland • T. Rahman
Department of Computing, Mathematics, and Physics, Bergen University College, Nygårdsgaten 112, N-5020 Bergen, Norway
e-mail: Atle.Loneland@hib.no; Talal.Rahman@hib.no

T. Dickopf et al. (eds.), *Domain Decomposition Methods in Science and Engineering XXII*, Lecture Notes in Computational Science and Engineering 104, DOI 10.1007/978-3-319-18827-0_61

number of results focused on iterative methods for the CR finite element for second order problems; cf. [1, 9, 11].

The purpose of this paper is to construct two parallel algorithms based on the edge based discrete space decomposition in the abstract Schwarz scheme. The algorithms are very similar in application.

We present almost optimal estimates for the convergence of the GMRES iteration applied to the preconditioned system, showing that the minimum eigenvalue of the preconditioned operator in the estimate, grows like $(1+\log(H/h))^{-2}$, where H is the maximal diameter of the subdomains and h is the fine mesh size parameter. Some preliminary results of numerical tests are also presented.

2 Discrete Problem

In this section we present the Crouzeix-Raviart finite element (CRFE) and finite volume (CRFV) discretizations of a model second order elliptic problem with discontinuous coefficients across prescribed substructures boundaries.

Let Ω be a polygonal domain in the plane. We assume that there exists a partition of Ω into disjoint polygonal subdomains Ω_k such that $\overline{\Omega} = \bigcup_{k=1}^{N} \overline{\Omega}_k$ with $\overline{\Omega}_k \cap \overline{\Omega}_l$ being an empty set, an edge or a vertex (crosspoint). We also assume that these subdomains form a coarse triangulation of the domain which is shape regular as in [2]. We introduce a global interface $\Gamma = \bigcup_i \partial\Omega_i \setminus \partial\Omega$ which plays an important role in our study.

Our model differential problem is to find u^* such that

$$\begin{aligned} -\nabla A(x)\nabla u^*(x) &= f(x) \qquad x \in \Omega \\ u^*(s) &= 0 \qquad s \in \partial\Omega, \end{aligned} \tag{1}$$

where $A(x)$ is the symmetric coefficients matrix.

The standard variational (weak) formulation is to find $u^* \in H_0^1(\Omega)$ such that $a(u^*, v) = \int_\Omega f v \, dx$ for all $v \in H_0^1(\Omega)$, where $f \in L^2(\Omega)$, and $a(u, v) = \sum_{k=1}^{N} \int_{\Omega_k} \nabla u^T A(x) \nabla v \, dx$. We assume that the restriction of the symmetric coefficients matrices to Ω_k: $A_k = A_{|\Omega_k}$ is in $W^{1,\infty}(\Omega_k)$ and bounded and positive definite, i.e.

$$\exists \alpha_k > 0 \ \forall x \in \Omega_k \ \forall \xi \in \mathbb{R}^2 \quad \xi^T A(x) \xi \geq \alpha_k |\xi|^2, \tag{2}$$

$$\exists M_k > 0 \ \forall x \in \Omega_k \ \forall \xi, \mu \in \mathbb{R}^2 \quad \mu^T A(x) \xi \leq M_k |\mu||\xi|. \tag{3}$$

Here $|\xi| = \sqrt{\xi^T \xi}$. We can always scale the matrix functions A in such a way that all $\alpha_k \geq 1$. Thus we assume that the restriction of the coefficient matrices to Ω_k: $A_k = A_{|\Omega_k}$ is in $W^{1,\infty}(\Omega_k)$ with the following bounds: $\|A_k\|_{W^{1,\infty}(\Omega_k)} \leq C$, and $M_k \leq C_e \alpha_k$, i.e. we assume that the coefficient matrix locally is smooth, isotropic and not too much varying.

We assume that there exists a sequence of quasiuniform triangulations: $T_h = T_h(\Omega) = \{\tau\}$, of Ω such that any element τ of T_h is contained in only one subdomain, as a consequence any subdomain Ω_k inherits a sequence of local triangulations: $T_h(\Omega_k) = \{\tau\}_{\tau\subset\Omega_k,\tau\in T_h}$.

Let $h = \max_{\tau\in T_h(\Omega)} \operatorname{diam}(\tau)$ be the mesh size parameter of the triangulation. We introduce the following sets of Crouzeix-Raviart (CR) nodal points or nodes: let $\Omega_h^{CR}, \partial\Omega_h^{CR}, \Omega_{k,h}^{CR}, \partial\Omega_{k,h}^{CR}, \Gamma_h^{CR}$, and $\Gamma_{kl,h}^{CR}$ be the midpoints of edges of elements in T_h which are on $\Omega, \partial\Omega, \Omega_k, \partial\Omega_k, \Gamma$, and Γ_{kl}, respectively. Here Γ_{kl} is an interface, an open edge, which is shared by the two subdomains, Ω_k and Ω_l. Note that $\Gamma_h^{CR} = \bigcup_{\Gamma_{kl}\subset\Gamma} \Gamma_{kl,h}^{CR}$. Now we define a dual triangulation T_h^* to the initial one. For an edge e of an element not on $\partial\Omega$ i.e. being the common edge of two elements τ_1 and τ_2 i.e. $e = \partial\tau_1 \cap \partial\tau_2$ we introduce two triangles: $V_k \subset \tau_k$ obtained by connecting the ends of e to the centroid (barycenter) of τ_k for $k = 1, 2$. Then, let the control volume $b_e = V_1 \cup e \cup V_2$, cf. Fig. 1. For an edge of an element τ contained in $\partial\Omega$ let the control volume be the triangle V obtained analogously i.e. by connecting the ends of e with the centroid of τ. Then let $T_h^* = \{b_e\}_{e\in E_h}$, where E_h is the set of all edges of elements in T_h.

Next we introduce two discrete spaces contained in $L^2(\Omega)$:

$$V_h := \{v \in L^2(\Omega) : v_{|\tau} \in P_1, \quad \tau \in T_h \quad v(m) = 0 \quad m \in \partial\Omega_h^{CR}\},$$
$$V_h^* := \{v \in L^2(\Omega) : v_{|b_e} \in P_0, \quad b_e \in T_h^* \quad v(m) = 0 \quad m \in \partial\Omega_h^{CR}\}.$$

The first space is the classical nonconforming Crouzeix-Raviart finite element space, cf. Fig. 2, and the second space is the space of piecewise constant functions which are zero on the boundary of the domain.

Let $\{\phi_m\}_{m\in\Omega_h^{CR}}$ be the standard CR nodal basis of V^h and $\{\psi_m\}_{m\in\Omega_h^{CR}}$ be the standard basis of V_h^* consisting of characteristic functions of the control volumes.

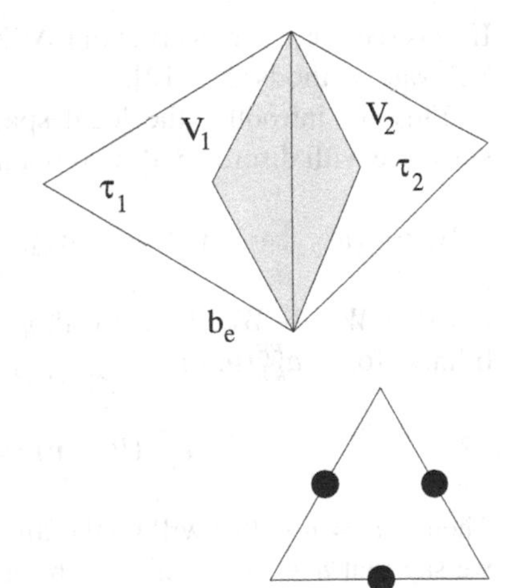

Fig. 1 The control volume b_e for an edge e which is the common edge to the triangles τ_1 and τ_2

Fig. 2 Edge midpoint corresponding to the degrees of freedom of the nonconforming Crouzeix-Raviart element

We also introduce two interpolation operators, I_h and I_h^*, defined for any function that has properly defined and unique values at each midpoint $m \in \Omega_h^{CR}$:

$$I_h(u) = \sum_{m \in \Omega_h^{CR}} u(m)\phi_m, \qquad I_h^*(u) = \sum_{m \in \Omega_h^{CR}} u(m)\psi_m.$$

Note that $I_h I_h^* u = u$ for any $u \in V_h$ and $I_h^* I_h u = u$ for any $u \in V_h^*$. We also define a nonsymmetric in general bilinear form $a_h : V_h \times V_h^* \to \mathbb{R}$:

$$a_h^{CRFV}(u, v) = -\sum_{e \in E_h^{in}} v(m_e) \int_{\partial b_e} \mathbf{n}^T A(s) \nabla u \, ds, \tag{4}$$

where $\mathbf{n}$ is a normal unit vector outer to ∂b_e, m_e is the median (midpoint) of the edge e and $E_h^{in} \subset E_h$ is the set of all interior edges, i.e. those which are not on $\partial\Omega$.

Then our discrete CRFV problem is to find $u^* \in V_h$ such that:

$$a_h^{FV}(u^*, v) = f(I_h^* v) \qquad \forall v \in V_h \tag{5}$$

for $a_h^{FV}(u, v) := a_h^{CRFV}(u, I_h^* v)$. In general the problem is nonsymmetric unless the coefficients matrix is a piecewise constant matrix over T_h. One can prove that there exists $h_0 > 0$ such that for all $h \leq h_0$ the form $a_h^{FV}(u, v)$ is positive definite over V_h. Thus this problem has a unique solution. Some error estimates are also proven, cf. [7] or [4] in the case of the smooth coefficients.

3 Additive Schwarz Method

In this section, we construct our ASM based on the abstract framework for additive Schwarz methods, see [12].

First we introduce the local spaces being the restriction of V_h to $\overline{\Omega}_k$ and its subspace with discrete CR zero boundary conditions:

$$W_k := \{v_{|\overline{\Omega}_k} : v \in V_h\}, \qquad W_{k,0} := \{v \in W_k : w(m) = 0 \quad m \in \partial\Omega_{k,h}^{CR}\} \subset W_k.$$

Let $P_k : W_k \to W_{k,0}$ be the orthogonal projection onto $W_{k,0}$ in terms of the local bilinear form: $a_{k,h}^{FE}(u, v) = \sum_{\tau \in T_h(\Omega_k)} \int_\tau \nabla u^T A \nabla v \, dx$, i.e.

$$a_{k,h}^{FE}(P_k u, v) = a_{k,h}^{FE}(u, v) \quad \forall v \in W_{k,0}.$$

Then $H_k u = u - P_k u$ will be the discrete harmonic part of $u \in W_k$. If $u = H_k u$ then we say that $u \in W_k$ is discrete harmonic. A function $u \in V_h$ is discrete harmonic if its all restrictions to subdomains are discrete harmonic i.e. $u_{|\Omega_k} = H_k u_{|\Omega_k}$ for $k = 1, \ldots, N$. We also define an edge function $\theta_{\Gamma_{kl}} \in V_h$ as a discrete harmonic function such that it is equal to one at CR nodes interior to Γ_{kl} and zero at all other CR nodes on the interface.

We now define the decomposition of V_h. Let $V_0 = \text{Span}(\theta_{\Gamma_{kl}})_{\Gamma_{kl}\subset\Gamma}$ be the coarse space, V_{kl} be the edge space associated with the interface Γ_{kl} formed by discrete harmonic functions that are zero at each $x \in \Gamma_h^{CR} \setminus \Gamma_{kl,h}^{CR}$. Finally let V_k be the space $W_{k,0}$ extended by zero to all remaining subdomains. Thus we have the following decomposition: $V_h = V_0 + \sum_{\Gamma_{kl}\subset\Gamma} V_{kl} + \sum_{k=1}^{N} V_k$. Note that this is a direct sum and that the subspace $V_0 + \sum_{\Gamma_{kl}\subset\Gamma} V_{kl}$ is $a_h^{FE}(u,v) = \sum_k a_{k,h}^{FE}(u,v)$ orthogonal to $\sum_{k=1}^{N} V_k$. Now we define the first type of projection like operators: the coarse and the local operators, $T_k^{sym} : V_h \to V_k$, as

$$a_h^{FE}(T_k^{sym}u, v) = a_h^{FV}(u, v) \qquad \forall v \in V_k, \quad k = 0, 1, \ldots, N,$$

the edge related operators, $T_{kl}^{sym} : V_h \to V_{kl}$, as

$$a_h^{FE}(T_{kl}^{sym}u, v) = a_h^{FV}(u, v) \qquad \forall v \in V_{kl}, \quad \Gamma_{kl} \subset \Gamma.$$

Note that $T_k^{sym}u$ can be computed by solving a symmetric local discrete CRFE Dirichlet problem and then extended by zero to the other subdomains.

The second type of operators is based solely on the nonsymmetric bilinear form $a_h^{FV}(u, v)$. We define the coarse and the local operators, $T_k^{nsym} : V_h \to V_k$, as

$$a_h^{FV}(T_k^{nsym}u, v) = a_h^{FV}(u, v) \qquad \forall v \in V_k, \quad k = 0, 1, \ldots, N,$$

and the edge related operators, $T_{kl}^{nsym} : V_h \to V_{kl}$, as

$$a_h^{FV}(T_{kl}^{nsym}u, v) = a_h^{FV}(u, v) \qquad \forall v \in V_{kl}, \quad \Gamma_{kl} \subset \Gamma.$$

We define the two ASM operators as follows:

$$T^{type} := \sum_{\Gamma_{kl}\subset\Gamma} T_{kl}^{type} + \sum_{k=0}^{N} T_k^{type},$$

where the super-index *type* is either *sym* or *nsym*. We can replace our discrete CRFV equation (5) by the following system:

$$T^{type}u_h^* = g^{type}, \tag{6}$$

where $g^{type} = g_0^{type} + \sum_{\Gamma_{kl}\subset\Gamma} g_{kl}^{type} + \sum_{k=1}^{N} g_k^{type}$, $g_0^{type} = T_0^{type}u_h^*$, $g_{kl}^{type} = T_{kl}^{type}u_h^*$ $g_k^{type} = T_k^{type}u_h^*$, and $type \in \{sym, nsym\}$.

We apply the GMRES method in the inner product $a_h^{FE}(u, v)$, to the new system (6), and get the the following estimate (see [6] for the case of standard l_2 inner product, and [3] for the general case):

$$\|g - T^{type} u_j\|_a \leq \left(1 - \frac{\alpha_{min}^2}{\alpha_{max}^2}\right)^{j/2} \|g - T^{type} u_0\|_a. \tag{7}$$

where $\alpha_{min} = \min_{u \in V_h \setminus \{0\}} \frac{a_h^{FE}(T^{type} u, u)}{\|u\|_a^2}$ and $\alpha_{max} = \max_{u \in V_h \setminus \{0\}} \frac{\|T^{type} u\|_a}{\|u\|_a}$, $\|v\|_a := \sqrt{a_h^{FE}(v, v)}$, and T^{type} is either T^{sym} or T^{nsym}.

Next, we present the main theoretical result of this paper, namely an estimate of the convergence rate of the GMRES method, which is the same for both preconditioned systems (6). The proof of this theorem is an extension of the proof in [10] to the case of CRFV and will be published in [8].

Theorem 1 *There exists $h_0 > 0$ such that for all $h < h_0$ and $u \in V_h$*

$$\|T^{type} u\|_a \leq C\|u\|_a, \qquad a^{FE}(T^{type} u, u) \geq c\left(1 + \log\left(\frac{H}{h}\right)\right)^{-2} \|u\|_a^2$$

where T^{type} is either T^{sym} or T^{nsym}, C and c are positive constants independent of h, $H = \max_{k=1,\dots,N} diam(\Omega_k)$, and the magnitudes of α_k and M_k, but they depend on $\frac{M_k}{\alpha_k} \leq C_e$, cf. (2)–(3).

This theorem together with (7) gives as an estimate of the rate of convergence of the GMRES iteration for the two cases showing that the rates slow down very slowly—poly-logarithmically.

4 Numerical Results

In this section, we present some preliminary numerical results for the proposed method. All experiments are done for the symmetric preconditioner, that is for T^{sym}, but we expect a similar performance for T^{nsym}. In all cases Ω is a unit square domain. The coefficient A is equal to $2 + \sin(100\pi x)\sin(100\pi y)$, except for areas (subdomains) marked with red where A equals $\alpha_1(2 + \sin(100\pi x)\sin(100\pi y))$ with α_1 being a parameter (cf. Fig. 3 and Table 1). The right hand side is chosen as $f = 1$. The numerical solution is found by using the generalized minimal residual method (GMRES) (Fig. 4).

For the paper, we consider two test problems as shown in Fig. 3. We run the method until the l_2 norm of the residual is reduced by a factor of 10^6, that is when $\|r_i\|_2/\|r_0\|_2 \leq 10^{-6}$. Number of iterations, for the problems under consideration,

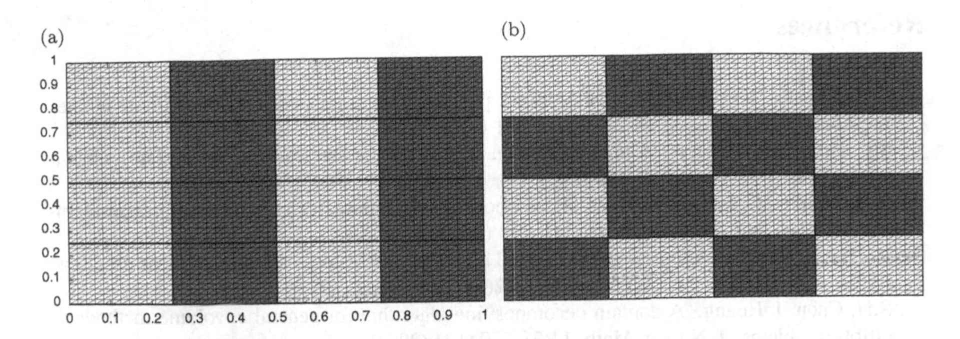

Fig. 3 Test problems (**a**) 1 and (**b**) 2. Regions (subdomains) marked with *red* are where A depends on α_1. Fine mesh consists of 48×48 *rectangular blocks*, while coarse mesh consists of 4×4 rectangular subdomains

Table 1 Number of GMRES iterations until convergence for the solution of (5), with different values of α_1 describing the coefficient A in the red regions, cf. Fig. 3a, b

α_1	1e0	1e1	1e2	1e3	1e4	1e5	1e6
Problem 1	18	26	26	27	27	27	28
Problem 2	18	21	22	22	22	23	24

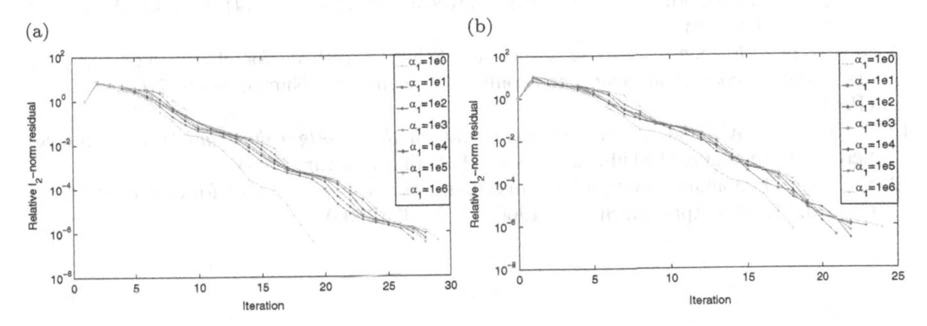

Fig. 4 Relative residual norms for GMRES minimizing the A-norm for different values of α_1. (**a**) Problem 1. (**b**) Problem 2

for different values of α_1, are shown in Table 1. The results show that the methods are robust for the present distribution of the coefficients, and supports our theory.

Acknowledgements This work was partially supported by Polish Scientific Grant 2011/01/B/ST1/01179 and Chinese Academy of Science Project: 2013FFGA0009 - GJHS20140901004635677.

References

1. S.C. Brenner, Two-level additive Schwarz preconditioners for nonconforming finite element methods. Math. Comput. **65**(215), 897–921 (1996)
2. S.C. Brenner, L.-Y. Sung, Balancing domain decomposition for nonconforming plate elements. Numer. Math. **83**(1), 25–52 (1999)
3. X.-C. Cai, O.B. Widlund, Domain decomposition algorithms for indefinite elliptic problems. SIAM J. Sci. Stat. Comput. **13**(1), 243–258 (1992)
4. P. Chatzipantelidis, A finite volume method based on the Crouzeix-Raviart element for elliptic PDE's in two dimensions. Numer. Math. **82**(3), 409–432 (1999)
5. S.H. Chou, J. Huang, A domain decomposition algorithm for general covolume methods for elliptic problems. J. Numer. Math. **11**(3), 179–194 (2003)
6. S.C. Eisenstat, H.C. Elman, M.H. Schultz, Variational iterative methods for nonsymmetric systems of linear equations. SIAM J. Numer. Anal. **20**(2), 345–357 (1983)
7. A. Loneland, L. Marcinkowski, T. Rahman, Additive average Schwarz method for the Crouzeix-Raviart finite volume element discretization of elliptic problems (2014a, submitted)
8. A. Loneland, L. Marcinkowski, T. Rahman, Edge based Schwarz methods for the Crouzeix-Raviart finite volume element discretization of elliptic problems (2014b, to appear in ETNA in 2015)
9. L. Marcinkowski, T. Rahman, Neumann-Neumann algorithms for a mortar Crouzeix-Raviart element for 2nd order elliptic problems. BIT Numer. Math. **48**(3), 607–626 (2008)
10. L. Marcinkowski, T. Rahman, J. Valdman, Additive Schwarz preconditioner for the general finite volume element discretization of symmetric elliptic problems. Tech. Report 204, Institute of Applied Mathematics and Mechanics, University of Warsaw (2014) [Published online in arXiv:1405.0185] [math.NA]
11. M. Sarkis, Nonstandard coarse spaces and Schwarz methods for elliptic problems with discontinuous coefficients using non-conforming elements. Numer. Math. **77**(3), 383–406 (1997)
12. A. Toselli, O. Widlund, *Domain Decomposition Methods—Algorithms and Theory*. Springer Series in Computational Mathematics, vol. 34 (Springer, Berlin, 2005)
13. S. Zhang, On domain decomposition algorithms for covolume methods for elliptic problems. Comput. Methods Appl. Mech. Eng. **196**(1–3), 24–32 (2006)

Preconditioning of the Reduced System Associated with the Restricted Additive Schwarz Method

François Pacull and Damien Tromeur-Dervout

It is of interest to solve large scale sparse linear systems on distributed computers, using Krylov subspace methods along with domain decomposition methods. If accurate subdomain solutions are used, the restricted additive Schwarz preconditioner allows a reduction to the interface via the Schur complement, which leads to an unpreconditioned reduced operator for the interface unknowns. Our purpose is to form a preconditioner for this interface operator by approximating it as a low-rank correction of the identity matrix. To this end, we use a sequence of orthogonal vectors and their image under the interface operator, which are both available after some iterations of the generalized minimal residual method.

The framework of study is purely algebraic and general real sparse nonsymmetric and indefinite matrices are considered. The linear system to solve is:

$$Au = f \tag{1}$$

with $A \in \mathbb{R}^{n \times n}$, $u \in \mathbb{R}^n$ and $f \in \mathbb{R}^n$.

Next, we set up the classical notations and terminologies from the algebraic Schwarz literature.

F. Pacull (✉) • D. Tromeur-Dervout
Institut Camille Jordan CNRS UMR 5208, Université Lyon 1, Université de Lyon, 43 boulevard du 11 novembre 1918, 69622 Villeurbanne Cedex, France
e-mail: francois.pacull@univ-lyon1.fr; fpacull@hotmail.com; damien.tromeur-dervout@univ-lyon1.fr

T. Dickopf et al. (eds.), *Domain Decomposition Methods in Science and Engineering XXII*, Lecture Notes in Computational Science and Engineering 104, DOI 10.1007/978-3-319-18827-0_62

1 Notations

We denote by $\mathcal{V} = \{1, \cdots, n\}$ the set of vertices and by $\mathcal{E}$ the set of edges of the connectivity graph of A: $\mathcal{G} \equiv \mathcal{G}(A) = (\mathcal{V}, \mathcal{E})$. In the present work, we assume that the structure of A is not too far from being symmetric, which is common for matrices issued from partial differential equations. For this reason, edges from $\mathcal{E}$ are made of unordered pairs of vertices from $\mathcal{V}$, and the graph $\mathcal{G}$ is said to be unoriented: given two vertices i and j from $\mathcal{V}$, the edge (i,j) belongs to $\mathcal{E}$ if and only if $A_{i,j} \neq 0$ or $A_{j,i} \neq 0$.

Given a subset $\mathcal{S} \subset \mathcal{V}$, the induced subgraph $\mathcal{G}|_{\mathcal{S}}$ consists of the vertices $\mathcal{S}$ and the edges $\mathcal{E}|_{\mathcal{S}} = \left\{(i,j) \in \mathcal{E} \;/\; (i,j) \in \mathcal{S}^2\right\} \subset \mathcal{E}$.

Two vertices are said to be adjacent if they share an edge in $\mathcal{E}$. Given a subset $S \subset \mathcal{V}$, the adjacent set $adj(S)$ contains all the vertices that are adjacent to at least one vertex of $\mathcal{S}$, but which do not belong to $\mathcal{S}$.

This allows the definition of overlapping and non-overlapping partitions of $\mathcal{V}$, as used by the algebraic Schwarz preconditioners.

A set $\mathcal{P}_0 = \{\mathcal{V}_{i,0}\}_{1 \leqslant i \leqslant p}$ of subsets $\mathcal{V}_{i,0} \subset \mathcal{V}$ is called a *non-overlapping partition* of $\mathcal{V}$ if:

- no element of $\mathcal{P}_0$ is empty,
- the elements of $\mathcal{P}_0$ are pairwise disjoint,
- the union of the elements of $\mathcal{P}_0$ is equal to $\mathcal{V}$.

A set $\mathcal{P}_\Delta = \{\mathcal{V}_{i,\Delta}\}_{1 \leqslant i \leqslant p}$ of subsets $\mathcal{V}_{i,\Delta} \subset \mathcal{V}$ is called an *overlapping partition* of $\mathcal{V}$ associated with the non-overlapping partition $\mathcal{P}_0$ if, for $1 \leqslant i \leqslant p$:

- $\mathcal{V}_{i,0}$ is a subset of $\mathcal{V}_{i,\Delta}$,
- each vertex from the overlap subset $\mathcal{V}_{i,\Delta} \setminus \mathcal{V}_{i,0}$ is connected to at least one vertex of $\mathcal{V}_{i,0}$ within the subgraph $\mathcal{G}|_{\mathcal{V}_{i,\Delta}}$.

By *connected*, it is meant that there exists a path made of successive adjacent vertices.

The parts or subsets of the partitions are referred to as *subdomains*. The techniques for partitioning a graph and growing subdomains from $\mathcal{P}_0$ to $\mathcal{P}_\Delta$ are beyond the scope of this paper. Let us only say that a common strategy for partitioning a graph is to minimize the number of edges that straddle across the non-overlapping subdomains, while creating p equal size subsets. Also, a basic technique to get an overlapping partition is to add the adjacent vertices of each subdomain: $\mathcal{V}_{i,\Delta} = \mathcal{V}_{i,0} \cup adj\,(\mathcal{V}_{i,0})$. This is what is used for all the computations presented hereafter. Note that this process could be performed recursively in order to further extend the overlap: $\mathcal{V}_{i,\Delta} \leftarrow \mathcal{V}_{i,\Delta} \cup adj\,(\mathcal{V}_{i,\Delta})$.

While all the notations and definitions given above allow the description of the Restricted Additive Schwarz (RAS) preconditioner, another subset of vertices is required in order to study the reduction of the unknowns: $\mathcal{V}_0^E$, the *external interface vertices* (we follow the terminology of [6]) of $\mathcal{P}_0$, which is the union of the p subdomain adjacency sets: $\mathcal{V}_0^E = \cup_{i=1}^{p} adj\,(\mathcal{V}_{i,0})$.

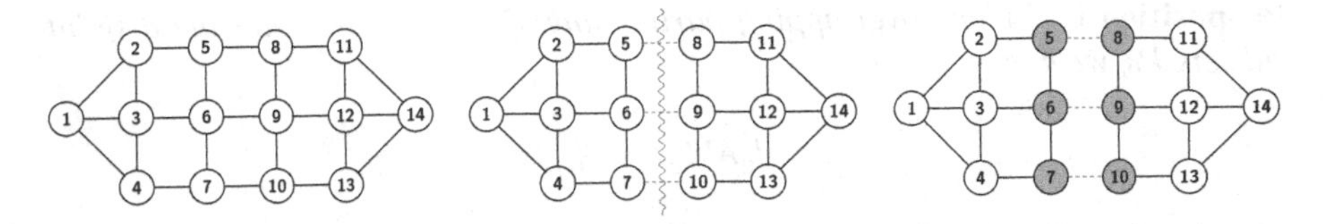

Fig. 1 Example of a connectivity graph (*left*); non-overlapping partition with two subdomains (*middle*); external interface vertices of the non-overlapping partition (*right*)

All the other vertices are referred to as the *irrelevant* vertices (see [1]): $\mathcal{V}_0^I = \mathcal{V} \setminus \mathcal{V}_0^E$. A simple example of a non-overlapping partition, with marked external interface vertices, is shown on Fig. 1.

If $n_{i,0} = |\mathcal{V}_{i,0}|$, for $1 \leqslant i \leqslant p$, we denote by $R_{i,0} \in \mathbb{R}^{n_{i,0} \times n}$ the restriction operator from $\mathbb{R}^n$ onto the subspace associated with $\mathcal{V}_{i,0}$. Similarly, if $n_{i,\Delta} = |\mathcal{V}_{i,\Delta}|$, we denote by $R_{i,\Delta} \in \mathbb{R}^{n_{i,\Delta} \times n}$ the restriction operator from $\mathbb{R}^n$ onto the subspace associated with $\mathcal{V}_{i,\Delta}$. The special restriction operator $\tilde{R}_{i,\Delta}$ is defined as follows: $\tilde{R}_{i,\Delta} = R_{i,\Delta} R_{i,0}^T R_{i,0}$.

If $n_0^E = |\mathcal{V}_0^E|$ and $n_0^I \equiv n - n_0^E$, we denote respectively by $R_0^E \in \mathbb{R}^{n_0^E \times n}$ and $R_0^I \in \mathbb{R}^{n_0^I \times n}$ the restriction operators from $\mathbb{R}^n$ onto the subspaces associated with $\mathcal{V}_0^E$ and $\mathcal{V}_0^I$.

Also, the diagonal operator I_0^I stands for $R_0^{I^T} R_0^I$. Note that the transpose of the restriction operators are the corresponding prolongation operators. Finally, the subdomain operators, assumed to be non-singular, are denoted by: $A_{i,\Delta} = R_{i,\Delta} A R_{i,\Delta}^T$.

Despite the fact that the methods related to the reduction to the interface are well-known in the community, we are not aware of a detailed description of this reduction in the right RAS preconditioning case: the next section provides this. We refer to [5] for the left RAS preconditioning case. Anyhow, convergence behaviors are alike when setting the preconditioner either on the left or the right side of A, for Krylov subspace methods, since $M_\Delta^{-1} A$ and $A M_\Delta^{-1}$ are similar operators.

2 Reduction to the Interface of the Right Preconditioned System

The preconditioned operator writes, by definition (see [2]):

$$A M_\Delta^{-1} = A \sum_{i=1}^{p} \tilde{R}_{i,\Delta}^T A_{i,\Delta}^{-1} R_{i,\Delta}$$

Let us start by stating the following proposition.

Proposition 1 *For any overlapping partitioning $\mathcal{P}_\Delta$ associated with the disjoint subsets $\mathcal{P}_0$, we have:*

$$I_0^I A M_\Delta^{-1} = I_0^I$$

The proof is rather lengthy compared to this paper format. Because it only involves classical algebra, it is left to the reader.

If we come back to the system (1), the right preconditioned version is the following:

$$AM_\Delta^{-1}\hat{u} = f, \qquad u = M_\Delta^{-1}\hat{u}$$

By introducing a global permutation matrix $P_0^T = [R_0^{I^T} R_0^{E^T}]$, which reorders the unknowns such that those from the external interface of the non-overlapping partition $\mathcal{V}_0^E$ are second, and by permuting this latter system, we get: $P_0 A M_\Delta^{-1} P_0^T P_0 \hat{u} = P_0 f$.

We denote respectively by $\hat{x}_0 = R_0^I \hat{u}$ and $\hat{y}_0 = R_0^E \hat{u}$ the irrelevant and external interface unknowns. Proposition 1 yields:

$$\begin{bmatrix} I & 0 \\ R_0^E A M_\Delta^{-1} R_0^{I^T} & R_0^E A M_\Delta^{-1} R_0^{E^T} \end{bmatrix} \begin{Bmatrix} \hat{x}_0 \\ \hat{y}_0 \end{Bmatrix} = \begin{Bmatrix} R_0^I f \\ R_0^E f \end{Bmatrix} \tag{2}$$

Let G_Δ denote the operator $I - R_0^E A M_\Delta^{-1} R_0^{E^T}$. With $\hat{x}_0 = R_0^I f$ and h_Δ standing for $R_0^E \left(I - AM_\Delta^{-1} I_0^I\right) f$, Eq. (2) can be reduced to the following system:

$$(I - G_\Delta)\,\hat{y}_0 = h_\Delta \tag{3}$$

In order to solve the unpreconditioned system (3) with the Generalized Minimal RESidual (GMRES) method, given an initial guess $\hat{y}_0^{(0)} = \hat{y}_0^{init}$, the evaluation of the initial residual is needed:

$$r_0 = h_\Delta - (I - G_\Delta)\, y_0^{(0)} = R_0^E \left(f - AM_\Delta^{-1}\left(R_0^{E^T} \hat{y}_0^{init} + I_0^I f\right)\right) \tag{4}$$

Then, in the GMRES outer loop, the following matrix-vector product is required: $w \leftarrow (I - G_\Delta)\, v_j = R_0^E A M_\Delta^{-1} R_0^{E^T} v_j$. This implies solving local problems on each $\mathcal{V}_{i,\Delta}$ subdomain with a right-hand side that is not zero-valued only in $\mathcal{V}_{i,\Delta} \cap \mathcal{V}_0^E$.

```
1: Initial guess: u_0 ∈ R^n
2: r_0 ← f − A u_0
3: β ← ||r_0||_2, v_1 ← r_0/β
4: for j = 1, ··· , m do
5:     w ← A M_Δ^{-1} v_j
6:     for i = 1, ··· , j do
7:         H_{i,j} ← (w, v_i)
8:         w ← w − H_{i,j} v_i
9:     end for
10:    H_{j+1,j} ← ||w||_2
11:    v_{j+1} ← w/H_{j+1,j}
12: end for
13: V_m = [v_1, ··· , v_m]
14: H̄_m = {H_{i,j} / 1 ⩽ i ⩽ m+1, 1 ⩽ j ⩽ m}
15: z_m ← argmin_z ||β e_1 − H̄_m z||_2
16: u ← u_0 + M_Δ^{-1} V_m z_m
```

```
1: Initial guess: ŷ_0 ∈ R^{n_0^E}
2: r_0 ← R_0^E (f − A M_Δ^{-1} (R_0^{E T} ŷ_0 + I_0^I f))
3: β ← ||r_0||_2, v_1 ← r_0/β
4: for j = 1, ··· , m do
5:     w ← R_0^E A M_Δ^{-1} R_0^{E T} v_j
6:     for i = 1, ··· , j do
7:         H_{i,j} ← (w, v_i)
8:         w ← w − H_{i,j} v_i
9:     end for
10:    H_{j+1,j} ← ||w||_2
11:    v_{j+1} ← w/H_{j+1,j}
12: end for
13: V_m = [v_1, ··· , v_m]
14: H̄_m = {H_{i,j} / 1 ⩽ i ⩽ m+1, 1 ⩽ j ⩽ m}
15: z_m ← argmin_z ||β e_1 − H̄_m z||_2
16: u ← M_Δ^{-1} (R_0^{E T} (ŷ_0 + V_m z_m) + I_0^I f)
```

Fig. 2 GMRES solvers for the global (*left*) and the interface (*right*) unknowns

Finally, once the iterative method converged to $\hat{y}_0^{(\infty)}$, the solution of the system (1) is recovered as follows:

$$u^{(\infty)} = M_\Delta^{-1} \hat{u}^{(\infty)} = M_\Delta^{-1} \left(R_0^{E^T} \hat{y}_0^{(\infty)} + I_0^I f \right)$$

The algorithm is described on the right of Fig. 2. At each iteration, we can monitor the global system's residual norm from the interface residual norm. Using Proposition 1, it is easy to check that:

$$\| f - AM_\Delta^{-1} \left(R_0^{E^T} \hat{y}_0 + I_0^I f \right) \|_2 = \| h_\Delta - (I - G_\Delta) \hat{y}_0 \|_2$$

If solving the interface system instead of the global one represents only a slight modification of the GMRES algorithm (described on the left of Fig. 2), the advantage lies in the size of the system, n_0^E against n, and thus the floating point operation count and memory usage of the GMRES method. The difference between the respective convergence behaviors is not significant, as shown on an example in Fig. 3. Indeed, we can see from Eq. (2) that the spectrum of AM_Δ^{-1} is equal to the spectrum of $I - G_\Delta$ augmented with n_0^I one-valued eigenvalues.

The next section is devoted to the preconditioning applied to the reduced system (3).

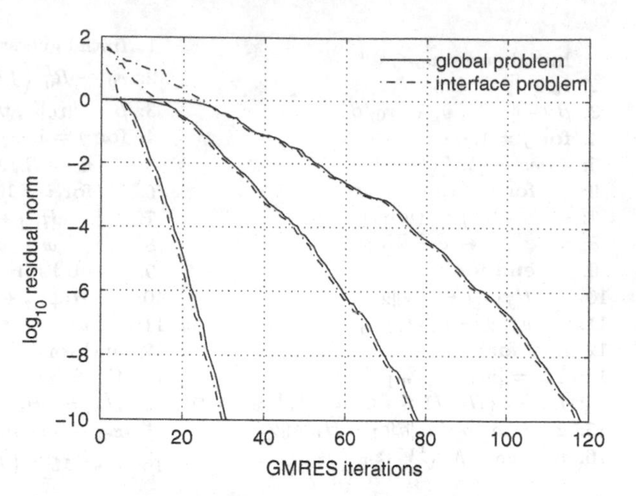

Fig. 3 Full GMRES convergence of the global and interface systems. The GT01R matrix and RHS from the UF sparse matrix collection [3] is used, with a zero initial guess. From *left to right*, the domain is divided into two, four and eight parts. The number of primary unknowns is 7980, while the number of interface unknowns is 420, 1260, and 2940 respectively from *left to right*

3 Preconditioning the Reduced System

The main difficulty with the preconditioning of $I - G_\Delta$ is that G_Δ is rather dense, as shown with an example in Table 1.

The cost of an approximate inverse approach appears to be prohibitive regarding computational time and memory. Our motivation is to only use the matrix-vector product $(I - G_\Delta)v$ in order to build the preconditioning strategy (Fig. 4).

If we have a set V_q of q orthonormal vectors of size n_0^E, we can approach G_Δ using the orthogonal projection matrix $V_q V_q^T$: $\tilde{G}_\Delta \equiv V_q V_q^T G_\Delta V_q V_q^T$. If we note W_q the image of V_q under G_Δ, and $\hat{G}_\Delta \equiv V_q^T G_\Delta V_q = V_q^T W_q \in \mathbb{R}^{q \times q}$, we get by Woodbury matrix identity:

$$(I - \tilde{G}_\Delta)^{-1} = I - V_q(I - \hat{G}_\Delta^{-1})^{-1} V_q^T = I + V_q((I - \hat{G}_\Delta)^{-1} - I)V_q^T \tag{5}$$

$$= I + V_q((I - V_q^T W_q)^{-1} - I)V_q^T \tag{6}$$

As we can see on Fig. 2, we already have an orthonormal basis $V_q = [1, \cdots, v_q]$ after q iterations of the GMRES algorithm. Also, the image of each vector of V_q under $(I - G_\Delta)$ is computed on line 5 of the algorithm on the right side. Thus we need to store theses images $W_q = (I - G_\Delta)V_q$ in order to build the preconditioner (6) and use it subsequently. Some results of this strategy are shown on Fig. 5: in this approach, the preconditioner is build first and then kept throughout the GMRES process. On the whole, we observe a trade-off between the number of GMRES iterations used to build the preconditioner q and those saved thanks to the preconditioning. If we note m_{conv} the number of GMRES iterations required to reach a given tolerance, we observe that $q + m_{conv}$ remains constant for all values of q.

Table 1 Density percentage of matrices A and G_Δ

Matrix	A	G_Δ		
Number of parts p	N.A.	2	4	8
Size	7980	420	1260	2940
Density (%)	0.68	66.06	51.10	25.72

The GT01R matrix and RHS from the UF sparse matrix collection [3] is used

1: Choose $tol > 0$
2: Initial guess: $\hat{y}_0 \in \mathbb{R}^{n_0^E}$
3: $restart \leftarrow 0$
4: $convergence \leftarrow false$
5: **repeat**
6: $\quad r_0 \leftarrow R_0^E(f - AM_\Delta^{-1}(R_0^{E^T}\hat{y}_0 + I_0^I f))$
7: $\quad$ **if** $restart = 1$ **then**
8: $\quad\quad r_0 \leftarrow (I + V_q(Q - I)V_q^T)r_0$
9: $\quad$ **end if**
10: $\quad \beta \leftarrow \|r_0\|_2,\ v_1 \leftarrow r_0/\beta$
11: $\quad$ **for** $j = 1, \cdots, q$ **do**
12: $\quad\quad w \leftarrow R_0^E A M_\Delta^{-1} R_0^{E^T} v_j$
13: $\quad\quad w_j \leftarrow w$
14: $\quad\quad$ **if** $restart = 1$ **then**
15: $\quad\quad\quad w \leftarrow (I + V_q(Q - I)V_q^T)w$
16: $\quad\quad$ **end if**
17: $\quad\quad$ **for** $i = 1, \cdots, j$ **do**
18: $\quad\quad\quad H_{i,j} \leftarrow (w, v_i)$
19: $\quad\quad\quad w \leftarrow w - H_{i,j}v_i$
20: $\quad\quad$ **end for**
21: $\quad\quad H_{j+1,j} \leftarrow \|w\|_2$
22: $\quad\quad v_{j+1} \leftarrow w/H_{j+1,j}$
23: $\quad$ **end for**
24: $\quad V_q = [v_1, \cdots, v_q]$
25: $\quad W_q = [w_1, \cdots, w_q]$
26: $\quad \overline{H}_q = \{H_{i,j} \ /\ 1 \leqslant i \leqslant q+1, 1 \leqslant j \leqslant q\}$
27: $\quad z_q = argmin_z \|\beta e_1 - \overline{H}_q z\|_2$
28: $\quad \hat{y}_q \leftarrow \hat{y}_0 + V_q z_q$
29: $\quad$ **if** $\|\beta e_1 - \overline{H}_q z_q\|_2 < tol$ **then**
30: $\quad\quad convergence \leftarrow true$
31: $\quad$ **else**
32: $\quad\quad restart \leftarrow 1$
33: $\quad\quad Q \leftarrow (I - V_q^T W_q)^{-1}$
34: $\quad\quad \hat{y}_0 \leftarrow \hat{y}_q$
35: $\quad$ **end if**
36: **until** $convergence$
37: $u \leftarrow M_\Delta^{-1}(R_0^{E^T}\hat{y}_q + I_0^I f)$

Fig. 4 GMRES(q) solver for the interface unknowns with a variable left preconditioner

Indeed, this interface preconditioner appears to be a cheap and efficient way to avoid stagnation when restarting by keeping some of the most recent convergence information. This is why we tested it on the GMRES(q) technique. At each restart, a new preconditioner is built using the just computed V_q basis. This preconditioner is only used for the subsequent q GMRES iterations. The left-preconditioned GMRES(q) algorithm for the interface unknowns is described on Fig. 4.

Some results are shown on Fig. 6: this preconditioned restarted GMRES method appears to be robust, while avoiding the growth of memory and orthogonalization time of the full GMRES approach.

Actually, by plugging the equality $I - \hat{G}_\Delta = V_q^T(I - G_\Delta)V_q$ into Eq. (5), it appears that this preconditioner is related to the preconditioner by deflation from [4], but with a fixed-size approximate invariant subspace that is fully renewed at each restart.

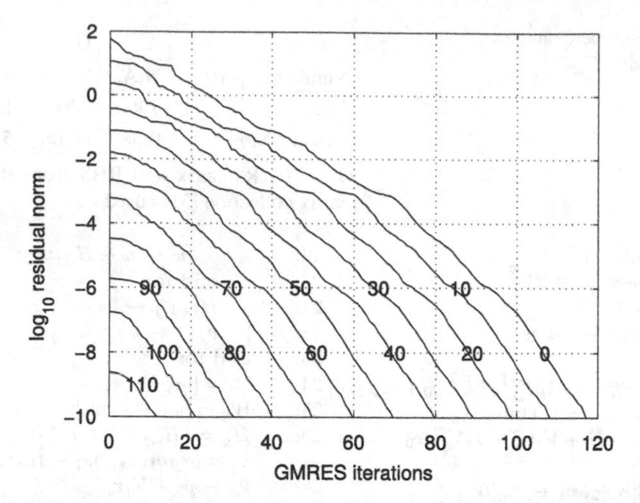

Fig. 5 Full GMRES convergence of the interface system preconditioned on the *right* side. The GT01R matrix and RHS from the UF sparse matrix collection [3] is used. The initial guess is the outcome of the preconditioner building process. The domain is divided into eight parts. The value of q, the size of the approximation space for $\widetilde{G}_\Delta$, ranges from 0 (*right*) to 110 (*left*)

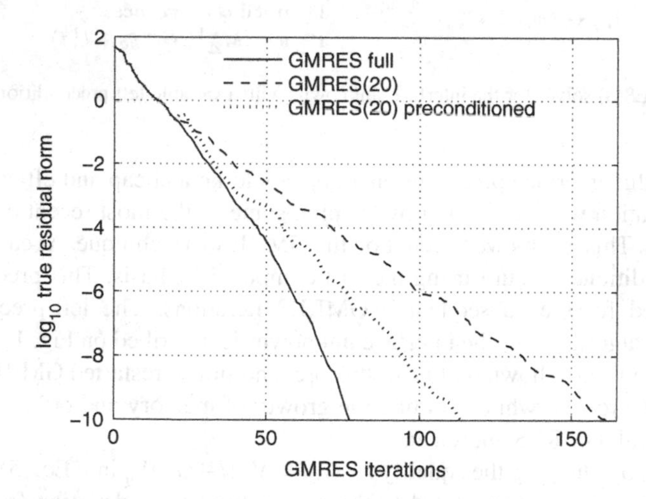

Fig. 6 GMRES convergence of the interface system for three different strategies: full GMRES, GMRES(20) and a left-preconditioned GMRES(20). The GT01R matrix and RHS from the UF sparse matrix collection [3] is used. The domain is divided into four parts

4 Conclusion

At first, we saw that if accurate subdomain solutions are employed, the right RAS preconditioned system can be reduced to a system of interface equations. The interface unknowns are located at the external interface vertices of the non-overlapping partition. Then, our purpose was to approximate the reduced interface operator as a low-rank correction of the identity matrix, using a sequence of Arnoldi vectors and their image. As might be expected, it is observed that the total cost of the linear solver regarding the number of applications of the Schwarz preconditioner remains approximately constant: what is gained by using an unvarying interface preconditioner is counterbalanced with its building cost. However, this technique becomes beneficial when the restarted variant of the Krylov subspace method is used along with a new interface preconditioner at each restart. A link with the deflation preconditioner was also presented.

Acknowledgement This work has been supported by the French National Agency of Research, through the ANR-MONU12-0012 H2MNO4 project. The authors wish to express their thanks to Stéphane Aubert for some stimulating conversations and suggestions.

References

1. E. Brakkee, P. Wilders, A domain decomposition method for the advection-diffusion equation. Technical Report, Delft University of Technology (1994)
2. X.-C. Cai, M. Sarkis, A restricted additive Schwarz preconditioner for general sparse linear systems. SIAM J. Sci. Comput. **21**(2), 792–797 (electronic) (1999)
3. T.A. Davis, Y. Hu, The University of Florida sparse matrix collection. ACM Trans. Math. Softw. **38**(1), Art. 1, 25 (2011)
4. J. Erhel, K. Burrage, B. Pohl, Restarted GMRES preconditioned by deflation. J. Comput. Appl. Math. **69**(2), 303–318 (1996)
5. F. Pacull, S. Aubert, GMRES acceleration of restricted Schwarz iterations, in *Domain Decomposition Methods in Science and Engineering XXI*. Lecture Notes in Computational Science and Engineering, vol. 98 (Springer, Berlin, 2014), pp. 725–732
6. Y. Saad, M. Sosonkina, Distributed Schur complement techniques for general sparse linear systems. SIAM J. Sci. Comput. **21**(4), 1337–1356 (electronic) (1999/2000)

Decoupled Schemes for Free Flow and Porous Medium Systems

Iryna Rybak and Jim Magiera

A comparison study of different decoupled schemes for the evolutionary Stokes/Darcy problem is carried out. Stability and error estimates of a mass conservative multiple-time-step algorithm are provided under a time step restriction which depends on the physical parameters of the flow system and the ratio between the time steps applied in the free flow and porous medium domains. Numerical results are presented and the advantage of multirate time integration is demonstrated.

1 Introduction

Modeling coupled porous medium and free flow systems is of interest for a wide spectrum of industrial and environmental applications. Physical processes in these systems evolve on different scales in space and time that require different models for each flow domain and an accurate treatment of transitions between them at the interface. In the free flow region, the Navier–Stokes or Stokes equations are typically applied to describe momentum conservation while Darcy's law is used in the porous medium. To couple these flow models, which are of different orders, the Beavers–Joseph–Saffman condition [1, 11] is usually applied together with restrictions that arise due to mass conservation and balance of normal forces across the interface.

Over the last decade, work has been carried out mainly for stationary flow systems aimed at providing rigorous problem formulations and numerical methods

I. Rybak (✉) • J. Magiera
Institute of Applied Analysis and Numerical Simulation, University of Stuttgart, Pfaffenwaldring 57, 70569 Stuttgart, Germany
e-mail: rybak@ians.uni-stuttgart.de; magierjm@mathematik.uni-stuttgart.de

T. Dickopf et al. (eds.), *Domain Decomposition Methods in Science and Engineering XXII*, Lecture Notes in Computational Science and Engineering 104, DOI 10.1007/978-3-319-18827-0_63

for solving such coupled flow problems [3–5, 9]. Recent advances in coupling techniques for nonstationary flow problems are presented in [2, 6–8], where the same time step is used in both domains.

Since the free flow velocity is usually much higher than the velocity of fluids through porous media, it is reasonable to apply a multiple-time-step technique: to compute fast/slow solutions using a small/large time step. First results on multirate time integration for the coupled Stokes/Darcy problem are presented in [10, 12]. Multiple-time-stepping pays off for single-fluid-phase systems when the free flow domain is smaller than the porous medium (modelling karst aquifers, flows in fractured porous media, flows in blood vessels and biological tissues) and it is especially efficient when a second fluid phase is present in the subsurface and the porous medium model is nonlinear and expensive (overland flow interactions with unsaturated groundwater aquifers).

The overall goal of this work is to investigate different multiple-time-step techniques for solving coupled free flow and porous medium flow problems.

2 Flow System Description

The system of interest includes a free flow region Ω_{ff}, containing a single fluid phase, and a porous medium layer Ω_{pm}, which contains a fluid and a solid phase (Fig. 1, left). At the macroscale, the system is described as two different continuum flow domains separated by the interface Γ (Fig. 1, right).

We deal with isothermal processes and consider the same incompressible fluid in both flow domains. The mass conservation equation reads

$$\nabla \cdot \mathbf{v} = 0 \qquad \text{in } \Omega_{\text{ff}} \times (0, T]. \tag{1}$$

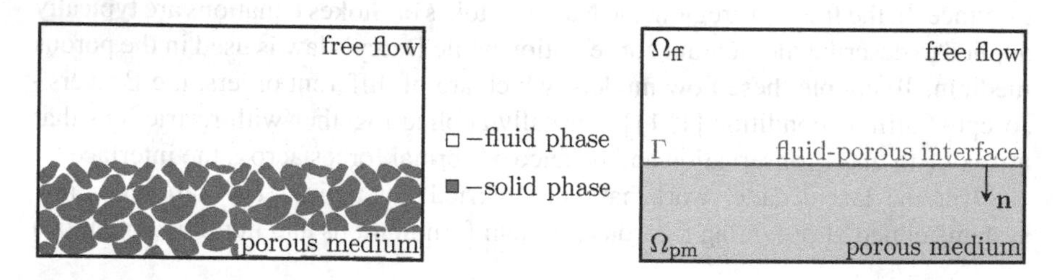

Fig. 1 Schematic representation of the coupled free flow and porous medium flow system

Considering laminar flows and neglecting the inertia term, the momentum balance in the free flow domain reduces to the Stokes equation

$$\rho \frac{\partial \mathbf{v}}{\partial t} - \nabla \cdot \mathbf{T}(\mathbf{v}, p) - \rho \mathbf{g} = 0 \qquad \text{in } \Omega_{\mathrm{ff}} \times (0, T], \tag{2}$$

where ρ is the density, $\mathbf{v}$ is the velocity, p is the pressure, $\mathbf{g}$ is the gravitational acceleration, $\mathbf{T}(\mathbf{v}, p) = 2\mu \mathbf{D}(\mathbf{v}) - p\mathbf{I}$ is the stress tensor, μ is the viscosity, $\mathbf{D}(\mathbf{v}) = \frac{1}{2}\left(\nabla \mathbf{v} + (\nabla \mathbf{v})^{\mathrm{T}}\right)$ is the strain tensor, and $\mathbf{I}$ is the identity tensor.

Fluid flows through the porous medium are usually described by Darcy's law $\mathbf{v} = -\mu^{-1}\mathbf{K}(\nabla p - \rho \mathbf{g})$, which, together with the mass conservation equation for compressible soils, yields the porous medium flow formulation

$$\beta \frac{\partial p}{\partial t} - \nabla \cdot \left(\frac{\mathbf{K}}{\mu} (\nabla p - \rho \mathbf{g}) \right) = 0 \qquad \text{in } \Omega_{\mathrm{pm}} \times (0, T], \tag{3}$$

where $\mathbf{K}$ is the intrinsic permeability tensor and β is the soil compressibility.

The *mass conservation* across the interface reads

$$\mathbf{v}_{\mathrm{ff}} \cdot \mathbf{n} = \mathbf{v}_{\mathrm{pm}} \cdot \mathbf{n} \qquad \text{on } \Gamma \times (0, T], \tag{4}$$

and the *balance of normal forces* is given by

$$-\mathbf{n} \cdot \mathbf{T}(\mathbf{v}_{\mathrm{ff}}, p_{\mathrm{ff}}) \cdot \mathbf{n} = p_{\mathrm{pm}} \qquad \text{on } \Gamma \times (0, T]. \tag{5}$$

The *Beavers–Joseph–Saffman* interface condition can be written as follows

$$\mathbf{v}_{\mathrm{ff}} \cdot \boldsymbol{\tau}_i + 2\alpha_{\mathrm{BJ}}^{-1} \sqrt{\mathbf{K}} \mathbf{n} \cdot \mathbf{D}(\mathbf{v}_{\mathrm{ff}}) \cdot \boldsymbol{\tau}_i = 0, \quad i = 1, \ldots, d-1 \qquad \text{on } \Gamma \times (0, T], \tag{6}$$

where $\mathbf{n}$ and $\boldsymbol{\tau}$ are the unit normal and tangential vectors to the interface (Fig. 1), $\alpha_{\mathrm{BJ}} > 0$ is the Beavers–Joseph parameter, and d is the number of space dimensions.

Problem (1)–(6) is subject to initial and boundary conditions at the external boundary of the coupled domain.

3 Decoupled Schemes

Multiphysics problems can be solved using the monolithic approach when the systems of linear algebraical equations resulting from the discretization of two models are assembled together with the interface conditions into one matrix, or applying partitioning techniques when each subdomain is treated separately.

For nonstationary problems where the processes run on different time scales, different time steps can be applied in each subdomain. Typically, fluid velocity in the free flow domain is much higher than through the porous medium, therefore it is

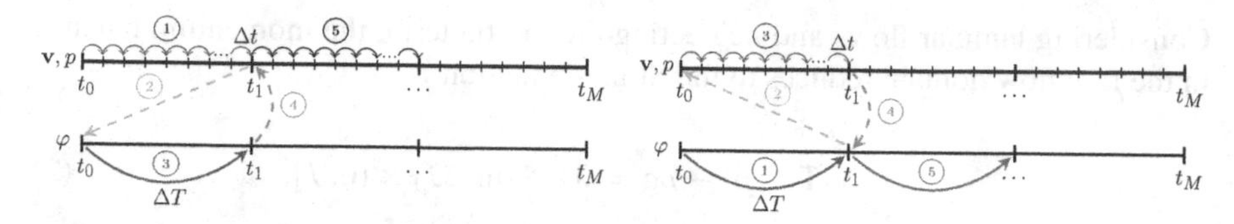

Fig. 2 Stokes–Darcy (*left*) and Darcy–Stokes (*right*) decoupled multistep schemes

Algorithm 1 (Stokes–Darcy)

for $k = 0$ **to** $M-1$ **do**

  for $m = m_k$ **to** $m_{k+1}-1$ **do**

$$\rho\frac{\mathbf{v}_h^{m+1}-\mathbf{v}_h^m}{\Delta t} + \mathrm{A}_{\mathrm{ff}}\left(\mathbf{v}_h^{m+1}, p_h^{m+1}\right) + \mathrm{A}_{\mathrm{ffpm}}\left(\mathbf{v}_h^{m+1}, p_h^{m+1}, \varphi_h^{m_k}\right) = \mathbf{f}_{\mathrm{ff}}^{m+1}$$

  end for

$$\beta\frac{\varphi_h^{m_{k+1}}-\varphi_h^{m_k}}{\Delta T} + \mathrm{A}_{\mathrm{pm}}\left(\varphi_h^{m_{k+1}}\right) + \mathrm{A}_{\mathrm{pmff}}\left(\mathbf{v}_h^{m_{k+1}}, \varphi_h^{m_{k+1}}\right) = f_{\mathrm{pm}}^{m_{k+1}}$$

end for

Algorithm 2 (Darcy–Stokes)

for $k = 0$ **to** $M-1$ **do**

$$\beta\frac{\varphi_h^{m_{k+1}}-\varphi_h^{m_k}}{\Delta T} + \mathrm{A}_{\mathrm{pm}}\left(\varphi_h^{m_{k+1}}\right) + \mathrm{A}_{\mathrm{pmff}}\left(\mathbf{v}_h^{m_k}, \varphi_h^{m_{k+1}}\right) = f_{\mathrm{pm}}^{m_{k+1}}$$

  for $m = m_k$ **to** $m_{k+1}-1$ **do**

$$\rho\frac{\mathbf{v}_h^{m+1}-\mathbf{v}_h^m}{\Delta t} + \mathrm{A}_{\mathrm{ff}}\left(\mathbf{v}_h^{m+1}, p_h^{m+1}\right) + \mathrm{A}_{\mathrm{ffpm}}\left(\mathbf{v}_h^{m+1}, p_h^{m+1}, \varphi_h^{m_{k+1}}\right) = \mathbf{f}_{\mathrm{ff}}^{m+1}$$

  end for

end for

reasonable to compute free flow solutions on a fine time mesh and porous medium solutions on a coarse time mesh. Different decoupled schemes can be developed: first the free flow problem is solved and then the porous medium one (Fig. 2, left), or vice versa (Fig. 2, right).

In Algorithms 1–3, A_{ff} and A_{pm} are the space discretization operators for the free flow problem (1)–(2) and the porous medium problem (3), $\mathrm{A}_{\mathrm{ffpm}}$ is responsible for the coupling conditions (5)–(6), $\mathrm{A}_{\mathrm{pmff}}$ stands for the interface condition (4), φ is the porous medium pressure, m_k and m are indices for the coarse and fine time grids, r is the ratio between the large and small time steps $\Delta T = r\Delta t$. In both domains, uniform rectangular meshes matching at the interface are considered and second order finite volume schemes [13, Chap. 4.4, 6.3] are applied. We will compare Algorithms 1–3 numerically and provide stability and error estimates for the most accurate Algorithm 1.

Algorithm 3 (Stokes–Darcy, averaged velocity, [12])

for $k = 0$ **to** $M - 1$ **do**

for $m = m_k$ **to** $m_{k+1} - 1$ **do**

$$\rho \frac{\mathbf{v}_h^{m+1} - \mathbf{v}_h^m}{\Delta t} + \mathrm{A}_{\mathrm{ff}}\left(\mathbf{v}_h^{m+1}, p_h^{m+1}\right) + \mathrm{A}_{\mathrm{ffpm}}\left(\mathbf{v}_h^{m+1}, p_h^{m+1}, \varphi_h^{m_k}\right) = \mathbf{f}_{\mathrm{ff}}^{m+1}$$

end for

$$\beta \frac{\varphi_h^{m_{k+1}} - \varphi_h^{m_k}}{\Delta T} + \mathrm{A}_{\mathrm{pm}}\left(\varphi_h^{m_{k+1}}\right) + \mathrm{A}_{\mathrm{pmff}}\left(\frac{1}{r}\sum_{m=m_k}^{m_{k+1}-1} \mathbf{v}_h^m, \varphi_h^{m_{k+1}}\right) = f_{\mathrm{pm}}^{m_{k+1}}$$

end for

4 Stability and Error Estimates

In this section, we provide the long time stability and the *a priori* error estimates for the multiple-time-step scheme (Algorithm 1) in case of homogeneous Dirichlet boundary conditions. The proofs can be found in [10].

Theorem 1 (Long Time Stability) *Under the restriction*

$$\Delta t \le \min\left\{\frac{k_{\min}\rho}{2\mu(r-1)^2 C^2}, \frac{2k_{\min}\mu\beta}{r\overline{C}}\right\}, \tag{7}$$

Algorithm 1 is stable for $t \in [0, +\infty)$ and the a priori estimate

$$\begin{aligned}
&\rho \left\|\mathbf{v}_h^{m_M}\right\|_{L^2(\Omega_{\mathrm{ff}})}^2 + \beta \left\|\varphi_h^{m_M}\right\|_{L^2(\Omega_{\mathrm{pm}})}^2 + 2\Delta t \sum_{k=0}^{M-1}\sum_{m=m_k}^{m_{k+1}-1}\sum_{j=1}^{d-1}\int_\Gamma \frac{\alpha_{BJ}}{\sqrt{\mathbf{K}}}\left(\mathbf{v}_h^{m+1}\cdot\boldsymbol{\tau}_j\right)^2 \\
&\le \rho \left\|\mathbf{v}_0\right\|_{L^2(\Omega_{\mathrm{ff}})}^2 + \beta \left\|\varphi_0\right\|_{L^2(\Omega_{\mathrm{pm}})}^2 + \frac{k_{\min}^2 \beta}{2\overline{C}} \left\|\nabla\varphi_0\right\|_{L^2(\Omega_{\mathrm{pm}})}^2 \\
&+ \Delta t \frac{C_v^2}{2\mu} \sum_{k=0}^{M-1}\sum_{m=m_k}^{m_{k+1}-1} \left\|\mathbf{f}_{\mathrm{ff}}^{m+1}\right\|_{L^2(\Omega_{\mathrm{ff}})}^2 + \Delta T \frac{C_\varphi^2 \mu}{k_{\min}} \sum_{k=0}^{M-1} \left\|f_{\mathrm{pm}}^{m_{k+1}}\right\|_{L^2(\Omega_{\mathrm{pm}})}^2
\end{aligned}$$

is valid, where $\mathbf{v}_0$, φ_0 are the initial data, $k_{\min}$ is the minimal permeability and the constants C, $\overline{C}$, C_v, $C_\varphi > 0$ are independent of the solution and the discretization parameters.

Theorem 2 (Convergence) *Let condition (7) be satisfied, then the solution of Algorithm 1 converges to the exact solution of problem (1)–(6) and the a priori error estimate $\rho \left\|\mathbf{e}_{\mathbf{v}}^{m_M}\right\|_{L^2(\Omega_{\mathrm{ff}})}^2 + \beta \left\|e_\varphi^{m_M}\right\|_{L^2(\Omega_{\mathrm{pm}})}^2 \le \tilde{C}\left(|h|^4 + \Delta T^2\right)$ holds true, where $\mathbf{e}_{\mathbf{v}}$ and e_φ are the errors of the discrete free flow velocity and the porous medium pressure, and the constant $\tilde{C} > 0$ does not depend on the grid steps.*

5 Numerical Experiments

Consider $\Omega_{\text{ff}} = [0, 1] \times [1, 2]$, $\Omega_{\text{pm}} = [0, 1] \times [0, 1]$, $\Gamma = (0, 1) \times \{1\}$, and choose model parameters $\rho = 1$, $\mu = 1$, $\beta = 1$, $\alpha_{\text{BJ}} = 1$, $\mathbf{K} = \mathbf{I}$, $\mathbf{g} = \mathbf{0}$. The exact solution $u(x, y, t) = -\cos(\pi x)\sin(\pi y)\exp(t)$, $v(x, y, t) = \sin(\pi x)\cos(\pi y)\exp(t)$, $p(x, y, t) = \frac{y^2}{2}\sin(\pi x)\exp(t)$, $\varphi(x, y, t) = \frac{y}{2}\sin(\pi x)\exp(t)$ satisfies the interface conditions (4)–(6).

Comparison of Algorithm 1, using the same time steps in both subdomains and a larger time step in the porous medium, with the monolithic approach is presented in Fig. 3. At each level of space grid refinement, the time step is reduced by the factor of four starting with $\Delta t = 10^{-2}$. The errors are defined as $\boldsymbol{\varepsilon}_{\mathbf{v}} = \|\mathbf{v} - \mathbf{v}_h\|_{L^2(\Omega_{\text{ff}})} / \|\mathbf{v}\|_{L^2(\Omega_{\text{ff}})}$, and $\varepsilon_\varphi = \|\varphi - \varphi_h\|_{L^2(\Omega_{\text{pm}})} / \|\varphi\|_{L^2(\Omega_{\text{pm}})}$. Numerical results confirm second order convergence in space and first order in time for all the schemes. The multistep algorithm is slightly less accurate due to a larger time step applied in the porous medium domain.

Comparison of Algorithms 1–3 for the same parameters is presented in Fig. 4. All methods demonstrate second order convergence in space and first order in time. Algorithm 1 is more accurate than Algorithms 2–3.

We note that restriction (7) is fulfilled for this model problem. For realistic applications this restriction is severe. However, numerical simulations show that the multiple-time-step algorithm is stable and convergent even when this restriction is not fulfilled [10, Sec. 6.2].

We also present numerical simulations for a realistic setup. Consider a coupled domain of size $5\,\text{m} \times 1.2\,\text{m}$ with the interface $\Gamma = (0, 5\,\text{m}) \times \{1\,\text{m}\}$. In the porous medium, there are two inclusions $[0, 2\,\text{m}] \times [0.4\,\text{m}, 0.8\,\text{m}]$ and $[2.5\,\text{m}, 3.5\,\text{m}] \times [0.1\,\text{m}, 0.8\,\text{m}]$. The fluid is water with density $\rho = 10^3\,[\text{kg/m}^3]$ and dynamic viscosity $\mu = 10^{-3}\,[\text{Pa s}]$. The soil is isotropic with permeability $k = 10^{-8}\,[\text{m}^2]$ except for the inclusions, where $k_1 = 10^{-6}\,[\text{m}^2]$ and $k_2 = 10^{-10}\,[\text{m}^2]$ (Fig. 5, top),

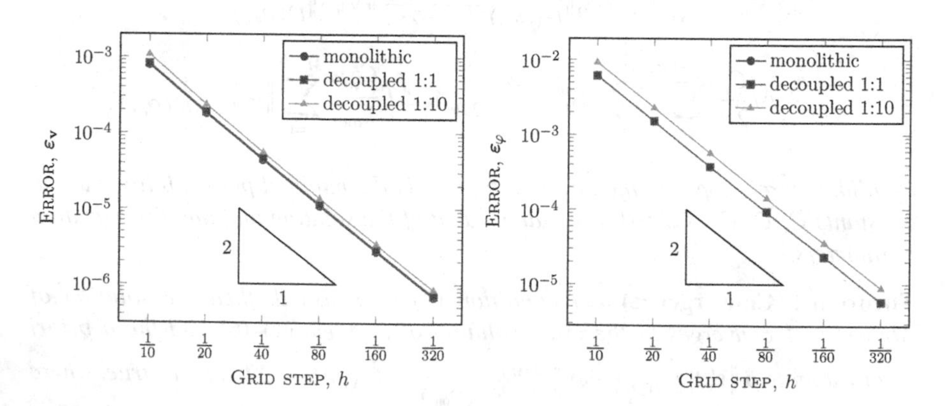

Fig. 3 Comparison of Algorithm 1 and the monolithic approach. Time step ratio $1 : r$

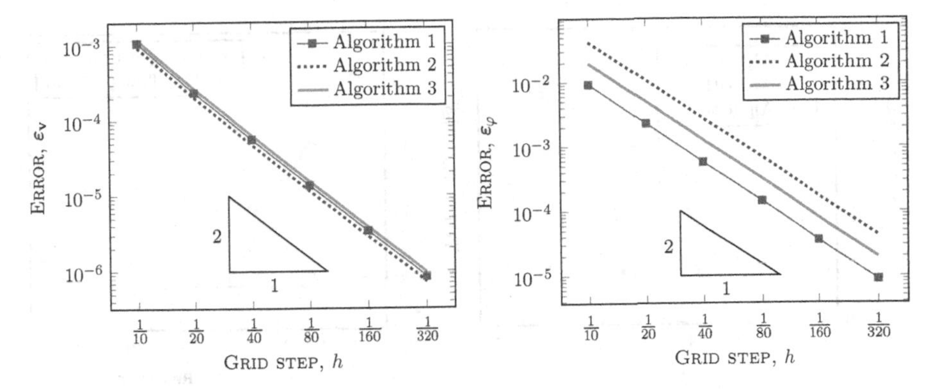

Fig. 4 Comparison of Algorithms 1–3. Time step ratio 1 : 10

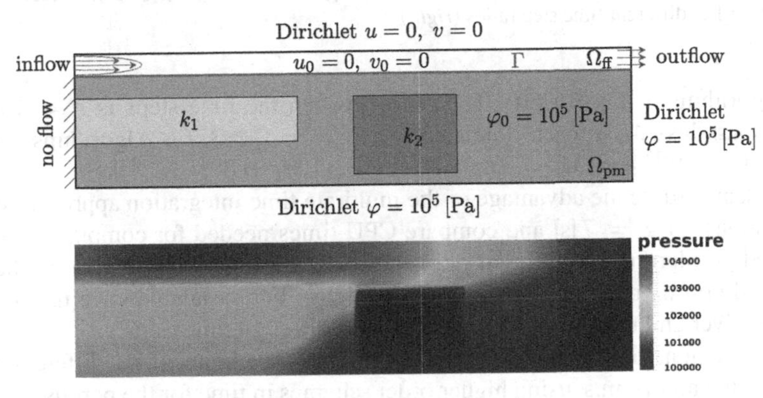

Fig. 5 Initial and boundary conditions (*top*) and pressure (*bottom*) for the realistic setup

and compressibility $\beta = 10^{-4}$ [1/Pa]. The Beavers–Joseph coefficient is $\alpha_{BJ} = 1$. Gravitational effects are neglected.

Initial and boundary conditions are prescribed in Fig. 5 (top), where the inflow conditions in the free flow domain are defined as $u = \left(2 - 190(y - 1.1)^2\right) \times (1 - \cos(\pi t/2))$ [m/s], $v = 0$, the no-flow condition in the porous medium is given by $\partial p/\partial x = 0$, and the outflow conditions in the free flow region are $\partial u/\partial x = 0$, $\partial v/\partial x = 0$.

The following discretization parameters are used $h = 10^{-2}$ [m], $\Delta t = 10^{-3}$ [s], and $r = 10$ except for the results presented in Fig. 6 (right), where r is varying. Numerical simulation results for the pressure distribution in the coupled domain at time $t = 2.4$ [s] are presented in Fig. 5 (bottom).

The finite volume method on staggered grids, used to discretize the free flow and the porous medium problems, is locally mass conservative. The only place where the mass can be lost is the interface Γ. Algorithm 1 is constructed in such a way that guarantees no mass loss across Γ. However, Algorithms 2–3 are not mass conservative. The overall mass loss through the interface is presented for all

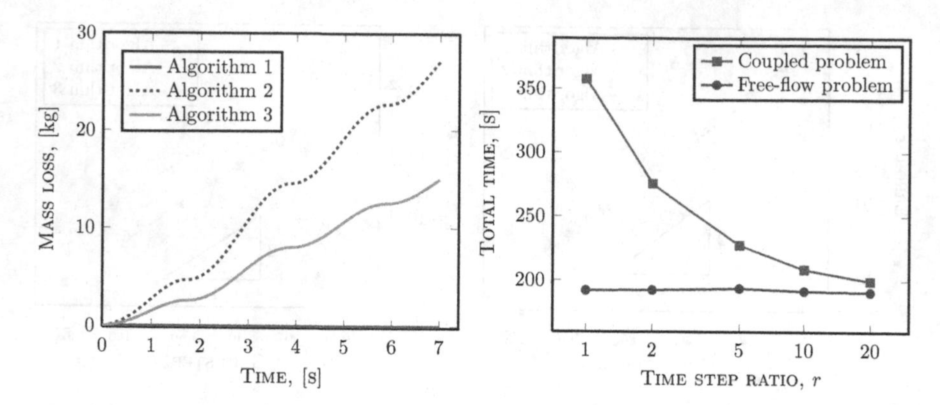

Fig. 6 Overall mass loss through the interface for Algorithms 1–3 (*left*). CPU time reduction for Algorithm 1 at different time step ratios (*right*)

the algorithms in Fig. 6 (left). The ratio between the time steps is $r = 10$. The mean mass loss M_i at each time step, where $i = 1, 2, 3$, for Algorithms 1–3 are $M_1 = 3.5 \cdot 10^{-14}$ [kg], $M_2 = 3.9 \cdot 10^{-2}$ [kg], and $M_3 = 2.1 \cdot 10^{-2}$ [kg].

To demonstrate the advantage of the multirate time integration approach we run simulations for $T = 7$ [s] and compare CPU times needed for computation of the coupled problem using different ratios between the time steps applied in the free flow and porous medium domains (Fig. 6, right). For simulations we use a direct sparse solver and reuse factorizations between different time steps.

Many extensions to this work are possible: development of different time-partitioning algorithms, using higher order schemes in time for the porous medium, application of various space discretizations in both domains, considering two fluid phases, and application of different flow models in the free flow and porous medium domains.

Acknowledgement This work was supported by the German Research Foundation (DFG) project RY 126/2-1.

References

1. G. Beavers, D. Joseph, Boundary conditions at a naturally permeable wall. J. Fluid Mech. **30**, 197–207 (1967)
2. A. Çeşmelioğlu, B. Rivière, Primal discontinuous Galerkin methods for time-dependent coupled surface and subsurface flow. J. Sci. Comput. **40**, 115–140 (2009)
3. M. Discacciati, A. Quarteroni, Navier–Stokes/Darcy coupling: modeling, analysis, and numerical approximation. Rev. Mat. Complut. **22**, 315–426 (2009)
4. M. Discacciati, E. Miglio, A. Quarteroni, Mathematical and numerical models for coupling surface and groundwater flows. Appl. Num. Math. **43**, 57–74 (2002)
5. W. Jäger, A. Mikelić, Modeling effective interface laws for transport phenomena between an unconfined fluid and a porous medium using homogenization. Transp. Porous Media **78**, 489–508 (2009)

6. W. Layton, H. Tran, X. Xiong, Long time stability of four methods for splitting the evolutionary Stokes–Darcy problem into Stokes and Darcy subproblems. J. Comput. Appl. Math. **236**, 3198–3217 (2012)
7. K. Mosthaf, K. Baber, B. Flemisch, R. Helmig, A. Leijnse, I. Rybak, B. Wohlmuth, A coupling concept for two-phase compositional porous-medium and single-phase compositional free flow. *Water Resour. Res.* **47**, W10522 (2011)
8. M. Mu, X. Zhu, Decoupled schemes for a non-stationary mixed Stokes–Darcy model. Math. Comp. **79**, 707–731 (2010)
9. B. Rivière, I. Yotov, Locally conservative coupling of Stokes and Darcy flow. SIAM J. Numer. Anal. **42**, 1959–1977 (2005)
10. I. Rybak, J. Magiera, A multiple-time-step technique for coupled free flow and porous medium systems. J. Comput. Phys. **272**, 327–342 (2014)
11. R. Saffman, On the boundary condition at the surface of a porous medium. Stud. Appl. Math. **50**, 93–101 (1971)
12. L. Shan, H. Zheng, W. Layton, A decoupling method with different subdomain time steps for the nonstationary Stokes–Darcy model. Numer. Methods Partial Differ. Eq. **29**, 549–583 (2013)
13. H. Versteeg, W. Malalasekra, *An Introduction to Computational Fluid Dynamics: The Finite Volume Method* (Prentice Hall, New Jersey, 2007)

Schwarz Waveform Relaxation for a Class of Non-dissipative Problems

Shu-Lin Wu

In this paper, we introduce the results for the Schwarz waveform relaxation (SWR) algorithm applied to a class of non-dissipative reaction diffusion equations. Both the Dirichlet and Robin transmission conditions (TCs) are considered. For the Dirichlet TC, we consider the algorithm for the nonlinear problem $\partial_t u = \mu\partial_{xx}u + f(u)$, in the case of many subdomains. For the Robin TC, we consider the linear problem $\partial_t u = \mu\partial_{xx}u + au$ with $a \geq 0$. We focus on the analysis of finding the optimal parameter involved in the Robin TC. For small overlap size $l = \mathcal{O}(\Delta x)$ and $\Delta t = \mathcal{O}(\Delta x^r)$ with $r < \frac{4}{3}$, we show that the *equioscillation* principle which works for $a < 0$ does not hold for $a \geq 0$. We show numerical results to support our theoretical conclusions.

1 Introduction

We are concerned with the SWR algorithm to compute solutions $u = u(x,t)$: $(0,L) \times (0,T) \to \mathbb{R}$ of the following problem

$$\partial_t u = \mu\partial_{xx}u + f(u), \ (x,t) \in (0,L) \times (0,T), \tag{1}$$

where $\mu > 0$ and $f \in \mathbb{C}^1(\mathbb{R})$ denotes a function which in general depends nonlinearly on u. For the case of two subdomains, Gander [3] proved that the classical SWR algorithm converges linearly on unbounded time intervals, if $f'(u)$ satisfies $f'(u) < \left(\sqrt{\mu}\pi/L\right)^2$. For the case $f'(u) \leq 0$, the analysis by Gander and

S.-L. Wu (✉)
School of Science, Sichuan University of Science and Engineering, Zigong, Sichuan 643000, PR China
e-mail: wushulin84@hotmail.com

T. Dickopf et al. (eds.), *Domain Decomposition Methods in Science and Engineering XXII*, Lecture Notes in Computational Science and Engineering 104, DOI 10.1007/978-3-319-18827-0_64

Stuart [5] and Wu et al. [8] can be used to prove convergence for the classical SWR algorithm in the case of N subdomains ($N \geq 3$). However, there are no results for the case $N \geq 3$ and $f'(u) < d$ with $d > 0$. In this paper, we show that the classical SWR algorithm is convergent for $N \geq 3$, provided $f'(u) \leq \left(\omega^* \sqrt{\mu}\pi/L\right)^2$, where $\omega^* \in (0, 1)$ depends on N.

For the purpose of fast convergence, one should use the Robin TC for the SWR algorithm, instead of the Dirichlet TC. For the linear model problem

$$\partial_t u = \mu \partial_{xx} u + au, \ (x, t) \in \mathbb{R} \times (0, T), \tag{2}$$

a key step for the convergence analysis is to solve a special min-max problem, whose solution corresponds to the best choice of the parameter p involved in the Robin TC. For $a < 0$, the optimization procedure has been deeply analyzed by Gander and Halpern [4] in the 1-D case, and by Bennequin et al. [1] in the 2-D case. Other related work also requires $a < 0$; see, e.g., [6]. For the case $a > 0$, the existing research always employs a variable transform, like $v(x, t) = e^{-\sigma t} u(x, t)$, and then the original equation is transformed to $\partial_t v = \mu \partial_{xx} v + (a - \sigma)v$ with negative coefficient, $a - \sigma < 0$. However, this trick is not advisable for practical computing. Roughly speaking, for σ large, we find numerically that even though $\max_j \|v_j^k - v\|_{L^\infty([0,T]\times\Omega_j)}$ is very small, $\max_j \|u_j^k - u\|_{L^\infty([0,T]\times\Omega_j)}$ is still a huge quantity, where j is the subdomain index, v_j^k denotes the k-th iterate of the optimized SWR algorithm applied to the transformed problem and u_j^k is obtained from the inverse transform $u_j^k = e^{\sigma t} v_j^k$ (see Fig. 3).

The parameter obtained for the linear problem (2) serves the optimized SWR algorithm for the nonlinear problem (1), by using the *'linearization'* idea introduced by Caetano et al. [2]. For the nonlinear problem (1) with $f'(u) \geq 0$, we first need to know the optimal parameter for (2) with $a \geq 0$ and to the best of our knowledge there are no results up to now. Here, we introduce our analysis of finding the best parameter for the Robin TC in the case $a \geq 0$. We show that, for overlap size $l = \mathcal{O}(\Delta x)$ small and $\Delta t = \mathcal{O}(\Delta x^r)$, the equioscillation principle established recently by Bennequin et al. [1] still holds, when $r \geq \frac{4}{3}$. But for $r < \frac{4}{3}$, this principle does not hold.

2 Main Results

In this section, we present the main results about the classical and optimized SWR algorithms. The concrete proof of the four propositions are given in our forthcoming paper [7].

2.1 Dirichlet Transmission Condition

The nonlinear IVP consists of the governing equation (1) and the initial and boundary conditions

$$u(x,0) = u_0(x), x \in [0,L];\ u(0,t) = g_1(t), u(L,t) = g_2(t), t > 0. \tag{3}$$

The domain $[0,L]$ is decomposed into N subdomains: $\Omega_j = [\alpha_j L, \beta_j L], j = 1,2,\ldots,N$, where $\alpha_1 = 0, \beta_N = 1$ and $0 < \alpha_{j+1} < \beta_j < 1$ for $j = 1,2,\ldots,N-1$. We assume that $\beta_j < \alpha_{j+2}$ so that all the subdomains overlap but domains which are not adjacent do not overlap. Then, the N-subdomain SWR algorithm with Dirichlet TC for the IVP (1) and (3) is

$$\begin{cases} \frac{\partial u_j^k(x,t)}{\partial t} = \mu \frac{\partial^2 u_j^k(x,t)}{\partial x^2} + f(u_j^k(x,t)), & (x,t) \in \Omega_j \times \mathbb{R}^+, \\ u_j^k(\alpha_j L, t) = u_{j-1}^{k-1}(\alpha_j L, t), u_j^k(\beta_j L, t) = u_{j+1}^{k-1}(\beta_j L, t), & t \in \mathbb{R}^+, \end{cases}$$

where k is the iteration index, $u_j^k(x,0) = u_0(x)$ for $x \in \Omega_j$, $\alpha_0 = \beta_0 = 0$, $\alpha_{N+1} = \beta_{N+1} = 1$, $u_0^k = g_1(t)$ and $u_{N+1}^k = g_2(t)$ for all $k \geq 0$. We assume that the overlapping domains and the subdomains are all of the same sizes.

Proposition 1 *Let l be the overlap size, N be the number of subdomains, $\phi = \frac{l}{L}\pi$ and $\varphi = \frac{L-l}{NL}\pi$. Then, suppose the function f in (1) satisfies $f'(u) \leq \left(\frac{\sqrt{\mu}\pi}{L}\omega^*\right)^2$ ($\forall u \in \mathbb{R}$), the classical SWR algorithm with $N \geq 2$ is convergent. Here, $\omega^* \in (0,1)$ is the unique solution of $\mathbf{r}(\omega) = 1$ and $\mathbf{r}(\omega)$ is defined by*

$$\mathbf{r}(\omega) = \frac{\min\{1, N-2\} \sin^2(\phi\omega) + \sin^2(\varphi\omega) + 2\cos\left(\frac{\pi}{N}\right)\sin(\phi\omega)\sin(\varphi\omega)}{\sin^2((\phi+\varphi)\omega)}.$$

2.2 Robin Transmission Condition

For the initial value problem (2) with $a > 0$, we decompose the spatial domain $\mathbb{R}$ into two subdomains $\Omega_1 = (-\infty, l]$ and $\Omega_2 = [0, +\infty)$, where $l \geq 0$. The SWR algorithm with Robin TC is given by

$$\begin{cases} \partial_t u_j^k = \mu \partial_{xx} u_j^k + a u_j^k,\ x \in \Omega_j, \\ (\partial_x + (-1)^{3-j} p) u_j^k((2-j)l, t) = (\partial_x + (-1)^{3-j} p) u_{3-j}^{k-1} j((2-j)l, t), \end{cases}$$

where $j = 1, 2$, $u_j^k(x,0) = u_0(x)$, k is the iteration index and p is a free parameter. Based on Laplace transform and maximum principle of analytic functions, we obtain the following results.

Proposition 2 (Overlapping Case $l > 0$) *Let $l > 0$ and $a \geq 0$. Then, the best performance of the SWR algorithm with Robin TC is obtained for $p = p_{opt} = \frac{q_{opt}}{2l}$. The argument q_{opt} is solution of the min-max problem*

$$\min_{q>0} \max_{y\in[y_0,y_1]} \mathcal{R}(y,q), \text{ with } \mathcal{R}(y,q) = \frac{(y-q)^2+y^2+z_0^2}{(y+q)^2+y^2+z_0^2}e^{-y}, z_0 = 2l\sqrt{\frac{a}{\mu}}. \quad (4)$$

where $y_j = 2l\sqrt{\left(\sqrt{a^2+(\pi/[j\Delta t+(1-j)T])^2}-a\right)/(2\mu)}$, $j = 0, 1$. *Define* $q_{\min} = \sqrt{2y_0^2+z_0^2}$, $q_{\max} = \sqrt{2y_1^2+z_0^2}$, $\mathcal{R}^\dagger(q) = \max\{\mathcal{R}(y_0,q), \mathcal{R}(y_1,q)\}$, $\tilde{q}_{\min} = \max\left\{q_1(z_0), q_{\min}, \frac{q_{\min}^2}{2}\right\}$, $\tilde{q}_{\max} = \min\left\{q_2(z_0), q_{\max}, \frac{q_{\max}^2}{2}\right\}$ *and*

$$q^\dagger = \begin{cases} q_{\min}, & \text{if } \mathcal{R}(y_0,q_{\min}) \geq \mathcal{R}(y_1,q_{\min}), \\ q_{\max}, & \text{if } \mathcal{R}(y_1,q_{\max}) \geq \mathcal{R}(y_0,q_{\max}), \\ q_0^\dagger, & \text{otherwise}, \end{cases} \quad (5)$$

where $q_0^\dagger \in (q_{\min}, q_{\max})$ is the unique root of $\mathcal{R}(y_0,q) = \mathcal{R}(y_1,q)$, and $q_1(z_0)$ and $q_2(z_0)$ are two different positive roots of the cubic polynomial $S(q,z_0) = q^3+4q^2-2q(2-z_0^2)+8z_0^2$ for $z_0 \in (0,z_0^*)$ with $z_0^* = 0.31920496942508$. Then, the solution of the min-max problem (4) *is given by*

$$q_{opt} = \begin{cases} q_{\max}, & \text{if } \mathcal{R}^\dagger(\tilde{q}_{\max}) < \bar{\mathcal{R}}(\tilde{q}_{\max}), \\ q_0^*, & \text{if } \mathcal{R}^\dagger(\tilde{q}_{\max}) \geq \bar{\mathcal{R}}(\tilde{q}_{\max}), \end{cases} \quad (6)$$

provided $z_0 < z_0^*$, $\tilde{q}_{\min} < \tilde{q}_{\max}$, $q^\dagger \in [\tilde{q}_{\min}, \tilde{q}_{\max}]$ and $\mathcal{R}^\dagger(q^\dagger) < \bar{\mathcal{R}}(q^\dagger)$, where $q_0^* \in [q^\dagger, q_{\max}]$ is the unique root of $\bar{\mathcal{R}}(q) = \mathcal{R}(y_0,q)$; otherwise $q_{opt} = q^\dagger$. Here, $\bar{\mathcal{R}}(q) = \mathcal{R}(\bar{y}(q),q)$ and $\bar{y}(q) = \sqrt{\frac{2q-z_0^2+\sqrt{-qS(q,z_0)}}{2}}$.

Proposition 3 (Non-overlapping Case: $l = 0$) *For $l = 0$ and $a \geq 0$, the best parameter p_{opt} for the Robin TC is given by*

$$p_{opt} = \begin{cases} \sqrt{z_{\min}^2+a_0}, & \text{if } \mathcal{R}_0(z_{\min},\sqrt{z_{\min}^2+a_0}) \geq \mathcal{R}_0(z_{\max},\sqrt{z_{\min}^2+a_0}), \\ \sqrt{z_{\max}^2+a_0}, & \text{if } \mathcal{R}_0(z_{\max},\sqrt{z_{\max}^2+a_0}) \geq \mathcal{R}_0(z_{\min},\sqrt{z_{\max}^2+a_0}), \\ p_0^*, & \text{otherwise}, \end{cases} \quad (7)$$

where $a_0 = \frac{a}{\mu}$ and p_0^* is the unique root of $\mathcal{R}_0(z_{\min},p) = \mathcal{R}_0(z_{\max},p)$.

Proposition 4 (Asymptotic Properties) *Let $\Delta t = C\Delta x^r$ with some positive constants C and r. Then, for Δx small and fixed length of the time interval, the convergence factor ρ_{Robin} of the SWR algorithms with Robin TC satisfies the* *following asymptotic properties:*

$$l = 0 : \rho_{Robin} \approx 1 - \mathcal{O}(\Delta t^{\frac{1}{4}});\ l = C_l \Delta x : \rho_{Robin} \approx \begin{cases} 1 - \mathcal{O}(\Delta x^{\frac{r}{4}}), & \text{if } r \leq \frac{4}{3}, \\ 1 - \mathcal{O}(\Delta x^{\frac{1}{3}}), & \text{otherwise.} \end{cases}$$

For T sufficiently large and fixed Δx, we have the asymptotic properties:

$$\rho_{Robin} \approx \begin{cases} 1 - \mathcal{O}(T^{-1}), & \text{if } l \geq 0 \text{ and } a > 0, \\ 1 - \mathcal{O}(T^{-\frac{1}{6}}), & \text{if } l > 0 \text{ and } a = 0, \\ 1 - \mathcal{O}(T^{-\frac{1}{4}}), & \text{if } l = 0 \text{ and } a = 0. \end{cases}$$

Remark 1 For the initial value problem (2) with $a < 0$, the min-max problem concerning the best choice of the parameter is

$$\min_{q>0} \max_{y \in [y_0, y_1]} \mathcal{R}(y, q), \text{ with } \mathcal{R}(y, q) = \frac{(y-q)^2 + y^2 - y_0^2}{(y+q)^2 + y^2 - y_0^2} e^{-y},$$

where $y_0 = 2l\sqrt{\left(\sqrt{a^2 + (\pi/T)^2} - a\right)/(2\mu)}$. We see that, this min-max problem is different from the one given by (4). For $a < 0$, Δx small and $\Delta t = \mathcal{O}(\Delta x^r)$, the solution q_{opt} is determined by the *equioscillation* principle [4]; an illustration is shown in Fig. 1 on the left. However, this principle does not always hold for the case $a \geq 0$; in particular, we have shown that for $\Delta t = \mathcal{O}(\Delta x^r)$ with $r < \frac{4}{3}$, it does not hold [7]. A concrete example is shown in Fig. 1 on the right, where we see that, based on the optimal parameter q_{opt}, the local maximum of the objective function $\mathcal{R}$ defined by (4) is smaller than $\mathcal{R}(y_0, q_{opt})$.

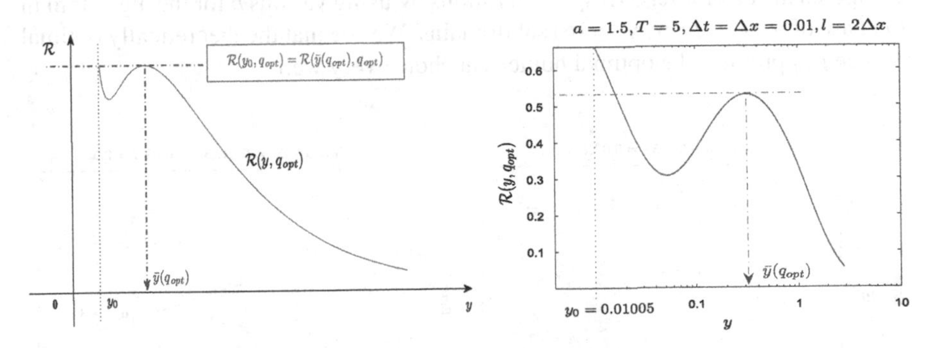

Fig. 1 *Left*: illustration of the *equioscillation* principle for the case $a < 0$. *Right*: an example for $a \geq 0$ and $\Delta t = \mathcal{O}(\Delta x^r)$ with $r < \frac{4}{3}$, where the *equioscillation* does not hold

3 Numerical Results

We consider the following linear problem with homogeneous initial and boundary conditions:

$$u_t = u_{xx} + au + t^2 \sin(xt), \quad (x,t) \in (0,4) \times (0,T), \tag{8}$$

The Laplace operator ∂_{xx} is treated by the centered finite difference scheme and then the derived system of ODEs is solved by the backward Euler method.

Example 1 (Dirichlet Transmission Condition) For (8), we choose $a > 0$ and $T = 60$. Let $\Delta t = 0.02$, $\Delta x = 0.01$ and $l = 2\Delta x$ (overlap size). Then, from Proposition 1 we know that the allowed maximal a is 0.5814 for $N = 4$ and 0.4312 for $N = 16$. In Fig. 2, we show the measured error corresponding to several choices of a. By "error" we denote here the discrete L_∞ norm in time and space of the difference between the converged solution and the iterate. We see that when a tends to its allowed upper bound, the SWR algorithm converges slowly.

Example 2 (Robin Transmission Condition) We now choose $T = 5$ for (8), and $\Delta t = \Delta x = \frac{1}{2^5}$ for the discretization parameters. For $a > 0$, by employing a changed variables $v(x,t) = e^{-\sigma t}u(x,t)$ the linear problem (8) can be transformed to $\partial_t v = \partial_{xx} v + (a - \sigma)v + e^{-\sigma t}t^2 \sin(xt)$ with homogeneous initial and boundary conditions. Then, by choosing a large σ we will have $a - \sigma < 0$. The SWR algorithm with negative coefficient $a - \sigma$ can converge very fast, while the error $\max_j \left\| e^{\sigma t} v_j^k - u_j \right\|_{\infty,\infty}$ diminishes slowly. By letting $l = 5\Delta x$ and $a = 1.5$, we illustrate this point in Fig. 3 for $\sigma = 2$ (left) and $\sigma = 3.5$ (right).

We next investigated how close the parameter p_{opt} given by Proposition 1 is to the best possible one for the numerical code. In Fig. 4 on the left (resp. right), we computed the error after 5 (resp. 7) iterations by using various p for the algorithm in the case of $N = 2$ (resp. $N = 16$) subdomains. We see that the theoretically optimal choice p_{opt} predicts the optimal numerical choice very well.

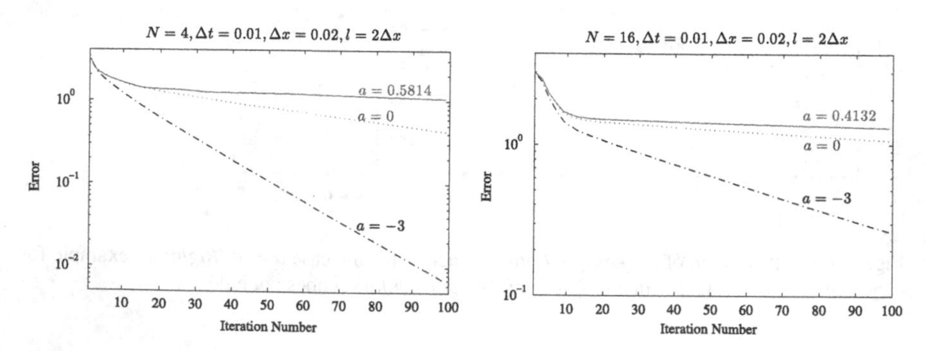

Fig. 2 Measured error of the classical SWR algorithm for different choices of a

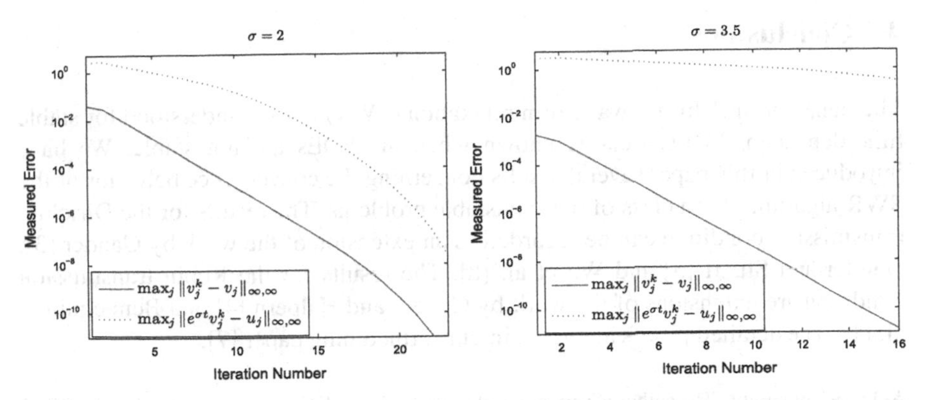

Fig. 3 Measured diminishing rate of $\max_j \|v_j^k - v_j\|_{\infty,\infty}$ and $\max_j \|e^{\sigma t} v_j^k - u_j\|_{\infty,\infty}$, with two choices of σ: $\sigma = 2$ (*left*) and $\sigma = 3.5$ (*right*)

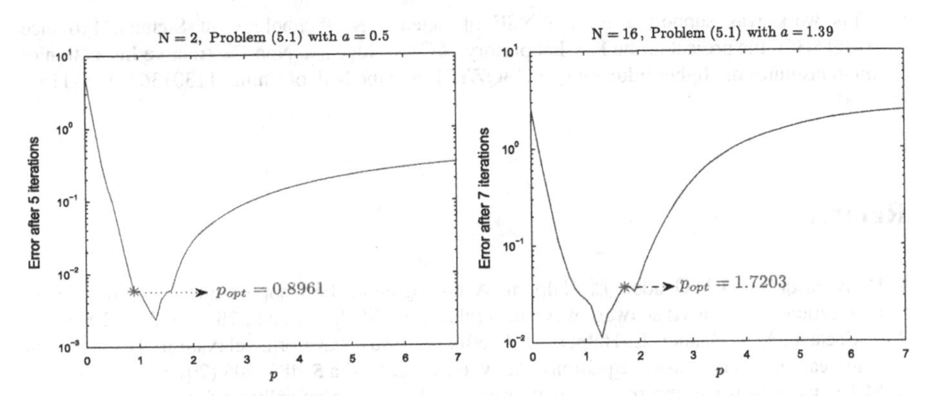

Fig. 4 Comparison of the numerical and analytical optimal parameter. *Left*: 2 subdomains and $l = 5\Delta x$. *Right*: 16 subdomains and $l = 4\Delta x$

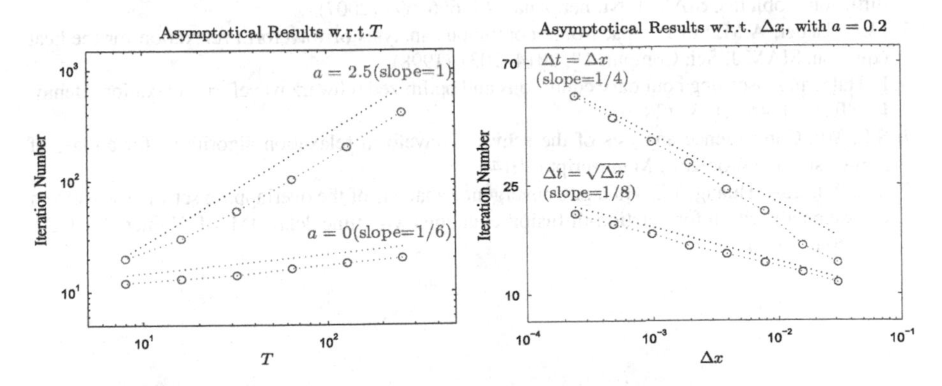

Fig. 5 Asymptotic behavior of the optimized SWR algorithm in the two subdomain case

The asymptotic behavior of the optimized SWR algorithm is shown in Fig. 5, and we see that the results coincide with Proposition 4.

4 Conclusions

The behavior of Schwarz waveform relaxation (SWR) is well understood for stable time-dependent PDEs. Less is known when the PDEs are not stable. We have introduced in this paper several results concerning the convergence behavior of the SWR algorithm for a class of these unstable problems. The results for the Dirichlet transmission condition can be regarded as an extension of the work by Gander [3], Gander and Stuart [5] and Wu et al. [8]. The results for the Robin transmission condition are extensions of the work by Gander and Halpern [4] and Bennequin et al. [1]. The detailed proofs are given in our forthcoming paper [7].

Acknowledgement The author is very grateful to Prof. Martin J. Gander, for his fund of the DD22 conference, his careful reading and revision of this paper, and his professional instructions in many fields.

This work was supported by the NSF of Science & Technology of Sichuan Province (2014JQ0035), the project of the Key Laboratory of Cambridge and Non-Destructive Inspection of Sichuan Institutes of Higher Education (2013QZY01) and the NSF of China (11301362, 11371157, 91130003).

References

1. D. Bennequin, M.J. Gander, L. Halpern, A homographic best approximation problem with application to optimized schwarz waveform relaxation. Math. Comput. **78**, 185–223 (2009)
2. F. Caetano, M.J. Gander, L. Halpern, J. Szeftel, Schwarz waveform relaxation algorithms for smilinear reaction-diffusion equations. Netw. Heterog. Media **5**, 487–505 (2010)
3. M.J. Gander, A waveform relaxation algorithm with overlapping splitting for reaction diffusion equations. Numer. Linear Algebra Appl. **6**, 125–145 (1998)
4. M.J. Gander, L. Halpern, Optimized schwarz waveform relaxation for advection reaction diffusion problems. SIAM J. Numer. Anal. **45**, 666–697 (2007)
5. M.J. Gander, A.M. Stuart, Space-time continuous analysis of waveform relaxation for the heat equation. SIAM J. Sci. Comput. **19**, 2014–2031 (1998)
6. L. Halpern, Absorbing boundary conditions and optimized schwarz waveform relaxation. Behav. Inf. Technol. **46**, 21–34 (2006)
7. S.L. Wu, Convergence analysis of the schwarz waveform relaxation algorithms for a class of non-dissipative problems. Manuscript (2014)
8. S.L. Wu, C.M. Huang, T.Z. Huang, Convergence analysis of the overlapping schwarz waveform relaxation algorithm for reaction-diffusion equations with time delay. IMA J. Numer. Anal. **32**, 632–671 (2012)

Optimized Schwarz Method with Two-Sided Transmission Conditions in an Unsymmetric Domain Decomposition

Martin J. Gander and Yingxiang Xu

1 Introduction

Domain decomposition (DD) methods are important techniques for designing parallel algorithms for solving partial differential equations. Since the decomposition is often performed using automatic mesh partitioning tools, one can in general not make any assumptions on the shape or physical size of the subdomains, especially if local mesh refinement is used. In many of the popular domain decomposition methods, neighboring subdomains are not using the same type of boundary conditions, e.g. the Dirichlet-Neumann methods invented by Bjørstad and Widlund [2], or the two-sided optimized Schwarz methods proposed in [3], and one has to decide which subdomain uses which boundary condition. A similar question also arises in mortar methods, see [1], where one has to decide on the master and slave side at the interfaces. In [4], it was found that for optimized Schwarz methods, the subdomain geometry and problem boundary conditions influence the optimized Robin parameters for symmetrical finite domain decompositions, and in [5], it was observed numerically that swapping the optimized two-sided Robin parameters can accelerate the convergence for a circular domain decomposition.

We study in this paper two-sided optimized Schwarz methods for a model decomposition into a larger and a smaller subdomain, to investigate which Robin parameter should be used on which subdomain in order to get fast convergence.

M.J. Gander
Section de Mathématiques, Université de Genève, 2-4 rue du Lièvre, CP 64, 1211 Genève 4, Switzerland
e-mail: Martin.Gander@unige.ch

Y. Xu (✉)
School of Mathematics and Statistics, Northeast Normal University, Changchun 130024, China
e-mail: yxxu@nenu.edu.cn

T. Dickopf et al. (eds.), *Domain Decomposition Methods in Science and Engineering XXII*, Lecture Notes in Computational Science and Engineering 104, DOI 10.1007/978-3-319-18827-0_65

We consider the model problem

$$\Delta u - \eta u = f \quad \text{in } \Omega, \qquad u|_{\partial\Omega} = 0, \tag{1}$$

where $\Omega = \{(x,y) \in \mathbb{R}^2 | -a \leq x \leq b\}$ is decomposed into two subdomains $\Omega = \Omega_1 \cup \Omega_2$, with $\Omega_1 = \{(x,y) \in \mathbb{R}^2 | -a \leq x \leq L\}$, $\Omega_2 = \{(x,y) \in \mathbb{R}^2 | 0 \leq x \leq b\}$, and $L \geq 0$ is the overlap between subdomains, $a + L < b$. Note here in the y-direction, the domain Ω is still infinite, but this will not affect our theoretical findings, since in numerical computations the Fourier frequency lies in between $k_{\min}$ and $k_{\max}$, the lowest and the highest frequencies involved in the computation, and we will use this in our analysis.

We focus in this short paper on the parallel Schwarz method

$$\begin{aligned} &\Delta u_1^n - \eta u_1^n = f, \text{ in } \Omega_1, \quad &&\Delta u_2^n - \eta u_2^n = f, \text{ in } \Omega_2, \\ &u_1^n(-a,y) = 0, &&u_2^n(b,y) = 0, \end{aligned} \tag{2}$$

with two-sided Robin transmission conditions

$$\begin{aligned} (\partial_x + p_1)u_1^n(L,\cdot) &= (\partial_x + p_1)u_2^{n-1}(L,\cdot), \\ (\partial_x - p_2)u_2^n(0,\cdot) &= (\partial_x - p_2)u_1^{n-1}(0,\cdot), \end{aligned} \tag{3}$$

where p_1, p_2 are positive constants.

2 Optimized Two-Sided Robin Transmission Conditions

Inserting a Fourier expansion of the iterates, $u_i^n(x,y) = \sum_{k=-\infty}^{\infty} \hat{u}_i^n(x,k)e^{iky}$, into (2) and iterating the solutions between subdomains through the transmission condition (3), see for example [3], we obtain for each Fourier mode k the contraction factor

$$\begin{aligned} \rho(k,\eta,L,p_1,p_2,a,b) = &\frac{\sqrt{\eta+k^2}(1+e^{-2\sqrt{\eta+k^2}(b-L)}) - p_1(1-e^{-2\sqrt{\eta+k^2}(b-L)})}{\sqrt{\eta+k^2}(1+e^{-2\sqrt{\eta+k^2}(a+L)}) + p_1(1-e^{-2\sqrt{\eta+k^2}(a+L)})} \cdot \\ &\frac{\sqrt{\eta+k^2}(1+e^{-2\sqrt{\eta+k^2}a}) - p_2(1-e^{-2\sqrt{\eta+k^2}a})}{\sqrt{\eta+k^2}(1+e^{-2\sqrt{\eta+k^2}b}) + p_2(1-e^{-2\sqrt{\eta+k^2}b})} \cdot e^{-2\sqrt{\eta+k^2}L}. \end{aligned} \tag{4}$$

To obtain the fastest method for all relevant Fourier modes k, we have to solve the optimization problem

$$\min_{p_1,p_2>0} \rho_{\max}(L,p_1,p_2), \tag{5}$$

where $\rho_{\max}(L, p_1, p_2) := \max_{k_{\min} \le k \le k_{\max}} |\rho(k, \eta, L, p_1, p_2, a, b)|$ and $k_{\min}$, $k_{\max}$ are estimates of the lowest and the highest frequencies involved in the computation. If h is the mesh size along the interface, and the interface length is c, one can estimate $k_{\min} = \pi/c$ and $k_{\max} = \pi/h$, see [3].

Since the frequency k is involved in the contraction factor in a complicated fashion, (5) can not be solved analytically. We show here a new idea, namely to approximate ρ for large k asymptotically accurately in order to solve the optimization problem (5). To this end, we introduce

$$\rho_{app}(k, \eta, L, p_1, p_2) = \frac{\sqrt{\eta + k^2} - p_1}{\sqrt{\eta + k^2} + p_1} \cdot \frac{\sqrt{\eta + k^2} - p_2}{\sqrt{\eta + k^2} + p_2} \cdot e^{-2\sqrt{\eta+k^2}L}, \tag{6}$$

which is the contraction factor obtained by Gander [3] in the infinite, symmetric domain decomposition analysis.

Theorem 1 (Approximation to the Contraction Factor) *The difference between the exact and approximate contraction factor satisfies the estimate*

$$|\rho(k, \eta, L, p_1, p_2, a, b) - \rho_{app}(k, \eta, L, p_1, p_2)| \le 4e^{-2\sqrt{\eta+k^2}(a+L)}. \tag{7}$$

Proof The contraction factor ρ can be rewritten in the form

$$\rho = \rho_{app} + (1 - \rho_{app}) \left(\frac{\sqrt{\eta+k^2} - p_2}{\sqrt{\eta+k^2} + p_2} e^{-2\sqrt{\eta+k^2}b} + \frac{\sqrt{\eta+k^2} - p_1}{\sqrt{\eta+k^2} + p_1} e^{-2\sqrt{\eta+k^2}(a+L)} \right), \tag{8}$$

and the result then follows by the triangle inequality and using that $-1 \le \rho_{app} \le 1$.
□

Theorem 1 shows that ρ_{app} is a good approximation for k large, but not for k small. We thus propose to only use the approximation for k large, and the exact ρ for k small, in order to solve the min-max problem (5) asymptotically. We obtain the following theorems, whose proofs are beyond the scope of this short paper, see our forthcoming paper [6].

Theorem 2 (Optimized Parameters, Overlapping Case) *With the overlap $L > 0$, the parameters $p_1^* = G^{\frac{4}{5}} L^{-\frac{1}{5}}$, $p_2^* = G^{\frac{2}{5}} L^{-\frac{3}{5}}$ solve asymptotically the equioscillation equations*

$$\rho(k_{\min}, \eta, L, p_1^*, p_2^*, a, b) = -\rho_{app}(\bar{k}_1, \eta, L, p_1^*, p_2^*) = \rho_{app}(\bar{k}_2, \eta, L, p_1^*, p_2^*), \tag{9}$$

where $G = G(k_{\min}, \eta, a, b) := \frac{\sqrt{\eta + k_{\min}^2}}{2} \frac{1 - e^{-2\sqrt{\eta + k_{\min}^2}(a+b)}}{(1 - e^{-2\sqrt{\eta + k_{\min}^2}a})(1 - e^{-2\sqrt{\eta + k_{\min}^2}b})}$, *and* $\bar{k}_1 = G^{\frac{3}{5}} L^{-\frac{2}{5}}$ *and* $\bar{k}_2 = G^{\frac{1}{5}} L^{-\frac{4}{5}}$ *are the locations of the interior maxima of* ρ_{app}. *Furthermore,* p_1^*, p_2^* *approximately solve the min-max problem (5) as* $L \to 0$ *with* *the error estimate*

$$|\rho_{\max}(L, p_1^*, p_2^*) - \min_{p_1,p_2>0} \rho_{\max}(L, p_1, p_2)| \leq 4e^{-2\sqrt{\eta+\underline{k}_1^2}(a+L)}, \quad (10)$$

where $\underline{k}_1 = cL^{-\frac{1}{5}}$, *and c is some constant. The associated contraction factor is*

$$\rho_{\max}(L, p_1^*, p_2^*) = 1 - 4G^{\frac{1}{5}}L^{\frac{1}{5}} + O(L^{\frac{2}{5}}). \quad (11)$$

Theorem 3 (Optimized Parameters, Nonoverlapping Case) *When* $L = 0$, *the parameters* $\bar{p}_1 = 2^{\frac{1}{4}}G^{\frac{3}{4}}k_{\max}^{\frac{1}{4}}$, $\bar{p}_2 = 2^{\frac{3}{4}}G^{\frac{1}{4}}k_{\max}^{\frac{3}{4}}$ *solve asymptotically the equioscillation equations*

$$\rho(k_{\min}, \eta, 0, \bar{p}_1, \bar{p}_2, a, b) = -\rho_{app}(\bar{k}, \eta, 0, \bar{p}_1, \bar{p}_2) = \rho_{app}(k_{\max}, \eta, 0, \bar{p}_1, \bar{p}_2), \quad (12)$$

where $\bar{k} = (2G)^{\frac{1}{2}}k_{\max}^{\frac{1}{2}}$ *is the location of the interior maximum of* ρ_{app}. *Furthermore,* $\bar{p}_1$, $\bar{p}_2$ *solve approximately the min-max problem (5) as* $k_{\max} \to \infty$ *with the error estimate*

$$|\rho_{\max}(0, \bar{p}_1, \bar{p}_2) - \min_{p_1,p_2>0} \rho_{\max}(0, p_1, p_2)| \leq 4e^{-2\sqrt{\eta+\underline{k}_0^2}a}, \quad (13)$$

where $\underline{k}_0 = ck_{\max}^{\frac{1}{4}}$, *and c is some constant. The associated contraction factor is*

$$\rho_{\max}(0, \bar{p}_1, \bar{p}_2) = 1 - 2^{\frac{7}{4}}G^{\frac{1}{4}}k_{\max}^{-\frac{1}{4}} + O(k_{\max}^{-\frac{1}{2}}). \quad (14)$$

3 Swapping the Robin Parameters

Theorems 2 and 3 do not allow us to see which Robin parameter of the two should be used on which subdomain, swapping them leads to the same asymptotic results. To see the influence of the domain size, we have to push the asymptotic analysis further.

The Overlapping Case To get further insight, we compute one more term in the asymptotic expansions of the equioscillation equations (9) both for the parameter ordering given, and swapped. We obtain at the interior maximum points $\bar{k}_1$ and $\bar{k}_2$ the same result $\rho_{\max} = 1 - 4G^{\frac{1}{5}}L^{\frac{1}{5}} + 8G^{\frac{2}{5}}L^{\frac{2}{5}} + O(L^{\frac{3}{5}})$, while at $k_{\min}$

$$\rho(k_{\min}, \eta, L, p_1^*, p_2^*, a, b) = 1 - 4G^{\frac{1}{5}}L^{\frac{1}{5}} + 8G^{\frac{2}{5}}L^{\frac{2}{5}}(1+d) + O(L^{\frac{3}{5}}), \quad (15)$$

$$\rho(k_{\min}, \eta, L, p_2^*, p_1^*, a, b) = 1 - 4G^{\frac{1}{5}}L^{\frac{1}{5}} + 8G^{\frac{2}{5}}L^{\frac{2}{5}}(1-d) + O(L^{\frac{3}{5}}), \quad (16)$$

where the additional term

$$d := \frac{e^{-2\sqrt{\eta+k_{\min}^2}a} - e^{-2\sqrt{\eta+k_{\min}^2}b}}{1 - e^{-2\sqrt{\eta+k_{\min}^2}(a+b)}} \tag{17}$$

appears. Hence, if $d > 0$, i.e. $a < b$, one should swap the parameters to get a uniform contraction factor bounded by $\rho_{\max}$, since $G > 0$, and we get

Theorem 4 *If $a < b$ and L is small, swapping the transmission parameters p_1^* and p_2^* improves the performance of the optimized Schwarz method (2), and the bigger the value of d in (17) is, the larger the improvement becomes.*

The natural next question is: from which overlap on should one swap the transmission parameters to get better performance? Notice that $|\rho|$ has the same asymptotic expansions at $\bar{k}_1$ and $\bar{k}_2$ up to $O(L^{\frac{3}{5}})$. We should thus look for an $L^* > 0$ such that when $L < L^*$

$$|\rho(k_{\min}, \eta, L, p_2^*, p_1^*, a, b)| < |\rho(k_{\min}, \eta, L, p_1^*, p_2^*, a, b)|. \tag{18}$$

Though it is hard to obtain an explicit expression of such an L^*, the inequality (18) can be used numerically as a necessary condition for judging when we should swap the optimized transmission parameters. A sufficient condition can be obtained as follows: if

$$p_1^* > \sqrt{\eta + k_{\min}^2}\,\frac{1 + e^{-2\sqrt{\eta+k_{min}^2}a}}{1 - e^{-2\sqrt{\eta+k_{\min}^2}a}}, \tag{19}$$

then (4) implies $\rho(k_{\min}, \eta, L, p_2^*, p_1^*, a, b) > 0$. Now using (8), we obtain with a direct comparison after a short calculation $\rho(k_{\min}, \eta, L, p_2^*, p_1^*, a, b) < \rho(k_{\min}, \eta, L, p_1^*, p_2^*, a, b)$, which together with the positivity implies (18). Solving (19) asymptotically yields

$$L < \frac{1}{16\sqrt{\eta + k_{\min}^2}} \frac{(1 - e^{-2\sqrt{\eta+k_{\min}^2}(a+b)})^4 (1 - e^{-2\sqrt{\eta+k_{\min}^2}a})}{(1 + e^{-2\sqrt{\eta+k_{\min}^2}a})^5 (1 - e^{-2\sqrt{\eta+k_{\min}^2}b})^4} =: L^*. \tag{20}$$

Noting that $GL^* < 1$, all the above mentioned asymptotic expansions converge, and we arrive at

Theorem 5 *For $a < b$, with an overlap $L < L^*$, where L^* is defined in (20), swapping the transmission parameters p_1^* and p_2^* in the optimized Schwarz method (2) improves the performance.*

The Nonoverlapping Case We again compute one more term in the expansions of the equioscillation equations (12), both for the parameter ordering given, and swapped. We obtain as in the overlapping case at $\bar{k}$ and $k_{\max}$ the same result,

$1 - 2^{\frac{7}{4}} G^{\frac{1}{4}} k_{\max}^{-\frac{1}{4}} + 2^{\frac{5}{2}} G^{\frac{1}{2}} k_{\max}^{-\frac{1}{2}} + O(k_{\max}^{-\frac{3}{4}})$, while at $k_{\min}$

$$\rho(k_{\min}, \eta, 0, \bar{p}_1, \bar{p}_2, a, b) = 1 - 2^{\frac{7}{4}} G^{\frac{1}{4}} k_{\max}^{-\frac{1}{4}} + 2^{\frac{5}{2}} G^{\frac{1}{2}} k_{\max}^{-\frac{1}{2}}(1+d) + O(k_{\max}^{-\frac{3}{4}}), \quad (21)$$

$$\rho(k_{\min}, \eta, 0, \bar{p}_2, \bar{p}_1, a, b) = 1 - 2^{\frac{7}{4}} G^{\frac{1}{4}} k_{\max}^{-\frac{1}{4}} + 2^{\frac{5}{2}} G^{\frac{1}{2}} k_{\max}^{-\frac{1}{2}}(1-d) + O(k_{\max}^{-\frac{3}{4}}), \quad (22)$$

where the same term d from (17) appears. Hence, as in the overlapping case, if

$$|\rho(k_{\min}, \eta, 0, \bar{p}_2, \bar{p}_1, a, b)| < |\rho(k_{\min}, \eta, 0, \bar{p}_1, \bar{p}_2, a, b)|, \quad (23)$$

swapping the transmission parameters in the optimized Schwarz method (2) improves the performance. Solving $\bar{p}_1 > \sqrt{\eta + k_{\min}^2}\frac{1+e^{-2\sqrt{\eta+k_{\min}^2}a}}{1-e^{-2\sqrt{\eta+k_{\min}^2}a}}$ with $k_{\max} = \pi/h$ gives an $\bar{h} = 2G^3\pi(\frac{1}{\sqrt{\eta+k_{\min}^2}}\frac{1-e^{-2\sqrt{\eta+k_{\min}^2}a}}{1+e^{-2\sqrt{\eta+k_{\min}^2}a}})^4$ such that for any $h < \bar{h}$ inequality (23) holds, and we get

Theorem 6 *If $a < b$ and there is no overlap, and if $h < \bar{h}$, swapping the transmission parameters $\bar{p}_1$ and $\bar{p}_2$ of the optimized Schwarz method (2) improves the performance.*

4 Numerical Experiments

We consider the model problem (1), where $\eta = 2$, and the domain $\Omega = (-a, b) \times (0, 1)$ is decomposed into $\Omega_1 = (-a, L) \times (0, 1)$, $\Omega_2 = (0, b) \times (0, 1)$, with $a = 0.1$, and $b = 0.5$. We discretize (2) with the classical five-point finite difference scheme on a uniform mesh with mesh parameter h, and simulate directly the error equations, i.e. $f = 0$. The initial guesses on the interfaces are chosen randomly so that all frequencies are present. We count the number of iterations required to reach an error reduction of $1e-6$, and compare the results obtained with our parameters to those obtained with parameters from the infinite domain decomposition analysis in [3], denoted by the subscript "inf". Table 1 shows the corresponding results, both for

Table 1 Number of iterations required by the various optimized Schwarz methods

		h				
L	Transmission parameters	1/50	1/100	1/200	1/400	1/800
h	$p_1 = p_1^*(p_2^*), p_2 = p_2^*(p_1^*)$	6(9)	7(9)	8(9)	10(10)	12(11)
h	$p_1 = p_{1,\mathtt{inf}}^*(p_{2,\mathtt{inf}}^*), p_2 = p_{2,\mathtt{inf}}^*(p_{1,\mathtt{inf}}^*)$	7(12)	8(15)	10(18)	11(18)	13(15)
0	$p_1 = \bar{p}_1(\bar{p}_2), p_2 = \bar{p}_2(\bar{p}_1)$	11(10)	14(12)	17(14)	19(16)	24(20)
0	$p_1 = \bar{p}_{1,\mathtt{inf}}(\bar{p}_{2,\mathtt{inf}}), p_2 = \bar{p}_{2,\mathtt{inf}}(\bar{p}_{1,\mathtt{inf}})$	13(14)	16(14)	18(17)	22(20)	27(24)

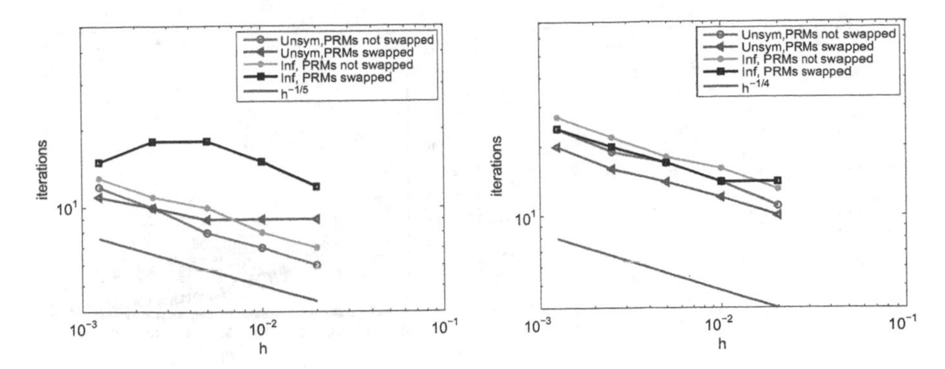

Fig. 1 Number of iterations required by the optimized Schwarz methods: overlapping case on the *left*, nonoverlapping case on the *right*

Table 2 Error reduction of each optimized Schwarz method

L	$2h$	$3h$	$4h$	$5h$
$p_1 = p_1^*, p_2 = p_2^*$	$1.5467e-06$	$1.6942e-07$	$1.6390e-08$	$1.0579e-08$
$p_1 = p_2^*, p_2 = p_1^*$	$1.4941e-07$	$3.7717e-08$	$3.4429e-08$	$2.9584e-08$

the overlapping case, $L = h$, and the nonoverlapping case, $L = 0$, with the results after parameter swapping in parentheses. In both cases, our parameters require less iterations than those from the infinite domain decomposition analysis. For the new parameters in the nonoverlapping case, the swapped transmission parameters perform better, which is in agreement with Theorem 6, since all the mesh sizes involved in this computation are less than $\bar{h} \approx 0.0234$. For the overlapping case, we see that swapping for the larger mesh sizes is not advantageous, but as soon as the mesh size becomes small, the swapped parameters catch up to give lower iteration numbers. The situation is similar for the parameters from the infinite domain decomposition analysis, but a more refined mesh would be required. We also plot all the results in Fig. 1, on the left for the overlapping case and on the right for the nonoverlapping case. We observe that each method performs as predicted by the asymptotic analysis, except in the case of the infinite domain decomposition analysis with overlap where a more refined mesh would be needed to reach the asymptotic regime.

We next illustrate numerically that there is indeed a critical value L^* so that when the overlap $L < L^*$, swapping the parameters can improve the performance, as predicted by Theorem 5. Table 2 shows the error after 10 iterations of the optimized Schwarz method with varying overlap and $h = 1/800$. We see that in this case L^* lies in between $3h$ and $4h$.

To finally test how well our analysis predicts the optimal parameters to be used in a numerical setting, we vary the parameters p_1 and p_2 with 51 equidistant samples of each for a fixed problem with $h = 1/200$ and count for each parameter pair (p_1, p_2) the number of iterations to reach a residual of $1e-6$. In the left column of Fig. 2 we

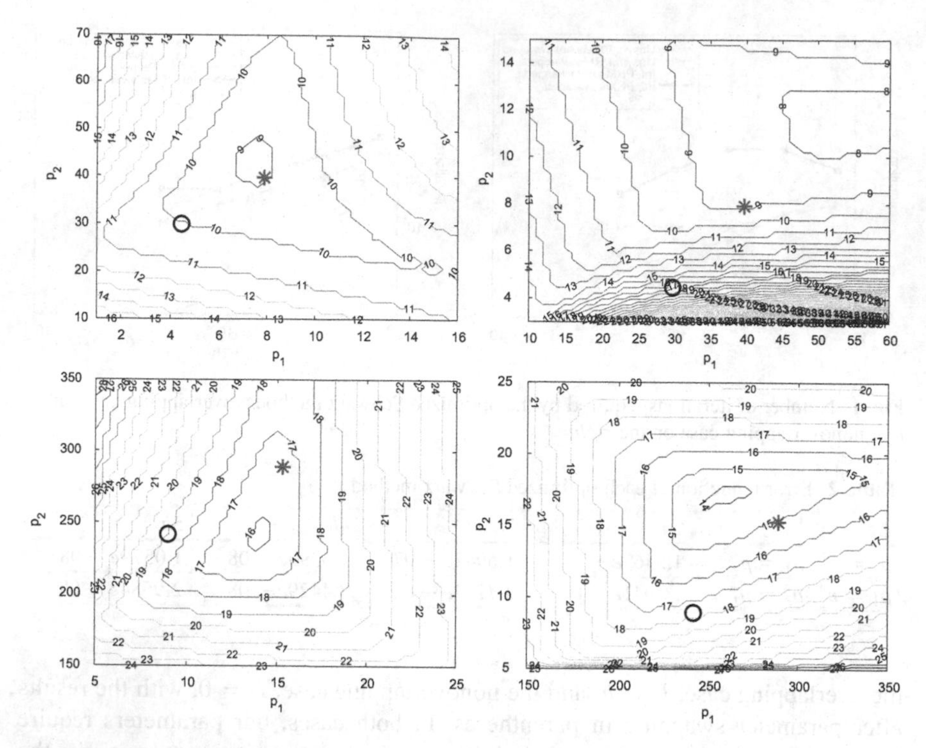

Fig. 2 Optimized parameters found by our analysis (*asterisk*), as well as by the infinite domain decomposition analysis (*circle*), compared to the performance of other values of the parameters: *first row* for the overlapping case, *second row* for the nonoverlapping case, with the parameters before parameter swapping in the *left column* and after swapping in the *right column*

show a contour plot before transmission parameter swapping, the overlapping case (on the top) and the nonoverlapping case (at the bottom), and in the right column the corresponding contour plots with transmission parameters swapped. We see that the transmission parameters obtained by our analysis ($*$) are always closer to the numerical optimum than those from the infinite domain decomposition analysis ($\circ$). The left and right columns of Fig. 2 also show numerically that there exist at least 2 local minimizers and the swapped parameters are close to the one resulting in the smaller contraction factor.

5 Conclusion

We have shown that when there are two different transmission conditions to be imposed between subdomains, the geometry, in our case the size of the subdomain, can indicate which subdomain should use which transmission condition. Using asymptotic analysis for a two subdomain model problem, we developed a necessary and a sufficient condition on the overlap or mesh size for when transmission conditions should be swapped between neighboring subdomains of different size to get better performance. Numerical experiments confirm well our theoretical findings. We also observed numerically that the min-max problem (5) has at least two local minimizers, but a more refined pre-asymptotic study of this problem is needed for a complete understanding of (5).

Acknowledgements The author "Yingxiang Xu" was partly supported by NSFC-11201061, 11471047, 11271065, CPSF-2012M520657 and the Science and Technology Development Planning of Jilin Province 20140520058JH.

References

1. C. Bernardi, Y. Maday, A.T. Patera, Domain decomposition by the mortar element method, in *Asymptotic and Numerical Methods for Partial Differential Equations with Critical Parameters* (Springer, New York, 1993), pp. 269–286
2. P.E. Bjørstad, O.B. Widlund, Iterative methods for the solution of elliptic problems on regions partitioned into substructures. SIAM J. Numer. Anal. **23**(6), 1097–1120 (1986)
3. M.J. Gander, Optimized Schwarz methods. SIAM J. Numer. Anal. **44**(2):699–731 (2006)
4. M.J. Gander, On the influence of geometry on optimized Schwarz methods. SeMA J. **53**(1), 71–78 (2011)
5. M.J. Gander, Y. Xu, Optimized Schwarz methods for circular domain decompositions with overlap. SIAM J. Numer. Anal. **52**(4), 1981–2004 (2014)
6. Y. Xu, Optimized Schwarz methods with unsymmetric domain decomposition. SIAM J. Numer. Anal. (submitted, 2015)

A Domain Decomposition Approach in the Electrocardiography Inverse Problem

Nejib Zemzemi

1 Introduction

The inverse problem in cardiac electrophysiology also known as electrocardiography imaging (ECGI) is a new and a powerful diagnosis technique. It allows the reconstruction of the electrical potential on the heart surface from electrical potentials measured on the body surface. This non-invasive technology and other similar techniques like the electroencephalography imaging interest more and more medical industries. The success of these technology would be considered as a breakthrough in the cardiac and brain diagnosis. However, in many cases the quality of reconstructed electrical potential is not sufficiently accurate. The difficulty comes from the fact that the inverse problem in cardiac electrophysiology is well known as a mathematically ill-posed problem. Different methods based on Thikhnov regularization [4] have been used in order to regularize the problem, but still the reconstructed electrical potential is not sufficiently satisfactory. In this study we present a domain decomposition approach to solve the inverse problem.

2 Methods

The domain decomposition method that will be presented in this paper is tested on synthetical data. This data is generated by solving the forward problem of ECGs. In the following paragraphs we will present first, the forward problem then the domain decomposition method for solving the inverse problem.

N. Zemzemi (✉)
INRIA Bordeaux Sud-Ouest, Carmen team, 200 Avenue de la Vieille Tour, 33405 Talence, France

Electrophysiology and Heart Modeling Institute (IHU LIRYC), Bordeaux, France
e-mail: nejib.zemzemi@inria.fr

T. Dickopf et al. (eds.), *Domain Decomposition Methods in Science and Engineering XXII*, Lecture Notes in Computational Science and Engineering 104, DOI 10.1007/978-3-319-18827-0_66

2.1 Forward Problem

The bidomain equations were used to simulate the electrical activity of the heart and extracellular potentials in the whole body (see e.g. [8, 9, 11]). These equations in the heart domain Ω_{H} are given by:

$$\begin{cases} A_{\mathrm{m}}\big(C_{\mathrm{m}}\dot{V}_{\mathrm{m}} + I_{\mathrm{ion}}(V_{\mathrm{m}},\boldsymbol{w})\big) - \operatorname{div}\big(\boldsymbol{\sigma}_{\mathrm{i}}\nabla V_{\mathrm{m}}\big) & \\ \qquad\qquad = \operatorname{div}\big(\boldsymbol{\sigma}_{\mathrm{i}}\nabla u_{\mathrm{e}}\big) + I_{\mathrm{stim}}, & \text{in } \Omega_{\mathrm{H}}, \\ -\operatorname{div}\big((\boldsymbol{\sigma}_{\mathrm{i}} + \boldsymbol{\sigma}_{\mathrm{e}})\nabla u_{\mathrm{e}}\big) = \operatorname{div}(\boldsymbol{\sigma}_{\mathrm{i}}\nabla V_{\mathrm{m}}), & \text{in } \Omega_{\mathrm{H}}, \\ \dot{\boldsymbol{w}} + \boldsymbol{g}(V_{\mathrm{m}}, \boldsymbol{w}) = 0, & \text{in } \Omega_{\mathrm{H}}, \\ \boldsymbol{\sigma}_{\mathrm{i}}\nabla V_{\mathrm{m}} \cdot \boldsymbol{n} = -\boldsymbol{\sigma}_{\mathrm{i}}\nabla u_{\mathrm{e}} \cdot \boldsymbol{n}, & \text{on } \Sigma. \end{cases} \tag{1}$$

The state variables V_{m} and u_{e} stand for the transmembrane and the extra-cellular potentials. Constants A_{m} and C_{m} represent the rate of membrane surface per unit of volume and the membrane capacitance, respectively. I_{stim} and I_{ion} are the stimulation and the transmembrane ionic currents. The heart-torso interface is denoted by Σ. The intra- and extracellular (anisotropic) conductivity tensors, $\boldsymbol{\sigma}_{\mathrm{i}}$ and $\boldsymbol{\sigma}_{\mathrm{e}}$, are given by $\boldsymbol{\sigma}_{\mathrm{i},e} = \sigma^{\mathrm{t}}_{\mathrm{i},e}\boldsymbol{I} + (\sigma^{\mathrm{l}}_{\mathrm{i},e} - \sigma^{\mathrm{t}}_{\mathrm{i},e})\boldsymbol{a} \otimes \boldsymbol{a}$, where $\boldsymbol{a}$ is a unit vector parallel to the local fibre direction and $\sigma^{\mathrm{l}}_{\mathrm{i},e}$ and $\sigma^{\mathrm{t}}_{\mathrm{i},e}$ are, respectively, the longitudinal and transverse conductivities of the intra- and extra-cellular media. The field of variables $\boldsymbol{w}$ is a vector containing different chemical concentrations and various gate variables. Its time derivative is given by the vector of functions $\boldsymbol{g}$.

The precise definition of $\boldsymbol{g}$ and I_{ion} depend on the electrophysiological transmembrane ionic model. In the present work we make use of one of the biophysically detailed human ventricular myocyte model [10]. The ion channels and transporters have been modeled on the basis of the most recent experimental data from human ventricular myocytes.

Figure 1 provides a geometrical representation of the domains considered to compute extracellular potentials in the human body. In the torso domain Ω_{T}, the electrical potential u_{T} is described by the Laplace equation.

$$\begin{cases} \operatorname{div}(\boldsymbol{\sigma}_{\mathrm{T}}\nabla u_{\mathrm{T}}) = 0, & \text{in} \quad \Omega_{\mathrm{T}}, \\ \boldsymbol{\sigma}_{\mathrm{T}}\nabla u_{\mathrm{T}} \cdot \boldsymbol{n}_{\mathrm{T}} = 0, & \text{on} \quad \Gamma_{\mathrm{ext}}. \end{cases} \tag{2}$$

where $\boldsymbol{\sigma}_{\mathrm{T}}$ stands for the torso conductivity tensor and $\boldsymbol{n}_{\mathrm{T}}$ is the outward unit normal to the torso external boundary Γ_{ext}.

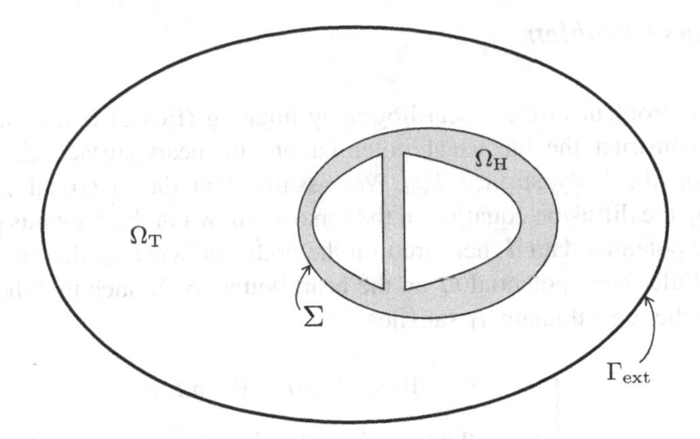

Fig. 1 Two-dimensional geometrical description: heart domain Ω_{H}, torso domain Ω_{T} (extramyocardial regions), heart-torso interface Σ and torso external boundary Γ_{ext}

The heart-torso interface Σ is supposed to be a perfect conductor. Then we have a continuity of current and potential between the extra-cellular myocardial region and the torso region.

$$\begin{cases} u_{\mathrm{e}} = u_{\mathrm{T}}, & \text{on} \quad \Sigma, \\ (\boldsymbol{\sigma}_{\mathrm{e}} + \boldsymbol{\sigma}_{\mathrm{i}})\nabla u_{\mathrm{e}} \cdot \boldsymbol{n}_{\mathrm{T}} = \boldsymbol{\sigma}_{\mathrm{T}} \nabla u_{\mathrm{T}} \cdot \boldsymbol{n}_{\mathrm{T}}, & \text{on} \quad \Gamma_{\mathrm{ext}}. \end{cases} \tag{3}$$

Other works [3, 6, 7] consider that the electrical current does not flow from the heart to the torso by assuming that the heart is isolated from torso. This approximation is appealing in terms of computational cost because it uncouples the Laplace equation (2) in the torso from the bidomain equations in the heart (1), which allows to reduce the size of the linear system to solve. It is even more appealing when the interest is only on the ECG computation, in that case the ECG solution could be an "off line" matrix vector multiplication after solving the bidomain equation, details about computing the transfer matrix could be found in [12]. Although this approach is very appealing in terms of computational cost, numerical evidence has shown that it can compromise the accuracy of the ECG signals (see e.g. [2, 6, 8]). Thus, in order to accurately compute ECGs we consider the state-of-the-art heart-torso full coupled electrophysiological problem (1)–(3) representing the cardiac electrical activity from the cell to the human body surface.

2.2 Inverse Problem

The inverse problem in electrocardiography imaging (ECGI) is a technique that allows to construct the electrical potential on the heart surface Σ from data measured on the body surface Γ_{ext}. We assume that the electrical potential is governed by the diffusion equation in the torso as shown in the previous paragraph. For a given potential data d measured on the body surface Γ_{ext}, the goal is to find the extracellular heart potential u_e on the heart boundary Σ such that the electrical potential in the torso domain u_T satisfies

$$\begin{cases} \operatorname{div}(\sigma_T \nabla u_T) = 0, \text{ in } \Omega_T, \\ u_T = d \text{ and } \sigma_T \nabla u_T.\boldsymbol{n} = 0, \text{ on } \Gamma_{\text{ext}}, \\ u_T = u_e, \text{ on } \Sigma. \end{cases} \tag{4}$$

In order to find u_T and the flux $\sigma_T \nabla u_T.\boldsymbol{n}$ on the boundary Σ we propose to decompose the problem into two auxiliary problems based on a mirror like boundary conditions in a fictitious domain as shown in Fig. 2. We consider $-\Omega_T$ (respectively, $-\Gamma_{\text{ext}}$) as the image of Ω_T (respectively, Γ_{ext}) through the interface boundary Σ. Then, we propose to define two functions u and v as follows,

$$\begin{cases} \operatorname{div}(\sigma_T \nabla u = 0, \text{ in } \Omega_T, \\ u = d, \text{ on } \Gamma_{\text{ext}}, \\ \sigma_T \nabla u.\boldsymbol{n} = -\sigma_T \nabla v.\boldsymbol{n}, \text{ on } \Sigma. \end{cases} \tag{5}$$

The function v is defined in the fictitious domain that we denoted $-\Omega_T$ as shown in Fig. 2.

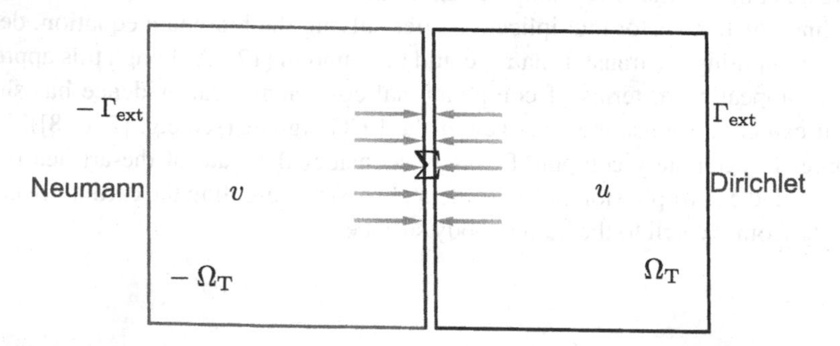

Fig. 2 Two-dimensional geometrical description of the physical torso domain Ω_T and its image $(-\Omega_T)$ through the boundary Σ. The *arrows* on the interface Σ show the opposite fluxes $\sigma_T \nabla u.\boldsymbol{n} = -\sigma_T \nabla v.\boldsymbol{n}$

$$\begin{cases} \operatorname{div}(\boldsymbol{\sigma}_{\mathrm{T}} \nabla v) = 0, \text{ in } -\Omega_{\mathrm{T}}, \\ \boldsymbol{\sigma}_{\mathrm{T}} \nabla v.\boldsymbol{n} = 0, \text{ on } -\Gamma_{\mathrm{ext}}, \\ v = u, \text{ on } \Sigma. \end{cases} \tag{6}$$

In order to solve the coupled problem (5)–(6), we use the domain decomposition technique. For a given initial guess $u^0 = 0$, we compute the incomplete boundary condition following the Algorithm 1.

Algorithm 1

for $i =$ *first time step* to *last time step* **do**
  $u^0 = 0$, on Γ_{ext}.
  load the known boundary data $u_{\mathrm{T}}(t_i)_{/\Gamma_{\mathrm{ext}}}$
  load the known boundary flux $\boldsymbol{\sigma}_{\mathrm{T}} \nabla u_{\mathrm{T}}(t_i)_{/\Gamma_{\mathrm{ext}}}.\boldsymbol{n}$
  error = 2 x tolerance.
  while error > tolerance) **do**
    Set $v^{k+1} = u^k$ on Σ
    compute v^{k+1} in $-\Omega_{\mathrm{T}}$
    Set $\boldsymbol{\sigma}_{\mathrm{T}} \nabla u^{k+1}.\boldsymbol{n} = -\boldsymbol{\sigma}_{\mathrm{T}} \nabla v^{k+1}.\boldsymbol{n}$ on Σ
    compute u^{k+1} in Ω_{T}
    compute error and relative error: norm($u^{k+1} - u^k$)
  end while
  save $u_{\mathrm{T}}(t_i)_{/\Sigma} = u^{k+1}_{/\Sigma}$
end for

The domain decomposition algorithm described in Algorithm 1 is accelerated using Aitken algorithm [5]. At the end of the while loop we have that $v^{k+1} = u^{k+1}$ up to the defined tolerance and $\boldsymbol{\sigma}_{\mathrm{T}} \nabla u^{k+1}.\boldsymbol{n} = -\boldsymbol{\sigma}_{\mathrm{T}} \nabla v^{k+1}.\boldsymbol{n}$ on Σ. If we define $\bar{v}^{k+1}$ as the symmetrical image of v^{k+1} through the interface Σ, we obtain $\boldsymbol{\sigma}_{\mathrm{T}} \nabla u^{k+1}.\boldsymbol{n} = \boldsymbol{\sigma}_{\mathrm{T}} \nabla \bar{v}^{k+1}.\boldsymbol{n}$ and $u^{k+1} = \bar{v}^{k+1}$ on Σ. Following [1, 13], this allows us to conclude that $u^{k+1} = \bar{v}^{k+1}$ in the whole torso domain Ω_{T}. Consequently, u^{k+1} satisfies both Dirichlet and Neumann boundary conditions on the external boundary Γ_{ext} and the Laplace equation in the torso domain, which solves the inverse problem.

3 Numerical Results

In order to test the domain decomposition approach for solving the inverse problem in electrocardiography, we generate synthetical data. We conducted numerical simulations solving the forward problem. We use the finite element LifeV[1] library for the numerical implementation of the method. For the sake of simplicity, we perform simulations on the volumes between three concentric spheres. The volume

[1]www.lifev.org.

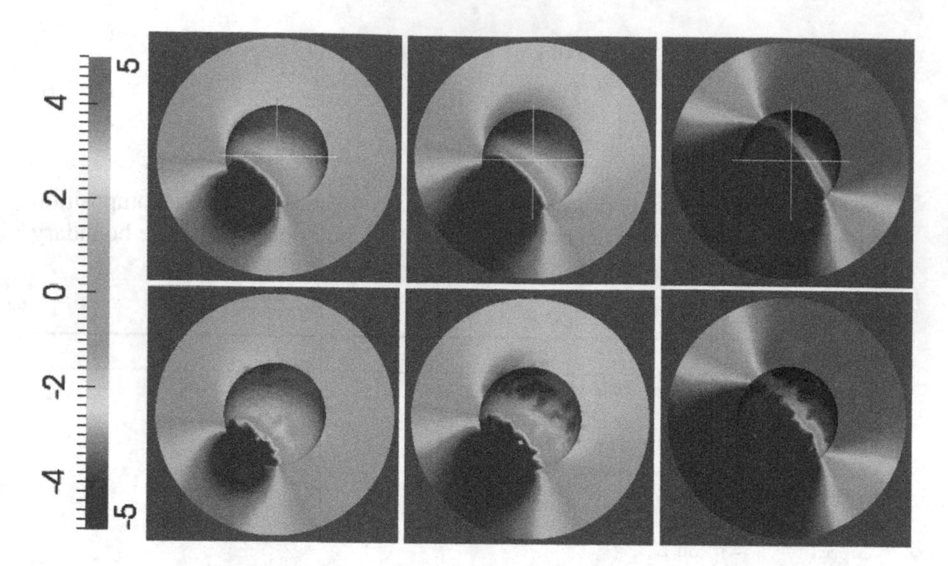

Fig. 3 *Top* (respectively, *bottom*): snapshots of numerical solution of the forward (respectively, inverse) problem at times 5, 10 and 25 ms (from *left to right*). The unit in the color bar is mV

between the small and the medium sphere represents the heart domain and the volume between the medium and the large spheres represents the torso. We show the torso potential in Fig. 3. The epicardium Σ is given by the internal sphere in Fig. 3 while the body surface Γ_{ext} is given by the external sphere in the same figure. In Fig. 3 (top) we show the forward problem solution at times 5, 10 and 25 ms (from left to right). We extract the electrical potential at the boundary Γ_{ext} and we use it as an input for the inverse problem after adding 10 % of white noise. We show the solution of the inverse problem in Fig. 3 (bottom). The three panels represent the inverse problem solution at times 5, 10 and 25 ms (from left to right). We observe that the space distribution of the electrical wave is well captured mainly in the internal sphere. This allows to construct the activation times on the heart surface with a good accuracy.

4 Conclusion

In this paper we presented a domain decomposition approach to solve the inverse problem in electrocardiography. The problem was formulated using a mirror-like boundary conditions at the heart torso interface, where we have continuity of potential and opposite fluxes. The preliminary results presented in this work show the capability of this method to capture the spatial distribution of the electrical wave. In particular the wave front is well captured even with a relatively high level of noise. In future work we will test this method on clinical data including CT-scans of real geometry of the torso and measurements of the electrical potential on the body surface.

References

1. F. Ben Belgacem, H. El Fekih, On cauchy's problem: I. A variational steklov–poincaré theory. Inverse Prob. **21**(6), 1915 (2005)
2. M. Boulakia, S. Cazeau, M.A. Fernández, J.F. Gerbeau, N. Zemzemi, Mathematical modeling of electrocardiograms: a numerical study. Ann. Biomed. Eng. **38**(3), 1071–1097 (2010) [ISSN 0090-6964]
3. J. Clements, J. Nenonen, P.K.J. Li, B.M. Horacek, Acivation dynamics in anisotropic cardiac tissue via decoupling. Ann. Biomed. Eng. **2**(32), 984–990 (2004)
4. S. Ghosh, Y. Rudy, Application of l1-norm regularization to epicardial potential solution of the inverse electrocardiography problem. Ann. Biomed. Eng. **37**(5), 902–912 (2009)
5. B. Irons, R.C. Tuck, A version of the aitken accelerator for computer implementation. Int. J. Numer. Methods Eng. **1**, 275–277 (1969)
6. G.T. Lines, M.L. Buist, P. Grottum, A.J. Pullan, J. Sundnes, A. Tveito, Mathematical models and numerical methods for the forward problem in cardiac electrophysiology. Comput. Visual. Sci. **5**(4), 215–239 (2003)
7. M. Potse, B. Dubé, M. Gulrajani, ECG simulations with realistic human membrane, heart, and torso models, in *Proceedings of the 25th Annual Intemational Conference of the IEEE EMBS* (2003), pp. 70–73
8. A.J. Pullan, M.L. Buist, L.K. Cheng, *Mathematically Modelling the Electrical Activity of the Heart. From Cell to Body Surface and Back Again* (World Scientific, Singapore, 2005)
9. J. Sundnes, G.T. Lines, X. Cai, B.F. Nielsen, K.-A. Mardal, A. Tveito, *Computing the Electrical Activity in the Heart* (Springer, Heidelberg, 2006)
10. K. Ten Tusscher, A.V. Panfilov, Cell model for efficient simulation of wave propagation in human ventricular tissue under normal and pathological conditions. Phys. Med. Biol. **51**, 6141 (2006)
11. L. Tung, *A bi-domain model for describing ischemic myocardial D–C potentials*. Ph.D. Thesis, MIT, 1978
12. N. Zemzemi, *Étude théorique et numérique de l'activité électrique du cœur: Applications aux électrocardiogrammes*. Ph.D. Thesis, Université Paris XI, 2009. http://tel.archives-ouvertes.fr/tel-00470375/en/
13. N. Zemzemi, A steklov-poincaré approach to solve the inverse problem in electrocardiography, in *Computing in Cardiology Conference (CinC), 2013* (IEEE, New York, 2013), pp. 703–706

Editorial Policy

1. Volumes in the following three categories will be published in LNCSE:

i) Research monographs
ii) Tutorials
iii) Conference proceedings

Those considering a book which might be suitable for the series are strongly advised to contact the publisher or the series editors at an early stage.

2. Categories i) and ii). Tutorials are lecture notes typically arising via summer schools or similar events, which are used to teach graduate students. These categories will be emphasized by Lecture Notes in Computational Science and Engineering. **Submissions by interdisciplinary teams of authors are encouraged.** The goal is to report new developments – quickly, informally, and in a way that will make them accessible to non-specialists. In the evaluation of submissions timeliness of the work is an important criterion. Texts should be well-rounded, well-written and reasonably self-contained. In most cases the work will contain results of others as well as those of the author(s). In each case the author(s) should provide sufficient motivation, examples, and applications. In this respect, Ph.D. theses will usually be deemed unsuitable for the Lecture Notes series. Proposals for volumes in these categories should be submitted either to one of the series editors or to Springer-Verlag, Heidelberg, and will be refereed. A provisional judgement on the acceptability of a project can be based on partial information about the work: a detailed outline describing the contents of each chapter, the estimated length, a bibliography, and one or two sample chapters – or a first draft. A final decision whether to accept will rest on an evaluation of the completed work which should include

– at least 100 pages of text;
– a table of contents;
– an informative introduction perhaps with some historical remarks which should be accessible to readers unfamiliar with the topic treated;
– a subject index.

3. Category iii). Conference proceedings will be considered for publication provided that they are both of exceptional interest and devoted to a single topic. One (or more) expert participants will act as the scientific editor(s) of the volume. They select the papers which are suitable for inclusion and have them individually refereed as for a journal. Papers not closely related to the central topic are to be excluded. Organizers should contact the Editor for CSE at Springer at the planning stage, see *Addresses* below.

In exceptional cases some other multi-author-volumes may be considered in this category.

4. Only works in English will be considered. For evaluation purposes, manuscripts may be submitted in print or electronic form, in the latter case, preferably as pdf- or zipped ps-files. Authors are requested to use the LaTeX style files available from Springer at http://www.springer.com/gp/authors-editors/book-authors-editors/manuscript-preparation/5636 (Click on LaTeX Template → monographs or contributed books).

For categories ii) and iii) we strongly recommend that all contributions in a volume be written in the same LaTeX version, preferably LaTeX2e. Electronic material can be included if appropriate. Please contact the publisher.

Careful preparation of the manuscripts will help keep production time short besides ensuring satisfactory appearance of the finished book in print and online.

5. The following terms and conditions hold. Categories i), ii) and iii):

Authors receive 50 free copies of their book. No royalty is paid.
Volume editors receive a total of 50 free copies of their volume to be shared with authors, but no royalties.

Authors and volume editors are entitled to a discount of 33.3 % on the price of Springer books purchased for their personal use, if ordering directly from Springer.

6. Springer secures the copyright for each volume.

Addresses:

Timothy J. Barth
NASA Ames Research Center
NAS Division
Moffett Field, CA 94035, USA
barth@nas.nasa.gov

Michael Griebel
Institut für Numerische Simulation
der Universität Bonn
Wegelerstr. 6
53115 Bonn, Germany
griebel@ins.uni-bonn.de

David E. Keyes
Mathematical and Computer Sciences
and Engineering
King Abdullah University of Science
and Technology
P.O. Box 55455
Jeddah 21534, Saudi Arabia
david.keyes@kaust.edu.sa

and

Department of Applied Physics
and Applied Mathematics
Columbia University
500 W. 120 th Street
New York, NY 10027, USA
kd2112@columbia.edu

Risto M. Nieminen
Department of Applied Physics
Aalto University School of Science
and Technology
00076 Aalto, Finland
risto.nieminen@aalto.fi

Dirk Roose
Department of Computer Science
Katholieke Universiteit Leuven
Celestijnenlaan 200A
3001 Leuven-Heverlee, Belgium
dirk.roose@cs.kuleuven.be

Tamar Schlick
Department of Chemistry
and Courant Institute
of Mathematical Sciences
New York University
251 Mercer Street
New York, NY 10012, USA
schlick@nyu.edu

Editor for Computational Science
and Engineering at Springer:
Martin Peters
Springer-Verlag
Mathematics Editorial IV
Tiergartenstrasse 17
69121 Heidelberg, Germany
martin.peters@springer.com

Lecture Notes in Computational Science and Engineering

1. D. Funaro, *Spectral Elements for Transport-Dominated Equations.*
2. H.P. Langtangen, *Computational Partial Differential Equations.* Numerical Methods and Diffpack Programming.
3. W. Hackbusch, G. Wittum (eds.), *Multigrid Methods V.*
4. P. Deuflhard, J. Hermans, B. Leimkuhler, A.E. Mark, S. Reich, R.D. Skeel (eds.), *Computational Molecular Dynamics: Challenges, Methods, Ideas.*
5. D. Kröner, M. Ohlberger, C. Rohde (eds.), *An Introduction to Recent Developments in Theory and Numerics for Conservation Laws.*
6. S. Turek, *Efficient Solvers for Incompressible Flow Problems.* An Algorithmic and Computational Approach.
7. R. von Schwerin, ***M**ulti **B**ody **S**ystem **SIM**ulation.* Numerical Methods, Algorithms, and Software.
8. H.-J. Bungartz, F. Durst, C. Zenger (eds.), *High Performance Scientific and Engineering Computing.*
9. T.J. Barth, H. Deconinck (eds.), *High-Order Methods for Computational Physics.*
10. H.P. Langtangen, A.M. Bruaset, E. Quak (eds.), *Advances in Software Tools for Scientific Computing.*
11. B. Cockburn, G.E. Karniadakis, C.-W. Shu (eds.), *Discontinuous Galerkin Methods.* Theory, Computation and Applications.
12. U. van Rienen, *Numerical Methods in Computational Electrodynamics.* Linear Systems in Practical Applications.
13. B. Engquist, L. Johnsson, M. Hammill, F. Short (eds.), *Simulation and Visualization on the Grid.*
14. E. Dick, K. Riemslagh, J. Vierendeels (eds.), *Multigrid Methods VI.*
15. A. Frommer, T. Lippert, B. Medeke, K. Schilling (eds.), *Numerical Challenges in Lattice Quantum Chromodynamics.*
16. J. Lang, *Adaptive Multilevel Solution of Nonlinear Parabolic PDE Systems.* Theory, Algorithm, and Applications.
17. B.I. Wohlmuth, *Discretization Methods and Iterative Solvers Based on Domain Decomposition.*
18. U. van Rienen, M. Günther, D. Hecht (eds.), *Scientific Computing in Electrical Engineering.*
19. I. Babuška, P.G. Ciarlet, T. Miyoshi (eds.), *Mathematical Modeling and Numerical Simulation in Continuum Mechanics.*
20. T.J. Barth, T. Chan, R. Haimes (eds.), *Multiscale and Multiresolution Methods.* Theory and Applications.
21. M. Breuer, F. Durst, C. Zenger (eds.), *High Performance Scientific and Engineering Computing.*
22. K. Urban, *Wavelets in Numerical Simulation.* Problem Adapted Construction and Applications.
23. L.F. Pavarino, A. Toselli (eds.), *Recent Developments in Domain Decomposition Methods.*

24. T. Schlick, H.H. Gan (eds.), *Computational Methods for Macromolecules: Challenges and Applications.*

25. T.J. Barth, H. Deconinck (eds.), *Error Estimation and Adaptive Discretization Methods in Computational Fluid Dynamics.*

26. M. Griebel, M.A. Schweitzer (eds.), *Meshfree Methods for Partial Differential Equations.*

27. S. Müller, *Adaptive Multiscale Schemes for Conservation Laws.*

28. C. Carstensen, S. Funken, W. Hackbusch, R.H.W. Hoppe, P. Monk (eds.), *Computational Electromagnetics.*

29. M.A. Schweitzer, *A Parallel Multilevel Partition of Unity Method for Elliptic Partial Differential Equations.*

30. T. Biegler, O. Ghattas, M. Heinkenschloss, B. van Bloemen Waanders (eds.), *Large-Scale PDE-Constrained Optimization.*

31. M. Ainsworth, P. Davies, D. Duncan, P. Martin, B. Rynne (eds.), *Topics in Computational Wave Propagation.* Direct and Inverse Problems.

32. H. Emmerich, B. Nestler, M. Schreckenberg (eds.), *Interface and Transport Dynamics.* Computational Modelling.

33. H.P. Langtangen, A. Tveito (eds.), *Advanced Topics in Computational Partial Differential Equations.* Numerical Methods and Diffpack Programming.

34. V. John, *Large Eddy Simulation of Turbulent Incompressible Flows.* Analytical and Numerical Results for a Class of LES Models.

35. E. Bänsch (ed.), *Challenges in Scientific Computing - CISC 2002.*

36. B.N. Khoromskij, G. Wittum, *Numerical Solution of Elliptic Differential Equations by Reduction to the Interface.*

37. A. Iske, *Multiresolution Methods in Scattered Data Modelling.*

38. S.-I. Niculescu, K. Gu (eds.), *Advances in Time-Delay Systems.*

39. S. Attinger, P. Koumoutsakos (eds.), *Multiscale Modelling and Simulation.*

40. R. Kornhuber, R. Hoppe, J. Périaux, O. Pironneau, O. Wildlund, J. Xu (eds.), *Domain Decomposition Methods in Science and Engineering.*

41. T. Plewa, T. Linde, V.G. Weirs (eds.), *Adaptive Mesh Refinement – Theory and Applications.*

42. A. Schmidt, K.G. Siebert, *Design of Adaptive Finite Element Software.* The Finite Element Toolbox ALBERTA.

43. M. Griebel, M.A. Schweitzer (eds.), *Meshfree Methods for Partial Differential Equations II.*

44. B. Engquist, P. Lötstedt, O. Runborg (eds.), *Multiscale Methods in Science and Engineering.*

45. P. Benner, V. Mehrmann, D.C. Sorensen (eds.), *Dimension Reduction of Large-Scale Systems.*

46. D. Kressner, *Numerical Methods for General and Structured Eigenvalue Problems.*

47. A. Boriçi, A. Frommer, B. Joó, A. Kennedy, B. Pendleton (eds.), *QCD and Numerical Analysis III.*

48. F. Graziani (ed.), *Computational Methods in Transport.*

49. B. Leimkuhler, C. Chipot, R. Elber, A. Laaksonen, A. Mark, T. Schlick, C. Schütte, R. Skeel (eds.), *New Algorithms for Macromolecular Simulation.*

50. M. Bücker, G. Corliss, P. Hovland, U. Naumann, B. Norris (eds.), *Automatic Differentiation: Applications, Theory, and Implementations.*

51. A.M. Bruaset, A. Tveito (eds.), *Numerical Solution of Partial Differential Equations on Parallel Computers.*

52. K.H. Hoffmann, A. Meyer (eds.), *Parallel Algorithms and Cluster Computing.*

53. H.-J. Bungartz, M. Schäfer (eds.), *Fluid-Structure Interaction.*

54. J. Behrens, *Adaptive Atmospheric Modeling.*

55. O. Widlund, D. Keyes (eds.), *Domain Decomposition Methods in Science and Engineering XVI.*

56. S. Kassinos, C. Langer, G. Iaccarino, P. Moin (eds.), *Complex Effects in Large Eddy Simulations.*

57. M. Griebel, M.A Schweitzer (eds.), *Meshfree Methods for Partial Differential Equations III.*

58. A.N. Gorban, B. Kégl, D.C. Wunsch, A. Zinovyev (eds.), *Principal Manifolds for Data Visualization and Dimension Reduction.*

59. H. Ammari (ed.), *Modeling and Computations in Electromagnetics: A Volume Dedicated to Jean-Claude Nédélec.*

60. U. Langer, M. Discacciati, D. Keyes, O. Widlund, W. Zulehner (eds.), *Domain Decomposition Methods in Science and Engineering XVII.*

61. T. Mathew, *Domain Decomposition Methods for the Numerical Solution of Partial Differential Equations.*

62. F. Graziani (ed.), *Computational Methods in Transport: Verification and Validation.*

63. M. Bebendorf, *Hierarchical Matrices.* A Means to Efficiently Solve Elliptic Boundary Value Problems.

64. C.H. Bischof, H.M. Bücker, P. Hovland, U. Naumann, J. Utke (eds.), *Advances in Automatic Differentiation.*

65. M. Griebel, M.A. Schweitzer (eds.), *Meshfree Methods for Partial Differential Equations IV.*

66. B. Engquist, P. Lötstedt, O. Runborg (eds.), *Multiscale Modeling and Simulation in Science.*

67. I.H. Tuncer, Ü. Gülcat, D.R. Emerson, K. Matsuno (eds.), *Parallel Computational Fluid Dynamics 2007.*

68. S. Yip, T. Diaz de la Rubia (eds.), *Scientific Modeling and Simulations.*

69. A. Hegarty, N. Kopteva, E. O'Riordan, M. Stynes (eds.), *BAIL* 2008 – *Boundary and Interior Layers.*

70. M. Bercovier, M.J. Gander, R. Kornhuber, O. Widlund (eds.), *Domain Decomposition Methods in Science and Engineering XVIII.*

71. B. Koren, C. Vuik (eds.), *Advanced Computational Methods in Science and Engineering.*

72. M. Peters (ed.), *Computational Fluid Dynamics for Sport Simulation.*

73. H.-J. Bungartz, M. Mehl, M. Schäfer (eds.), *Fluid Structure Interaction II - Modelling, Simulation, Optimization.*

74. D. Tromeur-Dervout, G. Brenner, D.R. Emerson, J. Erhel (eds.), *Parallel Computational Fluid Dynamics 2008.*

75. A.N. Gorban, D. Roose (eds.), *Coping with Complexity: Model Reduction and Data Analysis.*

76. J.S. Hesthaven, E.M. Rønquist (eds.), *Spectral and High Order Methods for Partial Differential Equations.*
77. M. Holtz, *Sparse Grid Quadrature in High Dimensions with Applications in Finance and Insurance.*
78. Y. Huang, R. Kornhuber, O.Widlund, J. Xu (eds.), *Domain Decomposition Methods in Science and Engineering XIX.*
79. M. Griebel, M.A. Schweitzer (eds.), *Meshfree Methods for Partial Differential Equations V.*
80. P.H. Lauritzen, C. Jablonowski, M.A. Taylor, R.D. Nair (eds.), *Numerical Techniques for Global Atmospheric Models.*
81. C. Clavero, J.L. Gracia, F.J. Lisbona (eds.), *BAIL 2010 – Boundary and Interior Layers, Computational and Asymptotic Methods.*
82. B. Engquist, O. Runborg, Y.R. Tsai (eds.), *Numerical Analysis and Multiscale Computations.*
83. I.G. Graham, T.Y. Hou, O. Lakkis, R. Scheichl (eds.), *Numerical Analysis of Multiscale Problems.*
84. A. Logg, K.-A. Mardal, G. Wells (eds.), *Automated Solution of Differential Equations by the Finite Element Method.*
85. J. Blowey, M. Jensen (eds.), *Frontiers in Numerical Analysis - Durham 2010.*
86. O. Kolditz, U.-J. Gorke, H. Shao, W. Wang (eds.), *Thermo-Hydro-Mechanical-Chemical Processes in Fractured Porous Media - Benchmarks and Examples.*
87. S. Forth, P. Hovland, E. Phipps, J. Utke, A. Walther (eds.), *Recent Advances in Algorithmic Differentiation.*
88. J. Garcke, M. Griebel (eds.), *Sparse Grids and Applications.*
89. M. Griebel, M.A. Schweitzer (eds.), *Meshfree Methods for Partial Differential Equations VI.*
90. C. Pechstein, *Finite and Boundary Element Tearing and Interconnecting Solvers for Multiscale Problems.*
91. R. Bank, M. Holst, O. Widlund, J. Xu (eds.), *Domain Decomposition Methods in Science and Engineering XX.*
92. H. Bijl, D. Lucor, S. Mishra, C. Schwab (eds.), *Uncertainty Quantification in Computational Fluid Dynamics.*
93. M. Bader, H.-J. Bungartz, T. Weinzierl (eds.), *Advanced Computing.*
94. M. Ehrhardt, T. Koprucki (eds.), *Advanced Mathematical Models and Numerical Techniques for Multi-Band Effective Mass Approximations.*
95. M. Azaïez, H. El Fekih, J.S. Hesthaven (eds.), *Spectral and High Order Methods for Partial Differential Equations ICOSAHOM 2012.*
96. F. Graziani, M.P. Desjarlais, R. Redmer, S.B. Trickey (eds.), *Frontiers and Challenges in Warm Dense Matter.*
97. J. Garcke, D. Pflüger (eds.), *Sparse Grids and Applications – Munich 2012.*
98. J. Erhel, M. Gander, L. Halpern, G. Pichot, T. Sassi, O. Widlund (eds.), *Domain Decomposition Methods in Science and Engineering XXI.*
99. R. Abgrall, H. Beaugendre, P.M. Congedo, C. Dobrzynski, V. Perrier, M. Ricchiuto (eds.), *High Order Nonlinear Numerical Methods for Evolutionary PDEs - HONOM 2013.*
100. M. Griebel, M.A. Schweitzer (eds.), *Meshfree Methods for Partial Differential Equations VII.*

101. R. Hoppe (ed.), *Optimization with PDE Constraints - OPTPDE 2014.*

102. S. Dahlke, W. Dahmen, M. Griebel, W. Hackbusch, K. Ritter, R. Schneider, C. Schwab, H. Yserentant (eds.), *Extraction of Quantifiable Information from Complex Systems.*

103. A. Abdulle, S. Deparis, D. Kressner, F. Nobile, M. Picasso (eds.), *Numerical Mathematics and Advanced Applications - ENUMATH 2013.*

104. T. Dickopf, M.J. Gander, L. Halpern, R. Krause, L.F. Pavarino (eds.), *Domain Decomposition Methods in Science and Engineering XXII.*

For further information on these books please have a look at our mathematics catalogue at the following URL: `www.springer.com/series/3527`

Monographs in Computational Science and Engineering

1. J. Sundnes, G.T. Lines, X. Cai, B.F. Nielsen, K.-A. Mardal, A. Tveito, *Computing the Electrical Activity in the Heart.*

For further information on this book, please have a look at our mathematics catalogue at the following URL: `www.springer.com/series/7417`

Texts in Computational Science and Engineering

1. H. P. Langtangen, *Computational Partial Differential Equations.* Numerical Methods and Diffpack Programming. 2nd Edition
2. A. Quarteroni, F. Saleri, P. Gervasio, *Scientific Computing with MATLAB and Octave.* 4th Edition
3. H. P. Langtangen, *Python Scripting for Computational Science.* 3rd Edition
4. H. Gardner, G. Manduchi, *Design Patterns for e-Science.*
5. M. Griebel, S. Knapek, G. Zumbusch, *Numerical Simulation in Molecular Dynamics.*
6. H. P. Langtangen, *A Primer on Scientific Programming with Python.* 4th Edition
7. A. Tveito, H. P. Langtangen, B. F. Nielsen, X. Cai, *Elements of Scientific Computing.*
8. B. Gustafsson, *Fundamentals of Scientific Computing.*
9. M. Bader, *Space-Filling Curves.*
10. M. Larson, F. Bengzon, *The Finite Element Method: Theory, Implementation and Applications.*
11. W. Gander, M. Gander, F. Kwok, *Scientific Computing: An Introduction using Maple and MATLAB.*
12. P. Deuflhard, S. Röblitz, *A Guide to Numerical Modelling in Systems Biology.*

For further information on these books please have a look at our mathematics catalogue at the following URL: `www.springer.com/series/5151`

Printed in the United States
By Bookmasters